FUNDAMENTALS OF BUSINESS STATISTICS, 6E

FUNDAMENTALS OF BUSINESS STATISTICS, 6E

Dennis J. Sweeney
University of Cincinnati

Thomas A. Williams
Rochester Institute of Technology

David R. Anderson
University of Cincinnati

SOUTH-WESTERN
CENGAGE Learning™

Australia · Brazil · Canada · Mexico · Singapore · Spain · United Kingdom · United States

Fundamentals of Business Statistics, Sixth Edition
Dennis J. Sweeney, Thomas A. Williams, David R. Anderson

VP/Editorial Director: Jack W. Calhoun

Publisher: Joe Sabatino

Senior Acquisitions Editor:
  Charles McCormick, Jr.

Developmental Editor: Maggie Kubale

Editorial Assistant: Nora Heink

Marketing Communications Manager:
  Libby Shipp

Marketing Manager: Adam Marsh

Content Project Manager:
  Jacquelyn K. Featherly

Media Editor: Chris Valentine

Manufacturing Buyer: Miranda Klapper

Print Buyer: Arethea Thomas

Production House/Compositor:
  MPS Limited, a Macmillan Company

Senior Rights Specialist:
  Mardell Glinski-Schultz

Senior Art Director: Stacy Jenkins Shirley

Internal Designer:
  Michael Stratton/Chris Miller Design

Cover Designer: Patti Hudepohl

Cover Images: © Shutterstock

Photo Credits:
  B/W Image: Getty Images/PhotoDisc/
              Chad Baker
  Cover Image: Shutterstock Images/
               Joel Kempson

© 2011, 2009 South-Western, a part of Cengage Learning

ALL RIGHTS RESERVED. No part of this work covered by the copyright herein may be reproduced, transmitted, stored or used in any form or by any means graphic, electronic, or mechanical, including but not limited to photocopying, recording, scanning, digitizing, taping, Web distribution, information networks, or information storage and retrieval systems, except as permitted under Section 107 or 108 of the 1976 United States Copyright Act, or applicable copyright law of another jurisdiction, without the prior written permission of the publisher.

> For product information and technology assistance, contact us at
> **Cengage Learning Customer & Sales Support, 1-800-354-9706**
> For permission to use material from this text or product, submit all requests online at **www.cengage.com/permissions**
> Further permissions questions can be emailed to
> **permissionrequest@cengage.com**

ExamView® and ExamView Pro® are registered trademarks of FSCreations, Inc. Windows is a registered trademark of the Microsoft Corporation used herein under license. Macintosh and Power Macintosh are registered trademarks of Apple Computer, Inc. used herein under license.

Library of Congress Control Number: 2010925381

International Edition:

ISBN 13: 978-1-111-22127-0

ISBN 10: 1-111-22127-8

Cengage Learning International Offices

Asia
www.cengageasia.com
tel: (65) 6410 1200

Brazil
www.cengage.com.br
tel: (55) 11 3665 9900

Latin America
www.cengage.com.mx
tel: (52) 55 1500 6000

Australia/New Zealand
www.cengage.com.au
tel: (61) 3 9685 4111

India
www.cengage.co.in
tel: (91) 11 4364 1111

UK/Europe/Middle East/Africa
www.cengage.co.uk
tel: (44) 0 1264 332 424

Represented in Canada by Nelson Education, Ltd.
tel: (416) 752 9100/(800) 668 0671
www.nelson.com

Cengage Learning is a leading provider of customized learning solutions with office locations around the globe, including Singapore, the United Kingdom, Australia, Mexico, Brazil, and Japan. Locate your local office at: **cengage.com/global**

For product information: **www.cengage.com/international**
Visit your local office: **www.cengage.com/global**
Visit our corporate website: **www.cengage.com**

Printed in China
2 3 4 5 6 7 14 13 12

Dedicated to
Marcia, Cherri, and Robbie

Brief Contents

Preface xxi
About the Authors xxxii

Chapter 1 Data and Statistics 1
Chapter 2 Descriptive Statistics: Tabular and Graphical Presentations 31
Chapter 3 Descriptive Statistics: Numerical Measures 86
Chapter 4 Introduction to Probability 148
Chapter 5 Discrete Probability Distributions 193
Chapter 6 Continuous Probability Distributions 232
Chapter 7 Sampling and Sampling Distributions 265
Chapter 8 Interval Estimation 304
Chapter 9 Hypothesis Tests 344
Chapter 10 Comparisons Involving Means, Experimental Design, and Analysis of Variance 392
Chapter 11 Comparisons Involving Proportions and a Test of Independence 448
Chapter 12 Simple Linear Regression 483
Chapter 13 Multiple Regression 552
Appendix A References and Bibliography 602
Appendix B Tables 604
Appendix C Summation Notation 631
Appendix D Self-Test Solutions and Answers to Exercises 633
Appendix E Using Excel Functions 665
Appendix F Computing p-Values Using Minitab and Excel 670
Index 674

Contents

Preface xxi
About the Authors xxxii

Chapter 1   Data and Statistics   1

Statistics in Practice: Businessweek 2
1.1   Applications in Business and Economics 3
      Accounting 3
      Finance 4
      Marketing 4
      Production 4
      Economics 4
1.2   Data 5
      Elements, Variables, and Observations 5
      Scales of Measurement 6
      Categorical and Quantitative Data 7
      Cross-Sectional and Time Series Data 7
1.3   Data Sources 10
      Existing Sources 10
      Statistical Studies 11
      Data Acquisition Errors 13
1.4   Descriptive Statistics 13
1.5   Statistical Inference 15
1.6   Computers and Statistical Analysis 17
1.7   Data Mining 17
1.8   Ethical Guidelines for Statistical Practice 18
Summary 20
Glossary 20
Supplementary Exercises 21
Appendix An Introduction to StatTools 28

Chapter 2 Descriptive Statistics: Tabular and Graphical Presentations 31

Statistics in Practice: Colgate-Palmolive Company 32

2.1 Summarizing Categorical Data 33
  Frequency Distribution 33
  Relative Frequency and Percent Frequency Distributions 34
  Bar Charts and Pie Charts 34

2.2 Summarizing Quantitative Data 39
  Frequency Distribution 39
  Relative Frequency and Percent Frequency Distributions 41
  Dot Plot 41
  Histogram 42
  Cumulative Distributions 44
  Ogive 44

2.3 Exploratory Data Analysis: The Stem-and-Leaf Display 49

2.4 Crosstabulations and Scatter Diagrams 54
  Crosstabulation 54
  Simpson's Paradox 57
  Scatter Diagram and Trendline 58

Summary 64
Glossary 65
Key Formulas 66
Supplementary Exercises 66
Case Problem 1: Pelican Stores 72
Case Problem 2: Motion Picture Industry 73
Appendix 2.1 Tabular and Graphical Presentations Using Minitab 74
Appendix 2.2 Tabular and Graphical Presentations Using Excel 76
Appendix 2.3 Tabular and Graphical Presentations Using StatTools 85

Chapter 3 Descriptive Statistics: Numerical Measures 86

Statistics in Practice: Small Fry Design 87

3.1 Measures of Location 88
  Mean 88
  Median 89
  Mode 90
  Percentiles 91
  Quartiles 92

3.2 Measures of Variability 96
  Range 97
  Interquartile Range 97

Variance 98
Standard Deviation 100
Coefficient of Variation 100

3.3 Measures of Distribution Shape, Relative Location, and Detection of Outliers 103
Distribution Shape 103
z-Scores 104
Chebyshev's Theorem 105
Empirical Rule 106
Detection of Outliers 107

3.4 Exploratory Data Analysis 110
Five-Number Summary 110
Box Plot 111

3.5 Measures of Association Between Two Variables 116
Covariance 116
Interpretation of the Covariance 118
Correlation Coefficient 120
Interpretation of the Correlation Coefficient 121

3.6 The Weighted Mean and Working with Grouped Data 125
Weighted Mean 125
Grouped Data 126

Summary 131
Glossary 131
Key Formulas 133
Supplementary Exercises 134
Case Problem 1: Pelican Stores 138
Case Problem 2: Motion Picture Industry 140
Case Problem 3: Heavenly Chocolates Website Transactions 140
Appendix 3.1 Descriptive Statistics Using Minitab 141
Appendix 3.2 Descriptive Statistics Using Excel 143
Appendix 3.3 Descriptive Statistics Using StatTools 146

Chapter 4   Introduction to Probability   148

Statistics in Practice: Oceanwide Seafood 149

4.1 Experiments, Counting Rules, and Assigning Probabilities 150
Counting Rules, Combinations, and Permutations 151
Assigning Probabilities 155
Probabilities for the KP&L Project 157

4.2 Events and Their Probabilities 160

4.3 Some Basic Relationships of Probability 164
      Complement of an Event 164
      Addition Law 165

4.4 Conditional Probability 171
      Independent Events 174
      Multiplication Law 174

4.5 Bayes' Theorem 178
      Tabular Approach 182

Summary 184
Glossary 184
Key Formulas 185
Supplementary Exercises 186
Case Problem: Hamilton County Judges 190

Chapter 5   Discrete Probability Distributions   193

Statistics in Practice: Citibank 194

5.1 Random Variables 194
      Discrete Random Variables 195
      Continuous Random Variables 196

5.2 Discrete Probability Distributions 197

5.3 Expected Value and Variance 202
      Expected Value 202
      Variance 203

5.4 Binomial Probability Distribution 207
      A Binomial Experiment 208
      Martin Clothing Store Problem 209
      Using Tables of Binomial Probabilities 213
      Expected Value and Variance for the Binomial Distribution 214

5.5 Poisson Probability Distribution 218
      An Example Involving Time Intervals 218
      An Example Involving Length or Distance Intervals 220

5.6 Hypergeometric Probability Distribution 221

Summary 225
Glossary 226
Key Formulas 226
Supplementary Exercises 227
Appendix 5.1 Discrete Probability Distributions Using Minitab 230
Appendix 5.2 Discrete Probability Distributions Using Excel 230

Contents

Chapter 6 Continuous Probability Distributions 232

Statistics in Practice: Procter & Gamble 233
6.1 Uniform Probability Distribution 234
    Area as a Measure of Probability 235
6.2 Normal Probability Distribution 238
    Normal Curve 238
    Standard Normal Probability Distribution 240
    Computing Probabilities for Any Normal Probability Distribution 245
    Grear Tire Company Problem 246
6.3 Normal Approximation of Binomial Probabilities 250
6.4 Exponential Probability Distribution 254
    Computing Probabilities for the Exponential Distribution 254
    Relationship Between the Poisson and Exponential Distributions 255
Summary 257
Glossary 258
Key Formulas 258
Supplementary Exercises 259
Case Problem: Specialty Toys 262
Appendix 6.1 Continuous Probability Distributions Using Minitab 263
Appendix 6.2 Continuous Probability Distributions Using Excel 263

Chapter 7 Sampling and Sampling Distributions 265

Statistics in Practice: Meadwestvaco Corporation 266
7.1 The Electronics Associates Sampling Problem 267
7.2 Selecting a Sample 268
    Sampling from a Finite Population 268
    Sampling from an Infinite Population 270
7.3 Point Estimation 273
    Practical Advice 275
7.4 Introduction to Sampling Distributions 276
7.5 Sampling Distribution of $\bar{x}$ 278
    Expected Value of $\bar{x}$ 279
    Standard Deviation of $\bar{x}$ 280
    Form of the Sampling Distribution of $\bar{x}$ 281
    Sampling Distribution of $\bar{x}$ for the EAI Problem 283
    Practical Value of the Sampling Distribution of $\bar{x}$ 283
    Relationship Between the Sample Size and the Sampling Distribution of $\bar{x}$ 285

7.6 **Sampling Distribution of $\bar{p}$** 289
　　Expected Value of $\bar{p}$ 289
　　Standard Deviation of $\bar{p}$ 290
　　Form of the Sampling Distribution of $\bar{p}$ 291
　　Practical Value of the Sampling Distribution of $\bar{p}$ 291

7.7 **Other Sampling Methods** 295
　　Stratified Random Sampling 295
　　Cluster Sampling 295
　　Systematic Sampling 296
　　Convenience Sampling 296
　　Judgment Sampling 297

Summary 297

Glossary 298

Key Formulas 299

Supplementary Exercises 299

Appendix 7.1 Random Sampling Using Minitab 301

Appendix 7.2 Random Sampling Using Excel 302

Appendix 7.3 Random Sampling Using StatTools 302

Chapter 8　Interval Estimation　304

Statistics in Practice: Food Lion 305

8.1 **Population Mean: σ Known** 306
　　Margin of Error and the Interval Estimate 306
　　Practical Advice 310

8.2 **Population Mean: σ Unknown** 312
　　Margin of Error and the Interval Estimate 313
　　Practical Advice 316
　　Using a Small Sample 316
　　Summary of Interval Estimation Procedures 318

8.3 **Determining the Sample Size** 321

8.4 **Population Proportion** 324
　　Determining the Sample Size 326

Summary 329

Glossary 330

Key Formulas 331

Supplementary Exercises 331

Case Problem 1: *Young Professional* Magazine 334
Case Problem 2: Gulf Real Estate Properties 335
Case Problem 3: Metropolitan Research, Inc. 337
Appendix 8.1 Interval Estimation Using Minitab 338
Appendix 8.2 Interval Estimation Using Excel 339
Appendix 8.3 Interval Estimation Using StatTools 341

Chapter 9 Hypothesis Tests 344

Statistics in Practice: John Morrell & Company 345

9.1 Developing Null and Alternative Hypotheses 346
  The Alternative Hypothesis as a Research Hypothesis 346
  The Null Hypothesis as an Assumption to Be Challenged 347
  Summary of Forms for Null and Alternative Hypotheses 348

9.2 Type I and Type II Errors 349

9.3 Population Mean: σ Known 352
  One-Tailed Test 352
  Two-Tailed Test 358
  Summary and Practical Advice 361
  Relationship Between Interval Estimation and Hypothesis Testing 362

9.4 Population Mean: σ Unknown 367
  One-Tailed Test 367
  Two-Tailed Test 368
  Summary and Practical Advice 370

9.5 Population Proportion 373
  Summary 375

Summary 378

Glossary 378

Key Formulas 379

Supplementary Exercises 379

Case Problem 1: Quality Associates, Inc. 382

Case Problem 2: Ethical Behavior of Business Students at Bayview Universtiy 383

Appendix 9.1 Hypothesis Testing Using Minitab 385

Appendix 9.2 Hypothesis Testing Using Excel 386

Appendix 9.3 Hypothesis Testing Using StatTools 391

Chapter 10 Comparisons Involving Means, Experimental Design, and Analysis of Variance 392

Statistics in Practice: U.S. Food and Drug Administration 393

10.1 Inferences About the Difference Between Two Population Means: σ_1 and σ_2 Known 394
Interval Estimation of $\mu_1 - \mu_2$ 394
Hypothesis Tests About $\mu_1 - \mu_2$ 397
Practical Advice 398

10.2 Inferences About the Difference Between Two Population Means: σ_1 and σ_2 Unknown 401
Interval Estimation of $\mu_1 - \mu_2$ 401
Hypothesis Tests About $\mu_1 - \mu_2$ 403
Practical Advice 405

10.3 Inferences About the Difference Between Two Population Means: Matched Samples 409

10.4 An Introduction to Experimental Design and Analysis of Variance 414
Data Collection 416
Assumptions for Analysis of Variance 417
Analysis of Variance: A Conceptual Overview 417

10.5 Analysis of Variance and the Completely Randomized Design 420
Between-Treatments Estimate of Population Variance 421
Within-Treatments Estimate of Population Variance 422
Comparing the Variance Estimates: The F Test 423
ANOVA Table 424
Computer Results for Analysis of Variance 425
Testing for the Equality of k Population Means: An Observational Study 427

Summary 431

Glossary 431

Key Formulas 431

Supplementary Exercises 433

Case Problem 1: Par, Inc. 438

Case Problem 2: Wentworth Medical Center 438

Case Problem 3 Compensation for Sales Professionals 439

Appendix 10.1 Inferences About Two Populations Using Minitab 440

Appendix 10.2 Analysis of Variance Using Minitab 442

Appendix 10.3 Inferences About Two Populations Using Excel 442

Appendix 10.4 Analysis of Variance Using Excel 443

Appendix 10.5 Inferences About Two Populations Using StatTools 444

Appendix 10.6 Analysis of Variance Using StatTools 446

Chapter 11 Comparisons Involving Proportions and a Test of Independence 448

Statistics in Practice: United Way 449

11.1 Inferences About the Difference Between Two Population Proportions 450
  Interval Estimation of $p_1 - p_2$ 450
  Hypothesis Tests About $p_1 - p_2$ 452

11.2 Hypothesis Test for Proportions of a Multinomial Population 456

11.3 Test of Independence 463

Summary 471

Glossary 471

Key Formulas 471

Supplementary Exercises 472

Case Problem: A Bipartisan Agenda for Change 477

Appendix 11.1 Inferences About Two Population Proportions Using Minitab 477

Appendix 11.2 Tests of Goodness of Fit and Independence Using Minitab 478

Appendix 11.3 Tests of Goodness of Fit and Independence Using Excel 479

Appendix 11.4 Inferences About Two Population Proportions Using StatTools 480

Appendix 11.5 Test of Independence Using StatTools 482

Chapter 12 Simple Linear Regression 483

Statistics in Practice: Alliance Data Systems 484

12.1 Simple Linear Regression Model 485
  Regression Model and Regression Equation 485
  Estimated Regression Equation 486

12.2 Least Squares Method 488

12.3 Coefficient of Determination 499
  Correlation Coefficient 502

12.4 Model Assumptions 506

12.5 Testing for Significance 508
  Estimate of σ^2 508
  t Test 509
  Confidence Interval for β_1 510
  F Test 511
  Some Cautions About the Interpretation of Significance Tests 513

12.6 Using the Estimated Regression Equation for Estimation and Prediction 517
Point Estimation 517
Interval Estimation 517
Confidence Interval for the Mean Value of y 518
Prediction Interval for an Individual Value of y 519

12.7 Computer Solution 523

12.8 Residual Analysis: Validating Model Assumptions 527
Residual Plot Against x 529
Residual Plot Against $\hat{y}$ 531

Summary 533

Glossary 534

Key Formulas 535

Supplementary Exercises 536

Case Problem 1: Measuring Stock Market Risk 543

Case Problem 2: U.S. Department of Transportation 544

Case Problem 3: Alumni Giving 545

Case Problem 4: PGA Tour Statistics 545

Appendix 12.1 Regression Analysis Using Minitab 547

Appendix 12.2 Regression Analysis Using Excel 548

Appendix 12.3 Regression Analysis Using StatTools 550

Chapter 13 Multiple Regression 552

Statistics in Practice: International Paper 553

13.1 Multiple Regression Model 554
Regression Model and Regression Equation 554
Estimated Multiple Regression Equation 554

13.2 Least Squares Method 555
An Example: Butler Trucking Company 556
Note on Interpretation of Coefficients 558

13.3 Multiple Coefficient of Determination 564

13.4 Model Assumptions 567

13.5 Testing for Significance 568
F Test 569
t Test 571
Multicollinearity 572

13.6 Using the Estimated Regression Equation for Estimation and Prediction 576

13.7 Categorical Independent Variables 578
An Example: Johnson Filtration, Inc. 578
Interpreting the Parameters 581
More Complex Categorical Variables 582

Summary 586

Glossary 587

Key Formulas 587
Supplementary Exercises 588
Case Problem 1: Consumer Research, Inc. 594
Case Problem 2: Alumni Giving 595
Case Problem 3: PGA Tour Statistics 597
Case Problem 4: Predicting Winning Percentage for the NFL 598
Appendix 13.1 Multiple Regression Using Minitab 599
Appendix 13.2 Multiple Regression Using Excel 599
Appendix 13.3 Multiple Regression Using StatTools 600

Appendix A References and Bibliography 602

Appendix B Tables 604

Appendix C Summation Notation 631

Appendix D Self-Test Solutions and Answers to Exercises 633

Appendix E Using Excel Functions 665

Appendix F Computing *p*-Values Using Minitab and Excel 670

Index 674

Preface

The purpose of *FUNDAMENTALS OF BUSINESS STATISTICS* is to give students, primarily those in the fields of business administration and economics, a conceptual introduction to the field of statistics and its many applications. The text is applications oriented and written with the needs of the nonmathematician in mind; the mathematical prerequisite is knowledge of algebra.

Applications of data analysis and statistical methodology are an integral part of the organization and presentation of the text material. The discussion and development of each technique is presented in an application setting, with the statistical results providing insights to decisions and solutions to problems.

Although the book is applications oriented, we have taken care to provide sound methodological development and to use notation that is generally accepted for the topic being covered. Hence, students will find that this text provides good preparation for the study of more advanced statistical material. A bibliography to guide further study is included as an appendix.

The text introduces the student to the software packages of Minitab 15 and Microsoft® Office Excel® 2007 and emphasizes the role of computer software in the application of statistical analysis. Minitab is illustrated as it is one of the leading statistical software packages for both education and statistical practice. Excel is not a statistical software package, but the wide availability and use of Excel make it important for students to understand the statistical capabilities of this package. With this edition, we are making available a commercial Excel add-in, StatTools, that extends the range of statistical options for Excel users. Minitab, Excel, and StatTools procedures are provided in chapter appendixes so that instructors have the flexibility of using as much computer emphasis as desired for the course.

It is likely there will be users of both Excel 2007 and Excel 2010 using this text. To accommodate both groups of users, the step-by-step procedures and the worksheets presented in our Excel appendixes were developed and tested using both Excel 2007 and the public beta versions of Excel 2010. For Excel 2007 users we have included on the website that accompanies the text a primer entitled Microsoft Excel 2007 and Tools for Statistical Analysis. A similar primer entitled Microsoft Excel 2010 and Tools for Statistical Analysis is provided on the website for Excel 2010 users.

AVAILABILITY OF RESOURCES MAY DIFFER BY REGION. Check with your local Cengage Learning representative for details.

Changes in the Sixth Edition

We appreciate the acceptance and positive response to the previous editions of *FUNDAMENTALS OF BUSINESS STATISTICS*. Accordingly, in making modifications for this new edition, we have maintained the presentation style and readability of those editions. The significant changes in the new edition are summarized here.

Content Revisions

- **StatTools Add-In for Excel.** Excel 2007 does not contain statistical functions or data analysis tools to perform all the statistical procedures discussed in the text. StatTools is a commercial Excel 2007 add-in, developed by Palisade Corporation, that provides additional statistical options for Excel users. In an appendix to Chapter 1 we show how to download and install StatTools, and most chapters include a chapter appendix that shows the steps required to implement a statistical procedure using StatTools.

    We have been very careful to make the use of StatTools completely optional so that instructors who want to teach using the standard tools available in Excel 2007 can continue to do so. But users who want additional statistical capabilities not available in standard Excel 2007 now have access to an industry standard statistics add-in that students will be able to continue to use in the workplace.

- **Change in Terminology for Data.** In the previous edition, nominal and ordinal data were classified as qualitative; interval and ratio data were classified as quantitative. In this edition, nominal and ordinal data are referred to as categorical data. Nominal and ordinal data use labels or names to identify categories of like items. Thus, we believe that the term *categorical* is more descriptive of this type of data.

- **Introducing Data Mining.** A new section in Chapter 1 introduces the relatively new field of data mining. We provide a brief overview of data mining and the concept of a data warehouse. We also describe how the fields of statistics and computer science join to make data mining operational and valuable.

- **Ethical Issues in Statistics.** Another new section in Chapter 1 provides a discussion of ethical issues when presenting and interpreting statistical information.

- **Updated Excel Appendix for Tabular and Graphical Descriptive Statistics.** The chapter-ending Excel appendix for Chapter 2 shows how the Chart Tools, PivotTable Report, and PivotChart Report can be used to enhance the capabilities for displaying tabular and graphical descriptive statistics.

- **Comparative Analysis with Box Plots.** The treatment of box plots in Chapter 2 has been expanded to include relatively quick and easy comparisons of two or more data sets. Typical starting salary data for accounting, finance, management, and marketing majors are used to illustrate box plot multigroup comparisons.

- **Revised Sampling Material.** The introduction of Chapter 7 has been revised and now includes the concepts of a sampled population and a frame. The distinction between sampling from a finite population and an infinite population has been clarified, with sampling from a process used to illustrate the selection of a random sample from an infinite population. A practical advice section stresses the importance of obtaining close correspondence between the sampled population and the target population.

AVAILABILITY OF RESOURCES MAY DIFFER BY REGION. Check with your local Cengage Learning representative for details.

- **Revised Introduction to Hypothesis Testing.** Section 9.1, Developing Null and Alternative Hypotheses, has been revised. A better set of guidelines has been developed for identifying the null and alternative hypotheses. The context of the situation and the purpose for taking the sample are key. In situations in which the focus is on finding evidence to support a research finding, the research hypothesis is the alternative hypothesis. In situations where the focus is on challenging an assumption, the assumption is the null hypothesis.
- **New Case Problems.** We have added 5 new case problems to this edition, bringing the total number of case problems to 31. A new case problem on descriptive statistics appears in Chapter 3 and a new case problem on hypothesis testing appears in Chapter 9. Two new case problems have been added to regression in Chapters 12 and 13. These case problems provide students with the opportunity to analyze larger data sets and prepare managerial reports based on the results of the analysis.
- **New Statistics in Practice Application.** Each chapter begins with a Statistics in Practice vignette that describes an application of the statistical methodology to be covered in the chapter. New to this edition is the Statistics in Practice article for Oceanwide Seafood in Chapter 4.
- **New Examples and Exercises Based on Real Data.** We continue to make a significant effort to update our text examples and exercises with the most current real data and referenced sources of statistical information. In this edition, we have added approximately 140 new examples and exercises based on real data and referenced sources. Using data from sources also used by *The Wall Street Journal*, *USA Today*, *Barron's*, and others, we have drawn from actual studies to develop explanations and to create exercises that demonstrate the many uses of statistics in business and economics. We believe that the use of real data helps generate more student interest in the material and enables the student to learn about both the statistical methodology and its application. The sixth edition of the text contains over 300 examples and exercises based on real data.

AVAILABILITY OF RESOURCES MAY DIFFER BY REGION. Check with your local Cengage Learning representative for details.

Features and Pedagogy

Authors Sweeney, Williams, and Anderson have continued many of the features that appeared in previous editions. Important ones for students are noted here.

Methods Exercises and Applications Exercises

The end-of-section exercises are split into two parts, Methods and Applications. The Methods exercises require students to use the formulas and make the necessary computations. The Applications exercises require students to use the chapter material in real-world situations. Thus, students first focus on the computational "nuts and bolts" and then move on to the subtleties of statistical application and interpretation.

Self-Test Exercises

Certain exercises are identified as "Self-Test Exercises." Completely worked-out solutions for these exercises are provided in Appendix D at the back of the book. Students can attempt the Self-Test Exercises and immediately check the solution to evaluate their understanding of the concepts presented in the chapter.

Margin Annotations and Notes and Comments

Margin annotations that highlight key points and provide additional insights for the student are a key feature of this text. These annotations are designed to emphasize and enhance understanding of the terms and concepts being presented in the text.

At the end of many sections, we provide Notes and Comments designed to give the student additional insights about the statistical methodology and its application. Notes and Comments include warnings about or limitations of the methodology, recommendations for application, brief descriptions of additional technical considerations, and other matters.

Data Files Accompany the Text

Approximately 250 data files are available on the website that accompanies the text. The data sets are available in both Minitab and Excel formats. File logos are used in the text to identify the data sets that are available on the website. Data sets for all case problems as well as data sets for larger exercises are included.

Innovative Technology

Aplia

Aplia's **online learning solution** makes business statistics relevant and engaging to students with interactive, automatically graded assignments. As students answer each question, they receive instant, detailed feedback—and their grades are automatically recorded in your Aplia gradebook.

Problem Sets

Students stay engaged in their coursework by regularly completing interactive problem sets. Aplia offers original, auto-graded problems—each question providing instant, detailed feedback.

Tutorials

Students prepare themselves to learn course concepts by using interactive tutorials that help them overcome deficiencies in necessary prerequisites. By assigning these tutorials, you no longer need to spend valuable class time reviewing these subjects.

Assessment & Grading

Aplia keeps you informed about student participation, progress, and performance through real-time graphical reports. You can easily download, save, manipulate, print, and import student grades into your current grading program.

Course Management System

You can post announcements, upload course materials, e-mail students, and manage your gradebook in Aplia's easy-to-use course management system. Aplia works independently or in conjunction with other course management systems.

AVAILABILITY OF RESOURCES MAY DIFFER BY REGION. Check with your local Cengage Learning representative for details.

CengageNOW

CengageNOW™ is an **online teaching and learning resource** that gives you **more control in less time** and **delivers better outcomes**—NOW! CengageNOW combines the best of current technology tools—including homework support, assessment and tutorial support, testing, and an eBook—to further your course goals and save you significant preparation and grading time!

CengageNOW is an Online Teaching and Learning Resource

CengageNOW offers all of your teaching and learning resources in one intuitive program organized around the essential activities you perform for class—lecturing, creating assignments, grading, quizzing, and tracking student progress and performance. CengageNOW provides students access to interactive tutorials, videos, animations, games, and other multimedia tools that help them get the most out of your course.

CengageNOW Provides More Control in Less Time

CengageNOW's flexible assignment and gradebook options provide you more control while saving you valuable time in planning and managing your course assignments. With CengageNOW, you can automatically grade all assignments, weigh grades, choose points or percentages, and set the number of attempts and due dates per problem to best suit your overall course plan.

CengageNOW Delivers Better Student Outcomes

The CengageNOW personalized study plans consist of a chapter-specific Pre-Test, Study Plan, and Post-Test that utilize valuable text-specific assets to empower students to master concepts, prepare for exams, and be more involved in class. Results to the personalized study plan materials provide immediate and ongoing feedback — *to both you and your students*. Best of all, the CengageNOW personalized study plan is designed to help your students get a better grade in your class.

For more information, visit www.cengage.com/now.

AVAILABILITY OF RESOURCES MAY DIFFER BY REGION. Check with your local Cengage Learning representative for details.

Supplements for Students

Student Access Card (Premium Online Content)

A useful access code, *provided at no extra cost with each new text*, provides access to data files to help students master key statistical software for success in today's classroom and tomorrow's business world. Students will have access to all of the templates and data sets for Excel® and Minitab® necessary to complete text exercises on the computer. Also posted are the "Microsoft Excel 2007 and Tools for Statistical Analysis" and "Microsoft Excel 2010 and Tools for Statistical Analysis" primers, and a link to download a textbook version of StatTools.

Supplements for Instructors

Test Bank

Efficiently assess your students' knowledge with a comprehensive selection of multiple-choice questions and problems for each chapter. *Available on the website*, www.cengage.com/international.

ExamView® Computerized Test Bank

This easy-to-use, test-creation program contains all of the questions from the Test Bank, making it easy for you to customize tests to your specific class needs, edit or create questions, and store customized exams. This is an ideal tool for online testing. *Available on the website*, www.cengage.com/international.

Solutions Manual

This manual, prepared by the text authors to ensure accuracy, provides the solutions for all problems in the text, For your convenience, this edition now shows the solution steps using the cummulative normal distrubution and offers more details in the explanations about how to compute the p-values. *Available on the website*, www.cengage.com/international.

Solutions for Case Problems

Prepared by the text authors to ensure accuracy, these solutions to the case problems from the text help you easily plan, assign, and efficiently grade case-problem assignments that are critical for student practice. *Available on the website*, www.cengage.com/international.

PowerPoint® Presentation Slides

Bring your lectures to life, clarify difficult concepts, and provide guides for student note-taking and study with these animated presentation slides. Step-by-step graphics that correspond with the text help ensure student attention and understanding. *Available on the website*, www.cengage.com/international.

AVAILABILITY OF RESOURCES MAY DIFFER BY REGION. Check with your local Cengage Learning representative for details.

Acknowledgments

We would like to acknowledge the work of our reviewers who provided comments and suggestions of ways to continue to improve our text. Thanks to:

Ahmad Saranjam
Bridgewater State College

Ahmad Syamil
Arkansas State University

Alan Olinsky
Bryant University

Amanda Felkey
Lake Forest College

Amy Schmidt
Saint Anselm College

Anirudh Ruhil
Ohio University

Asatar Bair
City College of San Francisco

Atul Gupta
Lynchburg College

Bedassa Tadesse
University of Minnesota, Duluth

Bill Swank
George Mason University

Billy L. Carson II
Itawamba Community College

Brad McDonald
Northern Illinois University

Bruce Gouldey
Shenandoah University

Carl Poch
Northern Illinois University

Carlton Scott
University of California, Irvine

Carol Jensen
Upper Iowa University

Carolyn Rochelle
East Tennessee State University

Ceyhun Ozgur
Valparaiso University

Charles Nicholas Gomersall
Luther College

Charles Vawter, Jr. Glendale Community College

Christopher Ball
Quinnipiac University

Chuck Parker
Wayne State College

Constance Lightner
Fayetteville State University

Dale Bails
Christian Brothers University

Dale DeBoer
University of Colorado, Colorado Springs

David Keswick
University of Michigan–Flint

Denise Robson
University of Wisconsin, Oshkosh

Doug Dotterweich
East Tennessee State University

Doug Morris
University of New Hampshire

Dwight Goehring
California State University–Monterey Bay

Edwin Shapiro
University of San Francisco

Elaine Zanutto
University of Pennsylvania

Emmanuelle Vaast
Long Island University

Eric B. Howington
Valdosta State University

Eric Huggins
Fort Lewis College

Gauri Shankar Guha
Arkansas State University

Geetha Vaidyanathan
University of North Carolina–Greensboro

George H. Jones
University of Wisconsin-Rock County

Gordon Stringer
University of Colorado, Colorado Springs

Greg Miller
U.S. Naval Academy

Harvey Singer
George Mason University

Helen Moshkovich
University of Montevallo Stephens' College of Business

Herbert Moskowitz
Purdue University

James Jozefowicz
Indiana University of Pennsylvania

James Perry
Owens State Community College

James Schmidt
University of Nebraska, Lincoln

James Thorson
Southern Connecticut State University

James Wright
Green Mountain College

Jan Stallaert
University of Connecticut

Janet Pol
University of Nebraska, Omaha

Jean Meyer
Xavier University of Louisiana

Jeffrey Bauer
University of Cincinnati, Clermont

Jeffrey Jarrett
University of Rhode Island

Jena Shafai
Bellevue University

Jennifer Kohn
Montclair State University

Jeremy Pittman
Coahoma Community College

Jerzy Kamburowski
The University of Toledo

Jigish Zaveri
Morgan State University

Jim Knudsen
Creighton University

Jim Kuchta
D'Youville College

Jim Zimmer
Chattanooga State Technical Community College

Jodey Lingg
City University

Joe Williams
Itawamba Community College

John Christiansen
Southwestern Oregon Community College

John Davis
University of the Incarnate Word

John Vangor
Fairfield University

Joseph Cavanaugh
Wright State University, Lake Campus

Joseph Williams
Itawamba Community College

Josh Kim
Quinnipiac University

Julie Szendrey
Malone College

Kazim Ruhi
University of Maryland

Ken Mayer
University of Nebraska at Omaha

Kevin Murphy
Oakland University

Kevin Nguyen
Montgomery College

Khosrow Moshirvaziri
California State University, Long Beach

Kiran R. Bhutani
The Catholic University of America

Kyle Vann Scott
Snead State Community College

Larry Corman
Fort Lewis College

Linda Sturges
SUNY Maritime College

Lyle Rupert
Hendrix College

Maggie Williams Flint
Northeast State Community College

Mark Gius
Quinnipiac University

Marvin Gonzalez
College of Charleston

Mary Lynn Engel
Saint Joseph's College of Maine

Maryanne Clifford
Eastern Connecticut State University

Melissa Miller
Meridian Community College

Michael Broida
Miami University of Ohio

Michael Gordinier
Washington University in St. Louis

Michael McKittrick
Santa Fe Community College

Michael Polomsky
Cleveland State University

Michael Sklar
Rutgers University

Mike Racer
University of Memphis

Minghe Sun
University of Texas–San Antonio

Molly Zimmer
University of Evansville

Nancy Brooks
University of Vermont

Omer Benli
California State University, Long Beach

Phuoc Huu Tran
Bellevue University

Phyllis Schumacher
Bryant University

Ranga Ramasesh
Texas Christian University

Robert Cochran
University of Wyoming

Robert Taylor
Mayland Community College

Robert Vokurka
Texas A&M University—Corpus Christi

Ronald Kizior
Loyola University Chicago

Ronnie Watson
Southern Arkansas University

Rosa Lemel
Kean University

Saiid Ganjalizadeh
The Catholic University of America

Scott Callan
Bentley College

Shauna L. Van Dewark
Humphreys College

Sheng-Kai Chang
Wayne State University

Shin-Ping Tucker
University of Wisconsin, Superior

Stephen Grubagh
Bentley University

Steven Eriksen
Babson College

Sue Umashankar
University of Arizona

Sunil Sapra
California State University, Los Angeles

Susan Emens
Kent State University, Trumbull Campus

Susan Sandblom
Scottsdale Community College

Tenpao Lee
Niagara University

Thomas R. Sexton
Stony Brook University

Toni Somers
Wayne State University

Vivek Shah
Texas State University

Wayne Bedford
University of West Alabama

William Pan
University of New Haven

Yongjing Zhang
Midwestern State University

Yuri Yatsenko
Houston Baptist University

We continue to owe a debt to our many colleagues and friends for their helpful comments and suggestions in the development of this and earlier editions of our text. Among them are:

Alan Smith
Robert Morris College

Ali Arshad
College of Santa Fe

Bennie Waller
Francis Marion University

Carlton Scott
University of California–Irvine

Charles Reichert
University of Wisconsin–Superior

Charles Zimmerman
Robert Morris College

Dale DeBoer
University of Colorado–Colorado Springs

Elaine Parks
Laramie County Community College

Gary Nelson
Central Community College–Columbus Campus

Gipsie Ranney
Belmont University

Habtu Braha
Coppin State College

Karen Gutermuth
Virginia Military Institute

Larry Scheuermann
University of Louisiana, Lafayette

Md. Mahbubul Kabir
Lyon College

Nader Ebrahimi
University of New Mexico

Raj Devasagayam
St. Norbert College

Robert Cochran
University of Wyoming

H. Robert Gadd
Southern Adventist University

Stephen Smith
Gordon College

Timothy Bergquist
Northwest Christian College

Wibawa Sutanto
Prairie View A&M University

Yan Yu
University of Cincinnati

Zhiwei Zhu
University of Louisiana at Lafayette

A special thanks goes to our associates from business and industry who supplied the Statistics in Practice features. We recognize them individually by a credit line in each of the articles. Finally, we are also indebted to our senior acquisitions editor, Charles McCormick, Jr.; our developmental editor, Maggie Kubale; our content project manager, Jacquelyn K Featherly; our Project Manager at MPS Content Services, Lynn Lustberg; our marketing manager, Adam Marsh, our media editor, Chris Valentine, and others at Cengage South-Western for their editorial counsel and support during the preparation of this text.

Dennis J. Sweeney
Thomas A. Williams
David R. Anderson

About the Authors

Dennis J. Sweeney. Dennis J. Sweeney is Professor of Quantitative Analysis and Founder of the Center for Productivity Improvement at the University of Cincinnati. Born in Des Moines, Iowa, he earned a B.S.B.A. degree from Drake University and his M.B.A. and D.B.A. degrees from Indiana University, where he was an NDEA Fellow. During 1978–79, Professor Sweeney worked in the management science group at Procter & Gamble; during 1981–82, he was a visiting professor at Duke University. Professor Sweeney served as Head of the Department of Quantitative Analysis and as Associate Dean of the College of Business Administration at the University of Cincinnati.

Professor Sweeney has published more than thirty articles and monographs in the area of management science and statistics. The National Science Foundation, IBM, Procter & Gamble, Federated Department Stores, Kroger, and Cincinnati Gas & Electric have funded his research, which has been published in *Management Science*, *Operations Research*, *Mathematical Programming*, *Decision Sciences*, and other journals.

Professor Sweeney has coauthored ten textbooks in the areas of statistics, management science, linear programming, and production and operations management.

Thomas A. Williams. Thomas A. Williams is Professor of Management Science in the College of Business at Rochester Institute of Technology. Born in Elmira, New York, he earned his B.S. degree at Clarkson University. He did his graduate work at Rensselaer Polytechnic Institute, where he received his M.S. and Ph.D. degrees.

Before joining the College of Business at RIT, Professor Williams served for seven years as a faculty member in the College of Business Administration at the University of Cincinnati, where he developed the undergraduate program in Information Systems and then served as its coordinator. At RIT he was the first chairman of the Decision Sciences Department. He teaches courses in management science and statistics, as well as graduate courses in regression and decision analysis.

Professor Williams is the coauthor of eleven textbooks in the areas of management science, statistics, production and operations management, and mathematics. He has been a consultant for numerous *Fortune* 500 companies and has worked on projects ranging from the use of data analysis to the development of large-scale regression models.

David R. Anderson. David R. Anderson is Professor of Quantitative Analysis in the College of Business Administration at the University of Cincinnati. Born in Grand Forks, North Dakota, he earned his B.S., M.S., and Ph.D. degrees from Purdue University. Professor Anderson has served as Head of the Department of Quantitative Analysis and Operations Management and as Associate Dean of the College of Business Administration. In addition, he was the coordinator of the College's first Executive Program.

At the University of Cincinnati, Professor Anderson has taught introductory statistics for business students as well as graduate-level courses in regression analysis, multivariate analysis, and management science. He has also taught statistical courses at the Department of Labor in Washington, D.C. He has been honored with nominations and awards for excellence in teaching and excellence in service to student organizations.

Professor Anderson has coauthored ten textbooks in the areas of statistics, management science, linear programming, and production and operations management. He is an active consultant in the field of sampling and statistical methods.

CHAPTER 1

Data and Statistics

CONTENTS

STATISTICS IN PRACTICE:
BUSINESSWEEK

1.1 APPLICATIONS IN BUSINESS AND ECONOMICS
Accounting
Finance
Marketing
Production
Economics

1.2 DATA
Elements, Variables, and Observations
Scales of Measurement
Categorical and Quantitative Data
Cross-Sectional and Time Series Data

1.3 DATA SOURCES
Existing Sources
Statistical Studies
Data Acquisition Errors

1.4 DESCRIPTIVE STATISTICS

1.5 STATISTICAL INFERENCE

1.6 COMPUTERS AND STATISTICAL ANALYSIS

1.7 DATA MINING

1.8 ETHICAL GUIDELINES FOR STATISTICAL PRACTICE

STATISTICS *in* PRACTICE

BUSINESSWEEK*
NEW YORK, NEW YORK

With a global circulation of more than 1 million, *BusinessWeek* is the most widely read business magazine in the world. More than 200 dedicated reporters and editors in 26 bureaus worldwide deliver a variety of articles of interest to the business and economic community. Along with feature articles on current topics, the magazine contains regular sections on International Business, Economic Analysis, Information Processing, and Science & Technology. Information in the feature articles and the regular sections helps readers stay abreast of current developments and assess the impact of those developments on business and economic conditions.

Most issues of *BusinessWeek* provide an in-depth report on a topic of current interest. Often, the in-depth reports contain statistical facts and summaries that help the reader understand the business and economic information. For example, the March 17, 2009 issue included a discussion of when the stock market would begin to recover, the May 4, 2009 issue had a special report on how to make pay cuts less painful, and the January 18, 2010 issue contained an article on the permanent temporary workforce. In addition, the weekly *BusinessWeek Investor* provides statistics about the state of the economy, including production indexes, stock prices, mutual funds, and interest rates.

BusinessWeek also uses statistics and statistical information in managing its own business. For example, an annual survey of subscribers helps the company learn about subscriber demographics, reading habits, likely purchases, lifestyles, and so on. *BusinessWeek* managers use statistical summaries from the survey to provide better

BusinessWeek uses statistical facts and summaries in many of its articles.

services to subscribers and advertisers. One recent North American subscriber survey indicated that 90% of *BusinessWeek* subscribers use a personal computer at home and that 64% of *BusinessWeek* subscribers are involved with computer purchases at work. Such statistics alert *BusinessWeek* managers to subscriber interest in articles about new developments in computers. The results of the survey are also made available to potential advertisers. The high percentage of subscribers using personal computers at home and the high percentage of subscribers involved with computer purchases at work would be an incentive for a computer manufacturer to consider advertising in *BusinessWeek*.

In this chapter, we discuss the types of data available for statistical analysis and describe how the data are obtained. We introduce descriptive statistics and statistical inference as ways of converting data into meaningful and easily interpreted statistical information.

---
*The authors are indebted to Charlene Trentham, Research Manager at *BusinessWeek*, for providing this Statistics in Practice.

Frequently, we see the following types of statements in newspapers and magazines:

- The National Association of Realtors reported that the median price paid by first-time home buyers is $165,000 (*The Wall Street Journal,* February 11, 2009).
- The National Collegiate Athletic Association (NCAA) reported that college athletes are earning degrees at record rates. Latest figures show that 79% of all men and women student-athletes graduate (Associated Press, October 15, 2008).
- The average one-way travel time to work is 25.3 minutes (U.S. Census Bureau, March 2009).

- A poll showed that 73% of the individuals surveyed expected the Dow Jones Industrial Average to gain 10% or more during the coming year (*Money Investor's Guide*, February 2010).
- The national average price for regular gasoline reached $4.00 per gallon for the first time in history (Cable News Network website, June 8, 2008).
- The New York Yankees have the highest salaries in major league baseball. The total payroll is $201,449,289 with a median salary of $5,000,000 (*USA Today Salary Data Base*, September 2009).
- The Dow Jones Industrial Average closed at 10,664 (*The Wall Street Journal*, January 12, 2010).

The numerical facts in the preceding statements ($165,000, 79%, 25.3, 73%, $4.00, $201,449,289, $5,000,000, and 10,664) are called statistics. In this usage, the term *statistics* refers to numerical facts such as averages, medians, percents, and index numbers that help us understand a variety of business and economic situations. However, as you will see, the field, or subject, of statistics involves much more than numerical facts. In a broader sense, **statistics** is defined as the art and science of collecting, analyzing, presenting, and interpreting data. Particularly in business and economics, the information provided by collecting, analyzing, presenting, and interpreting data gives managers and decision makers a better understanding of the business and economic environment and thus enables them to make more informed and better decisions. In this text, we emphasize the use of statistics for business and economic decision making.

Chapter 1 begins with some illustrations of the applications of statistics in business and economics. In Section 1.2 we define the term *data* and introduce the concept of a data set. This section also introduces key terms such as *variables* and *observations*, discusses the difference between quantitative and categorical data, and illustrates the uses of cross-sectional and time series data. Section 1.3 discusses how data can be obtained from existing sources or through survey and experimental studies designed to obtain new data. The important role that the Internet now plays in obtaining data is also highlighted. The uses of data in developing descriptive statistics and in making statistical inferences are described in Sections 1.4 and 1.5. The last three sections of Chapter 1 provide the role of the computer in statistical analysis, an introduction to the relative new field of data mining, and a discussion of ethical guidelines for statistical practice. A chapter-ending appendix includes an introduction to the add-in StatTools which can be used to extend the statistical options for users of Microsoft Excel.

1.1 Applications in Business and Economics

In today's global business and economic environment, anyone can access vast amounts of statistical information. The most successful managers and decision makers understand the information and know how to use it effectively. In this section, we provide examples that illustrate some of the uses of statistics in business and economics.

Accounting

Public accounting firms use statistical sampling procedures when conducting audits for their clients. For instance, suppose an accounting firm wants to determine whether the amount of accounts receivable shown on a client's balance sheet fairly represents the actual amount of accounts receivable. Usually the large number of individual accounts receivable makes reviewing and validating every account too time-consuming and expensive. As common practice in such situations, the audit staff selects a subset of the accounts called a sample. After reviewing the accuracy of the sampled accounts, the auditors draw a conclusion as to whether the accounts receivable amount shown on the client's balance sheet is acceptable.

Finance

Financial analysts use a variety of statistical information to guide their investment recommendations. In the case of stocks, the analysts review a variety of financial data including price/earnings ratios and dividend yields. By comparing the information for an individual stock with information about the stock market averages, a financial analyst can begin to draw a conclusion as to whether an individual stock is over- or underpriced. For example, *Barron's* (February 18, 2008) reported that the average dividend yield for the 30 stocks in the Dow Jones Industrial Average was 2.45%. Altria Group showed a dividend yield of 3.05%. In this case, the statistical information on dividend yield indicates a higher dividend yield for Altria Group than the average for the Dow Jones stocks. Therefore, a financial analyst might conclude that Altria Group was underpriced. This and other information about Altria Group would help the analyst make a buy, sell, or hold recommendation for the stock.

Marketing

Electronic scanners at retail checkout counters collect data for a variety of marketing research applications. For example, data suppliers such as ACNielsen and Information Resources, Inc., purchase point-of-sale scanner data from grocery stores, process the data, and then sell statistical summaries of the data to manufacturers. Manufacturers spend hundreds of thousands of dollars per product category to obtain this type of scanner data. Manufacturers also purchase data and statistical summaries on promotional activities such as special pricing and the use of in-store displays. Brand managers can review the scanner statistics and the promotional activity statistics to gain a better understanding of the relationship between promotional activities and sales. Such analyses often prove helpful in establishing future marketing strategies for the various products.

Production

Today's emphasis on quality makes quality control an important application of statistics in production. A variety of statistical quality control charts are used to monitor the output of a production process. In particular, an x-bar chart can be used to monitor the average output. Suppose, for example, that a machine fills containers with 12 ounces of a soft drink. Periodically, a production worker selects a sample of containers and computes the average number of ounces in the sample. This average, or x-bar value, is plotted on an x-bar chart. A plotted value above the chart's upper control limit indicates overfilling, and a plotted value below the chart's lower control limit indicates underfilling. The process is termed "in control" and allowed to continue as long as the plotted x-bar values fall between the chart's upper and lower control limits. Properly interpreted, an x-bar chart can help determine when adjustments are necessary to correct a production process.

Economics

Economists frequently provide forecasts about the future of the economy or some aspect of it. They use a variety of statistical information in making such forecasts. For instance, in forecasting inflation rates, economists use statistical information on such indicators as the Producer Price Index, the unemployment rate, and manufacturing capacity utilization. Often these statistical indicators are entered into computerized forecasting models that predict inflation rates.

Applications of statistics such as those described in this section are an integral part of this text. Such examples provide an overview of the breadth of statistical applications. To supplement these examples, practitioners in the fields of business and economics provided chapter-opening Statistics in Practice articles that introduce the material covered in each chapter. The Statistics in Practice applications show the importance of statistics in a wide variety of business and economic situations.

1.2 Data

Data are the facts and figures collected, analyzed, and summarized for presentation and interpretation. All the data collected in a particular study are referred to as the **data set** for the study. Table 1.1 shows a data set containing information for 25 mutual funds that are part of the *Morningstar Funds 500* for 2008. Morningstar is a company that tracks over 7000 mutual funds and prepares in-depth analyses of 2000 of these. Its recommendations are followed closely by financial analysts and individual investors.

Elements, Variables, and Observations

Elements are the entities on which data are collected. For the data set in Table 1.1, each individual mutual fund is an element: The element names appear in the first column. With 25 mutual funds, the data set contains 25 elements.

A **variable** is a characteristic of interest for the elements. The data set in Table 1.1 includes the following five variables:

- *Fund Type:* The type of mutual fund, labeled DE (Domestic Equity), IE (International Equity), and FI (Fixed Income)
- *Net Asset Value ($):* The closing price per share on December 31, 2007

TABLE 1.1 DATA SET FOR 25 MUTUAL FUNDS

Fund Name	Fund Type	Net Asset Value ($)	5-Year Average Return (%)	Expense Ratio (%)	Morningstar Rank
American Century Intl. Disc	IE	14.37	30.53	1.41	3-Star
American Century Tax-Free Bond	FI	10.73	3.34	0.49	4-Star
American Century Ultra	DE	24.94	10.88	0.99	3-Star
Artisan Small Cap	DE	16.92	15.67	1.18	3-Star
Brown Cap Small	DE	35.73	15.85	1.20	4-Star
DFA U.S. Micro Cap	DE	13.47	17.23	0.53	3-Star
Fidelity Contrafund	DE	73.11	17.99	0.89	5-Star
Fidelity Overseas	IE	48.39	23.46	0.90	4-Star
Fidelity Sel Electronics	DE	45.60	13.50	0.89	3-Star
Fidelity Sh-Term Bond	FI	8.60	2.76	0.45	3-Star
Gabelli Asset AAA	DE	49.81	16.70	1.36	4-Star
Kalmar Gr Val Sm Cp	DE	15.30	15.31	1.32	3-Star
Marsico 21st Century	DE	17.44	15.16	1.31	5-Star
Mathews Pacific Tiger	IE	27.86	32.70	1.16	3-Star
Oakmark I	DE	40.37	9.51	1.05	2-Star
PIMCO Emerg Mkts Bd D	FI	10.68	13.57	1.25	3-Star
RS Value A	DE	26.27	23.68	1.36	4-Star
T. Rowe Price Latin Am.	IE	53.89	51.10	1.24	4-Star
T. Rowe Price Mid Val	DE	22.46	16.91	0.80	4-Star
Thornburg Value A	DE	37.53	15.46	1.27	4-Star
USAA Income	FI	12.10	4.31	0.62	3-Star
Vanguard Equity-Inc	DE	24.42	13.41	0.29	4-Star
Vanguard Sht-Tm TE	FI	15.68	2.37	0.16	3-Star
Vanguard Sm Cp Idx	DE	32.58	17.01	0.23	3-Star
Wasatch Sm Cp Growth	DE	35.41	13.98	1.19	4-Star

Source: Morningstar Funds 500 (2008).

WEB file
Morningstar

Data sets such as Morningstar are available on the website for this text.

- *5-Year Average Return (%):* The average annual return for the fund over the past 5 years
- *Expense Ratio:* The percentage of assets deducted each fiscal year for fund expenses
- *Morningstar Rank:* The overall risk-adjusted star rating for each fund; Morningstar ranks go from a low of 1-Star to a high of 5-Stars

Measurements collected on each variable for every element in a study provide the data. The set of measurements obtained for a particular element is called an **observation**. Referring to Table 1.1, we see that the set of measurements for the first observation (American Century Intl. Disc) is IE, 14.37, 30.53, 1.41, and 3-Star. The set of measurements for the second observation (American Century Tax-Free Bond) is FI, 10.73, 3.34, 0.49, and 4-Star, and so on. A data set with 25 elements contains 25 observations.

Scales of Measurement

Data collection requires one of the following scales of measurement: nominal, ordinal, interval, or ratio. The scale of measurement determines the amount of information contained in the data and indicates the most appropriate data summarization and statistical analyses.

When the data for a variable consist of labels or names used to identify an attribute of the element, the scale of measurement is considered a **nominal scale**. For example, referring to the data in Table 1.1, we see that the scale of measurement for the Fund Type variable is nominal because DE, IE, and FI are labels used to identify the category or type of fund. In cases where the scale of measurement is nominal, a numerical code as well as nonnumerical labels may be used. For example, to facilitate data collection and to prepare the data for entry into a computer database, we might use a numerical code by letting 1 denote Domestic Equity, 2 denote International Equity, and 3 denote Fixed Income. In this case the numerical values 1, 2, and 3 identify the category of fund. The scale of measurement is nominal even though the data appear as numerical values.

The scale of measurement for a variable is called an **ordinal scale** if the data exhibit the properties of nominal data and the order or rank of the data is meaningful. For example, Eastside Automotive sends customers a questionnaire designed to obtain data on the quality of its automotive repair service. Each customer provides a repair service rating of excellent, good, or poor. Because the data obtained are the labels—excellent, good, or poor—the data have the properties of nominal data. In addition, the data can be ranked, or ordered, with respect to the service quality. Data recorded as excellent indicate the best service, followed by good and then poor. Thus, the scale of measurement is ordinal. As another example, note that the Morningstar Rank for the data in Table 1.1 is ordinal data. It provides a rank from 1 to 5-Stars based on Morningstar's assessment of the fund's risk-adjusted return. Ordinal data can also be provided using a numerical code, for example, your class rank in school.

The scale of measurement for a variable is an **interval scale** if the data have all the properties of ordinal data and the interval between values is expressed in terms of a fixed unit of measure. Interval data are always numerical. College admission SAT scores are an example of interval-scaled data. For example, three students with SAT math scores of 620, 550, and 470 can be ranked or ordered in terms of best performance to poorest performance in math. In addition, the differences between the scores are meaningful. For instance, student 1 scored $620 - 550 = 70$ points more than student 2, while student 2 scored $550 - 470 = 80$ points more than student 3.

The scale of measurement for a variable is a **ratio scale** if the data have all the properties of interval data and the ratio of two values is meaningful. Variables such as distance, height, weight, and time use the ratio scale of measurement. This scale requires that a zero value be included to indicate that nothing exists for the variable at the zero point.

For example, consider the cost of an automobile. A zero value for the cost would indicate that the automobile has no cost and is free. In addition, if we compare the cost of $30,000 for one automobile to the cost of $15,000 for a second automobile, the ratio property shows that the first automobile is $30,000/$15,000 = 2 times, or twice, the cost of the second automobile.

Categorical and Quantitative Data

Data can be classified as either categorical or quantitative. Data that can be grouped by specific categories are referred to as **categorical data**. Categorical data use either the nominal or ordinal scale of measurement. Data that use numerical values to indicate how much or how many are referred to as **quantitative data**. Quantitative data are obtained using either the interval or ratio scale of measurement.

The statistical method appropriate for summarizing data depends upon whether the data are categorical or quantitative.

A **categorical variable** is a variable with categorical data, and a **quantitative variable** is a variable with quantitative data. The statistical analysis appropriate for a particular variable depends upon whether the variable is categorical or quantitative. If the variable is categorical, the statistical analysis is limited. We can summarize categorical data by counting the number of observations in each category or by computing the proportion of the observations in each category. However, even when the categorical data are identified by a numerical code, arithmetic operations such as addition, subtraction, multiplication, and division do not provide meaningful results. Section 2.1 discusses ways for summarizing categorical data.

Arithmetic operations provide meaningful results for quantitative variables. For example, quantitative data may be added and then divided by the number of observations to compute the average value. This average is usually meaningful and easily interpreted. In general, more alternatives for statistical analysis are possible when data are quantitative. Section 2.2 and Chapter 3 provide ways of summarizing quantitative data.

Cross-Sectional and Time Series Data

For purposes of statistical analysis, distinguishing between cross-sectional data and time series data is important. **Cross-sectional data** are data collected at the same or approximately the same point in time. The data in Table 1.1 are cross-sectional because they describe the five variables for the 25 mutual funds at the same point in time. **Time series data** are data collected over several time periods. For example, the time series in Figure 1.1 shows the U.S. average price per gallon of conventional regular gasoline between 2006 and 2009. Note that higher gasoline prices have tended to occur in the summer months, with the all-time-high average of $4.05 per gallon occurring in July 2008. By January 2009, gasoline prices had taken a steep decline to a three-year low of $1.65 per gallon.

Graphs of time series data are frequently found in business and economic publications. Such graphs help analysts understand what happened in the past, identify any trends over time, and project future levels for the time series. The graphs of time series data can take on a variety of forms, as shown in Figure 1.2. With a little study, these graphs are usually easy to understand and interpret.

For example, Panel (A) in Figure 1.2 is a graph that shows the Dow Jones Industrial Average Index from 1997 to 2010. In April 1997, the popular stock market index was near 7000. Over the next 10 years the index rose to over 14,000 in July 2007. However, notice the sharp decline in the time series after the all-time high in 2007. By March 2009, poor economic conditions had caused the Dow Jones Industrial Average Index to return to the 7000 level of 1997. This was a scary and discouraging period for investors. By January 2010, the index was showing a recovery by reaching 10,600.

FIGURE 1.1 U.S. AVERAGE PRICE PER GALLON FOR CONVENTIONAL REGULAR GASOLINE

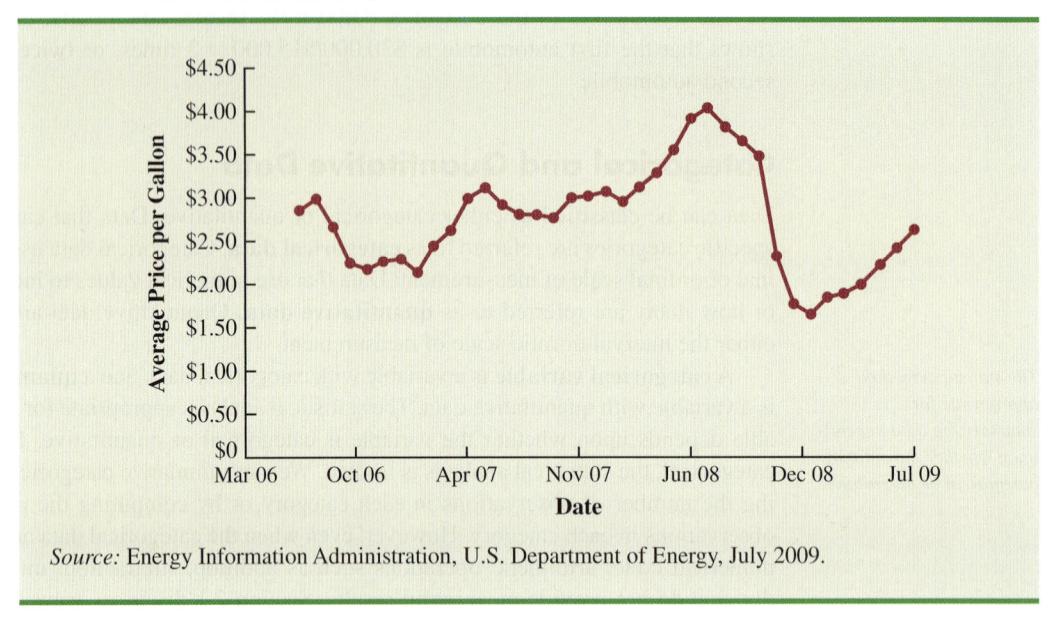

Source: Energy Information Administration, U.S. Department of Energy, July 2009.

The graph in Panel (B) shows the net income of McDonald's Inc. from 2003 to 2009. The declining economic conditions in 2008 and 2009 were actually beneficial to McDonald's as the company's net income rose to an all-time high. The growth in McDonald's net income showed that the company was thriving during the economic downturn as people were cutting back on the more expensive sit-down restaurants and seeking less-expensive alternatives offered by McDonald's.

Panel (C) shows the time series for the occupancy rate of hotels in South Florida over a one-year period. The highest occupancy rates, 95% and 98%, occur during the months of February and March when the climate of South Florida is attractive to tourists. In fact, January to April of each year is typically the high-occupancy season for South Florida hotels. On the other hand, note the low occupancy rates during the months of August to October, with the lowest occupancy rate of 50% occurring in September. High temperatures and the hurricane season are the primary reasons for the drop in hotel occupancy during this period.

NOTES AND COMMENTS

1. An observation is the set of measurements obtained for each element in a data set. Hence, the number of observations is always the same as the number of elements. The number of measurements obtained for each element equals the number of variables. Hence, the total number of data items can be determined by multiplying the number of observations by the number of variables.

2. Quantitative data may be discrete or continuous. Quantitative data that measure how many (e.g., number of telephone calls received in 15 minutes) are discrete. Quantitative data that measure how much (e.g., weight or time) are continuous because no separation occurs between the possible data values.

FIGURE 1.2 A VARIETY OF GRAPHS OF TIME SERIES DATA

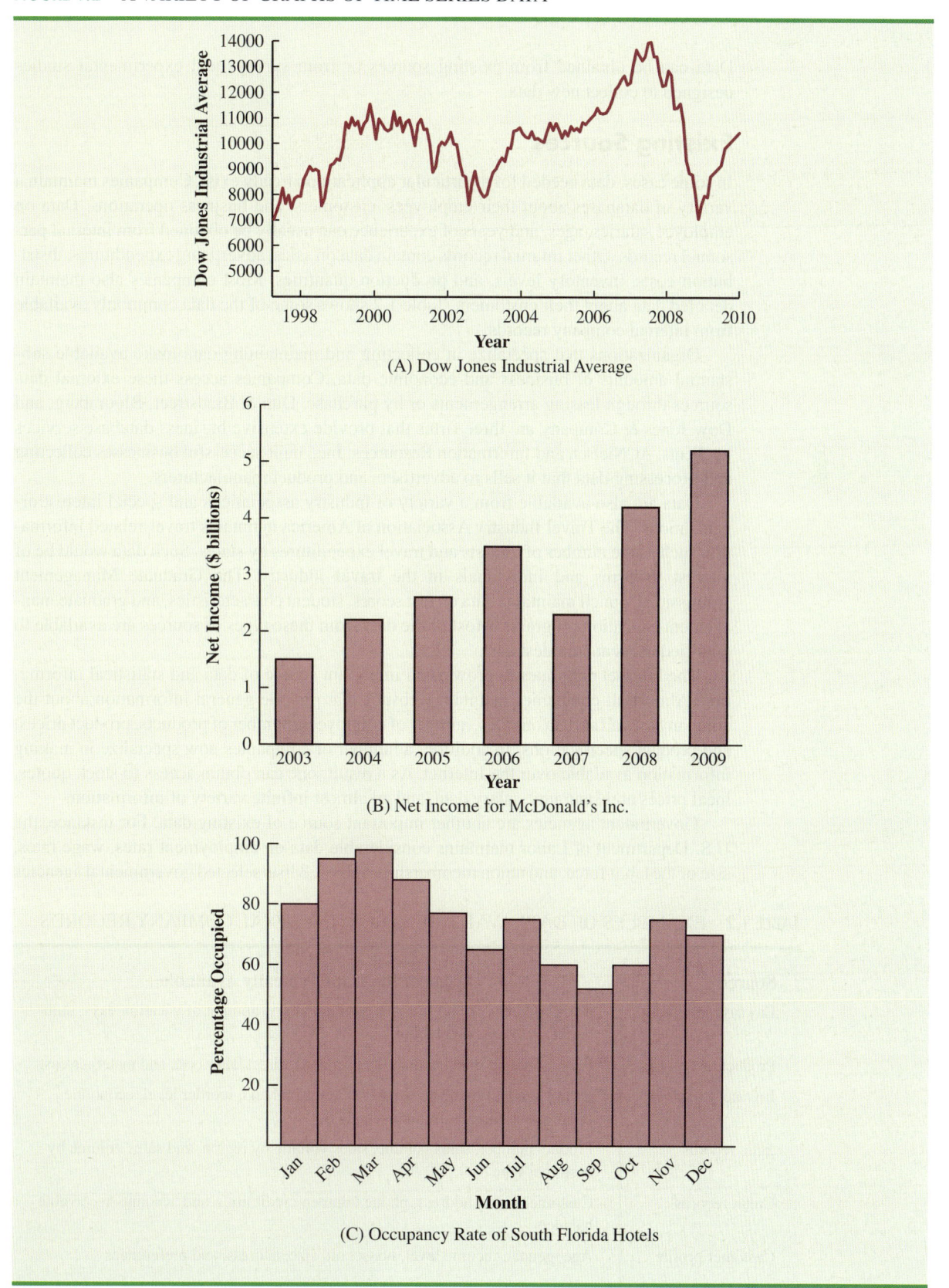

(A) Dow Jones Industrial Average

(B) Net Income for McDonald's Inc.

(C) Occupancy Rate of South Florida Hotels

1.3 Data Sources

Data can be obtained from existing sources or from surveys and experimental studies designed to collect new data.

Existing Sources

In some cases, data needed for a particular application already exist. Companies maintain a variety of databases about their employees, customers, and business operations. Data on employee salaries, ages, and years of experience can usually be obtained from internal personnel records. Other internal records contain data on sales, advertising expenditures, distribution costs, inventory levels, and production quantities. Most companies also maintain detailed data about their customers. Table 1.2 shows some of the data commonly available from internal company records.

Organizations that specialize in collecting and maintaining data make available substantial amounts of business and economic data. Companies access these external data sources through leasing arrangements or by purchase. Dun & Bradstreet, Bloomberg, and Dow Jones & Company are three firms that provide extensive business database services to clients. ACNielsen and Information Resources, Inc., built successful businesses collecting and processing data that it sells to advertisers and product manufacturers.

Data are also available from a variety of industry associations and special interest organizations. The Travel Industry Association of America maintains travel-related information such as the number of tourists and travel expenditures by states. Such data would be of interest to firms and individuals in the travel industry. The Graduate Management Admission Council maintains data on test scores, student characteristics, and graduate management education programs. Most of the data from these types of sources are available to qualified users at a modest cost.

The Internet continues to grow as an important source of data and statistical information. Almost all companies maintain websites that provide general information about the company as well as data on sales, number of employees, number of products, product prices, and product specifications. In addition, a number of companies now specialize in making information available over the Internet. As a result, one can obtain access to stock quotes, meal prices at restaurants, salary data, and an almost infinite variety of information.

Government agencies are another important source of existing data. For instance, the U.S. Department of Labor maintains considerable data on employment rates, wage rates, size of the labor force, and union membership. Table 1.3 lists selected governmental agencies

TABLE 1.2 EXAMPLES OF DATA AVAILABLE FROM INTERNAL COMPANY RECORDS

Source	Some of the Data Typically Available
Employee records	Name, address, social security number, salary, number of vacation days, number of sick days, and bonus
Production records	Part or product number, quantity produced, direct labor cost, and materials cost
Inventory records	Part or product number, number of units on hand, reorder level, economic order quantity, and discount schedule
Sales records	Product number, sales volume, sales volume by region, and sales volume by customer type
Credit records	Customer name, address, phone number, credit limit, and accounts receivable balance
Customer profile	Age, gender, income level, household size, address, and preferences

1.3 Data Sources

TABLE 1.3 EXAMPLES OF DATA AVAILABLE FROM SELECTED GOVERNMENT AGENCIES

Government Agency	Some of the Data Available
Census Bureau	Population data, number of households, and household income
Federal Reserve Board	Data on the money supply, installment credit, exchange rates, and discount rates
Office of Management and Budget	Data on revenue, expenditures, and debt of the federal government
Department of Commerce	Data on business activity, value of shipments by industry, level of profits by industry, and growing and declining industries
Bureau of Labor Statistics	Consumer spending, hourly earnings, unemployment rate, safety records, and international statistics

and some of the data they provide. Most government agencies that collect and process data also make the results available through a website. Figure 1.3 shows the homepage for the U.S. Census Bureau website.

Statistical Studies

Sometimes the data needed for a particular application are not available through existing sources. In such cases, the data can often be obtained by conducting a statistical study. Statistical studies can be classified as either *experimental* or *observational*.

In an experimental study, a variable of interest is first identified. Then one or more other variables are identified and controlled so that data can be obtained about how they influence the variable of interest. For example, a pharmaceutical firm might be interested in conducting an experiment to learn about how a new drug affects blood pressure. Blood pressure is the variable of interest in the study. The dosage level of the new drug is another variable that is hoped to have a causal effect on blood pressure. To obtain data about the effect of the

The largest experimental statistical study ever conducted is believed to be the 1954 Public Health Service experiment for the Salk polio vaccine. Nearly 2 million children in grades 1, 2, and 3 were selected from throughout the United States.

FIGURE 1.3 U.S. CENSUS BUREAU HOMEPAGE

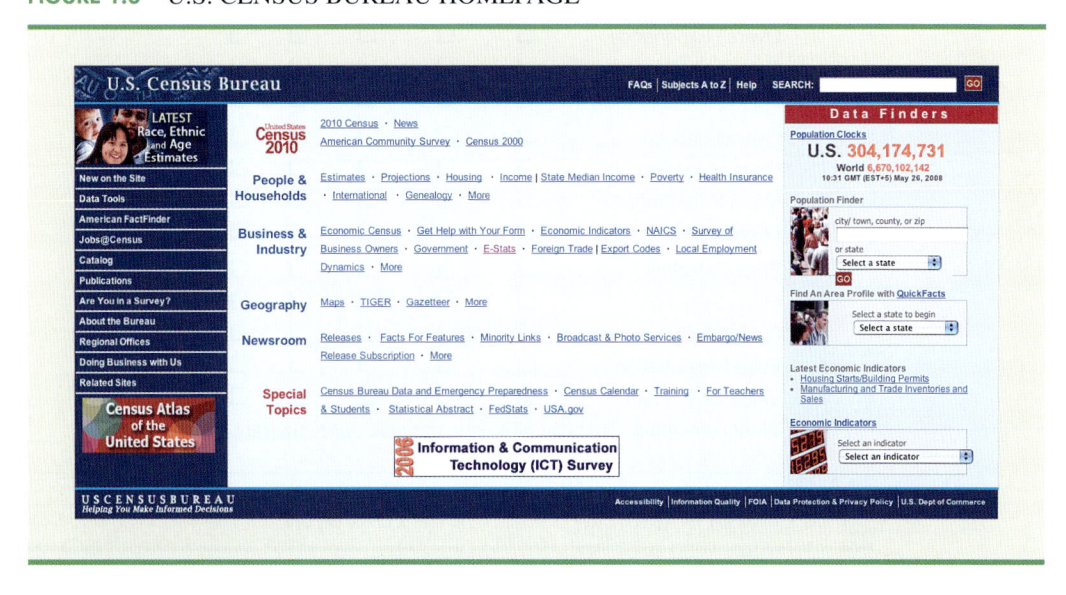

new drug, researchers select a sample of individuals. The dosage level of the new drug is controlled, as different groups of individuals are given different dosage levels. Before and after data on blood pressure are collected for each group. Statistical analysis of the experimental data can help determine how the new drug affects blood pressure.

Nonexperimental, or observational, statistical studies make no attempt to control the variables of interest. A survey is perhaps the most common type of observational study. For instance, in a personal interview survey, research questions are first identified. Then a questionnaire is designed and administered to a sample of individuals. Some restaurants use observational studies to obtain data about customer opinions on the quality of food, quality of service, atmosphere, and so on. A customer opinion questionnaire used by Chops City Grill in Naples, Florida, is shown in Figure 1.4. Note that the customers who fill out the questionnaire are asked to provide ratings for 12 variables, including overall experience, greeting by hostess, manager (table visit), overall service, and so on. The response categories of excellent, good, average, fair, and poor provide categorical data that enable Chops City Grill management to maintain high standards for the restaurant's food and service.

Anyone wanting to use data and statistical analysis as aids to decision making must be aware of the time and cost required to obtain the data. The use of existing data sources is desirable when data must be obtained in a relatively short period of time. If important data are not readily available from an existing source, the additional time and cost involved in obtaining the data must be taken into account. In all cases, the decision maker should

Studies of smokers and nonsmokers are observational studies because researchers do not determine or control who will smoke and who will not smoke.

FIGURE 1.4 CUSTOMER OPINION QUESTIONNAIRE USED BY CHOPS CITY GRILL RESTAURANT IN NAPLES, FLORIDA

Chop/s
CITY GRILL

Date: _____   Server Name: _____

*O*ur customers are our top priority. Please take a moment to fill out our survey card, so we can better serve your needs. You may return this card to the front desk or return by mail. Thank you!

SERVICE SURVEY	Excellent	Good	Average	Fair	Poor
Overall Experience	❑	❑	❑	❑	❑
Greeting by Hostess	❑	❑	❑	❑	❑
Manager (Table Visit)	❑	❑	❑	❑	❑
Overall Service	❑	❑	❑	❑	❑
Professionalism	❑	❑	❑	❑	❑
Menu Knowledge	❑	❑	❑	❑	❑
Friendliness	❑	❑	❑	❑	❑
Wine Selection	❑	❑	❑	❑	❑
Menu Selection	❑	❑	❑	❑	❑
Food Quality	❑	❑	❑	❑	❑
Food Presentation	❑	❑	❑	❑	❑
Value for $ Spent	❑	❑	❑	❑	❑

What comments could you give us to improve our restaurant?
_____
_____

Thank you, we appreciate your comments. —The staff of Chops City Grill.

Data Acquisition Errors

Managers should always be aware of the possibility of data errors in statistical studies. Using erroneous data can be worse than not using any data at all. An error in data acquisition occurs whenever the data value obtained is not equal to the true or actual value that would be obtained with a correct procedure. Such errors can occur in a number of ways. For example, an interviewer might make a recording error, such as a transposition in writing the age of a 24-year-old person as 42, or the person answering an interview question might misinterpret the question and provide an incorrect response.

Experienced data analysts take great care in collecting and recording data to ensure that errors are not made. Special procedures can be used to check for internal consistency of the data. For instance, such procedures would indicate that the analyst should review the accuracy of data for a respondent shown to be 22 years of age but reporting 20 years of work experience. Data analysts also review data with unusually large and small values, called outliers, which are candidates for possible data errors. In Chapter 3 we present some of the methods statisticians use to identify outliers.

Errors often occur during data acquisition. Blindly using any data that happen to be available or using data that were acquired with little care can result in misleading information and bad decisions. Thus, taking steps to acquire accurate data can help ensure reliable and valuable decision-making information.

1.4 Descriptive Statistics

Most of the statistical information in newspapers, magazines, company reports, and other publications consists of data that are summarized and presented in a form that is easy for the reader to understand. Such summaries of data, which may be tabular, graphical, or numerical, are referred to as **descriptive statistics**.

Refer again to the data set in Table 1.1 showing data on 25 mutual funds. Methods of descriptive statistics can be used to provide summaries of the information in this data set. For example, a tabular summary of the data for the categorical variable Fund Type is shown in Table 1.4. A graphical summary of the same data, called a bar chart, is shown in Figure 1.5. These types of tabular and graphical summaries generally make the data easier to interpret. Referring to Table 1.4 and Figure 1.5, we can see easily that the majority of the mutual funds are of the Domestic Equity type. On a percentage basis, 64% are of the Domestic Equity type, 16% are of the International Equity type, and 20% are of the Fixed Income type.

TABLE 1.4 FREQUENCIES AND PERCENT FREQUENCIES FOR MUTUAL FUND TYPE

Mutual Fund Type	Frequency	Percent Frequency
Domestic Equity	16	64
International Equity	4	16
Fixed Income	5	20
Totals	25	100

FIGURE 1.5 BAR CHART FOR MUTUAL FUND TYPE

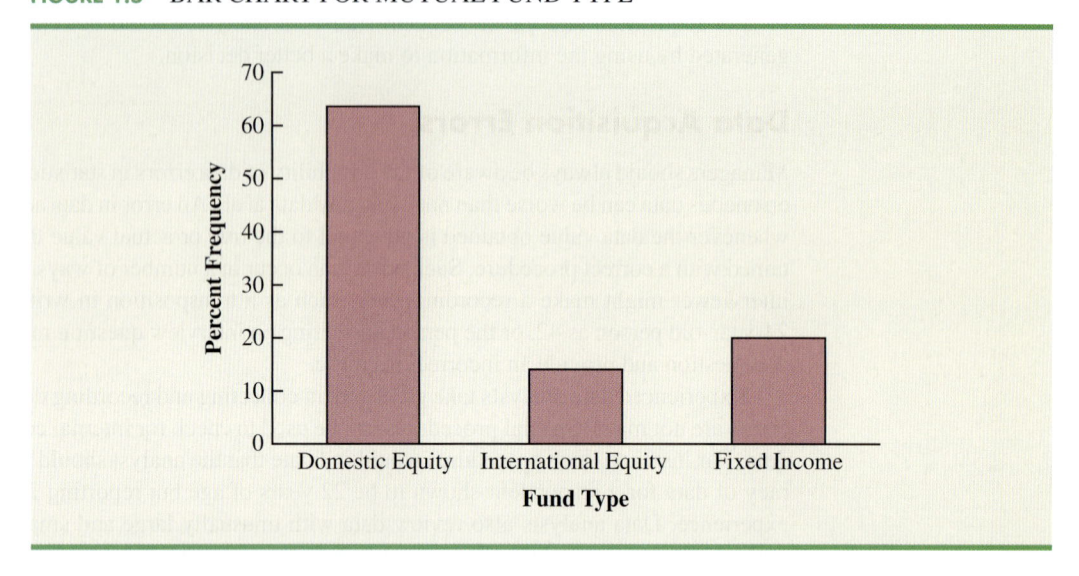

A graphical summary of the data for the quantitative variable Net Asset Value, called a histogram, is provided in Figure 1.6. The histogram makes it easy to see that the net asset values range from $0 to $75, with the highest concentration between $15 and $30. Only one of the net asset values is greater than $60.

In addition to tabular and graphical displays, numerical descriptive statistics are used to summarize data. The most common numerical descriptive statistic is the average, or

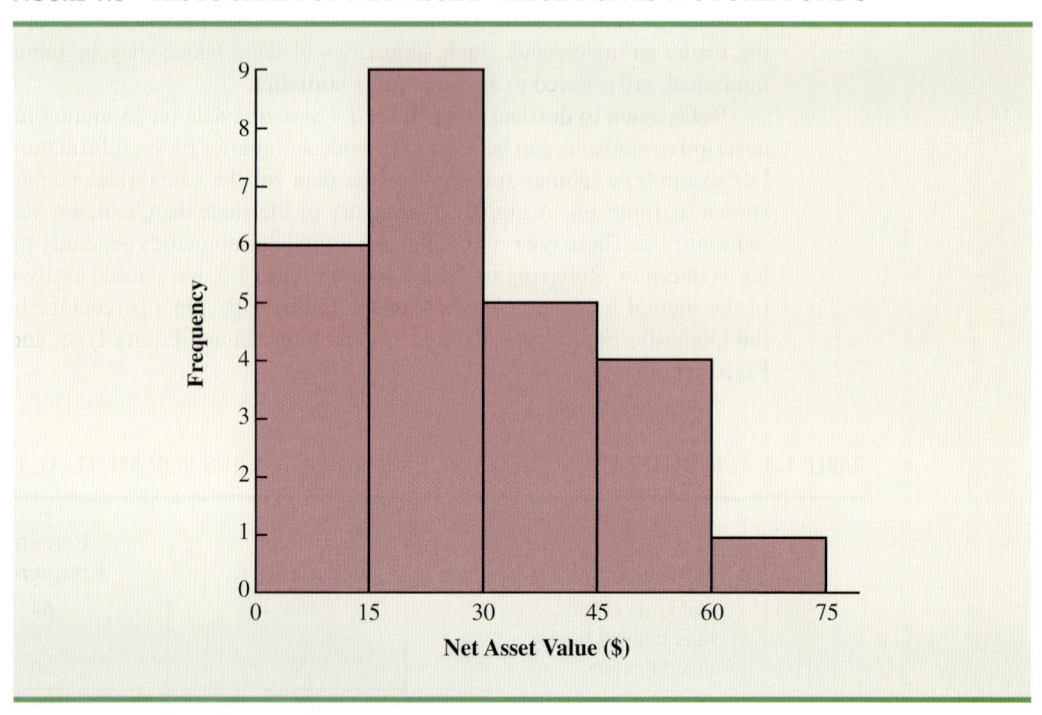

FIGURE 1.6 HISTOGRAM OF NET ASSET VALUE FOR 25 MUTUAL FUNDS

mean. Using the data on 5-Year Average Return for the mutual funds in Table 1.1, we can compute the average by adding the returns for all 25 mutual funds and dividing the sum by 25. Doing so provides a 5-year average return of 16.50%. This average demonstrates a measure of the central tendency, or central location, of the data for that variable.

There is a great deal of interest in effective methods for developing and presenting descriptive statistics. Chapters 2 and 3 devote attention to the tabular, graphical, and numerical methods of descriptive statistics.

1.5 Statistical Inference

Many situations require information about a large group of elements (individuals, companies, voters, households, products, customers, and so on). But, because of time, cost, and other considerations, data can be collected from only a small portion of the group. The larger group of elements in a particular study is called the **population**, and the smaller group is called the **sample**. Formally, we use the following definitions.

> **POPULATION**
>
> A population is the set of all elements of interest in a particular study.

> **SAMPLE**
>
> A sample is a subset of the population.

The U.S. government conducts a census every 10 years. Market research firms conduct sample surveys every day.

The process of conducting a survey to collect data for the entire population is called a **census**. The process of conducting a survey to collect data for a sample is called a **sample survey**. As one of its major contributions, statistics uses data from a sample to make estimates and test hypotheses about the characteristics of a population through a process referred to as **statistical inference**.

As an example of statistical inference, let us consider the study conducted by Norris Electronics. Norris manufactures a high-intensity lightbulb used in a variety of electrical products. In an attempt to increase the useful life of the lightbulb, the product design group developed a new lightbulb filament. In this case, the population is defined as all lightbulbs that could be produced with the new filament. To evaluate the advantages of the new filament, 200 bulbs with the new filament were manufactured and tested. Data collected from this sample showed the number of hours each lightbulb operated before filament burnout. See Table 1.5.

Suppose Norris wants to use the sample data to make an inference about the average hours of useful life for the population of all lightbulbs that could be produced with the new filament. Adding the 200 values in Table 1.5 and dividing the total by 200 provides the sample average lifetime for the lightbulbs: 76 hours. We can use this sample result to estimate that the average lifetime for the lightbulbs in the population is 76 hours. Figure 1.7 provides a graphical summary of the statistical inference process for Norris Electronics.

Whenever statisticians use a sample to estimate a population characteristic of interest, they usually provide a statement of the quality, or precision, associated with the estimate.

TABLE 1.5 HOURS UNTIL BURNOUT FOR A SAMPLE OF 200 LIGHTBULBS FOR THE NORRIS ELECTRONICS EXAMPLE

107	73	68	97	76	79	94	59	98	57
54	65	71	70	84	88	62	61	79	98
66	62	79	86	68	74	61	82	65	98
62	116	65	88	64	79	78	79	77	86
74	85	73	80	68	78	89	72	58	69
92	78	88	77	103	88	63	68	88	81
75	90	62	89	71	71	74	70	74	70
65	81	75	62	94	71	85	84	83	63
81	62	79	83	93	61	65	62	92	65
83	70	70	81	77	72	84	67	59	58
78	66	66	94	77	63	66	75	68	76
90	78	71	101	78	43	59	67	61	71
96	75	64	76	72	77	74	65	82	86
66	86	96	89	81	71	85	99	59	92
68	72	77	60	87	84	75	77	51	45
85	67	87	80	84	93	69	76	89	75
83	68	72	67	92	89	82	96	77	102
74	91	76	83	66	68	61	73	72	76
73	77	79	94	63	59	62	71	81	65
73	63	63	89	82	64	85	92	64	73

FIGURE 1.7 THE PROCESS OF STATISTICAL INFERENCE FOR THE NORRIS ELECTRONICS EXAMPLE

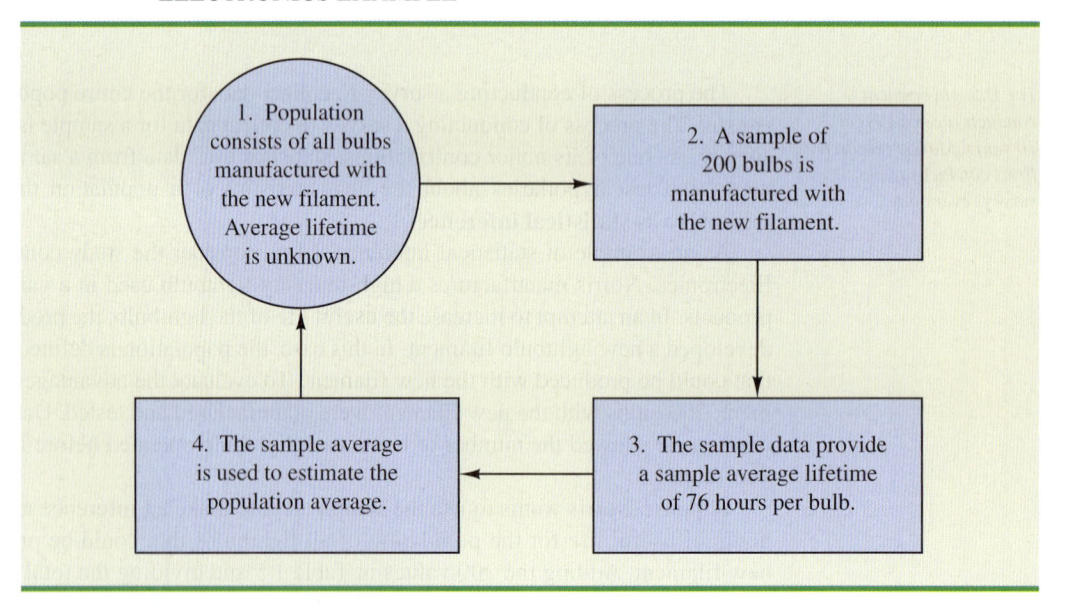

1. Population consists of all bulbs manufactured with the new filament. Average lifetime is unknown.

2. A sample of 200 bulbs is manufactured with the new filament.

3. The sample data provide a sample average lifetime of 76 hours per bulb.

4. The sample average is used to estimate the population average.

For the Norris example, the statistician might state that the point estimate of the average lifetime for the population of new lightbulbs is 76 hours with a margin of error of ±4 hours. Thus, an interval estimate of the average lifetime for all lightbulbs produced with the new filament is 72 hours to 80 hours. The statistician can also state how confident he or she is that the interval from 72 hours to 80 hours contains the population average.

1.6 Computers and Statistical Analysis

Statisticians frequently use computer software to perform the statistical computations required with large amounts of data. For example, computing the average lifetime for the 200 lightbulbs in the Norris Electronics example (see Table 1.5) would be quite tedious without a computer. To facilitate computer usage, many of the data sets in this book are available on the website that accompanies the text. The data files may be downloaded in either Minitab or Excel formats. In addition, the Excel add-in StatTools can be downloaded from the website. End-of-chapter appendixes cover the step-by-step procedures for using Minitab, Excel, and the Excel add-in StatTools to implement the statistical techniques presented in the chapter.

Minitab and Excel data sets and the Excel add-in StatTools are available on the website for this text.

1.7 Data Mining

With the aid of magnetic card readers, bar code scanners, and point-of-sale terminals, most organizations obtain large amounts of data on a daily basis. And, even for a small local restaurant that uses touch screen monitors to enter orders and handle billing, the amount of data collected can be significant. For large retail companies, the sheer volume of data collected is hard to conceptualize, and figuring out how to effectively use these data to improve profitability is a challenge. For example, mass retailers such as Walmart capture data on 20 to 30 million transactions every day, telecommunication companies such as France Telecom and AT&T generate over 300 million call records per day, and Visa processes 6800 payment transactions per second or approximately 600 million transactions per day. Storing and managing the transaction data is a significant undertaking.

The term *data warehousing* is used to refer to the process of capturing, storing, and maintaining the data. Computing power and data collection tools have reached the point where it is now feasible to store and retrieve extremely large quantities of data in seconds. Analysis of the data in the warehouse may result in decisions that will lead to new strategies and higher profits for the organization.

The subject of **data mining** deals with methods for developing useful decision-making information from large data bases. Using a combination of procedures from statistics, mathematics, and computer science, analysts "mine the data" in the warehouse to convert it into useful information, hence the name *data mining*. Dr. Kurt Thearling, a leading practitioner in the field, defines data mining as "the automated extraction of predictive information from large databases." The two key words in Dr. Thearling's definition are "automated" and "predictive." Data mining systems that are the most effective use automated procedures to extract information from the data using only the most general or even vague queries by the user. And data mining software automates the process of uncovering hidden predictive information that in the past required hands-on analysis.

The major applications of data mining have been made by companies with a strong consumer focus, such as retail businesses, financial organizations, and communication companies. Data mining has been successfully used to help retailers such as Amazon and Barnes & Noble determine one or more related products that customers who have already purchased a specific product are also likely to purchase. Then, when a customer logs on to the company's website and purchases a product, the website uses pop-ups to alert the customer about additional products that the customer is likely to purchase. In another application, data mining may be used to identify customers who are likely to spend more than $20 on a particular shopping trip. These customers may then be identified as the ones to receive special e-mail or regular mail discount offers to encourage them to make their next shopping trip before the discount termination date.

Data mining is a technology that relies heavily on statistical methodology such as multiple regression, logistic regression, and correlation. But it takes a creative integration of all

Statistical methods play an important role in data mining, both in terms of discovering relationships in the data and predicting future outcomes. However, a thorough coverage of data mining and the use of statistics in data mining is outside the scope of this text.

these methods and computer science technologies involving artificial intelligence and machine learning to make data mining effective. A significant investment in time and money is required to implement commercial data mining software packages developed by firms such as Oracle, Teradata, and SAS. The statistical concepts introduced in this text will be helpful in understanding the statistical methodology used by data mining software packages and enable you to better understand the statistical information that is developed.

Because statistical models play an important role in developing predictive models in data mining, many of the concerns that statisticians deal with in developing statistical models are also applicable. For instance, a concern in any statistical study involves the issue of model reliability. Finding a statistical model that works well for a particular sample of data does not necessarily mean that it can be reliably applied to other data. One of the common statistical approaches to evaluating model reliability is to divide the sample data set into two parts: a training data set and a test data set. If the model developed using the training data is able to accurately predict values in the test data, we say that the model is reliable. One advantage that data mining has over classical statistics is that the enormous amount of data available allows the data mining software to partition the data set so that a model developed for the training data set may be tested for reliability on other data. In this sense, the partitioning of the data set allows data mining to develop models and relationships and then quickly observe if they are repeatable and valid with new and different data. On the other hand, a warning for data mining applications is that with so much data available, there is a danger of overfitting the model to the point that misleading associations and cause/effect conclusions appear to exist. Careful interpretation of data mining results and additional testing will help avoid this pitfall.

1.8 Ethical Guidelines for Statistical Practice

Ethical behavior is something we should strive for in all that we do. Ethical issues arise in statistics because of the important role statistics plays in the collection, analysis, presentation, and interpretation of data. In a statistical study, unethical behavior can take a variety of forms including improper sampling, inappropriate analysis of the data, development of misleading graphs, use of inappropriate summary statistics, and/or a biased interpretation of the statistical results.

As you begin to do your own statistical work, we encourage you to be fair, thorough, objective, and neutral as you collect data, conduct analyses, make oral presentations, and present written reports containing information developed. As a consumer of statistics, you should also be aware of the possibility of unethical statistical behavior by others. When you see statistics in newspapers, on television, on the Internet, and so on, it is a good idea to view the information with some skepticism, always being aware of the source as well as the purpose and objectivity of the statistics provided.

The American Statistical Association, the nation's leading professional organization for statistics and statisticians, developed the report "Ethical Guidelines for Statistical Practice"[1] to help statistical practitioners make and communicate ethical decisions and assist students in learning how to perform statistical work responsibly. The report contains 67 guidelines organized into eight topic areas: Professionalism; Responsibilities to Funders, Clients, and Employers; Responsibilities in Publications and Testimony; Responsibilities to Research Subjects; Responsibilities to Research Team Colleagues; Responsibilities to Other Statisticians or Statistical Practitioners; Responsibilities Regarding Allegations of Misconduct; and Responsibilities of Employers Including Organizations, Individuals, Attorneys, or Other Clients Employing Statistical Practitioners.

---

[1] American Statistical Association "Ethical Guidelines for Statistical Practice," 1999.

One of the ethical guidelines in the professionalism area addresses the issue of running multiple tests until a desired result is obtained. Let us consider an example. In Section 1.5 we discussed a statistical study conducted by Norris Electronics involving a sample of 200 high-intensity lightbulbs manufactured with a new filament. The average lifetime for the sample, 76 hours, provided an estimate of the average lifetime for all lightbulbs produced with the new filament. However, consider this. Because Norris selected a sample of bulbs, it is reasonable to assume that another sample would have provided a different average lifetime.

Suppose Norris's management had hoped the sample results would enable it to claim that the average lifetime for the new lightbulbs was 80 hours or more. Suppose further that Norris's management decides to continue the study by manufacturing and testing repeated samples of 200 lightbulbs with the new filament until a sample mean of 80 hours or more is obtained. If the study is repeated enough times, a sample may eventually be obtained—by chance alone—that would provide the desired result and enable Norris to make such a claim. In this case, consumers would be misled into thinking the new product is better than it actually is. Clearly, this type of behavior is unethical and represents a gross misuse of statistics in practice.

Several ethical guidelines in the responsibilities and publications and testimony area deal with issues involving the handling of data. For instance, a statistician must account for all data considered in a study and explain the sample(s) actually used. In the Norris Electronics study the average lifetime for the 200 bulbs in the original sample is 76 hours; this is considerably less than the 80 hours or more that management hoped to obtain. Suppose now that after reviewing the results showing a 76 hour average lifetime, Norris discards all the observations with 70 or fewer hours until burnout, allegedly because these bulbs contain imperfections caused by startup problems in the manufacturing process. After discarding these lightbulbs, the average lifetime for the remaining lightbulbs in the sample turns out to be 82 hours. Would you be suspicious of Norris's claim that the lifetime for its lightbulbs is 82 hours?

If the Norris lightbulbs showing 70 or fewer hours until burnout were discarded to simply provide an average lifetime of 82 hours, there is no question that discarding the lightbulbs with 70 or fewer hours until burnout is unethical. But, even if the discarded lightbulbs contain imperfections due to startup problems in the manufacturing process—and, as a result, should not have been included in the analysis—the statistician who conducted the study must account for all the data that were considered and explain how the sample actually used was obtained. To do otherwise is potentially misleading and would constitute unethical behavior on the part of both the company and the statistician.

A guideline in the shared values section of the American Statistical Association report states that statistical practitioners should avoid any tendency to slant statistical work toward predetermined outcomes. This type of unethical practice is often observed when unrepresentative samples are used to make claims. For instance, in many areas of the country smoking is not permitted in restaurants. Suppose, however, a lobbyist for the tobacco industry interviews people in restaurants where smoking is permitted in order to estimate the percentage of people who are in favor of allowing smoking in restaurants. The sample results show that 90% of the people interviewed are in favor of allowing smoking in restaurants. Based upon these sample results, the lobbyist claims that 90% of all people who eat in restaurants are in favor of permitting smoking in restaurants. In this case we would argue that only sampling persons eating in restaurants that allow smoking has biased the results. If only the final results of such a study are reported, readers unfamiliar with the details of the study (i.e., that the sample was collected only in restaurants allowing smoking) can be misled.

The scope of the American Statistical Association's report is broad and includes ethical guidelines that are appropriate not only for a statistician, but also for consumers of statistical information. We encourage you to read the report to obtain a better perspective of ethical issues as you continue your study of statistics and to gain the background for determining how to ensure that ethical standards are met when you start to use statistics in practice.

Summary

Statistics is the art and science of collecting, analyzing, presenting, and interpreting data. Nearly every college student majoring in business or economics is required to take a course in statistics. We began the chapter by describing typical statistical applications for business and economics.

Data consist of the facts and figures that are collected and analyzed. Four scales of measurement used to obtain data on a particular variable include nominal, ordinal, interval, and ratio. The scale of measurement for a variable is nominal when the data are labels or names used to identify an attribute of an element. The scale is ordinal if the data demonstrate the properties of nominal data and the order or rank of the data is meaningful. The scale is interval if the data demonstrate the properties of ordinal data and the interval between values is expressed in terms of a fixed unit of measure. Finally, the scale of measurement is ratio if the data show all the properties of interval data and the ratio of two values is meaningful.

For purposes of statistical analysis, data can be classified as categorical or quantitative. Categorical data use labels or names to identify an attribute of each element. Categorical data use either the nominal or ordinal scale of measurement and may be nonnumerical or numerical. Quantitative data are numerical values that indicate how much or how many. Quantitative data use either the interval or ratio scale of measurement. Ordinary arithmetic operations are meaningful only if the data are quantitative. Therefore, statistical computations used for quantitative data are not always appropriate for categorical data.

In Sections 1.4 and 1.5 we introduced the topics of descriptive statistics and statistical inference. Descriptive statistics are the tabular, graphical, and numerical methods used to summarize data. The process of statistical inference uses data obtained from a sample to make estimates or test hypotheses about the characteristics of a population. The last three sections of the chapter provide information on the role of computers in statistical analysis, an introduction to the relative new field of data mining, and a summary of ethical guidelines for statistical practice.

Glossary

Statistics The art and science of collecting, analyzing, presenting, and interpreting data.

Data The facts and figures collected, analyzed, and summarized for presentation and interpretation.

Data set All the data collected in a particular study.

Elements The entities on which data are collected.

Variable A characteristic of interest for the elements.

Observation The set of measurements obtained for a particular element.

Nominal scale The scale of measurement for a variable when the data are labels or names used to identify an attribute of an element. Nominal data may be nonnumerical or numerical.

Ordinal scale The scale of measurement for a variable if the data exhibit the properties of nominal data and the order or rank of the data is meaningful. Ordinal data may be nonnumerical or numerical.

Interval scale The scale of measurement for a variable if the data demonstrate the properties of ordinal data and the interval between values is expressed in terms of a fixed unit of measure. Interval data are always numerical.

Ratio scale The scale of measurement for a variable if the data demonstrate all the properties of interval data and the ratio of two values is meaningful. Ratio data are always numerical.

Categorical data Labels or names used to identify an attribute of each element. Categorical data use either the nominal or ordinal scale of measurement and may be nonnumerical or numerical.

Quantitative data Numerical values that indicate how much or how many of something. Quantitative data are obtained using either the interval or ratio scale of measurement.
Categorical variable A variable with categorical data.
Quantitative variable A variable with quantitative data.
Cross-sectional data Data collected at the same or approximately the same point in time.
Time series data Data collected over several time periods.
Descriptive statistics Tabular, graphical, and numerical summaries of data.
Population The set of all elements of interest in a particular study.
Sample A subset of the population.
Census A survey to collect data on the entire population.
Sample survey A survey to collect data on a sample.
Statistical inference The process of using data obtained from a sample to make estimates or test hypotheses about the characteristics of a population.
Data mining The process of using procedures from statistics and computer science to extract useful information from extremely large databases.

Supplementary Exercises

1. *Foreign Affairs* magazine conducted a survey to develop a profile of its subscribers (Foreign Affairs website, February 23, 2008). The following questions were asked.
    a. How many nights have you stayed in a hotel in the past 12 months?
    b. Where do you purchase books? Three options were listed: Bookstore, Internet, and Book Club.
    c. Do you own or lease a luxury vehicle? (Yes or No)
    d. What is your age?
    e. For foreign trips taken in the past three years, what was your destination? Seven international destinations were listed.

    Comment on whether each question provides categorical or quantitative data.

2. The U.S. Department of Energy provides fuel economy information for a variety of motor vehicles. A sample of 10 automobiles is shown in Table 1.6 (Fuel Economy website, February 22, 2008). Data show the size of the automobile (compact, midsize, or large), the number of cylinders in the engine, the city driving miles per gallon, the highway driving miles per gallon, and the recommended fuel (diesel, premium, or regular).
    a. How many elements are in this data set?
    b. How many variables are in this data set?
    c. Which variables are categorical and which variables are quantitative?
    d. What type of measurement scale is used for each of the variables?

TABLE 1.6 FUEL ECONOMY INFORMATION FOR 10 AUTOMOBILES

Car	Size	Cylinders	City MPG	Highway MPG	Fuel
Audi A8	Large	12	13	19	Premium
BMW 328Xi	Compact	6	17	25	Premium
Cadillac CTS	Midsize	6	16	25	Regular
Chrysler 300	Large	8	13	18	Premium
Ford Focus	Compact	4	24	33	Regular
Hyundai Elantra	Midsize	4	25	33	Regular
Jeep Grand Cherokee	Midsize	6	17	26	Diesel
Pontiac G6	Compact	6	15	22	Regular
Toyota Camry	Midsize	4	21	31	Regular
Volkswagen Jetta	Compact	5	21	29	Regular

3. Refer to Table 1.6.
    a. What is the average miles per gallon for city driving?
    b. On average, how much higher is the miles per gallon for highway driving as compared to city driving?
    c. What percentage of the cars have four-cylinder engines?
    d. What percentage of the cars use regular fuel?

4. Consider the data set in Table 1.7.
    a. Compute the average endowment for the sample.
    b. Compute the average percentage of applicants admitted.
    c. What percentage of the schools have NCAA Division III varsity teams?
    d. What percentage of the schools have a City: Midsize campus setting?

5. Table 1.7 shows data for seven colleges and universities. The endowment (in billions of dollars) and the percentage of applicants admitted are shown (*USA Today,* February 3, 2008). The state each school is located in, the campus setting, and the NCAA Division for varsity teams were obtained from the National Center of Education Statistics website, February 22, 2008.
    a. How many elements are in the data set?
    b. How many variables are in the data set?
    c. Which of the variables are categorical and which are quantitative?

6. Discuss the differences between statistics as numerical facts and statistics as a discipline or field of study.

7. *The Wall Street Journal (WSJ)* subscriber survey (October 13, 2003) asked 46 questions about subscriber characteristics and interests. State whether each of the following questions provided categorical or quantitative data and indicate the measurement scale appropriate for each.
    a. What is your age?
    b. Are you male or female?
    c. When did you first start reading the *WSJ*? High school, college, early career, mid-career, late career, or retirement?
    d. How long have you been in your present job or position?
    e. What type of vehicle are you considering for your next purchase? Nine response categories include sedan, sports car, SUV, minivan, and so on.

8. The Commerce Department reported receiving the following applications for the Malcolm Baldrige National Quality Award: 23 from large manufacturing firms, 18 from large service firms, and 30 from small businesses.
    a. Is type of business a categorical or quantitative variable?
    b. What percentage of the applications came from small businesses?

TABLE 1.7 DATA FOR SEVEN COLLEGES AND UNIVERSITIES

School	State	Campus Setting	Endowment ($ billions)	% Applicants Admitted	NCAA Division
Amherst College	Massachusetts	Town: Fringe	1.7	18	III
Duke	North Carolina	City: Midsize	5.9	21	I-A
Harvard University	Massachusetts	City: Midsize	34.6	9	I-AA
Swarthmore College	Pennsylvania	Suburb: Large	1.4	18	III
University of Pennsylvania	Pennsylvania	City: Large	6.6	18	I-AA
Williams College	Massachusetts	Town: Fringe	1.9	18	III
Yale University	Connecticut	City: Midsize	22.5	9	I-AA

9. J. D. Power and Associates conducts vehicle quality surveys to provide automobile manufacturers with consumer satisfaction information about their products (Vehicle Quality Survey, January 2010). Using a sample of vehicle owners from recent vehicle purchase records, the survey asks the owners a variety of questions about their new vehicles such as those that follow. For each question, state whether the data collected are categorical or quantitative and indicate the measurement scale being used.
   a. What price did you pay for the vehicle?
   b. How did you pay for the vehicle? (Cash, Lease, or Finance)
   c. How likely would you be to recommend this vehicle to a friend? (Definitely Not, Probably Not, Probably Will, and Definitely Will)
   d. What is the current mileage?
   e. What is your overall rating of your new vehicle? A 10-point scale ranging from 1 for unacceptable to 10 for truly exceptional was used.

10. The *FinancialTimes*/Harris Poll is a monthly online poll of adults from six countries in Europe and the United States. A January poll included 1015 adults in the United States. One of the questions asked was, "How would you rate the Federal Bank in handling the credit problems in the financial markets?" Possible responses were Excellent, Good, Fair, Bad, and Terrible (Harris Interactive website, January 2008).
    a. What was the sample size for this survey?
    b. Are the data categorical or quantitative?
    c. Would it make more sense to use averages or percentages as a summary of the data for this question?
    d. Of the respondents in the United States, 10% said the Federal Bank is doing a good job. How many individuals provided this response?

11. The Hawaii Visitors Bureau collects data on visitors to Hawaii. The following questions were among 16 asked in a questionnaire handed out to passengers during incoming airline flights in June 2003.
    - This trip to Hawaii is my: 1st, 2nd, 3rd, 4th, and so on.
    - The primary reason for this trip is: (10 categories including vacation, convention, honeymoon)
    - Where I plan to stay: (11 categories including hotel, apartment, relatives, camping)
    - Total days in Hawaii
    a. What is the population being studied?
    b. Is the use of a questionnaire a good way to reach the population of passengers on incoming airline flights?
    c. Comment on each of the four questions in terms of whether it will provide categorical or quantitative data.

12. The Ritz-Carlton Hotel used a customer opinion questionnaire to obtain performance data about its dining and entertainment services (The Ritz-Carlton Hotel, Naples, Florida, February 2006). Customers were asked to rate six factors: Welcome, Service, Food, Menu Appeal, Atmosphere, and Overall Experience. Data were recorded for each factor with 1 for Fair, 2 for Average, 3 for Good, and 4 for Excellent.
    a. The customer responses provided data for six variables. Are the variables categorical or quantitative?
    b. What measurement scale is used?

13. Figure 1.8 provides a bar chart showing the amount of federal spending for the years 2002 to 2008 (*USA Today*, February 5, 2008).
    a. What is the variable of interest?
    b. Are the data categorical or quantitative?
    c. Are the data time series or cross-sectional?
    d. Comment on the trend in federal spending over time.

FIGURE 1.8 FEDERAL SPENDING

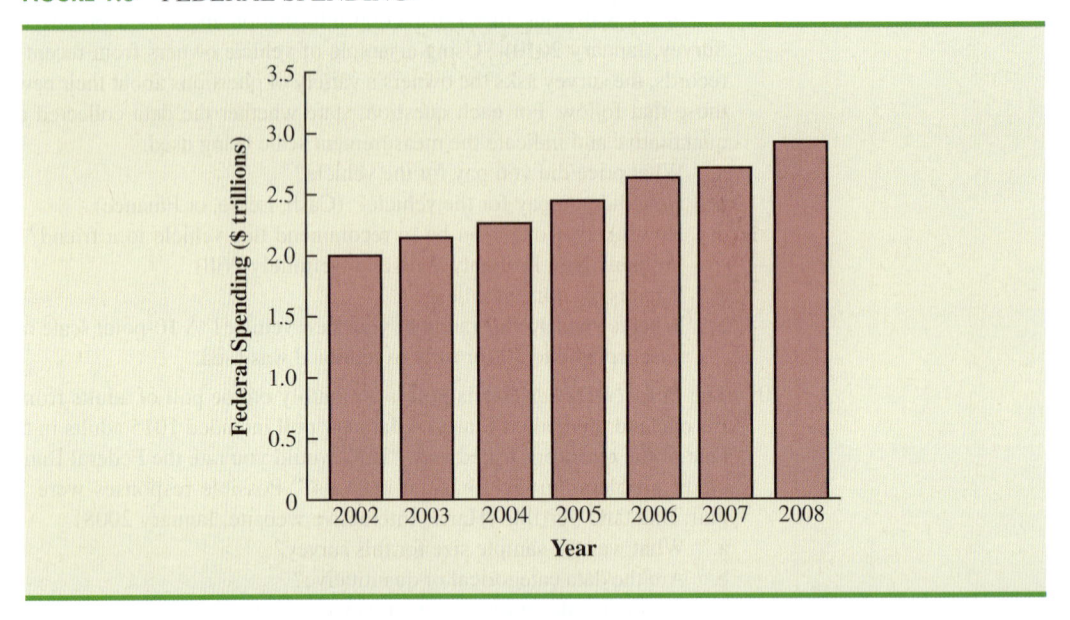

[Bar chart showing Federal Spending ($ trillions) by Year: 2002 ≈ 2.0, 2003 ≈ 2.15, 2004 ≈ 2.3, 2005 ≈ 2.45, 2006 ≈ 2.65, 2007 ≈ 2.75, 2008 ≈ 2.95]

14. The Energy Information Administration of the U.S. Department of Energy provided time series data for the U.S. average price per gallon of conventional regular gasoline between July 2006 and June 2009 (Energy Information Administration website, June 2009). Use the Internet to obtain the average price per gallon of conventional regular gasoline since June 2009.
    a. Extend the graph of the time series shown in Figure 1.1.
    b. What interpretations can you make about the average price per gallon of conventional regular gasoline since June 2009?
    c. Does the time series continue to show a summer increase in the average price per gallon? Explain.

15. CSM Worldwide forecasts global production for all automobile manufacturers. The following CSM data show the forecast of global auto production for General Motors, Ford, DaimlerChrysler, and Toyota for the years 2004 to 2007 (*USA Today*, December 21, 2007). Data are in millions of vehicles.

Manufacturer	2004	2005	2006	2007
General Motors	8.9	9.0	8.9	8.8
Ford	7.8	7.7	7.8	7.9
DaimlerChrysler	4.1	4.2	4.3	4.6
Toyota	7.8	8.3	9.1	9.6

a. Construct a time series graph for the years 2004 to 2007 showing the number of vehicles manufactured by each automotive company. Show the time series for all four manufacturers on the same graph.
b. General Motors has been the undisputed production leader of automobiles since 1931. What does the time series graph show about who is the world's biggest car company? Discuss.
c. Construct a bar graph showing vehicles produced by automobile manufacturer using the 2007 data. Is this graph based on cross-sectional or time series data?

FIGURE 1.9 NUMBER OF NEW DRUGS APPROVED BY THE FOOD AND DRUG ADMINISTRATION

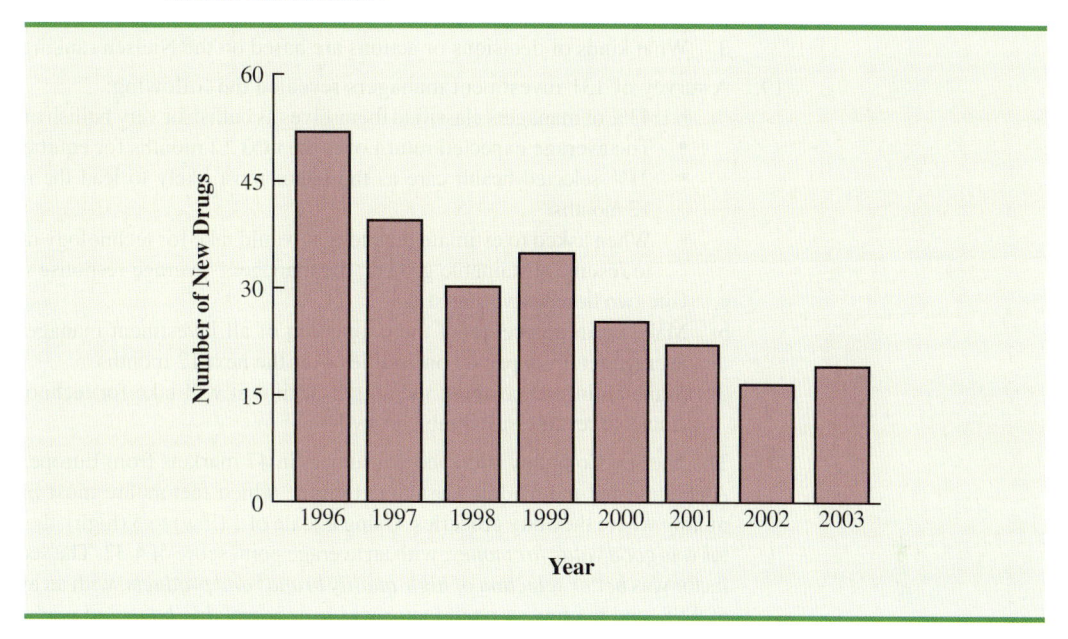

16. The Food and Drug Administration (FDA) reported the number of new drugs approved over an eight-year period (*The Wall Street Journal,* January 12, 2004). Figure 1.9 provides a bar chart summarizing the number of new drugs approved each year.
    a. Are the data categorical or quantitative?
    b. Are the data time series or cross-sectional?
    c. How many new drugs were approved in 2003?
    d. In what year were the fewest new drugs approved? How many?
    e. Comment on the trend in the number of new drugs approved by the FDA over the eight-year period.

17. A *BusinessWeek* North American subscriber study collected data from a sample of 2861 subscribers. Fifty-nine percent of the respondents indicated an annual income of $75,000 or more, and 50% reported having an American Express credit card.
    a. What is the population of interest in this study?
    b. Is annual income a categorical or quantitative variable?
    c. Is ownership of an American Express card a categorical or quantitative variable?
    d. Does this study involve cross-sectional or time series data?
    e. Describe any statistical inferences *BusinessWeek* might make on the basis of the survey.

18. Nielsen Media Research conducts weekly surveys of television viewing throughout the United States, publishing both rating and market share data. The Nielsen rating is the percentage of households with televisions watching a program, while the Nielsen share is the percentage of households watching a program among those households with televisions in use. For example, Nielsen Media Research results for the 2003 Baseball World Series between the New York Yankees and the Florida Marlins showed a rating of 12.8% and a share of 22% (Associated Press, October 27, 2003). Thus, 12.8% of households with televisions were watching the World Series and 22% of households with televisions in use were watching the World Series. Based on the rating and share data for major television programs, Nielsen publishes a weekly ranking of television programs as well as a weekly ranking of the four major networks: ABC, CBS, NBC, and Fox.

a. What is Nielsen Media Research attempting to measure?
b. What is the population?
c. Why would a sample be used in this situation?
d. What kinds of decisions or actions are based on the Nielsen rankings?

19. A survey of 131 investment managers revealed the following:
    - 43% of managers classified themselves as bullish or very bullish on the stock market.
    - The average expected return over the next 12 months for equities was 11.2%.
    - 21% selected health care as the sector most likely to lead the market in the next 12 months.
    - When asked to estimate how long it would take for technology and telecom stocks to resume sustainable growth, the managers' average response was 2.5 years.

    a. Cite two descriptive statistics.
    b. Make an inference about the population of all investment managers concerning the average return expected on equities over the next 12 months.
    c. Make an inference about the length of time it will take for technology and telecom stocks to resume sustainable growth.

20. The Nielsen Company surveyed consumers in 47 markets from Europe, Asia-Pacific, the Americas, and the Middle East to determine which factors are most important in determining where they buy groceries. Using a scale of 1 (low) to 5 (high), the highest rated factor was *good value for money*, with an average point score of 4.32. The second highest rated factor was *better selection of high-quality brands and products*, with an average point score of 3.78, and the lowest rated factor was *uses recyclable bags and packaging*, with an average point score of 2.71 (Nielsen website, February 24, 2008). Suppose that you have been hired by a grocery store chain to conduct a similar study to determine what factors customers at the chain's stores in Charlotte, North Carolina, think are most important in determining where they buy groceries.
    a. What is the population for the survey that you will be conducting?
    b. How would you collect the data for this study?

21. A survey of 430 business travelers found 155 used a travel agent to make travel arrangements (*USA Today*, November 20, 2003).
    a. Develop a descriptive statistic that can be used to estimate the percentage of all business travelers who use a travel agent to make travel arrangements.
    b. The survey reported that the most frequent way business travelers make travel arrangements is by using an online travel site. If 44% of business travelers surveyed made travel arrangements this way, how many of the 430 business travelers used an online travel site?
    c. Are the data on how travel arrangements are made categorical or quantitative?

22. A seven-year medical research study reported that women whose mothers took the drug DES during pregnancy were twice as likely to develop tissue abnormalities that might lead to cancer as were women whose mothers did not take the drug.
    a. This study involved the comparison of two populations. What were the populations?
    b. Do you suppose the data were obtained in a survey or an experiment?
    c. For the population of women whose mothers took the drug DES during pregnancy, a sample of 3980 women showed 63 developed tissue abnormalities that might lead to cancer. Provide a descriptive statistic that could be used to estimate the number of women out of 1000 in this population who have tissue abnormalities.
    d. For the population of women whose mothers did not take the drug DES during pregnancy, what is the estimate of the number of women out of 1000 who would be expected to have tissue abnormalities?
    e. Medical studies often use a relatively large sample (in this case, 3980). Why?

23. A manager of a large corporation recommends a $10,000 raise be given to keep a valued subordinate from moving to another company. What internal and external sources of data might be used to decide whether such a salary increase is appropriate?

TABLE 1.8 DATA SET FOR 25 SHADOW STOCKS

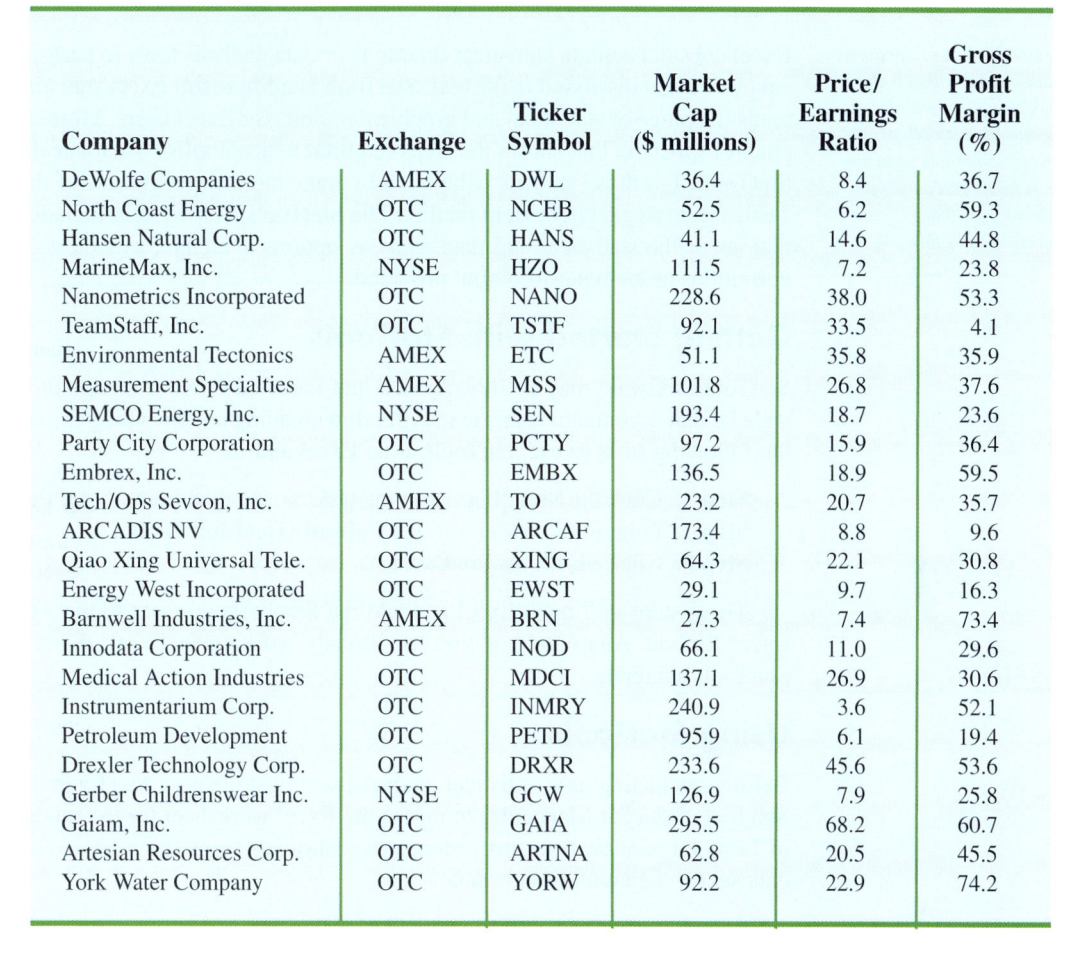

Company	Exchange	Ticker Symbol	Market Cap ($ millions)	Price/ Earnings Ratio	Gross Profit Margin (%)
DeWolfe Companies	AMEX	DWL	36.4	8.4	36.7
North Coast Energy	OTC	NCEB	52.5	6.2	59.3
Hansen Natural Corp.	OTC	HANS	41.1	14.6	44.8
MarineMax, Inc.	NYSE	HZO	111.5	7.2	23.8
Nanometrics Incorporated	OTC	NANO	228.6	38.0	53.3
TeamStaff, Inc.	OTC	TSTF	92.1	33.5	4.1
Environmental Tectonics	AMEX	ETC	51.1	35.8	35.9
Measurement Specialties	AMEX	MSS	101.8	26.8	37.6
SEMCO Energy, Inc.	NYSE	SEN	193.4	18.7	23.6
Party City Corporation	OTC	PCTY	97.2	15.9	36.4
Embrex, Inc.	OTC	EMBX	136.5	18.9	59.5
Tech/Ops Sevcon, Inc.	AMEX	TO	23.2	20.7	35.7
ARCADIS NV	OTC	ARCAF	173.4	8.8	9.6
Qiao Xing Universal Tele.	OTC	XING	64.3	22.1	30.8
Energy West Incorporated	OTC	EWST	29.1	9.7	16.3
Barnwell Industries, Inc.	AMEX	BRN	27.3	7.4	73.4
Innodata Corporation	OTC	INOD	66.1	11.0	29.6
Medical Action Industries	OTC	MDCI	137.1	26.9	30.6
Instrumentarium Corp.	OTC	INMRY	240.9	3.6	52.1
Petroleum Development	OTC	PETD	95.9	6.1	19.4
Drexler Technology Corp.	OTC	DRXR	233.6	45.6	53.6
Gerber Childrenswear Inc.	NYSE	GCW	126.9	7.9	25.8
Gaiam, Inc.	OTC	GAIA	295.5	68.2	60.7
Artesian Resources Corp.	OTC	ARTNA	62.8	20.5	45.5
York Water Company	OTC	YORW	92.2	22.9	74.2

WEB file
Shadow02

24. Table 1.8 shows a data set containing information for 25 of the shadow stocks tracked by the American Association of Individual Investors. Shadow stocks are common stocks of smaller companies that are not closely followed by Wall Street analysts. The data set is also on the website that accompanies the text in the file named Shadow02.
   a. How many variables are in the data set?
   b. Which of the variables are categorical and which are quantitative?
   c. For the Exchange variable, show the frequency and the percent frequency for AMEX, NYSE, and OTC. Construct a bar graph similar to Figure 1.5 for the Exchange variable.
   d. Show the frequency distribution for the Gross Profit Margin using the five intervals: 0–14.9, 15–29.9, 30–44.9, 45–59.9, and 60–74.9. Construct a histogram similar to Figure 1.6.
   e. What is the average price/earnings ratio?

25. A sample of midterm grades for five students showed the following results: 72, 65, 82, 90, 76. Which of the following statements are correct, and which should be challenged as being too generalized?
   a. The average midterm grade for the sample of five students is 77.
   b. The average midterm grade for all students who took the exam is 77.
   c. An estimate of the average midterm grade for all students who took the exam is 77.
   d. More than half of the students who take this exam will score between 70 and 85.
   e. If five other students are included in the sample, their grades will be between 65 and 90.

Appendix   An Introduction to StatTools

StatTools is a professional add-in that expands the statistical capabilities available with Microsoft Excel. StatTools software can be downloaded from the website that accompanies this text.

Excel does not contain statistical functions or data analysis tools to perform all the statistical procedures discussed in the text. StatTools is a Microsoft Excel statistics add-in that extends the range of statistical and graphical options for Excel users. Most chapters include a chapter appendix that shows the steps required to accomplish a statistical procedure using StatTools. For those students who want to make more extensive use of the software, StatTools offers an excellent Help facility. The StatTools Help system includes detailed explanations of the statistical and data analysis options available, as well as descriptions and definitions of the types of output provided.

Getting Started with StatTools

StatTools software may be downloaded and installed on your computer by accessing the website that accompanies this text. After downloading and installing the software, perform the following steps to use StatTools as an Excel add-in.

Step 1. Click the **Start** button on the taskbar and then point to **All Programs**
Step 2. Point to the folder entitled **Palisade Decision Tools**
Step 3. Click **StatTools for Excel**

These steps will open Excel and add the StatTools tab next to the Add-Ins tab on the Excel Ribbon. Alternately, if you are already working in Excel, these steps will make StatTools available.

Using StatTools

Before conducting any statistical analysis, we must create a StatTools data set using the StatTools Data Set Manager. Let us use the Excel worksheet for the mutual funds data set in Table 1.1 to show how this is done. The following steps show how to create a StatTools data set for the mutual funds data.

Step 1. Open the Excel file named Morningstar
Step 2. Select any cell in the data set (for example, cell A1)
Step 3. Click the **StatTools** tab on the Ribbon
Step 4. In the **Data** group, click **Data Set Manager**
Step 5. When StatTools asks if you want to add the range $A$1:$F$26 as a new StatTools data set, click **Yes**
Step 6. When the StatTools—Data Set Manager dialog box appears, click **OK**

Figure 1.10 shows the StatTools—Data Set Manager dialog box that appears in step 6. By default, the name of the new StatTools data set is Data Set #1. You can replace the name Data Set #1 in step 6 with a more descriptive name. And, if you select the Apply Cell Format option, the column labels will be highlighted in blue and the entire data set will have outside and inside borders. You can always select the Data Set Manager at any time in your analysis to make these types of changes.

Recommended Application Settings

StatTools allows the user to specify some of the application settings that control such things as where statistical output is displayed and how calculations are performed. The following steps show how to access the StatTools—Application Settings dialog box.

Step 1. Click the **StatTools** tab on the Ribbon
Step 2. In the **Tools Group**, click **Utilities**
Step 3. Choose **Application Settings** from the list of options

Appendix An Introduction to StatTools

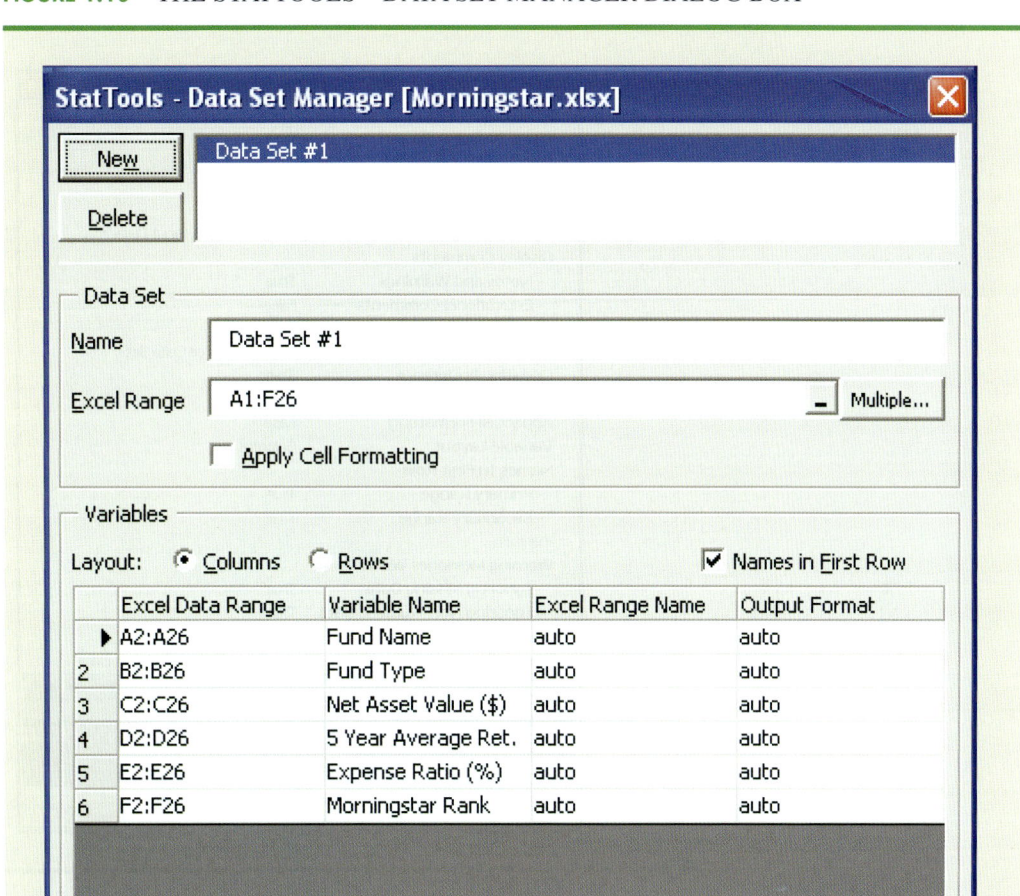

FIGURE 1.10 THE STATTOOLS—DATA SET MANAGER DIALOG BOX

Figure 1.11 shows that the StatTools—Application Settings dialog box has five sections: General Settings; Reports; Utilities; Data Set Defaults; and Analyses. Let us show how to make changes in the Reports section of the dialog box.

Figure 1.11 shows that the Placement option currently selected is **New Workbook**. Using this option, the StatTools output will be placed in a new workbook. But suppose you would like to place the StatTools output in the current (active) workbook. If you click the words **New Workbook**, a downward-pointing arrow will appear to the right. Clicking this arrow will display a list of all the placement options, including **Active Workbook**; we recommend using this option. Figure 1.11 also shows that the Updating Preferences option in the Reports section is currently **Live—Linked to Input Data**. With live updating, anytime one or more data values are changed StatTools will automatically change the output previously produced; we also recommend using this option. Note that there are two options available under Display Comments: **Notes and Warnings** and **Educational Comments**. Because these options provide useful notes and information regarding the output, we recommend using both options. Thus, to include educational

FIGURE 1.11 THE STATTOOLS—APPLICATION SETTINGS DIALOG BOX

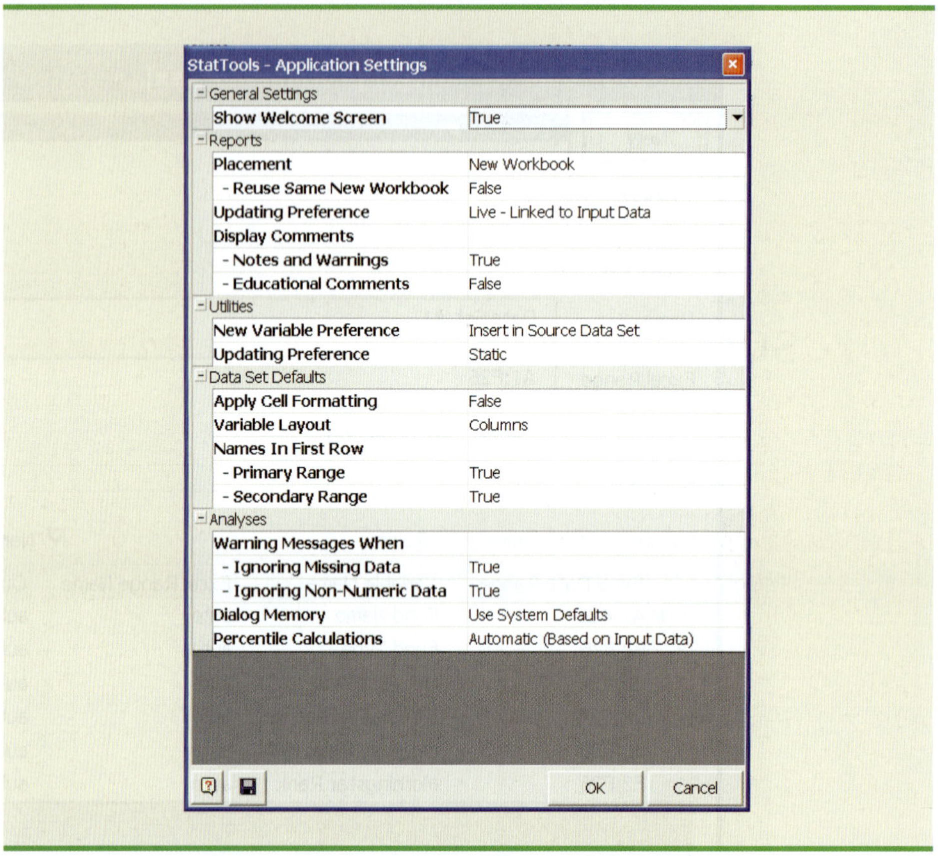

comments as part of the StatTools output, you will have to change the value of False for Educational Comments to True.

The StatTools—Settings dialog box contains numerous other features that enable you to customize the way that you want StatTools to operate. You can learn more about these features by selecting the Help option located in the Tools group, or by clicking the Help icon located in the lower left-hand corner of the dialog box. When you have finish making changes in the application settings, click OK at the bottom of the dialog box and then click Yes when StatTools asks you if you want to save the new application settings.

CHAPTER 2

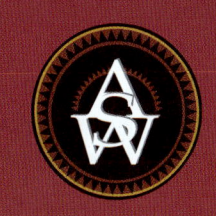

Descriptive Statistics: Tabular and Graphical Presentations

CONTENTS

STATISTICS IN PRACTICE:
COLGATE-PALMOLIVE COMPANY

2.1 SUMMARIZING CATEGORICAL DATA
Frequency Distribution
Relative Frequency and Percent Frequency Distributions
Bar Charts and Pie Charts

2.2 SUMMARIZING QUANTITATIVE DATA
Frequency Distribution
Relative Frequency and Percent Frequency Distributions
Dot Plot
Histogram
Cumulative Distributions
Ogive

2.3 EXPLORATORY DATA ANALYSIS: THE STEM-AND-LEAF DISPLAY

2.4 CROSSTABULATIONS AND SCATTER DIAGRAMS
Crosstabulation
Simpson's Paradox
Scatter Diagram and Trendline

STATISTICS *in* PRACTICE

COLGATE-PALMOLIVE COMPANY*
NEW YORK, NEW YORK

The Colgate-Palmolive Company started as a small soap and candle shop in New York City in 1806. Today, Colgate-Palmolive employs more than 40,000 people working in more than 200 countries and territories around the world. Although best known for its brand names of Colgate, Palmolive, Ajax, and Fab, the company also markets Mennen, Hill's Science Diet, and Hill's Prescription Diet products.

The Colgate-Palmolive Company uses statistics in its quality assurance program for home laundry detergent products. One concern is customer satisfaction with the quantity of detergent powder in a carton. Every carton in each size category is filled with the same amount of detergent by weight, but the volume of detergent is affected by the density of the detergent powder. For instance, if the powder density is on the heavy side, a smaller volume of detergent is needed to reach the carton's specified weight. As a result, the carton may appear to be underfilled when opened by the consumer.

To control the problem of heavy detergent powder, limits are placed on the acceptable range of powder density. Statistical samples are taken periodically, and the density of each powder sample is measured. Data summaries are then provided for operating personnel so that corrective action can be taken if necessary to keep the density within the desired quality specifications.

A frequency distribution for the densities of 150 samples taken over a one-week period and a histogram are shown in the accompanying table and figure. Density levels above .40 are unacceptably high. The frequency distribution and histogram show that the operation is meeting its quality guidelines with all of the densities less than or equal to .40. Managers viewing these statistical summaries would be pleased with the quality of the detergent production process.

In this chapter, you will learn about tabular and graphical methods of descriptive statistics such as frequency distributions, bar charts, histograms, stem-and-leaf displays, crosstabulations, and others. The goal of these methods is to summarize data so that the data can be easily understood and interpreted.

The Colgate-Palmolive Company uses statistical summaries to help maintain the quality of their products.

Frequency Distribution of Density Data

Density	Frequency
.29–.30	30
.31–.32	75
.33–.34	32
.35–.36	9
.37–.38	3
.39–.40	1
Total	150

Histogram of Density Data

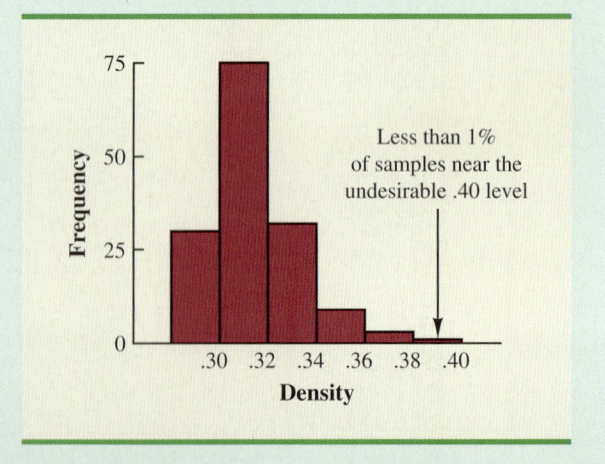

Less than 1% of samples near the undesirable .40 level

---
*The authors are indebted to William R. Fowle, Manager of Quality Assurance, Colgate-Palmolive Company, for providing this Statistics in Practice.

2.1 Summarizing Categorical Data

As indicated in Chapter 1, data can be classified as either categorical or quantitative. **Categorical data** use labels or names to identify categories of like items. **Quantitative data** are numerical values that indicate how much or how many.

This chapter introduces tabular and graphical methods commonly used to summarize both categorical and quantitative data. Tabular and graphical summaries of data can be found in annual reports, newspaper articles, and research studies. Everyone is exposed to these types of presentations. Hence, it is important to understand how they are prepared and how they should be interpreted. We begin with tabular and graphical methods for summarizing data concerning a single variable. The last section introduces methods for summarizing data when the relationship between two variables is of interest.

Modern statistical software packages provide extensive capabilities for summarizing data and preparing graphical presentations. Minitab and Excel are two packages that are widely available. In the chapter appendixes, we show some of their capabilities.

Summarizing Categorical Data

Frequency Distribution

We begin the discussion of how tabular and graphical methods can be used to summarize categorical data with the definition of a **frequency distribution**.

> **FREQUENCY DISTRIBUTION**
>
> A frequency distribution is a tabular summary of data showing the number (frequency) of items in each of several nonoverlapping classes.

Let us use the following example to demonstrate the construction and interpretation of a frequency distribution for categorical data. Coke Classic, Diet Coke, Dr. Pepper, Pepsi, and Sprite are five popular soft drinks. Assume that the data in Table 2.1 show the soft drink selected in a sample of 50 soft drink purchases.

TABLE 2.1   DATA FROM A SAMPLE OF 50 SOFT DRINK PURCHASES

Coke Classic	Sprite	Pepsi
Diet Coke	Coke Classic	Coke Classic
Pepsi	Diet Coke	Coke Classic
Diet Coke	Coke Classic	Coke Classic
Coke Classic	Diet Coke	Pepsi
Coke Classic	Coke Classic	Dr. Pepper
Dr. Pepper	Sprite	Coke Classic
Diet Coke	Pepsi	Diet Coke
Pepsi	Coke Classic	Pepsi
Pepsi	Coke Classic	Pepsi
Coke Classic	Coke Classic	Pepsi
Dr. Pepper	Pepsi	Pepsi
Sprite	Coke Classic	Coke Classic
Coke Classic	Sprite	Dr. Pepper
Diet Coke	Dr. Pepper	Pepsi
Coke Classic	Pepsi	Sprite
Coke Classic	Diet Coke	

TABLE 2.2

FREQUENCY DISTRIBUTION OF SOFT DRINK PURCHASES

Soft Drink	Frequency
Coke Classic	19
Diet Coke	8
Dr. Pepper	5
Pepsi	13
Sprite	5
Total	50

To develop a frequency distribution for these data, we count the number of times each soft drink appears in Table 2.1. Coke Classic appears 19 times, Diet Coke appears 8 times, Dr. Pepper appears 5 times, Pepsi appears 13 times, and Sprite appears 5 times. These counts are summarized in the frequency distribution in Table 2.2.

This frequency distribution provides a summary of how the 50 soft drink purchases are distributed across the five soft drinks. This summary offers more insight than the original data shown in Table 2.1. Viewing the frequency distribution, we see that Coke Classic is the leader, Pepsi is second, Diet Coke is third, and Sprite and Dr. Pepper are tied for fourth. The frequency distribution summarizes information about the popularity of the five soft drinks.

Relative Frequency and Percent Frequency Distributions

A frequency distribution shows the number (frequency) of items in each of several nonoverlapping classes. However, we are often interested in the proportion, or percentage, of items in each class. The *relative frequency* of a class equals the fraction or proportion of items belonging to a class. For a data set with n observations, the relative frequency of each class can be determined as follows:

RELATIVE FREQUENCY

$$\text{Relative frequency of a class} = \frac{\text{Frequency of the class}}{n} \quad (2.1)$$

The *percent frequency* of a class is the relative frequency multiplied by 100.

A **relative frequency distribution** gives a tabular summary of data showing the relative frequency for each class. A **percent frequency distribution** summarizes the percent frequency of the data for each class. Table 2.3 shows a relative frequency distribution and a percent frequency distribution for the soft drink data. In Table 2.3 we see that the relative frequency for Coke Classic is 19/50 = .38, the relative frequency for Diet Coke is 8/50 = .16, and so on. From the percent frequency distribution, we see that 38% of the purchases were Coke Classic, 16% of the purchases were Diet Coke, and so on. We can also note that 38% + 26% + 16% = 80% of the purchases were the top three soft drinks.

Bar Charts and Pie Charts

A **bar chart** is a graphical device for depicting categorical data summarized in a frequency, relative frequency, or percent frequency distribution. On one axis of the graph (usually the horizontal axis), we specify the labels that are used for the classes (categories). A frequency, relative frequency, or percent frequency scale can be used for the other axis of the chart

TABLE 2.3 RELATIVE FREQUENCY AND PERCENT FREQUENCY DISTRIBUTIONS OF SOFT DRINK PURCHASES

Soft Drink	Relative Frequency	Percent Frequency
Coke Classic	.38	38
Diet Coke	.16	16
Dr. Pepper	.10	10
Pepsi	.26	26
Sprite	.10	10
Total	1.00	100

2.1 Summarizing Categorical Data

FIGURE 2.1 BAR CHART OF SOFT DRINK PURCHASES

In quality control applications, bar charts are used to identify the most important causes of problems. When the bars are arranged in descending order of height from left to right with the most frequently occurring cause appearing first, the bar chart is called a Pareto *chart. The chart is named for its founder, Vilfredo Pareto, an Italian economist.*

(usually the vertical axis). Then, using a bar of fixed width drawn above each class label, we extend the length of the bar until we reach the frequency, relative frequency, or percent frequency of the class. For categorical data, the bars should be separated to emphasize the fact that each class is separate. Figure 2.1 shows a bar chart of the frequency distribution for the 50 soft drink purchases. Note how the graphical presentation shows Coke Classic, Pepsi, and Diet Coke to be the most preferred brands.

The **pie chart** provides another graphical device for presenting relative frequency and percent frequency distributions for categorical data. To construct a pie chart, we first draw a circle to represent all the data. Then we use the relative frequencies to subdivide the circle into sectors, or parts, that correspond to the relative frequency for each class. For example, because a circle contains 360 degrees and Coke Classic shows a relative frequency of .38, the sector of the pie chart labeled Coke Classic consists of .38(360) = 136.8 degrees. The sector of the pie chart labeled Diet Coke consists of .16(360) = 57.6 degrees. Similar calculations for the other classes yield the pie chart in Figure 2.2. The

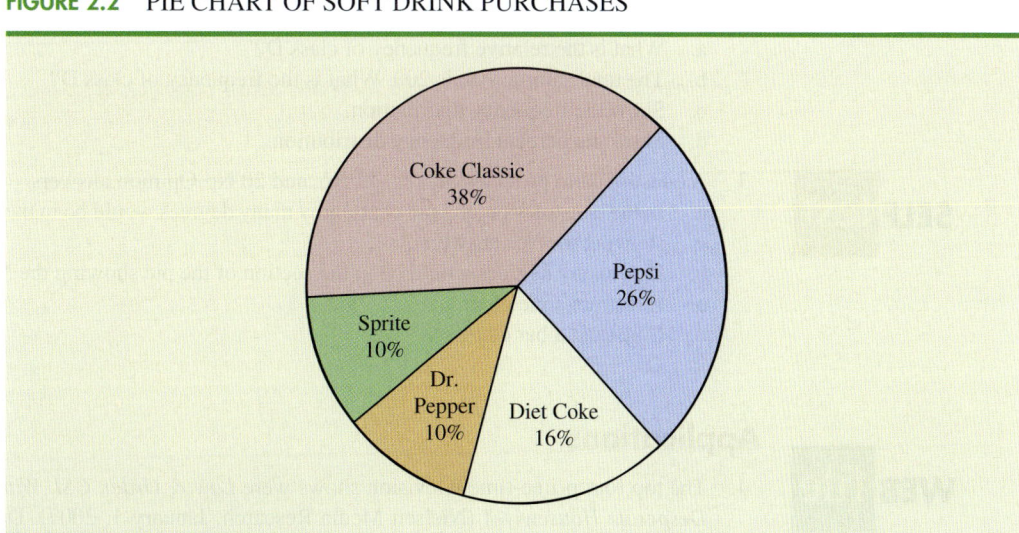

FIGURE 2.2 PIE CHART OF SOFT DRINK PURCHASES

numerical values shown for each sector can be frequencies, relative frequencies, or percent frequencies.

NOTES AND COMMENTS

1. Often the number of classes in a frequency distribution is the same as the number of categories found in the data, as is the case for the soft drink purchase data in this section. The data involve only five soft drinks, and a separate frequency distribution class was defined for each one. Data that included all soft drinks would require many categories, most of which would have a small number of purchases. Most statisticians recommend that classes with smaller frequencies be grouped into an aggregate class called "other." Classes with frequencies of 5% or less would most often be treated in this fashion.

2. The sum of the frequencies in any frequency distribution always equals the number of observations. The sum of the relative frequencies in any relative frequency distribution always equals 1.00, and the sum of the percentages in a percent frequency distribution always equals 100.

Exercises

Methods

1. The response to a question has three alternatives: A, B, and C. A sample of 120 responses provides 60 A, 24 B, and 36 C. Show the frequency and relative frequency distributions.

2. A partial relative frequency distribution is given.

Class	Relative Frequency
A	.22
B	.18
C	.40
D	

    a. What is the relative frequency of class D?
    b. The total sample size is 200. What is the frequency of class D?
    c. Show the frequency distribution.
    d. Show the percent frequency distribution.

3. A questionnaire provides 58 Yes, 42 No, and 20 No-Opinion answers.
    a. In the construction of a pie chart, how many degrees would be in the section of the pie showing the Yes answers?
    b. How many degrees would be in the section of the pie showing the No answers?
    c. Construct a pie chart.
    d. Construct a bar chart.

Applications

4. The top four prime-time television shows were *Law & Order, CSI, Without a Trace,* and *Desperate Housewives* (Nielsen Media Research, January 1, 2007). Data indicating the preferred shows for a sample of 50 viewers follow.

DH	CSI	DH	CSI	L&O
Trace	CSI	L&O	Trace	CSI
CSI	DH	Trace	CSI	DH
L&O	L&O	L&O	CSI	DH
CSI	DH	DH	L&O	CSI
DH	Trace	CSI	Trace	DH
DH	CSI	CSI	L&O	CSI
L&O	CSI	Trace	Trace	DH
L&O	CSI	CSI	CSI	DH
CSI	DH	Trace	Trace	L&O

a. Are these data categorical or quantitative?
b. Provide frequency and percent frequency distributions.
c. Construct a bar chart and a pie chart.
d. On the basis of the sample, which television show has the largest viewing audience? Which one is second?

5. In alphabetical order, the six most common last names in the United States are Brown, Davis, Johnson, Jones, Smith, and Williams (*The World Almanac,* 2006). Assume that a sample of 50 individuals with one of these last names provided the following data.

Brown	Williams	Williams	Williams	Brown
Smith	Jones	Smith	Johnson	Smith
Davis	Smith	Brown	Williams	Johnson
Johnson	Smith	Smith	Johnson	Brown
Williams	Davis	Johnson	Williams	Johnson
Williams	Johnson	Jones	Smith	Brown
Johnson	Smith	Smith	Brown	Jones
Jones	Jones	Smith	Smith	Davis
Davis	Jones	Williams	Davis	Smith
Jones	Johnson	Brown	Johnson	Davis

Summarize the data by constructing the following:
a. Relative and percent frequency distributions
b. A bar chart
c. A pie chart
d. Based on these data, what are the three most common last names?

6. The Nielsen Media Research television rating measures the percentage of television owners who are watching a particular television program. The highest-rated television program in television history was the *M*A*S*H Last Episode Special* shown on February 28, 1983. A 60.2 rating indicated that 60.2% of all television owners were watching this program. Nielsen Media Research provided the list of the 50 top-rated single shows in television history (*The New York Times Almanac,* 2006). The following data show the television network that produced each of these 50 top-rated shows.

ABC	ABC	ABC	NBC	CBS
ABC	CBS	ABC	ABC	NBC
NBC	NBC	CBS	ABC	NBC
CBS	ABC	CBS	NBC	ABC
CBS	NBC	NBC	CBS	NBC
CBS	CBS	CBS	NBC	NBC
FOX	CBS	CBS	ABC	NBC
ABC	ABC	CBS	NBC	NBC
NBC	CBS	NBC	CBS	CBS
ABC	CBS	ABC	NBC	ABC

a. Construct a frequency distribution, percent frequency distribution, and bar chart for the data.

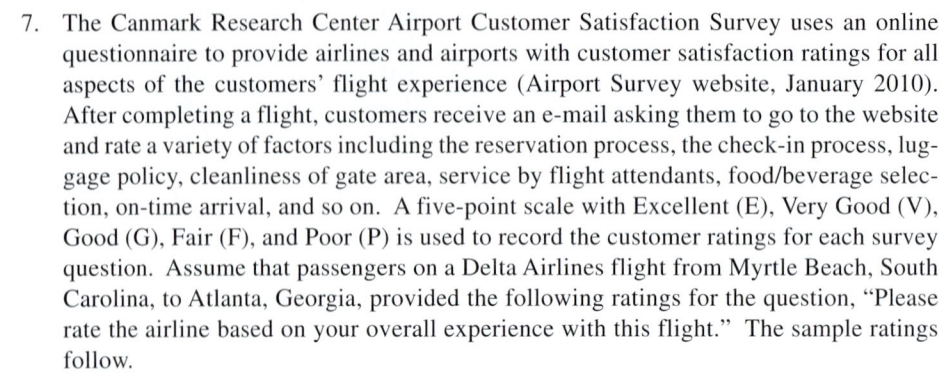

b. Which network or networks have done the best in terms of presenting top-rated television shows? Compare the performance of ABC, CBS, and NBC.

7. The Canmark Research Center Airport Customer Satisfaction Survey uses an online questionnaire to provide airlines and airports with customer satisfaction ratings for all aspects of the customers' flight experience (Airport Survey website, January 2010). After completing a flight, customers receive an e-mail asking them to go to the website and rate a variety of factors including the reservation process, the check-in process, luggage policy, cleanliness of gate area, service by flight attendants, food/beverage selection, on-time arrival, and so on. A five-point scale with Excellent (E), Very Good (V), Good (G), Fair (F), and Poor (P) is used to record the customer ratings for each survey question. Assume that passengers on a Delta Airlines flight from Myrtle Beach, South Carolina, to Atlanta, Georgia, provided the following ratings for the question, "Please rate the airline based on your overall experience with this flight." The sample ratings follow.

E	E	G	V	V	E	V	V	V	E
E	G	V	E	E	V	E	E	E	V
V	V	V	F	V	E	V	E	G	E
G	E	V	E	V	E	V	V	V	V
E	E	V	V	E	P	E	V	P	V

a. Use a percent frequency distribution and a bar chart to summarize these data. What do these summaries indicate about the overall customer satisfaction with the Delta flight?
b. The online survey questionnaire enabled respondents to explain any aspect of the flight that failed to meet expectations. Would this be helpful information to a manager looking for ways to improve the overall customer satisfaction on Delta flights? Explain.

8. Data for a sample of 55 members of the Baseball Hall of Fame in Cooperstown, New York, are shown here. Each observation indicates the primary position played by the Hall of Famers: pitcher (P), catcher (H), 1st base (1), 2nd base (2), 3rd base (3), shortstop (S), left field (L), center field (C), and right field (R).

L	P	C	H	2	P	R	1	S	S	1	L	P	R	P
P	P	P	R	C	S	L	R	P	C	C	P	P	R	P
2	3	P	H	L	P	1	C	P	P	S	1	L	R
R	1	2	H	S	3	H	2	L	P

a. Use frequency and relative frequency distributions to summarize the data.
b. What position provides the most Hall of Famers?
c. What position provides the fewest Hall of Famers?
d. What outfield position (L, C, or R) provides the most Hall of Famers?
e. Compare infielders (1, 2, 3, and S) to outfielders (L, C, and R).

9. The Pew Research Center's Social & Demographic Trends project found that 46% of U.S. adults would rather live in a different type of community than the one where they are living now (Pew Research Center, January 29, 2009). The national survey of 2260 adults asked, "Where do you live now?" and "What do you consider to be the ideal community?" Response options were City (C), Suburb (S), Small Town (T), or Rural (R). A representative portion of this survey for a sample of 100 respondents is as follows.

Where do you live now?

S	T	R	C	R	R	T	C	S	T	C	S	C	S	T
S	S	C	S	S	T	T	C	C	S	T	C	S	T	C
T	R	S	S	T	C	S	C	T	C	T	C	T	C	R
C	C	R	T	C	S	S	T	S	C	C	C	R	S	C
S	S	C	C	S	C	R	T	T	T	C	R	T	C	R
C	T	R	R	C	T	C	C	R	T	T	R	S	R	T
T	S	S	S	S	S	C	C	R	T

2.2 Summarizing Quantitative Data

What do you consider to be the ideal community?

S	C	R	R	R	S	T	S	S	T	T	S	C	S	T
C	C	R	T	R	S	T	T	S	S	C	C	T	T	S
S	R	C	S	C	C	S	C	R	C	T	S	R	R	R
C	T	S	T	T	T	R	R	S	C	C	R	R	S	S
S	T	C	T	T	C	R	T	T	T	C	T	T	R	R
C	S	R	T	C	T	C	C	T	T	T	R	C	R	T
T	C	S	S	C	S	T	S	S	R					

a. Provide a percent frequency distribution for each question.
b. Construct a bar chart for each question.
c. Where are most adults living now?
d. Where do most adults consider the ideal community?
e. What changes in living areas would you expect to see if people moved from where they currently live to their ideal community?

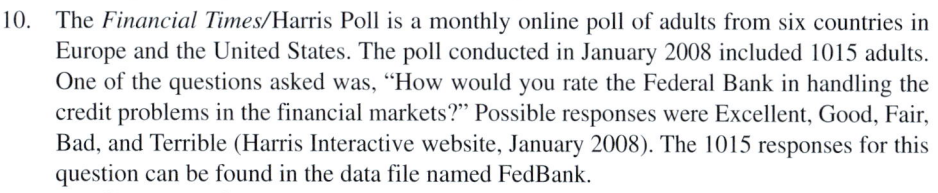

WEBfile
FedBank

10. The *Financial Times*/Harris Poll is a monthly online poll of adults from six countries in Europe and the United States. The poll conducted in January 2008 included 1015 adults. One of the questions asked was, "How would you rate the Federal Bank in handling the credit problems in the financial markets?" Possible responses were Excellent, Good, Fair, Bad, and Terrible (Harris Interactive website, January 2008). The 1015 responses for this question can be found in the data file named FedBank.
    a. Construct a frequency distribution.
    b. Construct a percent frequency distribution.
    c. Construct a bar chart for the percent frequency distribution.
    d. Comment on how adults in the United States think the Federal Bank is handling the credit problems in the financial markets.
    e. In Spain, 1114 adults were asked, "How would you rate the European Central Bank in handling the credit problems in the financial markets?" The percent frequency distribution obtained follows:

Rating	Percent Frequency
Excellent	0
Good	4
Fair	46
Bad	40
Terrible	10

Compare the results obtained in Spain with the results obtained in the United States.

Summarizing Quantitative Data

Frequency Distribution

As defined in Section 2.1, a frequency distribution is a tabular summary of data showing the number (frequency) of items in each of several nonoverlapping classes. This definition holds for quantitative as well as categorical data. However, with quantitative data we must be more careful in defining the nonoverlapping classes to be used in the frequency distribution.

For example, consider the quantitative data in Table 2.4. These data show the time in days required to complete year-end audits for a sample of 20 clients of Sanderson and Clifford, a small public accounting firm. The three steps necessary to define the classes for a frequency distribution with quantitative data are

1. Determine the number of nonoverlapping classes.
2. Determine the width of each class.
3. Determine the class limits.

TABLE 2.4

YEAR-END AUDIT TIMES (IN DAYS)

12	14	19	18
15	15	18	17
20	27	22	23
22	21	33	28
14	18	16	13

WEB file
Audit

Let us demonstrate these steps by developing a frequency distribution for the audit time data in Table 2.4.

Number of classes Classes are formed by specifying ranges that will be used to group the data. As a general guideline, we recommend using between 5 and 20 classes. For a small number of data items, as few as five or six classes may be used to summarize the data. For a larger number of data items, a larger number of classes is usually required. The goal is to use enough classes to show the variation in the data, but not so many classes that some contain only a few data items. Because the number of data items in Table 2.4 is relatively small ($n = 20$), we chose to develop a frequency distribution with five classes.

Width of the classes The second step in constructing a frequency distribution for quantitative data is to choose a width for the classes. As a general guideline, we recommend that the width be the same for each class. Thus the choices of the number of classes and the width of classes are not independent decisions. A larger number of classes means a smaller class width, and vice versa. To determine an approximate class width, we begin by identifying the largest and smallest data values. Then, with the desired number of classes specified, we can use the following expression to determine the approximate class width.

Making the classes the same width reduces the chance of inappropriate interpretations by the user.

$$\text{Approximate class width} = \frac{\text{Largest data value} - \text{Smallest data value}}{\text{Number of classes}} \tag{2.2}$$

The approximate class width given by equation (2.2) can be rounded to a more convenient value based on the preference of the person developing the frequency distribution. For example, an approximate class width of 9.28 might be rounded to 10 simply because 10 is a more convenient class width to use in presenting a frequency distribution.

For the data involving the year-end audit times, the largest data value is 33 and the smallest data value is 12. Because we decided to summarize the data with five classes, using equation (2.2) provides an approximate class width of $(33 - 12)/5 = 4.2$. We therefore decided to round up and use a class width of five days in the frequency distribution.

In practice, the number of classes and the appropriate class width are determined by trial and error. Once a possible number of classes is chosen, equation (2.2) is used to find the approximate class width. The process can be repeated for a different number of classes. Ultimately, the analyst uses judgment to determine the combination of the number of classes and class width that provides the best frequency distribution for summarizing the data.

No single frequency distribution is best for a data set. Different people may construct different, but equally acceptable, frequency distributions. The goal is to reveal the natural grouping and variation in the data.

For the audit time data in Table 2.4, after deciding to use five classes, each with a width of five days, the next task is to specify the class limits for each of the classes.

Class limits Class limits must be chosen so that each data item belongs to one and only one class. The *lower class limit* identifies the smallest possible data value assigned to the class. The *upper class limit* identifies the largest possible data value assigned to the class. In developing frequency distributions for qualitative data, we did not need to specify class limits because each data item naturally fell into a separate class. But with quantitative data, such as the audit times in Table 2.4, class limits are necessary to determine where each data value belongs.

Using the audit time data in Table 2.4, we selected 10 days as the lower class limit and 14 days as the upper class limit for the first class. This class is denoted 10–14 in Table 2.5. The smallest data value, 12, is included in the 10–14 class. We then selected 15 days as the lower class limit and 19 days as the upper class limit of the next class. We continued defining the lower and upper class limits to obtain a total of five classes: 10–14, 15–19, 20–24, 25–29, and 30–34. The largest data value, 33, is included in the 30–34 class. The difference between the lower class limits of adjacent classes is the class width. Using the first two lower class limits of 10 and 15, we see that the class width is $15 - 10 = 5$.

With the number of classes, class width, and class limits determined, a frequency distribution can be obtained by counting the number of data values belonging to each class.

TABLE 2.5

FREQUENCY DISTRIBUTION FOR THE AUDIT TIME DATA

Audit Time (days)	Frequency
10–14	4
15–19	8
20–24	5
25–29	2
30–34	1
Total	20

2.2 Summarizing Quantitative Data

TABLE 2.6 RELATIVE FREQUENCY AND PERCENT FREQUENCY DISTRIBUTIONS FOR THE AUDIT TIME DATA

Audit Time (days)	Relative Frequency	Percent Frequency
10–14	.20	20
15–19	.40	40
20–24	.25	25
25–29	.10	10
30–34	.05	5
Total	1.00	100

For example, the data in Table 2.4 show that four values—12, 14, 14, and 13—belong to the 10–14 class. Thus, the frequency for the 10–14 class is 4. Continuing this counting process for the 15–19, 20–24, 25–29, and 30–34 classes provides the frequency distribution in Table 2.5. Using this frequency distribution, we can observe the following:

1. The most frequently occurring audit times are in the class of 15–19 days. Eight of the twenty audit times belong to this class.
2. Only one audit required 30 or more days.

Other conclusions are possible, depending on the interests of the person viewing the frequency distribution. The value of a frequency distribution is that it provides insights about the data that are not easily obtained by viewing the data in their original unorganized form.

Class midpoint In some applications, we want to know the midpoints of the classes in a frequency distribution for quantitative data. The **class midpoint** is the value halfway between the lower and upper class limits. For the audit time data, the five class midpoints are 12, 17, 22, 27, and 32.

Relative Frequency and Percent Frequency Distributions

We define the relative frequency and percent frequency distributions for quantitative data in the same manner as for qualitative data. First, recall that the relative frequency is the proportion of the observations belonging to a class. With n observations,

$$\text{Relative frequency of class} = \frac{\text{Frequency of the class}}{n}$$

The percent frequency of a class is the relative frequency multiplied by 100.

Based on the class frequencies in Table 2.5 and with $n = 20$, Table 2.6 shows the relative frequency distribution and percent frequency distribution for the audit time data. Note that .40 of the audits, or 40%, required from 15 to 19 days. Only .05 of the audits, or 5%, required 30 or more days. Again, additional interpretations and insights can be obtained by using Table 2.6.

Dot Plot

One of the simplest graphical summaries of data is a **dot plot**. A horizontal axis shows the range for the data. Each data value is represented by a dot placed above the axis. Figure 2.3 is the dot plot for the audit time data in Table 2.4. The three dots located above 18 on the horizontal axis indicate that an audit time of 18 days occurred three times. Dot plots show the details of the data and are useful for comparing the distribution of the data for two or more variables.

FIGURE 2.3 DOT PLOT FOR THE AUDIT TIME DATA

Histogram

A common graphical presentation of quantitative data is a **histogram**. This graphical summary can be prepared for data previously summarized in either a frequency, relative frequency, or percent frequency distribution. A histogram is constructed by placing the variable of interest on the horizontal axis and the frequency, relative frequency, or percent frequency on the vertical axis. The frequency, relative frequency, or percent frequency of each class is shown by drawing a rectangle whose base is determined by the class limits on the horizontal axis and whose height is the corresponding frequency, relative frequency, or percent frequency.

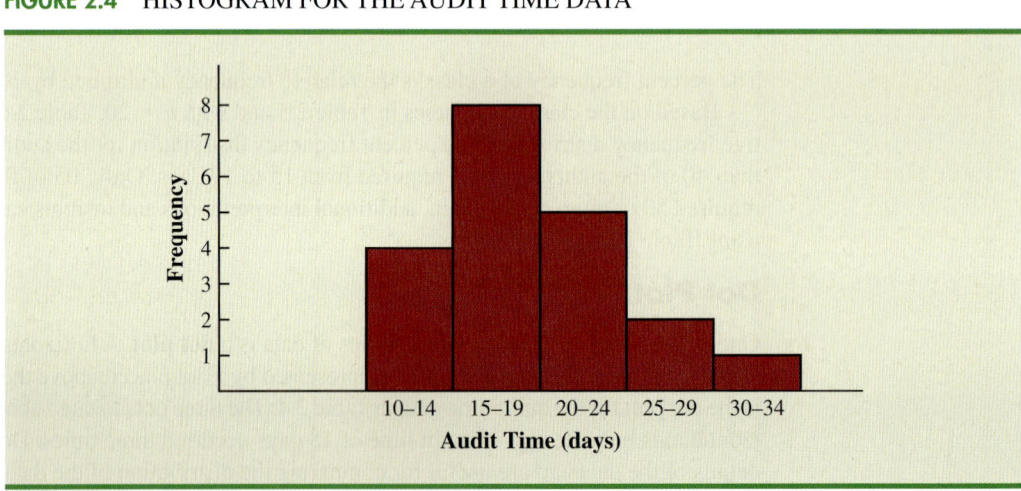

Figure 2.4 is a histogram for the audit time data. Note that the class with the greatest frequency is shown by the rectangle appearing above the class of 15–19 days. The height of the rectangle shows that the frequency of this class is 8. A histogram for the relative or percent frequency distribution of these data would look the same as the histogram in Figure 2.4 with the exception that the vertical axis would be labeled with relative or percent frequency values.

As Figure 2.4 shows, the adjacent rectangles of a histogram touch one another. Unlike a bar graph, a histogram contains no natural separation between the rectangles of adjacent classes. This format is the usual convention for histograms. Because the classes for the audit time data are stated as 10–14, 15–19, 20–24, 25–29, and 30–34, one-unit spaces of 14 to 15, 19 to 20, 24 to 25, and 29 to 30 would seem to be needed between the classes. These spaces are eliminated when constructing a histogram. Eliminating the spaces between classes in a histogram for the audit time data helps show that all values between the lower limit of the first class and the upper limit of the last class are possible.

FIGURE 2.4 HISTOGRAM FOR THE AUDIT TIME DATA

FIGURE 2.5 HISTOGRAMS SHOWING DIFFERING LEVELS OF SKEWNESS

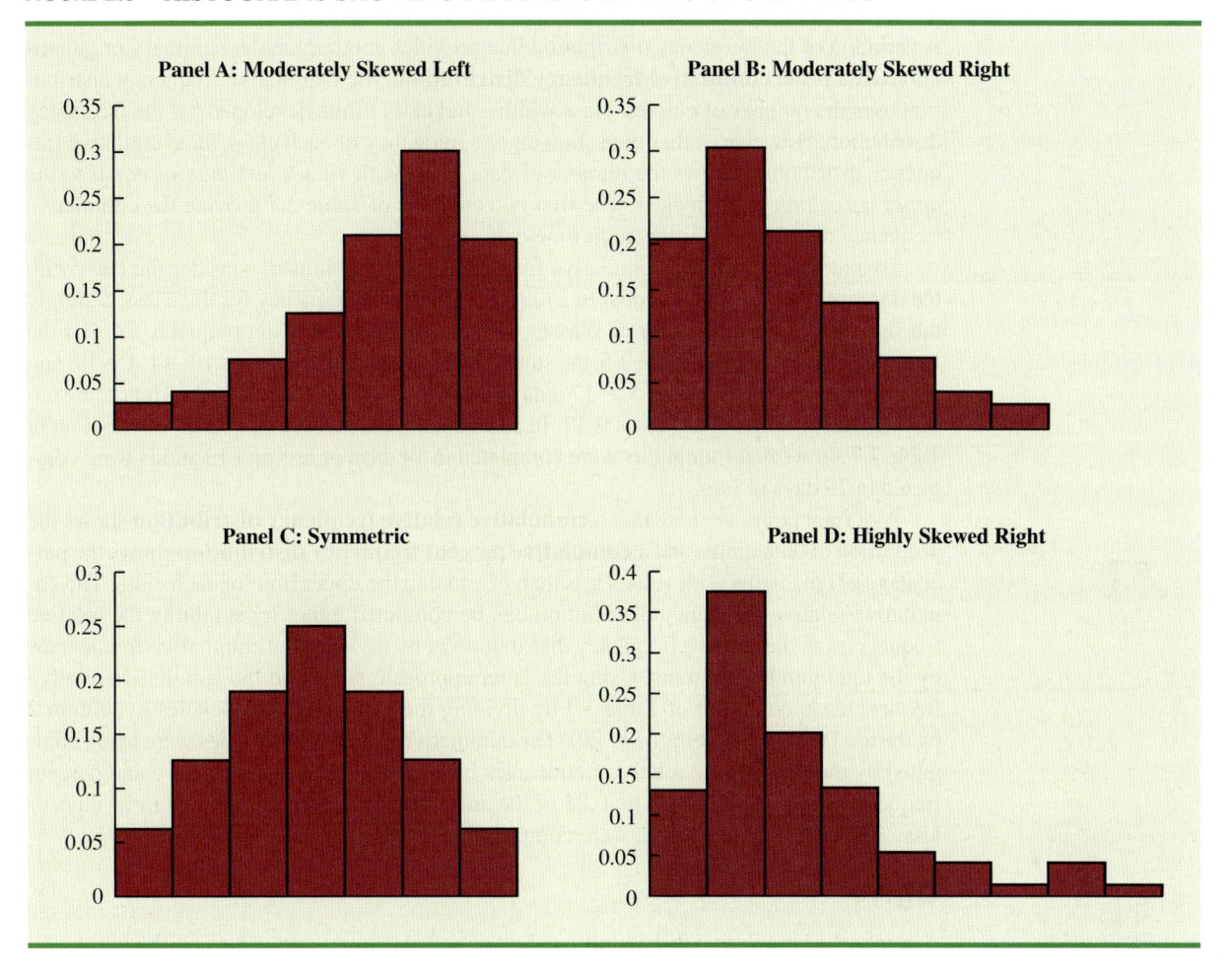

Panel A: Moderately Skewed Left

Panel B: Moderately Skewed Right

Panel C: Symmetric

Panel D: Highly Skewed Right

One of the most important uses of a histogram is to provide information about the shape, or form, of a distribution. Figure 2.5 contains four histograms constructed from relative frequency distributions. Panel A shows the histogram for a set of data moderately skewed to the left. A histogram is said to be skewed to the left if its tail extends farther to the left. This histogram is typical for exam scores, with no scores above 100%, most of the scores above 70%, and only a few really low scores. Panel B shows the histogram for a set of data moderately skewed to the right. A histogram is said to be skewed to the right if its tail extends farther to the right. An example of this type of histogram would be for data such as housing prices; a few expensive houses create the skewness in the right tail.

Panel C shows a symmetric histogram. In a symmetric histogram, the left tail mirrors the shape of the right tail. Histograms for data found in applications are never perfectly symmetric, but the histogram for many applications may be roughly symmetric. Data for SAT scores, heights and weights of people, and so on lead to histograms that are roughly symmetric. Panel D shows a histogram highly skewed to the right. This histogram was constructed from data on the amount of customer purchases over one day at a women's apparel store. Data from applications in business and economics often lead to histograms that are skewed to the right. For instance, data on housing prices, salaries, purchase amounts, and so on often result in histograms skewed to the right.

Cumulative Distributions

A variation of the frequency distribution that provides another tabular summary of quantitative data is the **cumulative frequency distribution**. The cumulative frequency distribution uses the number of classes, class widths, and class limits developed for the frequency distribution. However, rather than showing the frequency of each class, the cumulative frequency distribution shows the number of data items with values *less than or equal to the upper class limit* of each class. The first two columns of Table 2.7 provide the cumulative frequency distribution for the audit time data.

To understand how the cumulative frequencies are determined, consider the class with the description "less than or equal to 24." The cumulative frequency for this class is simply the sum of the frequencies for all classes with data values less than or equal to 24. For the frequency distribution in Table 2.5, the sum of the frequencies for classes 10–14, 15–19, and 20–24 indicates that 4 + 8 + 5 = 17 data values are less than or equal to 24. Hence, the cumulative frequency for this class is 17. In addition, the cumulative frequency distribution in Table 2.7 shows that four audits were completed in 14 days or less and 19 audits were completed in 29 days or less.

As a final point, we note that a **cumulative relative frequency distribution** shows the proportion of data items, and a **cumulative percent frequency distribution** shows the percentage of data items with values less than or equal to the upper limit of each class. The cumulative relative frequency distribution can be computed either by summing the relative frequencies in the relative frequency distribution or by dividing the cumulative frequencies by the total number of items. Using the latter approach, we found the cumulative relative frequencies in column 3 of Table 2.7 by dividing the cumulative frequencies in column 2 by the total number of items ($n = 20$). The cumulative percent frequencies were again computed by multiplying the relative frequencies by 100. The cumulative relative and percent frequency distributions show that .85 of the audits, or 85%, were completed in 24 days or less, .95 of the audits, or 95%, were completed in 29 days or less, and so on.

Ogive

A graph of a cumulative distribution, called an **ogive**, shows data values on the horizontal axis and either the cumulative frequencies, the cumulative relative frequencies, or the cumulative percent frequencies on the vertical axis. Figure 2.6 illustrates an ogive for the cumulative frequencies of the audit time data in Table 2.7.

The ogive is constructed by plotting a point corresponding to the cumulative frequency of each class. Because the classes for the audit time data are 10–14, 15–19, 20–24, and so on, one-unit gaps appear from 14 to 15, 19 to 20, and so on. These gaps are eliminated by plotting points halfway between the class limits. Thus, 14.5 is used for the 10–14 class, 19.5

TABLE 2.7 CUMULATIVE FREQUENCY, CUMULATIVE RELATIVE FREQUENCY, AND CUMULATIVE PERCENT FREQUENCY DISTRIBUTIONS FOR THE AUDIT TIME DATA

Audit Time (days)	Cumulative Frequency	Cumulative Relative Frequency	Cumulative Percent Frequency
Less than or equal to 14	4	.20	20
Less than or equal to 19	12	.60	60
Less than or equal to 24	17	.85	85
Less than or equal to 29	19	.95	95
Less than or equal to 34	20	1.00	100

2.2 Summarizing Quantitative Data

FIGURE 2.6 OGIVE FOR THE AUDIT TIME DATA

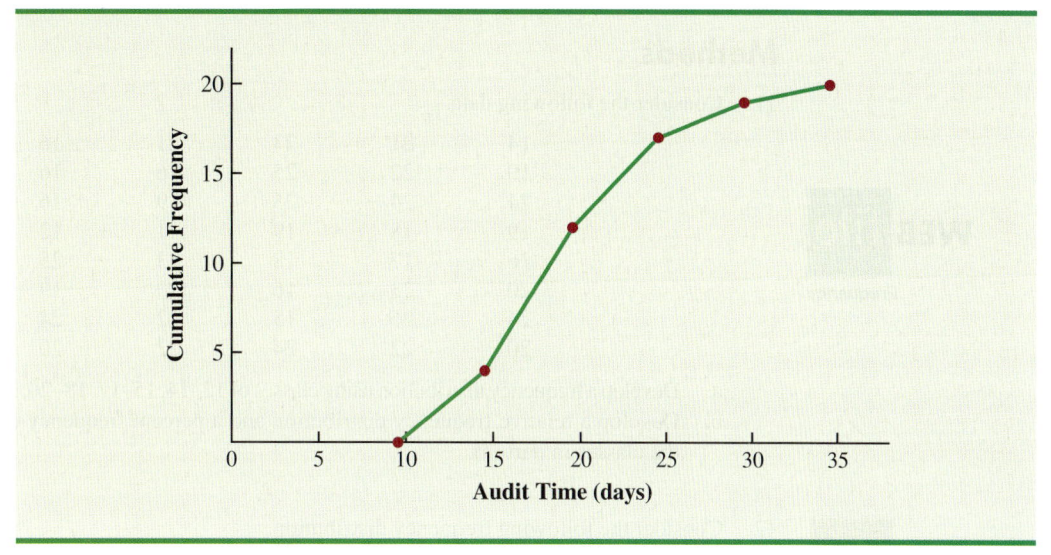

is used for the 15–19 class, and so on. The "less than or equal to 14" class with a cumulative frequency of 4 is shown on the ogive in Figure 2.6 by the point located at 14.5 on the horizontal axis and 4 on the vertical axis. The "less than or equal to 19" class with a cumulative frequency of 12 is shown by the point located at 19.5 on the horizontal axis and 12 on the vertical axis. Note that one additional point is plotted at the left end of the ogive. This point starts the ogive by showing that no data values fall below the 10–14 class. It is plotted at 9.5 on the horizontal axis and 0 on the vertical axis. The plotted points are connected by straight lines to complete the ogive.

NOTES AND COMMENTS

1. A bar chart and a histogram are essentially the same thing; both are graphical presentations of the data in a frequency distribution. A histogram is just a bar chart with no separation between bars. For some discrete quantitative data, a separation between bars is also appropriate. Consider, for example, the number of classes in which a college student is enrolled. The data may only assume integer values. Intermediate values such as 1.5, 2.73, and so on are not possible. With continuous quantitative data, however, such as the audit times in Table 2.4, a separation between bars is not appropriate.

2. The appropriate values for the class limits with quantitative data depend on the level of accuracy of the data. For instance, with the audit time data of Table 2.4 the limits used were integer values. If the data were rounded to the nearest tenth of a day (e.g., 12.3, 14.4, and so on), then the limits would be stated in tenths of days. For instance, the first class would be 10.0–14.9. If the data were recorded to the nearest hundredth of a day (e.g., 12.34, 14.45, and so on), the limits would be stated in hundredths of days. For instance, the first class would be 10.00–14.99.

3. An *open-ended* class requires only a lower class limit or an upper class limit. For example, in the audit time data of Table 2.4, suppose two of the audits had taken 58 and 65 days. Rather than continue with the classes of width 5 with classes 35–39, 40–44, 45–49, and so on, we could simplify the frequency distribution to show an open-end class of "35 or more." This class would have a frequency of 2. Most often the open-end class appears at the upper end of the distribution. Sometimes an open-end class appears at the lower end of the distribution, and occasionally such classes appear at both ends.

4. The last entry in a cumulative frequency distribution always equals the total number of observations. The last entry in a cumulative relative frequency distribution always equals 1.00 and the last entry in a cumulative percent frequency distribution always equals 100.

46 Chapter 2 Descriptive Statistics: Tabular and Graphical Presentations

Exercises

Methods

11. Consider the following data.

14	21	23	21	16
19	22	25	16	16
24	24	25	19	16
19	18	19	21	12
16	17	18	23	25
20	23	16	20	19
24	26	15	22	24
20	22	24	22	20

    a. Develop a frequency distribution using classes of 12–14, 15–17, 18–20, 21–23, and 24–26.
    b. Develop a relative frequency distribution and a percent frequency distribution using the classes in part (a).

12. Consider the following frequency distribution.

Class	Frequency
10–19	10
20–29	14
30–39	17
40–49	7
50–59	2

    Construct a cumulative frequency distribution and a cumulative relative frequency distribution.

13. Construct a histogram and an ogive for the data in exercise 12.

14. Consider the following data.

8.9	10.2	11.5	7.8	10.0	12.2	13.5	14.1	10.0	12.2
6.8	9.5	11.5	11.2	14.9	7.5	10.0	6.0	15.8	11.5

    a. Construct a dot plot.
    b. Construct a frequency distribution.
    c. Construct a percent frequency distribution.

Applications

15. A doctor's office staff studied the waiting times for patients who arrive at the office with a request for emergency service. The following data with waiting times in minutes were collected over a one-month period.

    2   5   10   12   4   4   5   17   11   8   9   8   12   21   6   8   7   13   18   3

    Use classes of 0–4, 5–9, and so on in the following:
    a. Show the frequency distribution.
    b. Show the relative frequency distribution.
    c. Show the cumulative frequency distribution.
    d. Show the cumulative relative frequency distribution.
    e. What proportion of patients needing emergency service wait nine minutes or less?

16. A shortage of candidates has required school districts to pay higher salaries and offer extras to attract and retain school district superintendents. The following data show the annual base

salary ($1000s) for superintendents in 20 districts in the greater Rochester, New York, area (*The Rochester Democrat and Chronicle,* February 10, 2008).

187	184	174	185
175	172	202	197
165	208	215	164
162	172	182	156
172	175	170	183

Use classes of 150–159, 160–169, and so on in the following.
a. Show the frequency distribution.
b. Show the percent frequency distribution.
c. Show the cumulative percent frequency distribution.
d. Develop a histogram for the annual base salary.
e. Do the data appear to be skewed? Explain.
f. What percentage of the superintendents make more than $200,000?

17. The Dow Jones Industrial Average (DJIA) underwent one of its infrequent reshufflings of companies when General Motors and Citigroup were replaced by Cisco Systems and Travelers (*The Wall Street Journal,* June 8, 2009). At the time, the prices per share for the 30 companies in the DJIA were as follows:

Company	$/Share	Company	$/Share
3M	61	IBM	107
Alcoa	11	Intel	16
American Express	25	JP Morgan Chase	35
AT&T	24	Johnson & Johnson	56
Bank of America	12	Kraft Foods	27
Boeing	52	McDonald's	59
Caterpillar	38	Merck	26
Chevron	69	Microsoft	22
Cisco Systems	20	Pfizer	14
Coca-Cola	49	Procter & Gamble	53
DuPont	27	Travelers	43
ExxonMobil	72	United Technologies	56
General Electric	14	Verizon	29
Hewlett-Packard	37	Walmart Stores	51
Home Depot	24	Walt Disney	25

a. What is the highest price per share? What is the lowest price per share?
b. Using a class width of 10, develop a frequency distribution for the data.
c. Prepare a histogram. Interpret the histogram, including a discussion of the general shape of the histogram, the midprice range, and the most frequent price range.
d. Use the *The Wall Street Journal* or another newspaper to find the current price per share for these companies. Prepare a histogram of the data and discuss any changes since June 2009. What company has had the largest increase in the price per share? What company has had the largest decrease in the price per share?

18. NRF/BIG research provided results of a consumer holiday spending survey (*USA Today,* December 20, 2005). The following data provide the dollar amount of holiday spending for a sample of 25 consumers.

1200	850	740	590	340
450	890	260	610	350
1780	180	850	2050	770
800	1090	510	520	220
1450	280	1120	200	350

a. What is the lowest holiday spending? The highest?
b. Use a class width of $250 to prepare a frequency distribution and a percent frequency distribution for the data.
c. Prepare a histogram and comment on the shape of the distribution.
d. What observations can you make about holiday spending?

19. *Fortune* provides a list of America's largest corporations based on annual revenue. The following table shows the 50 largest corporations' annual revenue expressed in billions of dollars (Money CNN website, January 15, 2010).

Corporation	Revenue	Corporation	Revenue
Amerisource Bergen	$ 71	Lowe's	$ 48
Archer Daniels Midland	70	Marathon Oil	74
AT&T	124	McKesson	102
Bank of America	113	Medco Health	51
Berkshire Hathaway	108	MetLife	55
Boeing	61	Microsoft	60
Cardinal Health	91	Morgan Stanley	62
Caterpillar	51	Pepsico	43
Chevron	263	Pfizer	48
Citigroup	112	Procter & Gamble	84
ConocoPhillips	231	Safeway	44
Costco Wholesale	72	Sears Holdings	47
CVS Caremark	87	State Farm Insurance	61
Dell	61	Sunoco	52
Dow Chemical	58	Target	65
Exxon Mobil	443	Time Warner	47
Ford Motors	146	United Parcel Service	51
General Electric	149	United Technologies	59
Goldman Sachs	54	UnitedHealth Group	81
Hewlett-Packard	118	Valero Energy	118
Home Depot	71	Verizon	97
IBM	104	Walgreen	59
JP Morgan Chase	101	Walmart	406
Johnson & Johnson	64	WellPoint	61
Kroger	76	Wells Fargo	52

a. Construct a frequency distribution (classes 0–49, 50–99, 100–149, and so on).
b. A relative frequency distribution
c. A cumulative frequency distribution
d. A cumulative relative frequency distribution
e. What do these distributions tell you about the annual revenue of the largest corporations in Amercia?
f. Show a histogram. Comment on the shape of the distribution.
g. What is the largest corporation in America and what is its annual revenue?

20. The *Golf Digest 50* lists the 50 professional golfers with the highest total annual income. Total income is the sum of both on-course and off-course earnings. Tiger Woods ranked first with a total annual income of $122 million. However, almost $100 million of this total was from off-course activities such as product endorsements and personal appearances. The 10 professional golfers with the highest *off-course* income are shown in the following table (Golf Digest website, February 2008).

Name	Off-Course Income ($1000s)
Tiger Woods	99,800
Phil Mickelson	40,200
Arnold Palmer	29,500
Vijay Singh	25,250
Ernie Els	24,500
Greg Norman	24,000
Jack Nicklaus	20,750
Sergio Garcia	14,500
Michelle Wie	12,500
Jim Furyk	11,000

The off-course income of all 50 professional golfers in the *Golf Digest 50* can be found on the website that accompanies the text. The income data are in $1000s. Use classes of 0–4999, 5000–9999, 10,000–14,999, and so on to answer the following questions. Include an open-ended class of 50,000 or more as the largest income class.

a. Construct a frequency distribution and percent frequency distribution of the annual off-course income of the 50 professional golfers.
b. Construct a histogram for these data.
c. Comment on the shape of the distribution of off-course income.
d. What is the most frequent off-course income class for the 50 professional golfers? Using your tabular and graphical summaries, what additional observations can you make about the off-course income of these 50 professional golfers?

21. The *Nielsen Home Technology Report* provides information about home technology and its usage. The following data are the hours of personal computer usage during one week for a sample of 50 persons.

4.1	1.5	10.4	5.9	3.4	5.7	1.6	6.1	3.0	3.7
3.1	4.8	2.0	14.8	5.4	4.2	3.9	4.1	11.1	3.5
4.1	4.1	8.8	5.6	4.3	3.3	7.1	10.3	6.2	7.6
10.8	2.8	9.5	12.9	12.1	0.7	4.0	9.2	4.4	5.7
7.2	6.1	5.7	5.9	4.7	3.9	3.7	3.1	6.1	3.1

Summarize the data by constructing the following:
a. A frequency distribution (use a class width of three hours)
b. A relative frequency distribution
c. A histogram
d. An ogive
e. Comment on what the data indicate about personal computer usage at home.

2.3 Exploratory Data Analysis: The Stem-and-Leaf Display

The techniques of **exploratory data analysis** consist of simple arithmetic and easy-to-draw graphs that can be used to summarize data quickly. One technique—referred to as a **stem-and-leaf display**—can be used to show both the rank order and shape of a data set simultaneously.

To illustrate the use of a stem-and-leaf display, consider the data in Table 2.8. These data result from a 150-question aptitude test given to 50 individuals recently interviewed

TABLE 2.8 NUMBER OF QUESTIONS ANSWERED CORRECTLY ON AN APTITUDE TEST

112	72	69	97	107
73	92	76	86	73
126	128	118	127	124
82	104	132	134	83
92	108	96	100	92
115	76	91	102	81
95	141	81	80	106
84	119	113	98	75
68	98	115	106	95
100	85	94	106	119

for a position at Haskens Manufacturing. The data indicate the number of questions answered correctly.

To develop a stem-and-leaf display, we first arrange the leading digits of each data value to the left of a vertical line. To the right of the vertical line, we record the last digit for each data value. Based on the top row of data in Table 2.8 (112, 72, 69, 97, and 107), the first five entries in constructing a stem-and-leaf display would be as follows:

```
 6 | 9
 7 | 2
 8 |
 9 | 7
10 | 7
11 | 2
12 |
13 |
14 |
```

For example, the data value 112 shows the leading digits 11 to the left of the line and the last digit 2 to the right of the line. Similarly, the data value 72 shows the leading digit 7 to the left of the line and last digit 2 to the right of the line. Continuing to place the last digit of each data value on the line corresponding to its leading digit(s) provides the following:

```
 6 | 9 8
 7 | 2 3 6 3 6 5
 8 | 6 2 3 1 1 0 4 5
 9 | 7 2 2 6 2 1 5 8 8 5 4
10 | 7 4 8 0 2 6 6 0 6
11 | 2 8 5 9 3 5 9
12 | 6 8 7 4
13 | 2 4
14 | 1
```

2.3 Exploratory Data Analysis: The Stem-and-Leaf Display

With this organization of the data, sorting the digits on each line into rank order is simple. Doing so provides the stem-and-leaf display shown here.

```
 6 | 8 9
 7 | 2 3 3 5 6 6
 8 | 0 1 1 2 3 4 5 6
 9 | 1 2 2 2 4 5 5 6 7 8 8
10 | 0 0 2 4 6 6 6 7 8
11 | 2 3 5 5 8 9 9
12 | 4 6 7 8
13 | 2 4
14 | 1
```

The numbers to the left of the vertical line (6, 7, 8, 9, 10, 11, 12, 13, and 14) form the *stem*, and each digit to the right of the vertical line is a *leaf*. For example, consider the first row with a stem value of 6 and leaves of 8 and 9.

```
6 | 8 9
```

This row indicates that two data values have a first digit of 6. The leaves show that the data values are 68 and 69. Similarly, the second row

```
7 | 2 3 3 5 6 6
```

indicates that six data values have a first digit of 7. The leaves show that the data values are 72, 73, 73, 75, 76, and 76.

To focus on the shape indicated by the stem-and-leaf display, let us use a rectangle to contain the leaves of each stem. Doing so, we obtain the following:

```
 6 | 8 9
 7 | 2 3 3 5 6 6
 8 | 0 1 1 2 3 4 5 6
 9 | 1 2 2 2 4 5 5 6 7 8 8
10 | 0 0 2 4 6 6 6 7 8
11 | 2 3 5 5 8 9 9
12 | 4 6 7 8
13 | 2 4
14 | 1
```

Rotating this page counterclockwise onto its side provides a picture of the data that is similar to a histogram with classes of 60–69, 70–79, 80–89, and so on.

Although the stem-and-leaf display may appear to offer the same information as a histogram, it has two primary advantages.

1. The stem-and-leaf display is easier to construct by hand.
2. Within a class interval, the stem-and-leaf display provides more information than the histogram because the stem-and-leaf shows the actual data.

Just as a frequency distribution or histogram has no absolute number of classes, neither does a stem-and-leaf display have an absolute number of rows or stems. If we believe that our original stem-and-leaf display condensed the data too much, we can easily stretch the display by using two or more stems for each leading digit. For example, to use two stems for each leading digit,

In a stretched stem-and-leaf display, whenever a stem value is stated twice, the first value corresponds to leaf values of 0–4, and the second value corresponds to leaf values of 5–9.

we would place all data values ending in 0, 1, 2, 3, and 4 in one row and all values ending in 5, 6, 7, 8, and 9 in a second row. The following stretched stem-and-leaf display illustrates this approach.

```
 6 | 8 9
 7 | 2 3 3
 7 | 5 6 6
 8 | 0 1 1 2 3 4
 8 | 5 6
 9 | 1 2 2 2 4
 9 | 5 5 6 7 8 8
10 | 0 0 2 4
10 | 6 6 6 7 8
11 | 2 3
11 | 5 5 8 9 9
12 | 4
12 | 6 7 8
13 | 2 4
13 |
14 | 1
```

Note that values 72, 73, and 73 have leaves in the 0–4 range and are shown with the first stem value of 7. The values 75, 76, and 76 have leaves in the 5–9 range and are shown with the second stem value of 7. This stretched stem-and-leaf display is similar to a frequency distribution with intervals of 65–69, 70–74, 75–79, and so on.

The preceding example showed a stem-and-leaf display for data with as many as three digits. Stem-and-leaf displays for data with more than three digits are possible. For example, consider the following data on the number of hamburgers sold by a fast-food restaurant for each of 15 weeks.

| 1565 | 1852 | 1644 | 1766 | 1888 | 1912 | 2044 | 1812 |
| 1790 | 1679 | 2008 | 1852 | 1967 | 1954 | 1733 |

A stem-and-leaf display of these data follows.

```
Leaf unit = 10
15 | 6
16 | 4 7
17 | 3 6 9
18 | 1 5 5 8
19 | 1 5 6
20 | 0 4
```

A single digit is used to define each leaf in a stem-and-leaf display. The leaf unit indicates how to multiply the stem-and-leaf numbers in order to approximate the original data. Leaf units may be 100, 10, 1, 0.1, and so on.

Note that a single digit is used to define each leaf and that only the first three digits of each data value have been used to construct the display. At the top of the display we have specified Leaf unit = 10. To illustrate how to interpret the values in the display, consider the first stem, 15, and its associated leaf, 6. Combining these numbers, we obtain 156. To reconstruct an approximation of the original data value, we must multiply this number by 10, the value of the *leaf unit.* Thus, 156 × 10 = 1560 is an approximation of the original data value used to construct the stem-and-leaf display. Although it is not possible to reconstruct the exact data value from this stem-and-leaf display, the convention of using a single digit for each leaf enables stem-and-leaf displays to be constructed for data having a large number of digits. For stem-and-leaf displays where the leaf unit is not shown, the leaf unit is assumed to equal 1.

2.3 Exploratory Data Analysis: The Stem-and-Leaf Display

Exercises

Methods

22. Construct a stem-and-leaf display for the following data.

70	72	75	64	58	83	80	82
76	75	68	65	57	78	85	72

23. Construct a stem-and-leaf display for the following data.

11.3	9.6	10.4	7.5	8.3	10.5	10.0
9.3	8.1	7.7	7.5	8.4	6.3	8.8

24. Construct a stem-and-leaf display for the following data. Use a leaf unit of 10.

1161	1206	1478	1300	1604	1725	1361	1422
1221	1378	1623	1426	1557	1730	1706	1689

Applications

25. A psychologist developed a new test of adult intelligence. The test was administered to 20 individuals, and the following data were obtained.

114	99	131	124	117	102	106	127	119	115
98	104	144	151	132	106	125	122	118	118

    Construct a stem-and-leaf display for the data.

26. *Money* magazine listed top career opportunities for work that is enjoyable, pays well, and will still be around 10 years from now (*Money*, November 2009). Shown in the following table are 20 top career opportunities with the median pay and top pay for workers with two to seven years of experience in the field. Data are shown in thousands of dollars.

Career	Median Pay	Top Pay
Account Executive	$ 81	$157
Certified Public Accountant	74	138
Computer Security Consultant	100	138
Director of Communications	78	135
Financial Analyst	80	109
Finance Director	121	214
Financial Research Analyst	66	155
Hotel General Manager	77	146
Human Resources Manager	72	111
Investment Banking	106	221
IT Business Analyst	83	119
IT Project Manager	99	140
Marketing Manager	77	126
Quality-Assurance Manager	80	122
Sales Representative	67	125
Senior Internal Auditor	76	106
Software Developer	79	116
Software Program Manager	110	152
Systems Engineer	87	130
Technical Writer	67	100

    Develop a stem-and-leaf display for both the median pay and the top pay. Comment on what you learn about the pay for these careers.

27. Most major ski resorts offer family programs that provide ski and snowboarding instruction for children. The typical classes provide four to six hours on the snow with a certified instructor. The daily rate for a group lesson at 15 ski resorts follows (*The Wall Street Journal*, January 20, 2006).

54 Chapter 2 Descriptive Statistics: Tabular and Graphical Presentations

Resort	Location	Daily Rate	Resort	Location	Daily Rate
Beaver Creek	Colorado	$137	Okemo	Vermont	$ 86
Deer Valley	Utah	115	Park City	Utah	145
Diamond Peak	California	95	Butternut	Massachusetts	75
Heavenly	California	145	Steamboat	Colorado	98
Hunter	New York	79	Stowe	Vermont	104
Mammoth	California	111	Sugar Bowl	California	100
Mount Sunapee	New Hampshire	96	Whistler-Blackcomb	British Columbia	104
Mount Bachelor	Oregon	83			

a. Develop a stem-and-leaf display for the data.
b. Interpret the stem-and-leaf display in terms of what it tells you about the daily rate for these ski and snowboarding instruction programs.

28. The 2004 Naples, Florida, minimarathon (13.1 miles) had 1228 registrants (*Naples Daily News,* January 17, 2004). Competition was held in six age groups. The following data show the ages for a sample of 40 individuals who participated in the marathon.

49	33	40	37	56
44	46	57	55	32
50	52	43	64	40
46	24	30	37	43
31	43	50	36	61
27	44	35	31	43
52	43	66	31	50
72	26	59	21	47

WEB file
Marathon

a. Show a stretched stem-and-leaf display.
b. What age group had the largest number of runners?
c. What age occurred most frequently?
d. A *Naples Daily News* feature article emphasized the number of runners who were "20-something." What percentage of the runners were in the 20-something age group? What do you suppose was the focus of the article?

2.4 Crosstabulations and Scatter Diagrams

Crosstabulations and scatter diagrams are used to summarize data in a way that reveals the relationship between two variables.

Thus far in this chapter, we have focused on tabular and graphical methods used to summarize the data for *one variable at a time.* Often a manager or decision maker requires tabular and graphical methods that will assist in the understanding of the *relationship between two variables.* Crosstabulation and scatter diagrams are two such methods.

Crosstabulation

A **crosstabulation** is a tabular summary of data for two variables. Let us illustrate the use of a crosstabulation by considering the following application based on data from *Zagat's Restaurant Review.* The quality rating and the meal price data were collected for a sample of 300 restaurants located in the Los Angeles area. Table 2.9 shows the data for the first 10 restaurants. Data on a restaurant's quality rating and typical meal price are reported. Quality rating is a categorical variable with rating categories of good, very good, and excellent. Meal price is a quantitative variable that ranges from $10 to $49.

A crosstabulation of the data for this application is shown in Table 2.10. The left and top margin labels define the classes for the two variables. In the left margin, the row labels (good, very good, and excellent) correspond to the three classes of the quality rating variable. In the top margin, the column labels ($10–19, $20–29, $30–39, and $40–49) correspond to

2.4 Crosstabulations and Scatter Diagrams

TABLE 2.9 QUALITY RATING AND MEAL PRICE FOR 300 LOS ANGELES RESTAURANTS

Restaurant	Quality Rating	Meal Price ($)
1	Good	18
2	Very Good	22
3	Good	28
4	Excellent	38
5	Very Good	33
6	Good	28
7	Very Good	19
8	Very Good	11
9	Very Good	23
10	Good	13
.	.	.
.	.	.
.	.	.

the four classes of the meal price variable. Each restaurant in the sample provides a quality rating and a meal price. Thus, each restaurant in the sample is associated with a cell appearing in one of the rows and one of the columns of the crosstabulation. For example, restaurant 5 is identified as having a very good quality rating and a meal price of $33. This restaurant belongs to the cell in row 2 and column 3 of Table 2.10. In constructing a crosstabulation, we simply count the number of restaurants that belong to each of the cells in the crosstabulation table.

In reviewing Table 2.10, we see that the greatest number of restaurants in the sample (64) have a very good rating and a meal price in the $20–29 range. Only two restaurants have an excellent rating and a meal price in the $10–19 range. Similar interpretations of the other frequencies can be made. In addition, note that the right and bottom margins of the crosstabulation provide the frequency distributions for quality rating and meal price separately. From the frequency distribution in the right margin, we see that data on quality ratings show 84 good restaurants, 150 very good restaurants, and 66 excellent restaurants. Similarly, the bottom margin shows the frequency distribution for the meal price variable.

Dividing the totals in the right margin of the crosstabulation by the total for that column provides a relative and percent frequency distribution for the quality rating variable.

Quality Rating	Relative Frequency	Percent Frequency
Good	.28	28
Very Good	.50	50
Excellent	.22	22
Total	1.00	100

TABLE 2.10 CROSSTABULATION OF QUALITY RATING AND MEAL PRICE FOR 300 LOS ANGELES RESTAURANTS

	Meal Price				
Quality Rating	$10–19	$20–29	$30–39	$40–49	Total
Good	42	40	2	0	84
Very Good	34	64	46	6	150
Excellent	2	14	28	22	66
Total	78	118	76	28	300

From the percent frequency distribution we see that 28% of the restaurants were rated good, 50% were rated very good, and 22% were rated excellent.

Dividing the totals in the bottom row of the crosstabulation by the total for that row provides a relative and percent frequency distribution for the meal price variable.

Meal Price	Relative Frequency	Percent Frequency
$10–19	.26	26
$20–29	.39	39
$30–39	.25	25
$40–49	.09	9
Total	1.00	100

Note that the sum of the values in each column does not add exactly to the column total, because the values being summed are rounded. From the percent frequency distribution we see that 26% of the meal prices are in the lowest price class ($10–19), 39% are in the next higher class, and so on.

The frequency and relative frequency distributions constructed from the margins of a crosstabulation provide information about each of the variables individually, but they do not shed any light on the relationship between the variables. The primary value of a crosstabulation lies in the insight it offers about the relationship between the variables. A review of the crosstabulation in Table 2.10 reveals that higher meal prices are associated with the higher quality restaurants, and the lower meal prices are associated with the lower quality restaurants.

Converting the entries in a crosstabulation into row percentages or column percentages can provide more insight into the relationship between the two variables. For row percentages, the results of dividing each frequency in Table 2.10 by its corresponding row total are shown in Table 2.11. Each row of Table 2.11 is a percent frequency distribution of meal price for one of the quality rating categories. Of the restaurants with the lowest quality rating (good), we see that the greatest percentages are for the less expensive restaurants (50% have $10–19 meal prices and 47.6% have $20–29 meal prices). Of the restaurants with the highest quality rating (excellent), we see that the greatest percentages are for the more expensive restaurants (42.4% have $30–39 meal prices and 33.4% have $40–49 meal prices). Thus, we continue to see that the more expensive meals are associated with the higher quality restaurants.

Crosstabulation is widely used for examining the relationship between two variables. In practice, the final reports for many statistical studies include a large number of crosstabulation tables. In the Los Angeles restaurant survey, the crosstabulation is based on one qualitative variable (quality rating) and one quantitative variable (meal price). Crosstabulations can also be developed when both variables are qualitative and when both variables are quantitative. When quantitative variables are used, however, we must first create classes for the values of the variable. For instance, in the restaurant example we grouped the meal prices into four classes ($10–19, $20–29, $30–39, and $40–49).

TABLE 2.11 ROW PERCENTAGES FOR EACH QUALITY RATING CATEGORY

Quality Rating	$10–19	$20–29	$30–39	$40–49	Total
Good	50.0	47.6	2.4	0.0	100
Very Good	22.7	42.7	30.6	4.0	100
Excellent	3.0	21.2	42.4	33.4	100

Simpson's Paradox

The data in two or more crosstabulations are often combined or aggregated to produce a summary crosstabulation showing how two variables are related. In such cases, we must be careful in drawing a conclusion because a conclusion based upon aggregate data can be reversed if we look at the unaggregated data. The reversal of conclusions based on aggregate and unaggregated data is called **Simpson's paradox**. To provide an illustration of Simpson's paradox, we consider an example involving the analysis of verdicts for two judges in two different courts.

Judges Ron Luckett and Dennis Kendall presided over cases in Common Pleas Court and Municipal Court during the past three years. Some of the verdicts they rendered were appealed. In most of these cases the appeals court upheld the original verdicts, but in some cases those verdicts were reversed. For each judge a crosstabulation was developed based upon two variables: Verdict (upheld or reversed) and Type of Court (Common Pleas and Municipal). Suppose that the two crosstabulations were then combined by aggregating the type of court data. The resulting aggregated crosstabulation contains two variables: Verdict (upheld or reversed) and Judge (Luckett or Kendall). This crosstabulation shows the number of appeals in which the verdict was upheld and the number in which the verdict was reversed for both judges. The following crosstabulation shows these results along with the column percentages in parentheses next to each value.

	Judge		
Verdict	Luckett	Kendall	Total
Upheld	129 (86%)	110 (88%)	239
Reversed	21 (14%)	15 (12%)	36
Total (%)	150 (100%)	125 (100%)	275

A review of the column percentages shows that 86% of the verdicts were upheld for Judge Luckett, while 88% of the verdicts were upheld for Judge Kendall. From this aggregated crosstabulation, we conclude that Judge Kendall is doing the better job because a greater percentage of Judge Kendall's verdicts are being upheld.

The following unaggregated crosstabulations show the cases tried by Judge Luckett and Judge Kendall in each court; column percentages are shown in parentheses next to each value.

Judge Luckett

Verdict	Common Pleas	Municipal Court	Total
Upheld	29 (91%)	100 (85%)	129
Reversed	3 (9%)	18 (15%)	21
Total (%)	32 (100%)	118 (100%)	150

Judge Kendall

Verdict	Common Pleas	Municipal Court	Total
Upheld	90 (90%)	20 (80%)	110
Reversed	10 (10%)	5 (20%)	15
Total (%)	100 (100%)	25 (100%)	125

From the crosstabulation and column percentages for Judge Luckett, we see that the verdicts were upheld in 91% of the Common Pleas Court cases and in 85% of the Municipal Court cases. From the crosstabulation and column percentages for Judge Kendall, we see that the verdicts were upheld in 90% of the Common Pleas Court cases and in 80% of the Municipal Court cases. Thus, when we unaggregate the data, we see that Judge Luckett has a better record because a greater percentage of Judge Luckett's verdicts are being upheld in both courts. This result contradicts the conclusion we reached with the aggregated data crosstabulation, which showed Judge Kendall had the better record. This reversal of conclusions based on aggregated and unaggregated data illustrates Simpson's paradox.

The original crosstabulation was obtained by aggregating the data in the separate crosstabulations for the two courts. Note that for both judges the percentage of appeals that resulted in reversals was much higher in Municipal Court than in Common Pleas Court. Because Judge Luckett tried a much higher percentage of his cases in Municipal Court, the aggregated data favored Judge Kendall. When we look at the crosstabulations for the two courts separately, however, Judge Luckett shows the better record. Thus, for the original crosstabulation, we see that the *type of court* is a hidden variable that cannot be ignored when evaluating the records of the two judges.

Because of the possibility of Simpson's paradox, realize that the conclusion or interpretation may be reversed depending upon whether you are viewing unaggregated or aggregate crosstabulation data. Before drawing a conclusion, you may want to investigate whether the aggregate or unaggregate form of the crosstabulation provides the better insight and conclusion. Especially when the crosstabulation involves aggreagrated data, you should investigate whether a hidden variable could affect the results such that separate or unaggregated crosstabulations provide a different and possibly better insight and conclusion.

Scatter Diagram and Trendline

A **scatter diagram** is a graphical presentation of the relationship between two quantitative variables, and a **trendline** is a line that provides an approximation of the relationship. As an illustration, consider the advertising/sales relationship for a stereo and sound equipment store in San Francisco. On 10 occasions during the past three months, the store used weekend television commercials to promote sales at its stores. The managers want to investigate whether a relationship exists between the number of commercials shown and sales at the store during the following week. Sample data for the 10 weeks with sales in hundreds of dollars are shown in Table 2.12.

Figure 2.7 shows the scatter diagram and the trendline[1] for the data in Table 2.12. The number of commercials (x) is shown on the horizontal axis and the sales (y) are shown on the vertical axis. For week 1, $x = 2$ and $y = 50$. A point with those coordinates is plotted on the scatter diagram. Similar points are plotted for the other nine weeks. Note that during two of the weeks one commercial was shown, during two of the weeks two commercials were shown, and so on.

The completed scatter diagram in Figure 2.7 indicates a positive relationship between the number of commercials and sales. Higher sales are associated with a higher number of commercials. The relationship is not perfect in that all points are not on a straight line. However, the general pattern of the points and the trendline suggest that the overall relationship is positive.

TABLE 2.12 SAMPLE DATA FOR THE STEREO AND SOUND EQUIPMENT STORE

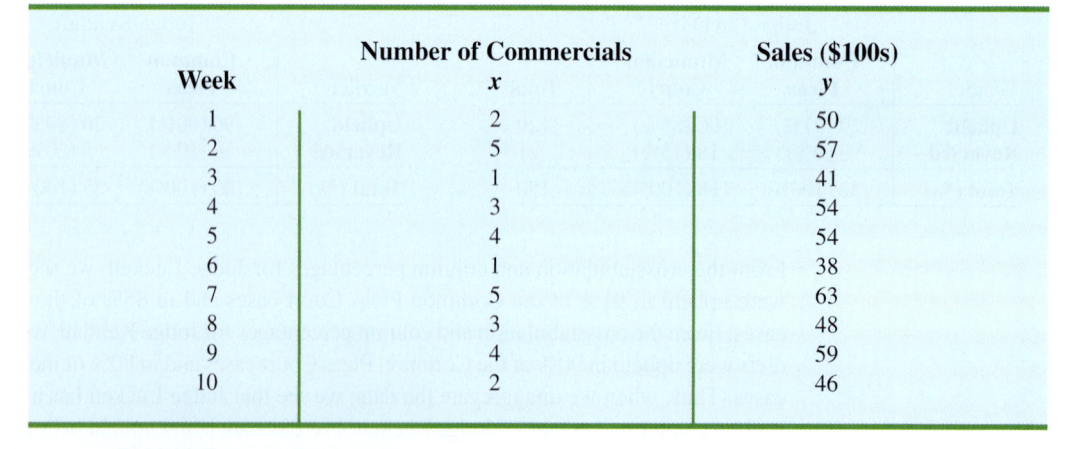

Week	Number of Commercials x	Sales ($100s) y
1	2	50
2	5	57
3	1	41
4	3	54
5	4	54
6	1	38
7	5	63
8	3	48
9	4	59
10	2	46

---

[1] The equation of the trendline is $y = 36.15 + 4.95x$. The slope of the trendline is 4.95 and the y-intercept (the point where the line intersects the y-axis) is 36.15. We will discuss in detail the interpretation of the slope and y-intercept for a linear trendline in Chapter 14 when we study simple linear regression.

2.4 Crosstabulations and Scatter Diagrams

FIGURE 2.7 SCATTER DIAGRAM AND TRENDLINE FOR THE STEREO AND SOUND EQUIPMENT STORE

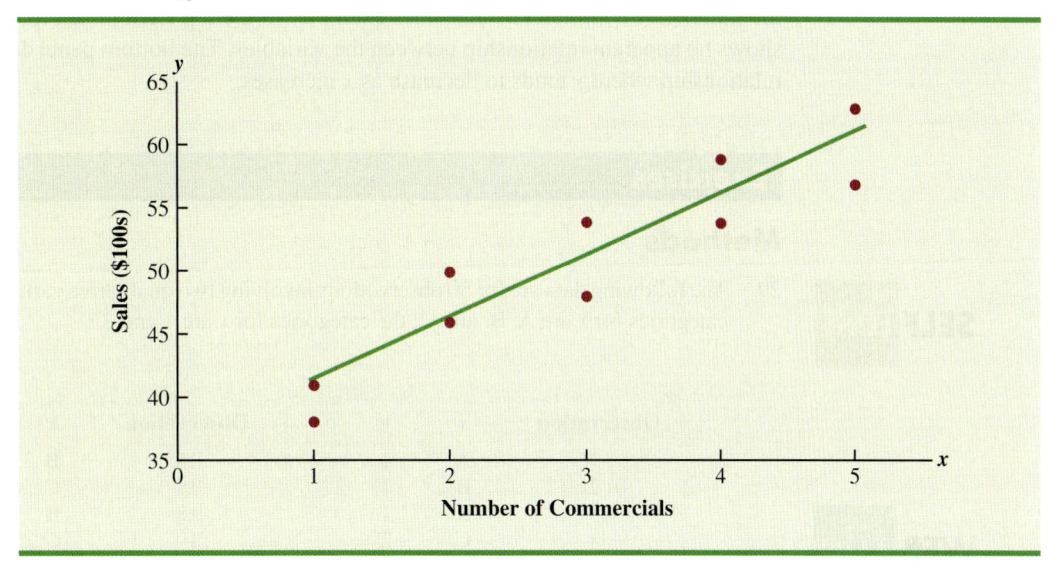

FIGURE 2.8 TYPES OF RELATIONSHIPS DEPICTED BY SCATTER DIAGRAMS

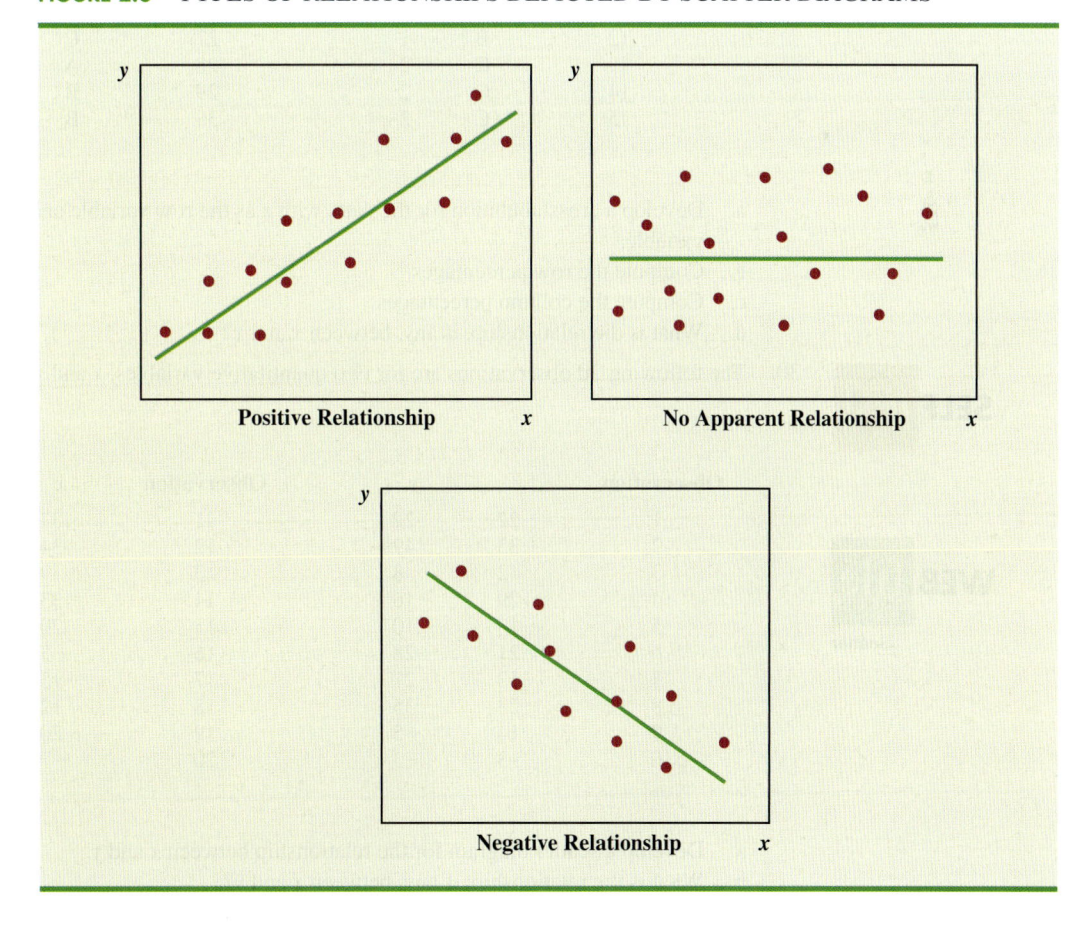

Some general scatter diagram patterns and the types of relationships they suggest are shown in Figure 2.8. The top left panel depicts a positive relationship similar to the one for the number of commercials and sales example. In the top right panel, the scatter diagram shows no apparent relationship between the variables. The bottom panel depicts a negative relationship where *y* tends to decrease as *x* increases.

Exercises

Methods

29. The following data are for 30 observations involving two qualitative variables, *x* and *y*. The categories for *x* are A, B, and C; the categories for *y* are 1 and 2.

Observation	x	y	Observation	x	y
1	A	1	16	B	2
2	B	1	17	C	1
3	B	1	18	B	1
4	C	2	19	C	1
5	B	1	20	B	1
6	C	2	21	C	2
7	B	1	22	B	1
8	C	2	23	C	2
9	A	1	24	A	1
10	B	1	25	B	1
11	A	1	26	C	2
12	B	1	27	C	2
13	C	2	28	A	1
14	C	2	29	B	1
15	C	2	30	B	2

a. Develop a crosstabulation for the data, with *x* as the row variable and *y* as the column variable.
b. Compute the row percentages.
c. Compute the column percentages.
d. What is the relationship, if any, between *x* and *y*?

30. The following 20 observations are for two quantitative variables, *x* and *y*.

Observation	x	y	Observation	x	y
1	−22	22	11	−37	48
2	−33	49	12	34	−29
3	2	8	13	9	−18
4	29	−16	14	−33	31
5	−13	10	15	20	−16
6	21	−28	16	−3	14
7	−13	27	17	−15	18
8	−23	35	18	12	17
9	14	−5	19	−20	−11
10	3	−3	20	−7	−22

a. Develop a scatter diagram for the relationship between *x* and *y*.
b. What is the relationship, if any, between *x* and *y*?

Applications

31. The following crosstabulation shows household income by educational level of the head of household (*Statistical Abstract of the United States: 2008*).

	\multicolumn{5}{c}{Household Income ($1000s)}	
Educational Level	Under 25	25.0–49.9	50.0–74.9	75.0–99.9	100 or more	Total
Not H.S. graduate	4,207	3,459	1,389	539	367	9,961
H.S. graduate	4,917	6,850	5,027	2,637	2,668	22,099
Some college	2,807	5,258	4,678	3,250	4,074	20,067
Bachelor's degree	885	2,094	2,848	2,581	5,379	13,787
Beyond bach. deg.	290	829	1,274	1,241	4,188	7,822
Total	13,106	18,490	15,216	10,248	16,676	73,736

   a. Compute the row percentages and identify the percent frequency distributions of income for households in which the head is a high school graduate and in which the head holds a bachelor's degree.
   b. What percentage of households headed by high school graduates earn $75,000 or more? What percentage of households headed by bachelor's degree recipients earn $75,000 or more?
   c. Construct percent frequency histograms of income for households headed by persons with a high school degree and for those headed by persons with a bachelor's degree. Is any relationship evident between household income and educational level?

32. Refer again to the crosstabulation of household income by educational level shown in exercise 31.
   a. Compute column percentages and identify the percent frequency distributions displayed. What percentage of the heads of households did not graduate from high school?
   b. What percentage of the households earning $100,000 or more were headed by a person having schooling beyond a bachelor's degree? What percentage of the households headed by a person with schooling beyond a bachelor's degree earned over $100,000? Why are these two percentages different?
   c. Compare the percent frequency distributions for those households earning "Under 25," "100 or more," and for "Total." Comment on the relationship between household income and educational level of the head of household.

33. Recently, management at Oak Tree Golf Course received a few complaints about the condition of the greens. Several players complained that the greens are too fast. Rather than react to the comments of just a few, the Golf Association conducted a survey of 100 male and 100 female golfers. The survey results are summarized here.

Male Golfers

	\multicolumn{2}{c}{Greens Condition}
Handicap	Too Fast	Fine
Under 15	10	40
15 or more	25	25

Female Golfers

	\multicolumn{2}{c}{Greens Condition}
Handicap	Too Fast	Fine
Under 15	1	9
15 or more	39	51

   a. Combine these two crosstabulations into one with Male and Female as the row labels and Too Fast and Fine as the column labels. Which group shows the highest percentage saying that the greens are too fast?

b. Refer to the initial crosstabulations. For those players with low handicaps (better players), which group (male or female) shows the highest percentage saying the greens are too fast?
c. Refer to the initial crosstabulations. For those players with higher handicaps, which group (male or female) shows the highest percentage saying the greens are too fast?
d. What conclusions can you draw about the preferences of men and women concerning the speed of the greens? Are the conclusions you draw from part (a) as compared with parts (b) and (c) consistent? Explain any apparent inconsistencies.

34. Table 2.13 shows a data set containing information for 45 mutual funds that are part of the *Morningstar Funds 500* for 2008. The data set includes the following five variables:

   Fund Type: The type of fund, labeled DE (Domestic Equity), IE (International Equity), and FI (Fixed Income)

   Net Asset Value ($):   The closing price per share

   Five-Year Average Return (%):   The average annual return for the fund over the past five years

   Expense Ratio (%):   The percentage of assets deducted each fiscal year for fund expenses

   Morningstar Rank: The risk adjusted star rating for each fund; Morningstar ranks go from a low of 1-Star to a high of 5-Stars

   a. Prepare a crosstabulation of the data on Fund Type (rows) and the average annual return over the past five years (columns). Use classes of 0–9.99, 10–19.99, 20–29.99, 30–39.99, 40–49.99, and 50–59.99 for the Five-Year Average Return (%).
   b. Prepare a frequency distribution for the data on Fund Type.
   c. Prepare a frequency distribution for the data on Five-Year Average Return (%).
   d. How has the crosstabulation helped in preparing the frequency distributions in parts (b) and (c)?
   e. What conclusions can you draw about the fund type and the average return over the past 5 years?

35. Refer to the data in Table 2.13.
   a. Prepare a crosstabulation of the data on Fund Type (rows) and the expense ratio (columns). Use classes of .25–.49, .50–.74, .75–.99, 1.00–1.24, and 1.25–1.49 for Expense Ratio (%).
   b. Prepare a percent frequency distribution for Expense Ratio (%).
   c. What conclusions can you draw about fund type and the expense ratio?

36. Refer to the data in Table 2.13.
   a. Prepare a scatter diagram with Five-Year Average Return (%) on the horizontal axis and Net Asset Value ($) on the vertical axis.
   b. Comment on the relationship, if any, between the variables.

37. The U.S. Department of Energy's Fuel Economy Guide provides fuel efficiency data for cars and trucks (Fuel Economy website, February 22, 2008). A portion of the data for 311 compact, midsize, and large cars is shown in Table 2.14. The data set contains the following variables:

   Size: Compact, Midsize, and Large

   Displacement: Engine size in liters

   Cylinders: Number of cylinders in the engine

   Drive: Front wheel (F), rear wheel (R), and four wheel (4)

   Fuel Type: Premium (P) or regular (R) fuel

   City MPG: Fuel efficiency rating for city driving in terms of miles per gallon

   Hwy MPG: Fuel efficiency rating for highway driving in terms of miles per gallon

The complete data set is contained in the file named FuelData08.
a. Prepare a crosstabulation of the data on Size (rows) and Hwy MPG (columns). Use classes of 15–19, 20–24, 25–29, 30–34, and 35–39 for Hwy MPG.
b. Comment on the relationship beween Size and Hwy MPG.

TABLE 2.13 FINANCIAL DATA FOR A SAMPLE OF 45 MUTUAL FUNDS

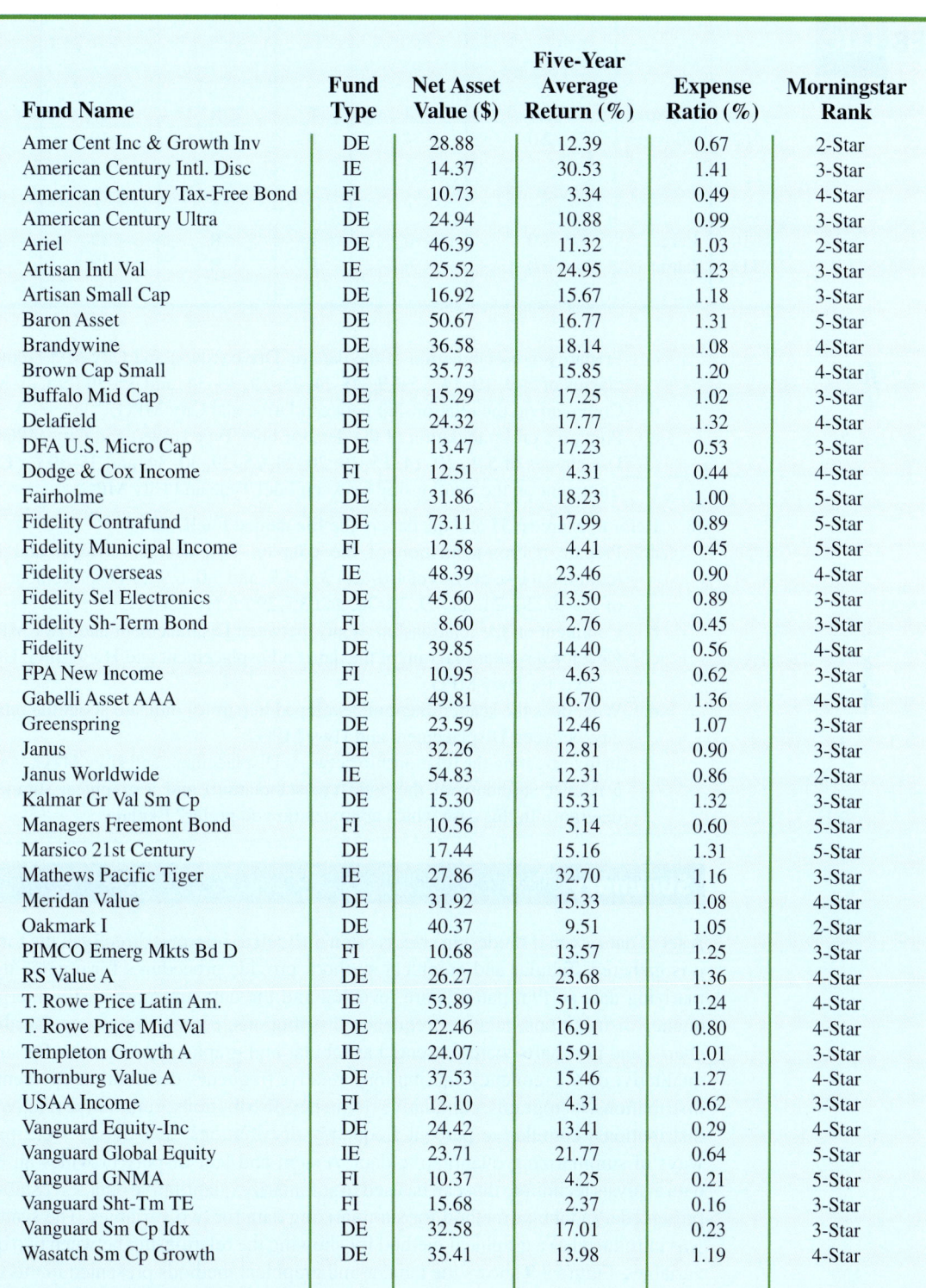

Fund Name	Fund Type	Net Asset Value ($)	Five-Year Average Return (%)	Expense Ratio (%)	Morningstar Rank
Amer Cent Inc & Growth Inv	DE	28.88	12.39	0.67	2-Star
American Century Intl. Disc	IE	14.37	30.53	1.41	3-Star
American Century Tax-Free Bond	FI	10.73	3.34	0.49	4-Star
American Century Ultra	DE	24.94	10.88	0.99	3-Star
Ariel	DE	46.39	11.32	1.03	2-Star
Artisan Intl Val	IE	25.52	24.95	1.23	3-Star
Artisan Small Cap	DE	16.92	15.67	1.18	3-Star
Baron Asset	DE	50.67	16.77	1.31	5-Star
Brandywine	DE	36.58	18.14	1.08	4-Star
Brown Cap Small	DE	35.73	15.85	1.20	4-Star
Buffalo Mid Cap	DE	15.29	17.25	1.02	3-Star
Delafield	DE	24.32	17.77	1.32	4-Star
DFA U.S. Micro Cap	DE	13.47	17.23	0.53	3-Star
Dodge & Cox Income	FI	12.51	4.31	0.44	4-Star
Fairholme	DE	31.86	18.23	1.00	5-Star
Fidelity Contrafund	DE	73.11	17.99	0.89	5-Star
Fidelity Municipal Income	FI	12.58	4.41	0.45	5-Star
Fidelity Overseas	IE	48.39	23.46	0.90	4-Star
Fidelity Sel Electronics	DE	45.60	13.50	0.89	3-Star
Fidelity Sh-Term Bond	FI	8.60	2.76	0.45	3-Star
Fidelity	DE	39.85	14.40	0.56	4-Star
FPA New Income	FI	10.95	4.63	0.62	3-Star
Gabelli Asset AAA	DE	49.81	16.70	1.36	4-Star
Greenspring	DE	23.59	12.46	1.07	3-Star
Janus	DE	32.26	12.81	0.90	3-Star
Janus Worldwide	IE	54.83	12.31	0.86	2-Star
Kalmar Gr Val Sm Cp	DE	15.30	15.31	1.32	3-Star
Managers Freemont Bond	FI	10.56	5.14	0.60	5-Star
Marsico 21st Century	DE	17.44	15.16	1.31	5-Star
Mathews Pacific Tiger	IE	27.86	32.70	1.16	3-Star
Meridan Value	DE	31.92	15.33	1.08	4-Star
Oakmark I	DE	40.37	9.51	1.05	2-Star
PIMCO Emerg Mkts Bd D	FI	10.68	13.57	1.25	3-Star
RS Value A	DE	26.27	23.68	1.36	4-Star
T. Rowe Price Latin Am.	IE	53.89	51.10	1.24	4-Star
T. Rowe Price Mid Val	DE	22.46	16.91	0.80	4-Star
Templeton Growth A	IE	24.07	15.91	1.01	3-Star
Thornburg Value A	DE	37.53	15.46	1.27	4-Star
USAA Income	FI	12.10	4.31	0.62	3-Star
Vanguard Equity-Inc	DE	24.42	13.41	0.29	4-Star
Vanguard Global Equity	IE	23.71	21.77	0.64	5-Star
Vanguard GNMA	FI	10.37	4.25	0.21	5-Star
Vanguard Sht-Tm TE	FI	15.68	2.37	0.16	3-Star
Vanguard Sm Cp Idx	DE	32.58	17.01	0.23	3-Star
Wasatch Sm Cp Growth	DE	35.41	13.98	1.19	4-Star

WEBfile
MutualFunds

TABLE 2.14 FUEL EFFICIENCY DATA FOR 311 CARS

Car	Size	Displacement	Cylinders	Drive	Fuel Type	City MPG	Hwy MPG
1	Compact	3.1	6	4	P	15	25
2	Compact	3.1	6	4	P	17	25
3	Compact	3.0	6	4	P	17	25
.	.	.	.	.	.	.	.
.	.	.	.	.	.	.	.
.	.	.	.	.	.	.	.
161	Midsize	2.4	4	F	R	22	30
162	Midsize	2.0	4	F	P	19	29
.	.	.	.	.	.	.	.
.	.	.	.	.	.	.	.
.	.	.	.	.	.	.	.
310	Large	3.0	6	F	R	17	25
311	Large	3.0	6	F	R	18	25

WEB file
FuelData08

    c. Prepare a crosstabulation of the data on Drive (rows) and City MPG (columns). Use classes of 5–9, 10–14, 15–19, 20–24, 25–29, 30–34, and 35–39 for City MPG.
    d. Comment on the relationship between Drive and City MPG.
    e. Prepare a crosstabulation of the data on Fuel Type (rows) and City MPG (columns). Use classes of 5–9, 10–14, 15–19, 20–24, 25–29, 30–34, and 35–39 for City MPG.
    f. Comment on the relationship between Fuel Type and City MPG.

38. Refer to exercise 37 and the data in the file named FuelData08.
    a. Prepare a crosstabulation of the data on Displacement (rows) and Hwy MPG (columns). Use classes of 1.0–2.9, 3.0–4.9, and 5.0–6.9 for Displacement. Use classes of 15–19, 20–24, 25–29, 30–34, and 35–39 for Hwy MPG.
    b. Comment on the relationship, if any, between Displacement and Hwy MPG.
    c. Develop a scatter diagram of the data on Displacement and Hwy MPG. Use the vertical axis for Hwy MPG.
    d. What does the scatter diagram developed in part (c) indicate about the relationship, if any, between Displacement and Hwy MPG?
    e. In investigating the relationship between Displacement and Hwy MPG, you developed a tabular summary of the data (crosstabulation) and a graphical summary (scatter diagram). In this case which approach do you prefer? Explain.

Summary

A set of data, even if modest in size, is often difficult to interpret directly in the form in which it is gathered. Tabular and graphical methods provide procedures for organizing and summarizing data so that patterns are revealed and the data are more easily interpreted. Frequency distributions, relative frequency distributions, percent frequency distributions, bar charts, and pie charts were presented as tabular and graphical procedures for summarizing qualitative data. Frequency distributions, relative frequency distributions, percent frequency distributions, histograms, cumulative frequency distributions, cumulative relative frequency distributions, cumulative percent frequency distributions, and ogives were presented as ways of summarizing quantitative data. A stem-and-leaf display provides an exploratory data analysis technique that can be used to summarize quantitative data. Crosstabulation was presented as a tabular method for summarizing data for two variables. The scatter diagram was introduced as a graphical method for showing the relationship between two quantitative variables. Figure 2.9 shows the tabular and graphical methods presented in this chapter.

FIGURE 2.9 TABULAR AND GRAPHICAL METHODS FOR SUMMARIZING DATA

```
                                  Data
                    ┌───────────────┴───────────────┐
              Categorical                      Quantitative
                 Data                              Data
           ┌───────┴───────┐              ┌───────────┴───────────┐
        Tabular         Graphical      Tabular                Graphical
        Methods         Methods        Methods                 Methods
```

- Frequency Distribution
- Relative Frequency Distribution
- Percent Frequency Distribution
- Crosstabulation

- Bar Chart
- Pie Chart

- Frequency Distribution
- Relative Frequency Distribution
- Percent Frequency Distribution
- Cumulative Frequency Distribution
- Cumulative Relative Frequency Distribution
- Cumulative Percent Frequency Distribution
- Crosstabulation

- Dot Plot
- Histogram
- Ogive
- Stem-and-Leaf Display
- Scatter Diagram

With large data sets, computer software packages are essential in constructing tabular and graphical summaries of data. In the chapter appendixes, we show how Minitab, Excel, and StatTools can be used for this purpose.

Glossary

Categorical data Labels or names used to identify categories of like items.

Quantitative data Numerical values that indicate how much or how many.

Frequency distribution A tabular summary of data showing the number (frequency) of data values in each of several nonoverlapping classes.

Relative frequency distribution A tabular summary of data showing the fraction or proportion of data values in each of several nonoverlapping classes.

Percent frequency distribution A tabular summary of data showing the percentage of data values in each of several nonoverlapping classes.

Bar chart A graphical device for depicting qualitative data that have been summarized in a frequency, relative frequency, or percent frequency distribution.

Pie chart A graphical device for presenting data summaries based on subdivision of a circle into sectors that correspond to the relative frequency for each class.

Class midpoint The value halfway between the lower and upper class limits.

Dot plot A graphical device that summarizes data by the number of dots above each data value on the horizontal axis.

Histogram A graphical presentation of a frequency distribution, relative frequency distribution, or percent frequency distribution of quantitative data constructed by placing the class intervals on the horizontal axis and the frequencies, relative frequencies, or percent frequencies on the vertical axis.

Cumulative frequency distribution A tabular summary of quantitative data showing the number of data values that are less than or equal to the upper class limit of each class.

Cumulative relative frequency distribution A tabular summary of quantitative data showing the fraction or proportion of data values that are less than or equal to the upper class limit of each class.

Cumulative percent frequency distribution A tabular summary of quantitative data showing the percentage of data values that are less than or equal to the upper class limit of each class.

Ogive A graph of a cumulative distribution.

Exploratory data analysis Methods that use simple arithmetic and easy-to-draw graphs to summarize data quickly.

Stem-and-leaf display An exploratory data analysis technique that simultaneously rank orders quantitative data and provides insight about the shape of the distribution.

Crosstabulation A tabular summary of data for two variables. The classes for one variable are represented by the rows; the classes for the other variable are represented by the columns.

Simpson's paradox Conclusions drawn from two or more separate crosstabulations that can be reversed when the data are aggregated into a single crosstabulation.

Scatter diagram A graphical presentation of the relationship between two quantitative variables. One variable is shown on the horizontal axis and the other variable is shown on the vertical axis.

Trendline A line that provides an approximation of the relationship between two variables.

Key Formulas

Relative Frequency

$$\frac{\text{Frequency of the class}}{n} \quad (2.1)$$

Approximate Class Width

$$\frac{\text{Largest data value} - \text{Smallest data value}}{\text{Number of classes}} \quad (2.2)$$

Supplementary Exercises

39. The Higher Education Research Institute at UCLA provides statistics on the most popular majors among incoming college freshmen. The five most popular majors are Arts and Humanities (A), Business Administration (B), Engineering (E), Professional (P), and Social Science (S) (*The New York Times Almanac*, 2006). A broad range of other (O) majors, including biological science, physical science, computer science, and education, is grouped together. The majors selected for a sample of 64 college freshmen follow.

S	P	P	O	B	E	O	E	P	O	O	B	O	O	O	A
O	E	E	B	S	O	B	O	A	O	E	O	E	O	B	P
B	A	S	O	E	A	B	O	S	S	O	O	E	B	O	B
A	E	B	E	A	A	P	O	O	E	O	B	B	O	P	B

    a. Show a frequency distribution and percent frequency distribution.
    b. Show a bar chart.

c. What percentage of freshmen select one of the five most popular majors?
d. What is the most popular major for incoming freshmen? What percentage of freshmen select this major?

40. In 2008 General Motors had a 23% share of the automobile industry with sales coming from eight divisions: Buick, Cadillac, Chevrolet, GMC, Hummer, Pontiac, Saab, and Saturn (*Forbes,* December 22, 2008). The data set GMSales shows the sales for a sample of 200 General Motors vehicles. The division for the vehicle is provided for each sale.
   a. Show the frequency distribution and the percent frequency distribution of sales by division for General Motors.
   b. Show a bar chart of the percent frequency distribution.
   c. Which General Motors division was the company leader in sales? What was the percentage of sales for this division? Was this General Motors' most important division? Explain.
   d. Due to the ongoing recession, high gasoline prices, and the decline in automobile sales, General Motors was facing bankruptcy in 2009. Expectations were that General Motors could not continue to operate all eight divisions. Based on the percentage of sales, which of the eight divisions looked to be the best candidates for General Motors to discontinue? Which divisions looked to be the least likely candidates for General Motors to discontinue?

41. Dividend yield is the annual dividend paid by a company expressed as a percentage of the price of the stock (Dividend/Stock Price × 100). The dividend yield for the Dow Jones Industrial Average companies is shown in Table 2.15 (*The Wall Street Journal,* June 8, 2009).
   a. Construct a frequency distribution and percent frequency distribution.
   b. Construct a histogram.
   c. Comment on the shape of the distribution.
   d. What do the tabular and graphical summaries tell about the dividend yields among the Dow Jones Industrial Average companies?
   e. What company has the highest dividend yield? If the stock for this company currently sells for $20 per share and you purchase 500 shares, how much dividend income will this investment generate in one year?

42. Approximately 1.5 million high school students take the SAT test each year and nearly 80% of the college and universities without open admissions policies use SAT scores in making admission decisions (College Board, March 2009). The current version of the SAT

TABLE 2.15 DIVIDEND YIELD FOR DOW JONES INDUSTRIAL AVERAGE COMPANIES

Company	Dividend Yield %	Company	Dividend Yield %
3M	3.6	IBM	2.1
Alcoa	1.3	Intel	3.4
American Express	2.9	JP Morgan Chase	0.5
AT&T	6.6	Johnson & Johnson	3.6
Bank of America	0.4	Kraft Foods	4.4
Boeing	3.8	McDonald's	3.4
Caterpillar	4.7	Merck	5.5
Chevron	3.9	Microsoft	2.5
Cisco Systems	0.0	Pfizer	4.2
Coca-Cola	3.3	Procter & Gamble	3.4
DuPont	5.8	Travelers	3.0
ExxonMobil	2.4	United Technologies	2.9
General Electric	9.2	Verizon	6.3
Hewlett-Packard	0.9	Walmart	2.2
Home Depot	3.9	Walt Disney	1.5

includes three parts: critical reading, mathematics, and writing. A perfect combined score for all three parts is 2400. A sample of SAT scores for the combined three-part SAT is as follows:

1665	1525	1355	1645	1780
1275	2135	1280	1060	1585
1650	1560	1150	1485	1990
1590	1880	1420	1755	1375
1475	1680	1440	1260	1730
1490	1560	940	1390	1175

WEB file NewSAT

a. Show a frequency distribution and histogram. Begin with the first class starting at 800 and use a class width of 200.
b. Comment on the shape of the distribution.
c. What other observations can be made about the SAT scores based on the tabular and graphical summaries?

43. The Pittsburgh Steelers defeated the Arizona Cardinals 27 to 23 in professional football's 43rd Super Bowl. With this win, its sixth championship, the Pittsburgh Steelers became the team with the most wins in the 43-year history of the event (*Tampa Tribune,* February 2, 2009). The Super Bowl has been played in eight different states: Arizona (AZ), California (CA), Florida (FL), Georgia (GA), Louisiana (LA), Michigan (MI), Minnesota (MN), and Texas (TX). Data in the following table show the state where the Super Bowls were played and the point margin of victory for the winning team.

WEB file SuperBowl

Super Bowl	State	Won By Points	Super Bowl	State	Won By Points	Super Bowl	State	Won By Points
1	CA	25	16	MI	5	31	LA	14
2	FL	19	17	CA	10	32	CA	7
3	FL	9	18	FL	19	33	FL	15
4	LA	16	19	CA	22	34	GA	7
5	FL	3	20	LA	36	35	FL	27
6	FL	21	21	CA	19	36	LA	3
7	CA	7	22	CA	32	37	CA	27
8	TX	17	23	FL	4	38	TX	3
9	LA	10	24	LA	45	39	FL	3
10	FL	4	25	FL	1	40	MI	11
11	CA	18	26	MN	13	41	FL	12
12	LA	17	27	CA	35	42	AZ	3
13	FL	4	28	GA	17	43	FL	4
14	CA	12	29	FL	23			
15	LA	17	30	AZ	10			

a. Show a frequency distribution and bar chart for the state where the Super Bowl was played.
b. What conclusions can you draw from your summary in part (a)? What percentage of Super Bowls were played in the states of Florida or California? What percentage of Super Bowls were played in northern or cold-weather states?
c. Show a stretched stem-and-leaf display for the point margin of victory for the winning team. Show a histogram.
d. What conclusions can you draw from your summary in part (c)? What percentage of Super Bowls have been close games with the margin of victory less than 5 points? What percentage of Super Bowls have been won by 20 or more points?
e. The closest Super Bowl occurred when the New York Giants beat the Buffalo Bills. Where was this game played and what was the winning margin of victory? The biggest point margin in Super Bowl history occurred when the San Francisco 49ers beat the Denver Broncos. Where was this game played and what was the winning margin of victory?

44. Data from the U.S. Census Bureau provide the population by state in millions of people (*The World Almanac*, 2006).

State	Population	State	Population	State	Population
Alabama	4.5	Louisiana	4.5	Ohio	11.5
Alaska	0.7	Maine	1.3	Oklahoma	3.5
Arizona	5.7	Maryland	5.6	Oregon	3.6
Arkansas	2.8	Massachusetts	6.4	Pennsylvania	12.4
California	35.9	Michigan	10.1	Rhode Island	1.1
Colorado	4.6	Minnesota	5.1	South Carolina	4.2
Connecticut	3.5	Mississippi	2.9	South Dakota	0.8
Delaware	0.8	Missouri	5.8	Tennessee	5.9
Florida	17.4	Montana	0.9	Texas	22.5
Georgia	8.8	Nebraska	1.7	Utah	2.4
Hawaii	1.3	Nevada	2.3	Vermont	0.6
Idaho	1.4	New Hampshire	1.3	Virginia	7.5
Illinois	12.7	New Jersey	8.7	Washington	6.2
Indiana	6.2	New Mexico	1.9	West Virginia	1.8
Iowa	3.0	New York	19.2	Wisconsin	5.5
Kansas	2.7	North Carolina	8.5	Wyoming	0.5
Kentucky	4.1	North Dakota	0.6		

   a. Develop a frequency distribution, a percent frequency distribution, and a histogram. Use a class width of 2.5 million.
   b. Discuss the skewness in the distribution.
   c. What observations can you make about the population of the 50 states?

45. *Drug Store News* (September 2002) provided data on annual pharmacy sales for the leading pharmacy retailers in the United States. The following data are annual sales in millions.

Retailer	Sales	Retailer	Sales
Ahold USA	$ 1700	Medicine Shoppe	$ 1757
CVS	12700	Rite-Aid	8637
Eckerd	7739	Safeway	2150
Kmart	1863	Walgreens	11660
Kroger	3400	Wal-Mart	7250

   a. Show a stem-and-leaf display.
   b. Identify the annual sales levels for the smallest, medium, and largest drug retailers.
   c. What are the two largest drug retailers?

46. The daily high and low temperatures for 20 cities follow (*USA Today*, March 3, 2006).

City	High	Low	City	High	Low
Albuquerque	66	39	Los Angeles	60	46
Atlanta	61	35	Miami	84	65
Baltimore	42	26	Minneapolis	30	11
Charlotte	60	29	New Orleans	68	50
Cincinnati	41	21	Oklahoma City	62	40
Dallas	62	47	Phoenix	77	50
Denver	60	31	Portland	54	38
Houston	70	54	St. Louis	45	27
Indianapolis	42	22	San Francisco	55	43
Las Vegas	65	43	Seattle	52	36

a. Prepare a stem-and-leaf display of the high temperatures.
b. Prepare a stem-and-leaf display of the low temperatures.
c. Compare the two stem-and-leaf displays and make comments about the difference between the high and low temperatures.
d. Provide a frequency distribution for both high and low temperatures.

47. Refer to the data set for high and low temperatures for 20 cities in exercise 46.
    a. Develop a scatter diagram to show the relationship between the two variables, high temperature and low temperature.
    b. Comment on the relationship between high and low temperatures.

48. Western University has only one women's softball scholarship remaining for the coming year. The final two players that Western is considering are Allison Fealey and Emily Janson. The coaching staff has concluded that the speed and defensive skills are virtually identical for the two players, and that the final decision will be based on which player has the best batting average. Crosstabulations of each player's batting performance in their junior and senior years of high school are as follows:

	Allison Fealey			Emily Janson	
Outcome	Junior	Senior	Outcome	Junior	Senior
Hit	15	75	Hit	70	35
No Hit	25	175	No Hit	130	85
Total At-Bats	40	250	Total At Bats	200	120

A player's batting average is computed by dividing the number of hits a player has by the total number of at-bats. Batting averages are represented as a decimal number with three places after the decimal.

a. Calculate the batting average for each player in her junior year. Then calculate the batting average of each player in her senior year. Using this analysis, which player should be awarded the scholarship? Explain.
b. Combine or aggregate the data for the junior and senior years into one crosstabulation as follows:

	Player	
Outcome	Fealey	Janson
Hit		
No Hit		
Total At-Bats		

Calculate each player's batting average for the combined two years. Using this analysis, which player should be awarded the scholarship? Explain.
c. Are the recommendations you made in parts (a) and (b) consistent? Explain any apparent inconsistencies.

49. A survey of commercial buildings served by the Cincinnati Gas & Electric Company asked what main heating fuel was used and what year the building was constructed. A partial crosstabulation of the findings follows.

Year Constructed	Fuel Type				
	Electricity	Natural Gas	Oil	Propane	Other
1973 or before	40	183	12	5	7
1974–1979	24	26	2	2	0
1980–1986	37	38	1	0	6
1987–1991	48	70	2	0	1

a. Complete the crosstabulation by showing the row totals and column totals.
b. Show the frequency distributions for year constructed and for fuel type.
c. Prepare a crosstabulation showing column percentages.
d. Prepare a crosstabulation showing row percentages.
e. Comment on the relationship between year constructed and fuel type.

50. One of the questions in a *Financial Times*/Harris Poll was, "How much do you favor or oppose a higher tax on higher carbon emission cars?" Possible responses were strongly favor, favor more than oppose, oppose more than favor, and strongly oppose. The following crosstabulation shows the responses obtained for 5372 adults surveyed in four countries in Europe and the United States (Harris Interactive website, February 27, 2008).

	Country					
Level of Support	**Great Britain**	**Italy**	**Spain**	**Germany**	**United States**	**Total**
Strongly favor	337	334	510	222	214	1617
Favor more than oppose	370	408	355	411	327	1871
Oppose more than favor	250	188	155	267	275	1135
Strongly oppose	130	115	89	211	204	749
Total	1087	1045	1109	1111	1020	5372

a. Construct a percent frequency distribution for the level of support variable. Do you think the results show support for a higher tax on higher carbon emission cars?
b. Construct a percent frequency distribution for the country variable.
c. Does the level of support among adults in the European countries appear to be different than the level of support among adults in the United States? Explain.

51. Table 2.16 contains a portion of the data in the file named Fortune. Data on stockholders' equity, market value, and profits for a sample of 50 *Fortune* 500 companies are shown.

TABLE 2.16 DATA FOR A SAMPLE OF 50 *FORTUNE* 500 COMPANIES

WEB file

Fortune

Company	Stockholders' Equity ($1000s)	Market Value ($1000s)	Profit ($1000s)
AGCO	982.1	372.1	60.6
AMP	2698.0	12017.6	2.0
Apple Computer	1642.0	4605.0	309.0
Baxter International	2839.0	21743.0	315.0
Bergen Brunswick	629.1	2787.5	3.1
Best Buy	557.7	10376.5	94.5
Charles Schwab	1429.0	35340.6	348.5
.	.	.	.
.	.	.	.
.	.	.	.
Walgreen	2849.0	30324.7	511.0
Westvaco	2246.4	2225.6	132.0
Whirlpool	2001.0	3729.4	325.0
Xerox	5544.0	35603.7	395.0

a. Prepare a crosstabulation for the variables Stockholders' Equity and Profit. Use classes of 0–200, 200–400, . . . , 1000–1200 for Profit, and classes of 0–1200, 1200–2400, . . . , 4800–6000 for Stockholders' Equity.
b. Compute the row percentages for your crosstabulation in part (a).
c. What relationship, if any, do you notice between Profit and Stockholders' Equity?

52. Refer to the data set in Table 2.16.
    a. Prepare a scatter diagram to show the relationship between the variables Profit and Stockholders' Equity.
    b. Comment on any relationship between the variables.

53. Refer to the data set in Table 2.16.
    a. Prepare a scatter diagram to show the relationship between the variables Market Value and Stockholders' Equity.
    b. Comment on any relationship between the variables.

54. Refer to the data set in Table 2.16.
    a. Prepare a crosstabulation for the variables Market Value and Profit.
    b. Compute the row percentages for your crosstabulation in part (a).
    c. Comment on any relationship between the variables.

Case Problem 1   Pelican Stores

Pelican Stores, a division of National Clothing, is a chain of women's apparel stores operating throughout the country. The chain recently ran a promotion in which discount coupons were sent to customers of other National Clothing stores. Data collected for a sample of 100 in-store credit card transactions at Pelican Stores during one day while the promotion was running are contained in the file named PelicanStores. Table 2.17 shows a portion of the data set. The Proprietary Card method of payment refers to charges made using a National Clothing charge card. Customers who made a purchase using a discount coupon are referred to as promotional customers and customers who made a purchase but did not use a discount coupon are referred to as regular customers. Because the promotional coupons were not sent to regular Pelican Stores customers, management considers the sales made to people presenting the promotional coupons as sales it would not otherwise make. Of course, Pelican also hopes that the promotional customers will continue to shop at its stores.

TABLE 2.17   DATA FOR A SAMPLE OF 100 CREDIT CARD PURCHASES AT PELICAN STORES

Customer	Type of Customer	Items	Net Sales	Method of Payment	Gender	Marital Status	Age
1	Regular	1	39.50	Discover	Male	Married	32
2	Promotional	1	102.40	Proprietary Card	Female	Married	36
3	Regular	1	22.50	Proprietary Card	Female	Married	32
4	Promotional	5	100.40	Proprietary Card	Female	Married	28
5	Regular	2	54.00	MasterCard	Female	Married	34
.	.	.	.	.	.	.	.
96	Regular	1	39.50	MasterCard	Female	Married	44
97	Promotional	9	253.00	Proprietary Card	Female	Married	30
98	Promotional	10	287.59	Proprietary Card	Female	Married	52
99	Promotional	2	47.60	Proprietary Card	Female	Married	30
100	Promotional	1	28.44	Proprietary Card	Female	Married	44

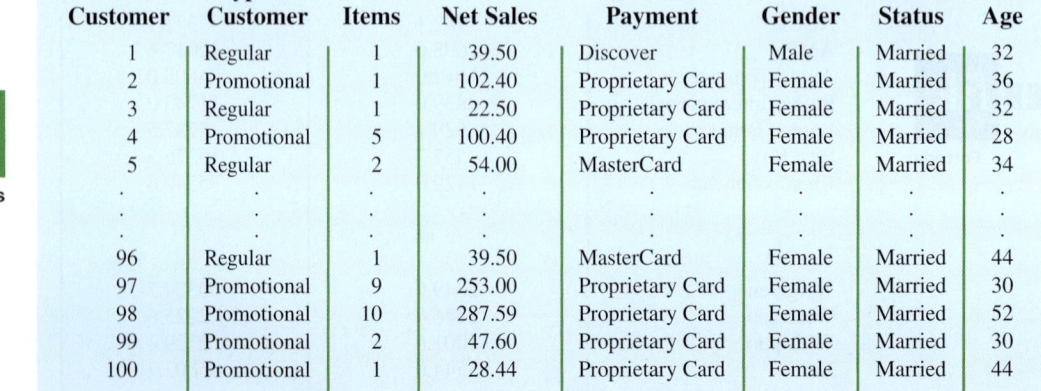

Most of the variables shown in Table 2.17 are self-explanatory, but two of the variables require some clarification.

Items   The total number of items purchased
Net Sales   The total amount ($) charged to the credit card

Pelican's management would like to use this sample data to learn about its customer base and to evaluate the promotion involving discount coupons.

Managerial Report

Use the tabular and graphical methods of descriptive statistics to help management develop a customer profile and to evaluate the promotional campaign. At a minimum, your report should include the following:

1. Percent frequency distribution for key variables
2. A bar chart or pie chart showing the number of customer purchases attributable to the method of payment
3. A crosstabulation of type of customer (regular or promotional) versus net sales. Comment on any similarities or differences present
4. A scatter diagram to explore the relationship between net sales and customer age

Case Problem 2 Motion Picture Industry

The motion picture industry is a competitive business. More than 50 studios produce a total of 300 to 400 new motion pictures each year, and the financial success of each motion picture varies considerably. The opening weekend gross sales ($ millions), the total gross sales ($ millions), the number of theaters the movie was shown in, and the number of weeks the motion picture was in the top 60 for gross sales are common variables used to measure the success of a motion picture. Data collected for a sample of 100 motion pictures produced in 2005 are contained in the file named Movies. Table 2.18 shows the data for the first 10 motion pictures in this file.

Managerial Report

Use the tabular and graphical methods of descriptive statistics to learn how these variables contribute to the success of a motion picture. Include the following in your report.

TABLE 2.18   PERFORMANCE DATA FOR 10 MOTION PICTURES

Motion Picture	Opening Gross Sales ($ millions)	Total Gross Sales ($ millions)	Number of Theaters	Weeks in Top 60
Coach Carter	29.17	67.25	2574	16
Ladies in Lavender	0.15	6.65	119	22
Batman Begins	48.75	205.28	3858	18
Unleashed	10.90	24.47	1962	8
Pretty Persuasion	0.06	0.23	24	4
Fever Pitch	12.40	42.01	3275	14
Harry Potter and the Goblet of Fire	102.69	287.18	3858	13
Monster-in-Law	23.11	82.89	3424	16
White Noise	24.11	55.85	2279	7
Mr. and Mrs. Smith	50.34	186.22	3451	21

1. Tabular and graphical summaries for each of the four variables along with a discussion of what each summary tells us about the motion picture industry.
2. A scatter diagram to explore the relationship between Total Gross Sales and Opening Weekend Gross Sales. Discuss.
3. A scatter diagram to explore the relationship between Total Gross Sales and Number of Theaters. Discuss.
4. A scatter diagram to explore the relationship between Total Gross Sales and Number of Weeks in the Top 60. Discuss.

Appendix 2.1 Tabular and Graphical Presentations Using Minitab

Minitab offers extensive capabilities for constructing tabular and graphical summaries of data. In this appendix we show how Minitab can be used to construct several graphical summaries and the tabular summary of a crosstabulation. The graphical methods presented include the dot plot, the histogram, the stem-and-leaf display, and the scatter diagram.

Dot Plot

We use the audit time data in Table 2.4 to demonstrate. The data are in column C1 of a Minitab worksheet. The following steps will generate a dot plot.

Step 1. Select the **Graph** menu and choose **Dotplot**
Step 2. Select **One Y, Simple** and click **OK**
Step 3. When the Dotplot-One Y, Simple dialog box appears:
   Enter C1 in the **Graph Variables** box
   Click **OK**

Histogram

We show how to construct a histogram with frequencies on the vertical axis using the audit time data in Table 2.4. The data are in column C1 of a Minitab worksheet. The following steps will generate a histogram for audit times.

Step 1. Select the **Graph** menu
Step 2. Choose **Histogram**
Step 3. Select **Simple** and click **OK**
Step 4. When the Histogram-Simple dialog box appears:
   Enter C1 in the **Graph Variables** box
   Click **OK**
Step 5. When the Histogram appears:
   Position the mouse pointer over any one of the bars
   Double-click
Step 6. When the Edit Bars dialog box appears:
   Click on the **Binning** tab
   Select **Cutpoint** for Interval Type
   Select **Midpoint/Cutpoint positions** for Interval Definition
   Enter 10:35/5 in the **Midpoint/Cutpoint positions** box*
   Click **OK**

---

*The entry 10:35/5 indicates that 10 is the starting value for the histogram, 35 is the ending value for the histogram, and 5 is the class width.

Note that Minitab also provides the option of scaling the *x*-axis so that the numerical values appear at the midpoints of the histogram rectangles. If this option is desired, modify step 6 to include Select **Midpoint** for Interval Type and Enter 12:32/5 in the **Midpoint/Cutpoint positions** box. These steps provide the same histogram with the midpoints of the histogram rectangles labeled 12, 17, 22, 27, and 32.

Stem-and-Leaf Display

We use the aptitude test data in Table 2.8 to demonstrate the construction of a stem-and-leaf display. The data are in column C1 of a Minitab worksheet. The following steps will generate the stretched stem-and-leaf display shown in Section 2.3.

Step 1. Select the **Graph** menu
Step 2. Choose **Stem-and-Leaf**
Step 3. When the Stem-and-Leaf dialog box appears:
   Enter C1 in the **Graph Variables** box
   Click **OK**

Scatter Diagram

We use the stereo and sound equipment store data in Table 2.12 to demonstrate the construction of a scatter diagram. The weeks are numbered from 1 to 10 in column C1, the data for number of commercials are in column C2, and the data for sales are in column C3 of a Minitab worksheet. The following steps will generate the scatter diagram shown in Figure 2.7.

Step 1. Select the **Graph** menu
Step 2. Choose **Scatterplot**
Step 3. Select **Simple** and click **OK**
Step 4. When the Scatterplot-Simple dialog box appears:
   Enter C3 under **Y variables** and C2 under **X variables**
   Click **OK**

Crosstabulation

We use the data from *Zagat's Restaurant Review*, part of which is shown in Table 2.9, to demonstrate. The restaurants are numbered from 1 to 300 in column C1 of the Minitab worksheet. The quality ratings are in column C2, and the meal prices are in column C3.

Minitab can only create a crosstabulation for qualitative variables and meal price is a quantitative variable. So we need to first code the meal price data by specifying the class to which each meal price belongs. The following steps will code the meal price data to create four classes of meal price in column C4: $10–19, $20–29, $30–39, and $40–49.

Step 1. Select the **Data** menu
Step 2. Choose **Code**
Step 3. Choose **Numerical to Text**
Step 4. When the Code-Numerical to Text dialog box appears:
   Enter C3 in the **Code data from columns** box
   Enter C4 in the **Store coded data in columns** box
   Enter 10:19 in the first **Original values** box and $10–19 in the adjacent **New** box
   Enter 20:29 in the second **Original values** box and $20–29 in the adjacent **New** box

Enter 30:39 in the third **Original values** box and $30–39 in the adjacent **New** box

Enter 40:49 in the fourth **Original values** box and $40–49 in the adjacent **New** box

Click **OK**

For each meal price in column C3, the associated meal price category will now appear in column C4. We can now develop a crosstabulation for quality rating and the meal price categories by using the data in columns C2 and C4. The following steps will create a crosstabulation containing the same information as shown in Table 2.10.

Step 1. Select the **Stat** menu
Step 2. Choose **Tables**
Step 3. Choose **Cross Tabulation and Chi-Square**
Step 4. When the Cross Tabulation and Chi-Square dialog box appears:

Enter C2 in the **For rows** box and C4 in the **For columns** box
Select **Counts** under Display
Click **OK**

Appendix 2.2 Tabular and Graphical Presentations Using Excel

Excel offers extensive capabilities for constructing tabular and graphical summaries of data. In this appendix, we show how Excel can be used to construct a frequency distribution, bar chart, pie chart, histogram, scatter diagram, and crosstabulation. We will demonstrate three of Excel's most powerful tools for data analysis: chart tools, PivotChart Report, and PivotTable Report.

Frequency Distribution and Bar Chart for Categorical Data

In this section we show how Excel can be used to construct a frequency distribution and a bar chart for categorical data. We illustrate each using the data on soft drink purchases in Table 2.1.

Frequency distribution We begin by showing how the COUNTIF function can be used to construct a frequency distribution for the data in Table 2.1. Refer to Figure 2.10 as we describe the steps involved. The formula worksheet (showing the functions and formulas used) is set in the background, and the value worksheet (showing the results obtained using the functions and formulas) appears in the foreground.

The label "Brand Purchased" and the data for the 50 soft drink purchases are in cells A1:A51. We also entered the labels "Soft Drink" and "Frequency" in cells C1:D1. The five soft drink names are entered into cells C2:C6. Excel's COUNTIF function can now be used to count the number of times each soft drink appears in cells A2:A51. The following steps are used.

Step 1. Select cell D2
Step 2. Enter =COUNTIF($A$2:$A$51,C2)
Step 3. Copy cell D2 to cells D3:D6

The formula worksheet in Figure 2.10 shows the cell formulas inserted by applying these steps. The value worksheet shows the values computed by the cell formulas. This worksheet shows the same frequency distribution that we developed in Table 2.2.

Appendix 2.2 Tabular and Graphical Presentations Using Excel

FIGURE 2.10 FREQUENCY DISTRIBUTION FOR SOFT DRINK PURCHASES CONSTRUCTED USING EXCEL'S COUNTIF FUNCTION

	A	B	C	D	E
1	Brand Purchased		Soft Drink	Frequency	
2	Coke Classic		Coke Classic	=COUNTIF($A$2:$A$51,C2)	
3	Diet Coke		Diet Coke	=COUNTIF($A$2:$A$51,C3)	
4	Pepsi		Dr. Pepper	=COUNTIF($A$2:$A$51,C4)	
5	Diet Coke		Pepsi	=COUNTIF($A$2:$A$51,C5)	
6	Coke Classic		Sprite	=COUNTIF($A$2:$A$51,C6)	
7	Coke Classic				
8	Dr. Pepper				
9	Diet Coke				
10	Pepsi				
45	Pepsi				
46	Pepsi				
47	Pepsi				
48	Coke Classic				
49	Dr. Pepper				
50	Pepsi				
51	Sprite				
52					

Note: Rows 11–44 are hidden.

	A	B	C	D	E
1	Brand Purchased		Soft Drink	Frequency	
2	Coke Classic		Coke Classic	19	
3	Diet Coke		Diet Coke	8	
4	Pepsi		Dr. Pepper	5	
5	Diet Coke		Pepsi	13	
6	Coke Classic		Sprite	5	
7	Coke Classic				
8	Dr. Pepper				
9	Diet Coke				
10	Pepsi				
45	Pepsi				
46	Pepsi				
47	Pepsi				
48	Coke Classic				
49	Dr. Pepper				
50	Pepsi				
51	Sprite				
52					

WEB file
SoftDrink

Bar chart Here we show how Excel's chart tools can be used to construct a bar chart for the soft drink data. Refer to the frequency distribution shown in the value worksheet of Figure 2.10. The bar chart that we are going to develop is an extension of this worksheet. The worksheet and the bar chart developed are shown in Figure 2.11. The steps are as follows:

Step 1. Select cells C2:D6
Step 2. Click the **Insert** tab on the Ribbon
Step 3. In the **Charts** group, click **Column**
Step 4. When the list of column chart subtypes appears:
   Go to the **2-D Column** section
   Click **Clustered Column** (the leftmost chart)
Step 5. In the **Chart Layouts** group, click the **More** button (the downward-pointing arrow with a line over it) to display all the options
Step 6. Choose **Layout 9**
Step 7. Select the **Chart Title** and replace it with **Bar Chart of Soft Drink Purchases**
Step 8. Select the **Horizontal (Category) Axis Title** and replace it with **Soft Drink**
Step 9. Select the **Vertical (Value) Axis Title** and replace it with **Frequency**
Step 10. Right-click the **Series 1 Legend Entry**
   Click **Delete**
Step 11. Right-click the vertical axis
   Click **Format Axis**

78 Chapter 2 Descriptive Statistics: Tabular and Graphical Presentations

FIGURE 2.11 BAR CHART OF SOFT DRINK PURCHASES CONSTRUCTED USING EXCEL'S CHART TOOLS

	A	B	C	D	E	F	G	H	I
1	Brand Purchased		Soft Drink	Frequency					
2	Coke Classic		Coke Classic	19					
3	Diet Coke		Diet Coke	8					
4	Pepsi		Dr. Pepper	5					
5	Diet Coke		Pepsi	13					
6	Coke Classic		Sprite	5					
7	Coke Classic								
8	Dr. Pepper								
9	Diet Coke								
10	Pepsi								
11	Pepsi								
12	Coke Classic								
13	Dr. Pepper								
14	Sprite								
15	Coke Classic								
16	Diet Coke								
17	Coke Classic								
18	Coke Classic								
19	Sprite								
20	Coke Classic								
50	Pepsi								
51	Sprite								
52									

Bar Chart of Soft Drink Purchases (chart embedded showing Frequency on y-axis with values for Coke Classic, Diet Coke, Dr. Pepper, Pepsi, Sprite on Soft Drink x-axis)

Step 12. When the Format Axis dialog box appears:
Go to the **Axis Options** section
Select **Fixed** for **Major Unit** and enter 5.0 in the corresponding box
Click **Close**

The resulting bar chart is shown in Figure 2.11.*

Excel can produce a pie chart for the soft drink data in a similar fashion. The major difference is that in step 3 we would click **Pie** in the **Charts** group. Several style pie charts are available.

Frequency Distribution and Histogram for Quantitative Data

In a later section of this appendix we describe how to use Excel's PivotTable Report to construct a crosstabulation.

WEB file
Audit

Excel's PivotTable Report is an interactive tool that allows you to quickly summarize data in a variety of ways, including developing a frequency distribution for quantitative data. Once a frequency distribution is created using the PivotTable Report, Excel's chart tools can then be used to construct the corresponding histogram. But, using Excel's PivotChart Report, we can construct a frequency distribution and a histogram simultaneously. We will illustrate this procedure using the audit time data in Table 2.4. The label "Audit Time" and the 20 audit time values are entered into cells A1:A21 of an Excel worksheet. The following steps describe how to use Excel's PivotChart Report to construct a frequency distribution and a histogram for the audit time data. Refer to Figure 2.12 as we describe the steps involved.

---
*The bar chart in Figure 2.11 can be resized. Resizing an Excel chart is not difficult. First, select the chart. Sizing handles will appear on the chart border. Click on the sizing handles and drag them to resize the figure to your preference.

Appendix 2.2 Tabular and Graphical Presentations Using Excel

FIGURE 2.12 USING EXCEL'S PIVOTCHART REPORT TO CONSTRUCT A FREQUENCY DISTRIBUTION AND HISTOGRAM FOR THE AUDIT TIME DATA

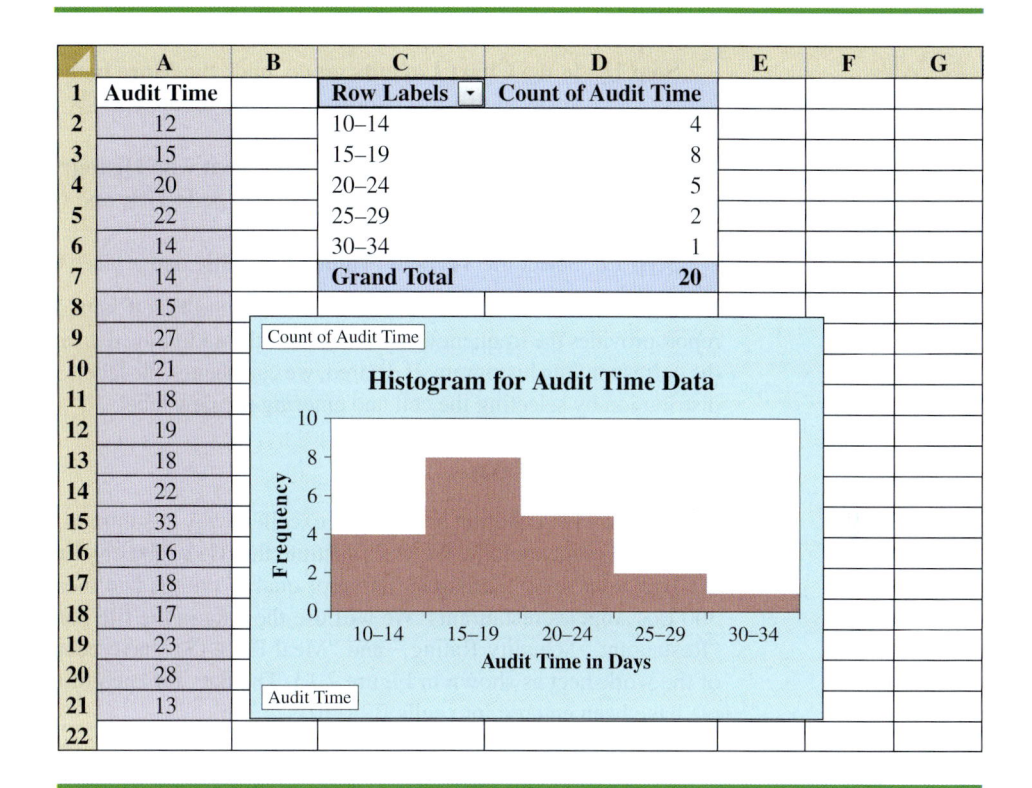

	A	B	C	D	E	F	G
1	Audit Time		Row Labels	Count of Audit Time			
2	12		10–14	4			
3	15		15–19	8			
4	20		20–24	5			
5	22		25–29	2			
6	14		30–34	1			
7	14		Grand Total	20			
8	15						
9	27						
10	21						
11	18						
12	19						
13	18						
14	22						
15	33						
16	16						
17	18						
18	17						
19	23						
20	28						
21	13						
22							

(Histogram for Audit Time Data shown as embedded PivotChart, with Frequency on the y-axis and Audit Time in Days on the x-axis.)

Step 1. Click the **Insert** tab on the Ribbon
Step 2. In the **Tables** group, click the word **PivotTable**
Step 3. Choose **PivotChart** from the options that appear
Step 4. When the Create PivotTable with PivotChart dialog box appears:
   Choose **Select a table or range**
   Enter A1:A21 in the **Table/Range** box
   Choose **Existing Worksheet** as the location for the PivotTable and PivotChart
   Enter C1 in the **Location** box
   Click **OK**
Step 5. In the **PivotTable Field List**, go to **Choose Fields to add to report**
   Drag the **Audit Time** field to the **Axis Fields (Categories)** area
   Drag the **Audit Time** field to the **Values** area
Step 6. Click **Sum of Audit Time** in the **Values** area
Step 7. Click **Value Field Settings** from the list of options that appears
Step 8. When the Value Field Settings dialog appears:
   Under **Summarize value field by**, choose **Count**
   Click **OK**
Step 9. Close the **PivotTable Field List**
Step 10. Right-click cell C2 in the PivotTable report or any other cell containing an audit time
Step 11. Choose **Group** from the list of options that appears
Step 12. When the Grouping dialog box appears:
   Enter 10 in the **Starting at** box

Enter 34 in the **Ending at** box
Enter 5 in the **By** box
Click **OK** (a PivotChart will appear)

Step 13. Click inside the resulting PivotChart
Step 14. Click the **Design** tab on the Ribbon
Step 15. In the **Chart Layouts** group, click the **More** button (the downward pointing arrow with a line over it) to display all the options
Step 16. Choose **Layout 8**
Step 17. Select the **Chart Title** and replace it with **Histogram for Audit Time Data**
Step 18. Select the **Horizontal (Category) Axis Title** and replace it with **Audit Time in Days**
Step 19. Select the **Vertical (Value) Axis Title** and replace it with **Frequency**

Figure 2.12 shows the resulting PivotTable and PivotChart. We see that the PivotTable report provides the frequency distribution for the audit time data and the PivotChart provides the corresponding histogram. If desired, we can change the labels in any cell in the frequency distribution by selecting the cell and entering the new label.

Crosstabulation

Excel's PivotTable Report provides an excellent way to summarize the data for two or more variables simultaneously. We will illustrate the use of Excel's PivotTable Report by showing how to develop a crosstabulation of quality ratings and meal prices for the sample of 300 Los Angeles restaurants. We will use the data in the file named Restaurant; the labels "Restaurant," "Quality Rating," and "Meal Price ($)" have been entered into cells A1:C1 of the worksheet as shown in Figure 2.13. The data for each of the restaurants in the sample have been entered into cells B2:C301.

FIGURE 2.13 EXCEL WORKSHEET CONTAINING RESTAURANT DATA

WEB file
Restaurant

Note: Rows 12–291 are hidden.

	A	B	C	D
1	Restaurant	Quality Rating	Meal Price ($)	
2	1	Good	18	
3	2	Very Good	22	
4	3	Good	28	
5	4	Excellent	38	
6	5	Very Good	33	
7	6	Good	28	
8	7	Very Good	19	
9	8	Very Good	11	
10	9	Very Good	23	
11	10	Good	13	
292	291	Very Good	23	
293	292	Very Good	24	
294	293	Excellent	45	
295	294	Good	14	
296	295	Good	18	
297	296	Good	17	
298	297	Good	16	
299	298	Good	15	
300	299	Very Good	38	
301	300	Very Good	31	
302				

Appendix 2.2 Tabular and Graphical Presentations Using Excel

In order to use the Pivot Table report to create a crosstabulation, we need to perform three tasks: Display the Initial PivotTable Field List and PivotTable Report; Set Up the PivotTable Field List; and Finalize the PivotTable Report. These tasks are described as follows.

Display the Initial PivotTable Field List and PivotTable Report: The following steps will display the initial PivotTable Field List and PivotTable report.

Step 1. Click the **Insert** tab on the Ribbon
Step 2. In the **Tables** group, click the icon above the word **PivotTable**
Step 3. When the Create PivotTable dialog box appears:
  Choose **Select a table or range**
  Enter A1:C301 in the **Table/Range** box
  Choose **New Worksheet** as the location for the PivotTable Report
  Click **OK**

The resulting initial PivotTable Field List and PivotTable Report are shown in Figure 2.14.

Set Up the PivotTable Field List: Each of the three columns in Figure 2.13 (labeled Restaurant, Quality Rating, and Meal Price ($)) is considered a field by Excel. Fields may be chosen to represent rows, columns, or values in the body of the PivotTable Report. The following steps show how to use Excel's PivotTable Field List to assign the Quality Rating field to the rows, the Meal Price ($) field to the columns, and the Restaurant field to the body of the PivotTable report.

Step 1. In the **PivotTable Field List,** go to **Choose Fields to add to report**
  Drag the **Quality Rating** field to the **Row Labels** area
  Drag the **Meal Price ($)** field to the **Column Labels** area
  Drag the **Restaurant** field to the **Values** area

FIGURE 2.14 INITIAL PIVOTTABLE FIELD LIST AND PIVOTTABLE FIELD REPORT FOR THE RESTAURANT DATA

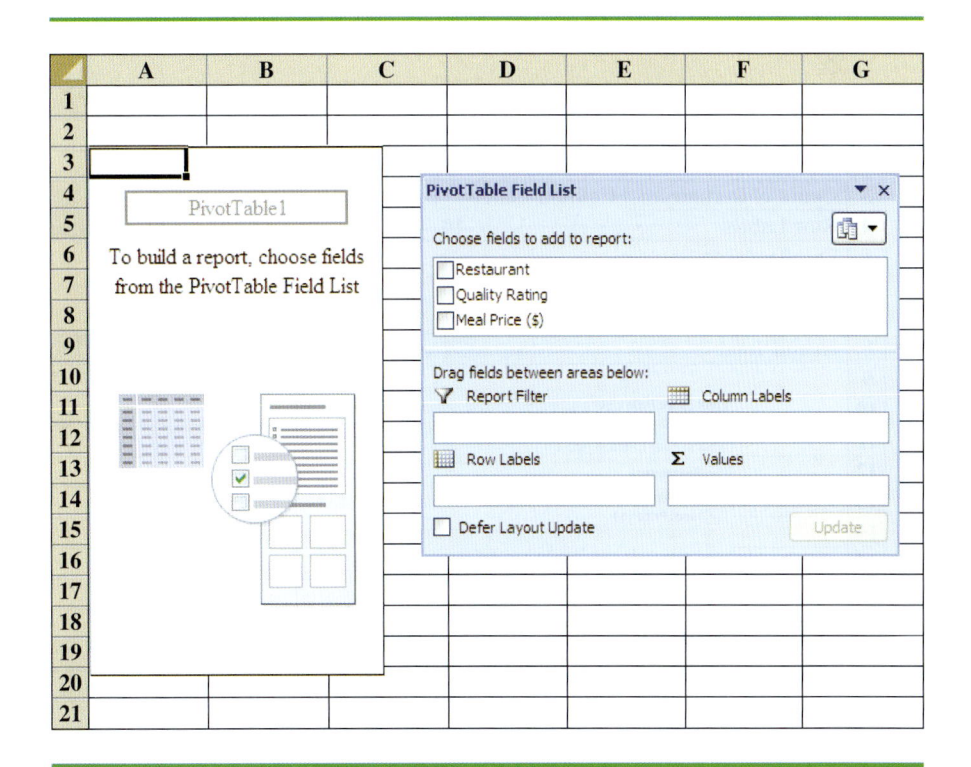

Step 2. Click on **Sum of Restaurant** in the **Values** area
Step 3. Click **Value Field Settings** from the list of options that appear
Step 4. When the Value Field Settings dialog appears:
　　Under **Summarize value field by**, choose **Count**
　　Click **OK**

Figure 2.15 shows the completed PivotTable Field List and a portion of the PivotTable worksheet as it now appears.

Finalize the PivotTable Report To complete the PivotTable Report, we need to group the columns representing meal prices and place the row labels for quality rating in the proper order. The following steps accomplish this.

Step 1. Right-click in cell B4 or any cell containing meal prices
Step 2. Choose **Group** from the list of options that appears
Step 3. When the Grouping dialog box appears:
　　Enter 10 in the **Starting at** box
　　Enter 49 in the **Ending at** box
　　Enter 10 in the **By** box
　　Click **OK**
Step 4. Right-click on **Excellent** in cell A5
Step 5. Choose **Move** and click **Move "Excellent" to End**

The final PivotTable Report is shown in Figure 2.16. Note that it provides the same information as the crosstabulation shown in Table 2.10.

Scatter Diagram

We can use Excel's chart tools to construct a scatter diagram and a trend line for the stereo and sound equipment store data presented in Table 2.12. Refer to Figures 2.17 and 2.18 as

FIGURE 2.15   COMPLETED PIVOTTABLE FIELD LIST AND A PORTION OF THE PIVOTTABLE REPORT FOR THE RESTAURANT DATA (COLUMNS H:AK ARE HIDDEN)

	A	B	C	D	E	F	G	AL	AM	AN	AO
1											
2											
3	Count of Restaurant	Column Labels									
4	Row Labels		10	11	12	13	14	15	47	48	Grand Total
5	Excellent					1			2	2	66
6	Good		6	4	3	3	2	4			84
7	Very Good		1	4	3	5	6	1		1	150
8	Grand Total		7	8	6	9	8	5	2	3	300

Appendix 2.2 Tabular and Graphical Presentations Using Excel

FIGURE 2.16 FINAL PIVOTTABLE REPORT FOR THE RESTAURANT DATA

	A	B	C	D	E	F
3	Count of Restaurant	Column Labels				
4	Row Labels	10–19	20–29	30–39	40–49	Grand Total
5	Good	42	40	2		84
6	Very Good	34	64	46	6	150
7	Excellent	2	14	28	22	66
8	Grand Total	78	118	76	28	300

PivotTable Field List

Choose fields to add to report:
- ☑ Restaurant
- ☑ Quality Rating
- ☑ Meal Price ($)

Drag fields between areas below:
- ▼ Report Filter
- ▥ Column Labels: Meal Price ($)
- ▤ Row Labels: Quality Rating
- Σ Values: Count of Restaurant

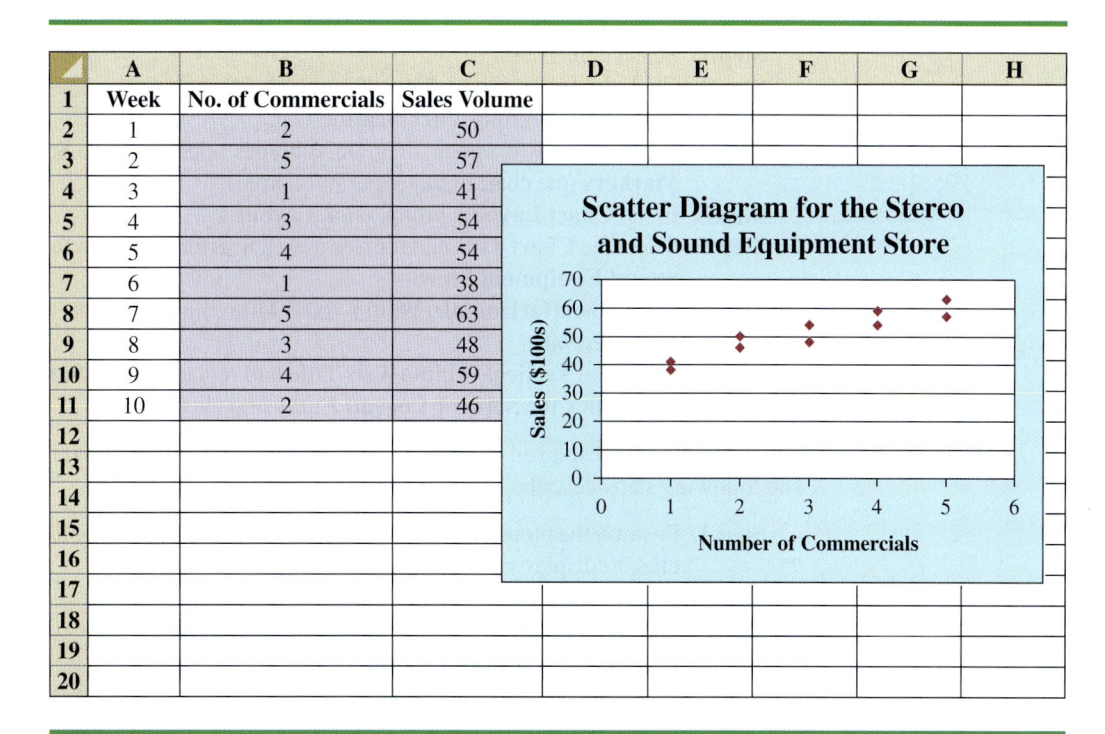

☐ Defer Layout Update   Update

FIGURE 2.17 SCATTER DIAGRAM FOR THE STEREO AND SOUND EQUIPMENT STORE USING EXCEL'S CHART TOOLS

	A	B	C
1	Week	No. of Commercials	Sales Volume
2	1	2	50
3	2	5	57
4	3	1	41
5	4	3	54
6	5	4	54
7	6	1	38
8	7	5	63
9	8	3	48
10	9	4	59
11	10	2	46

Scatter Diagram for the Stereo and Sound Equipment Store

84 Chapter 2 Descriptive Statistics: Tabular and Graphical Presentations

FIGURE 2.18 SCATTER DIAGRAM AND TRENDLINE FOR THE STEREO AND SOUND EQUIPMENT STORE USING EXCEL'S CHART TOOLS

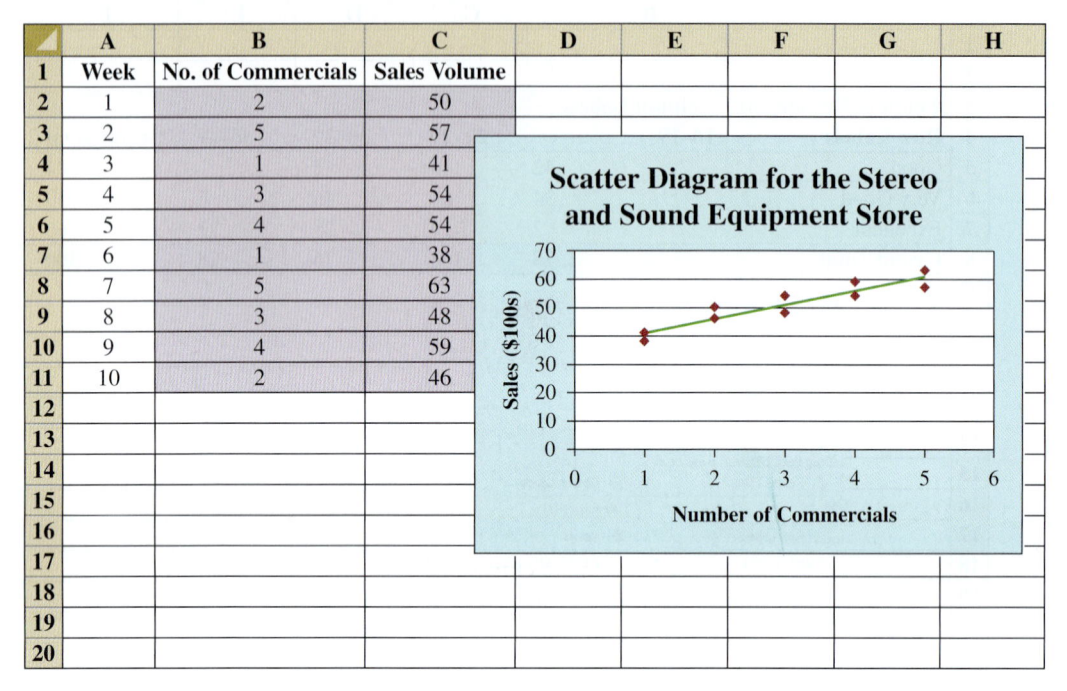

	A	B	C
1	Week	No. of Commercials	Sales Volume
2	1	2	50
3	2	5	57
4	3	1	41
5	4	3	54
6	5	4	54
7	6	1	38
8	7	5	63
9	8	3	48
10	9	4	59
11	10	2	46

Scatter Diagram for the Stereo and Sound Equipment Store — scatter plot with trendline, x-axis: Number of Commercials (0–6), y-axis: Sales ($100s) (0–70).

we describe the steps involved. We will use the data in the file named Stereo; the labels Week, No. of Commercials, and Sales Volume have been entered into cells A1:C1 of the worksheet. The data for each of the 10 weeks are entered into cells B2:C11. The following steps describe how to use Excel's chart tools to produce a scatter diagram for the data.

Step 1. Select cells B2:C11
Step 2. Click the **Insert** tab on the Ribbon
Step 3. In the **Charts** group, click **Scatter**
Step 4. When the list of scatter diagram subtypes appears, click **Scatter with only Markers** (the chart in the upper left corner)
Step 5. In the **Chart Layouts** group, click **Layout 1**
Step 6. Select the **Chart Title** and replace it with **Scatter Diagram for the Stereo and Sound Equipment Store**
Step 7. Select the **Horizontal (Value) Axis Title** and replace it with **Number of Commercials**
Step 8. Select the **Vertical (Value) Axis Title** and replace it with **Sales ($100s)**
Step 9. Right-click the **Series 1 Legend Entry** and click **Delete**

The worksheet displayed in Figure 2.17 shows the scatter diagram produced by Excel. The following steps describe how to add a trendline.

Step 1. Position the mouse pointer over any data point in the scatter diagram and right-click to display a list of options
Step 2. Choose **Add Trendline**
Step 3. When the Format Trendline dialog box appears:
  Select **Trendline Options**
  Choose **Linear** from the **Trend/Regression Type** list
  Click **Close**

The worksheet displayed in Figure 2.18 shows the scatter diagram with the trendline added.

Appendix 2.3 Tabular and Graphical Presentations Using StatTools

In this appendix we show how StatTools can be used to construct a histogram and a scatter diagram.

Histogram

We use the audit time data in Table 2.4 to illustrate. Begin by using the Data Set Manager to create a StatTools data set for these data using the procedure described in the appendix in Chapter 1. The following steps will generate a histogram.

WEB file
Audit

Step 1. Click the **StatTools** tab on the Ribbon
Step 2. In the **Analyses Group,** click **Summary Graphs**
Step 3. Choose the **Histogram** option
Step 4. When the StatTools—Histogram dialog box appears:
  In the **Variables** section, select **Audit Time**
  In the **Options** section,
    Enter 5 in the **Number of Bins** box
    Enter 9.5 in the **Histogram Minimum** box
    Enter 34.5 in the **Histogram Maximum** box
    Choose **Categorical** in the **X-Axis** box
    Choose **Frequency** in the **Y-Axis** box
  Click **OK**

A histogram for the audit time data similar to the histogram shown in Figure 2.12 will appear. The only difference is the histogram developed using StatTools shows the class midpoints on the horizontal axis.

Scatter Diagram

We use the stereo and sound equipment data in Table 2.12 to demonstrate the construction of a scatter diagram. Begin by using the Data Set Manager to create a StatTools data set for these data using the procedure described in the appendix in Chapter 1. The following steps will generate a scatter diagram.

WEB file
Stereo

Step 1. Click the **StatTools** tab on the Ribbon
Step 2. In the **Analyses Group,** click **Summary Graphs**
Step 3. Choose the **Scatterplot** option
Step 4. When the StatTools—Scatterplot dialog box appears:
  In the **Variables** section,
    In the column labeled **X**, select **No. of Commercials**
    In the column labeled **Y**, select **Sales Volume**
  Click **OK**

A scatter diagram similar to the one shown in Figure 2.17 will appear.

CHAPTER 3

Descriptive Statistics: Numerical Measures

CONTENTS

STATISTICS IN PRACTICE:
SMALL FRY DESIGN

3.1 MEASURES OF LOCATION
Mean
Median
Mode
Percentiles
Quartiles

3.2 MEASURES OF VARIABILITY
Range
Interquartile Range
Variance
Standard Deviation
Coefficient of Variation

3.3 MEASURES OF DISTRIBUTION SHAPE, RELATIVE LOCATION, AND DETECTION OF OUTLIERS
Distribution Shape
z-Scores
Chebyshev's Theorem
Empirical Rule
Detecting Outliers

3.4 EXPLORATORY DATA ANALYSIS
Five-Number Summary
Box Plot

3.5 MEASURES OF ASSOCIATION BETWEEN TWO VARIABLES
Covariance
Interpretation of the Covariance
Correlation Coefficient
Interpretation of the Correlation Coefficient

3.6 THE WEIGHTED MEAN AND WORKING WITH GROUPED DATA
Weighted Mean
Grouped Data

STATISTICS in PRACTICE

SMALL FRY DESIGN*
SANTA ANA, CALIFORNIA

Founded in 1997, Small Fry Design is a toy and accessory company that designs and imports products for infants. The company's product line includes teddy bears, mobiles, musical toys, rattles, and security blankets and features high-quality soft toy designs with an emphasis on color, texture, and sound. The products are designed in the United States and manufactured in China.

Small Fry Design uses independent representatives to sell the products to infant furnishing retailers, children's accessory and apparel stores, gift shops, upscale department stores, and major catalog companies. Currently, Small Fry Design products are distributed in more than 1000 retail outlets throughout the United States.

Cash flow management is one of the most critical activities in the day-to-day operation of this company. Ensuring sufficient incoming cash to meet both current and ongoing debt obligations can mean the difference between business success and failure. A critical factor in cash flow management is the analysis and control of accounts receivable. By measuring the average age and dollar value of outstanding invoices, management can predict cash availability and monitor changes in the status of accounts receivable. The company set the following goals: the average age for outstanding invoices should not exceed 45 days, and the dollar value of invoices more than 60 days old should not exceed 5% of the dollar value of all accounts receivable.

In a recent summary of accounts receivable status, the following descriptive statistics were provided for the age of outstanding invoices:

Mean	40 days
Median	35 days
Mode	31 days

*The authors are indebted to John A. McCarthy, President of Small Fry Design, for providing this Statistics in Practice.

Small Fry Design uses descriptive statistics to monitor its accounts receivable and incoming cash flow.

Interpretation of these statistics shows that the mean or average age of an invoice is 40 days. The median shows that half of the invoices remain outstanding 35 days or more. The mode of 31 days, the most frequent invoice age, indicates that the most common length of time an invoice is outstanding is 31 days. The statistical summary also showed that only 3% of the dollar value of all accounts receivable was more than 60 days old. Based on the statistical information, management was satisfied that accounts receivable and incoming cash flow were under control.

In this chapter, you will learn how to compute and interpret some of the statistical measures used by Small Fry Design. In addition to the mean, median, and mode, you will learn about other descriptive statistics such as the range, variance, standard deviation, percentiles, and correlation. These numerical measures will assist in the understanding and interpretation of data.

In Chapter 2 we discussed tabular and graphical presentations used to summarize data. In this chapter, we present several numerical measures that provide additional alternatives for summarizing data.

We start by developing numerical summary measures for data sets consisting of a single variable. When a data set contains more than one variable, the same numerical measures can be computed separately for each variable. However, in the two-variable case, we will also develop measures of the relationship between the variables.

Numerical measures of location, dispersion, shape, and association are introduced. If the measures are computed for data from a sample, they are called **sample statistics**. If the measures are computed for data from a population, they are called **population parameters**. In statistical inference, a sample statistic is referred to as the **point estimator** of the corresponding population parameter. In Chapter 7 we will discuss in more detail the process of point estimation.

In the three chapter appendixes we show how Minitab, Excel, and StatTools can be used to compute the numerical measures described in the chapter.

3.1 Measures of Location

Mean

Perhaps the most important measure of location is the **mean**, or average value, for a variable. The mean provides a measure of central location for the data. If the data are for a sample, the mean is denoted by $\bar{x}$; if the data are for a population, the mean is denoted by the Greek letter μ.

In statistical formulas, it is customary to denote the value of variable x for the first observation by x_1, the value of variable x for the second observation by x_2, and so on. In general, the value of variable x for the ith observation is denoted by x_i. For a sample with n observations, the formula for the sample mean is as follows.

The sample mean $\bar{x}$ is a sample statistic.

SAMPLE MEAN

$$\bar{x} = \frac{\Sigma x_i}{n} \tag{3.1}$$

In the preceding formula, the numerator is the sum of the values of the n observations. That is,

$$\Sigma x_i = x_1 + x_2 + \cdots + x_n$$

The Greek letter Σ is the summation sign.

To illustrate the computation of a sample mean, let us consider the following class size data for a sample of five college classes.

$$46 \quad 54 \quad 42 \quad 46 \quad 32$$

We use the notation x_1, x_2, x_3, x_4, x_5 to represent the number of students in each of the five classes.

$$x_1 = 46 \quad x_2 = 54 \quad x_3 = 42 \quad x_4 = 46 \quad x_5 = 32$$

Hence, to compute the sample mean, we can write

$$\bar{x} = \frac{\Sigma x_i}{n} = \frac{x_1 + x_2 + x_3 + x_4 + x_5}{5} = \frac{46 + 54 + 42 + 46 + 32}{5} = 44$$

The sample mean class size is 44 students.

Another illustration of the computation of a sample mean is given in the following situation. Suppose that a college placement office sent a questionnaire to a sample of business school graduates requesting information on monthly starting salaries. Table 3.1 shows the

3.1 Measures of Location

TABLE 3.1 MONTHLY STARTING SALARIES FOR A SAMPLE OF 12 BUSINESS SCHOOL GRADUATES

Graduate	Monthly Starting Salary ($)	Graduate	Monthly Starting Salary ($)
1	3450	7	3490
2	3550	8	3730
3	3650	9	3540
4	3480	10	3925
5	3355	11	3520
6	3310	12	3480

WEB file: StartSalary

collected data. The mean monthly starting salary for the sample of 12 business college graduates is computed as

$$\bar{x} = \frac{\Sigma x_i}{n} = \frac{x_1 + x_2 + \cdots + x_{12}}{12}$$

$$= \frac{3450 + 3550 + \cdots + 3480}{12}$$

$$= \frac{42{,}480}{12} = 3540$$

Equation (3.1) shows how the mean is computed for a sample with n observations. The formula for computing the mean of a population remains the same, but we use different notation to indicate that we are working with the entire population. The number of observations in a population is denoted by N and the symbol for a population mean is μ.

The sample mean $\bar{x}$ is a point estimator of the population mean μ.

POPULATION MEAN

$$\mu = \frac{\Sigma x_i}{N} \qquad (3.2)$$

Median

The **median** is another measure of central location. The median is the value in the middle when the data are arranged in ascending order (smallest value to largest value). With an odd number of observations, the median is the middle value. An even number of observations has no single middle value. In this case, we follow convention and define the median as the average of the values for the middle two observations. For convenience the definition of the median is restated as follows.

MEDIAN

Arrange the data in ascending order (smallest value to largest value).

(a) For an odd number of observations, the median is the middle value.
(b) For an even number of observations, the median is the average of the two middle values.

Let us apply this definition to compute the median class size for the sample of five college classes. Arranging the data in ascending order provides the following list:

$$32 \quad 42 \quad 46 \quad 46 \quad 54$$

Because $n = 5$ is odd, the median is the middle value. Thus the median class size is 46 students. Even though this data set contains two observations with values of 46, each observation is treated separately when we arrange the data in ascending order.

Suppose we also compute the median starting salary for the 12 business college graduates in Table 3.1. We first arrange the data in ascending order:

$$3310 \quad 3355 \quad 3450 \quad 3480 \quad 3480 \quad \underbrace{3490 \quad 3520}_{\text{Middle Two Values}} \quad 3540 \quad 3550 \quad 3650 \quad 3730 \quad 3925$$

Because $n = 12$ is even, we identify the middle two values: 3490 and 3520. The median is the average of these values.

$$\text{Median} = \frac{3490 + 3520}{2} = 3505$$

The median is the measure of location most often reported for annual income and property value data because a few extremely large incomes or property values can inflate the mean. In such cases, the median is the preferred measure of central location.

Although the mean is the more commonly used measure of central location, in some situations the median is preferred. The mean is influenced by extremely small and large data values. For instance, suppose that one of the graduates (see Table 3.1) had a starting salary of $10,000 per month (maybe the individual's family owns the company). If we change the highest monthly starting salary in Table 3.1 from $3925 to $10,000 and recompute the mean, the sample mean changes from $3540 to $4046. The median of $3505, however, is unchanged, because $3490 and $3520 are still the middle two values. With the extremely high starting salary included, the median provides a better measure of central location than the mean. We can generalize to say that whenever a data set contains extreme values, the median is often the preferred measure of central location.

Mode

A third measure of location is the **mode**. The mode is defined as follows.

> **MODE**
>
> The mode is the value that occurs with greatest frequency.

To illustrate the identification of the mode, consider the sample of five class sizes. The only value that occurs more than once is 46. Because this value, occurring with a frequency of 2, has the greatest frequency, it is the mode. As another illustration, consider the sample of starting salaries for the business school graduates. The only monthly starting salary that occurs more than once is $3480. Because this value has the greatest frequency, it is the mode.

Situations can arise for which the greatest frequency occurs at two or more different values. In these instances more than one mode exists. If the data contain exactly two modes, we say that the data are *bimodal*. If data contain more than two modes, we say that the data are *multimodal*. In multimodal cases the mode is almost never reported because listing three or more modes would not be particularly helpful in describing a location for the data.

Percentiles

A **percentile** provides information about how the data are spread over the interval from the smallest value to the largest value. For data that do not contain numerous repeated values, the pth percentile divides the data into two parts. Approximately p percent of the observations have values less than the pth percentile; approximately $(100 - p)$ percent of the observations have values greater than the pth percentile. The pth percentile is formally defined as follows.

PERCENTILE

The pth percentile is a value such that *at least* p percent of the observations are less than or equal to this value and *at least* $(100 - p)$ percent of the observations are greater than or equal to this value.

Colleges and universities frequently report admission test scores in terms of percentiles. For instance, suppose an applicant obtains a raw score of 54 on the verbal portion of an admission test. How this student performed in relation to other students taking the same test may not be readily apparent. However, if the raw score of 54 corresponds to the 70th percentile, we know that approximately 70% of the students scored lower than this individual and approximately 30% of the students scored higher than this individual.

The following procedure can be used to compute the pth percentile.

CALCULATING THE pTH PERCENTILE

Following these steps makes it easy to calculate percentiles.

Step 1. Arrange the data in ascending order (smallest value to largest value).
Step 2. Compute an index i

$$i = \left(\frac{p}{100}\right)n$$

where p is the percentile of interest and n is the number of observations.

Step 3. (a) If i *is not an integer, round up*. The next integer *greater* than i denotes the position of the pth percentile.
(b) If i *is an integer*, the pth percentile is the average of the values in positions i and $i + 1$.

As an illustration of this procedure, let us determine the 85th percentile for the starting salary data in Table 3.1.

Step 1. Arrange the data in ascending order.

3310   3355   3450   3480   3480   3490   3520   3540   3550   3650   3730   3925

Step 2.

$$i = \left(\frac{p}{100}\right)n = \left(\frac{85}{100}\right)12 = 10.2$$

Step 3. Because i is not an integer, *round up*. The position of the 85th percentile is the next integer greater than 10.2, the 11th position.

Returning to the data, we see that the 85th percentile is the data value in the 11th position, or 3730.

As another illustration of this procedure, let us consider the calculation of the 50th percentile for the starting salary data. Applying step 2, we obtain

$$i = \left(\frac{50}{100}\right)12 = 6$$

Because i is an integer, step 3(b) states that the 50th percentile is the average of the sixth and seventh data values; thus the 50th percentile is $(3490 + 3520)/2 = 3505$. Note that the *50th percentile is also the median.*

Quartiles

Quartiles are just specific percentiles; thus, the steps for computing percentiles can be applied directly in the computation of quartiles.

It is often desirable to divide data into four parts, with each part containing approximately one-fourth, or 25% of the observations. Figure 3.1 shows a data distribution divided into four parts. The division points are referred to as the **quartiles** and are defined as

Q_1 = first quartile, or 25th percentile

Q_2 = second quartile, or 50th percentile (also the median)

Q_3 = third quartile, or 75th percentile

The starting salary data are again arranged in ascending order. We already identified Q_2, the second quartile (median), as 3505.

3310 3355 3450 3480 3480 3490 3520 3540 3550 3650 3730 3925

The computations of quartiles Q_1 and Q_3 require the use of the rule for finding the 25th and 75th percentiles. These calculations follow.

For Q_1,

$$i = \left(\frac{p}{100}\right)n = \left(\frac{25}{100}\right)12 = 3$$

Because i is an integer, step 3(b) indicates that the first quartile, or 25th percentile, is the average of the third and fourth data values; thus, $Q_1 = (3450 + 3480)/2 = 3465$.

For Q_3,

$$i = \left(\frac{p}{100}\right)n = \left(\frac{75}{100}\right)12 = 9$$

Again, because i is an integer, step 3(b) indicates that the third quartile, or 75th percentile, is the average of the ninth and tenth data values; thus, $Q_3 = (3550 + 3650)/2 = 3600$.

FIGURE 3.1 LOCATION OF THE QUARTILES

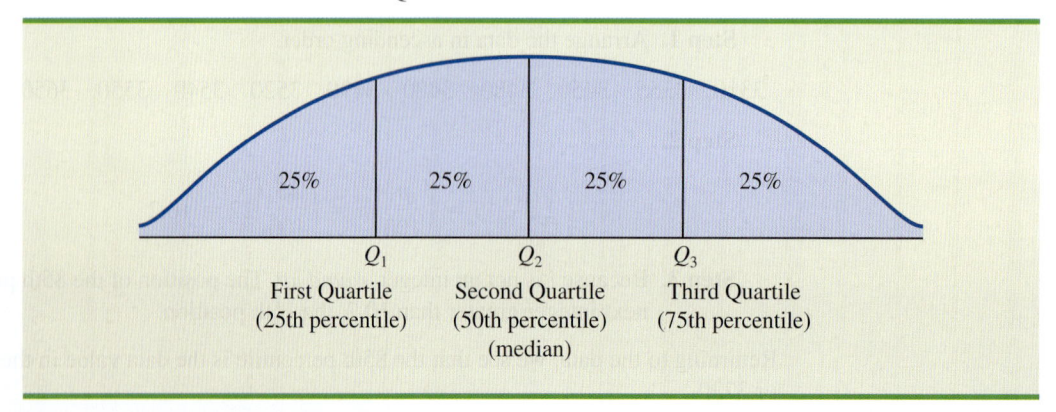

25% | 25% | 25% | 25%

Q_1
First Quartile
(25th percentile)

Q_2
Second Quartile
(50th percentile)
(median)

Q_3
Third Quartile
(75th percentile)

3.1 Measures of Location

The quartiles divide the starting salary data into four parts, with each part containing 25% of the observations.

$$3310 \quad 3355 \quad 3450 \mid 3480 \quad 3480 \quad 3490 \mid 3520 \quad 3540 \quad 3550 \mid 3650 \quad 3730 \quad 3925$$

$$Q_1 = 3465 \qquad Q_2 = 3505 \qquad Q_3 = 3600$$
$$\text{(Median)}$$

We defined the quartiles as the 25th, 50th, and 75th percentiles. Thus, we computed the quartiles in the same way as percentiles. However, other conventions are sometimes used to compute quartiles, and the actual values reported for quartiles may vary slightly depending on the convention used. Nevertheless, the objective of all procedures for computing quartiles is to divide the data into four equal parts.

NOTES AND COMMENTS

It is better to use the median than the mean as a measure of central location when a data set contains extreme values. Another measure, sometimes used when extreme values are present, is the *trimmed mean*. It is obtained by deleting a percentage of the smallest and largest values from a data set and then computing the mean of the remaining values. For example, the 5% trimmed mean is obtained by removing the smallest 5% and the largest 5% of the data values and then computing the mean of the remaining values. Using the sample with $n = 12$ starting salaries, $0.05(12) = 0.6$. Rounding this value to 1 indicates that the 5% trimmed mean would remove the 1 smallest data value and the 1 largest data value. The 5% trimmed mean using the 10 remaining observations is 3524.50.

Exercises

Methods

1. Consider a sample with data values of 10, 20, 12, 17, and 16. Compute the mean and median.
2. Consider a sample with data values of 10, 20, 21, 17, 16, and 12. Compute the mean and median.
3. Consider a sample with data values of 27, 25, 20, 15, 30, 34, 28, and 25. Compute the 20th, 25th, 65th, and 75th percentiles.
4. Consider a sample with data values of 53, 55, 70, 58, 64, 57, 53, 69, 57, 68, and 53. Compute the mean, median, and mode.

Applications

5. The Dow Jones Travel Index reported what business travelers pay for hotel rooms per night in major U.S. cities (*The Wall Street Journal*, January 16, 2004). The average hotel room rates for 20 cities are as follows:

City	Rate	City	Rate
Atlanta	$163	Minneapolis	$125
Boston	177	New Orleans	167
Chicago	166	New York	245
Cleveland	126	Orlando	146
Dallas	123	Phoenix	139
Denver	120	Pittsburgh	134
Detroit	144	San Francisco	167
Houston	173	Seattle	162
Los Angeles	160	St. Louis	145
Miami	192	Washington, D.C.	207

a. What is the mean hotel room rate?
b. What is the median hotel room rate?
c. What is the mode?
d. What is the first quartile?
e. What is the third quartile?

6. During the 2007–2008 NCAA college basketball season, men's basketball teams attempted an all-time high number of 3-point shots, averaging 19.07 shots per game (Associated Press Sports, January 24, 2009). In an attempt to discourage so many 3-point shots and encourage more inside play, the NCAA rules committee moved the 3-point line back from 19 feet, 9 inches to 20 feet, 9 inches at the beginning of the 2008–2009 basketball season. Shown in the following table are the 3-point shots taken and the 3-point shots made for a sample of 19 NCAA basketball games during the 2008–2009 season.

3-Point Shots	Shots Made	3-Point Shots	Shots Made
23	4	17	7
20	6	19	10
17	5	22	7
18	8	25	11
13	4	15	6
16	4	10	5
8	5	11	3
19	8	25	8
28	5	23	7
21	7		

a. What is the mean number of 3-point shots taken per game?
b. What is the mean number of 3-point shots made per game?
c. Using the closer 3-point line, players were making 35.2% of their shots. What percentage of shots were players making from the new 3-point line?
d. What was the impact of the NCAA rules change that moved the 3-point line back to 20 feet, 9 inches for the 2008–2009 season? Would you agree with the Associated Press Sports article that stated, "Moving back the 3-point line hasn't changed the game dramatically"? Explain.

7. Endowment income is a critical part of the annual budgets at colleges and universities. A study by the National Association of College and University Business Officers reported that the 435 colleges and universities surveyed held a total of $413 billion in endowments. The 10 wealthiest universities are shown in the following table (*The Wall Street Journal,* January 27, 2009). Amounts are in billions of dollars.

University	Endowment ($billion)	University	Endowment ($billion)
Columbia	7.2	Princeton	16.4
Harvard	36.6	Stanford	17.2
M.I.T.	10.1	Texas	16.1
Michigan	7.6	Texas A&M	6.7
Northwestern	7.2	Yale	22.9

a. What is the mean endowment for these universities?
b. What is the median endowment?
c. What is the mode endowment?
d. Compute the first and third quartiles.

e. What is the total endowment at these 10 universities? These universities represent 2.3% of the 435 colleges and universities surveyed. What percentage of the total $413 billion in endowments is held by these 10 universities?

f. *The Wall Street Journal* reported that over a recent five-month period, a downturn in the economy has caused endowments to decline 23%. What is the estimate of the dollar amount of the decline in the total endowments held by these 10 universities? Given this situation, what are some of the steps you would expect university administrators to be considering?

8. The cost of consumer purchases such as single-family housing, gasoline, Internet services, tax preparation, and hospitalization was provided in *The Wall-Street Journal* (January 2, 2007). Sample data typical of the cost of tax-return preparation by services such as H&R Block are as shown.

120	230	110	115	160
130	150	105	195	155
105	360	120	120	140
100	115	180	235	255

a. Compute the mean, median, and mode.
b. Compute the first and third quartiles.
c. Compute and interpret the 90th percentile.

9. The National Association of Realtors provided data showing that home sales were the slowest in 10 years (Associated Press, December 24, 2008). Sample data with representative sales prices for existing homes and new homes follow. Data are in thousands of dollars:

| *Existing Homes* | 315.5 | 202.5 | 140.2 | 181.3 | 470.2 | 169.9 | 112.8 | 230.0 | 177.5 |
| *New Homes* | 275.9 | 350.2 | 195.8 | 525.0 | 225.3 | 215.5 | 175.0 | 149.5 | |

a. What is the median sales price for existing homes?
b. What is the median sales price for new homes?
c. Do existing homes or new homes have the higher median sales price? What is the difference between the median sales prices?
d. A year earlier the median sales price for existing homes was $208.4 thousand and the median sales price for new homes was $249 thousand. Compute the percentage change in the median sales price of existing and new homes over the one-year period. Did existing homes or new homes have the larger percentage change in median sales price?

10. A panel of economists provided forecasts of the U.S. economy for the first six months of 2007 (*The Wall Street Journal,* January 2, 2007). The percent changes in the gross domestic product (GDP) forecasted by 30 economists are as follows.

2.6	3.1	2.3	2.7	3.4	0.9	2.6	2.8	2.0	2.4
2.7	2.7	2.7	2.9	3.1	2.8	1.7	2.3	2.8	3.5
0.4	2.5	2.2	1.9	1.8	1.1	2.0	2.1	2.5	0.5

a. What is the minimum forecast for the percent change in the GDP? What is the maximum?
b. Compute the mean, median, and mode.
c. Compute the first and third quartiles.
d. Did the economists provide an optimistic or pessimistic outlook for the U.S. economy? Discuss.

11. In automobile mileage and gasoline-consumption testing, 13 automobiles were road tested for 300 miles in both city and highway driving conditions. The following data were recorded for miles-per-gallon performance.

City: 16.2 16.7 15.9 14.4 13.2 15.3 16.8 16.0 16.1 15.3 15.2 15.3 16.2
Highway: 19.4 20.6 18.3 18.6 19.2 17.4 17.2 18.6 19.0 21.1 19.4 18.5 18.7

Use the mean, median, and mode to make a statement about the difference in performance for city and highway driving.

12. Walt Disney Company bought Pixar Animation Studios, Inc., in a deal worth $7.4 billion (CNN Money website, January 24, 2006). The animated movies produced by Disney and Pixar during the previous 10 years are listed in the following table. The box office revenues are in millions of dollars. Compute the total revenue, the mean, the median, and the quartiles to compare the box office success of the movies produced by both companies. Do the statistics suggest at least one of the reasons Disney was interested in buying Pixar? Discuss.

Disney Movies	Revenue ($millions)	Pixar Movies	Revenue ($millions)
Pocahontas	346	Toy Story	362
Hunchback of Notre Dame	325	A Bug's Life	363
Hercules	253	Toy Story 2	485
Mulan	304	Monsters, Inc.	525
Tarzan	448	Finding Nemo	865
Dinosaur	354	The Incredibles	631
The Emperor's New Groove	169		
Lilo & Stitch	273		
Treasure Planet	110		
The Jungle Book 2	136		
Brother Bear	250		
Home on the Range	104		
Chicken Little	249		

3.2 Measures of Variability

The variability in the delivery time creates uncertainty for production scheduling. Methods in this section help measure and understand variability.

In addition to measures of location, it is often desirable to consider measures of variability, or dispersion. For example, suppose that you are a purchasing agent for a large manufacturing firm and that you regularly place orders with two different suppliers. After several months of operation, you find that the mean number of days required to fill orders is 10 days for both of the suppliers. The histograms summarizing the number of working days required to fill orders from the suppliers are shown in Figure 3.2. Although the mean number of days is 10 for both suppliers, do the two suppliers demonstrate the same degree of reliability in terms of making deliveries on schedule? Note the dispersion, or variability, in delivery times indicated by the histograms. Which supplier would you prefer?

For most firms, receiving materials and supplies on schedule is important. The 7- or 8-day deliveries shown for J.C. Clark Distributors might be viewed favorably; however, a few of the slow 13- to 15-day deliveries could be disastrous in terms of keeping a workforce busy

3.2 Measures of Variability

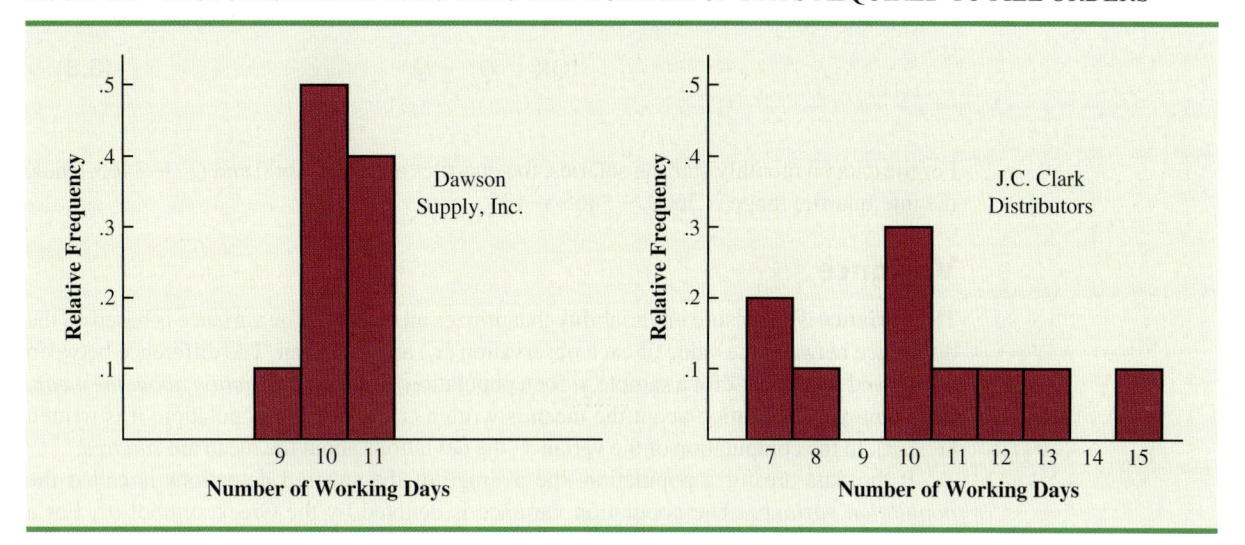

FIGURE 3.2 HISTORICAL DATA SHOWING THE NUMBER OF DAYS REQUIRED TO FILL ORDERS

and production on schedule. This example illustrates a situation in which the variability in the delivery times may be an overriding consideration in selecting a supplier. For most purchasing agents, the lower variability shown for Dawson Supply, Inc., would make Dawson the preferred supplier.

We turn now to a discussion of some commonly used measures of variability.

Range

The simplest measure of variability is the **range**.

> **RANGE**
>
> Range = Largest value − Smallest value

Let us refer to the data on starting salaries for business school graduates in Table 3.1. The largest starting salary is $3925 and the smallest is $3310. The range is 3925 − 3310 = 615.

Although the range is the easiest of the measures of variability to compute, it is seldom used as the only measure. The reason is that the range is based on only two of the observations and thus is highly influenced by extreme values. Suppose one of the graduates received a starting salary of $10,000 per month. In this case, the range would be 10,000 − 3310 = 6690 rather than 615. This large value for the range would not be especially descriptive of the variability in the data because 11 of the 12 starting salaries are closely grouped between 3310 and 3730.

Interquartile Range

A measure of variability that overcomes the dependency on extreme values is the **interquartile range (IQR)**. This measure of variability is the difference between the third quartile, Q_3, and the first quartile, Q_1. In other words, the interquartile range is the range for the middle 50% of the data.

INTERQUARTILE RANGE

$$IQR = Q_3 - Q_1 \qquad (3.3)$$

For the data on monthly starting salaries, the quartiles are $Q_3 = 3600$ and $Q_1 = 3465$. Thus, the interquartile range is $3600 - 3465 = 135$.

Variance

The **variance** is a measure of variability that utilizes all the data. The variance is based on the difference between the value of each observation (x_i) and the mean. The difference between each x_i and the mean ($\bar{x}$ for a sample, μ for a population) is called a *deviation about the mean*. For a sample, a deviation about the mean is written ($x_i - \bar{x}$); for a population, it is written ($x_i - \mu$). In the computation of the variance, the deviations about the mean are *squared*.

If the data are for a population, the average of the squared deviations is called the *population variance*. The population variance is denoted by the Greek symbol σ^2. For a population of N observations and with μ denoting the population mean, the definition of the population variance is as follows.

POPULATION VARIANCE

$$\sigma^2 = \frac{\Sigma(x_i - \mu)^2}{N} \qquad (3.4)$$

In most statistical applications, the data being analyzed are for a sample. When we compute a sample variance, we are often interested in using it to estimate the population variance σ^2. Although a detailed explanation is beyond the scope of this text, it can be shown that if the sum of the squared deviations about the sample mean is divided by $n - 1$, and not n, the resulting sample variance provides an unbiased estimate of the population variance. For this reason, the *sample variance,* denoted by s^2, is defined as follows.

The sample variance s^2 is the estimator of the population variance σ^2.

SAMPLE VARIANCE

$$s^2 = \frac{\Sigma(x_i - \bar{x})^2}{n - 1} \qquad (3.5)$$

To illustrate the computation of the sample variance, we will use the data on class size for the sample of five college classes as presented in Section 3.1. A summary of the data, including the computation of the deviations about the mean and the squared deviations about the mean, is shown in Table 3.2. The sum of squared deviations about the mean is $\Sigma(x_i - \bar{x})^2 = 256$. Hence, with $n - 1 = 4$, the sample variance is

$$s^2 = \frac{\Sigma(x_i - \bar{x})^2}{n - 1} = \frac{256}{4} = 64$$

Before moving on, let us note that the units associated with the sample variance often cause confusion. Because the values being summed in the variance calculation, $(x_i - \bar{x})^2$, are squared, the units associated with the sample variance are also *squared*. For instance, the

3.2 Measures of Variability

TABLE 3.2 COMPUTATION OF DEVIATIONS AND SQUARED DEVIATIONS ABOUT THE MEAN FOR THE CLASS SIZE DATA

Number of Students in Class (x_i)	Mean Class Size ($\bar{x}$)	Deviation About the Mean ($x_i - \bar{x}$)	Squared Deviation About the Mean ($x_i - \bar{x})^2$
46	44	2	4
54	44	10	100
42	44	−2	4
46	44	2	4
32	44	−12	144
		0	256
		$\Sigma(x_i - \bar{x})$	$\Sigma(x_i - \bar{x})^2$

The variance is useful in comparing the variability of two or more variables.

sample variance for the class size data is $s^2 = 64$ (students)^2. The squared units associated with variance make it difficult to obtain an intuitive understanding and interpretation of the numerical value of the variance. We recommend that you think of the variance as a measure useful in comparing the amount of variability for two or more variables. In a comparison of the variables, the one with the largest variance shows the most variability. Further interpretation of the value of the variance may not be necessary.

As another illustration of computing a sample variance, consider the starting salaries listed in Table 3.1 for the 12 business school graduates. In Section 3.1, we showed that the sample mean starting salary was $3540. The computation of the sample variance ($s^2 = 27,440.91$) is shown in Table 3.3.

TABLE 3.3 COMPUTATION OF THE SAMPLE VARIANCE FOR THE STARTING SALARY DATA

Monthly Salary (x_i)	Sample Mean ($\bar{x}$)	Deviation About the Mean ($x_i - \bar{x}$)	Squared Deviation About the Mean ($x_i - \bar{x})^2$
3450	3540	−90	8,100
3550	3540	10	100
3650	3540	110	12,100
3480	3540	−60	3,600
3355	3540	−185	34,225
3310	3540	−230	52,900
3490	3540	−50	2,500
3730	3540	190	36,100
3540	3540	0	0
3925	3540	385	148,225
3520	3540	−20	400
3480	3540	−60	3,600
		0	301,850
		$\Sigma(x_i - \bar{x})$	$\Sigma(x_i - \bar{x})^2$

Using equation (3.5),

$$s^2 = \frac{\Sigma(x_i - \bar{x})^2}{n-1} = \frac{301,850}{11} = 27,440.91$$

In Tables 3.2 and 3.3 we show both the sum of the deviations about the mean and the sum of the squared deviations about the mean. For any data set, the sum of the deviations about the mean will *always equal zero*. Note that in Tables 3.2 and 3.3, $\Sigma(x_i - \bar{x}) = 0$. The positive deviations and negative deviations cancel each other, causing the sum of the deviations about the mean to equal zero.

Standard Deviation

The **standard deviation** is defined to be the positive square root of the variance. Following the notation we adopted for a sample variance and a population variance, we use s to denote the sample standard deviation and σ to denote the population standard deviation. The standard deviation is derived from the variance in the following way.

The sample standard deviation s is the estimator of the population standard deviation σ.

STANDARD DEVIATION

$$\text{Sample standard deviation} = s = \sqrt{s^2} \tag{3.6}$$

$$\text{Population standard deviation} = \sigma = \sqrt{\sigma^2} \tag{3.7}$$

Recall that the sample variance for the sample of class sizes in five college classes is $s^2 = 64$. Thus, the sample standard deviation is $s = \sqrt{64} = 8$. For the data on starting salaries, the sample standard deviation is $s = \sqrt{27{,}440.91} = 165.65$.

The standard deviation is easier to interpret than the variance because the standard deviation is measured in the same units as the data.

What is gained by converting the variance to its corresponding standard deviation? Recall that the units associated with the variance are squared. For example, the sample variance for the starting salary data of business school graduates is $s^2 = 27{,}440.91$ (dollars)^2. Because the standard deviation is the square root of the variance, the units of the variance, dollars squared, are converted to dollars in the standard deviation. Thus, the standard deviation of the starting salary data is $165.65. In other words, the standard deviation is measured in the same units as the original data. For this reason the standard deviation is more easily compared to the mean and other statistics that are measured in the same units as the original data.

Coefficient of Variation

The coefficient of variation is a relative measure of variability; it measures the standard deviation relative to the mean.

In some situations we may be interested in a descriptive statistic that indicates how large the standard deviation is relative to the mean. This measure is called the **coefficient of variation** and is usually expressed as a percentage.

COEFFICIENT OF VARIATION

$$\left(\frac{\text{Standard deviation}}{\text{Mean}} \times 100 \right)\% \tag{3.8}$$

For the class size data, we found a sample mean of 44 and a sample standard deviation of 8. The coefficient of variation is $[(8/44) \times 100]\% = 18.2\%$. In words, the coefficient of variation tells us that the sample standard deviation is 18.2% of the value of the sample mean. For the starting salary data with a sample mean of 3540 and a sample standard deviation of 165.65, the coefficient of variation, $[(165.65/3540) \times 100]\% = 4.7\%$, tells us the sample standard deviation is only 4.7% of the value of the sample mean. In general, the coefficient of variation is a useful statistic for comparing the variability of variables that have different standard deviations and different means.

NOTES AND COMMENTS

1. Statistical software packages and spreadsheets can be used to develop the descriptive statistics presented in this chapter. After the data are entered into a worksheet, a few simple commands can be used to generate the desired output. In three chapter-ending appendixes we show how Minitab, Excel, and StatTools can be used to develop descriptive statistics.

2. The standard deviation is a commonly used measure of the risk associated with investing in stock and stock funds (*BusinessWeek*, January 17, 2000). It provides a measure of how monthly returns fluctuate around the long-run average return.

3. Rounding the value of the sample mean $\bar{x}$ and the values of the squared deviations $(x_i - \bar{x})^2$ may introduce errors when a calculator is used in the computation of the variance and standard deviation. To reduce rounding errors, we recommend carrying at least six significant digits during intermediate calculations. The resulting variance or standard deviation can then be rounded to fewer digits.

4. An alternative formula for the computation of the sample variance is

$$s^2 = \frac{\Sigma x_i^2 - n\bar{x}^2}{n-1}$$

where $\Sigma x_i^2 = x_1^2 + x_2^2 + \cdots + x_n^2$.

Exercises

Methods

13. Consider a sample with data values of 10, 20, 12, 17, and 16. Compute the range and interquartile range.

14. Consider a sample with data values of 10, 20, 12, 17, and 16. Compute the variance and standard deviation.

15. Consider a sample with data values of 27, 25, 20, 15, 30, 34, 28, and 25. Compute the range, interquartile range, variance, and standard deviation. **SELF test**

Applications

16. A bowler's scores for six games were 182, 168, 184, 190, 170, and 174. Using these data as a sample, compute the following descriptive statistics: **SELF test**
    a. Range
    b. Variance
    c. Standard deviation
    d. Coefficient of variation

17. A home theater in a box is the easiest and cheapest way to provide surround sound for a home entertainment center. A sample of prices is shown here (*Consumer Reports Buying Guide,* 2004). The prices are for models with a DVD player and for models without a DVD player.

Models with DVD Player	Price	Models without DVD Player	Price
Sony HT-1800DP	$450	Pioneer HTP-230	$300
Pioneer HTD-330DV	300	Sony HT-DDW750	300
Sony HT-C800DP	400	Kenwood HTB-306	360
Panasonic SC-HT900	500	RCA RT-2600	290
Panasonic SC-MTI	400	Kenwood HTB-206	300

a. Compute the mean price for models with a DVD player and the mean price for models without a DVD player. What is the additional price paid to have a DVD player included in a home theater unit?
b. Compute the range, variance, and standard deviation for the two samples. What does this information tell you about the prices for models with and without a DVD player?

18. Car rental rates per day for a sample of seven Eastern U.S. cities are as follows (*The Wall Street Journal,* January 16, 2004).

City	Daily Rate
Boston	$43
Atlanta	35
Miami	34
New York	58
Orlando	30
Pittsburgh	30
Washington, D.C.	36

    a. Compute the mean, variance, and standard deviation for the car rental rates.
    b. A similar sample of seven Western U.S. cities showed a sample mean car rental rate of $38 per day. The variance and standard deviation were 12.3 and 3.5, respectively. Discuss any difference between the car rental rates in Eastern and Western U.S. cities.

19. The *Los Angeles Times* regularly reports the air quality index for various areas of Southern California. A sample of air quality index values for Pomona provided the following data: 28, 42, 58, 48, 45, 55, 60, 49, and 50.
    a. Compute the range and interquartile range.
    b. Compute the sample variance and sample standard deviation.
    c. A sample of air quality index readings for Anaheim provided a sample mean of 48.5, a sample variance of 136, and a sample standard deviation of 11.66. What comparisons can you make between the air quality in Pomona and that in Anaheim on the basis of these descriptive statistics?

20. The following data were used to construct the histograms of the number of days required to fill orders for Dawson Supply, Inc., and J.C. Clark Distributors (see Figure 3.2).

    Dawson Supply Days for Delivery: 11 10 9 10 11 11 10 11 10 10
    Clark Distributors Days for Delivery: 8 10 13 7 10 11 10 7 15 12

    Use the range and standard deviation to support the previous observation that Dawson Supply provides the more consistent and reliable delivery times.

21. How do grocery costs compare across the country? Using a market basket of 10 items including meat, milk, bread, eggs, coffee, potatoes, cereal, and orange juice, *Where to Retire* magazine calculated the cost of the market basket in six cities and in six retirement areas across the country (*Where to Retire,* November/December 2003). The data with market basket cost to the nearest dollar are as follows:

City	Cost	Retirement Area	Cost
Buffalo, NY	$33	Biloxi-Gulfport, MS	$29
Des Moines, IA	27	Asheville, NC	32
Hartford, CT	32	Flagstaff, AZ	32
Los Angeles, CA	38	Hilton Head, SC	34
Miami, FL	36	Fort Myers, FL	34
Pittsburgh, PA	32	Santa Fe, NM	31

    a. Compute the mean, variance, and standard deviation for the sample of cities and the sample of retirement areas.
    b. What observations can be made based on the two samples?

22. The National Retail Federation reported that college freshman spend more on back-to-school items than any other college group (*USA Today*, August 4, 2006). Sample data comparing the back-to-school expenditures for 25 freshmen and 20 seniors are shown in the data file BackToSchool.
    a. What is the mean back-to-school expenditure for each group? Are the data consistent with the National Retail Federation's report?
    b. What is the range for the expenditures in each group?
    c. What is the interquartile range for the expenditures in each group?
    d. What is the standard deviation for expenditures in each group?
    e. Do freshmen or seniors have more variation in back-to-school expenditures?

23. Scores turned in by an amateur golfer at the Bonita Fairways Golf Course in Bonita Springs, Florida, during 2005 and 2006 are as follows:

    | 2005 Season: | 74 | 78 | 79 | 77 | 75 | 73 | 75 | 77 |
    | 2006 Season: | 71 | 70 | 75 | 77 | 85 | 80 | 71 | 79 |

    a. Use the mean and standard deviation to evaluate the golfer's performance over the two-year period.
    b. What is the primary difference in performance between 2005 and 2006? What improvement, if any, can be seen in the 2006 scores?

24. The following times were recorded by the quarter-mile and mile runners of a university track team (times are in minutes).

    | Quarter-Mile Times: | .92 | .98 | 1.04 | .90 | .99 |
    | Mile Times: | 4.52 | 4.35 | 4.60 | 4.70 | 4.50 |

    After viewing this sample of running times, one of the coaches commented that the quarter-milers turned in the more consistent times. Use the standard deviation and the coefficient of variation to summarize the variability in the data. Does the use of the coefficient of variation indicate that the coach's statement should be qualified?

3.3 Measures of Distribution Shape, Relative Location, and Detection of Outliers

We have described several measures of location and variability for data. In addition, it is often important to have a measure of the shape of a distribution. In Chapter 2 we noted that a histogram provides a graphical display showing the shape of a distribution. An important numerical measure of the shape of a distribution is called **skewness**.

Distribution Shape

Shown in Figure 3.3 are four histograms constructed from relative frequency distributions. The histograms in Panels A and B are moderately skewed. The one in Panel A is skewed to the left; its skewness is −.85. The histogram in Panel B is skewed to the right; its skewness is +.85. The histogram in Panel C is symmetric; its skewness is zero. The histogram in Panel D is highly skewed to the right; its skewness is 1.62. The formula used to compute skewness is somewhat complex.[1] However, the skewness can easily be

---
[1] The formula for the skewness of sample data:

$$\text{Skewness} = \frac{n}{(n-1)(n-2)} \sum \left(\frac{x_i - \bar{x}}{s}\right)^3$$

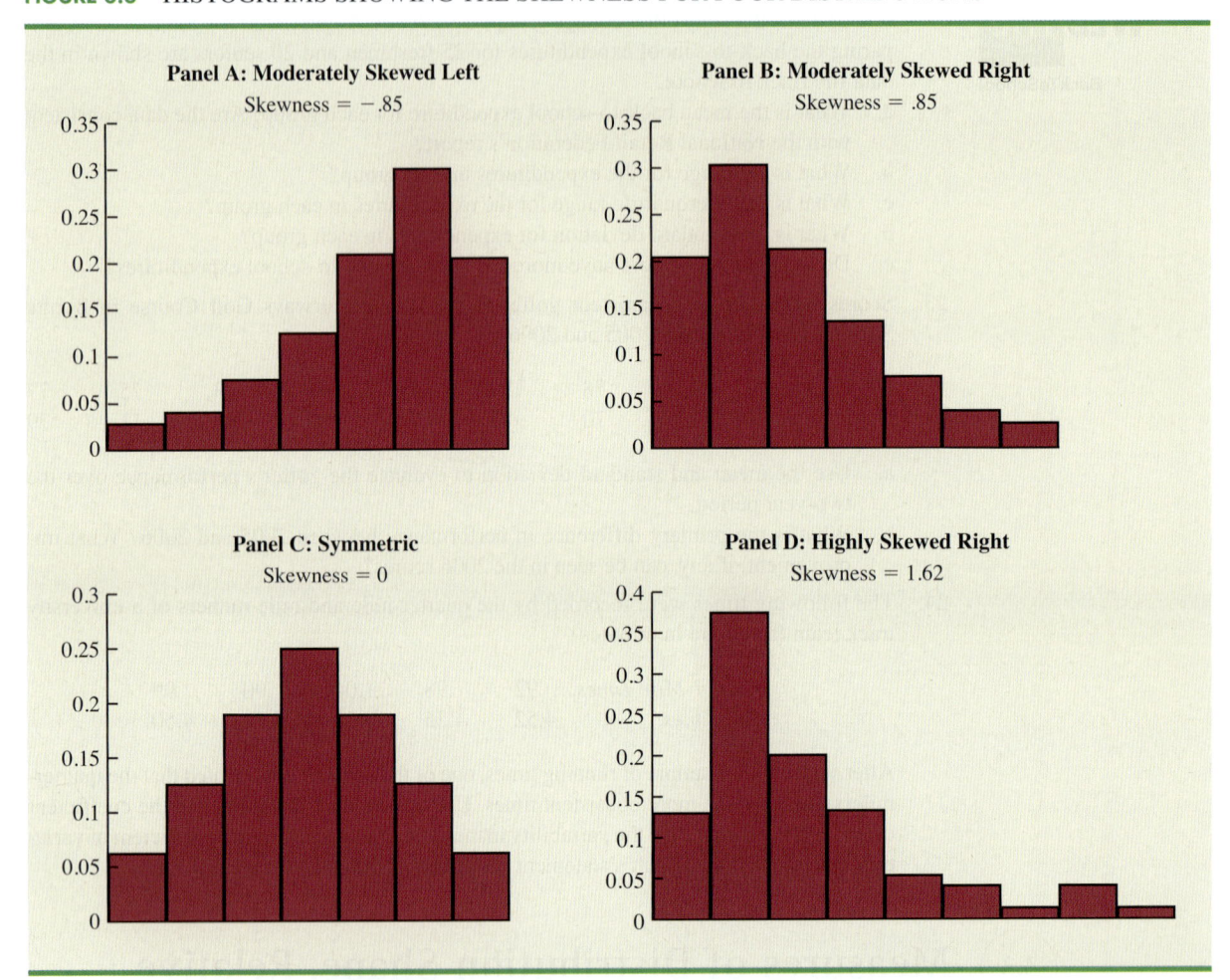

FIGURE 3.3 HISTOGRAMS SHOWING THE SKEWNESS FOR FOUR DISTRIBUTIONS

Panel A: Moderately Skewed Left
Skewness = −.85

Panel B: Moderately Skewed Right
Skewness = .85

Panel C: Symmetric
Skewness = 0

Panel D: Highly Skewed Right
Skewness = 1.62

computed using statistical software. For data skewed to the left, the skewness is negative; for data skewed to the right, the skewness is positive. If the data are symmetric, the skewness is zero.

For a symmetric distribution, the mean and the median are equal. When the data are positively skewed, the mean will usually be greater than the median; when the data are negatively skewed, the mean will usually be less than the median. The data used to construct the histogram in Panel D are customer purchases at a women's apparel store. The mean purchase amount is $77.60 and the median purchase amount is $59.70. The relatively few large purchase amounts tend to increase the mean, while the median remains unaffected by the large purchase amounts. The median provides the preferred measure of location when the data are highly skewed.

z-Scores

In addition to measures of location, variability, and shape, we are also interested in the relative location of values within a data set. Measures of relative location help us determine how far a particular value is from the mean.

By using both the mean and standard deviation, we can determine the relative location of any observation. Suppose we have a sample of n observations, with the values denoted

3.3 Measures of Distribution Shape, Relative Location, and Detection of Outliers

by $x_1, x_2, \ldots, x_n$. In addition, assume that the sample mean, $\bar{x}$, and the sample standard deviation, s, are already computed. Associated with each value, x_i, is another value called its **z-score**. Equation (3.9) shows how the z-score is computed for each x_i.

---

z-SCORE

$$z_i = \frac{x_i - \bar{x}}{s} \tag{3.9}$$

where

$z_i =$ the z-score for x_i
$\bar{x} =$ the sample mean
$s =$ the sample standard deviation

The z-score is often called the *standardized value*. The z-score, z_i, can be interpreted as the *number of standard deviations x_i is from the mean $\bar{x}$*. For example, $z_1 = 1.2$ would indicate that x_1 is 1.2 standard deviations greater than the sample mean. Similarly, $z_2 = -.5$ would indicate that x_2 is .5, or 1/2, standard deviation less than the sample mean. A z-score greater than zero occurs for observations with a value greater than the mean, and a z-score less than zero occurs for observations with a value less than the mean. A z-score of zero indicates that the value of the observation is equal to the mean.

The z-score for any observation can be interpreted as a measure of the relative location of the observation in a data set. Thus, observations in two different data sets with the same z-score can be said to have the same relative location in terms of being the same number of standard deviations from the mean.

The z-scores for the class size data are computed in Table 3.4. Recall the previously computed sample mean, $\bar{x} = 44$, and sample standard deviation, $s = 8$. The z-score of -1.50 for the fifth observation shows it is farthest from the mean; it is 1.50 standard deviations below the mean.

Chebyshev's Theorem

Chebyshev's theorem enables us to make statements about the proportion of data values that must be within a specified number of standard deviations of the mean.

TABLE 3.4 z-SCORES FOR THE CLASS SIZE DATA

Number of Students in Class (x_i)	Deviation About the Mean $(x_i - \bar{x})$	z-Score $\left(\dfrac{x_i - \bar{x}}{s}\right)$
46	2	2/8 = .25
54	10	10/8 = 1.25
42	−2	−2/8 = −.25
46	2	2/8 = .25
32	−12	−12/8 = −1.50

> **CHEBYSHEV'S THEOREM**
>
> At least $(1 - 1/z^2)$ of the data values must be within z standard deviations of the mean, where z is any value greater than 1.

Some of the implications of this theorem, with $z = 2, 3,$ and 4 standard deviations, follow.

- At least .75, or 75%, of the data values must be within $z = 2$ standard deviations of the mean.
- At least .89, or 89%, of the data values must be within $z = 3$ standard deviations of the mean.
- At least .94, or 94%, of the data values must be within $z = 4$ standard deviations of the mean.

For an example using Chebyshev's theorem, suppose that the midterm test scores for 100 students in a college business statistics course had a mean of 70 and a standard deviation of 5. How many students had test scores between 60 and 80? How many students had test scores between 58 and 82?

For the test scores between 60 and 80, we note that 60 is two standard deviations below the mean and 80 is two standard deviations above the mean. Using Chebyshev's theorem, we see that at least .75, or at least 75%, of the observations must have values within two standard deviations of the mean. Thus, at least 75% of the students must have scored between 60 and 80.

Chebyshev's theorem requires $z > 1$; but z need not be an integer.

For the test scores between 58 and 82, we see that $(58 - 70)/5 = -2.4$ indicates 58 is 2.4 standard deviations below the mean and that $(82 - 70)/5 = +2.4$ indicates 82 is 2.4 standard deviations above the mean. Applying Chebyshev's theorem with $z = 2.4$, we have

$$\left(1 - \frac{1}{z^2}\right) = \left(1 - \frac{1}{(2.4)^2}\right) = .826$$

At least 82.6% of the students must have test scores between 58 and 82.

Empirical Rule

The empirical rule is based on the normal probability distribution, which will be discussed in Chapter 6. The normal distribution is used extensively throughout the text.

One of the advantages of Chebyshev's theorem is that it applies to any data set regardless of the shape of the distribution of the data. Indeed, it could be used with any of the distributions in Figure 3.3. In many practical applications, however, data sets exhibit a symmetric mound-shaped or bell-shaped distribution like the one shown in Figure 3.4. When the data are believed to approximate this distribution, the **empirical rule** can be used to determine the percentage of data values that must be within a specified number of standard deviations of the mean.

> **EMPIRICAL RULE**
>
> For data having a bell-shaped distribution:
>
> - Approximately 68% of the data values will be within one standard deviation of the mean.
> - Approximately 95% of the data values will be within two standard deviations of the mean.
> - Almost all the data values will be within three standard deviations of the mean.

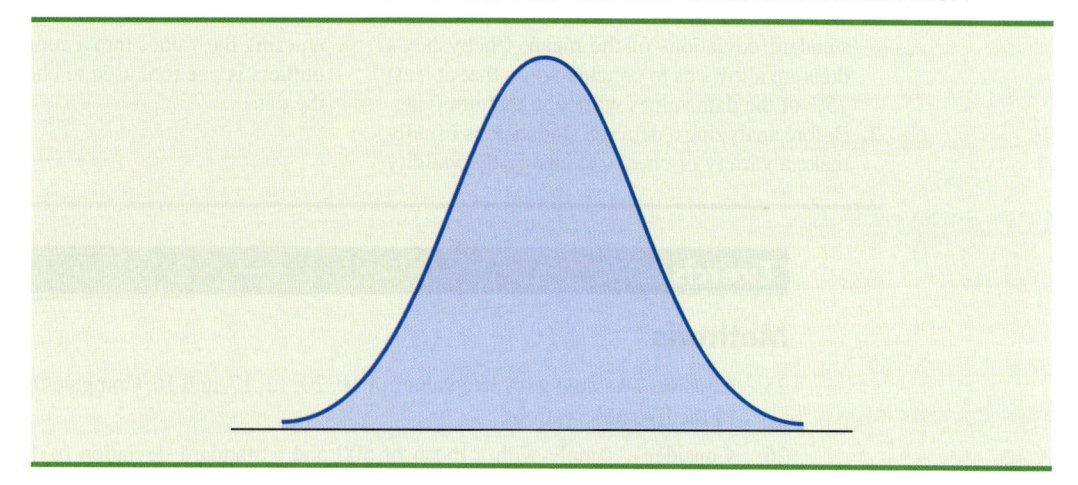

FIGURE 3.4 A SYMMETRIC MOUND-SHAPED OR BELL-SHAPED DISTRIBUTION

For example, liquid detergent bottles are filled automatically on a production line. Filling weights frequently have a bell-shaped distribution. If the mean filling weight is 16 ounces and the standard deviation is .25 ounce, we can use the empirical rule to draw the following conclusions.

- Approximately 68% of the filled cartons will have weights between 15.75 and 16.25 ounces (within one standard deviation of the mean).
- Approximately 95% of the filled cartons will have weights between 15.50 and 16.50 ounces (within two standard deviations of the mean).
- Almost all filled cartons will have weights between 15.25 and 16.75 ounces (within three standard deviations of the mean).

Detection of Outliers

Sometimes a data set will have one or more observations with unusually large or unusually small values. These extreme values are called **outliers**. Experienced statisticians take steps to identify outliers and then review each one carefully. An outlier may be a data value that has been incorrectly recorded. If so, it can be corrected before further analysis. An outlier may also be from an observation that was incorrectly included in the data set; if so, it can be removed. Finally, an outlier may be an unusual data value that has been recorded correctly and belongs in the data set. In such cases it should remain.

It is a good idea to check for outliers before making decisions based on data analysis. Errors are often made in recording data and entering data into the computer. Outliers should not necessarily be deleted, but their accuracy and appropriateness should be verified.

Standardized values (z-scores) can be used to identify outliers. Recall that the empirical rule allows us to conclude that for data with a bell-shaped distribution, almost all the data values will be within three standard deviations of the mean. Hence, in using z-scores to identify outliers, we recommend treating any data value with a z-score less than -3 or greater than $+3$ as an outlier. Such data values can then be reviewed for accuracy and to determine whether they belong in the data set.

Refer to the z-scores for the class size data in Table 3.4. The z-score of -1.50 shows that the fifth class size is farthest from the mean. However, this standardized value is well within the -3 to $+3$ guideline for outliers. Thus, the z-scores do not indicate that outliers are present in the class size data.

NOTES AND COMMENTS

1. Chebyshev's theorem is applicable for any data set and can be used to state the minimum number of data values that will be within a certain number of standard deviations of the mean. If the data are known to be approximately bell-shaped, more can be said. For instance, the

empirical rule allows us to say that *approximately* 95% of the data values will be within two standard deviations of the mean; Chebyshev's theorem allows us to conclude only that at least 75% of the data values will be in that interval.
2. Before analyzing a data set, statisticians usually make a variety of checks to ensure the validity of data. In a large study it is not uncommon for errors to be made in recording data values or in entering the values into a computer. Identifying outliers is one tool used to check the validity of the data.

Exercises

Methods

25. Consider a sample with data values of 10, 20, 12, 17, and 16. Compute the z-score for each of the five observations.

26. Consider a sample with a mean of 500 and a standard deviation of 100. What are the z-scores for the following data values: 520, 650, 500, 450, and 280?

27. Consider a sample with a mean of 30 and a standard deviation of 5. Use Chebyshev's theorem to determine the percentage of the data within each of the following ranges:
    a. 20 to 40
    b. 15 to 45
    c. 22 to 38
    d. 18 to 42
    e. 12 to 48

28. Suppose the data have a bell-shaped distribution with a mean of 30 and a standard deviation of 5. Use the empirical rule to determine the percentage of data within each of the following ranges:
    a. 20 to 40
    b. 15 to 45
    c. 25 to 35

Applications

29. The results of a national survey showed that on average, adults sleep 6.9 hours per night. Suppose that the standard deviation is 1.2 hours.
    a. Use Chebyshev's theorem to calculate the percentage of individuals who sleep between 4.5 and 9.3 hours.
    b. Use Chebyshev's theorem to calculate the percentage of individuals who sleep between 3.9 and 9.9 hours.
    c. Assume that the number of hours of sleep follows a bell-shaped distribution. Use the empirical rule to calculate the percentage of individuals who sleep between 4.5 and 9.3 hours per day. How does this result compare to the value that you obtained using Chebyshev's theorem in part (a)?

30. The Energy Information Administration reported that the mean retail price per gallon of regular grade gasoline was $2.05 (Energy Information Administration, May 2009). Suppose that the standard deviation was $.10 and that the retail price per gallon has a bell-shaped distribution.
    a. What percentage of regular grade gasoline sold between $1.95 and $2.15 per gallon?
    b. What percentage of regular grade gasoline sold between $1.95 and $2.25 per gallon?
    c. What percentage of regular grade gasoline sold for more than $2.25 per gallon?

31. The national average for the math portion of the College Board's SAT test is 515 (*The World Almanac,* 2009). The College Board periodically rescales the test scores such that the standard deviation is approximately 100. Answer the following questions using a bell-shaped distribution and the empirical rule for the math test scores.

a. What percentage of students have an SAT math score greater than 615?
b. What percentage of students have an SAT math score greater than 715?
c. What percentage of students have an SAT math score between 415 and 515?
d. What percentage of students have an SAT math score between 315 and 615?

32. The high costs in the California real estate market have caused families who cannot afford to buy bigger homes to consider backyard sheds as an alternative form of housing expansion. Many are using the backyard structures for home offices, art studios, and hobby areas as well as for additional storage. The mean price of a customized wooden, shingled backyard structure is $3100 (*Newsweek,* September 29, 2003). Assume that the standard deviation is $1200.
    a. What is the z-score for a backyard structure costing $2300?
    b. What is the z-score for a backyard structure costing $4900?
    c. Interpret the z-scores in parts (a) and (b). Comment on whether either should be considered an outlier.
    d. The *Newsweek* article described a backyard shed-office combination built in Albany, California, for $13,000. Should this structure be considered an outlier? Explain.

33. Florida Power & Light (FP&L) Company has enjoyed a reputation for quickly fixing its electric system after storms. However, during the hurricane seasons of 2004 and 2005, a new reality was that the company's historical approach to emergency electric system repairs was no longer good enough (*The Wall Street Journal,* January 16, 2006). Data showing the days required to restore electric service after seven hurricanes during 2004 and 2005 follow.

Hurricane	Days to Restore Service
Charley	13
Frances	12
Jeanne	8
Dennis	3
Katrina	8
Rita	2
Wilma	18

Based on this sample of seven, compute the following descriptive statistics:
    a. Mean, median, and mode
    b. Range and standard deviation
    c. Should Wilma be considered an outlier in terms of the days required to restore electric service?
    d. The seven hurricanes resulted in 10 million service interruptions to customers. Do the statistics show that FP&L should consider updating its approach to emergency electric system repairs? Discuss.

34. A sample of 10 NCAA college basketball game scores provided the following data (*USA Today,* January 26, 2004).

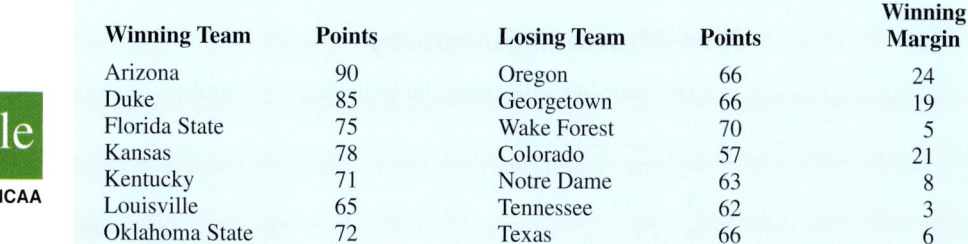

Winning Team	Points	Losing Team	Points	Winning Margin
Arizona	90	Oregon	66	24
Duke	85	Georgetown	66	19
Florida State	75	Wake Forest	70	5
Kansas	78	Colorado	57	21
Kentucky	71	Notre Dame	63	8
Louisville	65	Tennessee	62	3
Oklahoma State	72	Texas	66	6

Winning Team	Points	Losing Team	Points	Winning Margin
Purdue	76	Michigan State	70	6
Stanford	77	Southern Cal	67	10
Wisconsin	76	Illinois	56	20

a. Compute the mean and standard deviation for the points scored by the winning team.
b. Assume that the points scored by the winning teams for all NCAA games follow a bell-shaped distribution. Using the mean and standard deviation found in part (a), estimate the percentage of all NCAA games in which the winning team scores 84 or more points. Estimate the percentage of NCAA games in which the winning team scores more than 90 points.
c. Compute the mean and standard deviation for the winning margin. Do the data contain outliers? Explain.

35. The Associated Press Team Marketing Report listed the Dallas Cowboys as the team with the highest ticket prices in the National Football League (*USA Today*, October 20, 2009). Data showing the average ticket price for a sample of 14 teams in the National Football League are as follows.

Team	Ticket Price	Team	Ticket Price
Atlanta Falcons	$72	Green Bay Packers	$63
Buffalo Bills	51	Indianapolis Colts	83
Carolina Panthers	63	New Orleans Saints	62
Chicago Bears	88	New York Jets	87
Cleveland Browns	55	Pittsburgh Steelers	67
Dallas Cowboys	160	Seattle Seahawks	61
Denver Broncos	77	Tennessee Titans	61

a. What is the mean ticket price?
b. The previous year, the mean ticket price was $72.20. What was the percentage increase in the mean ticket price for the one-year period?
c. Compute the median ticket price.
d. Compute the first and third quartiles.
e. Compute the standard deviation.
f. What is the z-score for the Dallas Cowboys' ticket price? Should this price be considered an outlier? Explain.

3.4 Exploratory Data Analysis

In Chapter 2 we introduced the stem-and-leaf display as a technique of exploratory data analysis. Recall that exploratory data analysis enables us to use simple arithmetic and easy-to-draw pictures to summarize data. In this section we continue exploratory data analysis by considering five-number summaries and box plots.

Five-Number Summary

In a **five-number summary**, the following five numbers are used to summarize the data:

1. Smallest value
2. First quartile (Q_1)
3. Median (Q_2)
4. Third quartile (Q_3)
5. Largest value

3.4 Exploratory Data Analysis

The easiest way to develop a five-number summary is to first place the data in ascending order. Then it is easy to identify the smallest value, the three quartiles, and the largest value. The monthly starting salaries shown in Table 3.1 for a sample of 12 business school graduates are repeated here in ascending order.

$$3310 \quad 3355 \quad 3450 \mid 3480 \quad 3480 \quad 3490 \mid 3520 \quad 3540 \quad 3550 \mid 3650 \quad 3730 \quad 3925$$

$$Q_1 = 3465 \qquad\qquad Q_2 = 3505 \qquad\qquad Q_3 = 3600$$
$$\text{(Median)}$$

The median of 3505 and the quartiles $Q_1 = 3465$ and $Q_3 = 3600$ were computed in Section 3.1. Reviewing the data shows a smallest value of 3310 and a largest value of 3925. Thus the five-number summary for the salary data is 3310, 3465, 3505, 3600, 3925. Approximately one-fourth, or 25%, of the observations are between adjacent numbers in a five-number summary.

Box Plot

A **box plot** is a graphical summary of data that is based on a five-number summary. A key to the development of a box plot is the computation of the median and the quartiles, Q_1 and Q_3. The interquartile range, IQR = $Q_3 - Q_1$, is also used. Figure 3.5 is the box plot for the monthly starting salary data. The steps used to construct the box plot follow.

1. A box is drawn with the ends of the box located at the first and third quartiles. For the salary data, $Q_1 = 3465$ and $Q_3 = 3600$. This box contains the middle 50% of the data.
2. A vertical line is drawn in the box at the location of the median (3505 for the salary data).
3. By using the interquartile range, IQR = $Q_3 - Q_1$, limits are located. The limits for the box plot are 1.5(IQR) below Q_1 and 1.5(IQR) above Q_3. For the salary data, IQR = $Q_3 - Q_1 = 3600 - 3465 = 135$. Thus, the limits are $3465 - 1.5(135) = 3262.5$ and $3600 + 1.5(135) = 3802.5$. Data outside these limits are considered *outliers*.
4. The dashed lines in Figure 3.5 are called *whiskers*. The whiskers are drawn from the ends of the box to the smallest and largest values *inside the limits* computed in step 3. Thus, the whiskers end at salary values of 3310 and 3730.
5. Finally, the location of each outlier is shown with the symbol *. In Figure 3.5 we see one outlier, 3925.

Box plots provide another way to identify outliers. But they do not necessarily identify the same values as those with a z-score less than −3 or greater than +3. Either or both procedures may be used.

In Figure 3.5 we included lines showing the location of the upper and lower limits. These lines were drawn to show how the limits are computed and where they are located.

FIGURE 3.5 BOX PLOT OF THE STARTING SALARY DATA WITH LINES SHOWING THE LOWER AND UPPER LIMITS

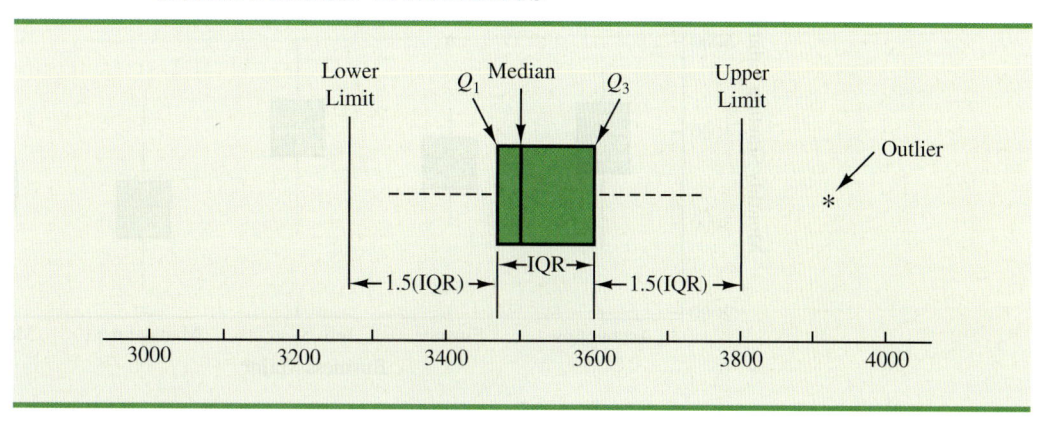

FIGURE 3.6 BOX PLOT OF MONTHLY STARTING SALARY DATA

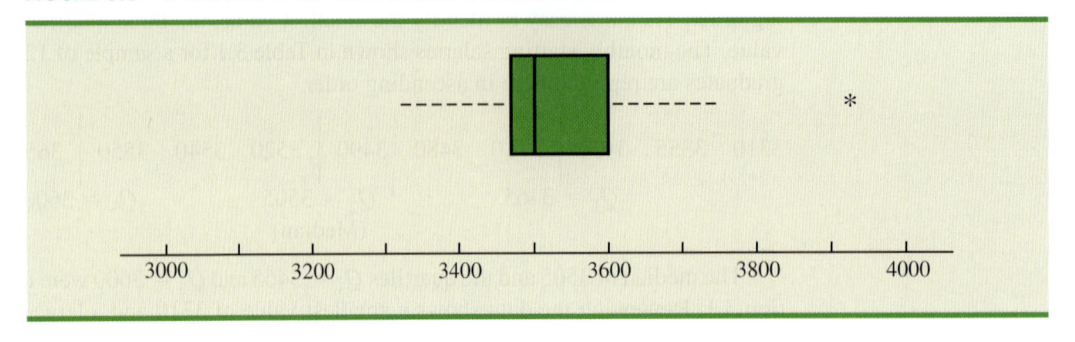

Although the limits are always computed, generally they are not drawn on the box plots. Figure 3.6 shows the usual appearance of a box plot for the salary data.

In order to compare monthly starting salaries for business school graduates by major, a sample of 111 recent graduates was selected. The major and the monthly starting salary were recorded for each graduate. Figure 3.7 shows the Minitab box plots for accounting, finance, information systems, management, and marketing majors. Note that the major is shown on the horizontal axis and each box plot is shown vertically above the corresponding major. Displaying box plots in this manner is an excellent graphical technique for making comparisons among two or more groups.

What observations can you make about monthly starting salaries by major using the box plots in Figured 3.7? Specifically, we note the following:

- The higher salaries are in accounting; the lower salaries are in management and marketing.
- Based on the medians, accounting and information systems have similar and higher median salaries. Finance is next with management and marketing showing lower median salaries.
- High salary outliers exist for accounting, finance, and marketing majors.
- Finance salaries appear to have the least variation, while accounting salaries appear to have the most variation.

Perhaps you can see additional interpretations based on these box plots.

FIGURE 3.7 MINITAB BOX PLOTS OF MONTLY STARTING SALARY BY MAJOR

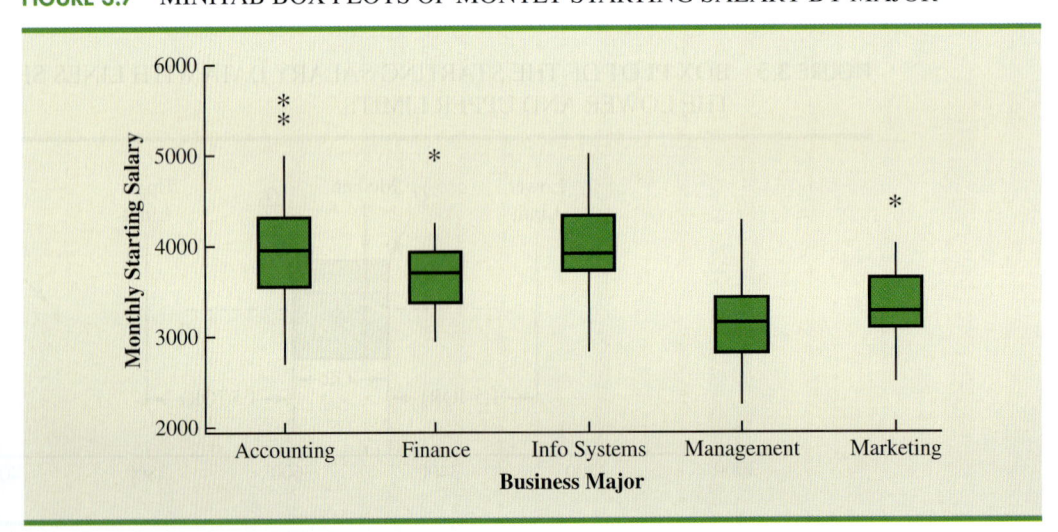

3.4 Exploratory Data Analysis

NOTES AND COMMENTS

1. An advantage of the exploratory data analysis procedures is that they are easy to use; few numerical calculations are necessary. We simply sort the data values into ascending order and identify the five-number summary. The box plot can then be constructed. It is not necessary to compute the mean and the standard deviation for the data.

2. In Appendix 3.1, we show how to construct a box plot for the starting salary data using Minitab. The box plot obtained looks just like the one in Figure 3.6, but turned on its side.

Exercises

Methods

36. Consider a sample with data values of 27, 25, 20, 15, 30, 34, 28, and 25. Provide the five-number summary for the data.

37. Show the box plot for the data in exercise 36.

38. Show the five-number summary and the box plot for the following data: 5, 15, 18, 10, 8, 12, 16, 10, 6.

39. A data set has a first quartile of 42 and a third quartile of 50. Compute the lower and upper limits for the corresponding box plot. Should a data value of 65 be considered an outlier?

Applications

40. Some of the best-known food franchises and the number of retail locations for each are shown in the following table (*The New York Times Almanac*, 2010).

Franchise	Locations	Franchise	Locations
Arby's	2,558	McDonald's	24,799
Baskin-Robbins	5,889	Papa John's	2,615
Dairy Queen	5,619	Pizza Hut	10,238
Domino's	8,053	Quiznos	5,110
Dunkin Donuts	8,082	Subway	29,612
Hardee's	1,397	Taco Bell	4,516
KFC Corp	11,553		

  a. What is the largest franchise? How many retail locations does it have?
  b. What is the median number of locations for the franchises?
  c. Provide a five-number summary.
  d. Are there any outliers?
  e. Show the box plot.

41. Naples, Florida, hosts a half-marathon (13.1-mile race) in January each year. The event attracts top runners from throughout the United States as well as from around the world. In January 2009, 22 men and 31 women entered the 19–24 age class. Finish times in minutes are as follows (*Naples Daily News*, January 19, 2009). Times are shown in order of finish.

Finish	Men	Women	Finish	Men	Women	Finish	Men	Women
1	65.30	109.03	11	109.05	123.88	21	143.83	136.75
2	66.27	111.22	12	110.23	125.78	22	148.70	138.20
3	66.52	111.65	13	112.90	129.52	23		139.00
4	66.85	111.93	14	113.52	129.87	24		147.18
5	70.87	114.38	15	120.95	130.72	25		147.35

Finish	Men	Women	Finish	Men	Women	Finish	Men	Women
6	87.18	118.33	16	127.98	131.67	26		147.50
7	96.45	121.25	17	128.40	132.03	27		147.75
8	98.52	122.08	18	130.90	133.20	28		153.88
9	100.52	122.48	19	131.80	133.50	29		154.83
10	108.18	122.62	20	138.63	136.57	30		189.27
						31		189.28

a. George Towett of Marietta, Georgia, finished in first place for the men and Lauren Wald of Gainesville, Florida, finished in first place for the women. Compare the first-place finish times for men and women. If the 53 men and women runners had competed as one group, in what place would Lauren have finished?
b. What is the median time for men and women runners? Compare men and women runners based on their median times.
c. Provide a five-number summary for both the men and the women.
d. Are there outliers in either group?
e. Show the box plots for the two groups. Did men or women have the most variation in finish times? Explain.

42. *Consumer Reports* provided overall customer satisfaction scores for AT&T, Sprint, T-Mobile, and Verizon cell-phone services in major metropolitan areas throughout the United States. The rating for each service reflects the overall customer satisfaction considering a variety of factors such as cost, connectivity problems, dropped calls, static interference, and customer support. A satisfaction scale from 0 to 100 was used with 0 indicating completely dissatisfied and 100 indicating completely satisfied. The ratings for the four cell-phone services in 20 metropolitan areas are as shown (*Consumer Reports,* January 2009).

Metropolitan Area	AT&T	Sprint	T-Mobile	Verizon
Atlanta	70	66	71	79
Boston	69	64	74	76
Chicago	71	65	70	77
Dallas	75	65	74	78
Denver	71	67	73	77
Detroit	73	65	77	79
Jacksonville	73	64	75	81
Las Vegas	72	68	74	81
Los Angeles	66	65	68	78
Miami	68	69	73	80
Minneapolis	68	66	75	77
Philadelphia	72	66	71	78
Phoenix	68	66	76	81
San Antonio	75	65	75	80
San Diego	69	68	72	79
San Francisco	66	69	73	75
Seattle	68	67	74	77
St. Louis	74	66	74	79
Tampa	73	63	73	79
Washington	72	68	71	76

a. Consider T-Mobile first. What is the median rating?
b. Develop a five-number summary for the T-Mobile service.
c. Are there outliers for T-Mobile? Explain.
d. Repeat parts (b) and (c) for the other three cell-phone services.

e. Show the box plots for the four cell-phone services on one graph. Discuss what a comparison of the box plots tells about the four services. Which service did *Consumer Reports* recommend as being best in terms of overall customer satisfaction?

43. The Philadelphia Phillies defeated the Tampa Bay Rays 4 to 3 to win the 2008 major league baseball World Series (*The Philadelphia Inquirer*, October 29, 2008). Earlier in the major league baseball playoffs, the Philadelphia Phillies defeated the Los Angeles Dodgers to win the National League Championship, while the Tampa Bay Rays defeated the Boston Red Sox to win the American League Championship. The file MLBSalaries contains the salaries for the 28 players on each of these four teams (USA Today Salary Database, October 2008). The data, shown in thousands of dollars, have been ordered from the highest salary to the lowest salary for each team.

   a. Analyze the salaries for the World Champion Philadelphia Phillies. What is the total payroll for the team? What is the median salary? What is the five-number summary?

   b. Were there salary outliers for the Philadelphia Phillies? If so, how many and what were the salary amounts?

   c. What is the total payroll for each of the other three teams? Develop the five-number summary for each team and identify any outliers.

   d. Show the box plots of the salaries for all four teams. What are your interpretations? Of these four teams, does it appear that the team with the higher salaries won the league championships and the World Series?

44. A listing of 46 mutual funds and their 12-month total return percentage is shown in Table 3.5 (*Smart Money*, February 2004).

   a. What are the mean and median return percentages for these mutual funds?
   b. What are the first and third quartiles?
   c. Provide a five-number summary.
   d. Do the data contain any outliers? Show a box plot.

TABLE 3.5 TWELVE-MONTH RETURN FOR MUTUAL FUNDS

Mutual Fund	Return (%)	Mutual Fund	Return (%)
Alger Capital Appreciation	23.5	Nations Small Company	21.4
Alger LargeCap Growth	22.8	Nations SmallCap Index	24.5
Alger MidCap Growth	38.3	Nations Strategic Growth	10.4
Alger SmallCap	41.3	Nations Value Inv	10.8
AllianceBernstein Technology	40.6	One Group Diversified Equity	10.0
Federated American Leaders	15.6	One Group Diversified Int'l	10.9
Federated Capital Appreciation	12.4	One Group Diversified Mid Cap	15.1
Federated Equity-Income	11.5	One Group Equity Income	6.6
Federated Kaufmann	33.3	One Group Int'l Equity Index	13.2
Federated Max-Cap Index	16.0	One Group Large Cap Growth	13.6
Federated Stock	16.9	One Group Large Cap Value	12.8
Janus Adviser Int'l Growth	10.3	One Group Mid Cap Growth	18.7
Janus Adviser Worldwide	3.4	One Group Mid Cap Value	11.4
Janus Enterprise	24.2	One Group Small Cap Growth	23.6
Janus High-Yield	12.1	PBHG Growth	27.3
Janus Mercury	20.6	Putnam Europe Equity	20.4
Janus Overseas	11.9	Putnam Int'l Capital Opportunity	36.6
Janus Worldwide	4.1	Putnam International Equity	21.5
Nations Convertible Securities	13.6	Putnam Int'l New Opportunity	26.3
Nations Int'l Equity	10.7	Strong Advisor Mid Cap Growth	23.7
Nations LargeCap Enhd. Core	13.2	Strong Growth 20	11.7
Nations LargeCap Index	13.5	Strong Growth Inv	23.2
Nation MidCap Index	19.5	Strong Large Cap Growth	14.5

3.5 Measures of Association Between Two Variables

Thus far we have examined numerical methods used to summarize the data for *one variable at a time*. Often a manager or decision maker is interested in the *relationship between two variables*. In this section we present covariance and correlation as descriptive measures of the relationship between two variables.

We begin by reconsidering the application concerning a stereo and sound equipment store in San Francisco as presented in Section 2.4. The store's manager wants to determine the relationship between the number of weekend television commercials shown and the sales at the store during the following week. Sample data with sales expressed in hundreds of dollars are provided in Table 3.6. It shows 10 observations ($n = 10$), one for each week. The scatter diagram in Figure 3.8 shows a positive relationship, with higher sales (y) associated with a greater number of commercials (x). In fact, the scatter diagram suggests that a straight line could be used as an approximation of the relationship. In the following discussion, we introduce **covariance** as a descriptive measure of the linear association between two variables.

Covariance

For a sample of size n with the observations (x_1, y_1), (x_2, y_2), and so on, the sample covariance is defined as follows:

> **SAMPLE COVARIANCE**
>
> $$s_{xy} = \frac{\Sigma(x_i - \bar{x})(y_i - \bar{y})}{n - 1} \qquad (3.10)$$

This formula pairs each x_i with a y_i. We then sum the products obtained by multiplying the deviation of each x_i from its sample mean $\bar{x}$ by the deviation of the corresponding y_i from its sample mean $\bar{y}$; this sum is then divided by $n - 1$.

TABLE 3.6 SAMPLE DATA FOR THE STEREO AND SOUND EQUIPMENT STORE

Week	Number of Commercials x	Sales Volume ($100s) y
1	2	50
2	5	57
3	1	41
4	3	54
5	4	54
6	1	38
7	5	63
8	3	48
9	4	59
10	2	46

3.5 Measures of Association Between Two Variables

FIGURE 3.8 SCATTER DIAGRAM FOR THE STEREO AND SOUND EQUIPMENT STORE

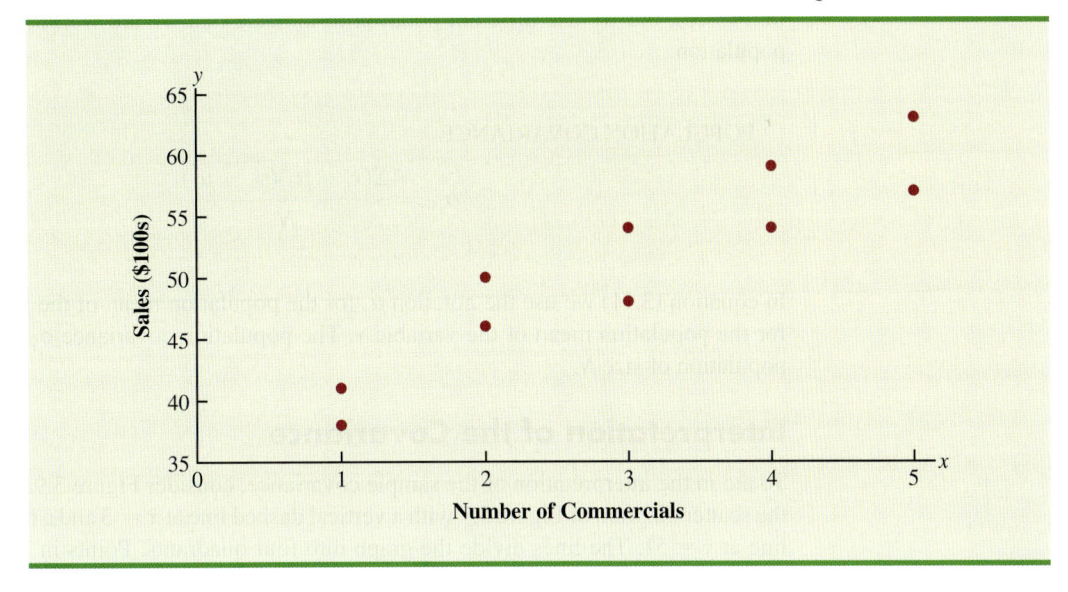

To measure the strength of the linear relationship between the number of commercials x and the sales volume y in the stereo and sound equipment store problem, we use equation (3.10) to compute the sample covariance. The calculations in Table 3.7 show the computation of $\Sigma(x_i - \bar{x})(y_i - \bar{y})$. Note that $\bar{x} = 30/10 = 3$ and $\bar{y} = 510/10 = 51$. Using equation (3.10), we obtain a sample covariance of

$$s_{xy} = \frac{\Sigma(x_i - \bar{x})(y_i - \bar{y})}{n - 1} = \frac{99}{9} = 11$$

TABLE 3.7 CALCULATIONS FOR THE SAMPLE COVARIANCE

x_i	y_i	$x_i - \bar{x}$	$y_i - \bar{y}$	$(x_i - \bar{x})(y_i - \bar{y})$
2	50	−1	−1	1
5	57	2	6	12
1	41	−2	−10	20
3	54	0	3	0
4	54	1	3	3
1	38	−2	−13	26
5	63	2	12	24
3	48	0	−3	0
4	59	1	8	8
2	46	−1	−5	5
Totals 30	510	0	0	99

$$s_{xy} = \frac{\Sigma(x_i - \bar{x})(y_i - \bar{y})}{n - 1} = \frac{99}{10 - 1} = 11$$

The formula for computing the covariance of a population of size N is similar to equation (3.10), but we use different notation to indicate that we are working with the entire population.

> **POPULATION COVARIANCE**
>
> $$\sigma_{xy} = \frac{\Sigma(x_i - \mu_x)(y_i - \mu_y)}{N} \quad (3.11)$$

In equation (3.11) we use the notation μ_x for the population mean of the variable x and μ_y for the population mean of the variable y. The population covariance σ_{xy} is defined for a population of size N.

Interpretation of the Covariance

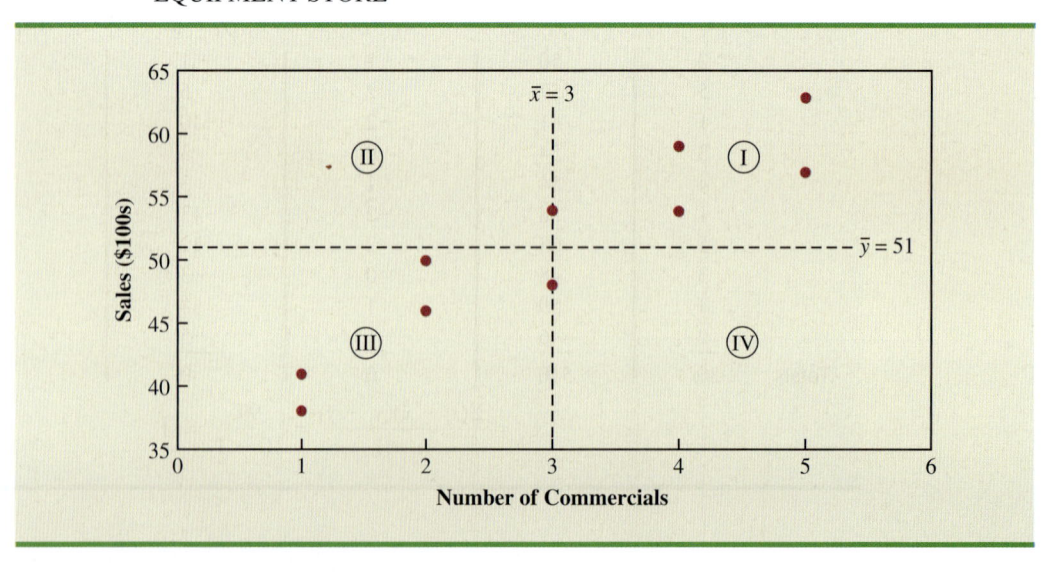

To aid in the interpretation of the sample covariance, consider Figure 3.9. It is the same as the scatter diagram of Figure 3.7 with a vertical dashed line at $\bar{x} = 3$ and a horizontal dashed line at $\bar{y} = 51$. The lines divide the graph into four quadrants. Points in quadrant I correspond to x_i greater than $\bar{x}$ and y_i greater than $\bar{y}$, points in quadrant II correspond to x_i less than $\bar{x}$ and y_i greater than $\bar{y}$, and so on. Thus, the value of $(x_i - \bar{x})(y_i - \bar{y})$ must be positive for points in quadrant I, negative for points in quadrant II, positive for points in quadrant III, and negative for points in quadrant IV.

The covariance is a measure of the linear association between two variables.

If the value of s_{xy} is positive, the points with the greatest influence on s_{xy} must be in quadrants I and III. Hence, a positive value for s_{xy} indicates a positive linear association between x and y; that is, as the value of x increases, the value of y increases. If the value of s_{xy} is negative, however, the points with the greatest influence on s_{xy} are in quadrants II and IV. Hence, a negative value for s_{xy} indicates a negative linear association between x and y; that is, as the value of x increases, the value of y decreases. Finally, if the points are evenly distributed across all four quadrants, the value of s_{xy} will be close to zero, indicating no linear association between x and y. Figure 3.10 shows the values of s_{xy} that can be expected with three different types of scatter diagrams.

FIGURE 3.9 PARTITIONED SCATTER DIAGRAM FOR THE STEREO AND SOUND EQUIPMENT STORE

3.5 Measures of Association Between Two Variables

FIGURE 3.10   INTERPRETATION OF SAMPLE COVARIANCE

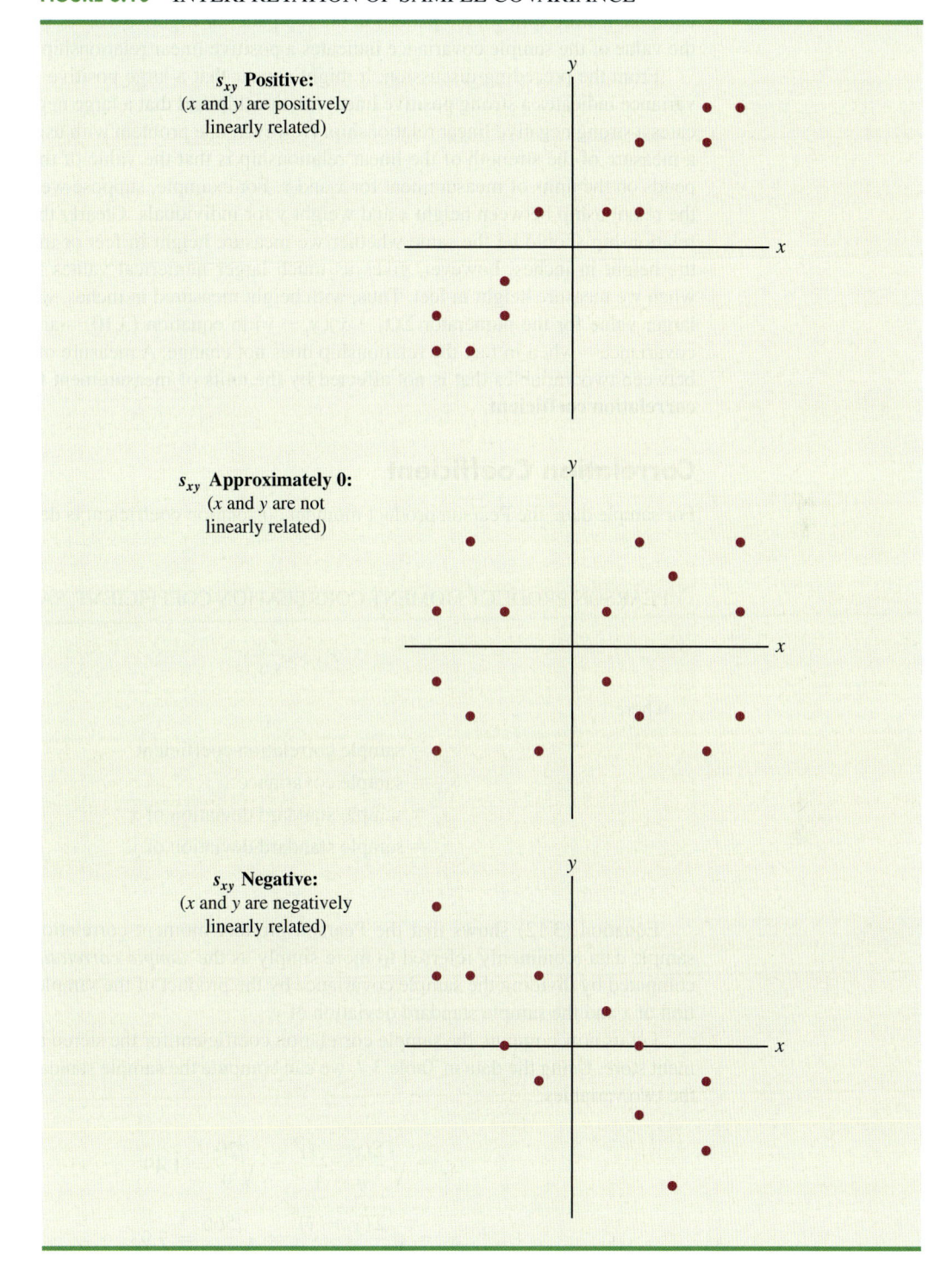

s_{xy} **Positive:**
(x and y are positively linearly related)

s_{xy} **Approximately 0:**
(x and y are not linearly related)

s_{xy} **Negative:**
(x and y are negatively linearly related)

Referring again to Figure 3.9, we see that the scatter diagram for the stereo and sound equipment store follows the pattern in the top panel of Figure 3.10. As we should expect, the value of the sample covariance indicates a positive linear relationship with $s_{xy} = 11$.

From the preceding discussion, it might appear that a large positive value for the covariance indicates a strong positive linear relationship and that a large negative value indicates a strong negative linear relationship. However, one problem with using covariance as a measure of the strength of the linear relationship is that the value of the covariance depends on the units of measurement for x and y. For example, suppose we are interested in the relationship between height x and weight y for individuals. Clearly the strength of the relationship should be the same whether we measure height in feet or inches. Measuring the height in inches, however, gives us much larger numerical values for $(x_i - \bar{x})$ than when we measure height in feet. Thus, with height measured in inches, we would obtain a larger value for the numerator $\Sigma(x_i - \bar{x})(y_i - \bar{y})$ in equation (3.10)—and hence a larger covariance—when in fact the relationship does not change. A measure of the relationship between two variables that is not affected by the units of measurement for x and y is the **correlation coefficient**.

Correlation Coefficient

For sample data, the Pearson product moment correlation coefficient is defined as follows.

PEARSON PRODUCT MOMENT CORRELATION COEFFICIENT: SAMPLE DATA

$$r_{xy} = \frac{s_{xy}}{s_x s_y} \qquad (3.12)$$

where

r_{xy} = sample correlation coefficient
s_{xy} = sample covariance
s_x = sample standard deviation of x
s_y = sample standard deviation of y

Equation (3.12) shows that the Pearson product moment correlation coefficient for sample data (commonly referred to more simply as the *sample correlation coefficient*) is computed by dividing the sample covariance by the product of the sample standard deviation of x and the sample standard deviation of y.

Let us now compute the sample correlation coefficient for the stereo and sound equipment store. Using the data in Table 3.7, we can compute the sample standard deviations for the two variables:

$$s_x = \sqrt{\frac{\Sigma(x_i - \bar{x})^2}{n - 1}} = \sqrt{\frac{20}{9}} = 1.49$$

$$s_y = \sqrt{\frac{\Sigma(y_i - \bar{y})^2}{n - 1}} = \sqrt{\frac{566}{9}} = 7.93$$

Now, because $s_{xy} = 11$, the sample correlation coefficient equals

$$r_{xy} = \frac{s_{xy}}{s_x s_y} = \frac{11}{(1.49)(7.93)} = .93$$

3.5 Measures of Association Between Two Variables

The formula for computing the correlation coefficient for a population, denoted by the Greek letter ρ_{xy} (rho, pronounced "row"), follows.

> **PEARSON PRODUCT MOMENT CORRELATION COEFFICIENT: POPULATION DATA**
>
> $$\rho_{xy} = \frac{\sigma_{xy}}{\sigma_x \sigma_y} \tag{3.13}$$
>
> where
>
> ρ_{xy} = population correlation coefficient
> σ_{xy} = population covariance
> σ_x = population standard deviation for x
> σ_y = population standard deviation for y

The sample correlation coefficient r_{xy} is the estimator of the population correlation coefficient ρ_{xy}.

The sample correlation coefficient r_{xy} provides an estimate of the population correlation coefficient ρ_{xy}.

Interpretation of the Correlation Coefficient

First let us consider a simple example that illustrates the concept of a perfect positive linear relationship. The scatter diagram in Figure 3.11 depicts the relationship between x and y based on the following sample data.

x_i	y_i
5	10
10	30
15	50

FIGURE 3.11 SCATTER DIAGRAM DEPICTING A PERFECT POSITIVE LINEAR RELATIONSHIP

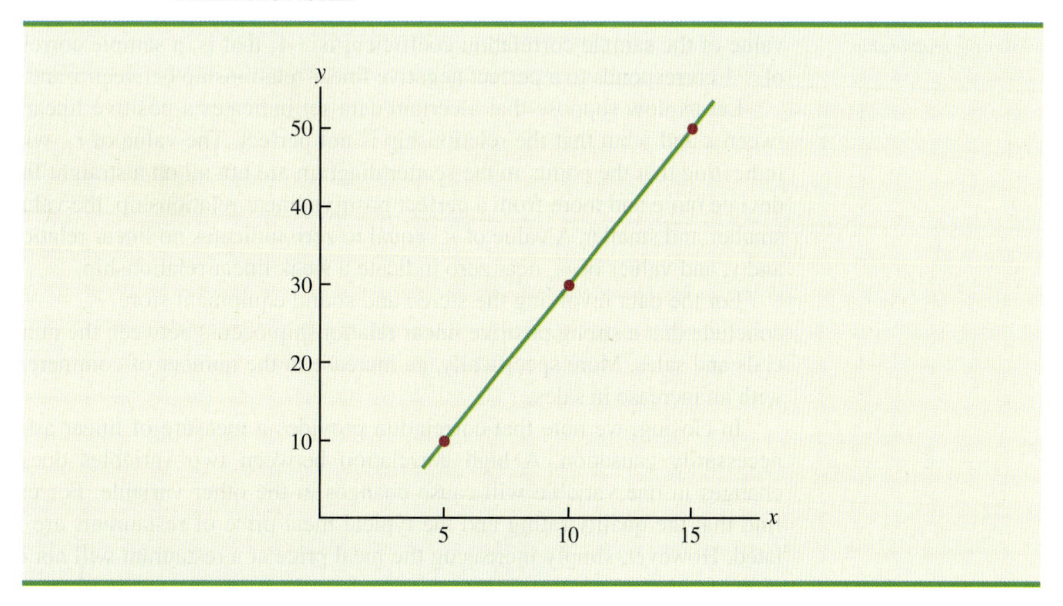

TABLE 3.8 COMPUTATIONS USED IN CALCULATING THE SAMPLE CORRELATION COEFFICIENT

	x_i	y_i	$x_i - \bar{x}$	$(x_i - \bar{x})^2$	$y_i - \bar{y}$	$(y_i - \bar{y})^2$	$(x_i - \bar{x})(y_i - \bar{y})$
	5	10	−5	25	−20	400	100
	10	30	0	0	0	0	0
	15	50	5	25	20	400	100
Totals	30	90	0	50	0	800	200

$\bar{x} = 10 \quad \bar{y} = 30$

The straight line drawn through each of the three points shows a perfect linear relationship between x and y. In order to apply equation (3.12) to compute the sample correlation, we must first compute s_{xy}, s_x, and s_y. Some of the computations are shown in Table 3.8. Using the results in this table, we find

$$s_{xy} = \frac{\Sigma(x_i - \bar{x})(y_i - \bar{y})}{n - 1} = \frac{200}{2} = 100$$

$$s_x = \sqrt{\frac{\Sigma(x_i - \bar{x})^2}{n - 1}} = \sqrt{\frac{50}{2}} = 5$$

$$s_y = \sqrt{\frac{\Sigma(y_i - \bar{y})^2}{n - 1}} = \sqrt{\frac{800}{2}} = 20$$

$$r_{xy} = \frac{s_{xy}}{s_x s_y} = \frac{100}{5(20)} = 1$$

The correlation coefficient ranges from −1 to +1. Values close to −1 or +1 indicate a strong linear relationship. The closer the correlation is to zero, the weaker the relationship.

Thus, we see that the value of the sample correlation coefficient is 1.

In general, it can be shown that if all the points in a data set fall on a positively sloped straight line, the value of the sample correlation coefficient is +1; that is, a sample correlation coefficient of +1 corresponds to a perfect positive linear relationship between x and y. Moreover, if the points in the data set fall on a straight line having negative slope, the value of the sample correlation coefficient is −1; that is, a sample correlation coefficient of −1 corresponds to a perfect negative linear relationship between x and y.

Let us now suppose that a certain data set indicates a positive linear relationship between x and y but that the relationship is not perfect. The value of r_{xy} will be less than 1, indicating that the points in the scatter diagram are not all on a straight line. As the points deviate more and more from a perfect positive linear relationship, the value of r_{xy} becomes smaller and smaller. A value of r_{xy} equal to zero indicates no linear relationship between x and y, and values of r_{xy} near zero indicate a weak linear relationship.

For the data involving the stereo and sound equipment store, $r_{xy} = .93$. Therefore, we conclude that a strong positive linear relationship occurs between the number of commercials and sales. More specifically, an increase in the number of commercials is associated with an increase in sales.

In closing, we note that correlation provides a measure of linear association and not necessarily causation. A high correlation between two variables does not mean that changes in one variable will cause changes in the other variable. For example, we may find that the quality rating and the typical meal price of restaurants are positively correlated. However, simply increasing the meal price at a restaurant will not cause the quality rating to increase.

3.5 Measures of Association Between Two Variables

Exercises

Methods

45. Five observations taken for two variables follow.

x_i	4	6	11	3	16
y_i	50	50	40	60	30

   a. Develop a scatter diagram with x on the horizontal axis.
   b. What does the scatter diagram developed in part (a) indicate about the relationship between the two variables?
   c. Compute and interpret the sample covariance.
   d. Compute and interpret the sample correlation coefficient.

46. Five observations taken for two variables follow.

x_i	6	11	15	21	27
y_i	6	9	6	17	12

   a. Develop a scatter diagram for these data.
   b. What does the scatter diagram indicate about a relationship between x and y?
   c. Compute and interpret the sample covariance.
   d. Compute and interpret the sample correlation coefficient.

Applications

47. Ten major college football bowl games were played in January 2010 with the University of Alabama beating the University of Texas 37 to 21 to become the national champion of college football. The results of the 10 bowl games are shown in the following table (*USA Today*, January 8, 2010). The predicted winning point margin was based on Las Vegas betting odds approximately one week before the bowl games were played. For example, Auburn was predicted to beat Northwestern in the Outback Bowl by 5 points. The actual winning point margin for Auburn was 3 points. A negative predicted winning point margin means that the team that won the bowl game was an underdog and expected to lose. For example, in the Rose Bowl, Ohio State was a 2-point underdog to Oregon and ended up winning by 9 points.

Bowl Game	Score	Predicted Point Margin	Actual Point Margin
Outback	Auburn 38 Northwestern 35	5	3
Gator	Florida State 33 West Virginia 21	1	12
Capital One	Penn State 19 LSU 17	3	2
Rose	Ohio State 26 Oregon 17	−2	9
Sugar	Florida 51 Cincinnati 24	14	27
Cotton	Mississippi State 21 Oklahoma State 7	3	14
Alamo	Texas Tech 41 Michigan State 31	9	10
Fiesta	Boise State 17 TCU 10	−4	7
Orange	Iowa 24 Georgia Tech 14	−3	10
Championship	Alabama 37 Texas 21	4	16

   a. Develop a scatter diagram with the predicted point margin on the horizontal axis.
   b. What is the relationship between predicted and actual point margins?
   c. Compute and interpret the sample covariance.
   d. Compute the sample correlation coefficient. What does this value indicate about the relationship between the Las Vegas predicted point margin and the actual point margin in college football bowl games?

48. A department of transportation's study on driving speed and miles per gallon for midsize automobiles resulted in the following data:

Speed (Miles per Hour)	30	50	40	55	30	25	60	25	50	55
Miles per Gallon	28	25	25	23	30	32	21	35	26	25

Compute and interpret the sample correlation coefficient.

49. At the beginning of 2009, the economic downturn resulted in the loss of jobs and an increase in delinquent loans for housing. The national unemployment rate was 6.5% and the percentage of delinquent loans was 6.12% (*The Wall Street Journal,* January 27, 2009). In projecting where the real estate market was headed in the coming year, economists studied the relationship between the jobless rate and the percentage of delinquent loans. The expectation was that if the jobless rate continued to increase, there would also be an increase in the percentage of delinquent loans. The following data show the jobless rate and the delinquent loan percentage for 27 major real estate markets.

Metro Area	Jobless Rate (%)	Delinquent Loan (%)	Metro Area	Jobless Rate (%)	Delinquent Loan (%)
Atlanta	7.1	7.02	New York	6.2	5.78
Boston	5.2	5.31	Orange County	6.3	6.08
Charlotte	7.8	5.38	Orlando	7.0	10.05
Chicago	7.8	5.40	Philadelphia	6.2	4.75
Dallas	5.8	5.00	Phoenix	5.5	7.22
Denver	5.8	4.07	Portland	6.5	3.79
Detroit	9.3	6.53	Raleigh	6.0	3.62
Houston	5.7	5.57	Sacramento	8.3	9.24
Jacksonville	7.3	6.99	St. Louis	7.5	4.40
Las Vegas	7.6	11.12	San Diego	7.1	6.91
Los Angeles	8.2	7.56	San Francisco	6.8	5.57
Miami	7.1	12.11	Seattle	5.5	3.87
Minneapolis	6.3	4.39	Tampa	7.5	8.42
Nashville	6.6	4.78			

 a. Compute the correlation coefficient. Is there a positive correlation between the jobless rate and the percentage of delinquent housing loans? What is your interpretation?
 b. Show a scatter diagram of the relationship between jobless rate and the percentage of delinquent housing loans.

50. The Dow Jones Industrial Average (DJIA) and the Standard & Poor's 500 Index (S&P 500) are both used to measure the performance of the stock market. The DJIA is based on the price of stocks for 30 large companies; the S&P 500 is based on the price of stocks for 500 companies. If both the DJIA and S&P 500 measure the performance of the stock market, how are they correlated? The following data show the daily percent increase or daily percent decrease in the DJIA and S&P 500 for a sample of nine days over a three-month period (*The Wall Street Journal,* January 15 to March 10, 2006).

DJIA	.20	.82	−.99	.04	−.24	1.01	.30	.55	−.25
S&P 500	.24	.19	−.91	.08	−.33	.87	.36	.83	−.16

 a. Show a scatter diagram.
 b. Compute the sample correlation coefficient for these data.
 c. Discuss the association between the DJIA and S&P 500. Do you need to check both before having a general idea about the daily stock market performance?

51. The daily high and low temperatures for 14 cities around the world are shown (The Weather Channel, April 22, 2009).

City	High	Low	City	High	Low
Athens	68	50	London	67	45
Beijing	70	49	Moscow	44	29
Berlin	65	44	Paris	69	44
Cairo	96	64	Rio de Janeiro	76	69
Dublin	57	46	Rome	69	51
Geneva	70	45	Tokyo	70	58
Hong Kong	80	73	Toronto	44	39

a. What is the sample mean high temperature?
b. What is the sample mean low temperature?
c. What is the correlation between the high and low temperatures? Discuss.

3.6 The Weighted Mean and Working with Grouped Data

In Section 3.1, we presented the mean as one of the most important measures of central location. The formula for the mean of a sample with n observations is restated as follows.

$$\bar{x} = \frac{\Sigma x_i}{n} = \frac{x_1 + x_2 + \cdots + x_n}{n} \quad (3.14)$$

In this formula, each x_i is given equal importance or weight. Although this practice is most common, in some instances, the mean is computed by giving each observation a weight that reflects its importance. A mean computed in this manner is referred to as a **weighted mean**.

Weighted Mean

The weighted mean is computed as follows:

WEIGHTED MEAN

$$\bar{x} = \frac{\Sigma w_i x_i}{\Sigma w_i} \quad (3.15)$$

where

x_i = value of observation i
w_i = weight for observation i

When the data are from a sample, equation (3.15) provides the weighted sample mean. When the data are from a population, μ replaces $\bar{x}$ and equation (3.15) provides the weighted population mean.

As an example of the need for a weighted mean, consider the following sample of five purchases of a raw material over the past three months.

Purchase	Cost per Pound ($)	Number of Pounds
1	3.00	1200
2	3.40	500
3	2.80	2750
4	2.90	1000
5	3.25	800

Note that the cost per pound varies from $2.80 to $3.40, and the quantity purchased varies from 500 to 2750 pounds. Suppose that a manager asked for information about the mean cost per pound of the raw material. Because the quantities ordered vary, we must use the formula for a weighted mean. The five cost-per-pound data values are $x_1 = 3.00$, $x_2 = 3.40$, $x_3 = 2.80$, $x_4 = 2.90$, and $x_5 = 3.25$. The weighted mean cost per pound is found by weighting each cost by its corresponding quantity. For this example, the weights are $w_1 = 1200$, $w_2 = 500$, $w_3 = 2750$, $w_4 = 1000$, and $w_5 = 800$. Based on equation (3.15), the weighted mean is calculated as follows:

$$\bar{x} = \frac{1200(3.00) + 500(3.40) + 2750(2.80) + 1000(2.90) + 800(3.25)}{1200 + 500 + 2750 + 1000 + 800}$$

$$= \frac{18{,}500}{6250} = 2.96$$

Thus, the weighted mean computation shows that the mean cost per pound for the raw material is $2.96. Note that using equation (3.14) rather than the weighted mean formula would have provided misleading results. In this case, the mean of the five cost-per-pound values is $(3.00 + 3.40 + 2.80 + 2.90 + 3.25)/5 = 15.35/5 = \$3.07$, which overstates the actual mean cost per pound purchased.

Computing a grade point average is a good example of the use of a weighted mean.

The choice of weights for a particular weighted mean computation depends upon the application. An example that is well known to college students is the computation of a grade point average (GPA). In this computation, the data values generally used are 4 for an A grade, 3 for a B grade, 2 for a C grade, 1 for a D grade, and 0 for an F grade. The weights are the number of credits hours earned for each grade. Exercise 54 at the end of this section provides an example of this weighted mean computation. In other weighted mean computations, quantities such as pounds, dollars, or volume are frequently used as weights. In any case, when observations vary in importance, the analyst must choose the weight that best reflects the importance of each observation in the determination of the mean.

Grouped Data

In most cases, measures of location and variability are computed by using the individual data values. Sometimes, however, data are available only in a grouped or frequency distribution form. In the following discussion, we show how the weighted mean formula can be used to obtain approximations of the mean, variance, and standard deviation for **grouped data**.

In Section 2.2 we provided a frequency distribution of the time in days required to complete year-end audits for the public accounting firm of Sanderson and Clifford. The frequency distribution of audit times is shown in Table 3.9. Based on this frequency distribution, what is the sample mean audit time?

To compute the mean using only the grouped data, we treat the midpoint of each class as being representative of the items in the class. Let M_i denote the midpoint for class i and let f_i denote the frequency of class i. The weighted mean formula (3.15) is then used with the data values denoted as M_i and the weights given by the frequencies f_i. In this case, the denominator of equation (3.15) is the sum of the frequencies, which is the

3.6 The Weighted Mean and Working with Grouped Data

TABLE 3.9 FREQUENCY DISTRIBUTION OF AUDIT TIMES

Audit Time (days)	Frequency
10–14	4
15–19	8
20–24	5
25–29	2
30–34	1
Total	20

sample size n. That is, $\Sigma f_i = n$. Thus, the equation for the sample mean for grouped data is as follows.

SAMPLE MEAN FOR GROUPED DATA

$$\bar{x} = \frac{\Sigma f_i M_i}{n} \quad (3.16)$$

where

M_i = the midpoint for class i
f_i = the frequency for class i
n = the sample size

With the class midpoints, M_i, halfway between the class limits, the first class of 10–14 in Table 3.9 has a midpoint at $(10 + 14)/2 = 12$. The five class midpoints and the weighted mean computation for the audit time data are summarized in Table 3.10. As can be seen, the sample mean audit time is 19 days.

To compute the variance for grouped data, we use a slightly altered version of the formula for the variance provided in equation (3.5). In equation (3.5), the squared deviations of the data about the sample mean $\bar{x}$ were written $(x_i - \bar{x})^2$. However, with grouped data, the values are not known. In this case, we treat the class midpoint, M_i, as being representative of the x_i values in the corresponding class. Thus, the squared deviations about the sample mean, $(x_i - \bar{x})^2$, are replaced by $(M_i - \bar{x})^2$. Then, just as we did with the sample mean calculations for grouped data, we weight each value by the frequency of the class, f_i. The sum of the squared deviations about the mean for all the data is approximated by $\Sigma f_i(M_i - \bar{x})^2$. The term $n - 1$ rather than n appears in the denominator in order to make the sample variance the estimate of the population variance. Thus, the following formula is used to obtain the sample variance for grouped data.

SAMPLE VARIANCE FOR GROUPED DATA

$$s^2 = \frac{\Sigma f_i(M_i - \bar{x})^2}{n - 1} \quad (3.17)$$

TABLE 3.10 COMPUTATION OF THE SAMPLE MEAN AUDIT TIME FOR GROUPED DATA

Audit Time (days)	Class Midpoint (M_i)	Frequency (f_i)	$f_i M_i$
10–14	12	4	48
15–19	17	8	136
20–24	22	5	110
25–29	27	2	54
30–34	32	1	32
		20	380

$$\text{Sample mean } \bar{x} = \frac{\Sigma f_i M_i}{n} = \frac{380}{20} = 19 \text{ days}$$

The calculation of the sample variance for audit times based on the grouped data is shown in Table 3.11. The sample variance is 30.

The standard deviation for grouped data is simply the square root of the variance for grouped data. For the audit time data, the sample standard deviation is $s = \sqrt{30} = 5.48$.

Before closing this section on computing measures of location and dispersion for grouped data, we note that formulas (3.16) and (3.17) are for a sample. Population summary measures are computed similarly. The grouped data formulas for a population mean and variance follow.

POPULATION MEAN FOR GROUPED DATA

$$\mu = \frac{\Sigma f_i M_i}{N} \quad (3.18)$$

POPULATION VARIANCE FOR GROUPED DATA

$$\sigma^2 = \frac{\Sigma f_i (M_i - \mu)^2}{N} \quad (3.19)$$

TABLE 3.11 COMPUTATION OF THE SAMPLE VARIANCE OF AUDIT TIMES FOR GROUPED DATA (SAMPLE MEAN $\bar{x} = 19$)

Audit Time (days)	Class Midpoint (M_i)	Frequency (f_i)	Deviation ($M_i - \bar{x}$)	Squared Deviation ($M_i - \bar{x}$)^2	$f_i(M_i - \bar{x})^2$
10–14	12	4	−7	49	196
15–19	17	8	−2	4	32
20–24	22	5	3	9	45
25–29	27	2	8	64	128
30–34	32	1	13	169	169
		20			570

$\Sigma f_i(M_i - \bar{x})^2$

$$\text{Sample variance } s^2 = \frac{\Sigma f_i(M_i - \bar{x})^2}{n-1} = \frac{570}{19} = 30$$

3.6 The Weighted Mean and Working with Grouped Data

NOTES AND COMMENTS

In computing descriptive statistics for grouped data, the class midpoints are used to approximate the data values in each class. As a result, the descriptive statistics for grouped data approximate the descriptive statistics that would result from using the original data directly. We therefore recommend computing descriptive statistics from the original data rather than from grouped data whenever possible.

Exercises

Methods

52. Consider the following data and corresponding weights.

x_i	Weight (w_i)
3.2	6
2.0	3
2.5	2
5.0	8

    a. Compute the weighted mean.
    b. Compute the sample mean of the four data values without weighting. Note the difference in the results provided by the two computations.

53. Consider the sample data in the following frequency distribution.

Class	Midpoint	Frequency
3–7	5	4
8–12	10	7
13–17	15	9
18–22	20	5

    a. Compute the sample mean.
    b. Compute the sample variance and sample standard deviation.

Applications

54. The grade point average for college students is based on a weighted mean computation. For most colleges, the grades are given the following data values: A (4), B (3), C (2), D (1), and F (0). After 60 credit hours of course work, a student at State University earned 9 credit hours of A, 15 credit hours of B, 33 credit hours of C, and 3 credit hours of D.
    a. Compute the student's grade point average.
    b. Students at State University must maintain a 2.5 grade point average for their first 60 credit hours of course work in order to be admitted to the business college. Will this student be admitted?

55. Morningstar tracks the total return for a large number of mutual funds. The following table shows the total return and the number of funds for four categories of mutual funds (*Morningstar Funds 500*, 2008).

Type of Fund	Number of Funds	Total Return (%)
Domestic Equity	9191	4.65
International Equity	2621	18.15
Specialty Stock	1419	11.36
Hybrid	2900	6.75

a. Using the number of funds as weights, compute the weighted average total return for the mutual funds covered by Morningstar.
b. Is there any difficulty associated with using the "number of funds" as the weights in computing the weighted average total return for Morningstar in part (a)? Discuss. What else might be used for weights?
c. Suppose you had invested $10,000 in mutual funds at the beginning of 2007 and diversified the investment by placing $2000 in Domestic Equity funds, $4000 in International Equity funds, $3000 in Specialty Stock funds, and $1000 in Hybrid funds. What is the expected return on the portfolio?

56. Based on a survey of 425 master's programs in business administration, *U.S. News & World Report* ranked the Indiana University Kelley Business School as the 20th best business program in the country (*America's Best Graduate Schools,* 2009). The ranking was based in part on surveys of business school deans and corporate recruiters. Each survey respondent was asked to rate the overall academic quality of the master's program on a scale from 1 "marginal" to 5 "outstanding." Use the following sample of responses to compute the weighted mean score for the business school deans and the corporate recruiters. Discuss.

Quality Assessment	Business School Deans	Corporate Recruiters
5	44	31
4	66	34
3	60	43
2	10	12
1	0	0

57. The following frequency distribution shows the price per share of the 30 companies in the Dow Jones Industrial Average (*Barron's,* February 2, 2009).

Price per Share	Number of Companies
$0–9	4
$10–19	5
$20–29	7
$30–39	3
$40–49	4
$50–59	4
$60–69	0
$70–79	2
$80–89	0
$90–99	1

a. Compute the mean price per share and the standard deviation of the price per share for the Dow Jones Industrial Average companies.
b. On January 16, 2006, the mean price per share was $45.83 and the standard deviation was $18.14. Comment on the changes in the price per share over the three-year period.

Summary

In this chapter we introduced several descriptive statistics that can be used to summarize the location, variability, and shape of a data distribution. Unlike the tabular and graphical procedures introduced in Chapter 2, the measures introduced in this chapter summarize the data in terms of numerical values. When the numerical values obtained are for a sample, they are called sample statistics. When the numerical values obtained are for a population, they are called population parameters. Some of the notation used for sample statistics and population parameters follow.

In statistical inference, the sample statistic is referred to as the point estimator of the population parameter.

	Sample Statistic	Population Parameter
Mean	$\bar{x}$	μ
Variance	s^2	σ^2
Standard deviation	s	σ
Covariance	s_{xy}	σ_{xy}
Correlation	r_{xy}	ρ_{xy}

As measures of central location, we defined the mean, median, and mode. Then the concept of percentiles was used to describe other locations in the data set. Next, we presented the range, interquartile range, variance, standard deviation, and coefficient of variation as measures of variability or dispersion. Our primary measure of the shape of a data distribution was the skewness. Negative values indicate a data distribution skewed to the left. Positive values indicate a data distribution skewed to the right. We then described how the mean and standard deviation could be used, applying Chebyshev's theorem and the empirical rule, to provide more information about the distribution of data and to identify outliers.

In Section 3.4 we showed how to develop a five-number summary and a box plot to provide simultaneous information about the location, variability, and shape of the distribution. In Section 3.5 we introduced covariance and the correlation coefficient as measures of association between two variables. In the final section, we showed how to compute a weighted mean and how to calculate a mean, variance, and standard deviation for grouped data.

The descriptive statistics we discussed can be developed using statistical software packages and spreadsheets. In the chapter-ending appendixes we show how to use Minitab, Excel, and StatTools to develop the descriptive statistics introduced in this chapter.

Glossary

Sample statistic A numerical value used as a summary measure for a sample (e.g., the sample mean, $\bar{x}$, the sample variance, s^2, and the sample standard deviation, s).

Population parameter A numerical value used as a summary measure for a population (e.g., the population mean, μ, the population variance, σ^2, and the population standard deviation, σ).

Point estimator The sample statistic, such as $\bar{x}$, s^2, and s, when used to estimate the corresponding population parameter.

Mean A measure of central location computed by summing the data values and dividing by the number of observations.

Median A measure of central location provided by the value in the middle when the data are arranged in ascending order.

Mode A measure of location, defined as the value that occurs with greatest frequency.

Percentile A value such that at least p percent of the observations are less than or equal to this value and at least $(100 - p)$ percent of the observations are greater than or equal to this value. The 50th percentile is the median.

Quartiles The 25th, 50th, and 75th percentiles, referred to as the first quartile, the second quartile (median), and third quartile, respectively. The quartiles can be used to divide a data set into four parts, with each part containing approximately 25% of the data.

Range A measure of variability, defined to be the largest value minus the smallest value.

Interquartile range (IQR) A measure of variability, defined to be the difference between the third and first quartiles.

Variance A measure of variability based on the squared deviations of the data values about the mean.

Standard deviation A measure of variability computed by taking the positive square root of the variance.

Coefficient of variation A measure of relative variability computed by dividing the standard deviation by the mean and multiplying by 100.

Skewness A measure of the shape of a data distribution. Data skewed to the left result in negative skewness; a symmetric data distribution results in zero skewness; and data skewed to the right result in positive skewness.

z-score A value computed by dividing the deviation about the mean $(x_i - \bar{x})$ by the standard deviation s. A z-score is referred to as a standardized value and denotes the number of standard deviations x_i is from the mean.

Chebyshev's theorem A theorem that can be used to make statements about the proportion of data values that must be within a specified number of standard deviations of the mean.

Empirical rule A rule that can be used to compute the percentage of data values that must be within one, two, and three standard deviations of the mean for data that exhibit a bell-shaped distribution.

Outlier An unusually small or unusually large data value.

Five-number summary An exploratory data analysis technique that uses five numbers to summarize the data: smallest value, first quartile, median, third quartile, and largest value.

Box plot A graphical summary of data based on a five-number summary.

Covariance A measure of linear association between two variables. Positive values indicate a positive relationship; negative values indicate a negative relationship.

Correlation coefficient A measure of linear association between two variables that takes on values between -1 and $+1$. Values near $+1$ indicate a strong positive linear relationship; values near -1 indicate a strong negative linear relationship; and values near zero indicate the lack of a linear relationship.

Weighted mean The mean obtained by assigning each observation a weight that reflects its importance.

Grouped data Data available in class intervals as summarized by a frequency distribution. Individual values of the original data are not available.

Key Formulas

Sample Mean

$$\bar{x} = \frac{\Sigma x_i}{n} \quad (3.1)$$

Population Mean

$$\mu = \frac{\Sigma x_i}{N} \quad (3.2)$$

Interquartile Range

$$\text{IQR} = Q_3 - Q_1 \quad (3.3)$$

Population Variance

$$\sigma^2 = \frac{\Sigma(x_i - \mu)^2}{N} \quad (3.4)$$

Sample Variance

$$s^2 = \frac{\Sigma(x_i - \bar{x})^2}{n - 1} \quad (3.5)$$

Standard Deviation

$$\text{Sample standard deviation} = s = \sqrt{s^2} \quad (3.6)$$
$$\text{Population standard deviation} = \sigma = \sqrt{\sigma^2} \quad (3.7)$$

Coefficient of Variation

$$\left(\frac{\text{Standard deviation}}{\text{Mean}} \times 100\right)\% \quad (3.8)$$

z-Score

$$z_i = \frac{x_i - \bar{x}}{s} \quad (3.9)$$

Sample Covariance

$$s_{xy} = \frac{\Sigma(x_i - \bar{x})(y_i - \bar{y})}{n - 1} \quad (3.10)$$

Population Covariance

$$\sigma_{xy} = \frac{\Sigma(x_i - \mu_x)(y_i - \mu_y)}{N} \quad (3.11)$$

Pearson Product Moment Correlation Coefficient: Sample Data

$$r_{xy} = \frac{s_{xy}}{s_x s_y} \quad (3.12)$$

Pearson Product Moment Correlation Coefficient: Population Data

$$\rho_{xy} = \frac{\sigma_{xy}}{\sigma_x \sigma_y} \tag{3.13}$$

Weighted Mean

$$\bar{x} = \frac{\Sigma w_i x_i}{\Sigma w_i} \tag{3.15}$$

Sample Mean for Grouped Data

$$\bar{x} = \frac{\Sigma f_i M_i}{n} \tag{3.16}$$

Sample Variance for Grouped Data

$$s^2 = \frac{\Sigma f_i (M_i - \bar{x})^2}{n-1} \tag{3.17}$$

Population Mean for Grouped Data

$$\mu = \frac{\Sigma f_i M_i}{N} \tag{3.18}$$

Population Variance for Grouped Data

$$\sigma^2 = \frac{\Sigma f_i (M_i - \mu)^2}{N} \tag{3.19}$$

Supplementary Exercises

58. According to an annual consumer spending survey, the average monthly Bank of America Visa credit card charge was $1838 (*U.S. Airways Attaché Magazine,* December 2003). A sample of monthly credit card charges provides the following data.

    | 236 | 1710 | 1351 | 825 | 7450 |
    | 316 | 4135 | 1333 | 1584 | 387 |
    | 991 | 3396 | 170 | 1428 | 1688 |

    a. Compute the mean and median.
    b. Compute the first and third quartiles.
    c. Compute the range and interquartile range.
    d. Compute the variance and standard deviation.
    e. The skewness measure for these data is 2.12. Comment on the shape of this distribution. Is it the shape you would expect? Why or why not?
    f. Do the data contain outliers?

59. The U.S. Census Bureau provides statistics on family life in the United States, including the age at the time of first marriage, current marital status, and size of household (U.S. Census Bureau website, March 20, 2006). The following data show the age at the time of first marriage for a sample of men and a sample of women.

Supplementary Exercises

Men	26	23	28	25	27	30	26	35	28
	21	24	27	29	30	27	32	27	25
Women	20	28	23	30	24	29	26	25	
	22	22	25	23	27	26	19		

a. Determine the median age at the time of first marriage for men and women.
b. Compute the first and third quartiles for both men and women.
c. Twenty-five years ago the median age at the time of first marriage was 25 for men and 22 for women. What insight does this information provide about the decision of when to marry among young people today?

60. The U.S. Department of Education reports that about 50% of all college students use a student loan to help cover college expenses (National Center for Educational Studies, January 2006). A sample of students who graduated with student loan debt is shown here. The data, in thousands of dollars, show typical amounts of debt upon graduation.

    10.1   14.8   5.0   10.2   12.4   12.2   2.0   11.5   17.8   4.0

a. For those students who use a student loan, what is the mean loan debt upon graduation?
b. What is the variance? Standard deviation?

61. Dividend yield is the annual dividend per share a company pays divided by the current market price per share expressed as a percentage. A sample of 10 large companies provided the following dividend yield data (*The Wall Street Journal*, January 16, 2004).

Company	Yield %	Company	Yield %
Altria Group	5.0	General Motors	3.7
American Express	0.8	JPMorgan Chase	3.5
Caterpillar	1.8	McDonald's	1.6
Eastman Kodak	1.9	United Technology	1.5
ExxonMobil	2.5	Wal-Mart Stores	0.7

a. What are the mean and median dividend yields?
b. What are the variance and standard deviation?
c. Which company provides the highest dividend yield?
d. What is the z-score for McDonald's? Interpret this z-score.
e. What is the z-score for General Motors? Interpret this z-score.
f. Based on z-scores, do the data contain any outliers?

62. Small business owners often look to payroll service companies to handle their employee payroll. Reasons are that small business owners face complicated tax regulations, and penalties for employment tax errors are costly. According to the Internal Revenue Service, 26% of all small business employment tax returns contained errors that resulted in a tax penalty to the owner (*The Wall Street Journal*, January 30, 2006). The tax penalty for a sample of 20 small business owners follows:

    820   270   450   1010   890   700   1350   350   300   1200
    390   730   2040   230   640   350   420   270   370   620

a. What is the mean tax penalty for improperly filed employment tax returns?
b. What is the standard deviation?
c. Is the highest penalty, $2040, an outlier?
d. What are some of the advantages of a small business owner hiring a payroll service company to handle employee payroll services, including the employment tax returns?

63. Public transportation and the automobile are two methods an employee can use to get to work each day. Samples of times recorded for each method are shown. Times are in minutes.

| Public Transportation: | 28 | 29 | 32 | 37 | 33 | 25 | 29 | 32 | 41 | 34 |
| Automobile: | | 29 | 31 | 33 | 32 | 34 | 30 | 31 | 32 | 35 | 33 |

a. Compute the sample mean time to get to work for each method.
b. Compute the sample standard deviation for each method.
c. On the basis of your results from parts (a) and (b), which method of transportation should be preferred? Explain.
d. Develop a box plot for each method. Does a comparison of the box plots support your conclusion in part (c)?

64. The National Association of Realtors reported the median home price in the United States and the increase in median home price over a five-year period (*The Wall Street Journal*, January 16, 2006). Use the sample home prices shown here to answer the following questions.

| 995.9 | 48.8 | 175.0 | 263.5 | 298.0 | 218.9 | 209.0 |
| 628.3 | 111.0 | 212.9 | 92.6 | 2325.0 | 958.0 | 212.5 |

a. What is the sample median home price?
b. In January 2001, the National Association of Realtors reported a median home price of $139,300 in the United States. What was the percentage increase in the median home price over the five-year period?
c. What are the first quartile and the third quartile for the sample data?
d. Provide a five-number summary for the home prices.
e. Do the data contain any outliers?
f. What is the mean home price for the sample? Why does the National Association of Realtors prefer to use the median home price in its reports?

65. The U.S. Census Bureau's American Community Survey reported the percentage of children under 18 years of age who had lived below the poverty level during the previous 12 months (U.S. Census Bureau website, August 2008). The region of the country, Northeast (NE), Southeast (SE), Midwest (MW), Southwest (SW), and West (W) and the percentage of children under 18 who had lived below the poverty level are shown for each state.

State	Region	Poverty %	State	Region	Poverty %
Alabama	SE	23.0	Montana	W	17.3
Alaska	W	15.1	Nebraska	MW	14.4
Arizona	SW	19.5	Nevada	W	13.9
Arkansas	SE	24.3	New Hampshire	NE	9.6
California	W	18.1	New Jersey	NE	11.8
Colorado	W	15.7	New Mexico	SW	25.6
Connecticut	NE	11.0	New York	NE	20.0
Delaware	NE	15.8	North Carolina	SE	20.2
Florida	SE	17.5	North Dakota	MW	13.0
Georgia	SE	20.2	Ohio	MW	18.7
Hawaii	W	11.4	Oklahoma	SW	24.3
Idaho	W	15.1	Oregon	W	16.8
Illinois	MW	17.1	Pennsylvania	NE	16.9
Indiana	MW	17.9	Rhode Island	NE	15.1
Iowa	MW	13.7	South Carolina	SE	22.1
Kansas	MW	15.6	South Dakota	MW	16.8
Kentucky	SE	22.8	Tennessee	SE	22.7
Louisiana	SE	27.8	Texas	SW	23.9
Maine	NE	17.6	Utah	W	11.9
Maryland	NE	9.7	Vermont	NE	13.2
Massachusetts	NE	12.4	Virginia	SE	12.2
Michigan	MW	18.3	Washington	W	15.4
Minnesota	MW	12.2	West Virginia	SE	25.2
Mississippi	SE	29.5	Wisconsin	MW	14.9
Missouri	MW	18.6	Wyoming	W	12.0

a. What is the median poverty level percentage for the 50 states?
b. What are the first and third quartiles? What is your interpretation of the quartiles?
c. Show a box plot for the data. Interpret the box plot in terms of what it tells you about the level of poverty for children in the United States. Are any states considered outliers? Discuss.
d. Identify the states in the lower quartile. What is your interpretation of this group and what region or regions are represented most in the lower quartile?

66. *Travel + Leisure* magazine presented its annual list of the 500 best hotels in the world (*Travel + Leisure,* January 2009). The magazine provides a rating for each hotel along with a brief description that includes the size of the hotel, amenities, and the cost per night for a double room. A sample of 12 of the top-rated hotels in the United States follows.

Hotel	Location	Rooms	Cost/Night
Boulders Resort & Spa	Phoenix, AZ	220	499
Disney's Wilderness Lodge	Orlando, FL	727	340
Four Seasons Hotel Beverly Hills	Los Angeles, CA	285	585
Four Seasons Hotel	Boston, MA	273	495
Hay-Adams	Washington, DC	145	495
Inn on Biltmore Estate	Asheville, NC	213	279
Loews Ventana Canyon Resort	Phoenix, AZ	398	279
Mauna Lani Bay Hotel	Island of Hawaii	343	455
Montage Laguna Beach	Laguna Beach, CA	250	595
Sofitel Water Tower	Chicago, IL	414	367
St. Regis Monarch Beach	Dana Point, CA	400	675
The Broadmoor	Colorado Springs, CO	700	420

a. What is the mean number of rooms?
b. What is the mean cost per night for a double room?
c. Develop a scatter diagram with the number of rooms on the horizontal axis and the cost per night on the vertical axis. Does there appear to be a relationship between the number of rooms and the cost per night? Discuss.
d. What is the sample correlation coefficient? What does it tell you about the relationship between the number of rooms and the cost per night for a double room? Does this appear reasonable? Discuss.

67. Morningstar tracks the performance of a large number of companies and publishes an evaluation of each. Along with a variety of financial data, Morningstar includes a Fair Value estimate for the price that should be paid for a share of the company's common stock. Data for 30 companies are available in the file named FairValue. The data include the Fair Value estimate per share of common stock, the most recent price per share, and the earning per share for the company (*Morningstar Stocks 500,* 2008).
    a. Develop a scatter diagram for the Fair Value and Share Price data with Share Price on the horizontal axis. What is the sample correlation coefficient, and what can you say about the relationship between the variables?
    b. Develop a scatter diagram for the Fair Value and Earnings per Share data with Earnings per Share on the horizontal axis. What is the sample correlation coefficient, and what can you say about the relationship between the variables?

68. Does a major league baseball team's record during spring training indicate how the team will play during the regular season? Over the last six years, the correlation coefficient between a team's winning percentage in spring training and its winning percentage in the regular season is .18 (*The Wall Street Journal,* March 30, 2009). Shown are the winning percentages for the 14 American League teams during the 2008 season.

Team	Spring Training	Regular Season	Team	Spring Training	Regular Season
Baltimore Orioles	.407	.422	Minnesota Twins	.500	.540
Boston Red Sox	.429	.586	New York Yankees	.577	.549
Chicago White Sox	.417	.546	Oakland A's	.692	.466
Cleveland Indians	.569	.500	Seattle Mariners	.500	.377
Detroit Tigers	.569	.457	Tampa Bay Rays	.731	.599
Kansas City Royals	.533	.463	Texas Rangers	.643	.488
Los Angeles Angels	.724	.617	Toronto Blue Jays	.448	.531

WEB file
SpringTraining

a. What is the correlation coefficient between the spring training and the regular season winning percentages?
b. What is your conclusion about a team's record during spring training indicating how the team will play during the regular season? What are some of the reasons why this occurs? Discuss.

69. Automobiles traveling on a road with a posted speed limit of 55 miles per hour are checked for speed by a state police radar system. Following is a frequency distribution of speeds.

Speed (miles per hour)	Frequency
45–49	10
50–54	40
55–59	150
60–64	175
65–69	75
70–74	15
75–79	10
Total	475

a. What is the mean speed of the automobiles traveling on this road?
b. Compute the variance and the standard deviation.

70. The days to maturity for a sample of five money market funds are shown here. The dollar amounts invested in the funds are provided. Use the weighted mean to determine the mean number of days to maturity for dollars invested in these five money market funds.

Days to Maturity	Dollar Value ($millions)
20	20
12	30
7	10
5	15
6	10

Case Problem 1 Pelican Stores

Pelican Stores, a division of National Clothing, is a chain of women's apparel stores operating throughout the country. The chain recently ran a promotion in which discount

Case Problem 1 Pelican Stores

TABLE 3.12 SAMPLE OF 100 CREDIT CARD PURCHASES AT PELICAN STORES

WEB file
PelicanStores

Customer	Type of Customer	Items	Net Sales	Method of Payment	Gender	Marital Status	Age
1	Regular	1	39.50	Discover	Male	Married	32
2	Promotional	1	102.40	Proprietary Card	Female	Married	36
3	Regular	1	22.50	Proprietary Card	Female	Married	32
4	Promotional	5	100.40	Proprietary Card	Female	Married	28
5	Regular	2	54.00	MasterCard	Female	Married	34
6	Regular	1	44.50	MasterCard	Female	Married	44
7	Promotional	2	78.00	Proprietary Card	Female	Married	30
8	Regular	1	22.50	Visa	Female	Married	40
9	Promotional	2	56.52	Proprietary Card	Female	Married	46
10	Regular	1	44.50	Proprietary Card	Female	Married	36
⋮	⋮	⋮	⋮	⋮	⋮	⋮	⋮
96	Regular	1	39.50	MasterCard	Female	Married	44
97	Promotional	9	253.00	Proprietary Card	Female	Married	30
98	Promotional	10	287.59	Proprietary Card	Female	Married	52
99	Promotional	2	47.60	Proprietary Card	Female	Married	30
100	Promotional	1	28.44	Proprietary Card	Female	Married	44

coupons were sent to customers of other National Clothing stores. Data collected for a sample of 100 in-store credit card transactions at Pelican Stores during one day while the promotion was running are contained in the file named PelicanStores. Table 3.12 shows a portion of the data set. The proprietary card method of payment refers to charges made using a National Clothing charge card. Customers who made a purchase using a discount coupon are referred to as promotional customers and customers who made a purchase but did not use a discount coupon are referred to as regular customers. Because the promotional coupons were not sent to regular Pelican Stores customers, management considers the sales made to people presenting the promotional coupons as sales it would not otherwise make. Of course, Pelican also hopes that the promotional customers will continue to shop at its stores.

Most of the variables shown in Table 3.12 are self-explanatory, but two of the variables require some clarification.

Items   The total number of items purchased
Net Sales   The total amount ($) charged to the credit card

Pelican's management would like to use this sample data to learn about its customer base and to evaluate the promotion involving discount coupons.

Managerial Report

Use the methods of descriptive statistics presented in this chapter to summarize the data and comment on your findings. At a minimum, your report should include the following:

1. Descriptive statistics on net sales and descriptive statistics on net sales by various classifications of customers.
2. Descriptive statistics concerning the relationship between age and net sales.

TABLE 3.13 PERFORMANCE DATA FOR 10 MOTION PICTURES

Motion Picture	Opening Gross Sales ($millions)	Total Gross Sales ($millions)	Number of Theaters	Weeks in Top 60
Coach Carter	29.17	67.25	2574	16
Ladies in Lavender	0.15	6.65	119	22
Batman Begins	48.75	205.28	3858	18
Unleashed	10.90	24.47	1962	8
Pretty Persuasion	0.06	0.23	24	4
Fever Pitch	12.40	42.01	3275	14
Harry Potter and the Goblet of Fire	102.69	287.18	3858	13
Monster-in-Law	23.11	82.89	3424	16
White Noise	24.11	55.85	2279	7
Mr. and Mrs. Smith	50.34	186.22	3451	21

Case Problem 2 Motion Picture Industry

The motion picture industry is a competitive business. More than 50 studios produce a total of 300 to 400 new motion pictures each year, and the financial success of each motion picture varies considerably. The opening weekend gross sales, the total gross sales, the number of theaters the movie was shown in, and the number of weeks the motion picture was in the top 60 for gross sales are common variables used to measure the success of a motion picture. Data collected for a sample of 100 motion pictures produced in 2005 are contained in the file named Movies. Table 3.13 shows the data for the first 10 motion pictures in the file.

Managerial Report

Use the numerical methods of descriptive statistics presented in this chapter to learn how these variables contribute to the success of a motion picture. Include the following in your report.

1. Descriptive statistics for each of the four variables along with a discussion of what the descriptive statistics tell us about the motion picture industry.
2. What motion pictures, if any, should be considered high-performance outliers? Explain.
3. Descriptive statistics showing the relationship between total gross sales and each of the other variables. Discuss.

Case Problem 3 Heavenly Chocolates Website Transactions

Heavenly Chocolates manufactures and sells quality chocolate products at its plant and retail store located in Saratoga Springs, New York. Two years ago the company developed a website and began selling its products over the Internet. Website sales have exceeded the company's expectations, and mangement is now considering stragegies to increase sales even further. To learn more about the website customers, a sample of 50 Heavenly Chocolate transactions was selected from the previous month's sales. Data showing the day of the week each transaction was made, the type of browser the customer used, the time spent on the website, the number of website pages viewed, and the amount spent by each of the 50 customers are contained in the file named Shoppers. A portion of the data are shown in Table 3.14.

Heavenly Chocolates would like to use the sample data to determine if online shoppers who spend more time and view more pages also spend more money during their visit to the

Appendix 3.1 Descriptive Statistics Using Minitab

TABLE 3.14 A SAMPLE OF 50 HEAVENLY CHOCOLATES WEBSITE TRANSACTIONS

Customer	Day	Browser	Time (min)	Pages Viewed	Amount Spent ($)
1	Mon	Internet Explorer	12.0	4	54.52
2	Wed	Other	19.5	6	94.90
3	Mon	Internet Explorer	8.5	4	26.68
4	Tue	Firefox	11.4	2	44.73
5	Wed	Internet Explorer	11.3	4	66.27
6	Sat	Firefox	10.5	6	67.80
7	Sun	Internet Explorer	11.4	2	36.04
.	.	.	.	.	.
.	.	.	.	.	.
.	.	.	.	.	.
48	Fri	Internet Explorer	9.7	5	103.15
49	Mon	Other	7.3	6	52.15
50	Fri	Internet Explorer	13.4	3	98.75

website. The company would also like to investigate the effect that the day of the week and the type of browser have on sales.

Managerial Report

Use the methods of descriptive statistics to learn about the customers who visit the Heavenly Chocolates website. Include the following in your report.

1. Graphical and numerical summaries for the length of time the shopper spends on the website, the number of pages viewed, and the mean amount spent per transaction. Discuss what you learn about Heavenly Cholcolates' online shoppers from these numerical summaries.
2. Summarize the frequency, the total dollars spent, and the mean amount spent per transaction for each day of week. What observations can you make about Hevenly Chocolates' business based on the day of the week? Discuss.
3. Summarize the frequency, the total dollars spent, and the mean amount spent per transaction for each type of browser. What observations can you make about Heavenly Chocolate's business based on the type of browser? Discuss.
4. Develop a scatter diagram and compute the sample correlation coefficient to explore the relationship between the time spent on the website and the dollar amount spent. Use the horizontal axis for the time spent on the website. Discuss.
5. Develop a scatter diagram and compute the sample correlation coefficient to explore the relationship between the the number of website pages viewed and the amount spent. Use the horizontal axis for the number of website pages viewed. Discuss.
6. Develop a scatter diagram and compute the sample correlation coefficient to explore the relationship between the time spent on the website and the number of pages viewed. Use the horizontal axis to represent the number of pages viewed. Discuss.

Appendix 3.1 Descriptive Statistics Using Minitab

In this appendix, we describe how Minitab can be used to compute a variety of descriptive statistics and display box plots. We then show how Minitab can be used to obtain covariance and correlation measures for two variables.

Descriptive Statistics

Table 3.1 provided the starting salaries for 12 business school graduates. These data are available in column C2 of the file StartSalary. The following steps can be used to generate descriptive statistics for the starting salaries.

Step 1. Select the **Stat** menu
Step 2. Choose **Basic Statistics**
Step 3. Choose **Display Descriptive Statistics**
Step 4. When the Display Descriptive Statistics dialog box appears:
Enter C2 in the **Variables** box
Click **OK**

Figure 3.12 shows the descriptive statistics for the salary data obtained by using Minitab. Definitions of the headings follow.

N	number of data values
N*	number of missing data values
Mean	mean
SE Mean	standard error of mean
StDev	standard deviation
Minimum	minimum data value
Q1	first quartile
Median	median
Q3	third quartile
Maximum	maximum data value

The label SE Mean refers to the *standard error of the mean*. It is computed by dividing the standard deviation by the square root of N. The interpretation and use of this measure are discussed in Chapter 7 when we introduce the topics of sampling and sampling distributions.

Note that Minitab's quartiles $Q_1 = 3457.5$ and $Q_3 = 3625$ are slightly different from the quartiles $Q_1 = 3465$ and $Q_3 = 3600$ computed in Section 3.1. The different conventions[†] used to identify the quartiles explain this variation. Hence, the values of Q_1 and Q_3 provided by one convention may not be identical to the values of Q_1 and Q_3 provided by another convention. Any differences tend to be negligible, however, and the results provided should not mislead the user in making the usual interpretations associated with quartiles.

FIGURE 3.12 DESCRIPTIVE STATISTICS PROVIDED BY MINITAB

N	N*	Mean	SEMean	StDev
12	0	3540.0	47.8	165.7

Minimum	Q1	Median	Q3	Maximum
3310.0	3457.5	3505.0	3625.0	3925.0

[†]With the n observations arranged in ascending order (smallest value to largest value), Minitab uses the positions given by $(n + 1)/4$ and $3(n + 1)/4$ to locate Q_1 and Q_3, respectively. When a position is fractional, Minitab interpolates between the two adjacent ordered data values to determine the corresponding quartile.

Minitab provides 15 additional descriptive statistics that may be selected at the option of the user. These additional descriptive statistics may be obtained by modifying step 4 as follows:

Step 4. When the Display Descriptive Statistics dialog box appears:
Select **Statistics**
When the Display Descriptive Statistics—Statistics dialog box appears:
Check the desired descriptive statistics
Click **OK**
Click **OK**

Common additional descriptive statistics that may be selected include the variance, coefficient of variation, interquartile range, mode, sum, range, and skewness.

Box Plot

The following steps use the file StartSalary to generate the box plot for the starting salary data.

Step 1. Select the **Graph** menu
Step 2. Choose **Boxplot**
Step 3. Select **Simple** and click **OK**
Step 4. When the Boxplot-**One Y**, Simple dialog box appears:
Enter C2 in the **Graph variables** box
Click **OK**

If you would like to show the box plots for two or more groups side-by-side on one graph, select **One Y With Groups** in step 3 and then enter the groups in step 4.

Covariance and Correlation

Table 3.6 provided for the number of commercials and the sales volume for a stereo and sound equipment store. These data are available in the file Stereo, with the number of commercials in column C2 and the sales volume in column C3. The following steps show how Minitab can be used to compute the covariance for the two variables.

Step 1. Select the **Stat** menu
Step 2. Choose **Basic Statistics**
Step 3. Choose **Covariance**
Step 4. When the Covariance dialog box appears:
Enter C2 C3 in the **Variables** box
Click **OK**

The Minitab output provides the variance of each variable in addition to the covariance.

To obtain the correlation coefficient for the number of commercials and the sales volume, only one change is necessary in the preceding procedure. In step 3, choose the **Correlation** option.

Appendix 3.2 Descriptive Statistics Using Excel

Excel can be used to generate the descriptive statistics discussed in this chapter. We show how Excel can be used to generate several measures of location and variability for a single variable and to generate the covariance and correlation coefficient as measures of association between two variables.

Using Excel Functions

Excel provides functions for computing the mean, median, mode, sample variance, and sample standard deviation. We illustrate the use of these Excel functions by computing the mean, median, mode, sample variance, and sample standard deviation for the starting salary data in Table 3.1. Refer to Figure 3.13 as we describe the steps involved. The data are entered in column B.

Excel's AVERAGE function can be used to compute the mean by entering the following formula into cell E1:

$$=\text{AVERAGE(B2:B13)}$$

Similarly, the formulas =MEDIAN(B2:B13), =MODE(B2:B13), =VAR(B2:B13), and =STDEV(B2:B13) are entered into cells E2:E5, respectively, to compute the median, mode, variance, and standard deviation. The worksheet in the foreground shows that the values computed using the Excel functions are the same as we computed earlier in the chapter.

Excel also provides functions that can be used to compute the covariance and correlation coefficient. You must be careful when using these functions because the covariance function treats the data as a population and the correlation function treats the data as a sample. Thus, the result obtained using Excel's covariance function must be adjusted to provide the sample covariance. We show here how these functions can be used to compute the sample covariance and the sample correlation coefficient for the stereo and sound equipment store data in Table 3.7. Refer to Figure 3.14 as we present the steps involved.

FIGURE 3.13 USING EXCEL FUNCTIONS FOR COMPUTING THE MEAN, MEDIAN, MODE, VARIANCE, AND STANDARD DEVIATION

	A	B	C	D	E	F
1	Graduate	Starting Salary		Mean	=AVERAGE(B2:B13)	
2	1	3450		Median	=MEDIAN(B2:B13)	
3	2	3550		Mode	=MODE(B2:B13)	
4	3	3650		Variance	=VAR(B2:B13)	
5	4	3480		Standard Deviation	=STDEV(B2:B13)	
6	5	3355				
7	6	3310				
8	7	3490				
9	8	3730				
10	9	3540				
11	10	3925				
12	11	3520				
13	12	3480				
14						

	A	B	C	D	E	F
1	Graduate	Starting Salary		Mean	3540	
2	1	3450		Median	3505	
3	2	3550		Mode	3480	
4	3	3650		Variance	27440.91	
5	4	3480		Standard Deviation	165.65	
6	5	3355				
7	6	3310				
8	7	3490				
9	8	3730				
10	9	3540				
11	10	3925				
12	11	3520				
13	12	3480				
14						

FIGURE 3.14 USING EXCEL FUNCTIONS FOR COMPUTING COVARIANCE AND CORRELATION

	A	B	C	D	E	F	G
1	Week	Commercials	Sales		Population Covariance	=COVAR(B2:B11,C2:C11)	
2	1	2	50		Sample Correlation	=CORREL(B2:B11,C2:C11)	
3	2	5	57				
4	3	1	41				
5	4	3	54				
6	5	4	54				
7	6	1	38				
8	7	5	63				
9	8	3	48				
10	9	4	59				
11	10	2	46				
12							

	A	B	C	D	E	F	G
1	Week	Commercials	Sales		Population Covariance	9.90	
2	1	2	50		Sample Correlation	0.93	
3	2	5	57				
4	3	1	41				
5	4	3	54				
6	5	4	54				
7	6	1	38				
8	7	5	63				
9	8	3	48				
10	9	4	59				
11	10	2	46				
12							

Excel's covariance function, COVAR, can be used to compute the population covariance by entering the following formula into cell F1:

$$=\text{COVAR}(B2:B11,C2:C11)$$

Similarly, the formula =CORREL(B2:B11,C2:C11) is entered into cell F2 to compute the sample correlation coefficient. The worksheet in the foreground shows the values computed using the Excel functions. Note that the value of the sample correlation coefficient (.93) is the same as computed using equation (3.12). However, the result provided by the Excel COVAR function, 9.9, was obtained by treating the data as a population. Thus, we must adjust the Excel result of 9.9 to obtain the sample covariance. The adjustment is rather simple. First, note that the formula for the population covariance, equation (3.11), requires dividing by the total number of observations in the data set. But the formula for the sample covariance, equation (3.10), requires dividing by the total number of observations minus 1. So, to use the Excel result of 9.9 to compute the sample covariance, we simply multiply 9.9 by $n/(n-1)$. Because $n = 10$, we obtain

$$s_{xy} = \left(\frac{10}{9}\right)9.9 = 11$$

Thus, the sample covariance for the stereo and sound equipment data is 11.

Using Excel's Descriptive Statistics Tool

As we already demonstrated, Excel provides statistical functions to compute descriptive statistics for a data set. These functions can be used to compute one statistic at a time (e.g., mean, variance, etc.). Excel also provides a variety of Data Analysis Tools. One of these tools, called Descriptive Statistics, allows the user to compute a variety of descriptive statistics at once. We show here how it can be used to compute descriptive statistics for the starting salary data in Table 3.1.

FIGURE 3.15 EXCEL'S DESCRIPTIVE STATISTICS TOOL OUTPUT

	A	B	C	D	E	F
1	Graduate	Starting Salary		*Starting Salary*		
2	1	3450				
3	2	3550		Mean	3540	
4	3	3650		Standard Error	47.82	
5	4	3480		Median	3505	
6	5	3355		Mode	3480	
7	6	3310		Standard Deviation	165.65	
8	7	3490		Sample Variance	27440.91	
9	8	3730		Kurtosis	1.7189	
10	9	3540		Skewness	1.0911	
11	10	3925		Range	615	
12	11	3520		Minimum	3310	
13	12	3480		Maximum	3925	
14				Sum	42480	
15				Count	12	
16						

Step 1. Click the **Data** tab on the Ribbon
Step 2. In the **Analysis** group, click **Data Analysis**
Step 3. When the Data Analysis dialog box appears:
    Choose **Descriptive Statistics**
    Click **OK**
Step 4. When the Descriptive Statistics dialog box appears:
    Enter B1:B13 in the **Input Range** box
    Select **Grouped By Columns**
    Select **Labels in First Row**
    Select **Output Range**
    Enter D1 in the **Output Range** box (to identify the upper left-hand corner of the section of the worksheet where the descriptive statistics will appear)
    Select **Summary statistics**
    Click **OK**

Cells D1:E15 of Figure 3.15 show the descriptive statistics provided by Excel. The boldface entries are the descriptive statistics we covered in this chapter. The descriptive statistics that are not boldface are either covered subsequently in the text or discussed in more advanced texts.

Appendix 3.3 Descriptive Statistics Using StatTools

In this appendix, we describe how StatTools can be used to compute a variety of descriptive statistics and also display box plots. We then show how StatTools can be used to obtain covariance and correlation measures for two variables.

Descriptive Statistics

We use the starting salary data in Table 3.1 to illustrate. Begin by using the Data Set Manager to create a StatTools data set for these data using the procedure described in the appendix in Chapter 1. The following steps will generate a variety of descriptive statistics.

> **Step 1.** Click the **StatTools** tab on the Ribbon
> **Step 2.** In the **Analyses Group,** click **Summary Statistics**
> **Step 3.** Choose the **One-Variable Summary** option
> **Step 4.** When the One-Variable Summary Statistics dialog box appears:
> In the **Variables** section, select **Starting Salary**
> Click **OK**

A variety of descriptive statistics will appear.

Box Plots

We use the starting salary data in Table 3.1 to illustrate. Begin by using the Data Set Manager to create a StatTools data set for these data using the procedure described in the appendix in Chapter 1. The following steps will create a box plot for these data.

> **Step 1.** Click the **StatTools** tab on the Ribbon
> **Step 2.** In the **Analyses Group,** click **Summary Graphs**
> **Step 3.** Choose the **Box-Whisker Plot** option
> **Step 4.** When the StatTools—Box-Whisker Plot dialog box appears:
> In the Variables section, select **Starting Salary**
> Click **OK**

The symbol □ is used to identify an outlier and x is used to identify the mean.

Covariance and Correlation

We use the stereo and sound equipment data in Table 3.7 to demonstrate the computation of the sample covariance and the sample correlation coefficient. Begin by using the Data Set Manager to create a StatTools data set for these data using the procedure described in the appendix in Chapter 1. The following steps will provide the sample covariance and sample correlation coefficient.

> **Step 1.** Click the **StatTools** tab on the Ribbon
> **Step 2.** In the **Analyses Group,** click **Summary Statistics**
> **Step 3.** Choose the **Correlation and Covariance** option
> **Step 4.** When the StatTools—Correlation and Covariance dialog box appears:
> In the **Variables** section
> Select **No. of Commercials**
> Select **Sales Volume**
> In the **Tables to Create** section
> Select **Table of Correlations**
> Select **Table of Covariances**
> In the **Table Structure** section select **Symmetric**
> Click **OK**

A table showing the correlation coefficient and the covariance will appear.

CHAPTER 4

Introduction to Probability

CONTENTS

STATISTICS IN PRACTICE:
OCEANWIDE SEAFOOD

4.1 EXPERIMENTS, COUNTING RULES, AND ASSIGNING PROBABILITIES
Counting Rules, Combinations, and Permutations
Assigning Probabilities
Probabilities for the KP&L Project

4.2 EVENTS AND THEIR PROBABILITIES

4.3 SOME BASIC RELATIONSHIPS OF PROBABILITY
Complement of an Event
Addition Law

4.4 CONDITIONAL PROBABILITY
Independent Events
Multiplication Law

4.5 BAYES' THEOREM
Tabular Approach

STATISTICS *in* PRACTICE

OCEANWIDE SEAFOOD*
SPRINGBORO, OHIO

Oceanwide Seafood is the leading provider of quality seafood in southwestern Ohio. The company stocks over 90 varieties of fresh and frozen seafood products from around the world and prepares specialty cuts according to customer specifications. Customers include major restaurants and retail food stores in Ohio, Kentucky, and Indiana. Established in 2005, the company has become successful by providing superior customer service and exceptional quality seafood.

Probability and statistical information are used for both operational and marketing decisions. For instance, a time series showing monthly sales is used to track the company's growth and to set future target sales levels. Statistics such as the mean customer order size and the mean number of days a customer takes to make payments help identify the firm's best customers as well as provide benchmarks for handling accounts receivable issues. In addition, data on monthly inventory levels are used in the analysis of operating profits and trends in product sales.

Probability analysis has helped Oceanwide determine reasonable and profitable prices for its products. For example, when Oceanwide receives a whole fresh fish from one of its suppliers, the fish must be processed and cut to fill individual customer orders. A fresh 100-pound whole tuna packed in ice might cost Oceanwide $500. At first glance, the company's cost for tuna appears to be $500/100 = $5 per pound. However, due to the loss in the processing and cutting operation, a 100-pound whole tuna will not provide 100 pounds of finished product. If the processing and cutting operation yields 75% of the whole tuna, the number of pounds of finished product available for sale to customers would be .75(100) = 75 pounds, not 100 pounds. In this case, the company's actual cost of tuna would be $500/75 = $6.67 per pound. Thus, Oceanwide would need to use a cost of $6.67 per pound to determine a profitable price to charge its customers.

---

*The authors are indebted to Dale Hartlage, president of Oceanwide Seafood Company, for providing this Statistics in Practice.

Oceanwide Seafood uses probability analysis to help determine reasonable and profitable prices for its products.

To help determine the yield percentage that is likely for processing and cutting whole tuna, data were collected on the yields from a sample of whole tunas. Let Y denote the yield percentage for whole tuna. Using the data, Oceanwide was able to determine that 5% of the time the yield for whole tuna was at least 90%. In conditional probability notation, this probability is written $P(Y \geq 90\% \mid \text{Tuna}) = .05$; in other words, the probability that the yield will be at least 90% given that the fish is a tuna is .05. If Oceanwide established the selling price for tuna based on a 90% yield, 95% of the time the company would realize a yield less than expected. As a result, the company would be understating its cost per pound and also understating the price of tuna for its customers. Additional conditional probability information for other yield percentages helped management select a 70% yield as the basis for determining the cost of tuna and the price to charge its customers. Similar conditional probabilities for other seafood products helped management establish pricing yield percentages for each type of seafood product. In this chapter, you will learn how to compute and interpret conditional probabilities and other probabilities that are helpful in the decision-making process.

Managers often base their decisions on an analysis of uncertainties such as the following:

1. What are the chances that sales will decrease if we increase prices?
2. What is the likelihood a new assembly method will increase productivity?
3. How likely is it that the project will be finished on time?
4. What is the chance that a new investment will be profitable?

Some of the earliest work on probability originated in a series of letters between Pierre de Fermat and Blaise Pascal in the 1650s.

Probability is a numerical measure of the likelihood that an event will occur. Thus, probabilities can be used as measures of the degree of uncertainty associated with the four events previously listed. If probabilities are available, we can determine the likelihood of each event occurring.

Probability values are always assigned on a scale from 0 to 1. A probability near zero indicates an event is unlikely to occur; a probability near 1 indicates an event is almost certain to occur. Other probabilities between 0 and 1 represent degrees of likelihood that an event will occur. For example, if we consider the event "rain tomorrow," we understand that when the weather report indicates "a near-zero probability of rain," it means almost no chance of rain. However, if a .90 probability of rain is reported, we know that rain is likely to occur. A .50 probability indicates that rain is just as likely to occur as not. Figure 4.1 depicts the view of probability as a numerical measure of the likelihood of an event occurring.

4.1 Experiments, Counting Rules, and Assigning Probabilities

In discussing probability, we define an **experiment** as a process that generates well-defined outcomes. On any single repetition of an experiment, one and only one of the possible experimental outcomes will occur. Several examples of experiments and their associated outcomes follow.

Experiment	Experimental Outcomes
Toss a coin	Head, tail
Select a part for inspection	Defective, nondefective
Conduct a sales call	Purchase, no purchase
Roll a die	1, 2, 3, 4, 5, 6
Play a football game	Win, lose, tie

By specifying all possible experimental outcomes, we identify the **sample space** for an experiment.

SAMPLE SPACE

The sample space for an experiment is the set of all experimental outcomes.

Experimental outcomes are also called sample points.

An experimental outcome is also called a **sample point** to identify it as an element of the sample space.

FIGURE 4.1 PROBABILITY AS A NUMERICAL MEASURE OF THE LIKELIHOOD OF AN EVENT OCCURRING

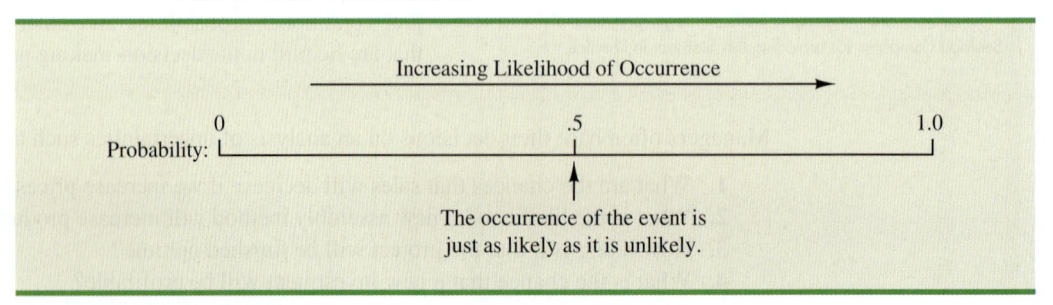

Increasing Likelihood of Occurrence

Probability: 0 ———————— .5 ———————— 1.0

The occurrence of the event is just as likely as it is unlikely.

4.1 Experiments, Counting Rules, and Assigning Probabilities

Consider the first experiment in the preceding table—tossing a coin. The upward face of the coin—a head or a tail—determines the experimental outcomes (sample points). If we let S denote the sample space, we can use the following notation to describe the sample space.

$$S = \{\text{Head, Tail}\}$$

The sample space for the second experiment in the table—selecting a part for inspection—can be described as follows:

$$S = \{\text{Defective, Nondefective}\}$$

Both of the experiments just described have two experimental outcomes (sample points). However, suppose we consider the fourth experiment listed in the table—rolling a die. The possible experimental outcomes, defined as the number of dots appearing on the upward face of the die, are the six points in the sample space for this experiment.

$$S = \{1, 2, 3, 4, 5, 6\}$$

Counting Rules, Combinations, and Permutations

Being able to identify and count the experimental outcomes is a necessary step in assigning probabilities. We now discuss three useful counting rules.

Multiple-step experiments The first counting rule applies to multiple-step experiments. Consider the experiment of tossing two coins. Let the experimental outcomes be defined in terms of the pattern of heads and tails appearing on the upward faces of the two coins. How many experimental outcomes are possible for this experiment? The experiment of tossing two coins can be thought of as a two-step experiment in which step 1 is the tossing of the first coin and step 2 is the tossing of the second coin. If we use H to denote a head and T to denote a tail, (H, H) indicates the experimental outcome with a head on the first coin and a head on the second coin. Continuing this notation, we can describe the sample space (S) for this coin-tossing experiment as follows:

$$S = \{(H, H), (H, T), (T, H), (T, T)\}$$

Thus, we see that four experimental outcomes are possible. In this case, we can easily list all the experimental outcomes.

The counting rule for multiple-step experiments makes it possible to determine the number of experimental outcomes without listing them.

> **COUNTING RULE FOR MULTIPLE-STEP EXPERIMENTS**
>
> If an experiment can be described as a sequence of k steps with n_1 possible outcomes on the first step, n_2 possible outcomes on the second step, and so on, then the total number of experimental outcomes is given by $(n_1)(n_2) \ldots (n_k)$.

Viewing the experiment of tossing two coins as a sequence of first tossing one coin $(n_1 = 2)$ and then tossing the other coin $(n_2 = 2)$, we can see from the counting rule that $(2)(2) = 4$ distinct experimental outcomes are possible. As shown, they are $S = \{(H, H), (H, T), (T, H), (T, T)\}$. The number of experimental outcomes in an experiment involving tossing six coins is $(2)(2)(2)(2)(2)(2) = 64$.

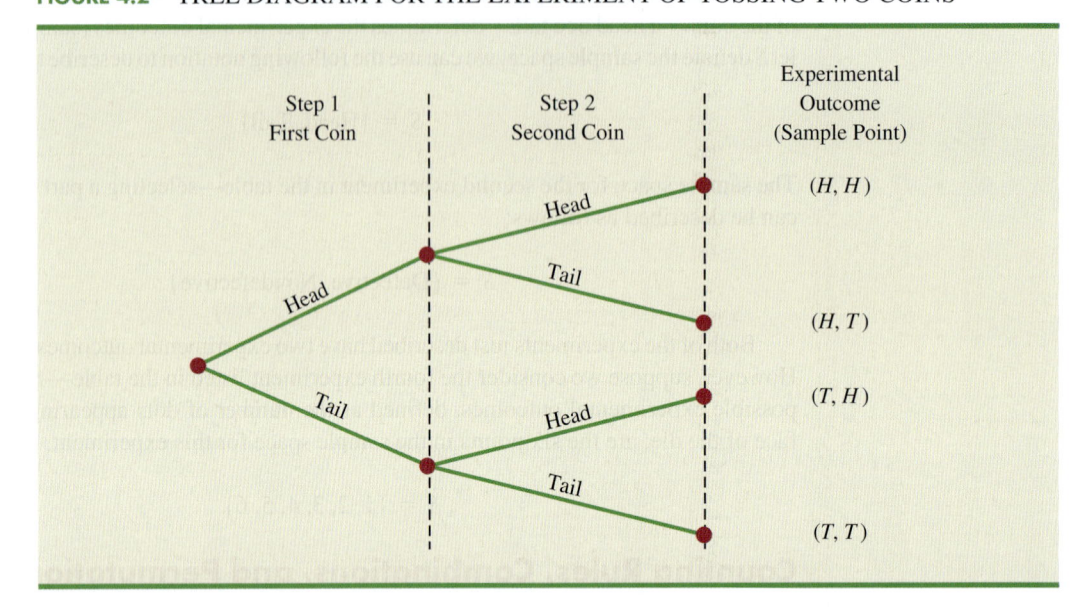

FIGURE 4.2 TREE DIAGRAM FOR THE EXPERIMENT OF TOSSING TWO COINS

	Step 1 First Coin	Step 2 Second Coin	Experimental Outcome (Sample Point)
		Head	(H, H)
	Head	Tail	(H, T)
	Tail	Head	(T, H)
		Tail	(T, T)

Without the tree diagram, one might think only three experimental outcomes are possible for two tosses of a coin: 0 heads, 1 head, and 2 heads.

A **tree diagram** is a graphical representation that helps in visualizing a multiple-step experiment. Figure 4.2 shows a tree diagram for the experiment of tossing two coins. The sequence of steps moves from left to right through the tree. Step 1 corresponds to tossing the first coin, and step 2 corresponds to tossing the second coin. For each step, the two possible outcomes are head or tail. Note that for each possible outcome at step 1 two branches correspond to the two possible outcomes at step 2. Each of the points on the right end of the tree corresponds to an experimental outcome. Each path through the tree from the leftmost node to one of the nodes at the right side of the tree corresponds to a unique sequence of outcomes.

Let us now see how the counting rule for multiple-step experiments can be used in the analysis of a capacity expansion project for the Kentucky Power & Light Company (KP&L). KP&L is starting a project designed to increase the generating capacity of one of its plants in northern Kentucky. The project is divided into two sequential stages or steps: stage 1 (design) and stage 2 (construction). Even though each stage will be scheduled and controlled as closely as possible, management cannot predict beforehand the exact time required to complete each stage of the project. An analysis of similar construction projects revealed possible completion times for the design stage of 2, 3, or 4 months and possible completion times for the construction stage of 6, 7, or 8 months. In addition, because of the critical need for additional electrical power, management set a goal of 10 months for the completion of the entire project.

Because this project has three possible completion times for the design stage (step 1) and three possible completion times for the construction stage (step 2), the counting rule for multiple-step experiments can be applied here to determine a total of (3)(3) = 9 experimental outcomes. To describe the experimental outcomes, we use a two-number notation; for instance, (2, 6) indicates that the design stage is completed in 2 months and the construction stage is completed in 6 months. This experimental outcome results in a total of 2 + 6 = 8 months to complete the entire project. Table 4.1 summarizes the nine experimental outcomes for the KP&L problem. The tree diagram in Figure 4.3 shows how the nine outcomes (sample points) occur.

The counting rule and tree diagram help the project manager identify the experimental outcomes and determine the possible project completion times. From the information in

TABLE 4.1 EXPERIMENTAL OUTCOMES (SAMPLE POINTS) FOR THE KP&L PROJECT

Completion Time (months)		Notation for Experimental Outcome	Total Project Completion Time (months)
Stage 1 Design	Stage 2 Construction		
2	6	(2, 6)	8
2	7	(2, 7)	9
2	8	(2, 8)	10
3	6	(3, 6)	9
3	7	(3, 7)	10
3	8	(3, 8)	11
4	6	(4, 6)	10
4	7	(4, 7)	11
4	8	(4, 8)	12

FIGURE 4.3 TREE DIAGRAM FOR THE KP&L PROJECT

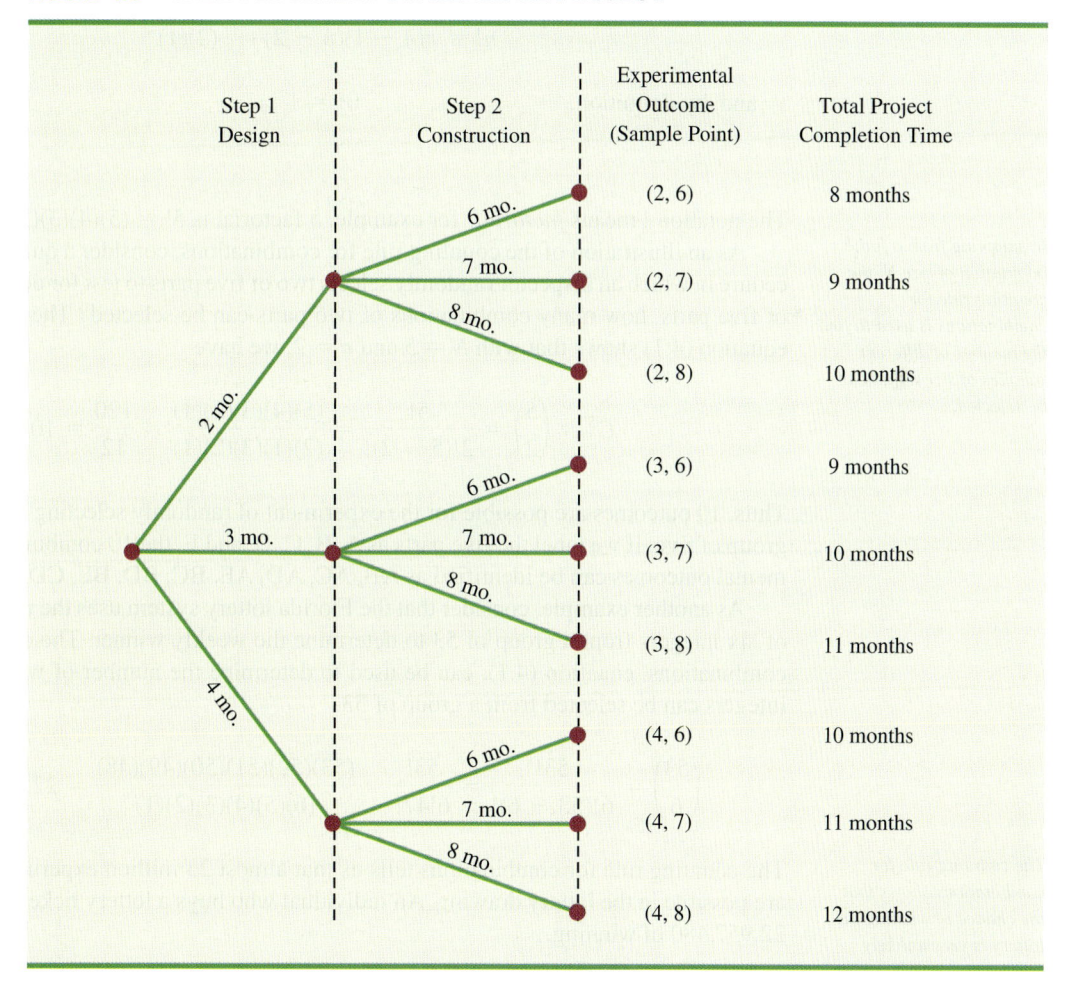

Step 1 Design	Step 2 Construction	Experimental Outcome (Sample Point)	Total Project Completion Time
2 mo.	6 mo.	(2, 6)	8 months
2 mo.	7 mo.	(2, 7)	9 months
2 mo.	8 mo.	(2, 8)	10 months
3 mo.	6 mo.	(3, 6)	9 months
3 mo.	7 mo.	(3, 7)	10 months
3 mo.	8 mo.	(3, 8)	11 months
4 mo.	6 mo.	(4, 6)	10 months
4 mo.	7 mo.	(4, 7)	11 months
4 mo.	8 mo.	(4, 8)	12 months

Figure 4.3, we see that the project will be completed in 8 to 12 months, with six of the nine experimental outcomes providing the desired completion time of 10 months or less. Even though identifying the experimental outcomes may be helpful, we need to consider how probability values can be assigned to the experimental outcomes before making an assessment of the probability that the project will be completed within the desired 10 months.

Combinations A second useful counting rule allows one to count the number of experimental outcomes when the experiment involves selecting n objects from a (usually larger) set of N objects. It is called the counting rule for combinations.

> COUNTING RULE FOR COMBINATIONS
>
> The number of combinations of N objects taken n at a time is
>
> $$C_n^N = \binom{N}{n} = \frac{N!}{n!(N-n)!} \quad (4.1)$$
>
> where
>
> $$N! = N(N-1)(N-2)\cdots(2)(1)$$
> $$n! = n(n-1)(n-2)\cdots(2)(1)$$
>
> and, by definition,
>
> $$0! = 1$$

The notation ! means *factorial*; for example, 5 factorial is $5! = (5)(4)(3)(2)(1) = 120$.

In sampling from a finite population of size N, the counting rule for combinations is used to find the number of different samples of size n that can be selected.

As an illustration of the counting rule for combinations, consider a quality control procedure in which an inspector randomly selects two of five parts to test for defects. In a group of five parts, how many combinations of two parts can be selected? The counting rule in equation (4.1) shows that with $N = 5$ and $n = 2$, we have

$$C_2^5 = \binom{5}{2} = \frac{5!}{2!(5-2)!} = \frac{(5)(4)(3)(2)(1)}{(2)(1)(3)(2)(1)} = \frac{120}{12} = 10$$

Thus, 10 outcomes are possible for the experiment of randomly selecting two parts from a group of five. If we label the five parts as A, B, C, D, and E, the 10 combinations or experimental outcomes can be identified as AB, AC, AD, AE, BC, BD, BE, CD, CE, and DE.

As another example, consider that the Florida lottery system uses the random selection of six integers from a group of 53 to determine the weekly winner. The counting rule for combinations, equation (4.1), can be used to determine the number of ways six different integers can be selected from a group of 53.

$$\binom{53}{6} = \frac{53!}{6!(53-6)!} = \frac{53!}{6!47!} = \frac{(53)(52)(51)(50)(49)(48)}{(6)(5)(4)(3)(2)(1)} = 22{,}957{,}480$$

The counting rule for combinations shows that the chance of winning the lottery is very unlikely.

The counting rule for combinations tells us that almost 23 million experimental outcomes are possible in the lottery drawing. An individual who buys a lottery ticket has 1 chance in 22,957,480 of winning.

Permutations A third counting rule that is sometimes useful is the counting rule for permutations. It allows one to compute the number of experimental outcomes when n objects are to be selected from a set of N objects where the order of selection is important. The same n objects selected in a different order are considered a different experimental outcome.

COUNTING RULE FOR PERMUTATIONS

The number of permutations of N objects taken n at a time is given by

$$P_n^N = n!\binom{N}{n} = \frac{N!}{(N-n)!} \tag{4.2}$$

The counting rule for permutations closely relates to the one for combinations; however, an experiment results in more permutations than combinations for the same number of objects because every selection of n objects can be ordered in $n!$ different ways.

As an example, consider again the quality control process in which an inspector selects two of five parts to inspect for defects. How many permutations may be selected? The counting rule in equation (4.2) shows that with $N = 5$ and $n = 2$, we have

$$P_2^5 = \frac{5!}{(5-2)!} = \frac{5!}{3!} = \frac{(5)(4)(3)(2)(1)}{(3)(2)(1)} = \frac{120}{6} = 20$$

Thus, 20 outcomes are possible for the experiment of randomly selecting two parts from a group of five when the order of selection must be taken into account. If we label the parts A, B, C, D, and E, the 20 permutations are AB, BA, AC, CA, AD, DA, AE, EA, BC, CB, BD, DB, BE, EB, CD, DC, CE, EC, DE, and ED.

Assigning Probabilities

Now let us see how probabilities can be assigned to experimental outcomes. The three approaches most frequently used are the classical, relative frequency, and subjective methods. Regardless of the method used, two **basic requirements for assigning probabilities** must be met.

BASIC REQUIREMENTS FOR ASSIGNING PROBABILITIES

1. The probability assigned to each experimental outcome must be between 0 and 1, inclusively. If we let E_i denote the ith experimental outcome and $P(E_i)$ its probability, then this requirement can be written as

$$0 \leq P(E_i) \leq 1 \text{ for all } i \tag{4.3}$$

2. The sum of the probabilities for all the experimental outcomes must equal 1.0. For n experimental outcomes, this requirement can be written as

$$P(E_1) + P(E_2) + \cdots + P(E_n) = 1 \tag{4.4}$$

The **classical method** of assigning probabilities is appropriate when all the experimental outcomes are equally likely. If n experimental outcomes are possible, a probability of $1/n$ is assigned to each experimental outcome. When using this approach, the two basic requirements for assigning probabilities are automatically satisfied.

For an example, consider the experiment of tossing a fair coin; the two experimental outcomes—head and tail—are equally likely. Because one of the two equally likely outcomes is a head, the probability of observing a head is 1/2, or .50. Similarly, the probability of observing a tail is also 1/2, or .50.

As another example, consider the experiment of rolling a die. It would seem reasonable to conclude that the six possible outcomes are equally likely, and hence each outcome is assigned a probability of 1/6. If $P(1)$ denotes the probability that one dot appears on the upward face of the die, then $P(1) = 1/6$. Similarly, $P(2) = 1/6$, $P(3) = 1/6$, $P(4) = 1/6$, $P(5) = 1/6$, and $P(6) = 1/6$. Note that these probabilities satisfy the two basic requirements of equations (4.3) and (4.4) because each of the probabilities is greater than or equal to zero and they sum to 1.0.

The **relative frequency method** of assigning probabilities is appropriate when data are available to estimate the proportion of the time the experimental outcome will occur if the experiment is repeated a large number of times. As an example, consider a study of waiting times in the X-ray department for a local hospital. A clerk recorded the number of patients waiting for service at 9:00 A.M. on 20 successive days and obtained the following results.

Number Waiting	Number of Days Outcome Occurred
0	2
1	5
2	6
3	4
4	3
Total	20

These data show that on 2 of the 20 days, zero patients were waiting for service; on 5 of the days, one patient was waiting for service; and so on. Using the relative frequency method, we would assign a probability of $2/20 = .10$ to the experimental outcome of zero patients waiting for service, $5/20 = .25$ to the experimental outcome of one patient waiting, $6/20 = .30$ to two patients waiting, $4/20 = .20$ to three patients waiting, and $3/20 = .15$ to four patients waiting. As with the classical method, using the relative frequency method automatically satisfies the two basic requirements of equations (4.3) and (4.4).

The **subjective method** of assigning probabilities is most appropriate when one cannot realistically assume that the experimental outcomes are equally likely and when little relevant data are available. When the subjective method is used to assign probabilities to the experimental outcomes, we may use any information available, such as our experience or intuition. After considering all available information, a probability value that expresses our *degree of belief* (on a scale from 0 to 1) that the experimental outcome will occur is specified. Because subjective probability expresses a person's degree of belief, it is personal. Using the subjective method, different people can be expected to assign different probabilities to the same experimental outcome.

The subjective method requires extra care to ensure that the two basic requirements of equations (4.3) and (4.4) are satisfied. Regardless of a person's degree of belief, the probability value assigned to each experimental outcome must be between 0 and 1, inclusive, and the sum of all the probabilities for the experimental outcomes must equal 1.0.

Consider the case in which Tom and Judy Elsbernd make an offer to purchase a house. Two outcomes are possible:

$$E_1 = \text{their offer is accepted}$$
$$E_2 = \text{their offer is rejected}$$

Judy believes that the probability that their offer will be accepted is .8; thus, Judy would set $P(E_1) = .8$ and $P(E_2) = .2$. Tom, however, believes that the probability that their offer will be accepted is .6; hence, Tom would set $P(E_1) = .6$ and $P(E_2) = .4$. Note that Tom's probability estimate for E_1 reflects a greater pessimism that their offer will be accepted.

Bayes' theorem (see Section 4.5) provides a means for combining subjectively determined prior probabilities with probabilities obtained by other means to obtain revised, or posterior, probabilities.

Both Judy and Tom assigned probabilities that satisfy the two basic requirements. The fact that their probability estimates are different emphasizes the personal nature of the subjective method.

Even in business situations where either the classical or the relative frequency approach can be applied, managers may want to provide subjective probability estimates. In such cases, the best probability estimates often are obtained by combining the estimates from the classical or relative frequency approach with subjective probability estimates.

Probabilities for the KP&L Project

To perform further analysis on the KP&L project, we must develop probabilities for each of the nine experimental outcomes listed in Table 4.1. On the basis of experience and judgment, management concluded that the experimental outcomes were not equally likely. Hence, the classical method of assigning probabilities could not be used. Management then decided to conduct a study of the completion times for similar projects undertaken by KP&L over the past three years. The results of a study of 40 similar projects are summarized in Table 4.2.

After reviewing the results of the study, management decided to employ the relative frequency method of assigning probabilities. Management could have provided subjective probability estimates but felt that the current project was quite similar to the 40 previous projects. Thus, the relative frequency method was judged best.

In using the data in Table 4.2 to compute probabilities, we note that outcome (2, 6)—stage 1 completed in 2 months and stage 2 completed in 6 months—occurred six times in the 40 projects. We can use the relative frequency method to assign a probability of $6/40 = .15$ to this outcome. Similarly, outcome (2, 7) also occurred in six of the 40 projects, providing a $6/40 = .15$ probability. Continuing in this manner, we obtain the probability assignments for the sample points of the KP&L project shown in Table 4.3. Note that $P(2, 6)$ represents the probability of the sample point (2, 6), $P(2, 7)$ represents the probability of the sample point (2, 7), and so on.

TABLE 4.2   COMPLETION RESULTS FOR 40 KP&L PROJECTS

Completion Time (months) Stage 1 Design	Stage 2 Construction	Sample Point	Number of Past Projects Having These Completion Times
2	6	(2, 6)	6
2	7	(2, 7)	6
2	8	(2, 8)	2
3	6	(3, 6)	4
3	7	(3, 7)	8
3	8	(3, 8)	2
4	6	(4, 6)	2
4	7	(4, 7)	4
4	8	(4, 8)	6
		Total	40

TABLE 4.3 PROBABILITY ASSIGNMENTS FOR THE KP&L PROJECT BASED ON THE RELATIVE FREQUENCY METHOD

Sample Point	Project Completion Time	Probability of Sample Point
(2, 6)	8 months	$P(2, 6) = 6/40 = .15$
(2, 7)	9 months	$P(2, 7) = 6/40 = .15$
(2, 8)	10 months	$P(2, 8) = 2/40 = .05$
(3, 6)	9 months	$P(3, 6) = 4/40 = .10$
(3, 7)	10 months	$P(3, 7) = 8/40 = .20$
(3, 8)	11 months	$P(3, 8) = 2/40 = .05$
(4, 6)	10 months	$P(4, 6) = 2/40 = .05$
(4, 7)	11 months	$P(4, 7) = 4/40 = .10$
(4, 8)	12 months	$P(4, 8) = 6/40 = .15$
		Total 1.00

NOTES AND COMMENTS

In statistics, the notion of an experiment differs somewhat from the notion of an experiment in the physical sciences. In the physical sciences, researchers usually conduct an experiment in a laboratory or a controlled environment in order to learn about cause and effect. In statistical experiments, probability determines outcomes. Even though the experiment is repeated in exactly the same way, an entirely different outcome may occur. Because of this influence of probability on the outcome, the experiments of statistics are sometimes called *random experiments*.

Exercises

Methods

1. An experiment has three steps with three outcomes possible for the first step, two outcomes possible for the second step, and four outcomes possible for the third step. How many experimental outcomes exist for the entire experiment?

2. How many ways can three items be selected from a group of six items? Use the letters A, B, C, D, E, and F to identify the items, and list each of the different combinations of three items.

3. How many permutations of three items can be selected from a group of six? Use the letters A, B, C, D, E, and F to identify the items, and list each of the permutations of items B, D, and F.

4. Consider the experiment of tossing a coin three times.
   a. Develop a tree diagram for the experiment.
   b. List the experimental outcomes.
   c. What is the probability for each experimental outcome?

5. Suppose an experiment has five equally likely outcomes: E_1, E_2, E_3, E_4, E_5. Assign probabilities to each outcome and show that the requirements in equations (4.3) and (4.4) are satisfied. What method did you use?

6. An experiment with three outcomes has been repeated 50 times, and it was learned that E_1 occurred 20 times, E_2 occurred 13 times, and E_3 occurred 17 times. Assign probabilities to the outcomes. What method did you use?

7. A decision maker subjectively assigned the following probabilities to the four outcomes of an experiment: $P(E_1) = .10$, $P(E_2) = .15$, $P(E_3) = .40$, and $P(E_4) = .20$. Are these probability assignments valid? Explain.

Applications

8. In the city of Milford, applications for zoning changes go through a two-step process: a review by the planning commission and a final decision by the city council. At step 1 the planning commission reviews the zoning change request and makes a positive or negative recommendation concerning the change. At step 2 the city council reviews the planning commission's recommendation and then votes to approve or to disapprove the zoning change. Suppose the developer of an apartment complex submits an application for a zoning change. Consider the application process as an experiment.
   a. How many sample points are there for this experiment? List the sample points.
   b. Construct a tree diagram for the experiment.

9. **SELF test** Simple random sampling uses a sample of size n from a population of size N to obtain data that can be used to make inferences about the characteristics of a population. Suppose that, from a population of 50 bank accounts, we want to take a random sample of 4 accounts in order to learn about the population. How many different random samples of 4 accounts are possible?

10. **SELF test** Many students accumulate debt by the time they graduate from college. Shown in the following table are the percentage of graduates with debt and the average amount of debt for these graduates at four universities and four liberal arts colleges (*U.S. News and World Report, America's Best Colleges,* 2008).

University	% with Debt	Amount($)	College	% with Debt	Amount($)
Pace	72	32,980	Wartburg	83	28,758
Iowa State	69	32,130	Morehouse	94	27,000
Massachusetts	55	11,227	Wellesley	55	10,206
SUNY–Albany	64	11,856	Wofford	49	11,012

   a. If you randomly choose a graduate of Morehouse College, what is the probability that this individual graduated with debt?
   b. If you randomly choose one of these eight institutions for a follow-up study on student loans, what is the probability that you will choose an institution with more than 60% of its graduates having debt?
   c. If you randomly choose one of these eight institutions for a follow-up study on student loans, what is the probability that you will choose an institution whose graduates with debts have an average debt of more than $30,000?
   d. What is the probability that a graduate of Pace University does not have debt?
   e. For graduates of Pace University with debt, the average amount of debt is $32,980. Considering all graduates from Pace University, what is the average debt per graduate?

11. The National Occupant Protection Use Survey (NOPUS) was conducted to provide probability-based data on motorcycle helmet use in the United States. The survey was conducted by sending observers to randomly selected roadway sites where they collected data on motorcycle helmet use, including the number of motorcyclists wearing a Department of Transportation (DOT) compliant helmet (National Highway Traffic Safety Administration website, January 7, 2010). Sample data consistent with the most recent NOPUS are as follows.

	Type of Helmet	
Region	**DOT-Compliant**	**Noncompliant**
Northeast	96	62
Midwest	86	43
South	92	49
West	76	16
Total	350	170

a. Use the sample data to compute an estimate of the probability that a motorcyclist wears a DOT-compliant helmet?
b. The probability that a motorcyclist wore a DOT-compliant helmet five years ago was .48, and last year this probability was .63. Would the National Highway Traffic Safety Administration be pleased with the most recent survey results?
c. What is the probability of DOT-compliant helmet use by region of the country? What region has the highest probability of DOT-compliant helmet use?

12. The Powerball lottery is played twice each week in 28 states, the Virgin Islands, and the District of Columbia. To play Powerball a participant must purchase a ticket and then select five numbers from the digits 1 through 55 and a Powerball number from the digits 1 through 42. To determine the winning numbers for each game, lottery officials draw 5 white balls out of a drum with 55 white balls, and 1 red ball out of a drum with 42 red balls. To win the jackpot, a participant's numbers must match the numbers on the 5 white balls in any order and the number on the red Powerball. Eight coworkers at the ConAgra Foods plant in Lincoln, Nebraska, claimed the record $365 million jackpot on February 18, 2006, by matching the numbers 15–17–43–44–49 and the Powerball number 29. A variety of other cash prizes are awarded each time the game is played. For instance, a prize of $200,000 is paid if the participant's five numbers match the numbers on the 5 white balls (Powerball website, March 19, 2006).
    a. Compute the number of ways the first five numbers can be selected.
    b. What is the probability of winning a prize of $200,000 by matching the numbers on the 5 white balls?
    c. What is the probability of winning the Powerball jackpot?

13. A company that manufactures toothpaste is studying five different package designs. Assuming that one design is just as likely to be selected by a consumer as any other design, what selection probability would you assign to each of the package designs? In an actual experiment, 100 consumers were asked to pick the design they preferred. The following data were obtained. Do the data confirm the belief that one design is just as likely to be selected as another? Explain.

Design	Number of Times Preferred
1	5
2	15
3	30
4	40
5	10

4.2 Events and Their Probabilities

In the introduction to this chapter we used the term *event* much as it would be used in everyday language. Then, in Section 4.1 we introduced the concept of an experiment and its associated experimental outcomes or sample points. Sample points and events provide the foundation for the study of probability. As a result, we must now introduce the formal definition of an **event** as it relates to sample points. Doing so will provide the basis for determining the probability of an event.

> **EVENT**
>
> An event is a collection of sample points.

4.2 Events and Their Probabilities

For an example, let us return to the KP&L project and assume that the project manager is interested in the event that the entire project can be completed in 10 months or less. Referring to Table 4.3, we see that six sample points—(2, 6), (2, 7), (2, 8), (3, 6), (3, 7), and (4, 6)—provide a project completion time of 10 months or less. Let C denote the event that the project is completed in 10 months or less; we write

$$C = \{(2, 6), (2, 7), (2, 8), (3, 6), (3, 7), (4, 6)\}$$

Event C is said to occur if *any one* of these six sample points appears as the experimental outcome.

Other events that might be of interest to KP&L management include the following.

L = The event that the project is completed in *less* than 10 months
M = The event that the project is completed in *more* than 10 months

Using the information in Table 4.3, we see that these events consist of the following sample points.

$$L = \{(2, 6), (2, 7), (3, 6)\}$$
$$M = \{(3, 8), (4, 7), (4, 8)\}$$

A variety of additional events can be defined for the KP&L project, but in each case the event must be identified as a collection of sample points for the experiment.

Given the probabilities of the sample points shown in Table 4.3, we can use the following definition to compute the probability of any event that KP&L management might want to consider.

PROBABILITY OF AN EVENT

The probability of any event is equal to the sum of the probabilities of the sample points in the event.

Using this definition, we calculate the probability of a particular event by adding the probabilities of the sample points (experimental outcomes) that make up the event. We can now compute the probability that the project will take 10 months or less to complete. Because this event is given by $C = \{(2, 6), (2, 7), (2, 8), (3, 6), (3, 7), (4, 6)\}$, the probability of event C, denoted $P(C)$, is given by

$$P(C) = P(2, 6) + P(2, 7) + P(2, 8) + P(3, 6) + P(3, 7) + P(4, 6)$$

Refer to the sample point probabilities in Table 4.3; we have

$$P(C) = .15 + .15 + .05 + .10 + .20 + .05 = .70$$

Similarly, because the event that the project is completed in less than 10 months is given by $L = \{(2, 6), (2, 7), (3, 6)\}$, the probability of this event is given by

$$P(L) = P(2, 6) + P(2, 7) + P(3, 6)$$
$$= .15 + .15 + .10 = .40$$

Finally, for the event that the project is completed in more than 10 months, we have $M = \{(3, 8), (4, 7), (4, 8)\}$ and thus

$$P(M) = P(3, 8) + P(4, 7) + P(4, 8)$$
$$= .05 + .10 + .15 = .30$$

Using these probability results, we can now tell KP&L management that there is a .70 probability that the project will be completed in 10 months or less, a .40 probability that the project will be completed in less than 10 months, and a .30 probability that the project will be completed in more than 10 months. This procedure of computing event probabilities can be repeated for any event of interest to the KP&L management.

Any time that we can identify all the sample points of an experiment and assign probabilities to each, we can compute the probability of an event using the definition. However, in many experiments the large number of sample points makes the identification of the sample points, as well as the determination of their associated probabilities, extremely cumbersome, if not impossible. In the remaining sections of this chapter, we present some basic probability relationships that can be used to compute the probability of an event without knowledge of all the sample point probabilities.

NOTES AND COMMENTS

1. The sample space, S, is an event. Because it contains all the experimental outcomes, it has a probability of 1; that is, $P(S) = 1$.
2. When the classical method is used to assign probabilities, the assumption is that the experimental outcomes are equally likely. In such cases, the probability of an event can be computed by counting the number of experimental outcomes in the event and dividing the result by the total number of experimental outcomes.

Exercises

Methods

14. An experiment has four equally likely outcomes: E_1, E_2, E_3, and E_4.
    a. What is the probability that E_2 occurs?
    b. What is the probability that any two of the outcomes occur (e.g., E_1 or E_3)?
    c. What is the probability that any three of the outcomes occur (e.g., E_1 or E_2 or E_4)?

15. Consider the experiment of selecting a playing card from a deck of 52 playing cards. Each card corresponds to a sample point with a 1/52 probability.
    a. List the sample points in the event an ace is selected.
    b. List the sample points in the event a club is selected.
    c. List the sample points in the event a face card (jack, queen, or king) is selected.
    d. Find the probabilities associated with each of the events in parts (a), (b), and (c).

16. Consider the experiment of rolling a pair of dice. Suppose that we are interested in the sum of the face values showing on the dice.
    a. How many sample points are possible? (*Hint:* Use the counting rule for multiple-step experiments.)
    b. List the sample points.
    c. What is the probability of obtaining a value of 7?
    d. What is the probability of obtaining a value of 9 or greater?
    e. Because each roll has six possible even values (2, 4, 6, 8, 10, and 12) and only five possible odd values (3, 5, 7, 9, and 11), the dice should show even values more often than odd values. Do you agree with this statement? Explain.
    f. What method did you use to assign the probabilities requested?

Applications

17. Refer to the KP&L sample points and sample point probabilities in Tables 4.2 and 4.3.
    a. The design stage (stage 1) will run over budget if it takes 4 months to complete. List the sample points in the event the design stage is over budget.
    b. What is the probability that the design stage is over budget?
    c. The construction stage (stage 2) will run over budget if it takes 8 months to complete. List the sample points in the event the construction stage is over budget.
    d. What is the probability that the construction stage is over budget?
    e. What is the probability that both stages are over budget?

18. To investigate how often families eat at home, Harris Interactive surveyed 496 adults living with children under the age of 18 (*USA Today*, January 3, 2007). The survey results are shown in the following table.

Number of Family Meals per Week	Number of Survey Responses
0	11
1	11
2	30
3	36
4	36
5	119
6	114
7 or more	139

    For a randomly selected family with children under the age of 18, compute the following.
    a. The probability that the family eats no meals at home during the week.
    b. The probability that the family eats at least four meals at home during the week.
    c. The probability that the family eats two or fewer meals at home during the week.

19. Do you think the government protects investors adequately? This question was part of an online survey of investors under age 65 living in the United States and Great Britain (*Financial Times*/Harris Poll, October 1, 2009). The number of investors from the United States and the number of investors from Great Britain who answered Yes, No, or Unsure to this question are provided as follows.

Response	United States	Great Britain
Yes	187	197
No	334	411
Unsure	256	213

    a. Estimate the probability that an investor living in the United States thinks the government is not protecting investors adequately.
    b. Estimate the probability that an investor living in Great Britain thinks the government is not protecting investors adequately or is unsure the government is protecting investors adequately.
    c. For a randomly selected investor from these two countries, estimate the probability that the investor thinks the government is not protecting investors adequately.
    d. Based upon the survey results, does there appear to be much difference between the perceptions of investors living in the United States and investors living in Great Britain regarding the issue of the government protecting investors adequately?

20. *Fortune* magazine publishes an annual list of the 500 largest companies in the United States. The following data show the five states with the largest number of *Fortune* 500 companies (*The New York Times Almanac, 2006*).

State	Number of Companies
New York	54
California	52
Texas	48
Illinois	33
Ohio	30

Suppose a *Fortune* 500 company is chosen for a follow-up questionnaire. What are the probabilities of the following events?
a. Let N be the event the company is headquartered in New York. Find $P(N)$.
b. Let T be the event the company is headquartered in Texas. Find $P(T)$.
c. Let B be the event the company is headquartered in one of these five states. Find $P(B)$.

21. The U.S. adult population by age is as follows (*The World Almanac, 2009*). The data are in millions of people.

Age	Number
18 to 24	29.8
25 to 34	40.0
35 to 44	43.4
45 to 54	43.9
55 to 64	32.7
65 and over	37.8

Assume that a person will be randomly chosen from this population.
a. What is the probability that the person is 18 to 24 years old?
b. What is the probability that the person is 18 to 34 years old?
c. What is the probability that the person is 45 or older?

4.3 Some Basic Relationships of Probability

Complement of an Event

Given an event A, the **complement of A** is defined to be the event consisting of all sample points that are *not* in A. The complement of A is denoted by A^c. Figure 4.4 is a diagram, known as a **Venn diagram**, which illustrates the concept of a complement. The rectangular area represents the sample space for the experiment and as such contains all possible sample points. The circle represents event A and contains only the sample points that belong to A. The shaded region of the rectangle contains all sample points not in event A and is by definition the complement of A.

In any probability application, either event A or its complement A^c must occur. Therefore, we have

$$P(A) + P(A^c) = 1$$

4.3 Some Basic Relationships of Probability

FIGURE 4.4 COMPLEMENT OF EVENT A IS SHADED

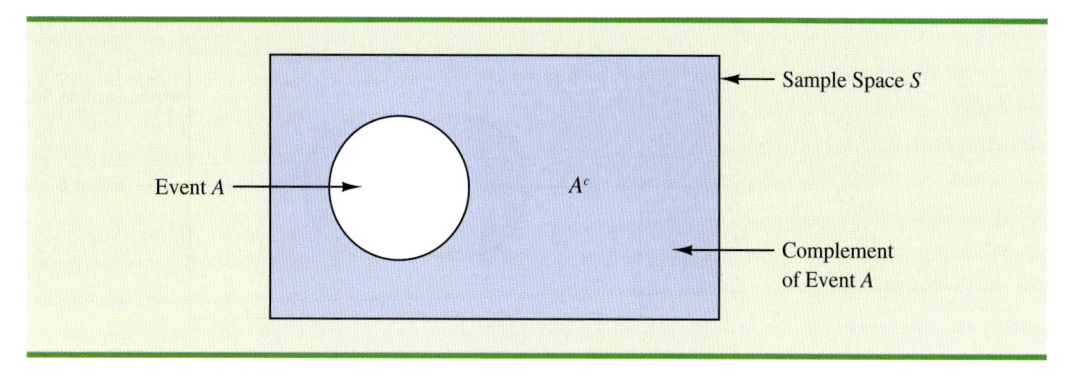

[Venn diagram showing Sample Space S as a rectangle, Event A as a white circle inside, and the shaded region outside the circle labeled A^c as the Complement of Event A.]

Solving for $P(A)$, we obtain the following result.

COMPUTING PROBABILITY USING THE COMPLEMENT

$$P(A) = 1 - P(A^c) \qquad (4.5)$$

Equation (4.5) shows that the probability of an event A can be computed easily if the probability of its complement, $P(A^c)$, is known.

As an example, consider the case of a sales manager who, after reviewing sales reports, states that 80% of new customer contacts result in no sale. By allowing A to denote the event of a sale and A^c to denote the event of no sale, the manager is stating that $P(A^c) = .80$. Using equation (4.5), we see that

$$P(A) = 1 - P(A^c) = 1 - .80 = .20$$

We can conclude that a new customer contact has a .20 probability of resulting in a sale.

In another example, a purchasing agent states a .90 probability that a supplier will send a shipment that is free of defective parts. Using the complement, we can conclude that there is a $1 - .90 = .10$ probability that the shipment will contain defective parts.

Addition Law

The addition law is helpful when we are interested in knowing the probability that at least one of two events occurs. That is, with events A and B we are interested in knowing the probability that event A or event B or both occur.

Before we present the addition law, we need to discuss two concepts related to the combination of events: the *union* of events and the *intersection* of events. Given two events A and B, the **union of A and B** is defined as follows.

UNION OF TWO EVENTS

The *union* of A and B is the event containing *all* sample points belonging to A or B or both. The union is denoted by $A \cup B$.

The Venn diagram in Figure 4.5 depicts the union of events A and B. Note that the two circles contain all the sample points in event A as well as all the sample points in event B.

FIGURE 4.5 UNION OF EVENTS *A* AND *B* IS SHADED

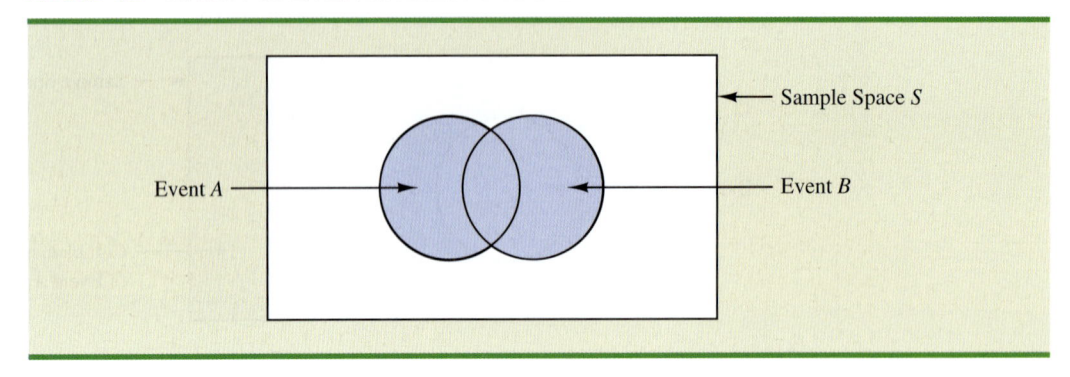

[Venn diagram: Sample Space S with Event A and Event B as overlapping circles, both shaded]

The fact that the circles overlap indicates that some sample points are contained in both *A* and *B*.

The definition of the **intersection of *A* and *B*** follows.

INTERSECTION OF TWO EVENTS

Given two events *A* and *B*, the *intersection* of *A* and *B* is the event containing the sample points belonging to *both A and B*. The intersection is denoted by $A \cap B$.

The Venn diagram depicting the intersection of events *A* and *B* is shown in Figure 4.6. The area where the two circles overlap is the intersection; it contains the sample points that are in both *A* and *B*.

Let us now continue with a discussion of the addition law. The **addition law** provides a way to compute the probability that event *A* or event *B* or both occur. In other words, the addition law is used to compute the probability of the union of two events. The addition law is written as follows.

ADDITION LAW

$$P(A \cup B) = P(A) + P(B) - P(A \cap B) \tag{4.6}$$

FIGURE 4.6 INTERSECTION OF EVENTS *A* AND *B* IS SHADED

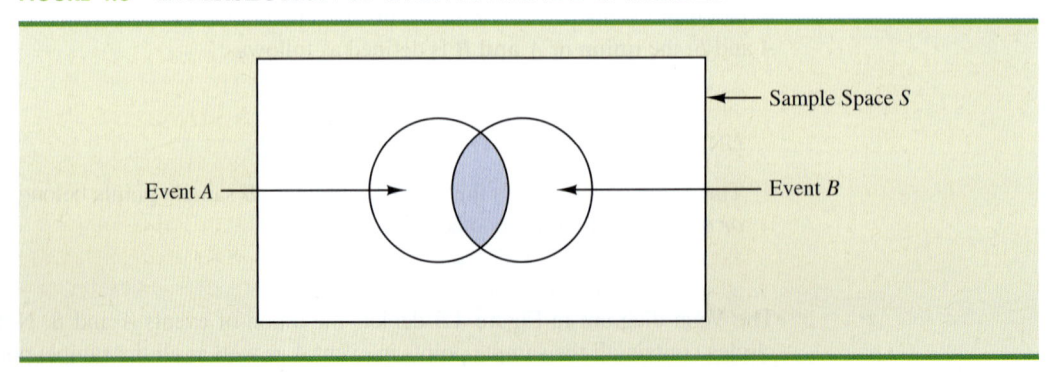

[Venn diagram: Sample Space S with Event A and Event B as overlapping circles, only the intersection shaded]

4.3 Some Basic Relationships of Probability

To understand the addition law intuitively, note that the first two terms in the addition law, $P(A) + P(B)$, account for all the sample points in $A \cup B$. However, because the sample points in the intersection $A \cap B$ are in both A and B, when we compute $P(A) + P(B)$, we are in effect counting each of the sample points in $A \cap B$ twice. We correct for this overcounting by subtracting $P(A \cap B)$.

As an example of an application of the addition law, let us consider the case of a small assembly plant with 50 employees. Each worker is expected to complete work assignments on time and in such a way that the assembled product will pass a final inspection. On occasion, some of the workers fail to meet the performance standards by completing work late or assembling a defective product. At the end of a performance evaluation period, the production manager found that 5 of the 50 workers completed work late, 6 of the 50 workers assembled a defective product, and 2 of the 50 workers both completed work late *and* assembled a defective product.

Let

$$L = \text{the event that the work is completed late}$$
$$D = \text{the event that the assembled product is defective}$$

The relative frequency information leads to the following probabilities.

$$P(L) = \frac{5}{50} = .10$$

$$P(D) = \frac{6}{50} = .12$$

$$P(L \cap D) = \frac{2}{50} = .04$$

After reviewing the performance data, the production manager decided to assign a poor performance rating to any employee whose work was either late or defective; thus the event of interest is $L \cup D$. What is the probability that the production manager assigned an employee a poor performance rating?

Note that the probability question is about the union of two events. Specifically, we want to know $P(L \cup D)$. Using equation (4.6), we have

$$P(L \cup D) = P(L) + P(D) - P(L \cap D)$$

Knowing values for the three probabilities on the right side of this expression, we can write

$$P(L \cup D) = .10 + .12 - .04 = .18$$

This calculation tells us that there is a .18 probability that a randomly selected employee received a poor performance rating.

As another example of the addition law, consider a recent study conducted by the personnel manager of a major computer software company. The study showed that 30% of the employees who left the firm within two years did so primarily because they were dissatisfied with their salary, 20% left because they were dissatisfied with their work assignments, and 12% of the former employees indicated dissatisfaction with *both* their salary and their work assignments. What is the probability that an employee who leaves within two years does so because of dissatisfaction with salary, dissatisfaction with the work assignment, or both?

Let

S = the event that the employee leaves because of salary
W = the event that the employee leaves because of work assignment

We have $P(S) = .30$, $P(W) = .20$, and $P(S \cap W) = .12$. Using equation (4.6), the addition law, we have

$$P(S \cup W) = P(S) + P(W) - P(S \cap W) = .30 + .20 - .12 = .38$$

We find a .38 probability that an employee leaves for salary or work assignment reasons.

Before we conclude our discussion of the addition law, let us consider a special case that arises for **mutually exclusive events**.

MUTUALLY EXCLUSIVE EVENTS

Two events are said to be mutually exclusive if the events have no sample points in common.

Events A and B are mutually exclusive if, when one event occurs, the other cannot occur. Thus, a requirement for A and B to be mutually exclusive is that their intersection must contain no sample points. The Venn diagram depicting two mutually exclusive events A and B is shown in Figure 4.7. In this case $P(A \cap B) = 0$ and the addition law can be written as follows.

ADDITION LAW FOR MUTUALLY EXCLUSIVE EVENTS

$$P(A \cup B) = P(A) + P(B)$$

FIGURE 4.7 MUTUALLY EXCLUSIVE EVENTS

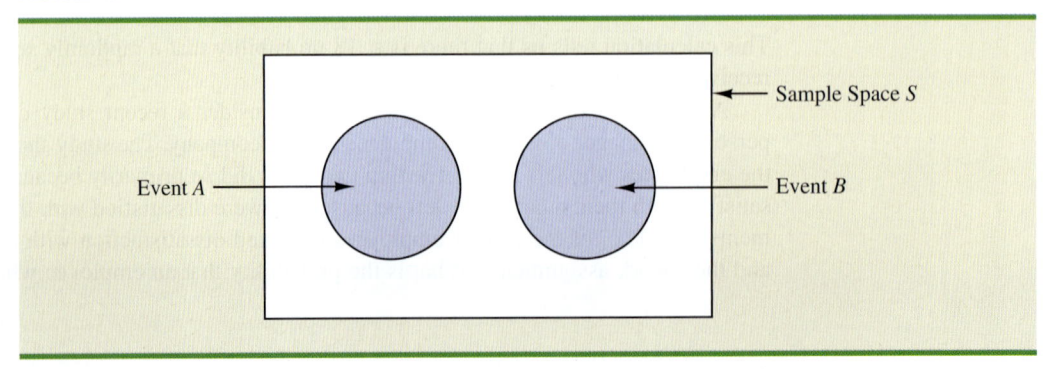

4.3 Some Basic Relationships of Probability

Exercises

Methods

22. Suppose that we have a sample space with five equally likely experimental outcomes: E_1, E_2, E_3, E_4, E_5. Let

$$A = \{E_1, E_2\}$$
$$B = \{E_3, E_4\}$$
$$C = \{E_2, E_3, E_5\}$$

  a. Find $P(A)$, $P(B)$, and $P(C)$.
  b. Find $P(A \cup B)$. Are A and B mutually exclusive?
  c. Find A^c, C^c, $P(A^c)$, and $P(C^c)$.
  d. Find $A \cup B^c$ and $P(A \cup B^c)$.
  e. Find $P(B \cup C)$.

SELF test

23. Suppose that we have a sample space $S = \{E_1, E_2, E_3, E_4, E_5, E_6, E_7\}$, where $E_1, E_2, \ldots, E_7$ denote the sample points. The following probability assignments apply: $P(E_1) = .05$, $P(E_2) = .20$, $P(E_3) = .20$, $P(E_4) = .25$, $P(E_5) = .15$, $P(E_6) = .10$, and $P(E_7) = .05$. Let

$$A = \{E_1, E_4, E_6\}$$
$$B = \{E_2, E_4, E_7\}$$
$$C = \{E_2, E_3, E_5, E_7\}$$

  a. Find $P(A)$, $P(B)$, and $P(C)$.
  b. Find $A \cup B$ and $P(A \cup B)$.
  c. Find $A \cap B$ and $P(A \cap B)$.
  d. Are events A and C mutually exclusive?
  e. Find B^c and $P(B^c)$.

Applications

24. Clarkson University surveyed alumni to learn more about what they think of Clarkson. One part of the survey asked respondents to indicate whether their overall experience at Clarkson fell short of expectations, met expectations, or surpassed expectations. The results showed that 4% of the respondents did not provide a response, 26% said that their experience fell short of expectations, and 65% of the respondents said that their experience met expectations.
  a. If we chose an alumnus at random, what is the probability that the alumnus would say their experience *surpassed* expectations?
  b. If we chose an alumnus at random, what is the probability that the alumnus would say their experience met or surpassed expectations?

25. The U.S. Census Bureau provides data on the number of young adults, ages 18–24, who are living in their parents' home.[1] Let

M = the event a male young adult is living in his parents' home
F = the event a female young adult is living in her parents' home

  If we randomly select a male young adult and a female young adult, the Census Bureau data enable us to conclude $P(M) = .56$ and $P(F) = .42$ (*The World Almanac*, 2006). The probability that both are living in their parents' home is .24.
  a. What is the probability at least one of the two young adults selected is living in his or her parents' home?
  b. What is the probability both young adults selected are living on their own (neither is living in their parents' home)?

---

[1] The data include single young adults who are living in college dormitories because it is assumed these young adults will return to their parents' home when school is not in session.

26. Information about mutual funds provided by Morningstar Investment Research includes the type of mutual fund (Domestic Equity, International Equity, or Fixed Income) and the Morningstar rating for the fund. The rating is expressed from 1-star (lowest rating) to 5-star (highest rating). A sample of 25 mutual funds was selected from *Morningstar Funds 500* (2008). The following counts were obtained:
    - Sixteen mutual funds were Domestic Equity funds.
    - Thirteen mutual funds were rated 3-star or less.
    - Seven of the Domestic Equity funds were rated 4-star.
    - Two of the Domestic Equity funds were rated 5-star.

    Assume that one of these 25 mutual funds will be randomly selected in order to learn more about the mutual fund and its investment strategy.
    a. What is the probability of selecting a Domestic Equity fund?
    b. What is the probability of selecting a fund with a 4-star or 5-star rating?
    c. What is the probability of selecting a fund that is both a Domestic Equity fund *and* a fund with a 4-star or 5-star rating?
    d. What is the probability of selecting a fund that is a Domestic Equity fund *or* a fund with a 4-star or 5-star rating?

27. What NCAA college basketball conferences have the higher probability of having a team play in college basketball's national championship game? Over the last 20 years, the Atlantic Coast Conference (ACC) ranks first by having a team in the championship game 10 times. The Southeastern Conference (SEC) ranks second by having a team in the championship game 8 times. However, these two conferences have both had teams in the championship game only one time, when Arkansas (SEC) beat Duke (ACC) 76–70 in 1994 (NCAA website, April 2009). Use these data to estimate the following probabilities.
    a. What is the probability that the ACC will have a team in the championship game?
    b. What is the probability that the SEC will have team in the championship game?
    c. What is the probability that the ACC and SEC will both have teams in the championship game?
    d. What is the probability that at least one team from these two conferences will be in the championship game? That is, what is the probability a team from the ACC or SEC will play in the championship game?
    e. What is the probability that the championship game will not a have team from one of these two conferences?

28. A survey of magazine subscribers showed that 45.8% rented a car during the past 12 months for business reasons, 54% rented a car during the past 12 months for personal reasons, and 30% rented a car during the past 12 months for both business and personal reasons.
    a. What is the probability that a subscriber rented a car during the past 12 months for business or personal reasons?
    b. What is the probability that a subscriber did not rent a car during the past 12 months for either business or personal reasons?

29. High school seniors with strong academic records apply to the nation's most selective colleges in greater numbers each year. Because the number of slots remains relatively stable, some colleges reject more early applicants. Suppose that for a recent admissions class, an Ivy League college received 2851 applications for early admission. Of this group, it admitted 1033 students early, rejected 854 outright, and deferred 964 to the regular admission pool for further consideration. In the past, this school has admitted 18% of the deferred early admission applicants during the regular admission process. Counting the students admitted early and the students admitted during the regular admission process, the total class size was 2375. Let E, R, and D represent the events that a student who applies for early admission is admitted early, rejected outright, or deferred to the regular admissions pool.
    a. Use the data to estimate $P(E)$, $P(R)$, and $P(D)$.
    b. Are events E and D mutually exclusive? Find $P(E \cap D)$.

c. For the 2375 students who were admitted, what is the probability that a randomly selected student was accepted during early admission?
d. Suppose a student applies for early admission. What is the probability that the student will be admitted for early admission or be deferred and later admitted during the regular admission process?

4.4 Conditional Probability

Often, the probability of an event is influenced by whether a related event already occurred. Suppose we have an event A with probability $P(A)$. If we obtain new information and learn that a related event, denoted by B, already occurred, we will want to take advantage of this information by calculating a new probability for event A. This new probability of event A is called a **conditional probability** and is written $P(A \mid B)$. We use the notation | to indicate that we are considering the probability of event A *given* the condition that event B has occurred. Hence, the notation $P(A \mid B)$ reads "the probability of A given B."

As an illustration of the application of conditional probability, consider the situation of the promotion status of male and female officers of a major metropolitan police force in the eastern United States. The police force consists of 1200 officers, 960 men and 240 women. Over the past two years, 324 officers on the police force received promotions. The specific breakdown of promotions for male and female officers is shown in Table 4.4.

After reviewing the promotion record, a committee of female officers raised a discrimination case on the basis that 288 male officers had received promotions but only 36 female officers had received promotions. The police administration argued that the relatively low number of promotions for female officers was due not to discrimination, but to the fact that relatively few females are members of the police force. Let us show how conditional probability could be used to analyze the discrimination charge.

Let

$$M = \text{event an officer is a man}$$
$$W = \text{event an officer is a woman}$$
$$A = \text{event an officer is promoted}$$
$$A^c = \text{event an officer is not promoted}$$

Dividing the data values in Table 4.4 by the total of 1200 officers enables us to summarize the available information with the following probability values.

$P(M \cap A) = 288/1200 = .24$ probability that a randomly selected officer is a man *and* is promoted

$P(M \cap A^c) = 672/1200 = .56$ probability that a randomly selected officer is a man *and* is not promoted

TABLE 4.4 PROMOTION STATUS OF POLICE OFFICERS OVER THE PAST TWO YEARS

	Men	Women	Total
Promoted	288	36	324
Not Promoted	672	204	876
Total	960	240	1200

TABLE 4.5 JOINT PROBABILITY TABLE FOR PROMOTIONS

	Men (*M*)	Women (*W*)	Total
Promoted (*A*)	.24	.03	.27
Not Promoted (*A^c*)	.56	.17	.73
Total	.80	.20	1.00

Joint probabilities appear in the body of the table.

Marginal probabilities appear in the margins of the table.

$$P(W \cap A) = 36/1200 = .03 \text{ probability that a randomly selected officer is a woman } and \text{ is promoted}$$

$$P(W \cap A^c) = 204/1200 = .17 \text{ probability that a randomly selected officer is a woman } and \text{ is not promoted}$$

Because each of these values gives the probability of the intersection of two events, the probabilities are called **joint probabilities**. Table 4.5, which provides a summary of the probability information for the police officer promotion situation, is referred to as a *joint probability table*.

The values in the margins of the joint probability table provide the probabilities of each event separately. That is, $P(M) = .80$, $P(W) = .20$, $P(A) = .27$, and $P(A^c) = .73$. These probabilities are referred to as **marginal probabilities** because of their location in the margins of the joint probability table. We note that the marginal probabilities are found by summing the joint probabilities in the corresponding row or column of the joint probability table. For instance, the marginal probability of being promoted is $P(A) = P(M \cap A) + P(W \cap A) = .24 + .03 = .27$. From the marginal probabilities, we see that 80% of the force is male, 20% of the force is female, 27% of all officers received promotions, and 73% were not promoted.

Let us begin the conditional probability analysis by computing the probability that an officer is promoted given that the officer is a man. In conditional probability notation, we are attempting to determine $P(A \mid M)$. To calculate $P(A \mid M)$, we first realize that this notation simply means that we are considering the probability of the event *A* (promotion) given that the condition designated as event *M* (the officer is a man) is known to exist. Thus $P(A \mid M)$ tells us that we are now concerned only with the promotion status of the 960 male officers. Because 288 of the 960 male officers received promotions, the probability of being promoted given that the officer is a man is $288/960 = .30$. In other words, given that an officer is a man, that officer had a 30% chance of receiving a promotion over the past two years.

This procedure was easy to apply because the values in Table 4.4 show the number of officers in each category. We now want to demonstrate how conditional probabilities such as $P(A \mid M)$ can be computed directly from related event probabilities rather than the frequency data of Table 4.4.

We have shown that $P(A \mid M) = 288/960 = .30$. Let us now divide both the numerator and denominator of this fraction by 1200, the total number of officers in the study.

$$P(A \mid M) = \frac{288}{960} = \frac{288/1200}{960/1200} = \frac{.24}{.80} = .30$$

We now see that the conditional probability $P(A \mid M)$ can be computed as $.24/.80$. Refer to the joint probability table (Table 4.5). Note in particular that .24 is the joint probability of

4.4 Conditional Probability

A and M; that is, $P(A \cap M) = .24$. Also note that .80 is the marginal probability that a randomly selected officer is a man; that is, $P(M) = .80$. Thus, the conditional probability $P(A \mid M)$ can be computed as the ratio of the joint probability $P(A \cap M)$ to the marginal probability $P(M)$.

$$P(A \mid M) = \frac{P(A \cap M)}{P(M)} = \frac{.24}{.80} = .30$$

The fact that conditional probabilities can be computed as the ratio of a joint probability to a marginal probability provides the following general formula for conditional probability calculations for two events A and B.

CONDITIONAL PROBABILITY

$$P(A \mid B) = \frac{P(A \cap B)}{P(B)} \tag{4.7}$$

or

$$P(B \mid A) = \frac{P(A \cap B)}{P(A)} \tag{4.8}$$

The Venn diagram in Figure 4.8 is helpful in obtaining an intuitive understanding of conditional probability. The circle on the right shows that event B has occurred; the portion of the circle that overlaps with event A denotes the event $(A \cap B)$. We know that once event B has occurred, the only way that we can also observe event A is for the event $(A \cap B)$ to occur. Thus, the ratio $P(A \cap B)/P(B)$ provides the conditional probability that we will observe event A given that event B has already occurred.

Let us return to the issue of discrimination against the female officers. The marginal probability in row 1 of Table 4.5 shows that the probability of promotion of an officer is $P(A) = .27$ (regardless of whether that officer is male or female). However, the critical issue in the discrimination case involves the two conditional probabilities $P(A \mid M)$ and $P(A \mid W)$. That is, what is the probability of a promotion *given* that the officer is a man, and what is the probability of a promotion *given* that the officer is a woman? If these two probabilities are equal, a discrimination argument has no basis because the chances of a promotion are the same for male and female officers. However, a difference in the two conditional probabilities will support the position that male and female officers are treated differently in promotion decisions.

FIGURE 4.8 CONDITIONAL PROBABILITY $P(A \mid B) = P(A \cap B)/P(B)$

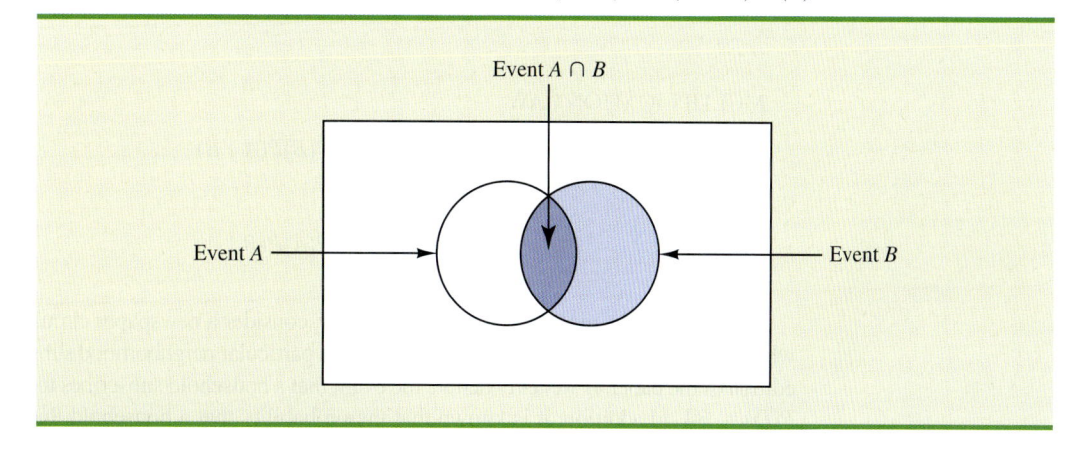

We already determined that $P(A \mid M) = .30$. Let us now use the probability values in Table 4.5 and the basic relationship of conditional probability in equation (4.7) to compute the probability that an officer is promoted given that the officer is a woman; that is, $P(A \mid W)$. Using equation (4.7), with W replacing B, we obtain

$$P(A \mid W) = \frac{P(A \cap W)}{P(W)} = \frac{.03}{.20} = .15$$

What conclusion do you draw? The probability of a promotion given that the officer is a man is .30, twice the .15 probability of a promotion given that the officer is a woman. Although the use of conditional probability does not in itself prove that discrimination exists in this case, the conditional probability values support the argument presented by the female officers.

Independent Events

In the preceding illustration, $P(A) = .27$, $P(A \mid M) = .30$, and $P(A \mid W) = .15$. We see that the probability of a promotion (event A) is affected or influenced by whether the officer is a man or a woman. Particularly, because $P(A \mid M) \neq P(A)$, we would say that events A and M are dependent events. That is, the probability of event A (promotion) is altered or affected by knowing that event M (the officer is a man) exists. Similarly, with $P(A \mid W) \neq P(A)$, we would say that events A and W are *dependent events*. However, if the probability of event A is not changed by the existence of event M—that is, $P(A \mid M) = P(A)$—we would say that events A and M are **independent events**. This situation leads to the following definition of the independence of two events.

INDEPENDENT EVENTS

Two events A and B are independent if

$$P(A \mid B) = P(A) \tag{4.9}$$

or

$$P(B \mid A) = P(B) \tag{4.10}$$

Otherwise, the events are dependent.

Multiplication Law

Whereas the addition law of probability is used to compute the probability of a union of two events, the multiplication law is used to compute the probability of the intersection of two events. The multiplication law is based on the definition of conditional probability. Using equations (4.7) and (4.8) and solving for $P(A \cap B)$, we obtain the **multiplication law**.

MULTIPLICATION LAW

$$P(A \cap B) = P(B)P(A \mid B) \tag{4.11}$$

or

$$P(A \cap B) = P(A)P(B \mid A) \tag{4.12}$$

To illustrate the use of the multiplication law, consider a newspaper circulation department where it is known that 84% of the households in a particular neighborhood subscribe to the daily edition of the paper. If we let D denote the event that a household subscribes to the daily edition, $P(D) = .84$. In addition, it is known that the probability that a household that already holds a daily subscription also subscribes to the Sunday edition (event S) is .75; that is, $P(S \mid D) = .75$. What is the probability that a household subscribes to both the Sunday and daily editions of the newspaper? Using the multiplication law, we compute the desired $P(S \cap D)$ as

$$P(S \cap D) = P(D)P(S \mid D) = .84(.75) = .63$$

We now know that 63% of the households subscribe to both the Sunday and daily editions.

Before concluding this section, let us consider the special case of the multiplication law when the events involved are independent. Recall that events A and B are independent whenever $P(A \mid B) = P(A)$ or $P(B \mid A) = P(B)$. Hence, using equations (4.11) and (4.12) for the special case of independent events, we obtain the following multiplication law.

> **MULTIPLICATION LAW FOR INDEPENDENT EVENTS**
>
> $$P(A \cap B) = P(A)P(B) \tag{4.13}$$

To compute the probability of the intersection of two independent events, we simply multiply the corresponding probabilities. Note that the multiplication law for independent events provides another way to determine whether A and B are independent. That is, if $P(A \cap B) = P(A)P(B)$, then A and B are independent; if $P(A \cap B) \neq P(A)P(B)$, then A and B are dependent.

As an application of the multiplication law for independent events, consider the situation of a service station manager who knows from past experience that 80% of the customers use a credit card when they purchase gasoline. What is the probability that the next two customers purchasing gasoline will each use a credit card? If we let

$A = $ the event that the first customer uses a credit card
$B = $ the event that the second customer uses a credit card

then the event of interest is $A \cap B$. Given no other information, we can reasonably assume that A and B are independent events. Thus,

$$P(A \cap B) = P(A)P(B) = (.80)(.80) = .64$$

To summarize this section, we note that our interest in conditional probability is motivated by the fact that events are often related. In such cases, we say the events are dependent and the conditional probability formulas in equations (4.7) and (4.8) must be used to compute the event probabilities. If two events are not related, they are independent; in this case neither event's probability is affected by whether the other event occurred.

NOTES AND COMMENTS

Do not confuse the notion of mutually exclusive events with that of independent events. Two events with nonzero probabilities cannot be both mutually exclusive and independent. If one mutually exclusive event is known to occur, the other cannot occur; thus, the probability of the other event occurring is reduced to zero. They are therefore dependent.

Exercises

Methods

30. Suppose that we have two events, A and B, with $P(A) = .50$, $P(B) = .60$, and $P(A \cap B) = .40$.
    a. Find $P(A \mid B)$.
    b. Find $P(B \mid A)$.
    c. Are A and B independent? Why or why not?

31. Assume that we have two events, A and B, that are mutually exclusive. Assume further that we know P(A) = .30 and P(B) = .40.
    a. What is P(A ∩ B)?
    b. What is P(A | B)?
    c. A student in statistics argues that the concepts of mutually exclusive events and independent events are really the same, and that if events are mutually exclusive they must be independent. Do you agree with this statement? Use the probability information in this problem to justify your answer.
    d. What general conclusion would you make about mutually exclusive and independent events given the results of this problem?

Applications

32. The automobile industry sold 657,000 vehicles in the United States during January 2009 (*The Wall Street Journal*, February 4, 2009). This volume was down 37% from January 2008 as economic conditions continued to decline. The Big Three U.S. automakers—General Motors, Ford, and Chrysler—sold 280,500 vehicles, down 48% from January 2008. A summary of sales by automobile manufacturer and type of vehicle sold is shown in the following table. Data are in thousands of vehicles. The non-U.S. manufacturers are led by Toyota, Honda, and Nissan. The category Light Truck includes pickup, minivan, SUV, and crossover models.

		Type of Vehicle	
		Car	Light Truck
Manufacturer	U.S.	87.4	193.1
	Non-U.S.	228.5	148.0

   a. Develop a joint probability table for these data and use the table to answer the remaining questions.
   b. What are the marginal probabilities? What do they tell you about the probabilities associated with the manufacturer and the type of vehicle sold?
   c. If a vehicle was manufactured by one of the U.S. automakers, what is the probability that the vehicle was a car? What is the probability that it was a light truck?
   d. If a vehicle was not manufactured by one of the U.S. automakers, what is the probability that the vehicle was a car? What is the probability that it was a light truck?
   e. If the vehicle was a light truck, what is the probability that it was manufactured by one of the U.S. automakers?
   f. What does the probability information tell you about sales?

33. In a survey of MBA students, the following data were obtained on "students' first reason for application to the school in which they matriculated."

		Reason for Application			
		School Quality	School Cost or Convenience	Other	Totals
Enrollment Status	Full Time	421	393	76	890
	Part Time	400	593	46	1039
	Totals	821	986	122	1929

   a. Develop a joint probability table for these data.
   b. Use the marginal probabilities of school quality, school cost or convenience, and other to comment on the most important reason for choosing a school.
   c. If a student goes full time, what is the probability that school quality is the first reason for choosing a school?

d. If a student goes part time, what is the probability that school quality is the first reason for choosing a school?
e. Let A denote the event that a student is full time and let B denote the event that the student lists school quality as the first reason for applying. Are events A and B independent? Justify your answer.

34. The U.S. Department of Transportation reported that during November, 83.4% of Southwest Airlines' flights, 75.1% of US Airways' flights, and 70.1% of JetBlue's flights arrived on time (*USA Today,* January 4, 2007). Assume that this on-time performance is applicable for flights arriving at concourse A of the Rochester International Airport, and that 40% of the arrivals at concourse A are Southwest Airlines flights, 35% are US Airways flights, and 25% are JetBlue flights.
    a. Develop a joint probability table with three rows (airlines) and two columns (on-time arrivals vs. late arrivals).
    b. An announcement has just been made that Flight 1424 will be arriving at gate 20 in concourse A. What is the most likely airline for this arrival?
    c. What is the probability that Flight 1424 will arrive on time?
    d. Suppose that an announcement is made saying that Flight 1424 will be arriving late. What is the most likely airline for this arrival? What is the least likely airline?

35. According to the Ameriprise Financial Money Across Generations study, 9 out of 10 parents with adult children ages 20 to 35 have helped their adult children with some type of financial assistance ranging from college, a car, rent, utilities, credit-card debt, and/or down payments for houses (*Money,* January 2009). The following table with sample data consistent with the study shows the number of times parents have given their adult children financial assistance to buy a car and to pay rent.

		Pay Rent	
		Yes	No
Buy a Car	Yes	56	52
	No	14	78

   a. Develop a joint probability table and use it to answer the remaining questions.
   b. Using the marginal probabilities for buy a car and pay rent, are parents more likely to assist their adult children with buying a car or paying rent? What is your interpretation of the marginal probabilities?
   c. If parents provided financial assistance to buy a car, what it the probability that the parents assisted with paying rent?
   d. If parents did not provide financial assistance to buy a car, what is the probability that the parents assisted with paying rent?
   e. Is financial assistance to buy a car independent of financial assistance to pay rent? Use probabilities to justify your answer.
   f. What is the probability that parents provided financial assistance for their adult children by either helping buy a car or paying rent?

36. Jerry Stackhouse of the National Basketball Association's Dallas Mavericks is the best free-throw shooter on the team, making 89% of his shots (ESPN website, July, 2008). Assume that late in a basketball game, Jerry Stackhouse is fouled and is awarded two shots.
    a. What is the probability that he will make both shots?
    b. What is the probability that he will make at least one shot?
    c. What is the probability that he will miss both shots?
    d. Late in a basketball game, a team often intentionally fouls an opposing player in order to stop the game clock. The usual strategy is to intentionally foul the other team's worst free-throw shooter. Assume that the Dallas Mavericks' center makes 58% of his free-throw shots. Calculate the probabilities for the center as shown in parts (a), (b), and (c), and show that intentionally fouling the Dallas Mavericks' center is a better strategy than intentionally fouling Jerry Stackhouse.

37. Visa Card USA studied how frequently young consumers, ages 18 to 24, use plastic (debit and credit) cards in making purchases (Associated Press, January 16, 2006). The results of the study provided the following probabilities.
    - The probability that a consumer uses a plastic card when making a purchase is .37.
    - Given that the consumer uses a plastic card, there is a .19 probability that the consumer is 18 to 24 years old.
    - Given that the consumer uses a plastic card, there is a .81 probability that the consumer is more than 24 years old.

    U.S. Census Bureau data show that 14% of the consumer population is 18 to 24 years old.
    a. Given the consumer is 18 to 24 years old, what is the probability that the consumer uses a plastic card?
    b. Given the consumer is over 24 years old, what is the probability that the consumer uses a plastic card?
    c. What is the interpretation of the probabilities shown in parts (a) and (b)?
    d. Should companies such as Visa, MasterCard, and Discover make plastic cards available to the 18 to 24 year old age group before these consumers have had time to establish a credit history? If no, why? If yes, what restrictions might the companies place on this age group?

38. Students in grades 3 through 8 in New York State are required to take a state mathematics exam. To meet the state's proficiency standards, a student must demonstrate an understanding of the mathematics expected at his or her grade level. The following data show the number of students tested in the New York City school system for grades 3 through 8 and the number who met and did not meet the proficiency standards on the exam (New York City Department of Education website, January 16, 2010).

	Met Proficiency Standards?	
Grade	Yes	No
3	47,401	23,975
4	35,020	34,740
5	36,062	33,540
6	36,361	32,929
7	40,945	29,768
8	40,720	31,931

a. Develop a joint probability table for these data.
b. What are the marginal probabilities? What do they tell about the probabilities of meeting or not meeting the proficiency standards on the exam?
c. If a randomly selected student is a third grader, what is the probability that the student met the proficiency standards? If the student is a fourth grader, what is the probability that the student met the proficiency standards?
d. If a randomly selected student is known to have met the proficiency standards on the exam, what it the probability that the student is a third grader? What is the probability if the student is a fourth grader?

4.5 Bayes' Theorem

In the discussion of conditional probability, we indicated that revising probabilities when new information is obtained is an important phase of probability analysis. Often, we begin the analysis with initial or **prior probability** estimates for specific events of interest. Then, from sources such as a sample, a special report, or a product test, we obtain additional information about the events. Given this new information, we update the prior probability values by calculating revised probabilities, referred to as **posterior probabilities**. **Bayes' theorem** provides a means for making these probability calculations. The steps in this probability revision process are shown in Figure 4.9.

4.5 Bayes' Theorem

FIGURE 4.9 PROBABILITY REVISION USING BAYES' THEOREM

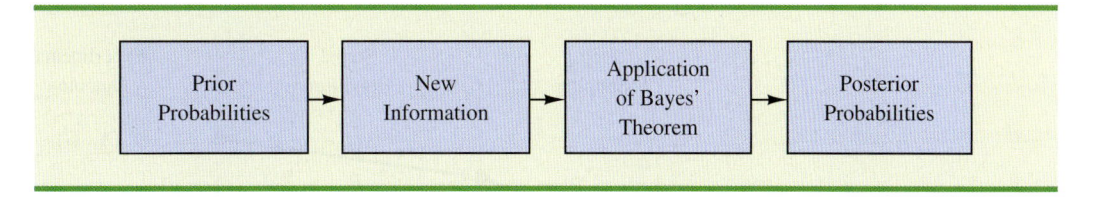

Prior Probabilities → New Information → Application of Bayes' Theorem → Posterior Probabilities

As an application of Bayes' theorem, consider a manufacturing firm that receives shipments of parts from two different suppliers. Let A_1 denote the event that a part is from supplier 1 and A_2 denote the event that a part is from supplier 2. Currently, 65% of the parts purchased by the company are from supplier 1 and the remaining 35% are from supplier 2. Hence, if a part is selected at random, we would assign the prior probabilities $P(A_1) = .65$ and $P(A_2) = .35$.

The quality of the purchased parts varies with the source of supply. Historical data suggest that the quality ratings of the two suppliers are as shown in Table 4.6. If we let G denote the event that a part is good and B denote the event that a part is bad, the information in Table 4.6 provides the following conditional probability values.

$$P(G \mid A_1) = .98 \quad P(B \mid A_1) = .02$$
$$P(G \mid A_2) = .95 \quad P(B \mid A_2) = .05$$

The tree diagram in Figure 4.10 depicts the process of the firm receiving a part from one of the two suppliers and then discovering that the part is good or bad as a two-step experiment. We see that four experimental outcomes are possible; two correspond to the part being good and two correspond to the part being bad.

Each of the experimental outcomes is the intersection of two events, so we can use the multiplication rule to compute the probabilities. For instance,

$$P(A_1, G) = P(A_1 \cap G) = P(A_1)P(G \mid A_1)$$

The process of computing these joint probabilities can be depicted in what is called a probability tree (see Figure 4.11). From left to right through the tree, the probabilities for each branch at step 1 are prior probabilities and the probabilities for each branch at step 2 are conditional probabilities. To find the probabilities of each experimental outcome, we simply multiply the probabilities on the branches leading to the outcome. Each of these joint probabilities is shown in Figure 4.11 along with the known probabilities for each branch.

Suppose now that the parts from the two suppliers are used in the firm's manufacturing process and that a machine breaks down because it attempts to process a bad part. Given the information that the part is bad, what is the probability that it came from supplier 1 and

TABLE 4.6 HISTORICAL QUALITY LEVELS OF TWO SUPPLIERS

	Percentage Good Parts	Percentage Bad Parts
Supplier 1	98	2
Supplier 2	95	5

FIGURE 4.10   TREE DIAGRAM FOR TWO-SUPPLIER EXAMPLE

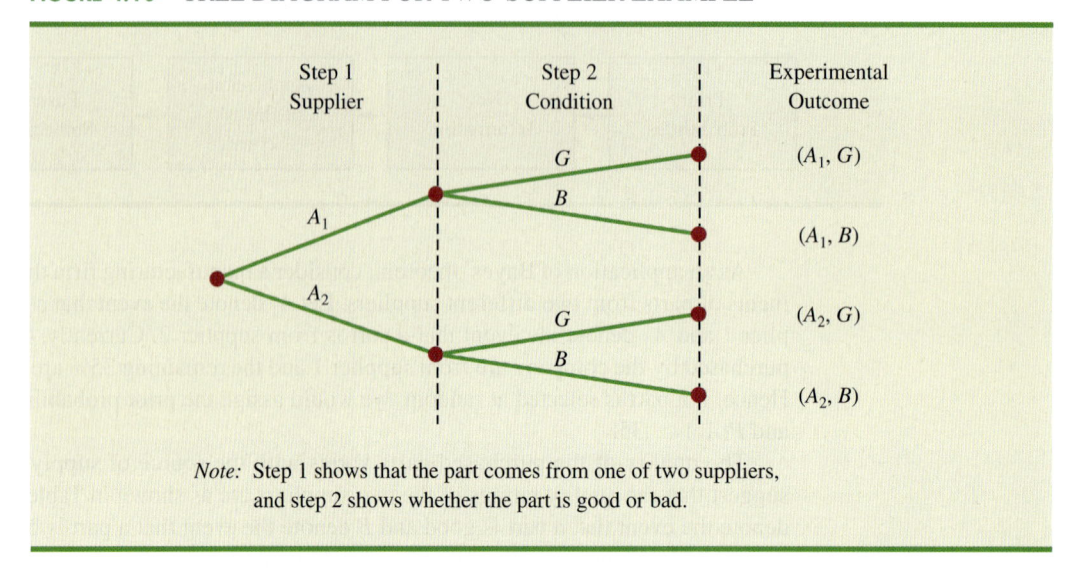

Note: Step 1 shows that the part comes from one of two suppliers, and step 2 shows whether the part is good or bad.

what is the probability that it came from supplier 2? With the information in the probability tree (Figure 4.11), Bayes' theorem can be used to answer these questions.

Letting B denote the event that the part is bad, we are looking for the posterior probabilities $P(A_1 \mid B)$ and $P(A_2 \mid B)$. From the law of conditional probability, we know that

$$P(A_1 \mid B) = \frac{P(A_1 \cap B)}{P(B)} \qquad (4.14)$$

Referring to the probability tree, we see that

$$P(A_1 \cap B) = P(A_1)P(B \mid A_1) \qquad (4.15)$$

FIGURE 4.11   PROBABILITY TREE FOR TWO-SUPPLIER EXAMPLE

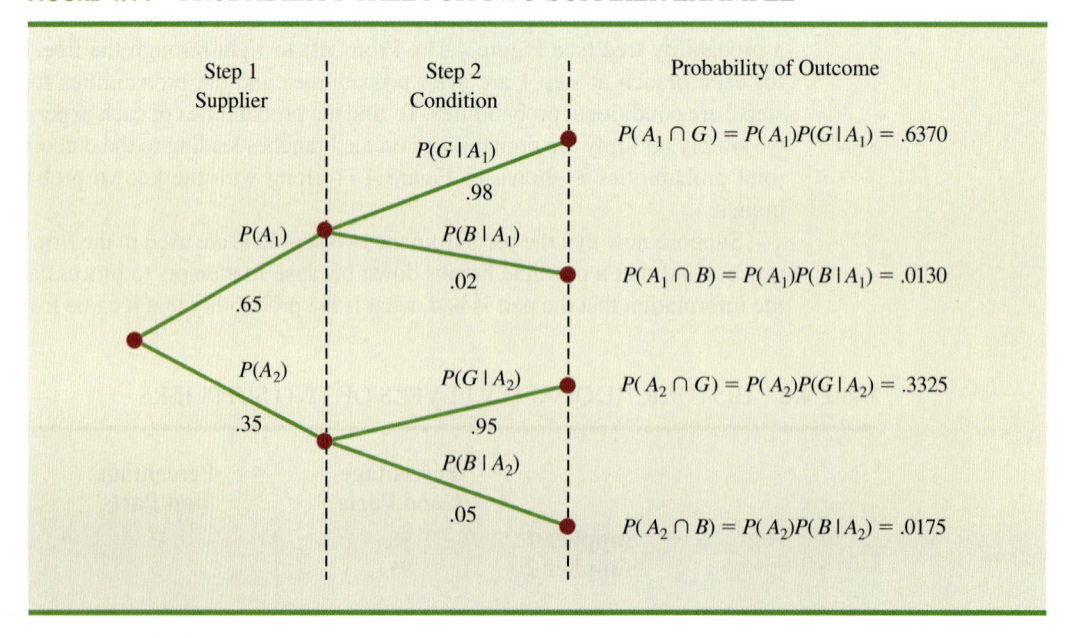

4.5 Bayes' Theorem

To find $P(B)$, we note that event B can occur in only two ways: $(A_1 \cap B)$ and $(A_2 \cap B)$. Therefore, we have

$$P(B) = P(A_1 \cap B) + P(A_2 \cap B) \qquad (4.16)$$
$$= P(A_1)P(B \mid A_1) + P(A_2)P(B \mid A_2)$$

Substituting from equations (4.15) and (4.16) into equation (4.14) and writing a similar result for $P(A_2 \mid B)$, we obtain Bayes' theorem for the case of two events.

The Reverend Thomas Bayes (1702–1761), a Presbyterian minister, is credited with the original work leading to the version of Bayes' theorem in use today.

BAYES' THEOREM (TWO-EVENT CASE)

$$P(A_1 \mid B) = \frac{P(A_1)P(B \mid A_1)}{P(A_1)P(B \mid A_1) + P(A_2)P(B \mid A_2)} \qquad (4.17)$$

$$P(A_2 \mid B) = \frac{P(A_2)P(B \mid A_2)}{P(A_1)P(B \mid A_1) + P(A_2)P(B \mid A_2)} \qquad (4.18)$$

Using equation (4.17) and the probability values provided in the example, we have

$$P(A_1 \mid B) = \frac{P(A_1)P(B \mid A_1)}{P(A_1)P(B \mid A_1) + P(A_2)P(B \mid A_2)}$$

$$= \frac{(.65)(.02)}{(.65)(.02) + (.35)(.05)} = \frac{.0130}{.0130 + .0175}$$

$$= \frac{.0130}{.0305} = .4262$$

In addition, using equation (4.18), we find $P(A_2 \mid B)$.

$$P(A_2 \mid B) = \frac{(.35)(.05)}{(.65)(.02) + (.35)(.05)}$$

$$= \frac{.0175}{.0130 + .0175} = \frac{.0175}{.0305} = .5738$$

Note that in this application we started with a probability of .65 that a part selected at random was from supplier 1. However, given information that the part is bad, the probability that the part is from supplier 1 drops to .4262. In fact, if the part is bad, it has better than a 50–50 chance that it came from supplier 2; that is, $P(A_2 \mid B) = .5738$.

Bayes' theorem is applicable when the events for which we want to compute posterior probabilities are mutually exclusive and their union is the entire sample space.[2] For the case of n mutually exclusive events $A_1, A_2, \ldots, A_n$, whose union is the entire sample space, Bayes' theorem can be used to compute any posterior probability $P(A_i \mid B)$ as shown here.

BAYES' THEOREM

$$P(A_i \mid B) = \frac{P(A_i)P(B \mid A_i)}{P(A_1)P(B \mid A_1) + P(A_2)P(B \mid A_2) + \cdots + P(A_n)P(B \mid A_n)} \qquad (4.19)$$

---

[2]If the union of events is the entire sample space, the events are said to be *collectively exhaustive*.

With prior probabilities $P(A_1), P(A_2), \ldots, P(A_n)$ and the appropriate conditional probabilities $P(B \mid A_1), P(B \mid A_2), \ldots, P(B \mid A_n)$, equation (4.19) can be used to compute the posterior probability of the events $A_1, A_2, \ldots, A_n$.

Tabular Approach

A tabular approach is helpful in conducting the Bayes' theorem calculations. Such an approach is shown in Table 4.7 for the parts supplier problem. The computations shown there are done in the following steps.

Step 1. Prepare the following three columns:
Column 1—The mutually exclusive events A_i for which posterior probabilities are desired
Column 2—The prior probabilities $P(A_i)$ for the events
Column 3—The conditional probabilities $P(B \mid A_i)$ of the new information B given each event

Step 2. In column 4, compute the joint probabilities $P(A_i \cap B)$ for each event and the new information B by using the multiplication law. These joint probabilities are found by multiplying the prior probabilities in column 2 by the corresponding conditional probabilities in column 3; that is, $P(A_i \cap B) = P(A_i)P(B \mid A_i)$.

Step 3. Sum the joint probabilities in column 4. The sum is the probability of the new information, $P(B)$. Thus we see in Table 4.7 that there is a .0130 probability that the part came from supplier 1 and is bad and a .0175 probability that the part came from supplier 2 and is bad. Because these are the only two ways in which a bad part can be obtained, the sum .0130 + .0175 shows an overall probability of .0305 of finding a bad part from the combined shipments of the two suppliers.

Step 4. In column 5, compute the posterior probabilities using the basic relationship of conditional probability.

$$P(A_i \mid B) = \frac{P(A_i \cap B)}{P(B)}$$

Note that the joint probabilities $P(A_i \cap B)$ are in column 4 and the probability $P(B)$ is the sum of column 4.

TABLE 4.7 TABULAR APPROACH TO BAYES' THEOREM CALCULATIONS FOR THE TWO-SUPPLIER PROBLEM

(1) Events A_i	(2) Prior Probabilities $P(A_i)$	(3) Conditional Probabilities $P(B \mid A_i)$	(4) Joint Probabilities $P(A_i \cap B)$	(5) Posterior Probabilities $P(A_i \mid B)$
A_1	.65	.02	.0130	.0130/.0305 = .4262
A_2	.35	.05	.0175	.0175/.0305 = .5738
	1.00		$P(B) = .0305$	1.0000

4.5 Bayes' Theorem

NOTES AND COMMENTS

1. Bayes' theorem is used extensively in decision analysis. The prior probabilities are often subjective estimates provided by a decision maker. Sample information is obtained and posterior probabilities are computed for use in choosing the best decision.

2. An event and its complement are mutually exclusive, and their union is the entire sample space. Thus, Bayes' theorem is always applicable for computing posterior probabilities of an event and its complement.

Exercises

Methods

39. The prior probabilities for events A_1 and A_2 are $P(A_1) = .40$ and $P(A_2) = .60$. It is also known that $P(A_1 \cap A_2) = 0$. Suppose $P(B \mid A_1) = .20$ and $P(B \mid A_2) = .05$.
    a. Are A_1 and A_2 mutually exclusive? Explain.
    b. Compute $P(A_1 \cap B)$ and $P(A_2 \cap B)$.
    c. Compute $P(B)$.
    d. Apply Bayes' theorem to compute $P(A_1 \mid B)$ and $P(A_2 \mid B)$.

40. The prior probabilities for events A_1, A_2, and A_3 are $P(A_1) = .20$, $P(A_2) = .50$, and $P(A_3) = .30$. The conditional probabilities of event B given A_1, A_2, and A_3 are $P(B \mid A_1) = .50$, $P(B \mid A_2) = .40$, and $P(B \mid A_3) = .30$.
    a. Compute $P(B \cap A_1)$, $P(B \cap A_2)$, and $P(B \cap A_3)$.
    b. Apply Bayes' theorem, equation (4.19), to compute the posterior probability $P(A_2 \mid B)$.
    c. Use the tabular approach to applying Bayes' theorem to compute $P(A_1 \mid B)$, $P(A_2 \mid B)$, and $P(A_3 \mid B)$.

Applications

41. A consulting firm submitted a bid for a large research project. The firm's management initially felt they had a 50–50 chance of getting the project. However, the agency to which the bid was submitted subsequently requested additional information on the bid. Past experience indicates that for 75% of the successful bids and 40% of the unsuccessful bids, the agency requested additional information.
    a. What is the prior probability of the bid being successful (that is, prior to the request for additional information)?
    b. What is the conditional probability of a request for additional information given that the bid will ultimately be successful?
    c. Compute the posterior probability that the bid will be successful given a request for additional information.

42. A local bank reviewed its credit card policy with the intention of recalling some of its credit cards. In the past approximately 5% of cardholders defaulted, leaving the bank unable to collect the outstanding balance. Hence, management established a prior probability of .05 that any particular cardholder will default. The bank also found that the probability of missing a monthly payment is .20 for customers who do not default. Of course, the probability of missing a monthly payment for those who default is 1.
    a. Given that a customer missed one or more monthly payments, compute the posterior probability that the customer will default.
    b. The bank would like to recall its card if the probability that a customer will default is greater than .20. Should the bank recall its card if the customer misses a monthly payment? Why or why not?

43. Two Wharton professors analyzed 1,613,234 putts by golfers on the Professional Golfers Association (PGA) Tour and found that 983,764 of the putts were made and 629,470 of the

putts were missed. Further analysis showed that for putts that were made, 64.0% of the time the player was attempting to make a par putt and 18.8% of the time the player was attempting to make a birdie putt. And, for putts that were missed, 20.3% of the time the player was attempting to make a par putt and 73.4% of the time the player was attempting to make a birdie putt (*Is Tiger Woods Loss Averse? Persistent Bias in the Face of Experience, Competition, and High Stakes*, D. G. Pope and M. E. Schweitzer, June 2009, The Wharton School, University of Pennsylvania).
   a. What is the probability that a PGA Tour player makes a putt?
   b. Suppose that a PGA Tour player has a putt for par. What is the probability that the player will make the putt?
   c. Suppose that a PGA Tour player has a putt for birdie. What is the probability that the player will make the putt?
   d. Comment on the differences in the probabilities computed in parts (b) and (c).

44. The American Council of Education reported that 47% of college freshmen earn a degree and graduate within five years. Assume that graduation records show women make up 50% of the students who graduated within five years, but only 45% of the students who did not graduate within five years. The students who had not graduated within five years either dropped out or were still working on their degrees.
   a. Let A_1 = the student graduated within five years
      A_2 = the student did not graduate within five years
      W = the student is a female student
      Using the given information, what are the values for $P(A_1)$, $P(A_2)$, $P(W \mid A_1)$, and $P(W \mid A_2)$?
   b. What is the probability that a female student will graduate within five years?
   c. What is the probability that a male student will graduate within five years?
   d. Given the preceding results, what are the percentage of women and the percentage of men in the entering freshman class?

45. In an article about investment alternatives, *Money* magazine reported that drug stocks provide a potential for long-term growth, with over 50% of the adult population of the United States taking prescription drugs on a regular basis. For adults age 65 and older, 82% take prescription drugs regularly. For adults age 18 to 64, 49% take prescription drugs regularly. The 18–64 age group accounts for 83.5% of the adult population (*Statistical Abstract of the United States*, 2008).
   a. What is the probability that a randomly selected adult is 65 or older?
   b. Given that an adult takes prescription drugs regularly, what is the probability that the adult is 65 or older?

Summary

In this chapter we introduced basic probability concepts and illustrated how probability analysis can be used to provide helpful information for decision making. We described how probability can be interpreted as a numerical measure of the likelihood that an event will occur. In addition, we saw that the probability of an event can be computed either by summing the probabilities of the experimental outcomes (sample points) comprising the event or by using the relationships established by the addition, conditional probability, and multiplication laws of probability. For cases in which additional information is available, we showed how Bayes' theorem can be used to obtain revised or posterior probabilities.

Glossary

Probability A numerical measure of the likelihood that an event will occur.
Experiment A process that generates well-defined outcomes.
Sample space The set of all experimental outcomes.

Sample point An element of the sample space. A sample point represents an experimental outcome.

Tree diagram A graphical representation that helps in visualizing a multiple-step experiment.

Basic requirements for assigning probabilities Two requirements that restrict the manner in which probability assignments can be made: (1) for each experimental outcome E_i we must have $0 \leq P(E_i) \leq 1$; (2) considering all experimental outcomes, we must have $P(E_1) + P(E_2) + \cdots + P(E_n) = 1.0$.

Classical method A method of assigning probabilities that is appropriate when all the experimental outcomes are equally likely.

Relative frequency method A method of assigning probabilities that is appropriate when data are available to estimate the proportion of the time the experimental outcome will occur if the experiment is repeated a large number of times.

Subjective method A method of assigning probabilities on the basis of judgment.

Event A collection of sample points.

Complement of A The event consisting of all sample points that are not in A.

Venn diagram A graphical representation for showing symbolically the sample space and operations involving events in which the sample space is represented by a rectangle and events are represented as circles within the sample space.

Union of A and B The event containing all sample points belonging to A or B or both. The union is denoted $A \cup B$.

Intersection of A and B The event containing the sample points belonging to both A and B. The intersection is denoted $A \cap B$.

Addition law A probability law used to compute the probability of the union of two events. It is $P(A \cup B) = P(A) + P(B) - P(A \cap B)$. For mutually exclusive events, $P(A \cap B) = 0$; in this case the addition law reduces to $P(A \cup B) = P(A) + P(B)$.

Mutually exclusive events Events that have no sample points in common; that is, $A \cap B$ is empty and $P(A \cap B) = 0$.

Conditional probability The probability of an event given that another event already occurred. The conditional probability of A given B is $P(A \mid B) = P(A \cap B)/P(B)$.

Joint probability The probability of two events both occurring; that is, the probability of the intersection of two events.

Marginal probability The values in the margins of a joint probability table that provide the probabilities of each event separately.

Independent events Two events A and B where $P(A \mid B) = P(A)$ or $P(B \mid A) = P(B)$; that is, the events have no influence on each other.

Multiplication law A probability law used to compute the probability of the intersection of two events. It is $P(A \cap B) = P(B)P(A \mid B)$ or $P(A \cap B) = P(A)P(B \mid A)$. For independent events it reduces to $P(A \cap B) = P(A)P(B)$.

Prior probabilities Initial estimates of the probabilities of events.

Posterior probabilities Revised probabilities of events based on additional information.

Bayes' theorem A method used to compute posterior probabilities.

Key Formulas

Counting Rule for Combinations

$$C_n^N = \binom{N}{n} = \frac{N!}{n!(N-n)!} \tag{4.1}$$

Counting Rule for Permutations

$$P_n^N = n!\binom{N}{n} = \frac{N!}{(N-n)!} \tag{4.2}$$

Computing Probability Using the Complement

$$P(A) = 1 - P(A^c) \quad (4.5)$$

Addition Law

$$P(A \cup B) = P(A) + P(B) - P(A \cap B) \quad (4.6)$$

Conditional Probability

$$P(A \mid B) = \frac{P(A \cap B)}{P(B)} \quad (4.7)$$

$$P(B \mid A) = \frac{P(A \cap B)}{P(A)} \quad (4.8)$$

Multiplication Law

$$P(A \cap B) = P(B)P(A \mid B) \quad (4.11)$$
$$P(A \cap B) = P(A)P(B \mid A) \quad (4.12)$$

Multiplication Law for Independent Events

$$P(A \cap B) = P(A)P(B) \quad (4.13)$$

Bayes' Theorem

$$P(A_i \mid B) = \frac{P(A_i)P(B \mid A_i)}{P(A_1)P(B \mid A_1) + P(A_2)P(B \mid A_2) + \cdots + P(A_n)P(B \mid A_n)} \quad (4.19)$$

Supplementary Exercises

46. An oil company purchased an option on land in Alaska. Preliminary geologic studies assigned the following prior probabilities.

    $P(\text{high-quality oil}) = .50$
    $P(\text{medium-quality oil}) = .20$
    $P(\text{no oil}) = .30$

    a. What is the probability of finding oil?
    b. After 200 feet of drilling on the first well, a soil test is taken. The probabilities of finding the particular type of soil identified by the test follow.

    $P(\text{soil} \mid \text{high-quality oil}) = .20$
    $P(\text{soil} \mid \text{medium-quality oil}) = .80$
    $P(\text{soil} \mid \text{no oil}) = .20$

    How should the firm interpret the soil test? What are the revised probabilities, and what is the new probability of finding oil?

47. Par Fore created a website to market golf equipment and apparel. Management would like a certain offer to appear for female visitors and a different offer to appear for male visitors. From a sample of past website visits, management learned that 60% of the visitors to the website ParFore are male and 40% are female.
    a. What is the prior probability that the next visitor to the website will be female?

b. Suppose you know that the current visitor to the website ParFore previously visited the Dillard's website, and that women are three times as likely to visit the Dillard's website as men. What is the revised probability that the current visitor to the website ParFore is female? Should you display the offer that appeals more to female visitors or the one that appeals more to male visitors?

48. An MBA new-matriculants survey provided the following data for 2018 students.

		Applied to More Than One School	
		Yes	No
Age Group	23 and under	207	201
	24–26	299	379
	27–30	185	268
	31–35	66	193
	36 and over	51	169

a. For a randomly selected MBA student, prepare a joint probability table for the experiment consisting of observing the student's age and whether the student applied to one or more schools.
b. What is the probability that a randomly selected applicant is 23 or under?
c. What is the probability that a randomly selected applicant is older than 26?
d. What is the probability that a randomly selected applicant applied to more than one school?

49. Refer again to the data from the MBA new-matriculants survey in exercise 48.
a. Given that a person applied to more than one school, what is the probability that the person is 24–26 years old?
b. Given that a person is in the 36-and-over age group, what is the probability that the person applied to more than one school?
c. What is the probability that a person is 24–26 years old or applied to more than one school?
d. Suppose a person is known to have applied to only one school. What is the probability that the person is 31 or more years old?
e. Is the number of schools applied to independent of age? Explain.

50. The following crosstabulation shows household income by educational level of the head of household (*Statistical Abstract of the United States*, 2008).

	Household Income ($1000s)					
Education Level	Under 25	25.0–49.9	50.0–74.9	75.0–99.9	100 or more	Total
Not H.S. Graduate	4,207	3,459	1,389	539	367	9,961
H.S. Graduate	4,917	6,850	5,027	2,637	2,668	22,099
Some College	2,807	5,258	4,678	3,250	4,074	20,067
Bachelor's Degree	885	2,094	2,848	2,581	5,379	13,787
Beyond Bach. Deg.	290	829	1,274	1,241	4,188	7,822
Total	13,106	18,490	15,216	10,248	16,676	73,736

a. Develop a joint probability table.
b. What is the probability of a head of household not being a high school graduate?
c. What is the probability of a head of household having a bachelor's degree or more education?

d. What is the probability of a household headed by someone with a bachelor's degree earning $100,000 or more?
e. What is the probability of a household having income below $25,000?
f. What is the probability of a household headed by someone with a bachelor's degree earning less than $25,000?
g. Is household income independent of educational level?

51. A survey showed that 8% of Internet users age 18 and older report keeping a blog. Referring to the 18–29 age group as young adults, the survey showed that for bloggers 54% are young adults and for nonbloggers 24% are young adults (Pew Internet & American Life Project, July 19, 2006).
    a. Develop a joint probability table for these data with two rows (bloggers vs. nonbloggers) and two columns (young adults vs. older adults).
    b. What is the probability that an Internet user is a young adult?
    c. What is the probability that an Internet user keeps a blog and is a young adult?
    d. Suppose that in a follow-up phone survey we contact someone who is 24 years old. What is the probability that this person keeps a blog?

52. A telephone survey to determine viewer response to a new television show obtained the following data.

Rating	Frequency
Poor	4
Below average	8
Average	11
Above average	14
Excellent	13

    a. What is the probability that a randomly selected viewer will rate the new show as average or better?
    b. What is the probability that a randomly selected viewer will rate the new show below average or worse?

53. A company studied the number of lost-time accidents occurring at its Brownsville, Texas, plant. Historical records show that 6% of the employees suffered lost-time accidents last year. Management believes that a special safety program will reduce such accidents to 5% during the current year. In addition, it estimates that 15% of employees who had lost-time accidents last year will experience a lost-time accident during the current year.
    a. What percentage of the employees will experience lost-time accidents in both years?
    b. What percentage of the employees will suffer at least one lost-time accident over the two-year period?

54. A study of 31,000 hospital admissions in New York State found that 4% of the admissions led to treatment-caused injuries. One-seventh of these treatment-caused injuries resulted in death, and one-fourth were caused by negligence. Malpractice claims were filed in one out of 7.5 cases involving negligence, and payments were made in one out of every two claims.
    a. What is the probability that a person admitted to the hospital will suffer a treatment-caused injury due to negligence?
    b. What is the probability that a person admitted to the hospital will die from a treatment-caused injury?
    c. In the case of a negligent treatment-caused injury, what is the probability that a malpractice claim will be paid?

55. Cooper Realty is a small real estate company located in Albany, New York, specializing primarily in residential listings. It recently became interested in determining the likelihood of one of its listings being sold within a certain number of days. An analysis of company sales of 800 homes in previous years produced the following data.

Supplementary Exercises

		Days Listed Until Sold			
		Under 30	31–90	Over 90	Total
Initial Asking Price	Under $150,000	50	40	10	100
	$150,000–$199,999	20	150	80	250
	$200,000–$250,000	20	280	100	400
	Over $250,000	10	30	10	50
	Total	100	500	200	800

a. If A is defined as the event that a home is listed for more than 90 days before being sold, estimate the probability of A.

b. If B is defined as the event that the initial asking price is under $150,000, estimate the probability of B.

c. What is the probability of $A \cap B$?

d. Assuming that a contract was just signed to list a home with an initial asking price of less than $150,000, what is the probability that the home will take Cooper Realty more than 90 days to sell?

e. Are events A and B independent?

56. Statistics from the 2009 Major League Baseball season show that there were 157 players who had at least 500 plate appearances. For this group, 42 players had a batting average of .300 or higher, 53 players hit 25 or more home runs, and 14 players had a batting average of .300 or higher and hit 25 or more home runs. Only four players had 200 or more hits (ESPN website, January 10, 2010). Use the 157 players who had at least 500 plate appearances to answer the following questions.

    a. What is the probability that a randomly selected player had a batting average of .300 or higher?

    b. What is the probability that a randomly selected player hit 25 or more home runs?

    c. Are the events having a batting average of .300 or higher and hitting 25 or more home runs mutually exclusive?

    d. What is the probability that a randomly selected player had a batting average of .300 or higher or hit 25 or more home runs?

    e. What is the probability that a randomly selected player had 200 or more hits? Does obtaining 200 or more hits appear to be more difficult than hitting 25 or more home runs? Explain.

57. A large consumer goods company ran a television advertisement for one of its soap products. On the basis of a survey that was conducted, probabilities were assigned to the following events.

    B = individual purchased the product
    S = individual recalls seeing the advertisement
    $B \cap S$ = individual purchased the product and recalls seeing the advertisement

    The probabilities assigned were $P(B) = .20$, $P(S) = .40$, and $P(B \cap S) = .12$.

    a. What is the probability of an individual's purchasing the product given that the individual recalls seeing the advertisement? Does seeing the advertisement increase the probability that the individual will purchase the product? As a decision maker, would you recommend continuing the advertisement (assuming that the cost is reasonable)?

    b. Assume that individuals who do not purchase the company's soap product buy from its competitors. What would be your estimate of the company's market share? Would you expect that continuing the advertisement will increase the company's market share? Why or why not?

c. The company also tested another advertisement and assigned it values of $P(S) = .30$ and $P(B \cap S) = .10$. What is $P(B \mid S)$ for this other advertisement? Which advertisement seems to have had the bigger effect on customer purchases?

58. A financial manager made two new investments—one in the oil industry and one in municipal bonds. After a one-year period, each of the investments will be classified as either successful or unsuccessful. Consider the making of the two investments as an experiment.
    a. How many sample points exist for this experiment?
    b. Show a tree diagram and list the sample points.
    c. Let O = the event that the oil industry investment is successful and M = the event that the municipal bond investment is successful. List the sample points in O and in M.
    d. List the sample points in the union of the events ($O \cup M$).
    e. List the sample points in the intersection of the events ($O \cap M$).
    f. Are events O and M mutually exclusive? Explain.

59. A poll conducted to learn about attitudes toward investment and retirement asked male and female respondents how important they felt level of risk was in choosing a retirement investment. The following joint probability table was constructed from the data provided. "Important" means the respondent said level of risk was either important or very important.

	Male	Female	Total
Important	.22	.27	.49
Not Important	.28	.23	.51
Total	.50	.50	1.00

a. What is the probability that a survey respondent will say level of risk is important?
b. What is the probability that a male respondent will say level of risk is important?
c. What is the probability that a female respondent will say level of risk is important?
d. Is the level of risk independent of the gender of the respondent? Why or why not?
e. Do male and female attitudes toward risk differ?

60. *The Wall Street Journal*/Harris Personal Finance poll asked 2082 adults if they owned a home (All Business website, January 23, 2008). A total of 1249 survey respondents answered Yes. Of the 450 respondents in the 18–34 age group, 117 responded Yes.
    a. What is the probability that a respondent to the poll owned a home?
    b. What is the probability that a respondent in the 18–34 age group owned a home?
    c. What is the probability that a respondent to the poll did not own a home?
    d. What is the probability that a respondent in the 18–34 age group did not own a home?

Case Problem Hamilton County Judges

Hamilton County judges try thousands of cases per year. In an overwhelming majority of the cases disposed, the verdict stands as rendered. However, some cases are appealed, and of those appealed, some of the cases are reversed. Kristen DelGuzzi of *The Cincinnati Enquirer* conducted a study of cases handled by Hamilton County judges over a three-year period. Shown in Table 4.8 are the results for 182,908 cases handled (disposed) by 38 judges in Common Pleas Court, Domestic Relations Court, and Municipal Court. Two of the judges (Dinkelacker and Hogan) did not serve in the same court for the entire three-year period.

The purpose of the newspaper's study was to evaluate the performance of the judges. Appeals are often the result of mistakes made by judges, and the newspaper wanted to know which judges were doing a good job and which were making too many mistakes. You are

Case Problem Hamilton County Judges

TABLE 4.8 TOTAL CASES DISPOSED, APPEALED, AND REVERSED IN HAMILTON COUNTY COURTS

Common Pleas Court

Judge	Total Cases Disposed	Appealed Cases	Reversed Cases
Fred Cartolano	3037	137	12
Thomas Crush	3372	119	10
Patrick Dinkelacker	1258	44	8
Timothy Hogan	1954	60	7
Robert Kraft	3138	127	7
William Mathews	2264	91	18
William Morrissey	3032	121	22
Norbert Nadel	2959	131	20
Arthur Ney, Jr.	3219	125	14
Richard Niehaus	3353	137	16
Thomas Nurre	3000	121	6
John O'Connor	2969	129	12
Robert Ruehlman	3205	145	18
J. Howard Sundermann	955	60	10
Ann Marie Tracey	3141	127	13
Ralph Winkler	3089	88	6
Total	43,945	1762	199

Domestic Relations Court

Judge	Total Cases Disposed	Appealed Cases	Reversed Cases
Penelope Cunningham	2729	7	1
Patrick Dinkelacker	6001	19	4
Deborah Gaines	8799	48	9
Ronald Panioto	12,970	32	3
Total	30,499	106	17

Municipal Court

Judge	Total Cases Disposed	Appealed Cases	Reversed Cases
Mike Allen	6149	43	4
Nadine Allen	7812	34	6
Timothy Black	7954	41	6
David Davis	7736	43	5
Leslie Isaiah Gaines	5282	35	13
Karla Grady	5253	6	0
Deidra Hair	2532	5	0
Dennis Helmick	7900	29	5
Timothy Hogan	2308	13	2
James Patrick Kenney	2798	6	1
Joseph Luebbers	4698	25	8
William Mallory	8277	38	9
Melba Marsh	8219	34	7
Beth Mattingly	2971	13	1
Albert Mestemaker	4975	28	9
Mark Painter	2239	7	3
Jack Rosen	7790	41	13
Mark Schweikert	5403	33	6
David Stockdale	5371	22	4
John A. West	2797	4	2
Total	108,464	500	104

WEB file
Judge

called in to assist in the data analysis. Use your knowledge of probability and conditional probability to help with the ranking of the judges. You also may be able to analyze the likelihood of appeal and reversal for cases handled by different courts.

Managerial Report

Prepare a report with your rankings of the judges. Also, include an analysis of the likelihood of appeal and case reversal in the three courts. At a minimum, your report should include the following:

1. The probability of cases being appealed and reversed in the three different courts.
2. The probability of a case being appealed for each judge.
3. The probability of a case being reversed for each judge.
4. The probability of reversal given an appeal for each judge.
5. Rank the judges within each court. State the criteria you used and provide a rationale for your choice.

CHAPTER 5

Discrete Probability Distributions

CONTENTS

STATISTICS IN PRACTICE:
CITIBANK

5.1 RANDOM VARIABLES
Discrete Random Variables
Continuous Random Variables

5.2 DISCRETE PROBABILITY DISTRIBUTIONS

5.3 EXPECTED VALUE AND VARIANCE
Expected Value
Variance

5.4 BINOMIAL PROBABILITY DISTRIBUTION
A Binomial Experiment
Martin Clothing Store Problem
Using Tables of Binomial Probabilities
Expected Value and Variance for the Binomial Distribution

5.5 POISSON PROBABILITY DISTRIBUTION
An Example Involving Time Intervals
An Example Involving Length or Distance Intervals

5.6 HYPERGEOMETRIC PROBABILITY DISTRIBUTION

STATISTICS *in* PRACTICE

CITIBANK*
LONG ISLAND CITY, NEW YORK

Citibank, the retail banking division of Citigroup, offers a wide range of financial services including checking and saving accounts, loans and mortgages, insurance, and investment services. It delivers these services through a unique system referred to as Citibanking.

Citibank was one of the first banks in the United States to introduce automatic teller machines (ATMs). Citibank's ATMs, located in Citicard Banking Centers (CBCs), let customers do all of their banking in one place with the touch of a finger, 24 hours a day, 7 days a week. More than 150 different banking functions—from deposits to managing investments—can be performed with ease. Citibank customers use ATMs for 80% of their transactions.

Each Citibank CBC operates as a waiting line system with randomly arriving customers seeking service at one of the ATMs. If all ATMs are busy, the arriving customers wait in line. Periodic CBC capacity studies are used to analyze customer waiting times and to determine whether additional ATMs are needed.

Data collected by Citibank showed that the random customer arrivals followed a probability distribution known as the Poisson distribution. Using the Poisson distribution, Citibank can compute probabilities for the number of customers arriving at a CBC during any time period and make decisions concerning the number of ATMs needed. For example, let $x =$ the number of

*The authors are indebted to Ms. Stacey Karter, Citibank, for providing this Statistics in Practice.

Periodic capacity studies are used to analyze customer waiting times and determine if additional ATMS are needed.

customers arriving during a one-minute period. Assuming that a particular CBC has a mean arrival rate of two customers per minute, the following table shows the probabilities for the number of customers arriving during a one-minute period.

x	Probability
0	.1353
1	.2707
2	.2707
3	.1804
4	.0902
5 or more	.0527

Discrete probability distributions, such as the one used by Citibank, are the topic of this chapter. In addition to the Poisson distribution, you will learn about the binomial and hypergeometric distributions and how they can be used to provide helpful probability information.

In this chapter we continue the study of probability by introducing the concepts of random variables and probability distributions. The focus of this chapter is discrete probability distributions. Three special discrete probability distributions—the binomial, Poisson, and hypergeometric—are covered.

5.1 Random Variables

In Chapter 4 we defined the concept of an experiment and its associated experimental outcomes. A random variable provides a means for describing experimental outcomes using numerical values. Random variables must assume numerical values.

5.1 Random Variables

Random variables must have numerical values.

> **RANDOM VARIABLE**
>
> A **random variable** is a numerical description of the outcome of an experiment.

In effect, a random variable associates a numerical value with each possible experimental outcome. The particular numerical value of the random variable depends on the outcome of the experiment. A random variable can be classified as being either *discrete* or *continuous* depending on the numerical values it assumes.

Discrete Random Variables

A random variable that may assume either a finite number of values or an infinite sequence of values such as 0, 1, 2, . . . is referred to as a **discrete random variable**. For example, consider the experiment of an accountant taking the certified public accountant (CPA) examination. The examination has four parts. We can define a random variable as x = the number of parts of the CPA examination passed. It is a discrete random variable because it may assume the finite number of values 0, 1, 2, 3, or 4.

As another example of a discrete random variable, consider the experiment of cars arriving at a tollbooth. The random variable of interest is x = the number of cars arriving during a one-day period. The possible values for x come from the sequence of integers 0, 1, 2, and so on. Hence, x is a discrete random variable assuming one of the values in this infinite sequence.

Although the outcomes of many experiments can naturally be described by numerical values, others cannot. For example, a survey question might ask an individual to recall the message in a recent television commercial. This experiment would have two possible outcomes: The individual cannot recall the message and the individual can recall the message. We can still describe these experimental outcomes numerically by defining the discrete random variable x as follows: let $x = 0$ if the individual cannot recall the message and $x = 1$ if the individual can recall the message. The numerical values for this random variable are arbitrary (we could use 5 and 10), but they are acceptable in terms of the definition of a random variable—namely, x is a random variable because it provides a numerical description of the outcome of the experiment.

Table 5.1 provides some additional examples of discrete random variables. Note that in each example the discrete random variable assumes a finite number of values or an infinite sequence of values such as 0, 1, 2, ... These types of discrete random variables are discussed in detail in this chapter.

TABLE 5.1 EXAMPLES OF DISCRETE RANDOM VARIABLES

Experiment	Random Variable (x)	Possible Values for the Random Variable
Contact five customers	Number of customers who place an order	0, 1, 2, 3, 4, 5
Inspect a shipment of 50 radios	Number of defective radios	0, 1, 2, $\cdots$, 49, 50
Operate a restaurant for one day	Number of customers	0, 1, 2, 3, $\cdots$
Sell an automobile	Gender of the customer	0 if male; 1 if female

Continuous Random Variables

A random variable that may assume any numerical value in an interval or collection of intervals is called a **continuous random variable**. Experimental outcomes based on measurement scales such as time, weight, distance, and temperature can be described by continuous random variables. For example, consider an experiment of monitoring incoming telephone calls to the claims office of a major insurance company. Suppose the random variable of interest is x = the time between consecutive incoming calls in minutes. This random variable may assume any value in the interval $x \geq 0$. Actually, an infinite number of values are possible for x, including values such as 1.26 minutes, 2.751 minutes, 4.3333 minutes, and so on. As another example, consider a 90-mile section of interstate highway I-75 north of Atlanta, Georgia. For an emergency ambulance service located in Atlanta, we might define the random variable as x = number of miles to the location of the next traffic accident along this section of I-75. In this case, x would be a continuous random variable assuming any value in the interval $0 \leq x \leq 90$. Additional examples of continuous random variables are listed in Table 5.2. Note that each example describes a random variable that may assume any value in an interval of values. Continuous random variables and their probability distributions will be the topic of Chapter 6.

TABLE 5.2   EXAMPLES OF CONTINUOUS RANDOM VARIABLES

Experiment	Random Variable (x)	Possible Values for the Random Variable
Operate a bank	Time between customer arrivals in minutes	$x \geq 0$
Fill a soft drink can (max = 12.1 ounces)	Number of ounces	$0 \leq x \leq 12.1$
Construct a new library	Percentage of project complete after six months	$0 \leq x \leq 100$
Test a new chemical process	Temperature when the desired reaction takes place (min 150° F; max 212° F)	$150 \leq x \leq 212$

NOTES AND COMMENTS

One way to determine whether a random variable is discrete or continuous is to think of the values of the random variable as points on a line segment. Choose two points representing values of the random variable. If the entire line segment between the two points also represents possible values for the random variable, then the random variable is continuous.

Exercises

Methods

1. Consider the experiment of tossing a coin twice.
    a. List the experimental outcomes.
    b. Define a random variable that represents the number of heads occurring on the two tosses.
    c. Show what value the random variable would assume for each of the experimental outcomes.
    d. Is this random variable discrete or continuous?

2. Consider the experiment of a worker assembling a product.
    a. Define a random variable that represents the time in minutes required to assemble the product.
    b. What values may the random variable assume?
    c. Is the random variable discrete or continuous?

Applications

3. Three students scheduled interviews for summer employment at the Brookwood Institute. In each case the interview results in either an offer for a position or no offer. Experimental outcomes are defined in terms of the results of the three interviews.
    a. List the experimental outcomes.
    b. Define a random variable that represents the number of offers made. Is the random variable continuous?
    c. Show the value of the random variable for each of the experimental outcomes.

4. In November the U.S. unemployment rate was 8.7% (*U.S. Department of Labor website*, January 10, 2010). The Census Bureau includes nine states in the Northeast region. Assume that the random variable of interest is the number of Northeastern states with an unemployment rate in November that was less than 8.7%. What values may this random variable have?

5. To perform a certain type of blood analysis, lab technicians must perform two procedures. The first procedure requires either one or two separate steps, and the second procedure requires either one, two, or three steps.
    a. List the experimental outcomes associated with performing the blood analysis.
    b. If the random variable of interest is the total number of steps required to do the complete analysis (both procedures), show what value the random variable will assume for each of the experimental outcomes.

6. Listed is a series of experiments and associated random variables. In each case, identify the values that the random variable can assume and state whether the random variable is discrete or continuous.

Experiment	Random Variable (x)
a. Take a 20-question examination	Number of questions answered correctly
b. Observe cars arriving at a tollbooth for one hour	Number of cars arriving at tollbooth
c. Audit 50 tax returns	Number of returns containing errors
d. Observe an employee's work	Number of nonproductive hours in an eight-hour workday
e. Weigh a shipment of goods	Number of pounds

5.2 Discrete Probability Distributions

The **probability distribution** for a random variable describes how probabilities are distributed over the values of the random variable. For a discrete random variable x, the probability distribution is defined by a **probability function**, denoted by $f(x)$. The probability function provides the probability for each value of the random variable.

As an illustration of a discrete random variable and its probability distribution, consider the sales of automobiles at DiCarlo Motors in Saratoga, New York. Over the past 300 days of operation, sales data show 54 days with no automobiles sold, 117 days with 1 automobile sold, 72 days with 2 automobiles sold, 42 days with 3 automobiles sold, 12 days with 4 automobiles sold, and 3 days with 5 automobiles sold. Suppose we consider the experiment of selecting a day of operation at DiCarlo Motors and define the random variable of interest as $x = $ the number of automobiles sold during a day. From historical data, we know

x is a discrete random variable that can assume the values 0, 1, 2, 3, 4, or 5. In probability function notation, $f(0)$ provides the probability of 0 automobiles sold, $f(1)$ provides the probability of 1 automobile sold, and so on. Because historical data show 54 of 300 days with 0 automobiles sold, we assign the value 54/300 = .18 to $f(0)$, indicating that the probability of 0 automobiles being sold during a day is .18. Similarly, because 117 of 300 days had 1 automobile sold, we assign the value 117/300 = .39 to $f(1)$, indicating that the probability of exactly 1 automobile being sold during a day is .39. Continuing in this way for the other values of the random variable, we compute the values for $f(2)$, $f(3)$, $f(4)$, and $f(5)$ as shown in Table 5.3, the probability distribution for the number of automobiles sold during a day at DiCarlo Motors.

A primary advantage of defining a random variable and its probability distribution is that once the probability distribution is known, it is relatively easy to determine the probability of a variety of events that may be of interest to a decision maker. For example, using the probability distribution for DiCarlo Motors as shown in Table 5.3, we see that the most probable number of automobiles sold during a day is 1 with a probability of $f(1) = .39$. In addition, there is an $f(3) + f(4) + f(5) = .14 + .04 + .01 = .19$ probability of selling 3 or more automobiles during a day. These probabilities, plus others the decision maker may ask about, provide information that can help the decision maker understand the process of selling automobiles at DiCarlo Motors.

In the development of a probability function for any discrete random variable, the following two conditions must be satisfied.

These conditions are the analogs to the two basic requirements for assigning probabilities to experimental outcomes presented in Chapter 4.

REQUIRED CONDITIONS FOR A DISCRETE PROBABILITY FUNCTION

$$f(x) \geq 0 \quad (5.1)$$
$$\Sigma f(x) = 1 \quad (5.2)$$

Table 5.3 shows that the probabilities for the random variable x satisfy equation (5.1); $f(x)$ is greater than or equal to 0 for all values of x. In addition, because the probabilities sum to 1, equation (5.2) is satisfied. Thus, the DiCarlo Motors probability function is a valid discrete probability function.

We can also present probability distributions graphically. In Figure 5.1 the values of the random variable x for DiCarlo Motors are shown on the horizontal axis and the probability associated with these values is shown on the vertical axis.

In addition to tables and graphs, a formula that gives the probability function, $f(x)$, for every value of x is often used to describe probability distributions. The simplest example of

TABLE 5.3 PROBABILITY DISTRIBUTION FOR THE NUMBER OF AUTOMOBILES SOLD DURING A DAY AT DICARLO MOTORS

x	$f(x)$
0	.18
1	.39
2	.24
3	.14
4	.04
5	.01
Total	1.00

5.2 Discrete Probability Distributions

FIGURE 5.1 GRAPHICAL REPRESENTATION OF THE PROBABILITY DISTRIBUTION FOR THE NUMBER OF AUTOMOBILES SOLD DURING A DAY AT DICARLO MOTORS

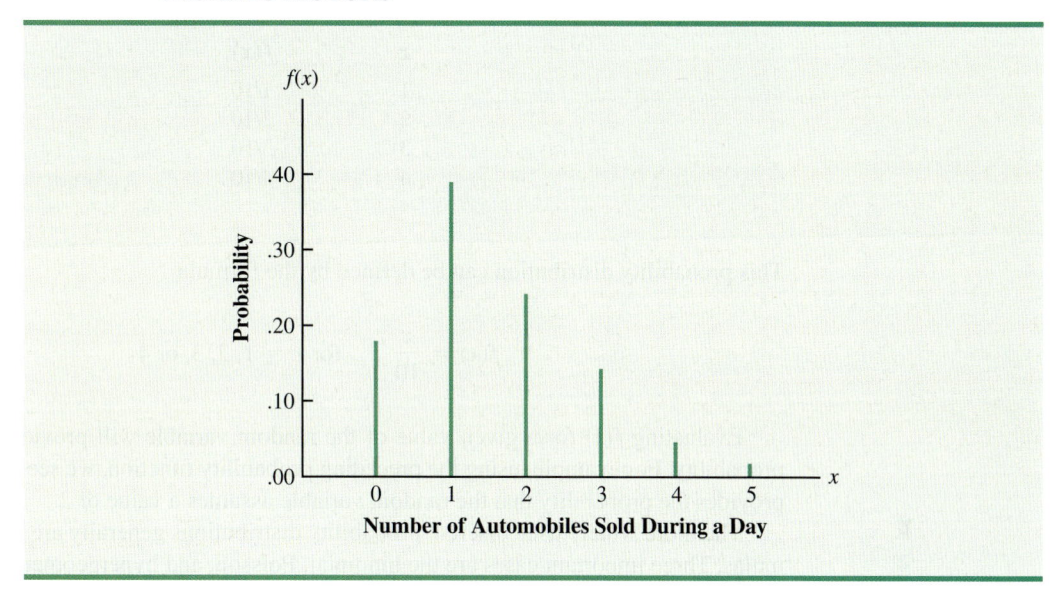

a discrete probability distribution given by a formula is the **discrete uniform probability distribution**. Its probability function is defined by equation (5.3).

DISCRETE UNIFORM PROBABILITY FUNCTION

$$f(x) = 1/n \quad (5.3)$$

where

n = the number of values the random variable may have

For example, suppose that for the experiment of rolling a die we define the random variable x to be the number of dots on the upward face. For this experiment, $n = 6$ values are possible for the random variable; $x = 1, 2, 3, 4, 5, 6$. Thus, the probability function for this discrete uniform random variable is

$$f(x) = 1/6 \quad x = 1, 2, 3, 4, 5, 6$$

The possible values of the random variable and the associated probabilities are shown.

x	$f(x)$
1	1/6
2	1/6
3	1/6
4	1/6
5	1/6
6	1/6

As another example, consider the random variable x with the following discrete probability distribution.

x	$f(x)$
1	1/10
2	2/10
3	3/10
4	4/10

This probability distribution can be defined by the formula

$$f(x) = \frac{x}{10} \quad \text{for } x = 1, 2, 3, \text{ or } 4$$

Evaluating $f(x)$ for a given value of the random variable will provide the associated probability. For example, using the preceding probability function, we see that $f(2) = 2/10$ provides the probability that the random variable assumes a value of 2.

The more widely used discrete probability distributions generally are specified by formulas. Three important cases are the binomial, Poisson, and hypergeometric distributions; these distributions are discussed later in the chapter.

Exercises

Methods

SELF test

7. The probability distribution for the random variable x follows.

x	$f(x)$
20	.20
25	.15
30	.25
35	.40

  a. Is this probability distribution valid? Explain.
  b. What is the probability that $x = 30$?
  c. What is the probability that x is less than or equal to 25?
  d. What is the probability that x is greater than 30?

Applications

SELF test

8. The following data were collected by counting the number of operating rooms in use at Tampa General Hospital over a 20-day period: On 3 of the days only one operating room was used, on 5 of the days two were used, on 8 of the days three were used, and on 4 days all four of the hospital's operating rooms were used.
  a. Use the relative frequency approach to construct a probability distribution for the number of operating rooms in use on any given day.
  b. Draw a graph of the probability distribution.
  c. Show that your probability distribution satisfies the required conditions for a valid discrete probability distribution.

9. For unemployed persons in the United States, the average number of months of unemployment at the end of December 2009 was approximately seven months (Bureau of Labor Statistics, January 2010). Suppose the following data are for a particular region in upstate New York. The values in the first column show the number of months unemployed and the values in the second column show the corresponding number of unemployed persons.

Months Unemployed	Number Unemployed
1	1029
2	1686
3	2269
4	2675
5	3487
6	4652
7	4145
8	3587
9	2325
10	1120

Let x be a random variable indicating the number of months a person is unemployed.
  a. Use the data to develop a probability distribution for x.
  b. Show that your probability distribution satisfies the conditions for a valid discrete probability distribution.
  c. What is the probability that a person is unemployed for two months or less? Unemployed for more than two months?
  d. What is the probability that a person is unemployed for more than six months?

10. The percent frequency distributions of job satisfaction scores for a sample of information systems (IS) senior executives and middle managers are as follows. The scores range from a low of 1 (very dissatisfied) to a high of 5 (very satisfied).

Job Satisfaction Score	IS Senior Executives (%)	IS Middle Managers (%)
1	5	4
2	9	10
3	3	12
4	42	46
5	41	28

  a. Develop a probability distribution for the job satisfaction score of a senior executive.
  b. Develop a probability distribution for the job satisfaction score of a middle manager.
  c. What is the probability that a senior executive will report a job satisfaction score of 4 or 5?
  d. What is the probability that a middle manager is very satisfied?
  e. Compare the overall job satisfaction of senior executives and middle managers.

11. A technician services mailing machines at companies in the Phoenix area. Depending on the type of malfunction, the service call can take one, two, three, or four hours. The different types of malfunctions occur at about the same frequency.
  a. Develop a probability distribution for the duration of a service call.
  b. Draw a graph of the probability distribution.
  c. Show that your probability distribution satisfies the conditions required for a discrete probability function.

d. What is the probability that a service call will take three hours?
e. A service call has just come in, but the type of malfunction is unknown. It is 3:00 P.M. and service technicians usually get off at 5:00 P.M. What is the probability that the service technician will have to work overtime to fix the machine today?

12. The two largest cable providers are Comcast Cable Communications, with 21.5 million subscribers, and Time Warner Cable, with 11.0 million subscribers (*The New York Times Almanac, 2007*). Suppose that the management of Time Warner Cable subjectively assesses a probability distribution for the number of new subscribers next year in the state of New York as follows.

x	f(x)
100,000	.10
200,000	.20
300,000	.25
400,000	.30
500,000	.10
600,000	.05

a. Is this probability distribution valid? Explain.
b. What is the probability that Time Warner will obtain more than 400,000 new subscribers?
c. What is the probability that Time Warner will obtain fewer than 200,000 new subscribers?

13. A psychologist determined that the number of sessions required to obtain the trust of a new patient is either 1, 2, or 3. Let x be a random variable indicating the number of sessions required to gain the patient's trust. The following probability function has been proposed.

$$f(x) = \frac{x}{6} \quad \text{for } x = 1, 2, \text{ or } 3$$

a. Is this probability function valid? Explain.
b. What is the probability that it takes exactly two sessions to gain the patient's trust?
c. What is the probability that it takes at least two sessions to gain the patient's trust?

14. The following table is a partial probability distribution for the MRA Company's projected profits (x = profit in $1000s) for the first year of operation (the negative value denotes a loss).

x	f(x)
−100	.10
0	.20
50	.30
100	.25
150	.10
200	

a. What is the proper value for $f(200)$? What is your interpretation of this value?
b. What is the probability that MRA will be profitable?
c. What is the probability that MRA will make at least $100,000?

5.3 Expected Value and Variance

Expected Value

The **expected value**, or mean, of a random variable is a measure of the central location for the random variable. The formula for the expected value of a discrete random variable x follows.

5.3 Expected Value and Variance

The expected value is a weighted average of the values of the random variable where the weights are the probabilities.

EXPECTED VALUE OF A DISCRETE RANDOM VARIABLE

$$E(x) = \mu = \Sigma x f(x) \tag{5.4}$$

Both the notations $E(x)$ and μ are used to denote the expected value of a random variable.

Equation (5.4) shows that to compute the expected value of a discrete random variable, we must multiply each value of the random variable by the corresponding probability $f(x)$ and then add the resulting products. Using the DiCarlo Motors automobile sales example from Section 5.2, we show the calculation of the expected value for the number of automobiles sold during a day in Table 5.4. The sum of the entries in the $xf(x)$ column shows that the expected value is 1.50 automobiles per day. We therefore know that although sales of 0, 1, 2, 3, 4, or 5 automobiles are possible on any one day, over time DiCarlo can anticipate selling an average of 1.50 automobiles per day. Assuming 30 days of operation during a month, we can use the expected value of 1.50 to forecast average monthly sales of 30(1.50) = 45 automobiles.

The expected value does not have to be a value the random variable can assume.

Variance

Even though the expected value provides the mean value for the random variable, we often need a measure of variability, or dispersion. Just as we used the variance in Chapter 3 to summarize the variability in data, we now use **variance** to summarize the variability in the values of a random variable. The formula for the variance of a discrete random variable follows.

The variance is a weighted average of the squared deviations of a random variable from its mean. The weights are the probabilities.

VARIANCE OF A DISCRETE RANDOM VARIABLE

$$\text{Var}(x) = \sigma^2 = \Sigma(x - \mu)^2 f(x) \tag{5.5}$$

As equation (5.5) shows, an essential part of the variance formula is the deviation, $x - \mu$, which measures how far a particular value of the random variable is from the expected value, or mean, μ. In computing the variance of a random variable, the deviations are squared and then weighted by the corresponding value of the probability function. The sum of these weighted squared deviations for all values of the random variable is referred to as the *variance*. The notations $\text{Var}(x)$ and σ^2 are both used to denote the variance of a random variable.

TABLE 5.4 CALCULATION OF THE EXPECTED VALUE FOR THE NUMBER OF AUTOMOBILES SOLD DURING A DAY AT DICARLO MOTORS

x	$f(x)$	$xf(x)$
0	.18	0(.18) = .00
1	.39	1(.39) = .39
2	.24	2(.24) = .48
3	.14	3(.14) = .42
4	.04	4(.04) = .16
5	.01	5(.01) = .05
		1.50

$$E(x) = \mu = \Sigma x f(x)$$

TABLE 5.5 CALCULATION OF THE VARIANCE FOR THE NUMBER OF AUTOMOBILES SOLD DURING A DAY AT DICARLO MOTORS

x	$x - \mu$	$(x - \mu)^2$	$f(x)$	$(x - \mu)^2 f(x)$
0	0 − 1.50 = −1.50	2.25	.18	2.25(.18) = .4050
1	1 − 1.50 = −.50	.25	.39	.25(.39) = .0975
2	2 − 1.50 = .50	.25	.24	.25(.24) = .0600
3	3 − 1.50 = 1.50	2.25	.14	2.25(.14) = .3150
4	4 − 1.50 = 2.50	6.25	.04	6.25(.04) = .2500
5	5 − 1.50 = 3.50	12.25	.01	12.25(.01) = .1225
				1.2500

$$\sigma^2 = \Sigma(x - \mu)^2 f(x)$$

The calculation of the variance for the probability distribution of the number of automobiles sold during a day at DiCarlo Motors is summarized in Table 5.5. We see that the variance is 1.25. The **standard deviation**, σ, is defined as the positive square root of the variance. Thus, the standard deviation for the number of automobiles sold during a day is

$$\sigma = \sqrt{1.25} = 1.118$$

The standard deviation is measured in the same units as the random variable ($\sigma = 1.118$ automobiles) and therefore is often preferred in describing the variability of a random variable. The variance σ^2 is measured in squared units and is thus more difficult to interpret.

Exercises

Methods

15. The following table provides a probability distribution for the random variable x.

x	$f(x)$
3	.25
6	.50
9	.25

 a. Compute $E(x)$, the expected value of x.
 b. Compute σ^2, the variance of x.
 c. Compute σ, the standard deviation of x.

16. The following table provides a probability distribution for the random variable y.

y	$f(y)$
2	.20
4	.30
7	.40
8	.10

 a. Compute $E(y)$.
 b. Compute Var(y) and σ.

5.3 Expected Value and Variance

Applications

17. The number of students taking the SAT has risen to an all-time high of more than 1.5 million (College Board, August 26, 2008). Students are allowed to repeat the test in hopes of improving the score that is sent to college and university admission offices. The number of times the SAT was taken and the number of students are as follows.

Number of Times	Number of Students
1	721,769
2	601,325
3	166,736
4	22,299
5	6,730

    a. Let x be a random variable indicating the number of times a student takes the SAT. Show the probability distribution for this random variable.
    b. What is the probability that a student takes the SAT more than one time?
    c. What is the probability that a student takes the SAT three or more times?
    d. What is the expected value of the number of times the SAT is taken? What is your interpretation of the expected value?
    e. What is the variance and standard deviation for the number of times the SAT is taken?

18. The American Housing Survey reported the following data on the number of bedrooms in owner-occupied and renter-occupied houses in central cities (U.S. Census Bureau website, March 31, 2003).

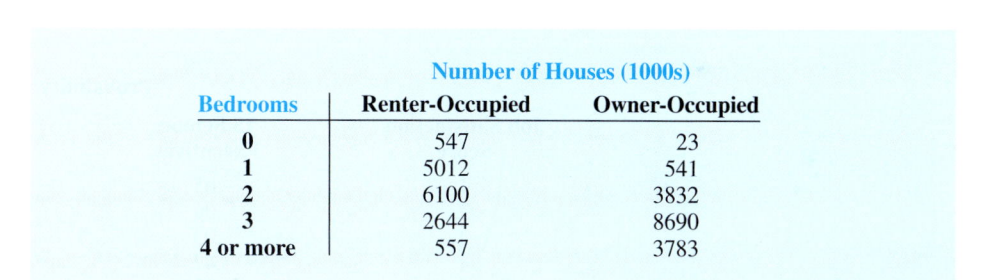

	Number of Houses (1000s)	
Bedrooms	Renter-Occupied	Owner-Occupied
0	547	23
1	5012	541
2	6100	3832
3	2644	8690
4 or more	557	3783

    a. Define a random variable x = number of bedrooms in renter-occupied houses and develop a probability distribution for the random variable. (Let $x = 4$ represent 4 or more bedrooms.)
    b. Compute the expected value and variance for the number of bedrooms in renter-occupied houses.
    c. Define a random variable y = number of bedrooms in owner-occupied houses and develop a probability distribution for the random variable. (Let $y = 4$ represent 4 or more bedrooms.)
    d. Compute the expected value and variance for the number of bedrooms in owner-occupied houses.
    e. What observations can you make from a comparison of the number of bedrooms in renter-occupied versus owner-occupied homes?

19. The National Basketball Association (NBA) records a variety of statistics for each team. Two of these statistics are the percentage of field goals made by the team and the percentage of three-point shots made by the team. For a portion of the 2004 season, the shooting records of the 29 teams in the NBA showed that the probability of scoring two points by

making a field goal was .44, and the probability of scoring three points by making a three-point shot was .34 (NBA website, January 3, 2004).

   a. What is the expected value of a two-point shot for these teams?
   b. What is the expected value of a three-point shot for these teams?
   c. If the probability of making a two-point shot is greater than the probability of making a three-point shot, why do coaches allow some players to shoot the three-point shot if they have the opportunity? Use expected value to explain your answer.

20. The probability distribution for damage claims paid by the Newton Automobile Insurance Company on collision insurance follows.

Payment ($)	Probability
0	.85
500	.04
1000	.04
3000	.03
5000	.02
8000	.01
10000	.01

   a. Use the expected collision payment to determine the collision insurance premium that would enable the company to break even.
   b. The insurance company charges an annual rate of $520 for the collision coverage. What is the expected value of the collision policy for a policyholder? (*Hint:* It is the expected payments from the company minus the cost of coverage.) Why does the policyholder purchase a collision policy with this expected value?

21. The following probability distributions of job satisfaction scores for a sample of information systems (IS) senior executives and middle managers range from a low of 1 (very dissatisfied) to a high of 5 (very satisfied).

	Probability	
Job Satisfaction Score	IS Senior Executives	IS Middle Managers
1	.05	.04
2	.09	.10
3	.03	.12
4	.42	.46
5	.41	.28

   a. What is the expected value of the job satisfaction score for senior executives?
   b. What is the expected value of the job satisfaction score for middle managers?
   c. Compute the variance of job satisfaction scores for executives and middle managers.
   d. Compute the standard deviation of job satisfaction scores for both probability distributions.
   e. Compare the overall job satisfaction of senior executives and middle managers.

22. The demand for a product of Carolina Industries varies greatly from month to month. The probability distribution in the following table, based on the past two years of data, shows the company's monthly demand.

Unit Demand	Probability
300	.20
400	.30
500	.35
600	.15

a. If the company bases monthly orders on the expected value of the monthly demand, what should Carolina's monthly order quantity be for this product?
b. Assume that each unit demanded generates $70 in revenue and that each unit ordered costs $50. How much will the company gain or lose in a month if it places an order based on your answer to part (a) and the actual demand for the item is 300 units?

23. The New York City Housing and Vacancy Survey showed a total of 59,324 rent-controlled housing units and 236,263 rent-stabilized units built in 1947 or later. For these rental units, the probability distributions for the number of persons living in the unit are given (U.S. Census Bureau website, January 12, 2004).

Number of Persons	Rent-Controlled	Rent-Stabilized
1	.61	.41
2	.27	.30
3	.07	.14
4	.04	.11
5	.01	.03
6	.00	.01

a. What is the expected value of the number of persons living in each type of unit?
b. What is the variance of the number of persons living in each type of unit?
c. Make some comparisons between the number of persons living in rent-controlled units and the number of persons living in rent-stabilized units.

24. The J. R. Ryland Computer Company is considering a plant expansion to enable the company to begin production of a new computer product. The company's president must determine whether to make the expansion a medium- or large-scale project. Demand for the new product is uncertain, which for planning purposes may be low demand, medium demand, or high demand. The probability estimates for demand are .20, .50, and .30, respectively. Letting x and y indicate the annual profit in thousands of dollars, the firm's planners developed the following profit forecasts for the medium- and large-scale expansion projects.

		Medium-Scale Expansion Profit		Large-Scale Expansion Profit	
		x	$f(x)$	y	$f(y)$
Demand	Low	50	.20	0	.20
	Medium	150	.50	100	.50
	High	200	.30	300	.30

a. Compute the expected value for the profit associated with the two expansion alternatives. Which decision is preferred for the objective of maximizing the expected profit?
b. Compute the variance for the profit associated with the two expansion alternatives. Which decision is preferred for the objective of minimizing the risk or uncertainty?

5.4 Binomial Probability Distribution

The binomial probability distribution is a discrete probability distribution that provides many applications. It is associated with a multiple-step experiment that we call the binomial experiment.

A Binomial Experiment

A **binomial experiment** exhibits the following four properties.

> PROPERTIES OF A BINOMIAL EXPERIMENT
>
> 1. The experiment consists of a sequence of n identical trials.
> 2. Two outcomes are possible on each trial. We refer to one outcome as a *success* and the other outcome as a *failure*.
> 3. The probability of a success, denoted by p, does not change from trial to trial. Consequently, the probability of a failure, denoted by $1 - p$, does not change from trial to trial.
> 4. The trials are independent.

Jakob Bernoulli (1654–1705), the first of the Bernoulli family of Swiss mathematicians, published a treatise on probability that contained the theory of permutations and combinations, as well as the binomial theorem.

If properties 2, 3, and 4 are present, we say the trials are generated by a Bernoulli process. If, in addition, property 1 is present, we say we have a binomial experiment. Figure 5.2 depicts one possible sequence of successes and failures for a binomial experiment involving eight trials.

In a binomial experiment, our interest is in the *number of successes occurring in the n trials*. If we let x denote the number of successes occurring in the n trials, we see that x can assume the values of 0, 1, 2, 3, . . . , n. Because the number of values is finite, x is a *discrete* random variable. The probability distribution associated with this random variable is called the **binomial probability distribution**. For example, consider the experiment of tossing a coin five times and on each toss observing whether the coin lands with a head or a tail on its upward face. Suppose we want to count the number of heads appearing over the five tosses. Does this experiment show the properties of a binomial experiment? What is the random variable of interest? Note that

1. The experiment consists of five identical trials; each trial involves the tossing of one coin.
2. Two outcomes are possible for each trial: a head or a tail. We can designate head a success and tail a failure.
3. The probability of a head and the probability of a tail are the same for each trial, with $p = .5$ and $1 - p = .5$.
4. The trials or tosses are independent because the outcome on any one trial is not affected by what happens on other trials or tosses.

FIGURE 5.2 ONE POSSIBLE SEQUENCE OF SUCCESSES AND FAILURES FOR AN EIGHT-TRIAL BINOMIAL EXPERIMENT

Property 1: The experiment consists of $n = 8$ identical trials.

Property 2: Each trial results in either success (*S*) or failure (*F*).

Trials →	1	2	3	4	5	6	7	8
Outcomes →	S	F	F	S	S	F	S	S

5.4 Binomial Probability Distribution

Thus, the properties of a binomial experiment are satisfied. The random variable of interest is $x =$ the number of heads appearing in the five trials. In this case, x can assume the values of 0, 1, 2, 3, 4, or 5.

As another example, consider an insurance salesperson who visits 10 randomly selected families. The outcome associated with each visit is classified as a success if the family purchases an insurance policy and a failure if the family does not. From past experience, the salesperson knows the probability that a randomly selected family will purchase an insurance policy is .10. Checking the properties of a binomial experiment, we observe that

1. The experiment consists of 10 identical trials; each trial involves contacting one family.
2. Two outcomes are possible on each trial: the family purchases a policy (success) or the family does not purchase a policy (failure).
3. The probabilities of a purchase and a nonpurchase are assumed to be the same for each sales call, with $p = .10$ and $1 - p = .90$.
4. The trials are independent because the families are randomly selected.

Because the four assumptions are satisfied, this example is a binomial experiment. The random variable of interest is the number of sales obtained in contacting the 10 families. In this case, x can assume the values of 0, 1, 2, 3, 4, 5, 6, 7, 8, 9, and 10.

Property 3 of the binomial experiment is called the *stationarity assumption* and is sometimes confused with property 4, independence of trials. To see how they differ, consider again the case of the salesperson calling on families to sell insurance policies. If, as the day wore on, the salesperson got tired and lost enthusiasm, the probability of success (selling a policy) might drop to .05, for example, by the tenth call. In such a case, property 3 (stationarity) would not be satisfied, and we would not have a binomial experiment. Even if property 4 held—that is, the purchase decisions of each family were made independently—it would not be a binomial experiment if property 3 was not satisfied.

In applications involving binomial experiments, a special mathematical formula, called the *binomial probability function,* can be used to compute the probability of x successes in the n trials. Using probability concepts introduced in Chapter 4, we will show in the context of an illustrative problem how the formula can be developed.

Martin Clothing Store Problem

Let us consider the purchase decisions of the next three customers who enter the Martin Clothing Store. On the basis of past experience, the store manager estimates the probability that any one customer will make a purchase is .30. What is the probability that two of the next three customers will make a purchase?

Using a tree diagram (Figure 5.3), we can see that the experiment of observing the three customers each making a purchase decision has eight possible outcomes. Using S to denote success (a purchase) and F to denote failure (no purchase), we are interested in experimental outcomes involving two successes in the three trials (purchase decisions). Next, let us verify that the experiment involving the sequence of three purchase decisions can be viewed as a binomial experiment. Checking the four requirements for a binomial experiment, we note that:

1. The experiment can be described as a sequence of three identical trials, one trial for each of the three customers who will enter the store.
2. Two outcomes—the customer makes a purchase (success) or the customer does not make a purchase (failure)—are possible for each trial.
3. The probability that the customer will make a purchase (.30) or will not make a purchase (.70) is assumed to be the same for all customers.
4. The purchase decision of each customer is independent of the decisions of the other customers.

FIGURE 5.3 TREE DIAGRAM FOR THE MARTIN CLOTHING STORE PROBLEM

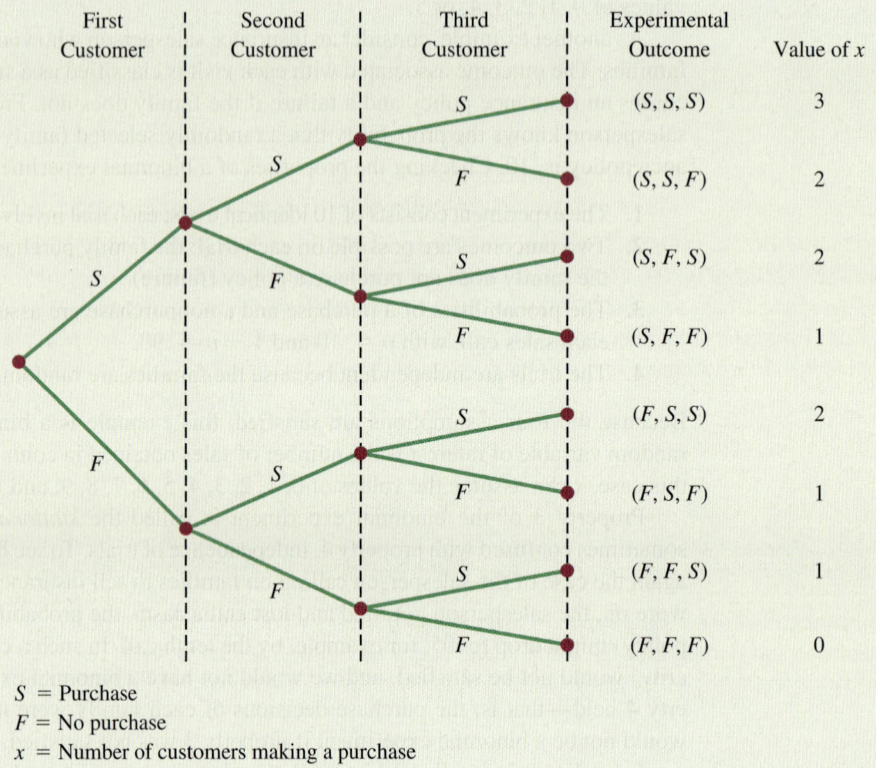

First Customer	Second Customer	Third Customer	Experimental Outcome	Value of x
S	S	S	(S, S, S)	3
S	S	F	(S, S, F)	2
S	F	S	(S, F, S)	2
S	F	F	(S, F, F)	1
F	S	S	(F, S, S)	2
F	S	F	(F, S, F)	1
F	F	S	(F, F, S)	1
F	F	F	(F, F, F)	0

S = Purchase
F = No purchase
x = Number of customers making a purchase

Hence, the properties of a binomial experiment are present.

The number of experimental outcomes resulting in exactly x successes in n trials can be computed using the following formula.[1]

NUMBER OF EXPERIMENTAL OUTCOMES PROVIDING EXACTLY x SUCCESSES IN n TRIALS

$$\binom{n}{x} = \frac{n!}{x!(n-x)!} \tag{5.6}$$

where

$$n! = n(n-1)(n-2) \cdots (2)(1)$$

and, by definition,

$$0! = 1$$

Now let us return to the Martin Clothing Store experiment involving three customer purchase decisions. Equation (5.6) can be used to determine the number of experimental

---

[1] This formula, introduced in Chapter 4, determines the number of combinations of n objects selected x at a time. For the binomial experiment, this combinatorial formula provides the number of experimental outcomes (sequences of n trials) resulting in x successes.

5.4 Binomial Probability Distribution

outcomes involving two purchases; that is, the number of ways of obtaining $x = 2$ successes in the $n = 3$ trials. From equation (5.6) we have

$$\binom{n}{x} = \binom{3}{2} = \frac{3!}{2!(3-2)!} = \frac{(3)(2)(1)}{(2)(1)(1)} = \frac{6}{2} = 3$$

Equation (5.6) shows that three of the experimental outcomes yield two successes. From Figure 5.3 we see that these three outcomes are denoted by (S, S, F), (S, F, S), and (F, S, S).

Using equation (5.6) to determine how many experimental outcomes have three successes (purchases) in the three trials, we obtain

$$\binom{n}{x} = \binom{3}{3} = \frac{3!}{3!(3-3)!} = \frac{3!}{3!0!} = \frac{(3)(2)(1)}{3(2)(1)(1)} = \frac{6}{6} = 1$$

From Figure 5.3 we see that the one experimental outcome with three successes is identified by (S, S, S).

We know that equation (5.6) can be used to determine the number of experimental outcomes that result in x successes. If we are to determine the probability of x successes in n trials, however, we must also know the probability associated with each of these experimental outcomes. Because the trials of a binomial experiment are independent, we can simply multiply the probabilities associated with each trial outcome to find the probability of a particular sequence of successes and failures.

The probability of purchases by the first two customers and no purchase by the third customer, denoted (S, S, F), is given by

$$pp(1-p)$$

With a .30 probability of a purchase on any one trial, the probability of a purchase on the first two trials and no purchase on the third is given by

$$(.30)(.30)(.70) = (.30)^2(.70) = .063$$

Two other experimental outcomes also result in two successes and one failure. The probabilities for all three experimental outcomes involving two successes follow.

Trial Outcomes				
1st Customer	2nd Customer	3rd Customer	Experimental Outcome	Probability of Experimental Outcome
Purchase	Purchase	No purchase	(S, S, F)	$pp(1-p) = p^2(1-p)$ $= (.30)^2(.70) = .063$
Purchase	No purchase	Purchase	(S, F, S)	$p(1-p)p = p^2(1-p)$ $= (.30)^2(.70) = .063$
No purchase	Purchase	Purchase	(F, S, S)	$(1-p)pp = p^2(1-p)$ $= (.30)^2(.70) = .063$

Observe that all three experimental outcomes with two successes have exactly the same probability. This observation holds in general. In any binomial experiment, all sequences of trial outcomes yielding x successes in n trials have the *same probability* of occurrence. The probability of each sequence of trials yielding x successes in n trials follows.

212 Chapter 5 Discrete Probability Distributions

$$\text{Probability of a particular sequence of trial outcomes with } x \text{ successes in } n \text{ trials} = p^x(1-p)^{(n-x)} \quad (5.7)$$

For the Martin Clothing Store, this formula shows that any experimental outcome with two successes has a probability of $p^2(1-p)^{(3-2)} = p^2(1-p)^1 = (.30)^2(.70)^1 = .063$.

Because equation (5.6) shows the number of outcomes in a binomial experiment with x successes and equation (5.7) gives the probability for each sequence involving x successes, we combine equations (5.6) and (5.7) to obtain the following **binomial probability function**.

BINOMIAL PROBABILITY FUNCTION

$$f(x) = \binom{n}{x} p^x (1-p)^{(n-x)} \quad (5.8)$$

where

$$x = \text{the number of successes}$$
$$p = \text{the probability of a success on one trial}$$
$$n = \text{the number of trials}$$
$$f(x) = \text{the probability of } x \text{ successes in } n \text{ trials}$$
$$\binom{n}{x} = \frac{n!}{x!(n-x)!}$$

For the binomial probability distribution, x is a discrete random variable with the probability function $f(x)$ applicable for values of $x = 0, 1, 2, \ldots, n$.

In the Martin Clothing Store example, let us use equation (5.8) to compute the probability that no customer makes a purchase, exactly one customer makes a purchase, exactly two customers make a purchase, and all three customers make a purchase. The calculations are summarized in Table 5.6, which gives the probability distribution of the number of customers making a purchase. Figure 5.4 is a graph of this probability distribution.

The binomial probability function can be applied to *any* binomial experiment. If we are satisfied that a situation demonstrates the properties of a binomial experiment and if we know the values of n and p, we can use equation (5.8) to compute the probability of x successes in the n trials.

TABLE 5.6 PROBABILITY DISTRIBUTION FOR THE NUMBER OF CUSTOMERS MAKING A PURCHASE

x	$f(x)$
0	$\frac{3!}{0!3!}(.30)^0(.70)^3 = .343$
1	$\frac{3!}{1!2!}(.30)^1(.70)^2 = .441$
2	$\frac{3!}{2!1!}(.30)^2(.70)^1 = .189$
3	$\frac{3!}{3!0!}(.30)^3(.70)^0 = \underline{.027}$
	1.000

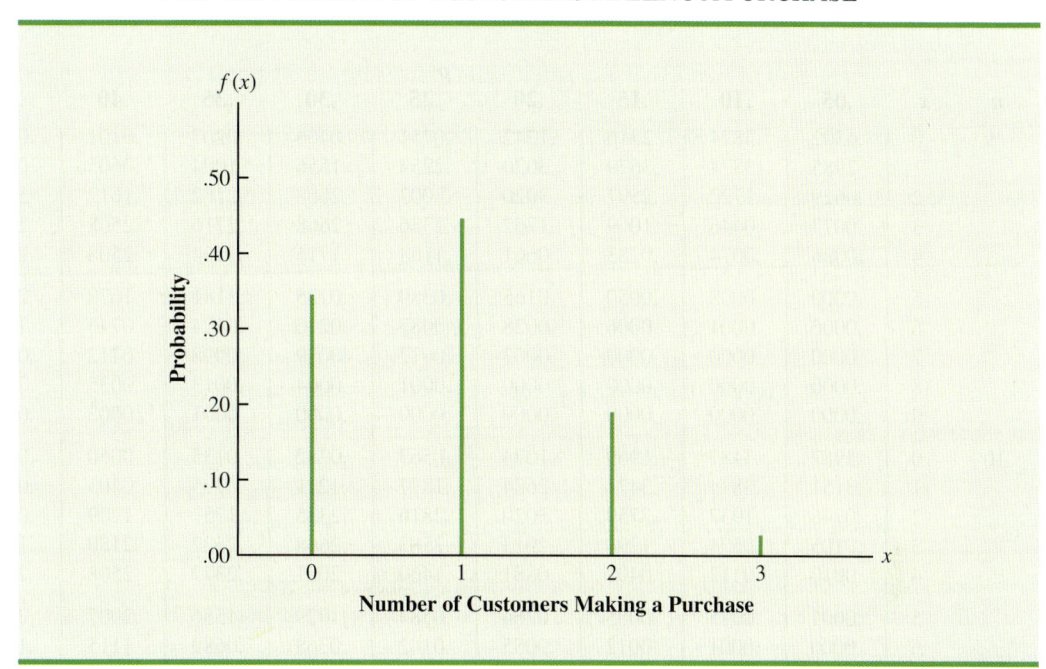

FIGURE 5.4 GRAPHICAL REPRESENTATION OF THE PROBABILITY DISTRIBUTION FOR THE NUMBER OF CUSTOMERS MAKING A PURCHASE

If we consider variations of the Martin experiment, such as 10 customers rather than 3 entering the store, the binomial probability function given by equation (5.8) is still applicable. Suppose we have a binomial experiment with $n = 10$, $x = 4$, and $p = .30$. The probability of making exactly four sales to 10 customers entering the store is

$$f(4) = \frac{10!}{4!6!} (.30)^4 (.70)^6 = .2001$$

Using Tables of Binomial Probabilities

Tables have been developed that give the probability of x successes in n trials for a binomial experiment. The tables are generally easy to use and quicker than equation (5.8). Table 5 of Appendix B provides such a table of binomial probabilities. A portion of this table appears in Table 5.7. To use this table, we must specify the values of n, p, and x for the binomial experiment of interest. In the example at the top of Table 5.7, we see that the probability of $x = 3$ successes in a binomial experiment with $n = 10$ and $p = .40$ is .2150. You can use equation (5.8) to verify that you would obtain the same answer if you used the binomial probability function directly.

Now let us use Table 5.7 to verify the probability of four successes in 10 trials for the Martin Clothing Store problem. Note that the value of $f(4) = .2001$ can be read directly from the table of binomial probabilities, with $n = 10$, $x = 4$, and $p = .30$.

With modern calculators, these tables are almost unnecessary. It is easy to evaluate equation (5.8) directly.

Even though the tables of binomial probabilities are relatively easy to use, it is impossible to have tables that show all possible values of n and p that might be encountered in a binomial experiment. However, with today's calculators, using equation (5.8) to calculate the desired probability is not difficult, especially if the number of trials is not large. In the exercises, you should practice using equation (5.8) to compute the binomial probabilities unless the problem specifically requests that you use the binomial probability table.

TABLE 5.7 SELECTED VALUES FROM THE BINOMIAL PROBABILITY TABLE
EXAMPLE: $n = 10$, $x = 3$, $p = .40$; $f(3) = .2150$

						p					
n	x	.05	.10	.15	.20	.25	.30	.35	.40	.45	.50
9	0	.6302	.3874	.2316	.1342	.0751	.0404	.0207	.0101	.0046	.0020
	1	.2985	.3874	.3679	.3020	.2253	.1556	.1004	.0605	.0339	.0176
	2	.0629	.1722	.2597	.3020	.3003	.2668	.2162	.1612	.1110	.0703
	3	.0077	.0446	.1069	.1762	.2336	.2668	.2716	.2508	.2119	.1641
	4	.0006	.0074	.0283	.0661	.1168	.1715	.2194	.2508	.2600	.2461
	5	.0000	.0008	.0050	.0165	.0389	.0735	.1181	.1672	.2128	.2461
	6	.0000	.0001	.0006	.0028	.0087	.0210	.0424	.0743	.1160	.1641
	7	.0000	.0000	.0000	.0003	.0012	.0039	.0098	.0212	.0407	.0703
	8	.0000	.0000	.0000	.0000	.0001	.0004	.0013	.0035	.0083	.0176
	9	.0000	.0000	.0000	.0000	.0000	.0000	.0001	.0003	.0008	.0020
10	0	.5987	.3487	.1969	.1074	.0563	.0282	.0135	.0060	.0025	.0010
	1	.3151	.3874	.3474	.2684	.1877	.1211	.0725	.0403	.0207	.0098
	2	.0746	.1937	.2759	.3020	.2816	.2335	.1757	.1209	.0763	.0439
	3	.0105	.0574	.1298	.2013	.2503	.2668	.2522	**.2150**	.1665	.1172
	4	.0010	.0112	.0401	.0881	.1460	.2001	.2377	.2508	.2384	.2051
	5	.0001	.0015	.0085	.0264	.0584	.1029	.1536	.2007	.2340	.2461
	6	.0000	.0001	.0012	.0055	.0162	.0368	.0689	.1115	.1596	.2051
	7	.0000	.0000	.0001	.0008	.0031	.0090	.0212	.0425	.0746	.1172
	8	.0000	.0000	.0000	.0001	.0004	.0014	.0043	.0106	.0229	.0439
	9	.0000	.0000	.0000	.0000	.0000	.0001	.0005	.0016	.0042	.0098
	10	.0000	.0000	.0000	.0000	.0000	.0000	.0000	.0001	.0003	.0010

Statistical software packages such as Minitab and spreadsheet packages such as Excel also provide a capability for computing binomial probabilities. Consider the Martin Clothing Store example with $n = 10$ and $p = .30$. Figure 5.5 shows the binomial probabilities generated by Minitab for all possible values of x. Note that these values are the same as those found in the $p = .30$ column of Table 5.7. Appendix 5.1 gives the step-by-step procedure for using Minitab to generate the output in Figure 5.5. Appendix 5.2 describes how Excel can be used to compute binomial probabilities.

Expected Value and Variance for the Binomial Distribution

In Section 5.3 we provided formulas for computing the expected value and variance of a discrete random variable. In the special case where the random variable has a binomial distribution with a known number of trials n and a known probability of success p, the general formulas for the expected value and variance can be simplified. The results follow.

EXPECTED VALUE AND VARIANCE FOR THE BINOMIAL DISTRIBUTION

$$E(x) = \mu = np \quad (5.9)$$

$$\text{Var}(x) = \sigma^2 = np(1 - p) \quad (5.10)$$

5.4 Binomial Probability Distribution

FIGURE 5.5 MINITAB OUTPUT SHOWING BINOMIAL PROBABILITIES FOR THE MARTIN CLOTHING STORE PROBLEM

x	P(X = x)
0	0.0282
1	0.1211
2	0.2335
3	0.2668
4	0.2001
5	0.1029
6	0.0368
7	0.0090
8	0.0014
9	0.0001
10	0.0000

For the Martin Clothing Store problem with three customers, we can use equation (5.9) to compute the expected number of customers who will make a purchase.

$$E(x) = np = 3(.30) = .9$$

Suppose that for the next month the Martin Clothing Store forecasts that 1000 customers will enter the store. What is the expected number of customers who will make a purchase? The answer is $\mu = np = (1000)(.3) = 300$. Thus, to increase the expected number of purchases, Martin's must induce more customers to enter the store and/or somehow increase the probability that any individual customer will make a purchase after entering.

For the Martin Clothing Store problem with three customers, we see that the variance and standard deviation for the number of customers who will make a purchase are

$$\sigma^2 = np(1 - p) = 3(.3)(.7) = .63$$
$$\sigma = \sqrt{.63} = .79$$

For the next 1000 customers entering the store, the variance and standard deviation for the number of customers who will make a purchase are

$$\sigma^2 = np(1 - p) = 1000(.3)(.7) = 210$$
$$\sigma = \sqrt{210} = 14.49$$

NOTES AND COMMENTS

1. The binomial table in Appendix B shows values of p up to and including $p = .95$. Some sources of the binomial table only show values of p up to and including $p = .50$. It would appear that such a table cannot be used when the probability of success exceeds $p = .50$. However, the table can be used by noting that the probability of $n - x$ failures is also the probability of x successes. Thus, when the probability of success is greater than $p = .50$, we can compute the probability of $n - x$ failures instead. The probability of failure, $1 - p$, will be less than .50 when $p > .50$.

2. Some sources present the binomial table in a cumulative form. In using such a table, one must subtract entries in the table to find the probability of exactly x success in n trials. For example, $f(2) = P(x \leq 2) - P(x \leq 1)$. The binomial table we provide in Appendix B provides $f(2)$ directly. To compute cumulative probabilities using the binomial table in Appendix B, sum the entries in the table. For example, to determine the cumulative probability $P(x \leq 2)$, compute the sum $f(0) + f(1) + f(2)$.

Exercises

Methods

25. Consider a binomial experiment with two trials and $p = .4$.
    a. Draw a tree diagram for this experiment (see Figure 5.3).
    b. Compute the probability of one success, $f(1)$.
    c. Compute $f(0)$.
    d. Compute $f(2)$.
    e. Compute the probability of at least one success.
    f. Compute the expected value, variance, and standard deviation.

26. Consider a binomial experiment with $n = 10$ and $p = .10$.
    a. Compute $f(0)$.
    b. Compute $f(2)$.
    c. Compute $P(x \leq 2)$.
    d. Compute $P(x \geq 1)$.
    e. Compute $E(x)$.
    f. Compute $Var(x)$ and σ.

27. Consider a binomial experiment with $n = 20$ and $p = .70$.
    a. Compute $f(12)$.
    b. Compute $f(16)$.
    c. Compute $P(x \geq 16)$.
    d. Compute $P(x \leq 15)$.
    e. Compute $E(x)$.
    f. Compute $Var(x)$ and σ.

Applications

28. A Harris Interactive survey for InterContinental Hotels & Resorts asked respondents, "When traveling internationally, do you generally venture out on your own to experience culture, or stick with your tour group and itineraries?" The survey found that 23% of the respondents stick with their tour group (*USA Today,* January 21, 2004).
    a. In a sample of six international travelers, what is the probability that two will stick with their tour group?
    b. In a sample of six international travelers, what is the probability that at least two will stick with their tour group?
    c. In a sample of 10 international travelers, what is the probability that none will stick with the tour group?

29. In San Francisco, 30% of workers take public transportation daily (*USA Today,* December 21, 2005).
    a. In a sample of 10 workers, what is the probability that exactly 3 workers take public transportation daily?
    b. In a sample of 10 workers, what is the probability that at least 3 workers take public transportation daily?

30. When a new machine is functioning properly, only 3% of the items produced are defective. Assume that we will randomly select two parts produced on the machine and that we are interested in the number of defective parts found.
    a. Describe the conditions under which this situation would be a binomial experiment.
    b. Draw a tree diagram similar to Figure 5.3 showing this problem as a two-trial experiment.
    c. How many experimental outcomes result in exactly one defect being found?
    d. Compute the probabilities associated with finding no defects, exactly one defect, and two defects.

5.4 Binomial Probability Distribution

31. A Randstad/Harris interactive survey reported that 25% of employees said their company is loyal to them (*USA Today*, November 11, 2009). Suppose 10 employees are selected randomly and will be interviewed about company loyalty.
    a. Is the selection of 10 employees a binomial experiment? Explain.
    b. What is the probability that none of the 10 employees will say their company is loyal to them?
    c. What is the probability that 4 of the 10 employees will say their company is loyal to them?
    d. What is the probability that at least 2 of the 10 employees will say their company is loyal to them?

32. Military radar and missile detection systems are designed to warn a country of an enemy attack. A reliability question is whether a detection system will be able to identify an attack and issue a warning. Assume that a particular detection system has a .90 probability of detecting a missile attack. Use the binomial probability distribution to answer the following questions.
    a. What is the probability that a single detection system will detect an attack?
    b. If two detection systems are installed in the same area and operate independently, what is the probability that at least one of the systems will detect the attack?
    c. If three systems are installed, what is the probability that at least one of the systems will detect the attack?
    d. Would you recommend that multiple detection systems be used? Explain.

33. Twelve of the top 20 finishers in the 2009 PGA Championship at Hazeltine National Golf Club in Chaska, Minnesota, used a Titleist brand golf ball (GolfBallTest website, November 12, 2009). Suppose these results are representative of the probability that a randomly selected PGA Tour player uses a Titleist brand golf ball. For a sample of 15 PGA Tour players, make the following calculations.
    a. Compute the probability that exactly 10 of the 15 PGA Tour players use a Titleist brand golf ball.
    b. Compute the probability that more than 10 of the 15 PGA Tour players use a Titleist brand golf ball.
    c. For a sample of 15 PGA Tour players, compute the expected number of players who use a Titleist brand golf ball.
    d. For a sample of 15 PGA Tour players, compute the variance and standard deviation of the number of players who use a Titleist brand golf ball.

34. The Census Bureau's Current Population Survey shows that 28% of individuals, ages 25 and older, have completed four years of college (*The New York Times Almanac*, 2006). For a sample of 15 individuals, ages 25 and older, answer the following questions:
    a. What is the probability that four will have completed four years of college?
    b. What is the probability that three or more will have completed four years of college?

35. A university found that 20% of its students withdraw without completing the introductory statistics course. Assume that 20 students registered for the course.
    a. Compute the probability that two or fewer will withdraw.
    b. Compute the probability that exactly four will withdraw.
    c. Compute the probability that more than three will withdraw.
    d. Compute the expected number of withdrawals.

36. According to a survey conducted by TD Ameritrade, one out of four investors have exchange-traded funds in their portfolios (*USA Today*, January 11, 2007). Consider a sample of 20 investors.
    a. Compute the probability that exactly four investors have exchange-traded funds in their portfolios.
    b. Compute the probability that at least two of the investors have exchange-traded funds in their portfolios.
    c. If you found that exactly 12 of the investors have exchange-traded funds in their portfolios, would you doubt the accuracy of the survey results?
    d. Compute the expected number of investors who have exchange-traded funds in their portfolios.

218 Chapter 5 Discrete Probability Distributions

37. Twenty-three percent of automobiles are not covered by insurance (CNN, February 23, 2006). On a particular weekend, 35 automobiles are involved in traffic accidents.
    a. What is the expected number of these automobiles that are not covered by insurance?
    b. What are the variance and standard deviation?

5.5 Poisson Probability Distribution

In this section we consider a discrete random variable that is often useful in estimating the number of occurrences over a specified interval of time or space. For example, the random variable of interest might be the number of arrivals at a car wash in one hour, the number of repairs needed in 10 miles of highway, or the number of leaks in 100 miles of pipeline. If the following two properties are satisfied, the number of occurrences is a random variable described by the **Poisson probability distribution**.

The Poisson probability distribution is often used to model random arrivals in waiting line situations.

> **PROPERTIES OF A POISSON EXPERIMENT**
>
> 1. The probability of an occurrence is the same for any two intervals of equal length.
> 2. The occurrence or nonoccurrence in any interval is independent of the occurrence or nonoccurrence in any other interval.

The **Poisson probability function** is defined by equation (5.11).

Siméon Poisson taught mathematics at the École Polytechnique in Paris from 1802 to 1808. In 1837, he published a work entitled, "Researches on the Probability of Criminal and Civil Verdicts," which includes a discussion of what later became known as the Poisson distribution.

> **POISSON PROBABILITY FUNCTION**
>
> $$f(x) = \frac{\mu^x e^{-\mu}}{x!} \quad (5.11)$$
>
> where
>
> $f(x)$ = the probability of x occurrences in an interval
> μ = expected value or mean number of occurrences in an interval
> e = 2.71828

For the Poisson probability distribution, x is a discrete random variable indicating the number of occurrences in the interval. Since there is no stated upper limit for the number of occurrences, the probability function $f(x)$ is applicable for values $x = 0, 1, 2, \ldots$ without limit. In practical applications, x will eventually become large enough so that $f(x)$ is approximately zero and the probability of any larger values of x becomes negligible.

An Example Involving Time Intervals

Suppose that we are interested in the number of arrivals at the drive-up teller window of a bank during a 15-minute period on weekday mornings. If we can assume that the probability of a car arriving is the same for any two time periods of equal length and that the arrival or nonarrival of a car in any time period is independent of the arrival or nonarrival in any other time period, the Poisson probability function is applicable. Suppose these assumptions are satisfied and an analysis of historical data shows that the average number of cars arriving in a 15-minute period of time is 10; in this case, the following probability function applies.

Bell Labs used the Poisson distribution to model the arrival of telephone calls.

$$f(x) = \frac{10^x e^{-10}}{x!}$$

5.5 Poisson Probability Distribution

The random variable here is x = number of cars arriving in any 15-minute period.

If management wanted to know the probability of exactly five arrivals in 15 minutes, we would set $x = 5$ and thus obtain

$$\text{Probability of exactly 5 arrivals in 15 minutes} = f(5) = \frac{10^5 e^{-10}}{5!} = .0378$$

Although this probability was determined by evaluating the probability function with $\mu = 10$ and $x = 5$, it is often easier to refer to a table for the Poisson distribution. The table provides probabilities for specific values of x and μ. We included such a table as Table 7 of Appendix B. For convenience, we reproduced a portion of this table as Table 5.8. Note that to use the table of Poisson probabilities, we need know only the values of x and μ. From Table 5.8 we see that the probability of five arrivals in a 15-minute period is found by locating the value in the row of the table corresponding to $x = 5$ and the column of the table corresponding to $\mu = 10$. Hence, we obtain $f(5) = .0378$.

In the preceding example, the mean of the Poisson distribution is $\mu = 10$ arrivals per 15-minute period. A property of the Poisson distribution is that the mean of the distribution and the variance of the distribution are *equal*. Thus, the variance for the number of arrivals during 15-minute periods is $\sigma^2 = 10$. The standard deviation is $\sigma = \sqrt{10} = 3.16$.

Our illustration involves a 15-minute period, but other time periods can be used. Suppose we want to compute the probability of one arrival in a 3-minute period. Because

A property of the Poisson distribution is that the mean and variance are equal.

TABLE 5.8 SELECTED VALUES FROM THE POISSON PROBABILITY TABLES
EXAMPLE: $\mu = 10, x = 5; f(5) = .0378$

					μ					
x	9.1	9.2	9.3	9.4	9.5	9.6	9.7	9.8	9.9	10
0	.0001	.0001	.0001	.0001	.0001	.0001	.0001	.0001	.0001	.0000
1	.0010	.0009	.0009	.0008	.0007	.0007	.0006	.0005	.0005	.0005
2	.0046	.0043	.0040	.0037	.0034	.0031	.0029	.0027	.0025	.0023
3	.0140	.0131	.0123	.0115	.0107	.0100	.0093	.0087	.0081	.0076
4	.0319	.0302	.0285	.0269	.0254	.0240	.0226	.0213	.0201	.0189
5	.0581	.0555	.0530	.0506	.0483	.0460	.0439	.0418	.0398	**.0378**
6	.0881	.0851	.0822	.0793	.0764	.0736	.0709	.0682	.0656	.0631
7	.1145	.1118	.1091	.1064	.1037	.1010	.0982	.0955	.0928	.0901
8	.1302	.1286	.1269	.1251	.1232	.1212	.1191	.1170	.1148	.1126
9	.1317	.1315	.1311	.1306	.1300	.1293	.1284	.1274	.1263	.1251
10	.1198	.1210	.1219	.1228	.1235	.1241	.1245	.1249	.1250	.1251
11	.0991	.1012	.1031	.1049	.1067	.1083	.1098	.1112	.1125	.1137
12	.0752	.0776	.0799	.0822	.0844	.0866	.0888	.0908	.0928	.0948
13	.0526	.0549	.0572	.0594	.0617	.0640	.0662	.0685	.0707	.0729
14	.0342	.0361	.0380	.0399	.0419	.0439	.0459	.0479	.0500	.0521
15	.0208	.0221	.0235	.0250	.0265	.0281	.0297	.0313	.0330	.0347
16	.0118	.0127	.0137	.0147	.0157	.0168	.0180	.0192	.0204	.0217
17	.0063	.0069	.0075	.0081	.0088	.0095	.0103	.0111	.0119	.0128
18	.0032	.0035	.0039	.0042	.0046	.0051	.0055	.0060	.0065	.0071
19	.0015	.0017	.0019	.0021	.0023	.0026	.0028	.0031	.0034	.0037
20	.0007	.0008	.0009	.0010	.0011	.0012	.0014	.0015	.0017	.0019
21	.0003	.0003	.0004	.0004	.0005	.0006	.0006	.0007	.0008	.0009
22	.0001	.0001	.0002	.0002	.0002	.0002	.0003	.0003	.0004	.0004
23	.0000	.0001	.0001	.0001	.0001	.0001	.0001	.0001	.0002	.0002
24	.0000	.0000	.0000	.0000	.0000	.0000	.0000	.0001	.0001	.0001

10 is the expected number of arrivals in a 15-minute period, we see that 10/15 = 2/3 is the expected number of arrivals in a 1-minute period and that (2/3)(3 minutes) = 2 is the expected number of arrivals in a 3-minute period. Thus, the probability of x arrivals in a 3-minute time period with $\mu = 2$ is given by the following Poisson probability function.

$$f(x) = \frac{2^x e^{-2}}{x!}$$

The probability of one arrival in a 3-minute period is calculated as follows:

$$\text{Probability of exactly 1 arrival in 3 minutes} = f(1) = \frac{2^1 e^{-2}}{1!} = .2707$$

Earlier we computed the probability of five arrivals in a 15-minute period; it was .0378. Note that the probability of one arrival in a 3-minute period (.2707) is not the same. When computing a Poisson probability for a different time interval, we must first convert the mean arrival rate to the time period of interest and then compute the probability.

An Example Involving Length or Distance Intervals

Let us illustrate an application not involving time intervals in which the Poisson distribution is useful. Suppose we are concerned with the occurrence of major defects in a highway one month after resurfacing. We will assume that the probability of a defect is the same for any two highway intervals of equal length and that the occurrence or nonoccurrence of a defect in any one interval is independent of the occurrence or nonoccurrence of a defect in any other interval. Hence, the Poisson distribution can be applied.

Suppose we learn that major defects one month after resurfacing occur at the average rate of two per mile. Let us find the probability of no major defects in a particular 3-mile section of the highway. Because we are interested in an interval with a length of 3 miles, μ = (2 defects/mile)(3 miles) = 6 represents the expected number of major defects over the 3-mile section of highway. Using equation (5.11), the probability of no major defects is $f(0) = 6^0 e^{-6}/0! = .0025$. Thus, it is unlikely that no major defects will occur in the 3-mile section. In fact, this example indicates a 1 − .0025 = .9975 probability of at least one major defect in the 3-mile highway section.

Exercises

Methods

38. Consider a Poisson distribution with $\mu = 3$.
    a. Write the appropriate Poisson probability function.
    b. Compute $f(2)$.
    c. Compute $f(1)$.
    d. Compute $P(x \geq 2)$.

39. Consider a Poisson distribution with a mean of two occurrences per time period.
    a. Write the appropriate Poisson probability function.
    b. What is the expected number of occurrences in three time periods?
    c. Write the appropriate Poisson probability function to determine the probability of x occurrences in three time periods.
    d. Compute the probability of two occurrences in one time period.
    e. Compute the probability of six occurrences in three time periods.
    f. Compute the probability of five occurrences in two time periods.

Applications

40. Phone calls arrive at the rate of 48 per hour at the reservation desk for Regional Airways.
    a. Compute the probability of receiving three calls in a 5-minute interval of time.
    b. Compute the probability of receiving exactly 10 calls in 15 minutes.
    c. Suppose no calls are currently on hold. If the agent takes 5 minutes to complete the current call, how many callers do you expect to be waiting by that time? What is the probability that none will be waiting?
    d. If no calls are currently being processed, what is the probability that the agent can take 3 minutes for personal time without being interrupted by a call?

41. During the period of time that a local university takes phone-in registrations, calls come in at the rate of one every two minutes.
    a. What is the expected number of calls in one hour?
    b. What is the probability of three calls in five minutes?
    c. What is the probability of no calls in a five-minute period?

42. More than 50 million guests stay at bed and breakfasts (B&Bs) each year. The website for the Bed and Breakfast Inns of North America, which averages seven visitors per minute, enables many B&Bs to attract guests (*Time*, September 2001).
    a. Compute the probability of no website visitors in a one-minute period.
    b. Compute the probability of two or more website visitors in a one-minute period.
    c. Compute the probability of one or more website visitors in a 30-second period.
    d. Compute the probability of five or more website visitors in a one-minute period.

43. Airline passengers arrive randomly and independently at the passenger-screening facility at a major international airport. The mean arrival rate is 10 passengers per minute.
    a. Compute the probability of no arrivals in a one-minute period.
    b. Compute the probability that three or fewer passengers arrive in a one-minute period.
    c. Compute the probability of no arrivals in a 15-second period.
    d. Compute the probability of at least one arrival in a 15-second period.

44. An average of 15 aircraft accidents occur each year (*The World Almanac and Book of Facts*, 2004).
    a. Compute the mean number of aircraft accidents per month.
    b. Compute the probability of no accidents during a month.
    c. Compute the probability of exactly one accident during a month.
    d. Compute the probability of more than one accident during a month.

45. The National Safety Council (NSC) estimates that off-the-job accidents cost U.S. businesses almost $200 billion annually in lost productivity (National Safety Council, March 2006). Based on NSC estimates, companies with 50 employees are expected to average three employee off-the-job accidents per year. Answer the following questions for companies with 50 employees.
    a. What is the probability of no off-the-job accidents during a one-year period?
    b. What is the probability of at least two off-the-job accidents during a one-year period?
    c. What is the expected number of off-the-job accidents during six months?
    d. What is the probability of no off-the-job accidents during the next six months?

5.6 Hypergeometric Probability Distribution

The **hypergeometric probability distribution** is closely related to the binomial distribution. The two probability distributions differ in two key ways. With the hypergeometric distribution, the trials are not independent; and the probability of success changes from trial to trial.

In the usual notation for the hypergeometric distribution, r denotes the number of elements in the population of size N labeled success, and $N - r$ denotes the number of elements in the population labeled failure. The **hypergeometric probability function** is used to compute the probability that in a random selection of n elements, selected without replacement, we obtain x elements labeled success and $n - x$ elements labeled failure. For this outcome to occur, we must obtain x successes from the r successes in the population and $n - x$ failures from the $N - r$ failures. The following hypergeometric probability function provides $f(x)$, the probability of obtaining x successes in n trials.

HYPERGEOMETRIC PROBABILITY FUNCTION

$$f(x) = \frac{\binom{r}{x}\binom{N-r}{n-x}}{\binom{N}{n}} \tag{5.12}$$

where

x = the number of successes
n = the number of trials
$f(x)$ = the probability of x successes in n trials
N = the number of elements in the population
r = the number of elements in the population labeled success

Note that $\binom{N}{n}$ represents the number of ways n elements can be selected from a population of size N; $\binom{r}{x}$ represents the number of ways that x successes can be selected from a total of r successes in the population; and $\binom{N-r}{n-x}$ represents the number of ways that $n - x$ failures can be selected from a total of $N - r$ failures in the population.

For the hypergeometric probability distribution, x is a discrete random variable and the probability function $f(x)$ given by equation (5.12) is usually applicable for values of $x = 0, 1, 2, \ldots, n$. However, only values of x where the number of observed successes is *less than or equal to* the number of successes in the population ($x \leq r$) and where the number of observed failures is *less than or equal to* the number of failures in the population ($n - x \leq N - r$) are valid. If these two conditions do not hold for one or more values of x, the corresponding value of $f(x) = 0$, indicating that the probability of this value of x is zero.

To illustrate the computations involved in using equation (5.12), let us consider the following quality control application. Electric fuses produced by Ontario Electric are packaged in boxes of 12 units each. Suppose an inspector randomly selects 3 of the 12 fuses in a box for testing. If the box contains exactly 5 defective fuses, what is the probability that the inspector will find exactly 1 of the 3 fuses defective? In this application, $n = 3$ and $N = 12$. With $r = 5$ defective fuses in the box the probability of finding $x = 1$ defective fuse is

$$f(1) = \frac{\binom{5}{1}\binom{7}{2}}{\binom{12}{3}} = \frac{\left(\frac{5!}{1!4!}\right)\left(\frac{7!}{2!5!}\right)}{\left(\frac{12!}{3!9!}\right)} = \frac{(5)(21)}{220} = .4773$$

5.6 Hypergeometric Probability Distribution

Now suppose that we wanted to know the probability of finding *at least* 1 defective fuse. The easiest way to answer this question is to first compute the probability that the inspector does not find any defective fuses. The probability of $x = 0$ is

$$f(0) = \frac{\binom{5}{0}\binom{7}{3}}{\binom{12}{3}} = \frac{\left(\frac{5!}{0!5!}\right)\left(\frac{7!}{3!4!}\right)}{\left(\frac{12!}{3!9!}\right)} = \frac{(1)(35)}{220} = .1591$$

With a probability of zero defective fuses $f(0) = .1591$, we conclude that the probability of finding at least 1 defective fuse must be $1 - .1591 = .8409$. Thus, there is a reasonably high probability that the inspector will find at least 1 defective fuse.

The mean and variance of a hypergeometric distribution are as follows.

$$E(x) = \mu = n\left(\frac{r}{N}\right) \qquad (5.13)$$

$$\text{Var}(x) = \sigma^2 = n\left(\frac{r}{N}\right)\left(1 - \frac{r}{N}\right)\left(\frac{N-n}{N-1}\right) \qquad (5.14)$$

In the preceding example $n = 3$, $r = 5$, and $N = 12$. Thus, the mean and variance for the number of defective fuses are

$$\mu = n\left(\frac{r}{N}\right) = 3\left(\frac{5}{12}\right) = 1.25$$

$$\sigma^2 = n\left(\frac{r}{N}\right)\left(1 - \frac{r}{N}\right)\left(\frac{N-n}{N-1}\right) = 3\left(\frac{5}{12}\right)\left(1 - \frac{5}{12}\right)\left(\frac{12-3}{12-1}\right) = .60$$

The standard deviation is $\sigma = \sqrt{.60} = .77$.

NOTES AND COMMENTS

Consider a hypergeometric distribution with n trials. Let $p = (r/N)$ denote the probability of a success on the first trial. If the population size is large, the term $(N - n)/(N - 1)$ in equation (5.14) approaches 1. As a result, the expected value and variance can be written $E(x) = np$ and $\text{Var}(x) = np(1 - p)$. Note that these expressions are the same as the expressions used to compute the expected value and variance of a binomial distribution, as in equations (5.9) and (5.10). When the population size is large, a hypergeometric distribution can be approximated by a binomial distribution with n trials and a probability of success $p = (r/N)$.

Exercises

Methods

46. Suppose $N = 10$ and $r = 3$. Compute the hypergeometric probabilities for the following values of n and x.
    a. $n = 4, x = 1$
    b. $n = 2, x = 2$
    c. $n = 2, x = 0$
    d. $n = 4, x = 2$
    e. $n = 4, x = 4$

47. Suppose $N = 15$ and $r = 4$. What is the probability of $x = 3$ for $n = 10$?

Applications

48. In a survey conducted by the Gallup Organization, respondents were asked, "What is your favorite sport to watch?" Football and basketball ranked number one and two in terms of preference (Gallup website, January 3, 2004). Assume that in a group of 10 individuals, 7 prefer football and 3 prefer basketball. A random sample of 3 of these individuals is selected.
    a. What is the probability that exactly 2 prefer football?
    b. What is the probability that the majority (either 2 or 3) prefer football?

49. Blackjack, or twenty-one as it is frequently called, is a popular gambling game played in Las Vegas casinos. A player is dealt two cards. Face cards (jacks, queens, and kings) and tens have a point value of 10. Aces have a point value of 1 or 11. A 52-card deck contains 16 cards with a point value of 10 (jacks, queens, kings, and tens) and four aces.
    a. What is the probability that both cards dealt are aces or 10-point cards?
    b. What is the probability that both of the cards are aces?
    c. What is the probability that both of the cards have a point value of 10?
    d. A blackjack is a 10-point card and an ace for a value of 21. Use your answers to parts (a), (b), and (c) to determine the probability that a player is dealt blackjack. (*Hint:* Part (d) is not a hypergeometric problem. Develop your own logical relationship as to how the hypergeometric probabilities from parts (a), (b), and (c) can be combined to answer this question.)

50. Axline Computers manufactures personal computers at two plants, one in Texas and the other in Hawaii. The Texas plant has 40 employees; the Hawaii plant has 20. A random sample of 10 employees is to be asked to fill out a benefits questionnaire.
    a. What is the probability that none of the employees in the sample work at the plant in Hawaii?
    b. What is the probability that one of the employees in the sample works at the plant in Hawaii?
    c. What is the probability that two or more of the employees in the sample work at the plant in Hawaii?
    d. What is the probability that nine of the employees in the sample work at the plant in Texas?

51. The Zagat Restaurant Survey provides food, decor, and service ratings for some of the top restaurants across the United States. For 15 restaurants located in Boston, the average price of a dinner, including one drink and tip, was $48.60. You are leaving for a business trip to Boston and will eat dinner at three of these restaurants. Your company will reimburse you for a maximum of $50 per dinner. Business associates familiar with these restaurants have told you that the meal cost at one-third of these restaurants will exceed $50. Suppose that you randomly select three of these restaurants for dinner.
    a. What is the probability that none of the meals will exceed the cost covered by your company?
    b. What is the probability that one of the meals will exceed the cost covered by your company?
    c. What is the probability that two of the meals will exceed the cost covered by your company?
    d. What is the probability that all three of the meals will exceed the cost covered by your company?

52. The Troubled Asset Relief Program (TARP), passed by the U.S. Congress in October 2008, provided $700 billion in assistance for the struggling U.S. economy. Over $200 billion was given to troubled financial institutions with the hope that there would be an increase in lending to help jump-start the economy. But three months later, a Federal Reserve survey found that two-thirds of the banks that had received TARP funds had tightened terms for business loans (*The Wall Street Journal*, February 3, 2009). Of the 10 banks that were the biggest recipients of TARP funds, only 3 had actually increased lending during this period.

Increased Lending	Decreased Lending
BB&T	Bank of America
Sun Trust Banks	Capital One
U.S. Bancorp	Citigroup
	Fifth Third Bancorp
	J.P. Morgan Chase
	Regions Financial
	U.S. Bancorp

For the purposes of this exercise, assume that you will randomly select 3 of these 10 banks for a study that will continue to monitor bank lending practices. Let x be a random variable indicating the number of banks in the study that had increased lending.

a. What is $f(0)$? What is your interpretation of this value?
b. What is $f(3)$? What is your interpretation of this value?
c. Compute $f(1)$ and $f(2)$. Show the probability distribution for the number of banks in the study that had increased lending. What value of x has the highest probability?
d. What is the probability that the study will have at least one bank that had increased lending?
e. Compute the expected value, variance, and standard deviation for the random variable.

Summary

A random variable provides a numerical description of the outcome of an experiment. The probability distribution for a random variable describes how the probabilities are distributed over the values the random variable can assume. For any discrete random variable x, the probability distribution is defined by a probability function, denoted by $f(x)$, which provides the probability associated with each value of the random variable. Once the probability function is defined, we can compute the expected value, variance, and standard deviation for the random variable.

The binomial distribution can be used to determine the probability of x successes in n trials whenever the experiment has the following properties:

1. The experiment consists of a sequence of n identical trials.
2. Two outcomes are possible on each trial, one called success and the other failure.
3. The probability of a success p does not change from trial to trial. Consequently, the probability of failure, $1 - p$, does not change from trial to trial.
4. The trials are independent.

When the four properties hold, the binomial probability function can be used to determine the probability of obtaining x successes in n trials. Formulas were also presented for the mean and variance of the binomial distribution.

The Poisson distribution is used when it is desirable to determine the probability of obtaining x occurrences over an interval of time or space. The following assumptions are necessary for the Poisson distribution to be applicable.

1. The probability of an occurrence of the event is the same for any two intervals of equal length.
2. The occurrence or nonoccurrence of the event in any interval is independent of the occurrence or nonoccurrence of the event in any other interval.

A third discrete probability distribution, the hypergeometric, was introduced in Section 5.6. Like the binomial, it is used to compute the probability of x successes in n trials. But, in contrast to the binomial, the probability of success changes from trial to trial.

Glossary

Random variable A numerical description of the outcome of an experiment.
Discrete random variable A random variable that may assume either a finite number of values or an infinite sequence of values.
Continuous random variable A random variable that may assume any numerical value in an interval or collection of intervals.
Probability distribution A description of how the probabilities are distributed over the values of the random variable.
Probability function A function, denoted by $f(x)$, that provides the probability that x assumes a particular value for a discrete random variable.
Discrete uniform probability distribution A probability distribution for which each possible value of the random variable has the same probability.
Expected value A measure of the central location of a random variable.
Variance A measure of the variability, or dispersion, of a random variable.
Standard deviation The positive square root of the variance.
Binomial experiment An experiment having the four properties stated at the beginning of Section 5.4.
Binomial probability distribution A probability distribution showing the probability of x successes in n trials of a binomial experiment.
Binomial probability function The function used to compute binomial probabilities.
Poisson probability distribution A probability distribution showing the probability of x occurrences of an event over a specified interval of time or space.
Poisson probability function The function used to compute Poisson probabilities.
Hypergeometric probability distribution A probability distribution showing the probability of x successes in n trials from a population with r successes and $N - r$ failures.
Hypergeometric probability function The function used to compute hypergeometric probabilities.

Key Formulas

Discrete Uniform Probability Function

$$f(x) = 1/n \tag{5.3}$$

Expected Value of a Discrete Random Variable

$$E(x) = \mu = \Sigma x f(x) \tag{5.4}$$

Variance of a Discrete Random Variable

$$\text{Var}(x) = \sigma^2 = \Sigma (x - \mu)^2 f(x) \tag{5.5}$$

Number of Experimental Outcomes Providing Exactly x Successes in n Trials

$$\binom{n}{x} = \frac{n!}{x!(n-x)!} \tag{5.6}$$

Binomial Probability Function

$$f(x) = \binom{n}{x} p^x (1-p)^{(n-x)} \tag{5.8}$$

Expected Value for the Binomial Distribution

$$E(x) = \mu = np \tag{5.9}$$

Variance for the Binomial Distribution

$$\text{Var}(x) = \sigma^2 = np(1-p) \tag{5.10}$$

Poisson Probability Function

$$f(x) = \frac{\mu^x e^{-\mu}}{x!} \tag{5.11}$$

Hypergeometric Probability Function

$$f(x) = \frac{\binom{r}{x}\binom{N-r}{n-x}}{\binom{N}{n}} \tag{5.12}$$

Expected Value for the Hypergeometric Distribution

$$E(x) = \mu = n\left(\frac{r}{N}\right) \tag{5.13}$$

Variance for the Hypergeometric Distribution

$$\text{Var}(x) = \sigma^2 = n\left(\frac{r}{N}\right)\left(1 - \frac{r}{N}\right)\left(\frac{N-n}{N-1}\right) \tag{5.14}$$

Supplementary Exercises

53. A poll conducted by Zogby International showed that of those Americans who said music plays a "very important" role in their lives, 30% said their local radio stations "always" play the kind of music they like (Zogby website, January 12, 2004). Suppose a sample of 800 people who say music plays an important role in their lives is taken.
    a. How many would you expect to say that their local radio stations always play the kind of music they like?
    b. What is the standard deviation of the number of respondents who think their local radio stations always play the kind of music they like?
    c. What is the standard deviation of the number of respondents who do not think their local radio stations always play the kind of music they like?

54. The *Barron's* Big Money Poll asked 131 investment managers across the United States about their short-term investment outlook (*Barron's,* October 28, 2002). Their responses showed that 4% were very bullish, 39% were bullish, 29% were neutral, 21% were bearish, and 7% were very bearish. Let x be the random variable reflecting the level of optimism about the market. Set $x = 5$ for very bullish down through $x = 1$ for very bearish.
    a. Develop a probability distribution for the level of optimism of investment managers.
    b. Compute the expected value for the level of optimism.
    c. Compute the variance and standard deviation for the level of optimism.
    d. Comment on what your results imply about the level of optimism and its variability.

55. Cars arrive at a car wash randomly and independently; the probability of an arrival is the same for any two time intervals of equal length. The mean arrival rate is 15 cars per hour. What is the probability that 20 or more cars will arrive during any given hour of operation?

56. The American Association of Individual Investors publishes an annual guide to the top mutual funds (*The Individual Investor's Guide to the Top Mutual Funds,* 22e, American Association of Individual Investors, 2003). The total risk ratings for 29 categories of mutual funds are as follows.

Total Risk	Number of Fund Categories
Low	7
Below Average	6
Average	3
Above Average	6
High	7

    a. Let $x = 1$ for low risk up through $x = 5$ for high risk, and develop a probability distribution for level of risk.
    b. What are the expected value and variance for total risk?
    c. It turns out that 11 of the fund categories were bond funds. For the bond funds, 7 categories were rated low and 4 were rated below average. Compare the total risk of the bond funds with the 18 categories of stock funds.

57. A new automated production process averages 1.5 breakdowns per day. Because of the cost associated with a breakdown, management is concerned about the possibility of having three or more breakdowns during a day. Assume that breakdowns occur randomly, that the probability of a breakdown is the same for any two time intervals of equal length, and that breakdowns in one period are independent of breakdowns in other periods. What is the probability of having three or more breakdowns during a day?

58. The budgeting process for a midwestern college resulted in expense forecasts for the coming year (in $ millions) of $9, $10, $11, $12, and $13. Because the actual expenses are unknown, the following respective probabilities are assigned: .3, .2, .25, .05, and .2.
    a. Show the probability distribution for the expense forecast.
    b. What is the expected value of the expense forecast for the coming year?
    c. What is the variance of the expense forecast for the coming year?
    d. If income projections for the year are estimated at $12 million, comment on the financial position of the college.

59. A regional director responsible for business development in the state of Pennsylvania is concerned about the number of small business failures. If the mean number of small business failures per month is 10, what is the probability that exactly four small businesses will fail during a given month? Assume that the probability of a failure is the same for any two months and that the occurrence or nonoccurrence of a failure in any month is independent of failures in any other month.

60. A survey showed that the average commuter spends about 26 minutes on a one-way door-to-door trip from home to work. In addition, 5% of commuters reported a one-way commute of more than one hour (Bureau of Transportation Statistics website, January 12, 2004).
    a. If 20 commuters are surveyed on a particular day, what is the probability that 3 will report a one-way commute of more than one hour?
    b. If 20 commuters are surveyed on a particular day, what is the probability that none will report a one-way commute of more than one hour?
    c. If a company has 2000 employees, what is the expected number of employees who have a one-way commute of more than one hour?
    d. If a company has 2000 employees, what are the variance and standard deviation of the number of employees who have a one-way commute of more than one hour?

61. Customer arrivals at a bank are random and independent; the probability of an arrival in any one-minute period is the same as the probability of an arrival in any other one-minute period. Answer the following questions, assuming a mean arrival rate of three customers per minute.

a. What is the probability of exactly three arrivals in a one-minute period?
b. What is the probability of at least three arrivals in a one-minute period?

62. A political action group is planning to interview home owners to assess the impact caused by a recent slump in housing prices. According to a *Wall Street Journal*/Harris Interactive Personal Finance poll, 26% of individuals aged 18–34, 50% of individuals aged 35–44, and 88% of individuals aged 55 and over are home owners (All Business website, January 23, 2008).
   a. How many people from the 18–34 age group must be sampled to find an expected number of at least 20 home owners?
   b. How many people from the 35–44 age group must be sampled to find an expected number of at least 20 home owners?
   c. How many people from the 55 and over age group must be sampled to find an expected number of at least 20 home owners?
   d. If the number of 18–34 year olds sampled is equal to the value identified in part (a), what is the standard deviation of the number who will be home owners?
   e. If the number of 35–44 year olds sampled is equal to the value identified in part (b), what is the standard deviation of the number who will be home owners?

63. A deck of playing cards contains 52 cards, four of which are aces. What is the probability that the deal of a five-card hand provides:
   a. A pair of aces?
   b. Exactly one ace?
   c. No aces?
   d. At least one ace?

64. Many companies use a quality control technique called acceptance sampling to monitor incoming shipments of parts, raw materials, and so on. In the electronics industry, component parts are commonly shipped from suppliers in large lots. Inspection of a sample of n components can be viewed as the n trials of a binomial experiment. The outcome for each component tested (trial) will be that the component is classified as good or defective. Reynolds Electronics accepts a lot from a particular supplier if the defective components in the lot do not exceed 1%. Suppose a random sample of five items from a recent shipment is tested.
   a. Assume that 1% of the shipment is defective. Compute the probability that no items in the sample are defective.
   b. Assume that 1% of the shipment is defective. Compute the probability that exactly one item in the sample is defective.
   c. What is the probability of observing one or more defective items in the sample if 1% of the shipment is defective?
   d. Would you feel comfortable accepting the shipment if one item was found to be defective? Why or why not?

65. *U.S. News & World Report*'s ranking of America's best graduate schools of business showed Harvard University and Stanford University in a tie for first place. In addition, 7 of the top 10 graduate schools of business showed students with an average undergraduate grade point average (GPA) of 3.50 or higher (*America's Best Graduate Schools, 2009 edition*, *U.S. News & World Report*). Suppose that we randomly select 2 of the top 10 graduate schools of business.
   a. What is the probability that exactly one school has students with an average undergraduate GPA of 3.50 or higher?
   b. What is the probability that both schools have students with an average undergraduate GPA of 3.50 or higher?
   c. What is the probability that neither school has students with an average undergraduate GPA of 3.50 or higher?

66. The unemployment rate in the state of Arizona is 4.1% (CNN Money website, May 2, 2007). Assume that 100 employable people in Arizona are selected randomly.
   a. What is the expected number of people who are unemployed?
   b. What are the variance and standard deviation of the number of people who are unemployed?

Appendix 5.1 Discrete Probability Distributions Using Minitab

Statistical packages such as Minitab offer a relatively easy and efficient procedure for computing binomial probabilities. In this appendix, we show the step-by-step procedure for determining the binomial probabilities for the Martin Clothing Store problem in Section 5.4. Recall that the desired binomial probabilities are based on $n = 10$ and $p = .30$. Before beginning the Minitab routine, the user must enter the desired values of the random variable x into a column of the worksheet. We entered the values 0, 1, 2, . . . , 10 in column 1 (see Figure 5.5) to generate the entire binomial probability distribution. The Minitab steps to obtain the desired binomial probabilities follow.

Step 1. Select the **Calc** menu
Step 2. Choose **Probability Distributions**
Step 3. Choose **Binomial**
Step 4. When the Binomial Distribution dialog box appears:
  Select **Probability**
  Enter 10 in the **Number of trials** box
  Enter .3 in the **Event probability** box
  Enter C1 in the **Input column** box
  Click **OK**

The Minitab output with the binomial probabilities will appear as shown in Figure 5.5.

Minitab provides Poisson and hypergeometric probabilities in a similar manner. For instance, to compute Poisson probabilities the only differences are in step 3, where the **Poisson** option would be selected, and step 4, where the **Mean** would be entered rather than the number of trials and the probability of success.

Appendix 5.2 Discrete Probability Distributions Using Excel

Excel provides functions for computing probabilities for the binomial, Poisson, and hypergeometric distributions introduced in this chapter. The Excel function for computing binomial probabilities is BINOMDIST. It has four arguments: x (the number of successes), n (the number of trials), p (the probability of success), and cumulative. FALSE is used for the fourth argument (cumulative) if we want the probability of x successes, and TRUE is used for the fourth argument if we want the cumulative probability of x or fewer successes. Here we show how to compute the probabilities of 0 through 10 successes for the Martin Clothing Store problem in Section 5.4 (see Figure 5.5).

As we describe the worksheet development, refer to Figure 5.6; the formula worksheet is set in the background, and the value worksheet appears in the foreground. We entered the number of trials (10) into cell B1, the probability of success into cell B2, and the values for the random variable into cells B5:B15. The following steps will generate the desired probabilities:

Step 1. Use the BINOMDIST function to compute the probability of $x = 0$ by entering the following formula into cell C5:

$$=\text{BINOMDIST(B5,\$B\$1,\$B\$2,FALSE)}$$

Step 2. Copy the formula in cell C5 into cells C6:C15.

Appendix 5.2 Discrete Probability Distributions Using Excel

FIGURE 5.6 EXCEL WORKSHEET FOR COMPUTING BINOMIAL PROBABILITIES

	A	B	C	D
1	Number of Trials (*n*)	10		
2	Probability of Success (*p*)	0.3		
3				
4		*x*	*f*(*x*)	
5		0	=BINOMDIST(B5,$B$1,$B$2,FALSE)	
6		1	=BINOMDIST(B6,$B$1,$B$2,FALSE)	
7		2	=BINOMDIST(B7,$B$1,$B$2,FALSE)	
8		3	=BINOMDIST(B8,$B$1,$B$2,FALSE)	
9		4	=BINOMDIST(B9,$B$1,$B$2,FALSE)	
10		5	=BINOMDIST(B10,$B$1,$B$2,FALSE)	
11		6	=BINOMDIST(B11,$B$1,$B$2,FALSE)	
12		7	=BINOMDIST(B12,$B$1,$B$2,FALSE)	
13		8	=BINOMDIST(B13,$B$1,$B$2,FALSE)	
14		9	=BINOMDIST(B14,$B$1,$B$2,FALSE)	
15		10	=BINOMDIST(B15,$B$1,$B$2,FALSE)	
16				

	A	B	C	D
1	Number of Trials (*n*)		10	
2	Probability of Success (*p*)		0.3	
3				
4		*x*	*f*(*x*)	
5		0	0.0282	
6		1	0.1211	
7		2	0.2335	
8		3	0.2668	
9		4	0.2001	
10		5	0.1029	
11		6	0.0368	
12		7	0.0090	
13		8	0.0014	
14		9	0.0001	
15		10	0.0000	
16				

The value worksheet in Figure 5.6 shows that the probabilities obtained are the same as in Figure 5.5. Poisson and hypergeometric probabilities can be computed in a similar fashion. The POISSON and HYPGEOMDIST functions are used. Excel's Insert Function dialog box can help the user in entering the proper arguments for these functions (see Appendix E).

CHAPTER 6

Continuous Probability Distributions

CONTENTS

STATISTICS IN PRACTICE: PROCTER & GAMBLE

6.1 UNIFORM PROBABILITY DISTRIBUTION
Area as a Measure of Probability

6.2 NORMAL PROBABILITY DISTRIBUTION
Normal Curve
Standard Normal Probability Distribution
Computing Probabilities for Any Normal Probability Distribution
Grear Tire Company Problem

6.3 NORMAL APPROXIMATION OF BINOMIAL PROBABILITIES

6.4 EXPONENTIAL PROBABILITY DISTRIBUTION
Computing Probabilities for the Exponential Distribution
Relationship Between the Poisson and Exponential Distributions

STATISTICS *in* PRACTICE

PROCTER & GAMBLE*
CINCINNATI, OHIO

Procter & Gamble (P&G) produces and markets such products as detergents, disposable diapers, over-the-counter pharmaceuticals, dentifrices, bar soaps, mouthwashes, and paper towels. Worldwide, it has the leading brand in more categories than any other consumer products company. Since its merger with Gillette, P&G also produces and markets razors, blades, and many other personal care products.

As a leader in the application of statistical methods in decision making, P&G employs people with diverse academic backgrounds: engineering, statistics, operations research, and business. The major quantitative technologies for which these people provide support are probabilistic decision and risk analysis, advanced simulation, quality improvement, and quantitative methods (e.g., linear programming, regression analysis, probability analysis).

The Industrial Chemicals Division of P&G is a major supplier of fatty alcohols derived from natural substances such as coconut oil and from petroleum-based derivatives. The division wanted to know the economic risks and opportunities of expanding its fatty-alcohol production facilities, so it called in P&G's experts in probabilistic decision and risk analysis to help. After structuring and modeling the problem, they determined that the key to profitability was the cost difference between the petroleum- and coconut-based raw materials. Future costs were unknown, but the analysts were able to approximate them with the following continuous random variables.

x = the coconut oil price per pound of fatty alcohol

and

y = the petroleum raw material price per pound of fatty alcohol

Because the key to profitability was the difference between these two random variables, a third random

Some of Procter & Gamble's many well-known products.

variable, $d = x - y$, was used in the analysis. Experts were interviewed to determine the probability distributions for x and y. In turn, this information was used to develop a probability distribution for the difference in prices d. This continuous probability distribution showed a .90 probability that the price difference would be $.0655 or less and a .50 probability that the price difference would be $.035 or less. In addition, there was only a .10 probability that the price difference would be $.0045 or less.[†]

The Industrial Chemicals Division thought that being able to quantify the impact of raw material price differences was key to reaching a consensus. The probabilities obtained were used in a sensitivity analysis of the raw material price difference. The analysis yielded sufficient insight to form the basis for a recommendation to management.

The use of continuous random variables and their probability distributions was helpful to P&G in analyzing the economic risks associated with its fatty-alcohol production. In this chapter, you will gain an understanding of continuous random variables and their probability distributions, including one of the most important probability distributions in statistics, the normal distribution.

---

*The authors are indebted to Joel Kahn of Procter & Gamble for providing this Statistics in Practice.

[†]The price differences stated here have been modified to protect proprietary data.

In the preceding chapter we discussed discrete random variables and their probability distributions. In this chapter we turn to the study of continuous random variables. Specifically, we discuss three continuous probability distributions: the uniform, the normal, and the exponential.

A fundamental difference separates discrete and continuous random variables in terms of how probabilities are computed. For a discrete random variable, the probability function $f(x)$ provides the probability that the random variable assumes a particular value. With continuous random variables, the counterpart of the probability function is the **probability density function**, also denoted by $f(x)$. The difference is that the probability density function does not directly provide probabilities. However, the area under the graph of $f(x)$ corresponding to a given interval does provide the probability that the continuous random variable x assumes a value in that interval. So when we compute probabilities for continuous random variables, we are computing the probability that the random variable assumes any value in an interval.

Because the area under the graph of $f(x)$ at any particular point is zero, one of the implications of the definition of probability for continuous random variables is that the probability of any particular value of the random variable is zero. In Section 6.1 we demonstrate these concepts for a continuous random variable that has a uniform distribution.

Much of the chapter is devoted to describing and showing applications of the normal distribution. The normal distribution is of major importance because of its wide applicability and its extensive use in statistical inference. The chapter closes with a discussion of the exponential distribution. The exponential distribution is useful in applications involving such factors as waiting times and service times.

6.1 Uniform Probability Distribution

Consider the random variable x representing the flight time of an airplane traveling from Chicago to New York. Suppose the flight time can be any value in the interval from 120 minutes to 140 minutes. Because the random variable x can assume any value in that interval, x is a continuous rather than a discrete random variable. Let us assume that sufficient actual flight data are available to conclude that the probability of a flight time within any 1-minute interval is the same as the probability of a flight time within any other 1-minute interval contained in the larger interval from 120 to 140 minutes. With every 1-minute interval being equally likely, the random variable x is said to have a **uniform probability distribution**. The probability density function, which defines the uniform distribution for the flight-time random variable, is

Whenever the probability is proportional to the length of the interval, the random variable is uniformly distributed.

$$f(x) = \begin{cases} 1/20 & \text{for } 120 \leq x \leq 140 \\ 0 & \text{elsewhere} \end{cases}$$

Figure 6.1 is a graph of this probability density function. In general, the uniform probability density function for a random variable x is defined by the following formula.

UNIFORM PROBABILITY DENSITY FUNCTION

$$f(x) = \begin{cases} \dfrac{1}{b-a} & \text{for } a \leq x \leq b \\ 0 & \text{elsewhere} \end{cases} \tag{6.1}$$

For the flight-time random variable, $a = 120$ and $b = 140$.

6.1 Uniform Probability Distribution

FIGURE 6.1 UNIFORM PROBABILITY DISTRIBUTION FOR FLIGHT TIME

As noted in the introduction, for a continuous random variable, we consider probability only in terms of the likelihood that a random variable assumes a value within a specified interval. In the flight-time example, an acceptable probability question is: What is the probability that the flight time is between 120 and 130 minutes? That is, what is $P(120 \leq x \leq 130)$? Because the flight time must be between 120 and 140 minutes and because the probability is described as being uniform over this interval, we feel comfortable saying $P(120 \leq x \leq 130) = .50$. In the following subsection we show that this probability can be computed as the area under the graph of $f(x)$ from 120 to 130 (see Figure 6.2).

Area as a Measure of Probability

Let us make an observation about the graph in Figure 6.2. Consider the area under the graph of $f(x)$ in the interval from 120 to 130. The area is rectangular, and the area of a rectangle is simply the width multiplied by the height. With the width of the interval equal to $130 - 120 = 10$ and the height equal to the value of the probability density function $f(x) = 1/20$, we have area = width × height = $10(1/20) = 10/20 = .50$.

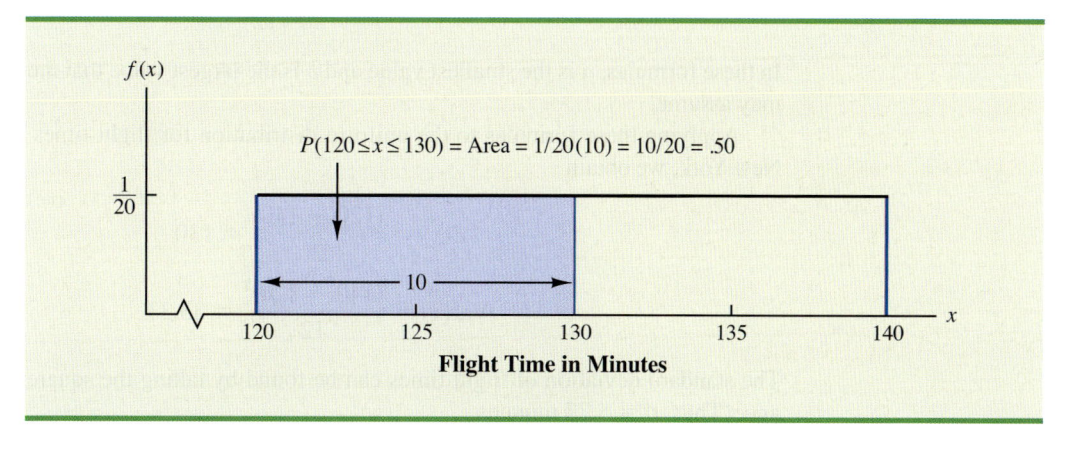

FIGURE 6.2 AREA PROVIDES PROBABILITY OF A FLIGHT TIME BETWEEN 120 AND 130 MINUTES

What observation can you make about the area under the graph of $f(x)$ and probability? They are identical! Indeed, this observation is valid for all continuous random variables. Once a probability density function $f(x)$ is identified, the probability that x takes a value between some lower value x_1 and some higher value x_2 can be found by computing the area under the graph of $f(x)$ over the interval from x_1 to x_2.

Given the uniform distribution for flight time and using the interpretation of area as probability, we can answer any number of probability questions about flight times. For example, what is the probability of a flight time between 128 and 136 minutes? The width of the interval is $136 - 128 = 8$. With the uniform height of $f(x) = 1/20$, we see that $P(128 \leq x \leq 136) = 8(1/20) = .40$.

Note that $P(120 \leq x \leq 140) = 20(1/20) = 1$; that is, the total area under the graph of $f(x)$ is equal to 1. This property holds for all continuous probability distributions and is the analog of the condition that the sum of the probabilities must equal 1 for a discrete probability function. For a continuous probability density function, we must also require that $f(x) \geq 0$ for all values of x. This requirement is the analog of the requirement that $f(x) \geq 0$ for discrete probability functions.

Two major differences stand out between the treatment of continuous random variables and the treatment of their discrete counterparts.

1. We no longer talk about the probability of the random variable assuming a particular value. Instead, we talk about the probability of the random variable assuming a value within some given interval.

2. The probability of a continuous random variable assuming a value within some given interval from x_1 to x_2 is defined to be the area under the graph of the probability density function between x_1 and x_2. Because a single point is an interval of zero width, this implies that the probability of a continuous random variable assuming any particular value exactly is zero. It also means that the probability of a continuous random variable assuming a value in any interval is the same whether or not the endpoints are included.

To see that the probability of any single point is 0, refer to Figure 6.2 and compute the probability of a single point, say, $x = 125$. $P(x = 125) = P(125 \leq x \leq 125) = 0(1/20) = 0$.

The calculation of the expected value and variance for a continuous random variable is analogous to that for a discrete random variable. However, because the computational procedure involves integral calculus, we leave the derivation of the appropriate formulas to more advanced texts.

For the uniform continuous probability distribution introduced in this section, the formulas for the expected value and variance are

$$E(x) = \frac{a+b}{2}$$

$$\text{Var}(x) = \frac{(b-a)^2}{12}$$

In these formulas, a is the smallest value and b is the largest value that the random variable may assume.

Applying these formulas to the uniform distribution for flight times from Chicago to New York, we obtain

$$E(x) = \frac{(120 + 140)}{2} = 130$$

$$\text{Var}(x) = \frac{(140 - 120)^2}{12} = 33.33$$

The standard deviation of flight times can be found by taking the square root of the variance. Thus, $\sigma = 5.77$ minutes.

NOTES AND COMMENTS

To see more clearly why the height of a probability density function is not a probability, think about a random variable with the following uniform probability distribution.

$$f(x) = \begin{cases} 2 & \text{for } 0 \leq x \leq .5 \\ 0 & \text{elsewhere} \end{cases}$$

The height of the probability density function, $f(x)$, is 2 for values of x between 0 and .5. However, we know probabilities can never be greater than 1. Thus, we see that $f(x)$ cannot be interpreted as the probability of x.

Exercises

Methods

1. The random variable x is known to be uniformly distributed between 1.0 and 1.5.
   a. Show the graph of the probability density function.
   b. Compute $P(x = 1.25)$.
   c. Compute $P(1.0 \leq x \leq 1.25)$.
   d. Compute $P(1.20 < x < 1.5)$.

2. The random variable x is known to be uniformly distributed between 10 and 20.
   a. Show the graph of the probability density function.
   b. Compute $P(x < 15)$.
   c. Compute $P(12 \leq x \leq 18)$.
   d. Compute $E(x)$.
   e. Compute $\text{Var}(x)$.

Applications

3. Delta Air Lines quotes a flight time of 2 hours, 5 minutes for its flights from Cincinnati to Tampa. Suppose we believe that actual flight times are uniformly distributed between 2 hours and 2 hours, 20 minutes.
   a. Show the graph of the probability density function for flight time.
   b. What is the probability that the flight will be no more than 5 minutes late?
   c. What is the probability that the flight will be more than 10 minutes late?
   d. What is the expected flight time?

4. Most computer languages include a function that can be used to generate random numbers. In Excel, the RAND function can be used to generate random numbers between 0 and 1. If we let x denote a random number generated using RAND, then x is a continuous random variable with the following probability density function.

$$f(x) = \begin{cases} 1 & \text{for } 0 \leq x \leq 1 \\ 0 & \text{elsewhere} \end{cases}$$

   a. Graph the probability density function.
   b. What is the probability of generating a random number between .25 and .75?
   c. What is the probability of generating a random number with a value less than or equal to .30?
   d. What is the probability of generating a random number with a value greater than .60?
   e. Generate 50 random numbers by entering =RAND() into 50 cells of an Excel worksheet.
   f. Compute the mean and standard deviation for the random numbers in part (e).

5. The driving distance for the top 100 golfers on the PGA tour is between 284.7 and 310.6 yards (*Golfweek,* March 29, 2003). Assume that the driving distance for these golfers is uniformly distributed over this interval.
   a. Give a mathematical expression for the probability density function of driving distance.
   b. What is the probability that the driving distance for one of these golfers is less than 290 yards?
   c. What is the probability that the driving distance for one of these golfers is at least 300 yards?
   d. What is the probability that the driving distance for one of these golfers is between 290 and 305 yards?
   e. How many of these golfers drive the ball at least 290 yards?

6. On average, 30-minute television sitcoms have 22 minutes of programming (CNBC, February 23, 2006). Assume that the probability distribution for minutes of programming can be approximated by a uniform distribution from 18 minutes to 26 minutes.
   a. What is the probability that a sitcom will have 25 or more minutes of programming?
   b. What is the probability that a sitcom will have between 21 and 25 minutes of programming?
   c. What is the probability that a sitcom will have more than 10 minutes of commercials or other nonprogramming interruptions?

7. Suppose we are interested in bidding on a piece of land and we know one other bidder is interested.[1] The seller announced that the highest bid in excess of $10,000 will be accepted. Assume that the competitor's bid x is a random variable that is uniformly distributed between $10,000 and $15,000.
   a. Suppose you bid $12,000. What is the probability that your bid will be accepted?
   b. Suppose you bid $14,000. What is the probability that your bid will be accepted?
   c. What amount should you bid to maximize the probability that you get the property?
   d. Suppose you know someone who is willing to pay you $16,000 for the property. Would you consider bidding less than the amount in part (c)? Why or why not?

6.2 Normal Probability Distribution

Abraham de Moivre, a French mathematician, published The Doctrine of Chances *in 1733. He derived the normal distribution.*

The most important probability distribution for describing a continuous random variable is the **normal probability distribution**. The normal distribution has been used in a wide variety of practical applications in which the random variables are heights and weights of people, test scores, scientific measurements, amounts of rainfall, and other similar values. It is also widely used in statistical inference, which is the major topic of the remainder of this book. In such applications, the normal distribution provides a description of the likely results obtained through sampling.

Normal Curve

The form, or shape, of the normal distribution is illustrated by the bell-shaped normal curve in Figure 6.3. The probability density function that defines the bell-shaped curve of the normal distribution follows.

---
[1]This exercise is based on a problem suggested to us by Professor Roger Myerson of Northwestern University.

6.2 Normal Probability Distribution

FIGURE 6.3 BELL-SHAPED CURVE FOR THE NORMAL DISTRIBUTION

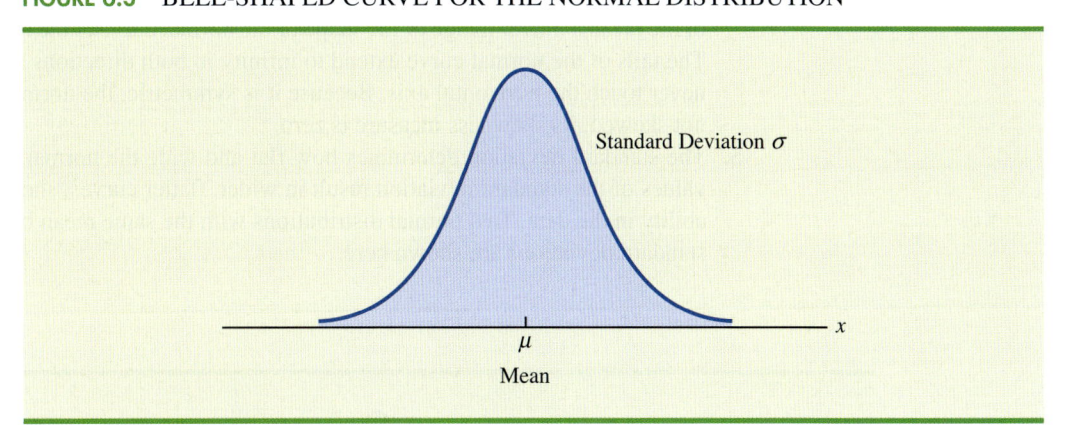

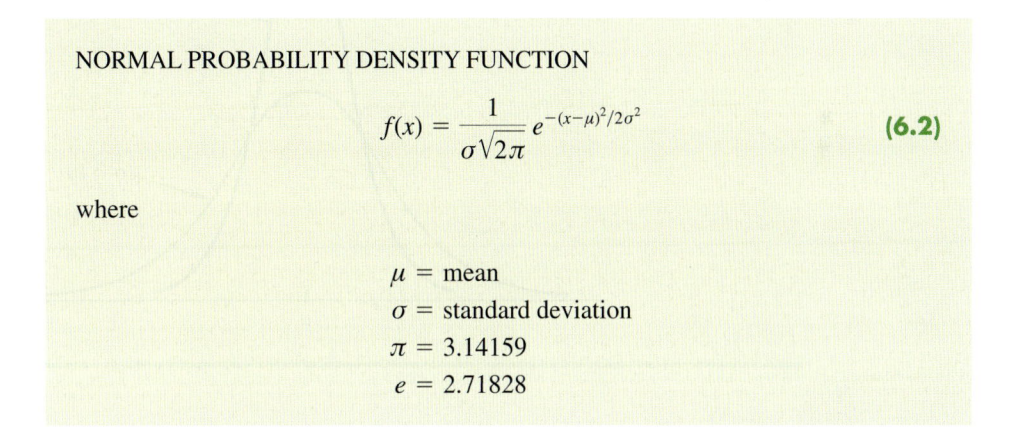

NORMAL PROBABILITY DENSITY FUNCTION

$$f(x) = \frac{1}{\sigma\sqrt{2\pi}} e^{-(x-\mu)^2/2\sigma^2} \tag{6.2}$$

where

μ = mean
σ = standard deviation
π = 3.14159
e = 2.71828

We make several observations about the characteristics of the normal distribution.

The normal curve has two parameters, μ and σ. They determine the location and shape of the normal distribution.

1. The entire family of normal distributions is differentiated by two parameters: the mean μ and the standard deviation σ.
2. The highest point on the normal curve is at the mean, which is also the median and mode of the distribution.
3. The mean of the distribution can be any numerical value: negative, zero, or positive. Three normal distributions with the same standard deviation but three different means (-10, 0, and 20) are shown here.

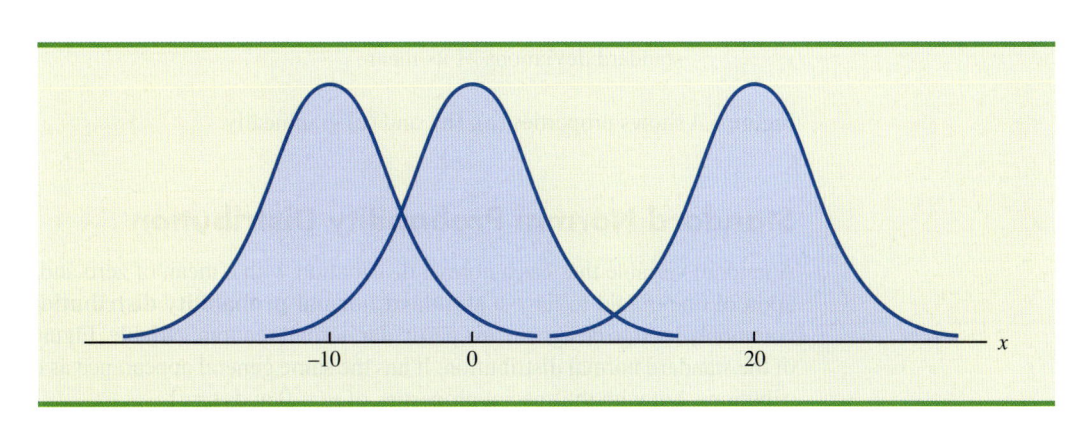

4. The normal distribution is symmetric, with the shape of the normal curve to the left of the mean a mirror image of the shape of the normal curve to the right of the mean. The tails of the normal curve extend to infinity in both directions and theoretically never touch the horizontal axis. Because it is symmetric, the normal distribution is not skewed; its skewness measure is zero.
5. The standard deviation determines how flat and wide the normal curve is. Larger values of the standard deviation result in wider, flatter curves, showing more variability in the data. Two normal distributions with the same mean but with different standard deviations are shown here.

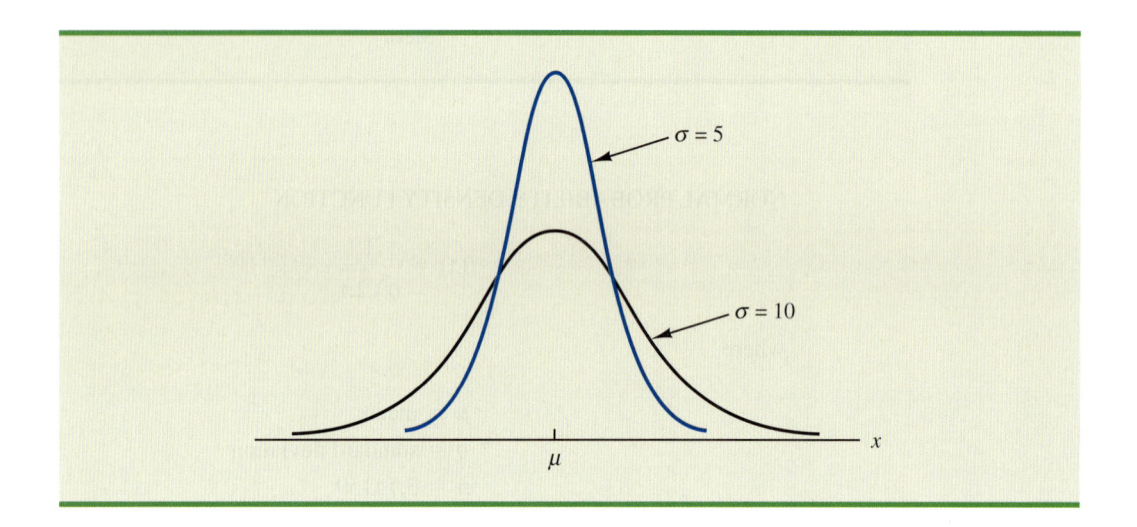

6. Probabilities for the normal random variable are given by areas under the normal curve. The total area under the curve for the normal distribution is 1. Because the distribution is symmetric, the area under the curve to the left of the mean is .50 and the area under the curve to the right of the mean is .50.
7. The percentage of values in some commonly used intervals are
   a. 68.3% of the values of a normal random variable are within plus or minus one standard deviation of its mean.
   b. 95.4% of the values of a normal random variable are within plus or minus two standard deviations of its mean.
   c. 99.7% of the values of a normal random variable are within plus or minus three standard deviations of its mean.

These percentages are the basis for the empirical rule introduced in Section 3.3.

Figure 6.4 shows properties (a), (b), and (c) graphically.

Standard Normal Probability Distribution

A random variable that has a normal distribution with a mean of zero and a standard deviation of one is said to have a **standard normal probability distribution**. The letter z is commonly used to designate this particular normal random variable. Figure 6.5 is the graph of the standard normal distribution. It has the same general appearance as other normal distributions, but with the special properties of $\mu = 0$ and $\sigma = 1$.

6.2 Normal Probability Distribution

FIGURE 6.4 AREAS UNDER THE CURVE FOR ANY NORMAL DISTRIBUTION

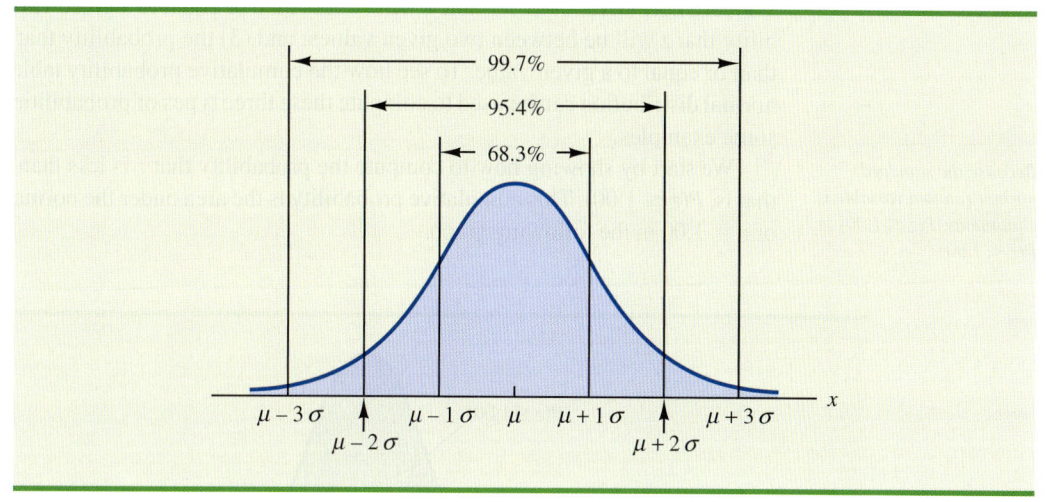

FIGURE 6.5 THE STANDARD NORMAL DISTRIBUTION

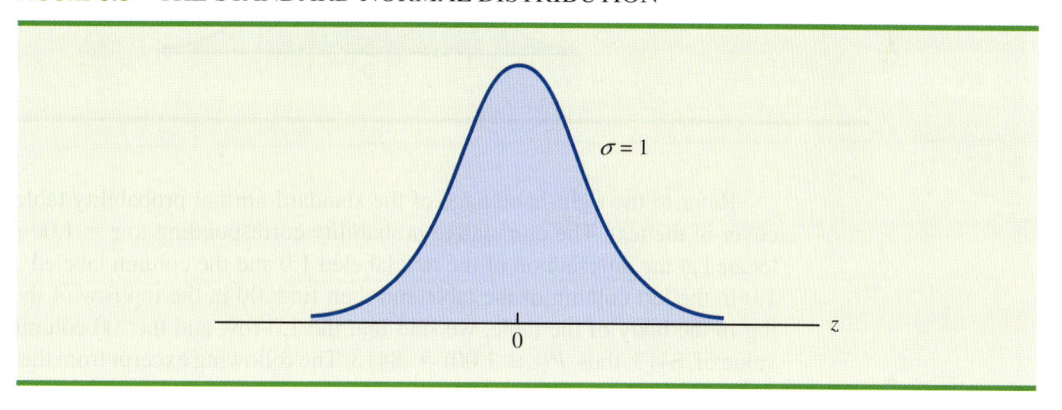

Because $\mu = 0$ and $\sigma = 1$, the formula for the standard normal probability density function is a simpler version of equation (6.2).

STANDARD NORMAL DENSITY FUNCTION

$$f(z) = \frac{1}{\sqrt{2\pi}} e^{-z^2/2}$$

As with other continuous random variables, probability calculations with any normal distribution are made by computing areas under the graph of the probability density function. Thus, to find the probability that a normal random variable is within any specific interval, we must compute the area under the normal curve over that interval.

For the standard normal distribution, areas under the normal curve have been computed and are available in tables that can be used to compute probabilities. Such a table appears on the two pages inside the front cover of the text. The table on the left-hand page contains areas, or cumulative probabilities, for z values less than or equal to the mean of zero. The table on the right-hand page contains areas, or cumulative probabilities, for z values greater than or equal to the mean of zero.

For the normal probability density function, the height of the normal curve varies and more advanced mathematics is required to compute the areas that represent probability.

Because the standard normal random variable is continuous, $P(z \leq 1.00) = P(z < 1.00)$.

The three types of probabilities we need to compute include (1) the probability that the standard normal random variable z will be less than or equal to a given value; (2) the probability that z will be between two given values; and (3) the probability that z will be greater than or equal to a given value. To see how the cumulative probability table for the standard normal distribution can be used to compute these three types of probabilities, let us consider some examples.

We start by showing how to compute the probability that z is less than or equal to 1.00; that is, $P(z \leq 1.00)$. This cumulative probability is the area under the normal curve to the left of $z = 1.00$ in the following graph.

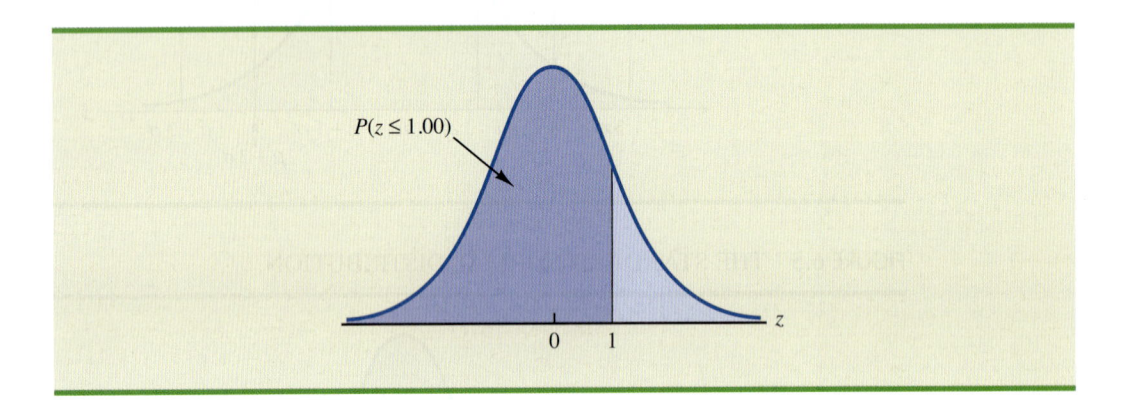

Refer to the right-hand page of the standard normal probability table inside the front cover of the text. The cumulative probability corresponding to $z = 1.00$ is the table value located at the intersection of the row labeled 1.0 and the column labeled .00. First we find 1.0 in the left column of the table and then find .00 in the top row of the table. By looking in the body of the table, we find that the 1.0 row and the .00 column intersect at the value of .8413; thus, $P(z \leq 1.00) = .8413$. The following excerpt from the probability table shows these steps.

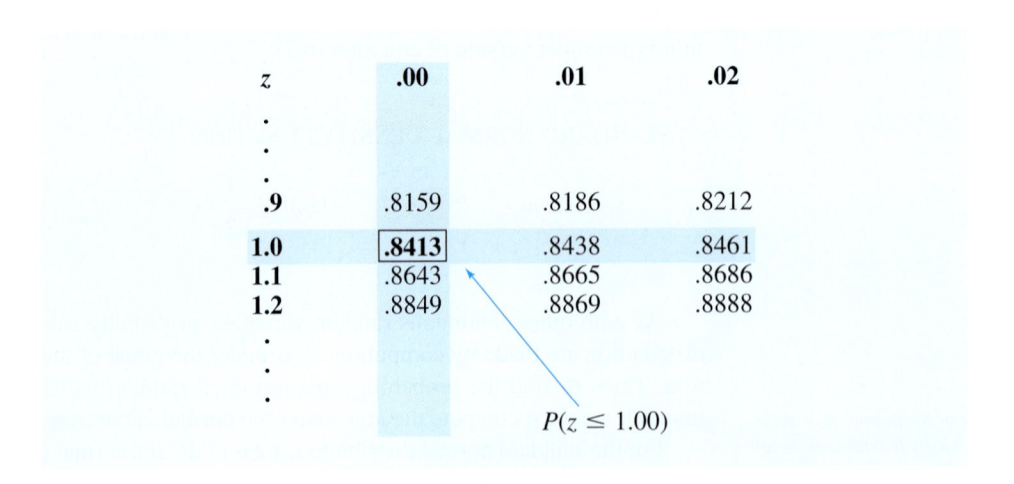

z	.00	.01	.02
.			
.			
.			
.9	.8159	.8186	.8212
1.0	.8413	.8438	.8461
1.1	.8643	.8665	.8686
1.2	.8849	.8869	.8888
.			
.			
.			

$P(z \leq 1.00)$

To illustrate the second type of probability calculation, we show how to compute the probability that z is in the interval between $-.50$ and 1.25; that is, $P(-.50 \leq z \leq 1.25)$. The following graph shows this area, or probability.

6.2 Normal Probability Distribution

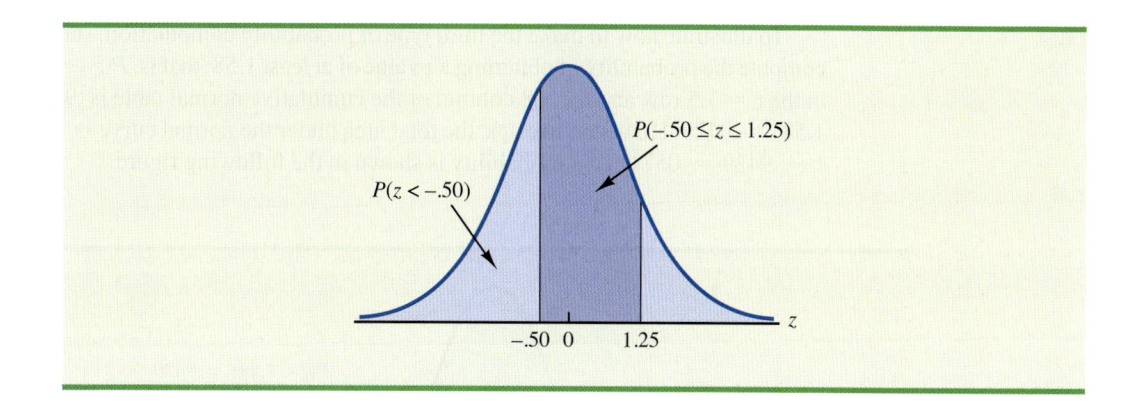

Three steps are required to compute this probability. First, we find the area under the normal curve to the left of $z = 1.25$. Second, we find the area under the normal curve to the left of $z = -.50$. Finally, we subtract the area to the left of $z = -.50$ from the area to the left of $z = 1.25$ to find $P(-.50 \leq z \leq 1.25)$.

To find the area under the normal curve to the left of $z = 1.25$, we first locate the 1.2 row in the standard normal probability table and then move across to the .05 column. Because the table value in the 1.2 row and the .05 column is .8944, $P(z \leq 1.25) = .8944$. Similarly, to find the area under the curve to the left of $z = -.50$, we use the left-hand page of the table to locate the table value in the $-.5$ row and the .00 column; with a table value of .3085, $P(z \leq -.50) = .3085$. Thus, $P(-.50 \leq z \leq 1.25) = P(z \leq 1.25) - P(z \leq -.50) = .8944 - .3085 = .5859$.

Let us consider another example of computing the probability that z is in the interval between two given values. Often it is of interest to compute the probability that a normal random variable assumes a value within a certain number of standard deviations of the mean. Suppose we want to compute the probability that the standard normal random variable is within one standard deviation of the mean; that is, $P(-1.00 \leq z \leq 1.00)$. To compute this probability we must find the area under the curve between -1.00 and 1.00. Earlier we found that $P(z \leq 1.00) = .8413$. Referring again to the table inside the front cover of the book, we find that the area under the curve to the left of $z = -1.00$ is .1587, so $P(z \leq -1.00) = .1587$. Therefore, $P(-1.00 \leq z \leq 1.00) = P(z \leq 1.00) - P(z \leq -1.00) = .8413 - .1587 = .6826$. This probability is shown graphically in the following figure.

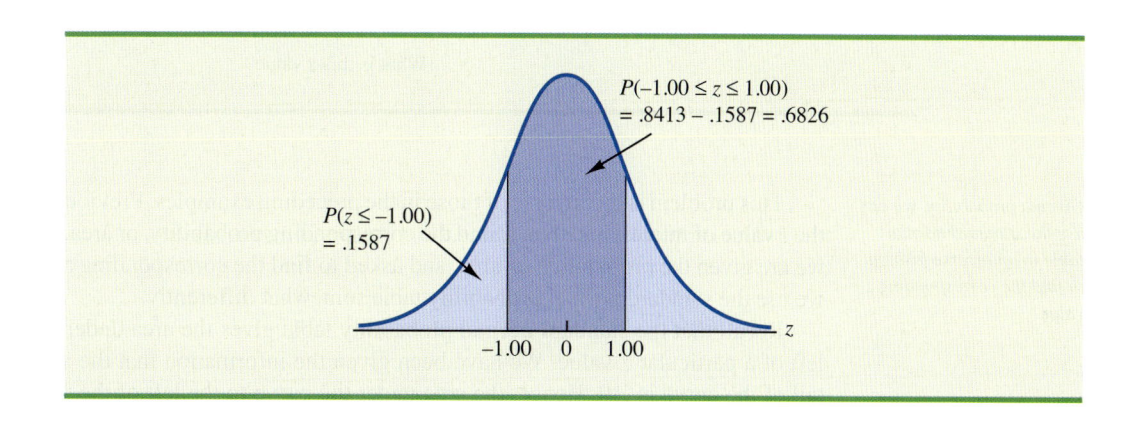

To illustrate how to make the third type of probability computation, suppose we want to compute the probability of obtaining a z value of at least 1.58; that is, $P(z \geq 1.58)$. The value in the $z = 1.5$ row and the .08 column of the cumulative normal table is .9429; thus, $P(z < 1.58) = .9429$. However, because the total area under the normal curve is 1, $P(z \geq 1.58) = 1 - .9429 = .0571$. This probability is shown in the following figure.

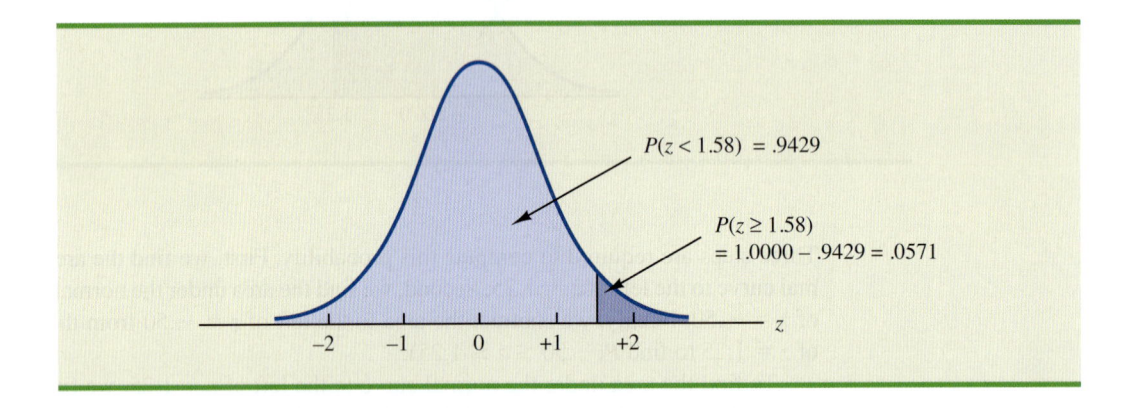

$P(z < 1.58) = .9429$

$P(z \geq 1.58)$
$= 1.0000 - .9429 = .0571$

In the preceding illustrations, we showed how to compute probabilities given specified z values. In some situations, we are given a probability and are interested in working backward to find the corresponding z value. Suppose we want to find a z value such that the probability of obtaining a larger z value is .10. The following figure shows this situation graphically.

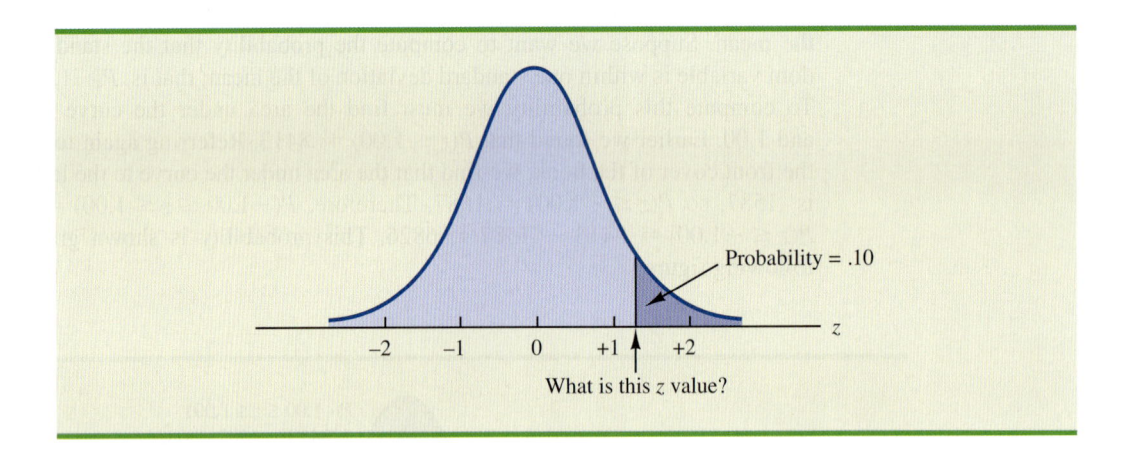

Probability = .10

What is this z value?

Given a probability, we can use the standard normal table in an inverse fashion to find the corresponding z value.

This problem is the inverse of those in the preceding examples. Previously, we specified the z value of interest and then found the corresponding probability, or area. In this example, we are given the probability, or area, and asked to find the corresponding z value. To do so, we use the standard normal probability table somewhat differently.

Recall that the standard normal probability table gives the area under the curve to the left of a particular z value. We have been given the information that the area in the upper tail of the curve is .10. Hence, the area under the curve to the left of the unknown z value must equal .9000. Scanning the body of the table, we find .8997 is the cumulative probability value closest to .9000. The section of the table providing this result follows.

6.2 Normal Probability Distribution

z	.06	.07	.08	.09
.				
.				
.				
1.0	.8554	.8577	.8599	.8621
1.1	.8770	.8790	.8810	.8830
1.2	.8962	.8980	.8997	.9015
1.3	.9131	.9147	.9162	.9177
1.4	.9279	.9292	.9306	.9319
.				
.				
.				

Cumulative probability value closest to .9000

Reading the z value from the leftmost column and the top row of the table, we find that the corresponding z value is 1.28. Thus, an area of approximately .9000 (actually .8997) will be to the left of $z = 1.28$.[2] In terms of the question originally asked, there is an approximately .10 probability of a z value larger than 1.28.

The examples illustrate that the table of cumulative probabilities for the standard normal probability distribution can be used to find probabilities associated with values of the standard normal random variable z. Two types of questions can be asked. The first type of question specifies a value, or values, for z and asks us to use the table to determine the corresponding areas or probabilities. The second type of question provides an area, or probability, and asks us to use the table to determine the corresponding z value. Thus, we need to be flexible in using the standard normal probability table to answer the desired probability question. In most cases, sketching a graph of the standard normal probability distribution and shading the appropriate area will help to visualize the situation and aid in determining the correct answer.

Computing Probabilities for Any Normal Probability Distribution

The reason for discussing the standard normal distribution so extensively is that probabilities for all normal distributions are computed by using the standard normal distribution. That is, when we have a normal distribution with any mean μ and any standard deviation σ, we answer probability questions about the distribution by first converting to the standard normal distribution. Then we can use the standard normal probability table and the appropriate z values to find the desired probabilities. The formula used to convert any normal random variable x with mean μ and standard deviation σ to the standard normal random variable z follows.

The formula for the standard normal random variable is similar to the formula we introduced in Chapter 3 for computing z-scores for a data set.

CONVERTING TO THE STANDARD NORMAL RANDOM VARIABLE

$$z = \frac{x - \mu}{\sigma} \tag{6.3}$$

---

[2] We could use interpolation in the body of the table to get a better approximation of the z value that corresponds to an area of .9000. Doing so to provide one more decimal place of accuracy would yield a z value of 1.282. However, in most practical situations, sufficient accuracy is obtained by simply using the table value closest to the desired probability.

A value of x equal to its mean μ results in $z = (\mu - \mu)/\sigma = 0$. Thus, we see that a value of x equal to its mean μ corresponds to $z = 0$. Now suppose that x is one standard deviation above its mean; that is, $x = \mu + \sigma$. Applying equation (6.3), we see that the corresponding z value is $z = [(\mu + \sigma) - \mu]/\sigma = \sigma/\sigma = 1$. Thus, an x value that is one standard deviation above its mean corresponds to $z = 1$. In other words, *we can interpret z as the number of standard deviations that the normal random variable x is from its mean μ.*

To see how this conversion enables us to compute probabilities for any normal distribution, suppose we have a normal distribution with $\mu = 10$ and $\sigma = 2$. What is the probability that the random variable x is between 10 and 14? Using equation (6.3), we see that at $x = 10$, $z = (x - \mu)/\sigma = (10 - 10)/2 = 0$ and that at $x = 14$, $z = (14 - 10)/2 = 4/2 = 2$. Thus, the answer to our question about the probability of x being between 10 and 14 is given by the equivalent probability that z is between 0 and 2 for the standard normal distribution. In other words, the probability that we are seeking is the probability that the random variable x is between its mean and two standard deviations above the mean. Using $z = 2.00$ and the standard normal probability table inside the front cover of the text, we see that $P(z \leq 2) = .9772$. Because $P(z \leq 0) = .5000$, we can compute $P(.00 \leq z \leq 2.00) = P(z \leq 2) - P(z \leq 0) = .9772 - .5000 = .4772$. Hence the probability that x is between 10 and 14 is .4772.

Grear Tire Company Problem

We turn now to an application of the normal probability distribution. Suppose the Grear Tire Company developed a new steel-belted radial tire to be sold through a national chain of discount stores. Because the tire is a new product, Grear's managers believe that the mileage guarantee offered with the tire will be an important factor in the acceptance of the product. Before finalizing the tire mileage guarantee policy, Grear's managers want probability information about x = number of miles the tires will last.

From actual road tests with the tires, Grear's engineering group estimated that the mean tire mileage is $\mu = 36{,}500$ miles and that the standard deviation is $\sigma = 5000$. In addition, the data collected indicate that a normal distribution is a reasonable assumption. What percentage of the tires can be expected to last more than 40,000 miles? In other words, what is the probability that the tire mileage, x, will exceed 40,000? This question can be answered by finding the area of the darkly shaded region in Figure 6.6.

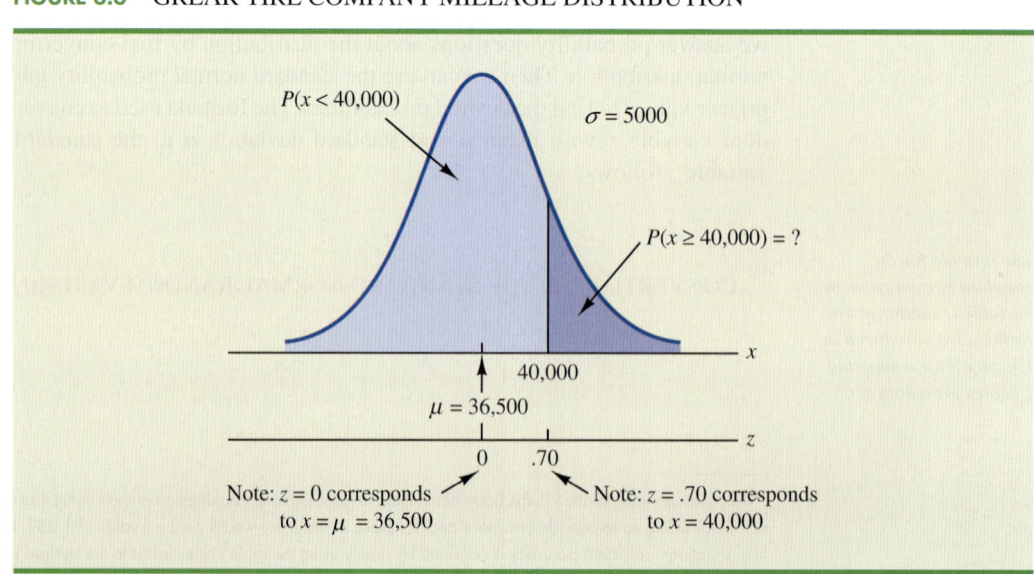

FIGURE 6.6   GREAR TIRE COMPANY MILEAGE DISTRIBUTION

6.2 Normal Probability Distribution

At $x = 40,000$, we have

$$z = \frac{x - \mu}{\sigma} = \frac{40,000 - 36,500}{5000} = \frac{3500}{5000} = .70$$

Refer now to the bottom of Figure 6.6. We see that a value of $x = 40,000$ on the Grear Tire normal distribution corresponds to a value of $z = .70$ on the standard normal distribution. Using the standard normal probability table, we see that the area under the standard normal curve to the left of $z = .70$ is .7580. Thus, $1.000 - .7580 = .2420$ is the probability that z will exceed .70 and hence x will exceed 40,000. We can conclude that about 24.2% of the tires will exceed 40,000 in mileage.

Let us now assume that Grear is considering a guarantee that will provide a discount on replacement tires if the original tires do not provide the guaranteed mileage. What should the guarantee mileage be if Grear wants no more than 10% of the tires to be eligible for the discount guarantee? This question is interpreted graphically in Figure 6.7.

According to Figure 6.7, the area under the curve to the left of the unknown guarantee mileage must be .10. So, we must first find the z value that cuts off an area of .10 in the left tail of a standard normal distribution. Using the standard normal probability table, we see that $z = -1.28$ cuts off an area of .10 in the lower tail. Hence, $z = -1.28$ is the value of the standard normal random variable corresponding to the desired mileage guarantee on the Grear Tire normal distribution. To find the value of x corresponding to $z = -1.28$, we have

The guarantee mileage we need to find is 1.28 standard deviations below the mean. Thus, $x = \mu - 1.28\sigma$.

$$z = \frac{x - \mu}{\sigma} = -1.28$$
$$x - \mu = -1.28\sigma$$
$$x = \mu - 1.28\sigma$$

With $\mu = 36,500$ and $\sigma = 5000$,

$$x = 36,500 - 1.28(5000) = 30,100$$

With the guarantee set at 30,000 miles, the actual percentage eligible for the guarantee will be 9.68%.

Thus, a guarantee of 30,100 miles will meet the requirement that approximately 10% of the tires will be eligible for the guarantee. Perhaps, with this information, the firm will set its tire mileage guarantee at 30,000 miles.

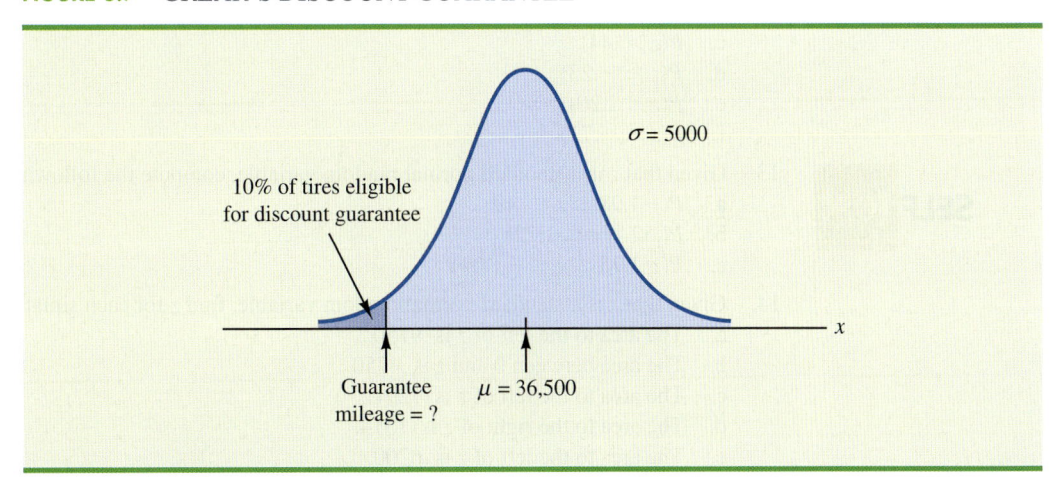

FIGURE 6.7 GREAR'S DISCOUNT GUARANTEE

EXERCISES

Methods

8. Using Figure 6.4 as a guide, sketch a normal curve for a random variable x that has a mean of $\mu = 100$ and a standard deviation of $\sigma = 10$. Label the horizontal axis with values of 70, 80, 90, 100, 110, 120, and 130.

9. A random variable is normally distributed with a mean of $\mu = 50$ and a standard deviation of $\sigma = 5$.
    a. Sketch a normal curve for the probability density function. Label the horizontal axis with values of 35, 40, 45, 50, 55, 60, and 65. Figure 6.4 shows that the normal curve almost touches the horizontal axis at three standard deviations below and at three standard deviations above the mean (in this case at 35 and 65).
    b. What is the probability that the random variable will assume a value between 45 and 55?
    c. What is the probability that the random variable will assume a value between 40 and 60?

10. Draw a graph for the standard normal distribution. Label the horizontal axis at values of −3, −2, −1, 0, 1, 2, and 3. Then use the table of probabilities for the standard normal distribution inside the front cover of the text to compute the following probabilities.
    a. $P(z \leq 1.5)$
    b. $P(z \leq 1)$
    c. $P(1 \leq z \leq 1.5)$
    d. $P(0 < z < 2.5)$

11. Given that z is a standard normal random variable, compute the following probabilities.
    a. $P(z \leq -1.0)$
    b. $P(z \geq -1)$
    c. $P(z \geq -1.5)$
    d. $P(-2.5 \leq z)$
    e. $P(-3 < z \leq 0)$

12. Given that z is a standard normal random variable, compute the following probabilities.
    a. $P(0 \leq z \leq .83)$
    b. $P(-1.57 \leq z \leq 0)$
    c. $P(z > .44)$
    d. $P(z \geq -.23)$
    e. $P(z < 1.20)$
    f. $P(z \leq -.71)$

13. Given that z is a standard normal random variable, compute the following probabilities.
    a. $P(-1.98 \leq z \leq .49)$
    b. $P(.52 \leq z \leq 1.22)$
    c. $P(-1.75 \leq z \leq -1.04)$

14. Given that z is a standard normal random variable, find z for each situation.
    a. The area to the left of z is .9750.
    b. The area between 0 and z is .4750.
    c. The area to the left of z is .7291.
    d. The area to the right of z is .1314.
    e. The area to the left of z is .6700.
    f. The area to the right of z is .3300.

6.2 Normal Probability Distribution

15. Given that z is a standard normal random variable, find z for each situation.
    a. The area to the left of z is .2119.
    b. The area between $-z$ and z is .9030.
    c. The area between $-z$ and z is .2052.
    d. The area to the left of z is .9948.
    e. The area to the right of z is .6915.

16. Given that z is a standard normal random variable, find z for each situation.
    a. The area to the right of z is .01.
    b. The area to the right of z is .025.
    c. The area to the right of z is .05.
    d. The area to the right of z is .10.

Applications

17. For borrowers with good credit scores, the mean debt for revolving and installment accounts is $15,015 (*BusinessWeek,* March 20, 2006). Assume the standard deviation is $3540 and that debt amounts are normally distributed.
    a. What is the probability that the debt for a borrower with good credit is more than $18,000?
    b. What is the probability that the debt for a borrower with good credit is less than $10,000?
    c. What is the probability that the debt for a borrower with good credit is between $12,000 and $18,000?
    d. What is the probability that the debt for a borrower with good credit is no more than $14,000?

18. The average stock price for companies making up the S&P 500 is $30, and the standard deviation is $8.20 (*BusinessWeek,* Special Annual Issue, Spring 2003). Assume the stock prices are normally distributed.
    a. What is the probability that a company will have a stock price of at least $40?
    b. What is the probability that a company will have a stock price no higher than $20?
    c. How high does a stock price have to be to put a company in the top 10%?

19. In an article about the cost of health care, *Money* magazine reported that a visit to a hospital emergency room for something as simple as a sore throat has a mean cost of $328 (*Money,* January 2009). Assume that the cost for this type of hospital emergency room visit is normally distributed with a standard deviation of $92. Answer the following questions about the cost of a hospital emergency room visit for this medical service.
    a. What is the probability that the cost will be more than $500?
    b. What is the probability that the cost will be less than $250?
    c. What is the probability that the cost will be between $300 and $400?
    d. If the cost to a patient is in the lower 8% of charges for this medical service, what was the cost of this patient's emergency room visit?

20. In January 2003, the American worker spent an average of 77 hours logged on to the Internet while at work (CNBC, March 15, 2003). Assume the population mean is 77 hours, the times are normally distributed, and that the standard deviation is 20 hours.
    a. What is the probability that in January 2003 a randomly selected worker spent fewer than 50 hours logged on to the Internet?
    b. What percentage of workers spent more than 100 hours in January 2003 logged on to the Internet?
    c. A person is classified as a heavy user if he or she is in the upper 20% of usage. In January 2003, how many hours did a worker have to be logged on to the Internet to be considered a heavy user?

21. A person must score in the upper 2% of the population on an IQ test to qualify for membership in Mensa, the international high-IQ society. If IQ scores are normally distributed with a mean of 100 and a standard deviation of 15, what score must a person have to qualify for Mensa?

250 Chapter 6 Continuous Probability Distributions

22. Television viewing reached a new high when the Nielsen Company reported a mean daily viewing time of 8.35 hours per household (*USA Today*, November 11, 2009). Use a normal probability distribution with a standard deviation of 2.5 hours to answer the following questions about daily television viewing per household.
    a. What is the probability that a household views television between 5 and 10 hours a day?
    b. How many hours of television viewing must a household have in order to be in the top 3% of all television viewing households?
    c. What is the probability that a household views television more than 3 hours a day?

23. The time needed to complete a final examination in a particular college course is normally distributed with a mean of 80 minutes and a standard deviation of 10 minutes. Answer the following questions.
    a. What is the probability of completing the exam in one hour or less?
    b. What is the probability that a student will complete the exam in more than 60 minutes but less than 75 minutes?
    c. Assume that the class has 60 students and that the examination period is 90 minutes in length. How many students do you expect will be unable to complete the exam in the allotted time?

24. Trading volume on the New York Stock Exchange is heaviest during the first half hour (early morning) and last half hour (late afternoon) of the trading day. The early morning trading volumes (millions of shares) for 13 days in January and February are shown here (*Barron's,* January 23, 2006; February 13, 2006; and February 27, 2006).

214	163	265	194	180
202	198	212	201	
174	171	211	211	

    The probability distribution of trading volume is approximately normal.
    a. Compute the mean and standard deviation to use as estimates of the population mean and standard deviation.
    b. What is the probability that, on a randomly selected day, the early morning trading volume will be less than 180 million shares?
    c. What is the probability that, on a randomly selected day, the early morning trading volume will exceed 230 million shares?
    d. How many shares would have to be traded for the early morning trading volume on a particular day to be among the busiest 5% of days?

25. According to the Sleep Foundation, the average night's sleep is 6.8 hours (*Fortune,* March 20, 2006). Assume the standard deviation is .6 hour and that the probability distribution is normal.
    a. What is the probability that a randomly selected person sleeps more than 8 hours?
    b. What is the probability that a randomly selected person sleeps 6 hours or less?
    c. Doctors suggest getting between 7 and 9 hour of sleep each night. What percentage of the population gets this much sleep?

6.3 Normal Approximation of Binomial Probabilities

In Section 5.4 we presented the discrete binomial distribution. Recall that a binomial experiment consists of a sequence of *n* identical independent trials with each trial having two possible outcomes, a success or a failure. The probability of a success on a trial is the same for all trials and is denoted by *p*. The binomial random variable is the number of successes in the *n* trials, and probability questions pertain to the probability of *x* successes in the *n* trials.

6.3 Normal Approximation of Binomial Probabilities

FIGURE 6.8 NORMAL APPROXIMATION TO A BINOMIAL PROBABILITY DISTRIBUTION WITH $n = 100$ AND $p = .10$ SHOWING THE PROBABILITY OF 12 ERRORS

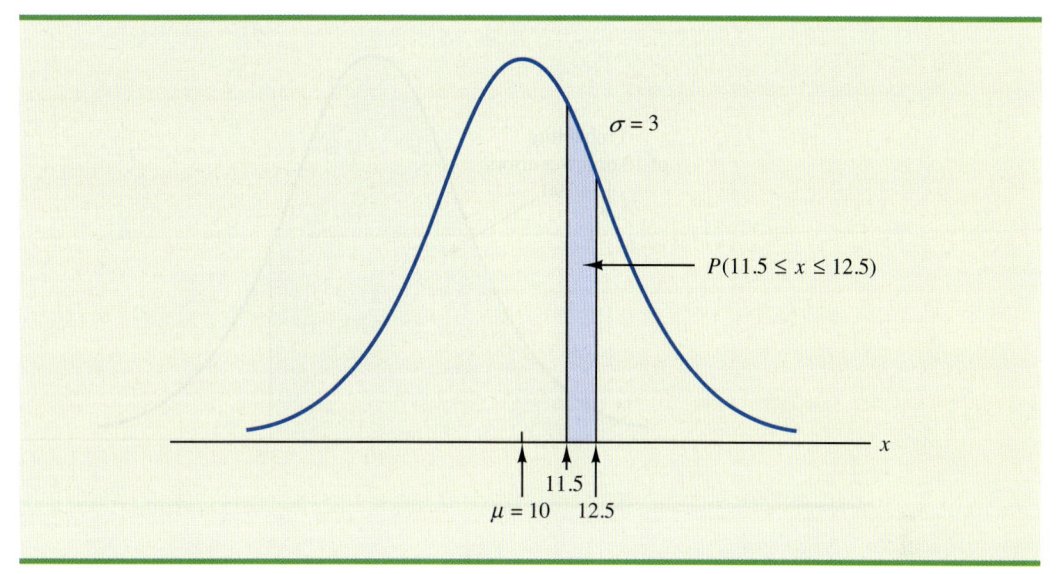

[Figure: Normal curve with $\sigma = 3$, $\mu = 10$, showing shaded region $P(11.5 \le x \le 12.5)$ between 11.5 and 12.5]

When the number of trials becomes large, evaluating the binomial probability function by hand or with a calculator is difficult. In cases where $np \ge 5$, and $n(1 - p) \ge 5$, the normal distribution provides an easy-to-use approximation of binomial probabilities. When using the normal approximation to the binomial, we set $\mu = np$ and $\sigma = \sqrt{np(1 - p)}$ in the definition of the normal curve.

Let us illustrate the normal approximation to the binomial by supposing that a particular company has a history of making errors in 10% of its invoices. A sample of 100 invoices has been taken, and we want to compute the probability that 12 invoices contain errors. That is, we want to find the binomial probability of 12 successes in 100 trials. In applying the normal approximation in this case, we set $\mu = np = (100)(.1) = 10$ and $\sigma = \sqrt{np(1-p)} = \sqrt{(100)(.1)(.9)} = 3$. A normal distribution with $\mu = 10$ and $\sigma = 3$ is shown in Figure 6.8.

Recall that, with a continuous probability distribution, probabilities are computed as areas under the probability density function. As a result, the probability of any single value for the random variable is zero. Thus to approximate the binomial probability of 12 successes, we compute the area under the corresponding normal curve between 11.5 and 12.5. The .5 that we add to and subtract from 12 is called a **continuity correction factor**. It is introduced because a continuous distribution is being used to approximate a discrete distribution. Thus, $P(x = 12)$ for the *discrete* binomial distribution is approximated by $P(11.5 \le x \le 12.5)$ for the *continuous* normal distribution.

Converting to the standard normal distribution to compute $P(11.5 \le x \le 12.5)$, we have

$$z = \frac{x - \mu}{\sigma} = \frac{12.5 - 10.0}{3} = .83 \quad \text{at } x = 12.5$$

and

$$z = \frac{x - \mu}{\sigma} = \frac{11.5 - 10.0}{3} = .50 \quad \text{at } x = 11.5$$

FIGURE 6.9 NORMAL APPROXIMATION TO A BINOMIAL PROBABILITY DISTRIBUTION WITH $n = 100$ AND $p = .10$ SHOWING THE PROBABILITY OF 13 OR FEWER ERRORS

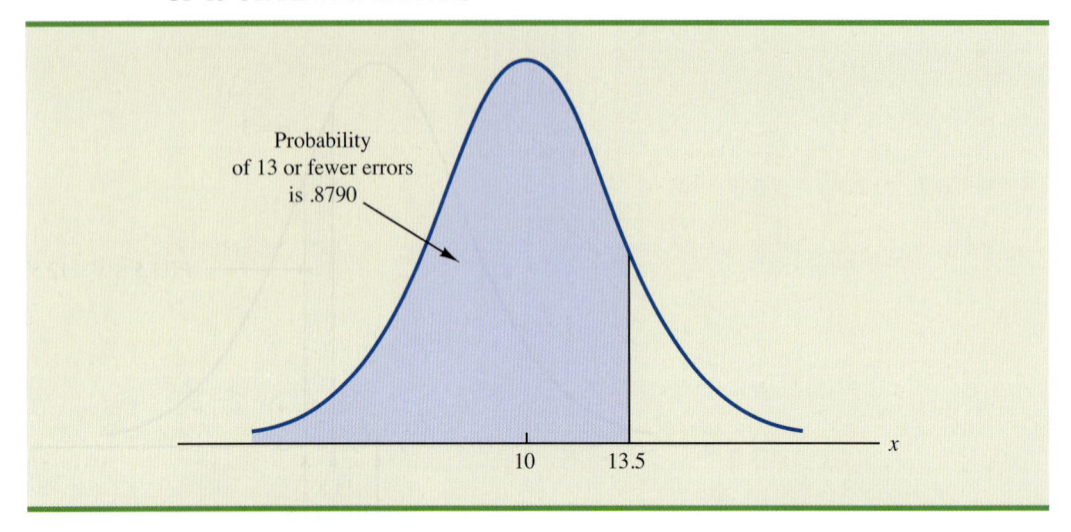

Using the standard normal probability table, we find that the area under the curve (in Figure 6.8) to the left of 12.5 is .7967. Similarly, the area under the curve to the left of 11.5 is .6915. Therefore, the area between 11.5 and 12.5 is .7967 − .6915 = .1052. The normal approximation to the probability of 12 successes in 100 trials is .1052.

For another illustration, suppose we want to compute the probability of 13 or fewer errors in the sample of 100 invoices. Figure 6.9 shows the area under the normal curve that approximates this probability. Note that the use of the continuity correction factor results in the value of 13.5 being used to compute the desired probability. The z value corresponding to $x = 13.5$ is

$$z = \frac{13.5 - 10.0}{3.0} = 1.17$$

The standard normal probability table shows that the area under the standard normal curve to the left of $z = 1.17$ is .8790. The area under the normal curve approximating the probability of 13 or fewer errors is given by the shaded portion of the graph in Figure 6.9.

Exercises

Methods

26. A binomial probability distribution has $p = .20$ and $n = 100$.
    a. What are the mean and standard deviation?
    b. Is this situation one in which binomial probabilities can be approximated by the normal probability distribution? Explain.
    c. What is the probability of exactly 24 successes?
    d. What is the probability of 18 to 22 successes?
    e. What is the probability of 15 or fewer successes?

27. Assume a binomial probability distribution has $p = .60$ and $n = 200$.
    a. What are the mean and standard deviation?
    b. Is this situation one in which binomial probabilities can be approximated by the normal probability distribution? Explain.

6.3 Normal Approximation of Binomial Probabilities

c. What is the probability of 100 to 110 successes?
d. What is the probability of 130 or more successes?
e. What is the advantage of using the normal probability distribution to approximate the binomial probabilities? Use part (d) to explain the advantage.

Applications

SELF test

28. Although studies continue to show smoking leads to significant health problems, 20% of adults in the United States smoke. Consider a group of 250 adults.
    a. What is the expected number of adults who smoke?
    b. What is the probability that fewer than 40 smoke?
    c. What is the probability that from 55 to 60 smoke?
    d. What is the probability that 70 or more smoke?

29. An Internal Revenue Oversight Board survey found that 82% of taxpayers said that it was very important for the Internal Revenue Service (IRS) to ensure that high-income taxpayers do not cheat on their tax returns (*The Wall Street Journal*, February 11, 2009).
    a. For a sample of eight taxpayers, what is the probability that at least six taxpayers say that it is very important to ensure that high-income taxpayers do not cheat on their tax returns? Use the binomial distribution probability function shown in Section 5.4 to answer this question.
    b. For a sample of 80 taxpayers, what is the probability that at least 60 taxpayers say that it is very important to ensure that high-income taxpayers do not cheat on their tax returns? Use the normal approximation of the binomial distribution to answer this question.
    c. As the number of trails in a binomial distribution application becomes large, what is the advantage of using the normal approximation of the binomial distribution to compute probabilities?
    d. When the number of trials for a binominal distribution application becomes large, would developers of statistical software packages prefer to use the binomial distribution probability function shown in Section 5.4 or the normal approximation of the binomial distribution shown in Section 6.3? Explain.

30. When you sign up for a credit card, do you read the contract carefully? In a FindLaw.com survey, individuals were asked, "How closely do you read a contract for a credit card?" (*USA Today,* October 16, 2003). The findings were that 44% read every word, 33% read enough to understand the contract, 11% just glance at it, and 4% don't read it at all.
    a. For a sample of 500 people, how many would you expect to say that they read every word of a credit card contract?
    b. For a sample of 500 people, what is the probability that 200 or fewer will say they read every word of a credit card contract?
    c. For a sample of 500 people, what is the probability that at least 15 say they don't read credit card contracts?

31. A Bureau of National Affairs survey found that 79% of employers provide their workers with a two-day paid Thanksgiving holiday with workers off both Thursday and Friday (*USA Today*, November 12, 2009). Nineteen percent of employers provide a one-day paid holiday with workers off Thanksgiving Day. Two percent of employers do not provide a paid Thanksgiving holiday. Consider a sample of 120 employers.
    a. What is the probability that at least 85 of the employers provide a two-day paid Thanksgiving holiday?
    b. What is the probability that between 90 and 100 employers provide a two-day paid Thanksgiving holiday? That is, what is $P(90 \leq x \leq 100)$?
    c. What is the probability that less than 20 employers provide a one-day paid Thanksgiving holiday?

6.4 Exponential Probability Distribution

The **exponential probability distribution** may be used for random variables such as the time between arrivals at a car wash, the time required to load a truck, the distance between major defects in a highway, and so on. The exponential probability density function follows.

EXPONENTIAL PROBABILITY DENSITY FUNCTION

$$f(x) = \frac{1}{\mu} e^{-x/\mu} \quad \text{for } x \geq 0 \qquad (6.4)$$

where μ = expected value or mean

As an example of the exponential distribution, suppose that x represents the loading time for a truck at the Schips loading dock and follows such a distribution. If the mean, or average, loading time is 15 minutes ($\mu = 15$), the appropriate probability density function for x is

$$f(x) = \frac{1}{15} e^{-x/15}$$

Figure 6.10 is the graph of this probability density function.

Computing Probabilities for the Exponential Distribution

As with any continuous probability distribution, the area under the curve corresponding to an interval provides the probability that the random variable assumes a value in that interval. In the Schips loading dock example, the probability that loading a truck will take 6 minutes or less $P(x \leq 6)$ is defined to be the area under the curve in Figure 6.10 from $x = 0$ to $x = 6$. Similarly, the probability that the loading time will be 18 minutes or less $P(x \leq 18)$ is the area under the curve from $x = 0$ to $x = 18$. Note also that the probability that the loading time will be between 6 minutes and 18 minutes $P(6 \leq x \leq 18)$ is given by the area under the curve from $x = 6$ to $x = 18$.

In waiting line applications, the exponential distribution is often used for service time.

FIGURE 6.10 EXPONENTIAL DISTRIBUTION FOR THE SCHIPS LOADING DOCK EXAMPLE

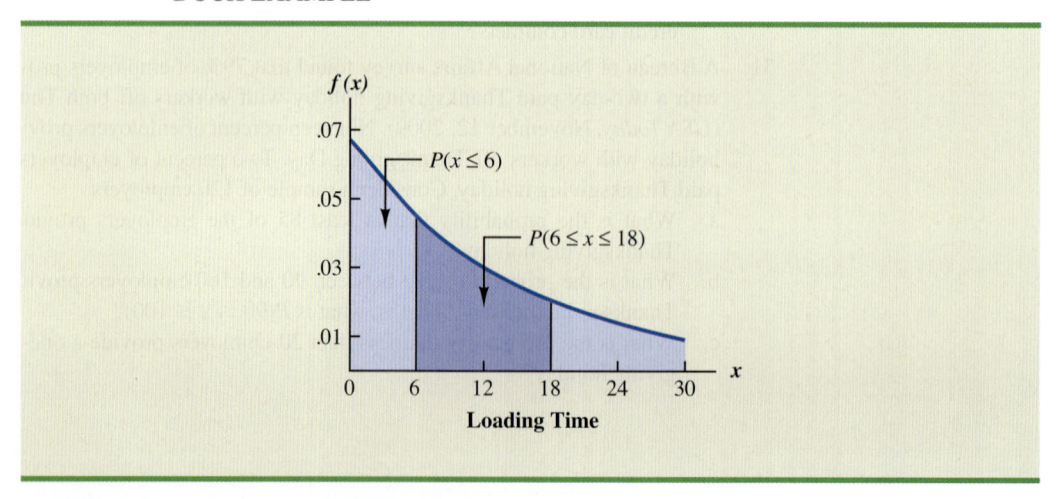

6.4 Exponential Probability Distribution

To compute exponential probabilities such as those just described, we use the following formula. It provides the cumulative probability of obtaining a value for the exponential random variable of less than or equal to some specific value denoted by x_0.

EXPONENTIAL DISTRIBUTION: CUMULATIVE PROBABILITIES

$$P(x \leq x_0) = 1 - e^{-x_0/\mu} \qquad (6.5)$$

For the Schips loading dock example, $x =$ loading time in minutes and $\mu = 15$ minutes. Using equation (6.5),

$$P(x \leq x_0) = 1 - e^{-x_0/15}$$

Hence, the probability that loading a truck will take 6 minutes or less is

$$P(x \leq 6) = 1 - e^{-6/15} = .3297$$

Again using equation (6.5), we calculate the probability of loading a truck in 18 minutes or less.

$$P(x \leq 18) = 1 - e^{-18/15} = .6988$$

Thus, the probability that loading a truck will take between 6 minutes and 18 minutes is equal to $.6988 - .3297 = .3691$. Probabilities for any other interval can be computed similarly.

A property of the exponential distribution is that the mean and standard deviation are equal.

In the preceding example, the mean time it takes to load a truck is $\mu = 15$ minutes. A property of the exponential distribution is that the mean of the distribution and the standard deviation of the distribution are *equal*. Thus, the standard deviation for the time it takes to load a truck is $\sigma = 15$ minutes. The variance is $\sigma^2 = (15)^2 = 225$.

Relationship Between the Poisson and Exponential Distributions

In Section 5.5 we introduced the Poisson distribution as a discrete probability distribution that is often useful in examining the number of occurrences of an event over a specified interval of time or space. Recall that the Poisson probability function is

$$f(x) = \frac{\mu^x e^{-\mu}}{x!}$$

where

$$\mu = \text{expected value or mean number of occurrences over a specified interval}$$

If arrivals follow a Poisson distribution, the time between arrivals must follow an exponential distribution.

The continuous exponential probability distribution is related to the discrete Poisson distribution. If the Poisson distribution provides an appropriate description of the number of occurrences per interval, the exponential distribution provides a description of the length of the interval between occurrences.

To illustrate this relationship, suppose the number of cars that arrive at a car wash during one hour is described by a Poisson probability distribution with a mean of 10 cars per hour. The Poisson probability function that gives the probability of x arrivals per hour is

$$f(x) = \frac{10^x e^{-10}}{x!}$$

Because the average number of arrivals is 10 cars per hour, the average time between cars arriving is

$$\frac{1 \text{ hour}}{10 \text{ cars}} = .1 \text{ hour/car}$$

Thus, the corresponding exponential distribution that describes the time between the arrivals has a mean of $\mu = .1$ hour per car; as a result, the appropriate exponential probability density function is

$$f(x) = \frac{1}{.1} e^{-x/.1} = 10e^{-10x}$$

NOTES AND COMMENTS

As we can see in Figure 6.10, the exponential distribution is skewed to the right. Indeed, the skewness measure for exponential distributions is 2. The exponential distribution gives us a good idea what a skewed distribution looks like.

Exercises

Methods

32. Consider the following exponential probability density function.

$$f(x) = \frac{1}{8} e^{-x/8} \quad \text{for } x \geq 0$$

   a. Find $P(x \leq 6)$.
   b. Find $P(x \leq 4)$.
   c. Find $P(x \geq 6)$.
   d. Find $P(4 \leq x \leq 6)$.

33. Consider the following exponential probability density function.

$$f(x) = \frac{1}{3} e^{-x/3} \quad \text{for } x \geq 0$$

   a. Write the formula for $P(x \leq x_0)$.
   b. Find $P(x \leq 2)$.
   c. Find $P(x \geq 3)$.
   d. Find $P(x \leq 5)$.
   e. Find $P(2 \leq x \leq 5)$.

Applications

34. The time required to pass through security screening at the airport can be annoying to travelers. The mean wait time during peak periods at Cincinnati/Northern Kentucky International Airport is 12.1 minutes (*The Cincinnati Enquirer*, February 2, 2006). Assume the time to pass through security screening follows an exponential distribution.
   a. What is the probability that it will take less than 10 minutes to pass through security screening during a peak period?
   b. What is the probability that it will take more than 20 minutes to pass through security screening during a peak period?

c. What is the probability that it will take between 10 and 20 minutes to pass through security screening during a peak period?
d. It is 8:00 A.M. (a peak period) and you just entered the security line. To catch your plane you must be at the gate within 30 minutes. If it takes 12 minutes from the time you clear security until you reach your gate, what is the probability that you will miss your flight?

35. The time between arrivals of vehicles at a particular intersection follows an exponential probability distribution with a mean of 12 seconds.
    a. Sketch this exponential probability distribution.
    b. What is the probability that the arrival time between vehicles is 12 seconds or less?
    c. What is the probability that the arrival time between vehicles is 6 seconds or less?
    d. What is the probability of 30 or more seconds between vehicle arrivals?

36. Comcast Corporation is the largest cable television company, the second largest Internet service provider, and the fourth largest telephone service provider in the United States. Generally known for quality and reliable service, the company periodically experiences unexpected service interruptions. On January 14, 2009, such an interruption occurred for the Comcast customers living in southwest Florida. When customers called the Comcast office, a recorded message told them that the company was aware of the service outage and that it was anticipated that service would be restored in two hours. Assume that two hours is the mean time to do the repair and that the repair time has an exponential probability distribution.
    a. What is the probability that the cable service will be repaired in one hour or less?
    b. What is the probability that the repair will take between one hour and two hours?
    c. For a customer who calls the Comcast office at 1:00 P.M., what is the probability that the cable service will not be repaired by 5:00 P.M.?

37. Collina's Italian Café in Houston, Texas, advertises that carryout orders take about 25 minutes (Collina's website, February 27, 2008). Assume that the time required for a carryout order to be ready for customer pickup has an exponential distribution with a mean of 25 minutes.
    a. What is the probability than a carryout order will be ready within 20 minutes?
    b. If a customer arrives 30 minutes after placing an order, what is the probability that the order will not be ready?
    c. A particular customer lives 15 minutes from Collina's Italian Café. If the customer places a telephone order at 5:20 P.M., what is the probability that the customer can drive to the café, pick up the order, and return home by 6:00 P.M.?

38. Do interruptions while you are working reduce your productivity? According to a University of California–Irvine study, businesspeople are interrupted at the rate of approximately $5\frac{1}{2}$ times per hour (*Fortune*, March 20, 2006). Suppose the number of interruptions follows a Poisson probability distribution.
    a. Show the probability distribution for the time between interruptions.
    b. What is the probability that a businessperson will have no interruptions during a 15-minute period?
    c. What is the probability that the next interruption will occur within 10 minutes for a particular businessperson?

Summary

This chapter extended the discussion of probability distributions to the case of continuous random variables. The major conceptual difference between discrete and continuous probability distributions involves the method of computing probabilities. With discrete distributions, the probability function $f(x)$ provides the probability that the random variable x assumes various values. With continuous distributions, the probability density function $f(x)$ does not provide probability values directly. Instead, probabilities are given by areas under the curve or graph of the probability density function $f(x)$. Because the area under the curve above a single point is zero, we observe that the probability of any particular value is zero for a continuous random variable.

Three continuous probability distributions—the uniform, normal, and exponential distributions—were treated in detail. The normal distribution is used widely in statistical inference and will be used extensively throughout the remainder of the text.

Glossary

Probability density function A function used to compute probabilities for a continuous random variable. The area under the graph of a probability density function over an interval represents probability.

Uniform probability distribution A continuous probability distribution for which the probability that the random variable will assume a value in any interval is the same for each interval of equal length.

Normal probability distribution A continuous probability distribution. Its probability density function is bell-shaped and determined by its mean μ and standard deviation σ.

Standard normal probability distribution A normal distribution with a mean of zero and a standard deviation of one.

Continuity correction factor A value of .5 that is added to or subtracted from a value of x when the continuous normal distribution is used to approximate the discrete binomial distribution.

Exponential probability distribution A continuous probability distribution that is useful in computing probabilities for the time it takes to complete a task.

Key Formulas

Uniform Probability Density Function

$$f(x) = \begin{cases} \dfrac{1}{b-a} & \text{for } a \leq x \leq b \\ 0 & \text{elsewhere} \end{cases} \tag{6.1}$$

Normal Probability Density Function

$$f(x) = \frac{1}{\sigma\sqrt{2\pi}} e^{-(x-\mu)^2/2\sigma^2} \tag{6.2}$$

Converting to the Standard Normal Random Variable

$$z = \frac{x - \mu}{\sigma} \tag{6.3}$$

Exponential Probability Density Function

$$f(x) = \frac{1}{\mu} e^{-x/\mu} \quad \text{for } x \geq 0 \tag{6.4}$$

Exponential Distribution: Cumulative Probabilities

$$P(x \leq x_0) = 1 - e^{-x_0/\mu} \tag{6.5}$$

Supplementary Exercises

39. The time (in minutes) between telephone calls at an insurance claims office has the following exponential probability distribution.

$$f(x) = .50e^{-.50x} \quad \text{for } x \geq 0$$

   a. What is the mean time between telephone calls?
   b. What is the probability of having 30 seconds or less between telephone calls?
   c. What is the probability of having 1 minute or less between telephone calls?
   d. What is the probability of having 5 or more minutes without a telephone call?

40. A business executive, transferred from Chicago to Atlanta, needs to sell her house in Chicago quickly. The executive's employer has offered to buy the house for $210,000, but the offer expires at the end of the week. The executive does not currently have a better offer but can afford to leave the house on the market for another month. From conversations with her realtor, the executive believes the price she will get by leaving the house on the market for another month is uniformly distributed between $200,000 and $225,000.
   a. If she leaves the house on the market for another month, what is the mathematical expression for the probability density function of the sales price?
   b. If she leaves it on the market for another month, what is the probability that she will get at least $215,000 for the house?
   c. If she leaves it on the market for another month, what is the probability that she will get less than $210,000?
   d. Should the executive leave the house on the market for another month? Why or why not?

41. The American Community Survey showed that residents of New York City have the longest travel times to get to work compared to residents of other cities in the United States (U.S. Census Bureau website, August 2008). According to the latest statistics available, the average travel time to work for residents of New York City is 38.3 minutes.
   a. Assume the exponential probability distribution is applicable and show the probability density function for the travel time to work for a resident of this city.
   b. What is the probability that it will take a resident of this city between 20 and 40 minutes to travel to work?
   c. What is the probability that it will take a resident of this city more than one hour to travel to work?

42. The U.S. Bureau of Labor Statistics reports that the average annual expenditure on food and drink for all families is $5700 (*Money*, December 2003). Assume that annual expenditure on food and drink is normally distributed and that the standard deviation is $1500.
   a. What is the range of expenditures of the 10% of families with the lowest annual spending on food and drink?
   b. What percentage of families spend more than $7000 annually on food and drink?
   c. What is the range of expenditures for the 5% of families with the highest annual spending on food and drink?

43. The website for the Bed and Breakfast Inns of North America gets approximately seven visitors per minute. Suppose the number of website visitors per minute follows a Poisson probability distribution.
   a. What is the mean time between visits to the website?
   b. Show the exponential probability density function for the time between website visits.
   c. What is the probability that no one will access the website in a 1-minute period?
   d. What is the probability that no one will access the website in a 12-second period?

44. Motorola used the normal distribution to determine the probability of defects and the number of defects expected in a production process. Assume a production process produces

items with a mean weight of 10 ounces. Calculate the probability of a defect and the expected number of defects for a 1000-unit production run in the following situations.
   a. The process standard deviation is .15, and the process control is set at plus or minus one standard deviation. Units with weights less than 9.85 or greater than 10.15 ounces will be classified as defects.
   b. Through process design improvements, the process standard deviation can be reduced to .05. Assume the process control remains the same, with weights less than 9.85 or greater than 10.15 ounces being classified as defects.
   c. What is the advantage of reducing process variation, thereby causing process control limits to be at a greater number of standard deviations from the mean?

45. The Information Systems Audit and Control Association surveyed office workers to learn about the anticipated usage of office computers for personal holiday shopping (*USA Today*, November 11, 2009). Assume that the number of hours a worker spends doing holiday shopping on an office computer follows an exponential distribution.
   a. The study reported that there is a .53 probability that a worker uses an office computer for holiday shopping 5 hours or less. Is the mean time spent using an office computer for holiday shopping closest to 5.8, 6.2, 6.6, or 7 hours?
   b. Using the mean time from part (a), what is the probability that a worker uses an office computer for holiday shopping more than 10 hours?
   c. What is the probability that a worker uses an office computer for holiday shopping between 4 and 8 hours?

46. The average annual amount American households spend for daily transportation is $6312 (*Money,* August 2001). Assume that the amount spent is normally distributed.
   a. Suppose you learn that 5% of American households spend less than $1000 for daily transportation. What is the standard deviation of the amount spent?
   b. What is the probability that a household spends between $4000 and $6000?
   c. What is the range of spending for the 3% of households with the highest daily transportation cost?

47. A blackjack player at a Las Vegas casino learned that the house will provide a free room if play is for four hours at an average bet of $50. The player's strategy provides a probability of .49 of winning on any one hand, and the player knows that there are 60 hands per hour. Suppose the player plays for four hours at a bet of $50 per hand.
   a. What is the player's expected payoff?
   b. What is the probability that the player loses $1000 or more?
   c. What is the probability that the player wins?
   d. Suppose the player starts with $1500. What is the probability of going broke?

48. *Condé Nast Traveler* publishes a Gold List of the top hotels all over the world. The Broadmoor Hotel in Colorado Springs contains 700 rooms and is on the 2004 Gold List (*Condé Nast Traveler,* January 2004). Suppose Broadmoor's marketing group forecasts a mean demand of 670 rooms for the coming weekend. Assume that demand for the upcoming weekend is normally distributed with a standard deviation of 30.
   a. What is the probability that all the hotel's rooms will be rented?
   b. What is the probability that 50 or more rooms will not be rented?
   c. Would you recommend the hotel consider offering a promotion to increase demand? What considerations would be important?

49. A machine fills containers with a particular product. The standard deviation of filling weights is known from past data to be .6 ounce. If only 2% of the containers hold less than 18 ounces, what is the mean filling weight for the machine? That is, what must μ equal? Assume the filling weights have a normal distribution.

50. Ward Doering Auto Sales is considering offering a special service contract that will cover the total cost of any service work required on leased vehicles. From experience, the

company manager estimates that yearly service costs are approximately normally distributed, with a mean of $150 and a standard deviation of $25.
   a. If the company offers the service contract to customers for a yearly charge of $200, what is the probability that any one customer's service costs will exceed the contract price of $200?
   b. What is Ward's expected profit per service contract?

51. Consider a multiple-choice examination with 50 questions. Each question has four possible answers. Assume that a student who has done the homework and attended lectures has a 75% probability of answering any question correctly.
   a. A student must answer 43 or more questions correctly to obtain a grade of A. What percentage of the students who have done their homework and attended lectures will obtain a grade of A on this multiple-choice examination?
   b. A student who answers 35 to 39 questions correctly will receive a grade of C. What percentage of students who have done their homework and attended lectures will obtain a grade of C on this multiple-choice examination?
   c. A student must answer 30 or more questions correctly to pass the examination. What percentage of the students who have done their homework and attended lectures will pass the examination?
   d. Assume that a student has not attended class and has not done the homework for the course. Furthermore, assume that the student will simply guess at the answer to each question. What is the probability that this student will answer 30 or more questions correctly and pass the examination?

52. Is lack of sleep causing traffic fatalities? A study conducted under the auspices of the National Highway Traffic Safety Administration found that the average number of fatal crashes caused by drowsy drivers each year was 1550 (*BusinessWeek*, January 26, 2004). Assume the annual number of fatal crashes per year is normally distributed with a standard deviation of 300.
   a. What is the probability of fewer than 1000 fatal crashes in a year?
   b. What is the probability that the number of fatal crashes will be between 1000 and 2000 for a year?
   c. For a year to be in the upper 5% with respect to the number of fatal crashes, how many fatal crashes would have to occur?

53. According to Salary Wizard, the average base salary for a brand manager in Houston, Texas, is $88,592 and the average base salary for a brand manager in Los Angeles, California, is $97,417 (Salary Wizard website, February 27, 2008). Assume that salaries are normally distributed, the standard deviation for brand managers in Houston is $19,900, and the standard deviation for brand managers in Los Angeles is $21,800.
   a. What is the probability that a brand manager in Houston has a base salary in excess of $100,000?
   b. What is the probability that a brand manager in Los Angeles has a base salary in excess of $100,000?
   c. What is the probability that a brand manager in Los Angeles has a base salary of less than $75,000?
   d. How much would a brand manager in Los Angeles have to make in order to have a higher salary than 99% of the brand managers in Houston?

54. Assume that the test scores from a college admissions test are normally distributed, with a mean of 450 and a standard deviation of 100.
   a. What percentage of the people taking the test score between 400 and 500?
   b. Suppose someone receives a score of 630. What percentage of the people taking the test score better? What percentage score worse?
   c. If a particular university will not admit anyone scoring below 480, what percentage of the persons taking the test would be acceptable to the university?

Case Problem: Specialty Toys

Specialty Toys, Inc., sells a variety of new and innovative children's toys. Management learned that the preholiday season is the best time to introduce a new toy, because many families use this time to look for new ideas for December holiday gifts. When Specialty discovers a new toy with good market potential, it chooses an October market entry date.

In order to get toys in its stores by October, Specialty places one-time orders with its manufacturers in June or July of each year. Demand for children's toys can be highly volatile. If a new toy catches on, a sense of shortage in the marketplace often increases the demand to high levels and large profits can be realized. However, new toys can also flop, leaving Specialty stuck with high levels of inventory that must be sold at reduced prices. The most important question the company faces is deciding how many units of a new toy should be purchased to meet anticipated sales demand. If too few are purchased, sales will be lost; if too many are purchased, profits will be reduced because of low prices realized in clearance sales.

For the coming season, Specialty plans to introduce a new product called Weather Teddy. This variation of a talking teddy bear is made by a company in Taiwan. When a child presses Teddy's hand, the bear begins to talk. A built-in barometer selects one of five responses that predict the weather conditions. The responses range from "It looks to be a very nice day! Have fun" to "I think it may rain today. Don't forget your umbrella." Tests with the product show that, even though it is not a perfect weather predictor, its predictions are surprisingly good. Several of Specialty's managers claimed Teddy gave predictions of the weather that were as good as many local television weather forecasters.

As with other products, Specialty faces the decision of how many Weather Teddy units to order for the coming holiday season. Members of the management team suggested order quantities of 15,000, 18,000, 24,000, or 28,000 units. The wide range of order quantities suggested indicates considerable disagreement concerning the market potential. The product management team asks you for an analysis of the stock-out probabilities for various order quantities, an estimate of the profit potential, and to help make an order quantity recommendation. Specialty expects to sell Weather Teddy for $24 based on a cost of $16 per unit. If inventory remains after the holiday season, Specialty will sell all surplus inventory for $5 per unit. After reviewing the sales history of similar products, Specialty's senior sales forecaster predicted an expected demand of 20,000 units with a .95 probability that demand would be between 10,000 units and 30,000 units.

Managerial Report

Prepare a managerial report that addresses the following issues and recommends an order quantity for the Weather Teddy product.

1. Use the sales forecaster's prediction to describe a normal probability distribution that can be used to approximate the demand distribution. Sketch the distribution and show its mean and standard deviation.
2. Compute the probability of a stock-out for the order quantities suggested by members of the management team.
3. Compute the projected profit for the order quantities suggested by the management team under three scenarios: worst case in which sales = 10,000 units, most likely case in which sales = 20,000 units, and best case in which sales = 30,000 units.
4. One of Specialty's managers felt that the profit potential was so great that the order quantity should have a 70% chance of meeting demand and only a 30% chance of any stock-outs. What quantity would be ordered under this policy, and what is the projected profit under the three sales scenarios?
5. Provide your own recommendation for an order quantity and note the associated profit projections. Provide a rationale for your recommendation.

Appendix 6.1 Continuous Probability Distributions Using Minitab

Let us demonstrate the Minitab procedure for computing continuous probabilities by referring to the Grear Tire Company problem where tire mileage was described by a normal distribution with $\mu = 36{,}500$ and $\sigma = 5000$. One question asked was, What is the probability that the tire mileage will exceed 40,000 miles?

For continuous probability distributions, Minitab provides a cumulative probability; that is, the probability that the random variable has a value less than or equal to a specified constant. For the Grear tire mileage question, Minitab can be used to determine the cumulative probability that the tire mileage will be less than or equal to 40,000 miles. After obtaining the cumulative probability, we can subtract it from 1 to determine the probability that the tire mileage will exceed 40,000 miles.

Prior to using Minitab to compute a cumulative probability, we enter the specified constant into a column of the worksheet. For the Grear tire mileage question we entered 40,000 into column C1. The steps in using Minitab to compute the cumulative probability of the normal random variable having a value less than or equal to 40,000 follow.

Step 1. Select the **Calc** menu
Step 2. Choose **Probability Distributions**
Step 3. Choose **Normal**
Step 4. When the Normal Distribution dialog box appears:
  Select **Cumulative probability**
  Enter 36500 in the **Mean** box
  Enter 5000 in the **Standard deviation** box
  Enter C1 in the **Input column** box (the column containing 40,000)
  Click **OK**

Minitab shows that this probability is .7580. Because we are interested in the probability that the tire mileage will be greater than 40,000, the desired probability is $1 - .7580 = .2420$.

A second question in the Grear Tire Company problem was, What mileage guarantee should Grear set to ensure that no more than 10% of the tires qualify for the guarantee? Here we are given a probability and want to find the corresponding value for the random variable. Minitab uses an inverse calculation routine to find the value of the random variable associated with a given cumulative probability. First, we enter the cumulative probability into a column of the Minitab worksheet. In this case, the desired cumulative probability is .10. Then, the first three steps of the Minitab procedure are as shown previously. In step 4, we select **Inverse cumulative probability** instead of **Cumulative probability** and complete the remaining parts of the step. Minitab then displays the mileage guarantee of 30,092 miles.

Minitab can be used to compute probabilities for other continuous probability distributions, including the exponential probability distribution. To compute exponential probabilities, follow the procedure shown previously for the normal probability distribution and choose the **Exponential** option in step 3. Step 4 is as shown, with the exception that entering the standard deviation is not required. Output for cumulative probabilities and inverse cumulative probabilities is identical to these described for the normal probability distribution.

Appendix 6.2 Continuous Probability Distributions Using Excel

Excel provides the capability for computing probabilities for several continuous probability distributions, including the normal probability distributions. In this appendix, we describe how Excel can be used to compute probabilities for any normal distribution. The procedures for other continuous distributions are similar to the one we describe for the normal distribution.

Let us return to the Grear Tire Company problem where the tire mileage was described by a normal distribution with $\mu = 36{,}500$ and $\sigma = 5000$. Assume we are interested in the probability that tire mileage will exceed 40,000 miles.

Excel's NORMDIST function provides cumulative probabilities for a normal distribution. The general form of the function is NORMDIST $(x,\mu,\sigma,\text{cumulative})$. For the fourth argument, TRUE is specified if a cumulative probability is desired. Thus, to compute the cumulative probability that the tire mileage will be less than or equal to 40,000 miles, we would enter the following formula into any cell of an Excel worksheet:

$$=\text{NORMDIST}(40000,36500,5000,\text{TRUE})$$

At this point, .7580 will appear in the cell where the formula was entered, indicating that the probability of tire mileage being less than or equal to 40,000 miles is .7580. Therefore, the probability that tire mileage will exceed 40,000 miles is $1 - .7580 = .2420$.

Excel's NORMINV function uses an inverse computation to find the x value corresponding to a given cumulative probability. For instance, suppose we want to find the guaranteed mileage Grear should offer so that no more than 10% of the tires will be eligible for the guarantee. We would enter the following formula into any cell of an Excel worksheet:

$$=\text{NORMINV}(.1,36500,5000)$$

At this point, 30092 will appear in the cell where the formula was entered, indicating that the probability of a tire lasting 30,092 miles or less is .10.

The Excel function for computing exponential probabilities is EXPONDIST. This function requires three arguments: x, the value of the variable; lamda, which is $1/\mu$, and TRUE if you would like the cumulative probability. For example, consider an exponential probability distribution with mean $\mu = 15$. The probability that the exponential variable is less than or equal to 6 can be computed by the Excel function

$$=\text{EXPONDIST}(6,1/15,\text{TRUE}).$$

If you need help inserting functions in a worksheet, Excel's Insert Function dialog box may be used. See Appendix E in the back of the text.

CHAPTER 7

Sampling and Sampling Distributions

CONTENTS

STATISTICS IN PRACTICE:
MEADWESTVACO CORPORATION

7.1 THE ELECTRONICS ASSOCIATES SAMPLING PROBLEM

7.2 SELECTING A SAMPLE
Sampling from a Finite Population
Sampling from an Infinite Population

7.3 POINT ESTIMATION
Practical Advice

7.4 INTRODUCTION TO SAMPLING DISTRIBUTIONS

7.5 SAMPLING DISTRIBUTION OF $\bar{x}$
Expected Value of $\bar{x}$
Standard Deviation of $\bar{x}$
Form of the Sampling Distribution of $\bar{x}$
Sampling Distribution of $\bar{x}$ for the EAI Problem
Practical Value of the Sampling Distribution of $\bar{x}$
Relationship Between the Sample Size and the Sampling Distribution of $\bar{x}$

7.6 SAMPLING DISTRIBUTION OF $\bar{p}$
Expected Value of $\bar{p}$
Standard Deviation of $\bar{p}$
Form of the Sampling Distribution of $\bar{p}$
Practical Value of the Sampling Distribution of $\bar{p}$

7.7 OTHER SAMPLING METHODS
Stratified Random Sampling
Cluster Sampling
Systematic Sampling
Convenience Sampling
Judgment Sampling

STATISTICS *in* PRACTICE

MEADWESTVACO CORPORATION*
STAMFORD, CONNECTICUT

MeadWestvaco Corporation, a leading producer of packaging, coated and specialty papers, consumer and office products, and specialty chemicals, employs more than 30,000 people. It operates worldwide in 29 countries and serves customers located in approximately 100 countries. MeadWestvaco holds a leading position in paper production, with an annual capacity of 1.8 million tons. The company's products include textbook paper, glossy magazine paper, beverage packaging systems, and office products. MeadWestvaco's internal consulting group uses sampling to provide a variety of information that enables the company to obtain significant productivity benefits and remain competitive.

For example, MeadWestvaco maintains large woodland holdings, which supply the trees, or raw material, for many of the company's products. Managers need reliable and accurate information about the timberlands and forests to evaluate the company's ability to meet its future raw material needs. What is the present volume in the forests? What is the past growth of the forests? What is the projected future growth of the forests? With answers to these important questions MeadWestvaco's managers can develop plans for the future, including long-term planting and harvesting schedules for the trees.

How does MeadWestvaco obtain the information it needs about its vast forest holdings? Data collected from sample plots throughout the forests are the basis for learning about the population of trees owned by the company. To identify the sample plots, the timberland holdings are first divided into three sections based on location and types of trees. Using maps and random numbers, MeadWestvaco analysts identify random samples of 1/5- to 1/7-acre plots in each section of the forest.

*The authors are indebted to Dr. Edward P. Winkofsky for providing this Statistics in Practice.

Random sampling of its forest holdings enables MeadWestvaco Corporation to meet future raw material needs.

MeadWestvaco foresters collect data from these sample plots to learn about the forest population.

Foresters throughout the organization participate in the field data collection process. Periodically, two-person teams gather information on each tree in every sample plot. The sample data are entered into the company's continuous forest inventory (CFI) computer system. Reports from the CFI system include a number of frequency distribution summaries containing statistics on types of trees, present forest volume, past forest growth rates, and projected future forest growth and volume. Sampling and the associated statistical summaries of the sample data provide the reports essential for the effective management of MeadWestvaco's forests and timberlands.

In this chapter you will learn about sampling and the sample selection process. In addition, you will learn how statistics such as the sample mean and sample proportion are used to estimate the population mean and population proportion. The important concept of a sampling distribution is also introduced.

In Chapter 1 we presented the following definitions of an element, a population, and a sample.

- An *element* is the entity on which data are collected.
- A *population* is the collection of all the elements of interest.
- A *sample* is a subset of the population.

The reason we select a sample is to collect data to make an inference and/or answer a research question about a population.

Let us begin by citing two examples in which sampling was used to answer a research question about a population.

1. Members of a political party in Texas were considering supporting a particular candidate for election to the U.S. Senate, and party leaders wanted to estimate the proportion of registered voters in the state favoring the candidate. A sample of 400 registered voters in Texas was selected and 160 of the 400 voters indicated a preference for the candidate. Thus, an estimate of the proportion of the population of registered voters favoring the candidate is 160/400 = .40.
2. A tire manufacturer is considering producing a new tire designed to provide an increase in mileage over the firm's current line of tires. To estimate the mean useful life of the new tires, the manufacturer produced a sample of 120 tires for testing. The test results provided a sample mean of 36,500 miles. Hence, an estimate of the mean useful life for the population of new tires was 36,500 miles.

It is important to realize that sample results provide only *estimates* of the values of the corresponding population characteristics. We do not expect exactly .40, or 40%, of the population of registered voters to favor the candidate, nor do we expect the sample mean of 36,500 miles to exactly equal the mean mileage for the population of all new tires produced. The reason is simply that the sample contains only a portion of the population. Some sampling error is to be expected. With proper sampling methods, the sample results will provide "good" estimates of the population parameters. But how good can we expect the sample results to be? Fortunately, statistical procedures are available for answering this question.

A sample mean provides an estimate of a population mean, and a sample proportion provides an estimate of a population proportion. With estimates such as these, some estimation error can be expected. This chapter provides the basis for determining how large that error might be.

Let us define some of the terms used in sampling. The **sampled population** is the population from which the sample is drawn, and a **frame** is a list of the elements that the sample will be selected from. In the first example, the sampled population is all registered voters in Texas, and the frame is a list of all the registered voters. Because the number of registered voters in Texas is a finite number, the first example is an illustration of sampling from a finite population. In Section 7.2, we discuss how a simple random sample can be selected when sampling from a finite population.

The sampled population for the tire mileage example is more difficult to define because the sample of 120 tires was obtained from a production process at a particular point in time. We can think of the sampled population as the conceptual population of all the tires that could have been made by the production process at that particular point in time. In this sense the sampled population is considered infinite, making it impossible to construct a frame to draw the sample from. In Section 7.2, we discuss how to select a random sample in such a situation.

In this chapter, we show how simple random sampling can be used to select a sample from a finite population and describe how a random sample can be taken from an infinite population that is generated by an ongoing process. We then show how data obtained from a sample can be used to compute estimates of a population mean, a population standard deviation, and a population proportion. In addition, we introduce the important concept of a sampling distribution. As we will show, knowledge of the appropriate sampling distribution enables us to make statements about how close the sample estimates are to the corresponding population parameters. The last section discusses some alternatives to simple random sampling that are often employed in practice.

7.1 The Electronics Associates Sampling Problem

The director of personnel for Electronics Associates, Inc. (EAI) has been assigned the task of developing a profile of the company's 2500 managers. The characteristics to be identified include the mean annual salary for the managers and the proportion of managers having completed the company's management training program.

WEB file

EAI

Using the 2500 managers as the population for this study, we can find the annual salary and the training program status for each individual by referring to the firm's personnel records. The data set containing this information for all 2500 managers in the population is in the file named EAI.

Using the EAI data and the formulas presented in Chapter 3, we compute the population mean and the population standard deviation for the annual salary data.

$$\text{Population mean:} \quad \mu = \$51{,}800$$
$$\text{Population standard deviation:} \quad \sigma = \$4{,}000$$

The data for the training program status show that 1500 of the 2500 managers completed the training program.

Numerical characteristics of a population are called **parameters**. Letting p denote the proportion of the population that completed the training program, we see that $p = 1500/2500 = .60$. The population mean annual salary ($\mu = \$51{,}800$), the population standard deviation of annual salary ($\sigma = \$4{,}000$), and the population proportion that completed the training program ($p = .60$) are parameters of the population of EAI managers.

Usually the cost of collecting information from a sample is substantially less than from a population, especially when personal interviews must be conducted to collect the information.

Now, suppose that the necessary information on all the EAI managers was not readily available in the company's database. The question we now consider is how the firm's director of personnel can obtain estimates of the population parameters by using a sample of managers rather than all 2500 managers in the population. Suppose that a sample of 30 managers will be used. Clearly, the time and the cost of developing a profile would be substantially less for 30 managers than for the entire population. If the personnel director could be assured that a sample of 30 managers would provide adequate information about the population of 2500 managers, working with a sample would be preferable to working with the entire population. Let us explore the possibility of using a sample for the EAI study by first considering how we can identify a sample of 30 managers.

7.2 Selecting a Sample

In this section we describe how to select a sample. We first describe how to sample from a finite population and then describe how to select a sample from an infinite population.

Sampling from a Finite Population

Statisticians recommend selecting a probability sample when sampling from a finite population because a probability sample allows them to make valid statistical inferences about the population. The simplest type of probability sample is one in which each sample of size n has the same probability of being selected. It is called a simple random sample. A simple random sample of size n from a finite population of size N is defined as follows.

Other methods of probability sampling are described in Section 7.8.

> **SIMPLE RANDOM SAMPLE (FINITE POPULATION)**
>
> A **simple random sample** of size n from a finite population of size N is a sample selected such that each possible sample of size n has the same probability of being selected.

Computer-generated random numbers can also be used to implement the random sample selection process. Excel provides a function for generating random numbers in its worksheets.

One procedure for selecting a simple random sample from a finite population is to choose the elements for the sample one at a time in such a way that, at each step, each of the elements remaining in the population has the same probability of being selected. Sampling n elements in this way will satisfy the definition of a simple random sample from a finite population.

To select a simple random sample from the finite population of EAI managers, we first construct a frame by assigning each manager a number. For example, we can assign the

7.2 Selecting a Sample

TABLE 7.1 RANDOM NUMBERS

63271	59986	71744	51102	15141	80714	58683	93108	13554	79945
88547	09896	95436	79115	08303	01041	20030	63754	08459	28364
55957	57243	83865	09911	19761	66535	40102	26646	60147	15702
46276	87453	44790	67122	45573	84358	21625	16999	13385	22782
55363	07449	34835	15290	76616	67191	12777	21861	68689	03263
69393	92785	49902	58447	42048	30378	87618	26933	40640	16281
13186	29431	88190	04588	38733	81290	89541	70290	40113	08243
17726	28652	56836	78351	47327	18518	92222	55201	27340	10493
36520	64465	05550	30157	82242	29520	69753	72602	23756	54935
81628	36100	39254	56835	37636	02421	98063	89641	64953	99337
84649	48968	75215	75498	49539	74240	03466	49292	36401	45525
63291	11618	12613	75055	43915	26488	41116	64531	56827	30825
70502	53225	03655	05915	37140	57051	48393	91322	25653	06543
06426	24771	59935	49801	11082	66762	94477	02494	88215	27191
20711	55609	29430	70165	45406	78484	31639	52009	18873	96927
41990	70538	77191	25860	55204	73417	83920	69468	74972	38712
72452	36618	76298	26678	89334	33938	95567	29380	75906	91807
37042	40318	57099	10528	09925	89773	41335	96244	29002	46453
53766	52875	15987	46962	67342	77592	57651	95508	80033	69828
90585	58955	53122	16025	84299	53310	67380	84249	25348	04332
32001	96293	37203	64516	51530	37069	40261	61374	05815	06714
62606	64324	46354	72157	67248	20135	49804	09226	64419	29457
10078	28073	85389	50324	14500	15562	64165	06125	71353	77669
91561	46145	24177	15294	10061	98124	75732	00815	83452	97355
13091	98112	53959	79607	52244	63303	10413	63839	74762	50289

managers the numbers 1 to 2500 in the order that their names appear in the EAI personnel file. Next, we refer to the table of random numbers shown in Table 7.1. Using the first row of the table, each digit, 6, 3, 2, ..., is a random digit having an equal chance of occurring. Because the largest number in the population list of EAI managers, 2500, has four digits, we will select random numbers from the table in sets or groups of four digits. Even though we may start the selection of random numbers anywhere in the table and move systematically in a direction of our choice, we will use the first row of Table 7.1 and move from left to right. The first 7 four-digit random numbers are

The random numbers in the table are shown in groups of five for readability.

   6327 1599 8671 7445 1102 1514 1807

Because the numbers in the table are random, these four-digit numbers are equally likely.

  We can now use these four-digit random numbers to give each manager in the population an equal chance of being included in the random sample. The first number, 6327, is greater than 2500. It does not correspond to one of the numbered managers in the population, and hence is discarded. The second number, 1599, is between 1 and 2500. Thus the first manager selected for the random sample is number 1599 on the list of EAI managers. Continuing this process, we ignore the numbers 8671 and 7445 before identifying managers number 1102, 1514, and 1807 to be included in the random sample. This process continues until the simple random sample of 30 EAI managers has been obtained.

  In implementing this simple random sample selection process, it is possible that a random number used previously may appear again in the table before the complete sample of 30 EAI managers has been selected. Because we do not want to select a manager more than one time, any previously used random numbers are ignored because the corresponding manager is already included in the sample. Selecting a sample in this manner is referred to as **sampling without replacement**. If we selected a sample such that previously used random numbers are acceptable and specific managers could be included in the sample two or more times, we would be **sampling with replacement**. Sampling with replacement is a valid way of identifying a simple random sample. However, sampling without replacement is the sampling procedure used most often. When we refer to simple random sampling, we will assume the sampling is without replacement.

Sampling from an Infinite Population

Sometimes we want to select a sample from a population, but the population is infinitely large or the elements of the population are being generated by an ongoing process for which there is no limit on the number of elements that can be generated. Thus, it is not possible to develop a list of all the elements in the population. This is considered the infinite population case. With an infinite population, we cannot select a simple random sample because we cannot construct a frame consisting of all the elements. In the infinite population case, statisticians recommend selecting what is called a random sample.

> RANDOM SAMPLE (INFINITE POPULATION)
>
> A **random sample** of size n from an infinite population is a sample selected such that the following conditions are satisfied.
>
> 1. Each element selected comes from the same population.
> 2. Each element is selected independently.

Care and judgment must be exercised in implementing the selection process for obtaining a random sample from an infinite population. Each case may require a different selection procedure. Let us consider two examples to see what we mean by the conditions (1) each element selected comes from the same population and (2) each element is selected independently.

A common quality control application involves a production process where there is no limit on the number of elements that can be produced. The conceptual population we are sampling from is all the elements that could be produced (not just the ones that are produced) by the ongoing production process. Because we cannot develop a list of all the elements that could be produced, the population is considered infinite. To be more specific, let us consider a production line designed to fill boxes of a breakfast cereal with a mean weight of 24 ounces of breakfast cereal per box. Samples of 12 boxes filled by this process are periodically selected by a quality control inspector to determine if the process is operating properly or if, perhaps, a machine malfunction has caused the process to begin underfilling or overfilling the boxes.

With a production operation such as this, the biggest concern in selecting a random sample is to make sure that condition 1, the sampled elements are selected from the same population, is satisfied. To ensure that this condition is satisfied, the boxes must be selected at approximately the same point in time. This way the inspector avoids the possibility of selecting some boxes when the process is operating properly and other boxes when the process is not operating properly and is underfilling or overfilling the boxes. With a production process such as this, the second condition, each element is selected independently, is satisfied by designing the production process so that each box of cereal is filled independently. With this assumption, the quality control inspector only needs to worry about satisfying the same population condition.

As another example of selecting a random sample from an infinite population, consider the population of customers arriving at a fast-food restaurant. Suppose an employee is asked to select and interview a sample of customers in order to develop a profile of customers who visit the restaurant. The customer arrival process is ongoing and there is no way to obtain a list of all customers in the population. So, for practical purposes, the population for this ongoing process is considered infinite. As long as a sampling procedure is designed so that all the elements in the sample are customers of the restaurant and they are selected independently, a random sample will be obtained. In this case, the employee collecting the sample needs to select the sample from people who come into the restaurant and make a purchase to ensure that the same population condition is satisfied. If, for instance, the employee selected someone for the sample who came into the restaurant just to use the restroom, that person would not be a customer and the same population condition would be violated. So, as long as the interviewer selects the sample from people making a purchase at the restaurant, condition 1 is satisfied. Ensuring that the customers are selected independently can be more difficult.

The purpose of the second condition of the random sample selection procedure (each element is selected independently) is to prevent selection bias. In this case, selection bias would occur if the interviewer were free to select customers for the sample arbitrarily. The interviewer might feel more comfortable selecting customers in a particular age group and might avoid customers in other age groups. Selection bias would also occur if the interviewer selected a group of five customers who entered the restaurant together and asked all of them to participate in the sample. Such a group of customers would be likely to exhibit similar characteristics, which might provide misleading information about the population of customers. Selection bias such as this can be avoided by ensuring that the selection of a particular customer does not influence the selection of any other customer. In other words, the elements (customers) are selected independently.

McDonald's, the fast-food restaurant leader, implemented a random sampling procedure for this situation. The sampling procedure was based on the fact that some customers presented discount coupons. Whenever a customer presented a discount coupon, the next customer served was asked to complete a customer profile questionnaire. Because arriving customers presented discount coupons randomly and independently of other customers, this sampling procedure ensured that customers were selected independently. As a result, the sample satisfied the requirements of a random sample from an infinite population.

Situations involving sampling from an infinite population are usually associated with a process that operates over time. Examples include parts being manufactured on a production line, repeated experimental trials in a laboratory, transactions occurring at a bank, telephone calls arriving at a technical support center, and customers entering a retail store. In each case, the situation may be viewed as a process that generates elements from an infinite population. As long as the sampled elements are selected from the same population and are selected independently, the sample is considered a random sample from an infinite population.

NOTES AND COMMENTS

1. In this section we have been careful to define two types of samples: a simple random sample from a finite population and a random sample from an infinite population. In the remainder of the text, we will generally refer to both of these as either a *random sample* or simply a *sample*. We will not make a distinction of the sample being a "simple" random sample unless it is necessary for the exercise or discussion.

2. Statisticians who specialize in sample surveys from finite populations use sampling methods that provide probability samples. With a probability sample, each possible sample has a known probability of selection and a random process is used to select the elements for the sample. Simple random sampling is one of these methods. In Section 7.8, we describe some other probability sampling methods: stratified random sampling, cluster sampling, and systematic sampling. We use the term *simple* in simple random sampling to clarify that this is the probability sampling method that assures each sample of size n has the same probability of being selected.

3. The number of different simple random samples of size n that can be selected from a finite population of size N is

$$\frac{N!}{n!(N-n)!}$$

In this formula, $N!$ and $n!$ are the factorial formulas discussed in Chapter 4. For the EAI problem with $N = 2500$ and $n = 30$, this expression can be used to show that approximately 2.75×10^{69} different simple random samples of 30 EAI managers can be obtained.

4. Computer software packages can be used to select a random sample. In the chapter appendixes, we show how Minitab and Excel can be used to select a simple random sample from a finite population.

Exercises

Methods

1. Consider a finite population with five elements labeled A, B, C, D, and E. Ten possible simple random samples of size 2 can be selected.
   a. List the 10 samples beginning with AB, AC, and so on.
   b. Using simple random sampling, what is the probability that each sample of size 2 is selected?
   c. Assume random number 1 corresponds to A, random number 2 corresponds to B, and so on. List the simple random sample of size 2 that will be selected by using the random digits 8 0 5 7 5 3 2.

2. Assume a finite population has 350 elements. Using the last three digits of each of the following five-digit random numbers (e.g., 601, 022, 448, . . .), determine the first four elements that will be selected for the simple random sample.

   98601   73022   83448   02147   34229   27553   84147   93289   14209

Applications

3. *Fortune* publishes data on sales, profits, assets, stockholders' equity, market value, and earnings per share for the 500 largest U.S. industrial corporations (*Fortune* 500, 2006). Assume that you want to select a simple random sample of 10 corporations from the *Fortune* 500 list. Use the last three digits in column 9 of Table 7.1, beginning with 554. Read down the column and identify the numbers of the 10 corporations that would be selected.

4. The 10 most active stocks on the New York Stock Exchange on March 6, 2006, are shown here (*The Wall Street Journal*, March 7, 2006).

   | AT&T | Lucent | Nortel | Qwest | Bell South |
   | Pfizer | Texas Instruments | Gen. Elect. | iShrMSJpn | LSI Logic |

   Exchange authorities decided to investigate trading practices using a sample of three of these stocks.
   a. Beginning with the first random digit in column 6 of Table 7.1, read down the column to select a simple random sample of three stocks for the exchange authorities.
   b. Using the information in the third Note and Comment, determine how many different simple random samples of size 3 can be selected from the list of 10 stocks.

5. A student government organization is interested in estimating the proportion of students who favor a mandatory "pass-fail" grading policy for elective courses. A list of names and addresses of the 645 students enrolled during the current quarter is available from the registrar's office. Using three-digit random numbers in row 10 of Table 7.1 and moving across the row from left to right, identify the first 10 students who would be selected using simple random sampling. The three-digit random numbers begin with 816, 283, and 610.

6. The *County and City Data Book*, published by the Census Bureau, lists information on 3139 counties throughout the United States. Assume that a national study will collect data from 30 randomly selected counties. Use four-digit random numbers from the last column of Table 7.1 to identify the numbers corresponding to the first five counties selected for the sample. Ignore the first digits and begin with the four-digit random numbers 9945, 8364, 5702, and so on.

7. Assume that we want to identify a simple random sample of 12 of the 372 doctors practicing in a particular city. The doctors' names are available from a local medical organization. Use the eighth column of five-digit random numbers in Table 7.1 to identify the 12 doctors for the sample. Ignore the first two random digits in each five-digit grouping of the random numbers. This process begins with random number 108 and proceeds down the column of random numbers.

8. The following stocks make up the Dow Jones Industrial Average (*Barron's*, March 23, 2009).

1. 3M	11. Disney	21. McDonald's
2. AT&T	12. DuPont	22. Merck
3. Alcoa	13. ExxonMobil	23. Microsoft
4. American Express	14. General Electric	24. J.P. Morgan
5. Bank of America	15. Hewlett-Packard	25. Pfizer
6. Boeing	16. Home Depot	26. Procter & Gamble
7. Caterpillar	17. IBM	27. Travelers
8. Chevron	18. Intel	28. United Technologies
9. Cisco Systems	19. Johnson & Johnson	29. Verizon
10. Coca-Cola	20. Kraft Foods	30. Walmart

   Suppose you would like to select a sample of six of these companies to conduct an in-depth study of management practices. Use the first two digits in each row of the ninth column of Table 7.1 to select a simple random sample of six companies.

9. *The Wall Street Journal* provides the net asset value, the year-to-date percent return, and the three-year percent return for 555 mutual funds (*The Wall Street Journal*, April 25, 2003). Assume that a simple random sample of 12 of the 555 mutual funds will be selected for a follow-up study on the size and performance of mutual funds. Use the fourth column of the random numbers in Table 7.1, beginning with 51102, to select the simple random sample of 12 mutual funds. Begin with mutual fund 102 and use the *last* three digits in each row of the fourth column for your selection process. What are the numbers of the 12 mutual funds in the simple random sample?

10. Indicate which of the following situations involve sampling from a finite population and which involve sampling from an infinite population. In cases where the sampled population is finite, describe how you would construct a frame.
    a. Obtain a sample of licensed drivers in the state of New York.
    b. Obtain a sample of boxes of cereal produced by the Breakfast Choice company.
    c. Obtain a sample of cars crossing the Golden Gate Bridge on a typical weekday.
    d. Obtain a sample of students in a statistics course at Indiana University.
    e. Obtain a sample of the orders that are processed by a mail-order firm.

7.3 Point Estimation

Now that we have described how to select a simple random sample, let us return to the EAI problem. A simple random sample of 30 managers and the corresponding data on annual salary and management training program participation are as shown in Table 7.2. The notation x_1, x_2, and so on is used to denote the annual salary of the first manager in the sample, the annual salary of the second manager in the sample, and so on. Participation in the management training program is indicated by Yes in the management training program column.

To estimate the value of a population parameter, we compute a corresponding characteristic of the sample, referred to as a **sample statistic**. For example, to estimate the population mean μ and the population standard deviation σ for the annual salary of EAI managers, we use the data in Table 7.2 to calculate the corresponding sample statistics: the

TABLE 7.2 ANNUAL SALARY AND TRAINING PROGRAM STATUS FOR A SIMPLE RANDOM SAMPLE OF 30 EAI MANAGERS

Annual Salary ($)	Management Training Program	Annual Salary ($)	Management Training Program
$x_1 = 49{,}094.30$	Yes	$x_{16} = 51{,}766.00$	Yes
$x_2 = 53{,}263.90$	Yes	$x_{17} = 52{,}541.30$	No
$x_3 = 49{,}643.50$	Yes	$x_{18} = 44{,}980.00$	Yes
$x_4 = 49{,}894.90$	Yes	$x_{19} = 51{,}932.60$	Yes
$x_5 = 47{,}621.60$	No	$x_{20} = 52{,}973.00$	Yes
$x_6 = 55{,}924.00$	Yes	$x_{21} = 45{,}120.90$	Yes
$x_7 = 49{,}092.30$	Yes	$x_{22} = 51{,}753.00$	Yes
$x_8 = 51{,}404.40$	Yes	$x_{23} = 54{,}391.80$	No
$x_9 = 50{,}957.70$	Yes	$x_{24} = 50{,}164.20$	No
$x_{10} = 55{,}109.70$	Yes	$x_{25} = 52{,}973.60$	No
$x_{11} = 45{,}922.60$	Yes	$x_{26} = 50{,}241.30$	No
$x_{12} = 57{,}268.40$	No	$x_{27} = 52{,}793.90$	No
$x_{13} = 55{,}688.80$	Yes	$x_{28} = 50{,}979.40$	Yes
$x_{14} = 51{,}564.70$	No	$x_{29} = 55{,}860.90$	Yes
$x_{15} = 56{,}188.20$	No	$x_{30} = 57{,}309.10$	No

sample mean and the sample standard deviation s. Using the formulas for a sample mean and a sample standard deviation presented in Chapter 3, the sample mean is

$$\bar{x} = \frac{\Sigma x_i}{n} = \frac{1{,}554{,}420}{30} = \$51{,}814$$

and the sample standard deviation is

$$s = \sqrt{\frac{\Sigma(x_i - \bar{x})^2}{n-1}} = \sqrt{\frac{325{,}009{,}260}{29}} = \$3{,}348$$

To estimate p, the proportion of managers in the population who completed the management training program, we use the corresponding sample proportion $\bar{p}$. Let x denote the number of managers in the sample who completed the management training program. The data in Table 7.2 show that $x = 19$. Thus, with a sample size of $n = 30$, the sample proportion is

$$\bar{p} = \frac{x}{n} = \frac{19}{30} = .63$$

By making the preceding computations, we perform the statistical procedure called *point estimation*. We refer to the sample mean $\bar{x}$ as the **point estimator** of the population mean μ, the sample standard deviation s as the point estimator of the population standard deviation σ, and the sample proportion $\bar{p}$ as the point estimator of the population proportion p. The numerical value obtained for $\bar{x}$, s, or $\bar{p}$ is called the **point estimate**. Thus, for the simple random sample of 30 EAI managers shown in Table 7.2, $\$51{,}814$ is the point estimate of μ, $\$3{,}348$ is the point estimate of σ, and .63 is the point estimate of p. Table 7.3 summarizes the sample results and compares the point estimates to the actual values of the population parameters.

As is evident from Table 7.3, the point estimates differ somewhat from the corresponding population parameters. This difference is to be expected because a sample, and not a census of the entire population, is being used to develop the point estimates. In the next chapter, we will show how to construct an interval estimate in order to provide information about how close the point estimate is to the population parameter.

7.3 Point Estimation

TABLE 7.3 SUMMARY OF POINT ESTIMATES OBTAINED FROM A SIMPLE RANDOM SAMPLE OF 30 EAI MANAGERS

Population Parameter	Parameter Value	Point Estimator	Point Estimate
μ = Population mean annual salary	$51,800	$\bar{x}$ = Sample mean annual salary	$51,814
σ = Population standard deviation for annual salary	$4,000	s = Sample standard deviation for annual salary	$3,348
p = Population proportion having completed the management training program	.60	$\bar{p}$ = Sample proportion having completed the management training program	.63

Practical Advice

The subject matter of most of the rest of the book is concerned with statistical inference. Point estimation is a form of statistical inference. We use a sample statistic to make an inference about a population parameter. When making inferences about a population based on a sample, it is important to have a close correspondence between the sampled population and the target population. The **target population** is the population we want to make inferences about, while the sampled population is the population from which the sample is actually taken. In this section, we have described the process of drawing a simple random sample from the population of EAI managers and making point estimates of characteristics of that same population. So the sampled population and the target population are identical, which is the desired situation. But in other cases, it is not as easy to obtain a close correspondence between the sampled and target populations.

Consider the case of an amusement park selecting a sample of its customers to learn about characteristics such as age and time spent at the park. Suppose all the sample elements were selected on a day when park attendance was restricted to employees of a large company. Then the sampled population would be composed of employees of that company and members of their families. If the target population we wanted to make inferences about were typical park customers over a typical summer, then we might encounter a significant difference between the sampled population and the target population. In such a case, we would question the validity of the point estimates being made. Park management would be in the best position to know whether a sample taken on a particular day was likely to be representative of the target population.

In summary, whenever a sample is used to make inferences about a population, we should make sure that the study is designed so that the sampled population and the target population are in close agreement. Good judgment is a necessary ingredient of sound statistical practice.

Exercises

Methods

11. The following data are from a simple random sample.

        5   8   10   7   10   14

    a. What is the point estimate of the population mean?
    b. What is the point estimate of the population standard deviation?

12. A survey question for a sample of 150 individuals yielded 75 Yes responses, 55 No responses, and 20 No Opinions.

    a. What is the point estimate of the proportion in the population who respond Yes?
    b. What is the point estimate of the proportion in the population who respond No?

Applications

13. A simple random sample of 5 months of sales data provided the following information:

Month:	1	2	3	4	5
Units Sold:	94	100	85	94	92

    a. Develop a point estimate of the population mean number of units sold per month.
    b. Develop a point estimate of the population standard deviation.

14. *BusinessWeek* published information on 283 equity mutual funds (*BusinessWeek*, January 26, 2004). A sample of 40 of those funds is contained in the data set MutualFund. Use the data set to answer the following questions.
    a. Develop a point estimate of the proportion of the *BusinessWeek* equity funds that are load funds.
    b. Develop a point estimate of the proportion of funds that are classified as high risk.
    c. Develop a point estimate of the proportion of funds that have a below-average risk rating.

15. Many drugs used to treat cancer are expensive. *BusinessWeek* reported on the cost per treatment of Herceptin, a drug used to treat breast cancer (*BusinessWeek*, January 30, 2006). Typical treatment costs (in dollars) for Herceptin are provided by a simple random sample of 10 patients.

4376	5578	2717	4920	4495
4798	6446	4119	4237	3814

    a. Develop a point estimate of the mean cost per treatment with Herceptin.
    b. Develop a point estimate of the standard deviation of the cost per treatment with Herceptin.

16. A sample of 50 *Fortune* 500 companies (*Fortune*, April 14, 2003) showed 5 were based in New York, 6 in California, 2 in Minnesota, and 1 in Wisconsin.
    a. Develop an estimate of the proportion of *Fortune* 500 companies based in New York.
    b. Develop an estimate of the number of *Fortune* 500 companies based in Minnesota.
    c. Develop an estimate of the proportion of *Fortune* 500 companies that are not based in these four states.

17. The American Association of Individual Investors (AAII) polls its subscribers on a weekly basis to determine the number who are bullish, bearish, or neutral on the short-term prospects for the stock market. Their findings for the week ending March 2, 2006, are consistent with the following sample results (AAII website, March 7, 2006).

    Bullish 409   Neutral 299   Bearish 291

    Develop a point estimate of the following population parameters.
    a. The proportion of all AAII subscribers who are bullish on the stock market.
    b. The proportion of all AAII subscribers who are neutral on the stock market.
    c. The proportion of all AAII subscribers who are bearish on the stock market.

7.4 Introduction to Sampling Distributions

In the preceding section we said that the sample mean $\bar{x}$ is the point estimator of the population mean μ, and the sample proportion $\bar{p}$ is the point estimator of the population proportion p. For the simple random sample of 30 EAI managers shown in Table 7.2, the point estimate of μ is $\bar{x} = \$51,814$ and the point estimate of p is $\bar{p} = .63$. Suppose we select another simple random sample of 30 EAI managers and obtain the following point estimates:

Sample mean: $\bar{x} = \$52,670$
Sample proportion: $\bar{p} = .70$

7.4 Introduction to Sampling Distributions

TABLE 7.4 VALUES OF $\bar{x}$ AND $\bar{p}$ FROM 500 SIMPLE RANDOM SAMPLES OF 30 EAI MANAGERS

Sample Number	Sample Mean ($\bar{x}$)	Sample Proportion ($\bar{p}$)
1	51,814	.63
2	52,670	.70
3	51,780	.67
4	51,588	.53
.	.	.
.	.	.
.	.	.
500	51,752	.50

Note that different values of $\bar{x}$ and $\bar{p}$ were obtained. Indeed, a second simple random sample of 30 EAI managers cannot be expected to provide the same point estimates as the first sample.

Now, suppose we repeat the process of selecting a simple random sample of 30 EAI managers over and over again, each time computing the values of $\bar{x}$ and $\bar{p}$. Table 7.4 contains a portion of the results obtained for 500 simple random samples, and Table 7.5 shows the frequency and relative frequency distributions for the 500 $\bar{x}$ values. Figure 7.1 shows the relative frequency histogram for the $\bar{x}$ values.

The ability to understand the material in subsequent chapters depends heavily on the ability to understand and use the sampling distributions presented in this chapter.

In Chapter 5 we defined a random variable as a numerical description of the outcome of an experiment. If we consider the process of selecting a simple random sample as an experiment, the sample mean $\bar{x}$ is the numerical description of the outcome of the experiment. Thus, the sample mean $\bar{x}$ is a random variable. As a result, just like other random variables, $\bar{x}$ has a mean or expected value, a standard deviation, and a probability distribution. Because the various possible values of $\bar{x}$ are the result of different simple random samples, the probability distribution of $\bar{x}$ is called the **sampling distribution** of $\bar{x}$. Knowledge of this sampling distribution and its properties will enable us to make probability statements about how close the sample mean $\bar{x}$ is to the population mean μ.

Let us return to Figure 7.1. We would need to enumerate every possible sample of 30 managers and compute each sample mean to completely determine the sampling distribution of $\bar{x}$. However, the histogram of 500 $\bar{x}$ values gives an approximation of this sampling distribution. From the approximation we observe the bell-shaped appearance of

TABLE 7.5 FREQUENCY AND RELATIVE FREQUENCY DISTRIBUTIONS OF $\bar{x}$ FROM 500 SIMPLE RANDOM SAMPLES OF 30 EAI MANAGERS

Mean Annual Salary ($)	Frequency	Relative Frequency
49,500.00–49,999.99	2	.004
50,000.00–50,499.99	16	.032
50,500.00–50,999.99	52	.104
51,000.00–51,499.99	101	.202
51,500.00–51,999.99	133	.266
52,000.00–52,499.99	110	.220
52,500.00–52,999.99	54	.108
53,000.00–53,499.99	26	.052
53,500.00–53,999.99	6	.012
Totals	500	1.000

FIGURE 7.1 RELATIVE FREQUENCY HISTOGRAM OF $\bar{x}$ VALUES FROM 500 SIMPLE RANDOM SAMPLES OF SIZE 30 EACH

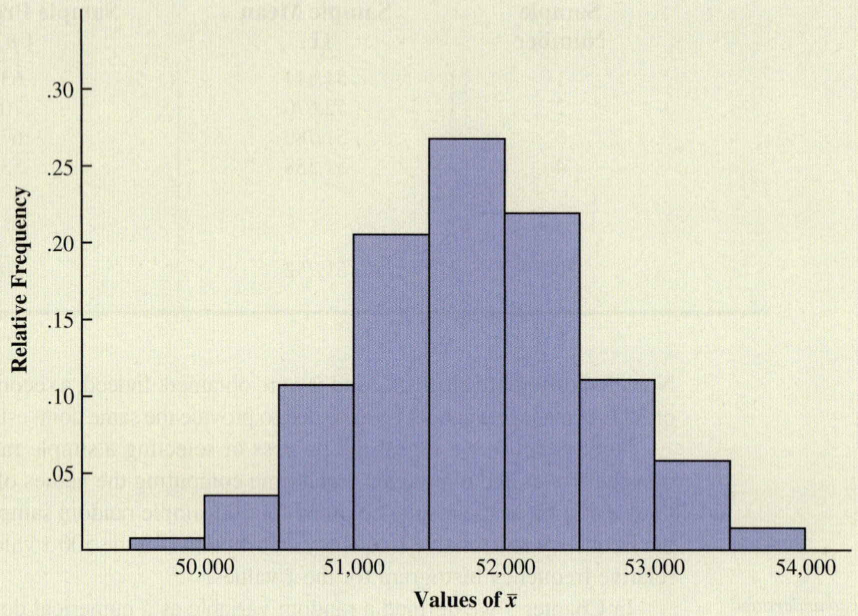

the distribution. We note that the largest concentration of the $\bar{x}$ values and the mean of the 500 $\bar{x}$ values is near the population mean $\mu = \$51{,}800$. We will describe the properties of the sampling distribution of $\bar{x}$ more fully in the next section.

The 500 values of the sample proportion $\bar{p}$ are summarized by the relative frequency histogram in Figure 7.2. As in the case of $\bar{x}$, $\bar{p}$ is a random variable. If every possible sample of size 30 were selected from the population and if a value of $\bar{p}$ were computed for each sample, the resulting probability distribution would be the sampling distribution of $\bar{p}$. The relative frequency histogram of the 500 sample values in Figure 7.2 provides a general idea of the appearance of the sampling distribution of $\bar{p}$.

In practice, we select only one simple random sample from the population. We repeated the sampling process 500 times in this section simply to illustrate that many different samples are possible and that the different samples generate a variety of values for the sample statistics $\bar{x}$ and $\bar{p}$. The probability distribution of any particular sample statistic is called the sampling distribution of the statistic. In Section 7.5 we show the characteristics of the sampling distribution of $\bar{x}$. In Section 7.6 we show the characteristics of the sampling distribution of $\bar{p}$.

7.5 Sampling Distribution of $\bar{x}$

In the previous section we said that the sample mean $\bar{x}$ is a random variable and its probability distribution is called the sampling distribution of $\bar{x}$.

SAMPLING DISTRIBUTION OF $\bar{x}$

The sampling distribution of $\bar{x}$ is the probability distribution of all possible values of the sample mean $\bar{x}$.

FIGURE 7.2   RELATIVE FREQUENCY HISTOGRAM OF $\bar{p}$ VALUES FROM 500 SIMPLE RANDOM SAMPLES OF SIZE 30 EACH

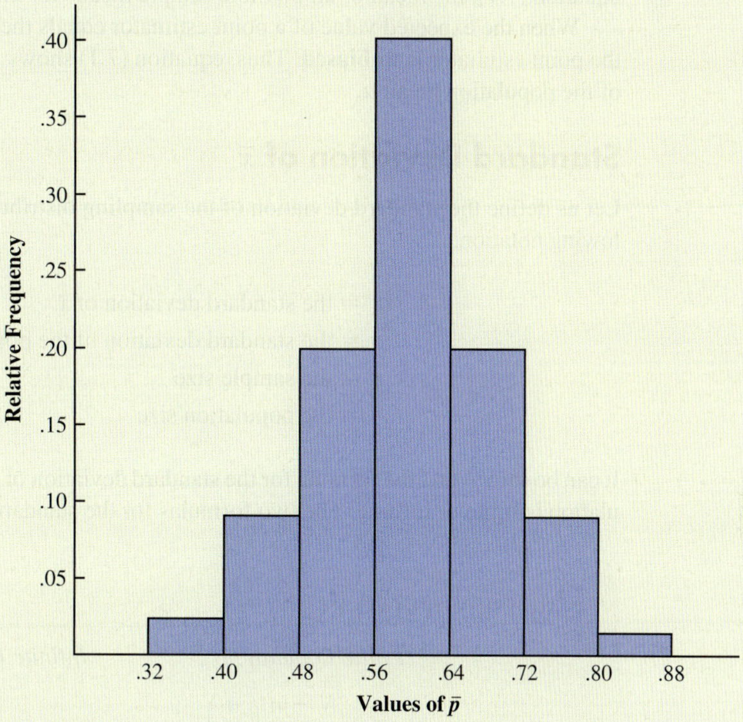

This section describes the properties of the sampling distribution of $\bar{x}$. Just as with other probability distributions we studied, the sampling distribution of $\bar{x}$ has an expected value or mean, a standard deviation, and a characteristic shape or form. Let us begin by considering the mean of all possible $\bar{x}$ values, which is referred to as the expected value of $\bar{x}$.

Expected Value of $\bar{x}$

In the EAI sampling problem we saw that different simple random samples result in a variety of values for the sample mean $\bar{x}$. Because many different values of the random variable $\bar{x}$ are possible, we are often interested in the mean of all possible values of $\bar{x}$ that can be generated by the various simple random samples. The mean of the $\bar{x}$ random variable is the expected value of $\bar{x}$. Let $E(\bar{x})$ represent the expected value of $\bar{x}$ and μ represent the mean of the population from which we are selecting a simple random sample. It can be shown that with simple random sampling, $E(\bar{x})$ and μ are equal.

The expected value of $\bar{x}$ equals the mean of the population from which the sample is selected.

EXPECTED VALUE OF $\bar{x}$

$$E(\bar{x}) = \mu \tag{7.1}$$

where

$E(\bar{x})$ = the expected value of $\bar{x}$
μ = the population mean

This result shows that with simple random sampling, the expected value or mean of the sampling distribution of $\bar{x}$ is equal to the mean of the population. In Section 7.1 we saw that the mean annual salary for the population of EAI managers is $\mu = \$51{,}800$. Thus, according to equation (7.1), the mean of all possible sample means for the EAI study is also $51,800.

When the expected value of a point estimator equals the population parameter, we say the point estimator is **unbiased**. Thus, equation (7.1) shows that $\bar{x}$ is an unbiased estimator of the population mean μ.

Standard Deviation of $\bar{x}$

Let us define the standard deviation of the sampling distribution of $\bar{x}$. We will use the following notation.

$$\sigma_{\bar{x}} = \text{the standard deviation of } \bar{x}$$
$$\sigma = \text{the standard deviation of the population}$$
$$n = \text{the sample size}$$
$$N = \text{the population size}$$

It can be shown that the formula for the standard deviation of $\bar{x}$ depends on whether the population is finite or infinite. The two formulas for the standard deviation of $\bar{x}$ follow.

STANDARD DEVIATION OF $\bar{x}$

Finite Population   *Infinite Population*

$$\sigma_{\bar{x}} = \sqrt{\frac{N-n}{N-1}}\left(\frac{\sigma}{\sqrt{n}}\right) \qquad \sigma_{\bar{x}} = \frac{\sigma}{\sqrt{n}} \qquad (7.2)$$

In comparing the two formulas in (7.2), we see that the factor $\sqrt{(N-n)/(N-1)}$ is required for the finite population case but not for the infinite population case. This factor is commonly referred to as the **finite population correction factor**. In many practical sampling situations, we find that the population involved, although finite, is "large," whereas the sample size is relatively "small." In such cases the finite population correction factor $\sqrt{(N-n)/(N-1)}$ is close to 1. As a result, the difference between the values of the standard deviation of $\bar{x}$ for the finite and infinite population cases becomes negligible. Then, $\sigma_{\bar{x}} = \sigma/\sqrt{n}$ becomes a good approximation to the standard deviation of $\bar{x}$ even though the population is finite. This observation leads to the following general guideline, or rule of thumb, for computing the standard deviation of $\bar{x}$.

USE THE FOLLOWING EXPRESSION TO COMPUTE THE STANDARD DEVIATION OF $\bar{x}$

$$\sigma_{\bar{x}} = \frac{\sigma}{\sqrt{n}} \qquad (7.3)$$

whenever

1. The population is infinite; or
2. The population is finite *and* the sample size is less than or equal to 5% of the population size; that is, $n/N \leq .05$.

7.5 Sampling Distribution of $\bar{x}$

Problem 21 shows that when $n/N \leq .05$, the finite population correction factor has little effect on the value of $\sigma_{\bar{x}}$.

In cases where $n/N > .05$, the finite population version of formula (7.2) should be used in the computation of $\sigma_{\bar{x}}$. Unless otherwise noted, throughout the text we will assume that the population size is "large," $n/N \leq .05$, and expression (7.3) can be used to compute $\sigma_{\bar{x}}$.

To compute $\sigma_{\bar{x}}$, we need to know σ, the standard deviation of the population. To further emphasize the difference between $\sigma_{\bar{x}}$ and σ, we refer to the standard deviation of $\bar{x}$, $\sigma_{\bar{x}}$, as the **standard error** of the mean. In general, the term *standard error* refers to the standard deviation of a point estimator. Later we will see that the value of the standard error of the mean is helpful in determining how far the sample mean may be from the population mean. Let us now return to the EAI example and compute the standard error of the mean associated with simple random samples of 30 EAI managers.

The term standard error is used throughout statistical inference to refer to the standard deviation of a point estimator.

In Section 7.1 we saw that the standard deviation of annual salary for the population of 2500 EAI managers is $\sigma = 4000$. In this case, the population is finite, with $N = 2500$. However, with a sample size of 30, we have $n/N = 30/2500 = .012$. Because the sample size is less than 5% of the population size, we can ignore the finite population correction factor and use equation (7.3) to compute the standard error.

$$\sigma_{\bar{x}} = \frac{\sigma}{\sqrt{n}} = \frac{4000}{\sqrt{30}} = 730.3$$

Form of the Sampling Distribution of $\bar{x}$

The preceding results concerning the expected value and standard deviation for the sampling distribution of $\bar{x}$ are applicable for any population. The final step in identifying the characteristics of the sampling distribution of $\bar{x}$ is to determine the form or shape of the sampling distribution. We will consider two cases: (1) The population has a normal distribution; and (2) the population does not have a normal distribution.

Population has a normal distribution. In many situations it is reasonable to assume that the population from which we are selecting a random sample has a normal, or nearly normal, distribution. When the population has a normal distribution, the sampling distribution of $\bar{x}$ is normally distributed for any sample size.

Population does not have a normal distribution. When the population from which we are selecting a random sample does not have a normal distribution, the **central limit theorem** is helpful in identifying the shape of the sampling distribution of $\bar{x}$. A statement of the central limit theorem as it applies to the sampling distribution of $\bar{x}$ follows.

> **CENTRAL LIMIT THEOREM**
>
> In selecting random samples of size n from a population, the sampling distribution of the sample mean $\bar{x}$ can be approximated by a *normal distribution* as the sample size becomes large.

Figure 7.3 shows how the central limit theorem works for three different populations; each column refers to one of the populations. The top panel of the figure shows that none of the populations are normally distributed. Population I follows a uniform distribution. Population II is often called the rabbit-eared distribution. It is symmetric, but the more likely values fall in the tails of the distribution. Population III is shaped like the exponential distribution; it is skewed to the right.

The bottom three panels of Figure 7.3 show the shape of the sampling distribution for samples of size $n = 2$, $n = 5$, and $n = 30$. When the sample size is 2, we see that the shape of each sampling distribution is different from the shape of the corresponding population

FIGURE 7.3 ILLUSTRATION OF THE CENTRAL LIMIT THEOREM FOR THREE POPULATIONS

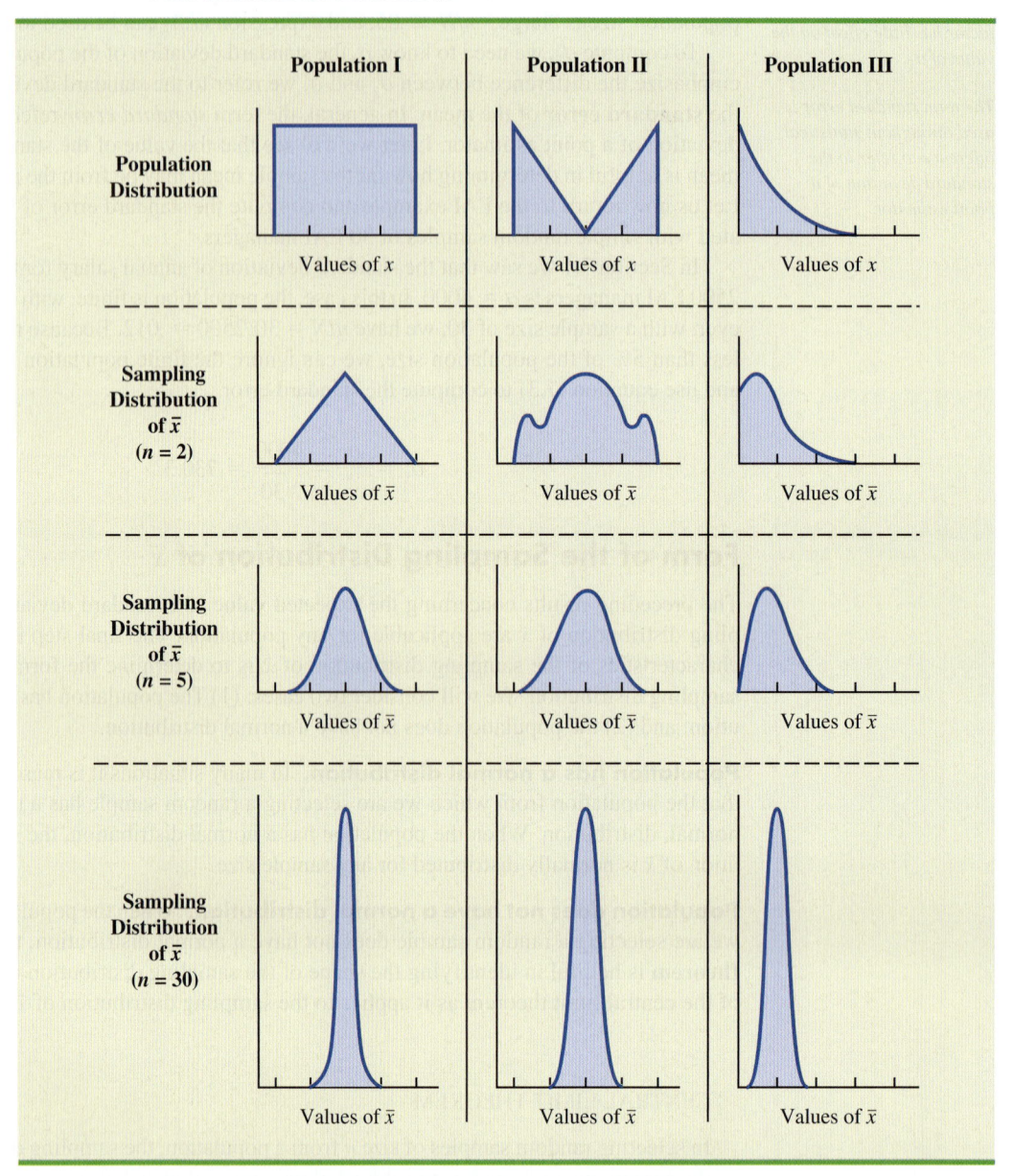

	Population I	Population II	Population III
Population Distribution	Values of x	Values of x	Values of x
Sampling Distribution of $\bar{x}$ ($n = 2$)	Values of $\bar{x}$	Values of $\bar{x}$	Values of $\bar{x}$
Sampling Distribution of $\bar{x}$ ($n = 5$)	Values of $\bar{x}$	Values of $\bar{x}$	Values of $\bar{x}$
Sampling Distribution of $\bar{x}$ ($n = 30$)	Values of $\bar{x}$	Values of $\bar{x}$	Values of $\bar{x}$

distribution. For samples of size 5, we see that the shapes of the sampling distributions for populations I and II begin to look similar to the shape of a normal distribution. Even though the shape of the sampling distribution for population III begins to look similar to the shape of a normal distribution, some skewness to the right is still present. Finally, for samples of size 30, the shapes of each of the three sampling distributions are approximately normal.

From a practitioner standpoint, we often want to know how large the sample size needs to be before the central limit theorem applies and we can assume that the shape of the sampling distribution is approximately normal. Statistical researchers have investigated this question by studying the sampling distribution of $\bar{x}$ for a variety of populations and a variety of sample sizes. General statistical practice is to assume that, for most applications, the sampling distribution of $\bar{x}$ can be approximated by a normal distribution whenever the sample is size 30 or more. In cases where the population is highly skewed or outliers are present, samples of size 50 may be needed. Finally, if the population is discrete, the sample size needed for a normal approximation often depends on the population proportion. We say more about this issue when we discuss the sampling distribution of $\bar{p}$ in Section 7.6.

Sampling Distribution of $\bar{x}$ for the EAI Problem

Let us return to the EAI problem where we previously showed that $E(\bar{x}) = \$51{,}800$ and $\sigma_{\bar{x}} = 730.3$. At this point, we do not have any information about the population distribution; it may or may not be normally distributed. If the population has a normal distribution, the sampling distribution of $\bar{x}$ is normally distributed. If the population does not have a normal distribution, the simple random sample of 30 managers and the central limit theorem enable us to conclude that the sampling distribution of $\bar{x}$ can be approximated by a normal distribution. In either case, we are comfortable proceeding with the conclusion that the sampling distribution of $\bar{x}$ can be described by the normal distribution shown in Figure 7.4.

Practical Value of the Sampling Distribution of $\bar{x}$

Whenever a simple random sample is selected and the value of the sample mean is used to estimate the value of the population mean μ, we cannot expect the sample mean to exactly equal the population mean. The practical reason we are interested in the sampling distribution of $\bar{x}$ is that it can be used to provide probability information about the difference between the sample mean and the population mean. To demonstrate this use, let us return to the EAI problem.

Suppose the personnel director believes the sample mean will be an acceptable estimate of the population mean if the sample mean is within $500 of the population mean. However, it is not possible to guarantee that the sample mean will be within $500 of the population mean. Indeed, Table 7.5 and Figure 7.1 show that some of the 500 sample means differed by more than $2,000 from the population mean. So we must think of the personnel director's request in probability terms. That is, the personnel director is concerned with the following question: What is the probability that the sample mean computed using a simple random sample of 30 EAI managers will be within $500 of the population mean?

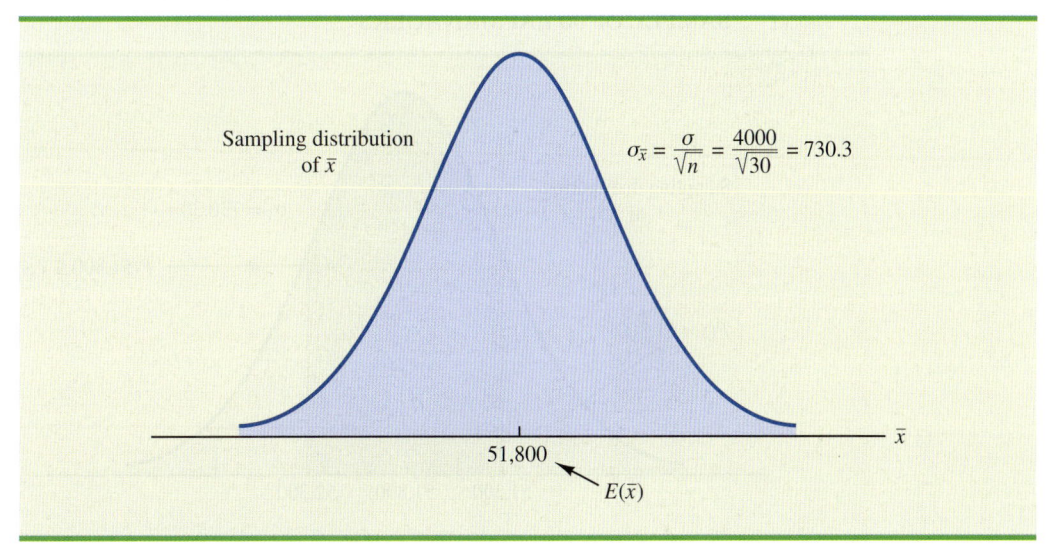

FIGURE 7.4 SAMPLING DISTRIBUTION OF $\bar{x}$ FOR THE MEAN ANNUAL SALARY OF A SIMPLE RANDOM SAMPLE OF 30 EAI MANAGERS

Sampling distribution of $\bar{x}$

$$\sigma_{\bar{x}} = \frac{\sigma}{\sqrt{n}} = \frac{4000}{\sqrt{30}} = 730.3$$

51,800

$E(\bar{x})$

Because we have identified the properties of the sampling distribution of $\bar{x}$ (see Figure 7.4), we will use this distribution to answer the probability question. Refer to the sampling distribution of $\bar{x}$ shown again in Figure 7.5. With a population mean of $51,800, the personnel director wants to know the probability that $\bar{x}$ is between $51,300 and $52,300. This probability is given by the darkly shaded area of the sampling distribution shown in Figure 7.5. Because the sampling distribution is normally distributed, with mean 51,800 and standard error of the mean 730.3, we can use the standard normal probability table to find the area or probability.

We first calculate the z value at the upper endpoint of the interval (52,300) and use the table to find the area under the curve to the left of that point (left tail area). Then we compute the z value at the lower endpoint of the interval (51,300) and use the table to find the area under the curve to the left of that point (another left tail area). Subtracting the second tail area from the first gives us the desired probability.

At $\bar{x} = 52,300$, we have

$$z = \frac{52{,}300 - 51{,}800}{730.30} = .68$$

Referring to the standard normal probability table, we find a cumulative probability (area to the left of $z = .68$) of .7517.

At $\bar{x} = 51,300$, we have

$$z = \frac{51{,}300 - 51{,}800}{730.30} = -.68$$

The area under the curve to the left of $z = -.68$ is .2483. Therefore, $P(51{,}300 \leq \bar{x} \leq 52{,}300) = P(z \leq .68) - P(z < -.68) = .7517 - .2483 = .5034$.

The preceding computations show that a simple random sample of 30 EAI managers has a .5034 probability of providing a sample mean $\bar{x}$ that is within $500 of the population mean. Thus, there is a $1 - .5034 = .4966$ probability that the difference between $\bar{x}$ and $\mu = \$51{,}800$ will be more than $500. In other words, a simple random sample of 30 EAI managers has roughly a 50/50 chance of providing a sample mean within the allowable

The sampling distribution of $\bar{x}$ can be used to provide probability information about how close the sample mean $\bar{x}$ is to the population mean μ.

FIGURE 7.5 PROBABILITY OF A SAMPLE MEAN BEING WITHIN $500 OF THE POPULATION MEAN FOR A SIMPLE RANDOM SAMPLE OF 30 EAI MANAGERS

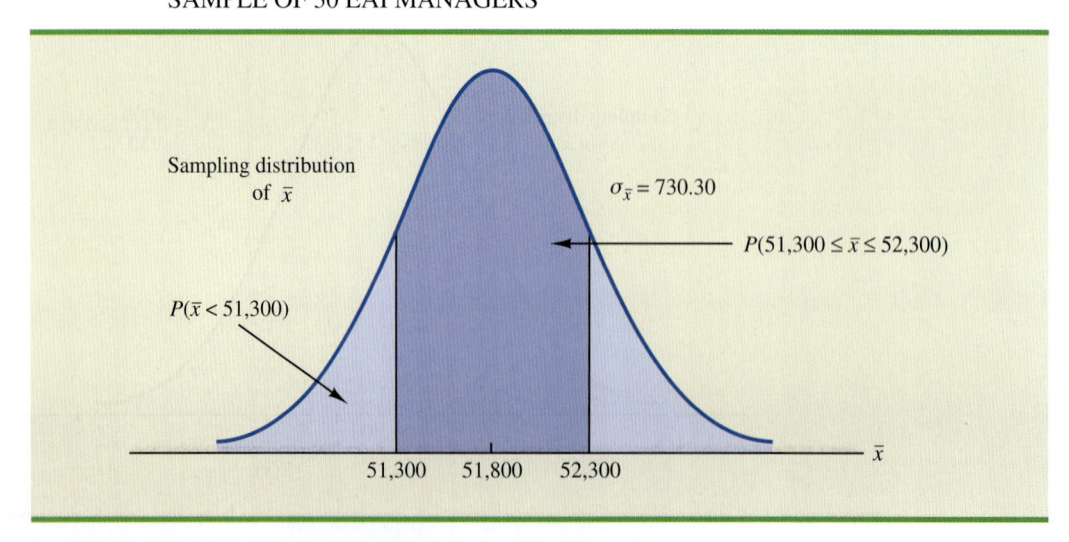

$500. Perhaps a larger sample size should be considered. Let us explore this possibility by considering the relationship between the sample size and the sampling distribution of $\bar{x}$.

Relationship Between the Sample Size and the Sampling Distribution of $\bar{x}$

Suppose that in the EAI sampling problem we select a simple random sample of 100 EAI managers instead of the 30 originally considered. Intuitively, it would seem that with more data provided by the larger sample size, the sample mean based on $n = 100$ should provide a better estimate of the population mean than the sample mean based on $n = 30$. To see how much better, let us consider the relationship between the sample size and the sampling distribution of $\bar{x}$.

First note that $E(\bar{x}) = \mu$ regardless of the sample size. Thus, the mean of all possible values of $\bar{x}$ is equal to the population mean μ regardless of the sample size n. However, note that the standard error of the mean, $\sigma_{\bar{x}} = \sigma/\sqrt{n}$, is related to the square root of the sample size. Whenever the sample size is increased, the standard error of the mean $\sigma_{\bar{x}}$ decreases. With $n = 30$, the standard error of the mean for the EAI problem is 730.3. However, with the increase in the sample size to $n = 100$, the standard error of the mean is decreased to

$$\sigma_{\bar{x}} = \frac{\sigma}{\sqrt{n}} = \frac{4000}{\sqrt{100}} = 400$$

The sampling distributions of $\bar{x}$ with $n = 30$ and $n = 100$ are shown in Figure 7.6. Because the sampling distribution with $n = 100$ has a smaller standard error, the values of $\bar{x}$ have less variation and tend to be closer to the population mean than the values of $\bar{x}$ with $n = 30$.

We can use the sampling distribution of $\bar{x}$ for the case with $n = 100$ to compute the probability that a simple random sample of 100 EAI managers will provide a sample mean that is within $500 of the population mean. Because the sampling distribution is normal, with mean 51,800 and standard error of the mean 400, we can use the standard normal probability table to find the area or probability.

At $\bar{x} = 52,300$ (see Figure 7.7), we have

$$z = \frac{52,300 - 51,800}{400} = 1.25$$

FIGURE 7.6 A COMPARISON OF THE SAMPLING DISTRIBUTIONS OF $\bar{x}$ FOR SIMPLE RANDOM SAMPLES OF $n = 30$ AND $n = 100$ EAI MANAGERS

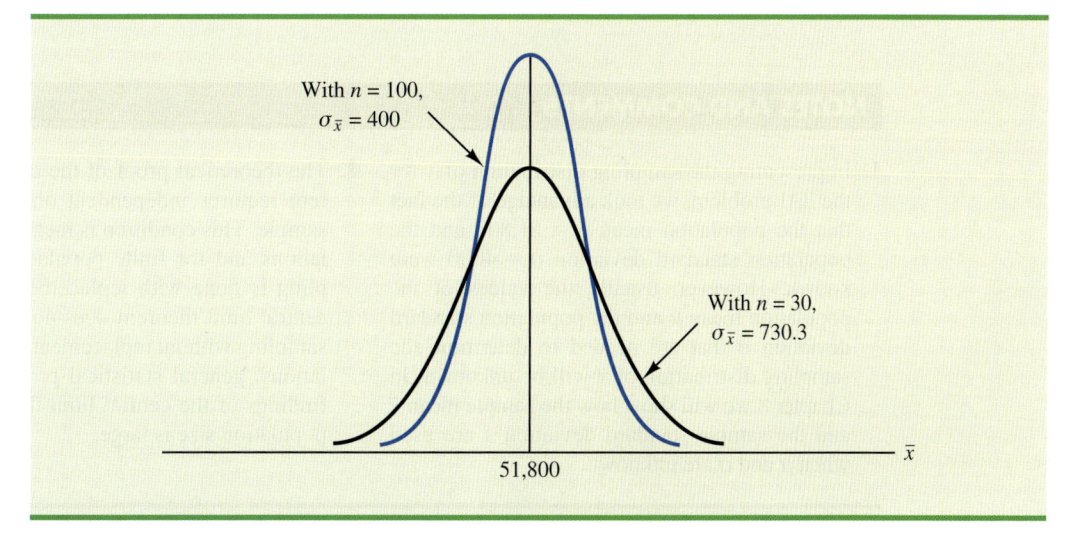

FIGURE 7.7 PROBABILITY OF A SAMPLE MEAN BEING WITHIN $500 OF THE POPULATION MEAN FOR A SIMPLE RANDOM SAMPLE OF 100 EAI MANAGERS

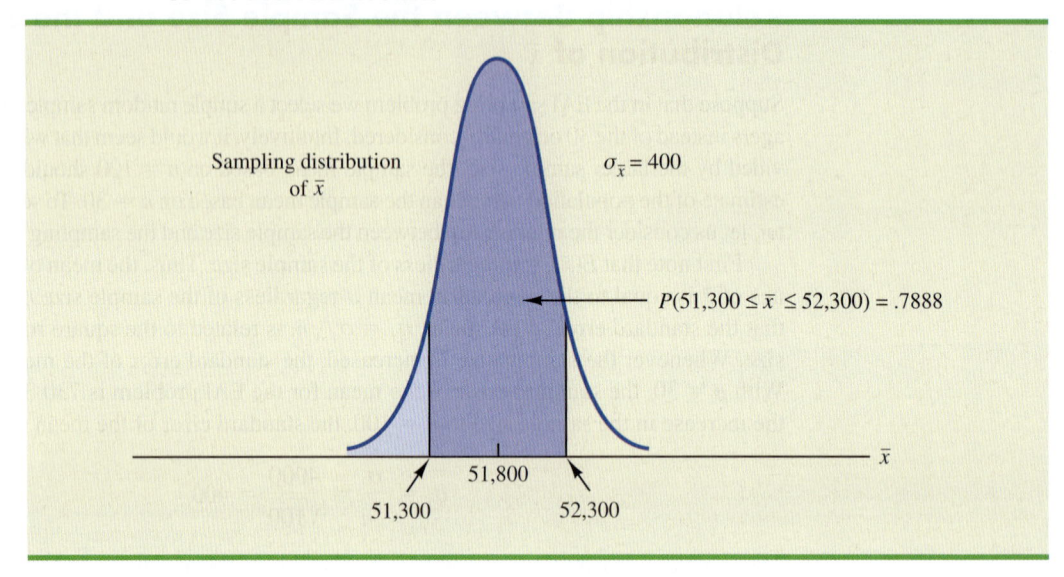

[Figure: Sampling distribution of $\bar{x}$ with $\sigma_{\bar{x}} = 400$, showing $P(51{,}300 \leq \bar{x} \leq 52{,}300) = .7888$, centered at 51,800 with bounds at 51,300 and 52,300.]

Referring to the standard normal probability table, we find a cumulative probability corresponding to $z = 1.25$ of .8944.

At $\bar{x} = 51{,}300$, we have

$$z = \frac{51{,}300 - 51{,}800}{400} = -1.25$$

The cumulative probability corresponding to $z = -1.25$ is .1056. Therefore, $P(51{,}300 \leq \bar{x} \leq 52{,}300) = P(z \leq 1.25) - P(z < -1.25) = .8944 - .1056 = .7888$. Thus, by increasing the sample size from 30 to 100 EAI managers, we increase the probability of obtaining a sample mean within $500 of the population mean from .5034 to .7888.

The important point in this discussion is that as the sample size is increased, the standard error of the mean decreases. As a result, the larger sample size provides a higher probability that the sample mean is within a specified distance of the population mean.

NOTES AND COMMENTS

1. In presenting the sampling distribution of $\bar{x}$ for the EAI problem, we took advantage of the fact that the population mean $\mu = 51{,}800$ and the population standard deviation $\sigma = 4000$ were known. However, usually the values of the population mean μ and the population standard deviation σ that are needed to determine the sampling distribution of $\bar{x}$ will be unknown. In Chapter 8 we will show how the sample mean $\bar{x}$ and the sample standard deviation s are used when μ and σ are unknown.

2. The theoretical proof of the central limit theorem requires independent observations in the sample. This condition is met for infinite populations and for finite populations where sampling is done with replacement. Although the central limit theorem does not directly address sampling without replacement from finite populations, general statistical practice applies the findings of the central limit theorem when the population size is large.

Exercises

Methods

18. A population has a mean of 200 and a standard deviation of 50. A simple random sample of size 100 will be taken and the sample mean $\bar{x}$ will be used to estimate the population mean.
    a. What is the expected value of $\bar{x}$?
    b. What is the standard deviation of $\bar{x}$?
    c. Show the sampling distribution of $\bar{x}$.
    d. What does the sampling distribution of $\bar{x}$ show?

19. A population has a mean of 200 and a standard deviation of 50. Suppose a simple random sample of size 100 is selected and $\bar{x}$ is used to estimate μ.
    a. What is the probability that the sample mean will be within ± 5 of the population mean?
    b. What is the probability that the sample mean will be within ± 10 of the population mean?

20. Assume the population standard deviation is $\sigma = 25$. Compute the standard error of the mean, $\sigma_{\bar{x}}$, for sample sizes of 50, 100, 150, and 200. What can you say about the size of the standard error of the mean as the sample size is increased?

21. Suppose a simple random sample of size 50 is selected from a population with $\sigma = 10$. Find the value of the standard error of the mean in each of the following cases (use the finite population correction factor if appropriate).
    a. The population size is infinite.
    b. The population size is $N = 50,000$.
    c. The population size is $N = 5000$.
    d. The population size is $N = 500$.

Applications

22. Refer to the EAI sampling problem. Suppose a simple random sample of 60 managers is used.
    a. Sketch the sampling distribution of $\bar{x}$ when simple random samples of size 60 are used.
    b. What happens to the sampling distribution of $\bar{x}$ if simple random samples of size 120 are used?
    c. What general statement can you make about what happens to the sampling distribution of $\bar{x}$ as the sample size is increased? Does this generalization seem logical? Explain.

23. In the EAI sampling problem (see Figure 7.5), we showed that for $n = 30$, there was .5034 probability of obtaining a sample mean within $\pm \$500$ of the population mean.
    a. What is the probability that $\bar{x}$ is within $\$500$ of the population mean if a sample of size 60 is used?
    b. Answer part (a) for a sample of size 120.

24. *Barron's* reported that the average number of weeks an individual is unemployed is 17.5 weeks (*Barron's*, February 18, 2008). Assume that for the population of all unemployed individuals the population mean length of unemployment is 17.5 weeks and that the population standard deviation is 4 weeks. Suppose you would like to select a random sample of 50 unemployed individuals for a follow-up study.
    a. Show the sampling distribution of $\bar{x}$, the sample mean average for a sample of 50 unemployed individuals.
    b. What is the probability that a simple random sample of 50 unemployed individuals will provide a sample mean within 1 week of the population mean?
    c. What is the probability that a simple random sample of 50 unemployed individuals will provide a sample mean within 1/2 week of the population mean?

25. The College Board reported the following mean scores for the three parts of the Scholastic Aptitude Test (SAT) (*The World Almanac*, 2009):

Critical Reading	502
Mathematics	515
Writing	494

Assume that the population standard deviation on each part of the test is $\sigma = 100$.

   a. What is the probability that a random sample of 90 test takers will provide a sample mean test score within 10 points of the population mean of 502 on the Critical Reading part of the test?
   b. What is the probability that a random sample of 90 test takers will provide a sample mean test score within 10 points of the population mean of 515 on the Mathematics part of the test? Compare this probability to the value computed in part (a).
   c. What is the probability that a random sample of 100 test takers will provide a sample mean test score within 10 of the population mean of 494 on the writing part of the test? Comment on the differences between this probability and the values computed in parts (a) and (b).

26. The mean annual cost of automobile insurance is $939 (CNBC, February 23, 2006). Assume that the standard deviation is $\sigma = \$245$.
   a. What is the probability that a simple random sample of automobile insurance policies will have a sample mean within $25 of the population mean for each of the following sample sizes: 30, 50, 100, and 400?
   b. What is the advantage of a larger sample size when attempting to estimate the population mean?

27. *BusinessWeek* conducted a survey of graduates from 30 top MBA programs (*BusinessWeek*, September 22, 2003). On the basis of the survey, assume that the mean annual salary for male and female graduates 10 years after graduation is $168,000 and $117,000, respectively. Assume the standard deviation for the male graduates is $40,000, and for the female graduates it is $25,000.
   a. What is the probability that a simple random sample of 40 male graduates will provide a sample mean within $10,000 of the population mean, $168,000?
   b. What is the probability that a simple random sample of 40 female graduates will provide a sample mean within $10,000 of the population mean, $117,000?
   c. In which of the preceding two cases, part (a) or part (b), do we have a higher probability of obtaining a sample estimate within $10,000 of the population mean? Why?
   d. What is the probability that a simple random sample of 100 male graduates will provide a sample mean more than $4000 below the population mean?

28. The average score for male golfers is 95 and the average score for female golfers is 106 (*Golf Digest*, April 2006). Use these values as the population means for men and women and assume that the population standard deviation is $\sigma = 14$ strokes for both. A simple random sample of 30 male golfers and another simple random sample of 45 female golfers will be taken.
   a. Show the sampling distribution of $\bar{x}$ for male golfers.
   b. What is the probability that the sample mean is within three strokes of the population mean for the sample of male golfers?
   c. What is the probability that the sample mean is within three strokes of the population mean for the sample of female golfers?
   d. In which case, part (b) or part (c), is the probability of obtaining a sample mean within three strokes of the population mean higher? Why?

29. The average price of a gallon of unleaded regular gasoline was reported to be $2.34 in northern Kentucky (*The Cincinnati Enquirer*, January 21, 2006). Use this price as the population mean, and assume the population standard deviation is $.20.

a. What is the probability that the mean price for a sample of 30 service stations is within $.03 of the population mean?
b. What is the probability that the mean price for a sample of 50 service stations is within $.03 of the population mean?
c. What is the probability that the mean price for a sample of 100 service stations is within $.03 of the population mean?
d. Which, if any, of the sample sizes in parts (a), (b), and (c) would you recommend to have at least a .95 probability that the sample mean is within $.03 of the population mean?

30. To estimate the mean age for a population of 4000 employees, a simple random sample of 40 employees is selected.
a. Would you use the finite population correction factor in calculating the standard error of the mean? Explain.
b. If the population standard deviation is $\sigma = 8.2$ years, compute the standard error both with and without the finite population correction factor. What is the rationale for ignoring the finite population correction factor whenever $n/N \leq .05$?
c. What is the probability that the sample mean age of the employees will be within ± 2 years of the population mean age?

7.6 Sampling Distribution of $\bar{p}$

The sample proportion $\bar{p}$ is the point estimator of the population proportion p. The formula for computing the sample proportion is

$$\bar{p} = \frac{x}{n}$$

where

x = the number of elements in the sample that possess the characteristic of interest
n = sample size

As noted in Section 7.4, the sample proportion $\bar{p}$ is a random variable and its probability distribution is called the sampling distribution of $\bar{p}$.

SAMPLING DISTRIBUTION OF $\bar{p}$

The sampling distribution of $\bar{p}$ is the probability distribution of all possible values of the sample proportion $\bar{p}$.

To determine how close the sample proportion $\bar{p}$ is to the population proportion p, we need to understand the properties of the sampling distribution of $\bar{p}$: the expected value of $\bar{p}$, the standard deviation of $\bar{p}$, and the shape or form of the sampling distribution of $\bar{p}$.

Expected Value of $\bar{p}$

The expected value of $\bar{p}$, the mean of all possible values of $\bar{p}$, is equal to the population proportion p.

EXPECTED VALUE OF $\bar{p}$

$$E(\bar{p}) = p \qquad (7.4)$$

where

$E(\bar{p})$ = the expected value of $\bar{p}$
p = the population proportion

Because $E(\bar{p}) = p$, $\bar{p}$ is an unbiased estimator of p. Recall from Section 7.1 that $p = .60$ for the EAI population, where p is the proportion of the population of managers who participated in the company's management training program. Thus, the expected value of $\bar{p}$ for the EAI sampling problem is .60.

Standard Deviation of $\bar{p}$

Just as we found for the standard deviation of $\bar{x}$, the standard deviation of $\bar{p}$ depends on whether the population is finite or infinite. The two formulas for computing the standard deviation of $\bar{p}$ follow.

STANDARD DEVIATION OF $\bar{p}$

Finite Population

$$\sigma_{\bar{p}} = \sqrt{\frac{N-n}{N-1}}\sqrt{\frac{p(1-p)}{n}}$$

Infinite Population

$$\sigma_{\bar{p}} = \sqrt{\frac{p(1-p)}{n}} \qquad (7.5)$$

Comparing the two formulas in (7.5), we see that the only difference is the use of the finite population correction factor $\sqrt{(N-n)/(N-1)}$.

As was the case with the sample mean $\bar{x}$, the difference between the expressions for the finite population and the infinite population becomes negligible if the size of the finite population is large in comparison to the sample size. We follow the same rule of thumb that we recommended for the sample mean. That is, if the population is finite with $n/N \leq .05$, we will use $\sigma_{\bar{p}} = \sqrt{p(1-p)/n}$. However, if the population is finite with $n/N > .05$, the finite population correction factor should be used. Again, unless specifically noted, throughout the text we will assume that the population size is large in relation to the sample size and thus the finite population correction factor is unnecessary.

In Section 7.5 we used the term *standard error of the mean* to refer to the standard deviation of $\bar{x}$. We stated that in general the term *standard error* refers to the standard deviation of a point estimator. Thus, for proportions we use *standard error of the proportion* to refer to the standard deviation of $\bar{p}$. Let us now return to the EAI example and compute the standard error of the proportion associated with simple random samples of 30 EAI managers.

For the EAI study we know that the population proportion of managers who participated in the management training program is $p = .60$. With $n/N = 30/2500 = .012$, we can ignore the finite population correction factor when we compute the standard error of the proportion. For the simple random sample of 30 managers, $\sigma_{\bar{p}}$ is

$$\sigma_{\bar{p}} = \sqrt{\frac{p(1-p)}{n}} = \sqrt{\frac{.60(1-.60)}{30}} = .0894$$

Form of the Sampling Distribution of $\bar{p}$

Now that we know the mean and standard deviation of the sampling distribution of $\bar{p}$, the final step is to determine the form or shape of the sampling distribution. The sample proportion is $\bar{p} = x/n$. For a simple random sample from a large population, the value of x is a binomial random variable indicating the number of elements in the sample with the characteristic of interest. Because n is a constant, the probability of x/n is the same as the binomial probability of x, which means that the sampling distribution of $\bar{p}$ is also a discrete probability distribution and that the probability for each value of x/n is the same as the probability of x.

In Chapter 6 we also showed that a binomial distribution can be approximated by a normal distribution whenever the sample size is large enough to satisfy the following two conditions:

$$np \geq 5 \quad \text{and} \quad n(1 - p) \geq 5$$

Assuming these two conditions are satisfied, the probability distribution of x in the sample proportion, $\bar{p} = x/n$, can be approximated by a normal distribution. And because n is a constant, the sampling distribution of $\bar{p}$ can also be approximated by a normal distribution. This approximation is stated as follows.

> The sampling distribution of $\bar{p}$ can be approximated by a normal distribution whenever $np \geq 5$ and $n(1 - p) \geq 5$.

In practical applications, when an estimate of a population proportion is desired, we find that sample sizes are almost always large enough to permit the use of a normal approximation for the sampling distribution of $\bar{p}$.

Recall that for the EAI sampling problem we know that the population proportion of managers who participated in the training program is $p = .60$. With a simple random sample of size 30, we have $np = 30(.60) = 18$ and $n(1 - p) = 30(.40) = 12$. Thus, the sampling distribution of $\bar{p}$ can be approximated by a normal distribution shown in Figure 7.8.

Practical Value of the Sampling Distribution of $\bar{p}$

The practical value of the sampling distribution of $\bar{p}$ is that it can be used to provide probability information about the difference between the sample proportion and the population proportion. For instance, suppose that in the EAI problem the personnel director wants to know the probability of obtaining a value of $\bar{p}$ that is within .05 of the population proportion of EAI managers who participated in the training program. That is, what is the probability of obtaining a sample with a sample proportion $\bar{p}$ between .55 and .65? The darkly shaded area in Figure 7.9 shows this probability. Using the fact that the sampling distribution of $\bar{p}$ can be approximated by a normal distribution with a mean of .60 and a standard error of the proportion of $\sigma_{\bar{p}} = .0894$, we find that the standard normal random variable corresponding to $\bar{p} = .65$ has a value of $z = (.65 - .60)/.0894 = .56$. Referring to the standard normal probability table, we see that the cumulative probability corresponding to $z = .56$ is .7123. Similarly, at $\bar{p} = .55$, we find $z = (.55 - .60)/.0894 = -.56$. From the standard normal probability table, we find the cumulative probability corresponding to $z = -.56$ is .2877. Thus, the probability of selecting a sample that provides a sample proportion $\bar{p}$ within .05 of the population proportion p is given by $.7123 - .2877 = .4246$.

FIGURE 7.8 SAMPLING DISTRIBUTION OF $\bar{p}$ FOR THE PROPORTION OF EAI MANAGERS WHO PARTICIPATED IN THE MANAGEMENT TRAINING PROGRAM

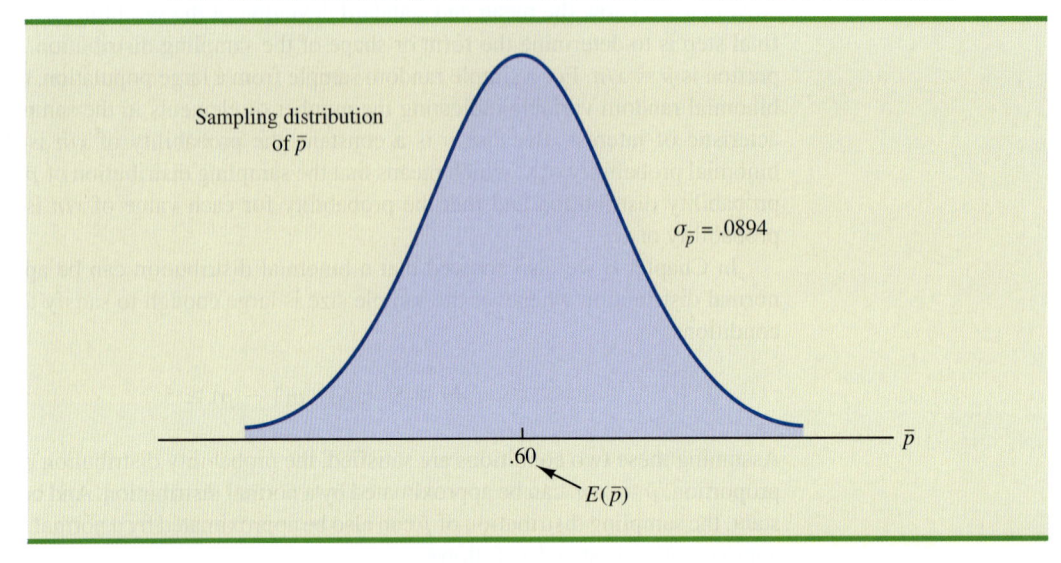

Sampling distribution of $\bar{p}$

$\sigma_{\bar{p}} = .0894$

.60

$E(\bar{p})$

If we consider increasing the sample size to $n = 100$, the standard error of the proportion becomes

$$\sigma_{\bar{p}} = \sqrt{\frac{.60(1 - .60)}{100}} = .049$$

With a sample size of 100 EAI managers, the probability of the sample proportion having a value within .05 of the population proportion can now be computed. Because the sampling distribution is approximately normal, with mean .60 and standard deviation .049, we can use the standard normal probability table to find the area or probability. At $\bar{p} = .65$, we have $z = (.65 - .60)/.049 = 1.02$. Referring to the standard normal probability table, we see that the cumulative probability corresponding to $z = 1.02$ is .8461. Similarly, at

FIGURE 7.9 PROBABILITY OF OBTAINING $\bar{p}$ BETWEEN .55 AND .65

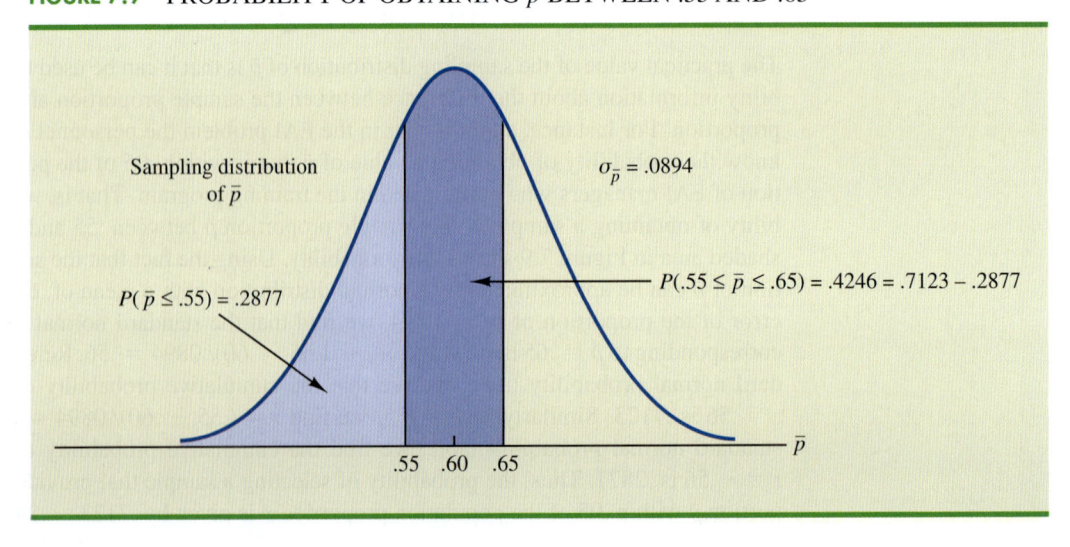

Sampling distribution of $\bar{p}$

$\sigma_{\bar{p}} = .0894$

$P(\bar{p} \leq .55) = .2877$

$P(.55 \leq \bar{p} \leq .65) = .4246 = .7123 - .2877$

.55 .60 .65

$\bar{p} = .55$, we have $z = (.55 - .60)/.049 = -1.02$. We find that the cumulative probability corresponding to $z = -1.02$ is .1539. Thus, if the sample size is increased from 30 to 100, the probability that the sample proportion $\bar{p}$ is within .05 of the population proportion p will increase to $.8461 - .1539 = .6922$.

Exercises

Methods

31. A simple random sample of size 100 is selected from a population with $p = .40$.
    a. What is the expected value of $\bar{p}$?
    b. What is the standard error of $\bar{p}$?
    c. Show the sampling distribution of $\bar{p}$.
    d. What does the sampling distribution of $\bar{p}$ show?

32. A population proportion is .40. A simple random sample of size 200 will be taken and the sample proportion $\bar{p}$ will be used to estimate the population proportion.
    a. What is the probability that the sample proportion will be within $\pm.03$ of the population proportion?
    b. What is the probability that the sample proportion will be within $\pm.05$ of the population proportion?

33. Assume that the population proportion is .55. Compute the standard error of the proportion, $\sigma_{\bar{p}}$, for sample sizes of 100, 200, 500, and 1000. What can you say about the size of the standard error of the proportion as the sample size is increased?

34. The population proportion is .30. What is the probability that a sample proportion will be within $\pm.04$ of the population proportion for each of the following sample sizes?
    a. $n = 100$
    b. $n = 200$
    c. $n = 500$
    d. $n = 1000$
    e. What is the advantage of a larger sample size?

Applications

35. The president of Doerman Distributors, Inc., believes that 30% of the firm's orders come from first-time customers. A random sample of 100 orders will be used to estimate the proportion of first-time customers.
    a. Assume that the president is correct and $p = .30$. What is the sampling distribution of $\bar{p}$ for this study?
    b. What is the probability that the sample proportion $\bar{p}$ will be between .20 and .40?
    c. What is the probability that the sample proportion will be between .25 and .35?

36. *The Cincinnati Enquirer* reported that, in the United States, 66% of adults and 87% of youths ages 12 to 17 use the Internet (*The Cincinnati Enquirer,* February 7, 2006). Use the reported numbers as the population proportions and assume that samples of 300 adults and 300 youths will be used to learn about attitudes toward Internet security.
    a. Show the sampling distribution of $\bar{p}$, where $\bar{p}$ is the sample proportion of adults using the Internet.
    b. What is the probability that the sample proportion of adults using the Internet will be within $\pm.04$ of the population proportion?
    c. What is the probability that the sample proportion of youths using the Internet will be within $\pm.04$ of the population proportion?

d. Is the probability different in parts (b) and (c)? If so, why?
e. Answer part (b) for a sample of size 600. Is the probability smaller? Why?

37. People end up tossing 12% of what they buy at the grocery store (*Reader's Digest*, March, 2009). Assume this is the true population proportion and that you plan to take a sample survey of 540 grocery shoppers to further investigate their behavior.
   a. Show the sampling distribution of $\bar{p}$, the proportion of groceries thrown out by your sample respondents.
   b. What is the probability that your survey will provide a sample proportion within $\pm.03$ of the population proportion?
   c. What is the probability that your survey will provide a sample proportion within $\pm.015$ of the population proportion?

38. Roper ASW conducted a survey to learn about American adults' attitudes toward money and happiness (*Money*, October 2003). Fifty-six percent of the respondents said they balance their checkbook at least once a month.
   a. Suppose a sample of 400 American adults were taken. Show the sampling distribution of the proportion of adults who balance their checkbook at least once a month.
   b. What is the probability that the sample proportion will be within $\pm.02$ of the population proportion?
   c. What is the probability that the sample proportion will be within $\pm.04$ of the population proportion?

39. In 2008 the Better Business Bureau settled 75% of complaints it received (*USA Today*, March 2, 2009). Suppose you have been hired by the Better Business Bureau to investigate the complaints it received this year involving new car dealers. You plan to select a sample of new car dealer complaints to estimate the proportion of complaints the Better Business Bureau is able to settle. Assume the population proportion of complaints settled for new car dealers is .75, the same as the overall proportion of complaints settled in 2008.
   a. Suppose you select a sample of 450 complaints involving new car dealers. Show the sampling distribution of $\bar{p}$.
   b. Based upon a sample of 450 complaints, what is the probability that the sample proportion will be within .04 of the population proportion?
   c. Suppose you select a sample of 200 complaints involving new car dealers. Show the sampling distribution of $\bar{p}$.
   d. Based upon the smaller sample of only 200 complaints, what is the probability that the sample proportion will be within .04 of the population proportion?
   e. As measured by the increase in probability, how much do you gain in precision by taking the larger sample in part (b)?

40. The Grocery Manufacturers of America reported that 76% of consumers read the ingredients listed on a product's label. Assume the population proportion is $p = .76$ and a sample of 400 consumers is selected from the population.
   a. Show the sampling distribution of the sample proportion $\bar{p}$, where $\bar{p}$ is the proportion of the sampled consumers who read the ingredients listed on a product's label.
   b. What is the probability that the sample proportion will be within $\pm.03$ of the population proportion?
   c. Answer part (b) for a sample of 750 consumers.

41. The Food Marketing Institute shows that 17% of households spend more than $100 per week on groceries. Assume the population proportion is $p = .17$ and a simple random sample of 800 households will be selected from the population.
   a. Show the sampling distribution of $\bar{p}$, the sample proportion of households spending more than $100 per week on groceries.
   b. What is the probability that the sample proportion will be within $\pm.02$ of the population proportion?
   c. Answer part (b) for a sample of 1600 households.

7.7 Other Sampling Methods

This section provides a brief introduction to survey sampling methods other than simple random sampling.

We described simple random sampling as a procedure for sampling from a finite population and discussed the properties of the sampling distributions of $\bar{x}$ and $\bar{p}$ when simple random sampling is used. Other methods such as stratified random sampling, cluster sampling, and systematic sampling provide advantages over simple random sampling in some of these situations. In this section we briefly introduce these alternative sampling methods.

Stratified Random Sampling

Stratified random sampling works best when the variance among elements in each stratum is relatively small.

In **stratified random sampling**, the elements in the population are first divided into groups called *strata,* such that each element in the population belongs to one and only one stratum. The basis for forming the strata, such as department, location, age, industry type, and so on, is at the discretion of the designer of the sample. However, the best results are obtained when the elements within each stratum are as much alike as possible. Figure 7.10 is a diagram of a population divided into H strata.

After the strata are formed, a simple random sample is taken from each stratum. Formulas are available for combining the results for the individual stratum samples into one estimate of the population parameter of interest. The value of stratified random sampling depends on how homogeneous the elements are within the strata. If elements within strata are alike, the strata will have low variances. Thus relatively small sample sizes can be used to obtain good estimates of the strata characteristics. If strata are homogeneous, the stratified random sampling procedure provides results just as precise as those of simple random sampling by using a smaller total sample size.

Cluster Sampling

Cluster sampling works best when each cluster provides a small-scale representation of the population.

In **cluster sampling**, the elements in the population are first divided into separate groups called *clusters*. Each element of the population belongs to one and only one cluster (see Figure 7.11). A simple random sample of the clusters is then taken. All elements within each sampled cluster form the sample. Cluster sampling tends to provide the best results when the elements within the clusters are not alike. In the ideal case, each cluster is a representative small-scale version of the entire population. The value of cluster sampling depends on how representative each cluster is of the entire population. If all clusters are alike in this regard, sampling a small number of clusters will provide good estimates of the population parameters.

One of the primary applications of cluster sampling is area sampling, where clusters are city blocks or other well-defined areas. Cluster sampling generally requires a larger total

FIGURE 7.10   DIAGRAM FOR STRATIFIED RANDOM SAMPLING

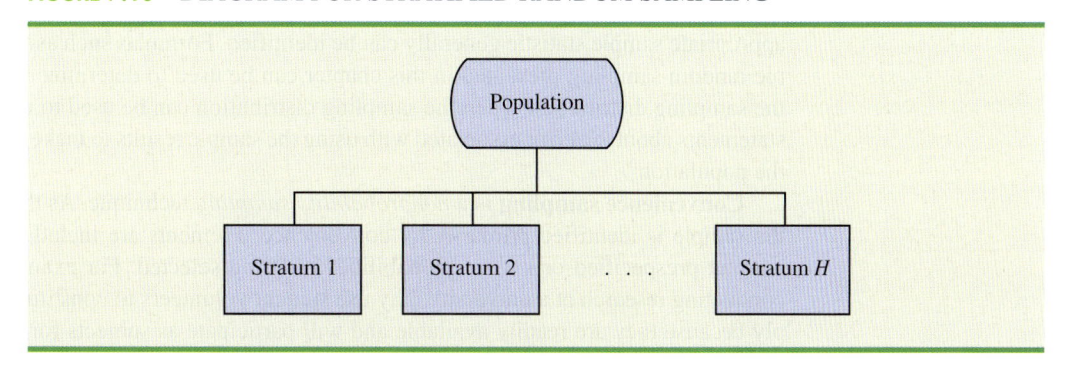

FIGURE 7.11 DIAGRAM FOR CLUSTER SAMPLING

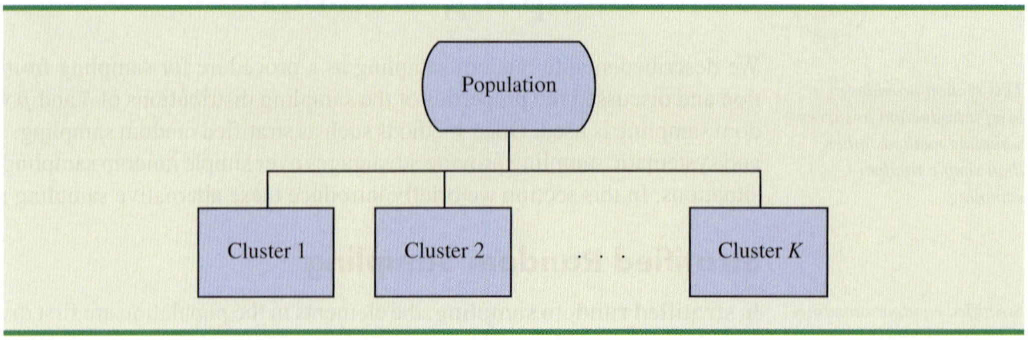

sample size than either simple random sampling or stratified random sampling. However, it can result in cost savings because of the fact that when an interviewer is sent to a sampled cluster (e.g., a city-block location), many sample observations can be obtained in a relatively short time. Hence, a larger sample size may be obtainable with a significantly lower total cost.

Systematic Sampling

In some sampling situations, especially those with large populations, it is time-consuming to select a simple random sample by first finding a random number and then counting or searching through the list of the population until the corresponding element is found. An alternative to simple random sampling is **systematic sampling**. For example, if a sample size of 50 is desired from a population containing 5000 elements, we will sample one element for every $5000/50 = 100$ elements in the population. A systematic sample for this case involves selecting randomly one of the first 100 elements from the population list. Other sample elements are identified by starting with the first sampled element and then selecting every 100th element that follows in the population list. In effect, the sample of 50 is identified by moving systematically through the population and identifying every 100th element after the first randomly selected element. The sample of 50 usually will be easier to identify in this way than it would be if simple random sampling were used. Because the first element selected is a random choice, a systematic sample is usually assumed to have the properties of a simple random sample. This assumption is especially applicable when the list of elements in the population is a random ordering of the elements.

Convenience Sampling

The sampling methods discussed thus far are referred to as *probability sampling* techniques. Elements selected from the population have a known probability of being included in the sample. The advantage of probability sampling is that the sampling distribution of the appropriate sample statistic generally can be identified. Formulas such as the ones for simple random sampling presented in this chapter can be used to determine the properties of the sampling distribution. Then the sampling distribution can be used to make probability statements about the error associated with using the sample results to make inferences about the population.

Convenience sampling is a *nonprobability sampling* technique. As the name implies, the sample is identified primarily by convenience. Elements are included in the sample without prespecified or known probabilities of being selected. For example, a professor conducting research at a university may use student volunteers to constitute a sample simply because they are readily available and will participate as subjects for little or no cost.

Similarly, an inspector may sample a shipment of oranges by selecting oranges haphazardly from among several crates. Labeling each orange and using a probability method of sampling would be impractical. Samples such as wildlife captures and volunteer panels for consumer research are also convenience samples.

Convenience samples have the advantage of relatively easy sample selection and data collection; however, it is impossible to evaluate the "goodness" of the sample in terms of its representativeness of the population. A convenience sample may provide good results or it may not; no statistically justified procedure allows a probability analysis and inference about the quality of the sample results. Sometimes researchers apply statistical methods designed for probability samples to a convenience sample, arguing that the convenience sample can be treated as though it were a probability sample. However, this argument cannot be supported, and we should be cautious in interpreting the results of convenience samples that are used to make inferences about populations.

Judgment Sampling

One additional nonprobability sampling technique is **judgment sampling**. In this approach, the person most knowledgeable on the subject of the study selects elements of the population that he or she feels are most representative of the population. Often this method is a relatively easy way of selecting a sample. For example, a reporter may sample two or three senators, judging that those senators reflect the general opinion of all senators. However, the quality of the sample results depends on the judgment of the person selecting the sample. Again, great caution is warranted in drawing conclusions based on judgment samples used to make inferences about populations.

NOTES AND COMMENTS

We recommend using probability sampling methods when sampling from finite populations: simple random sampling, stratified random sampling, cluster sampling, or systematic sampling. For these methods, formulas are available for evaluating the "goodness" of the sample results in terms of the closeness of the results to the population parameters being estimated. An evaluation of the goodness cannot be made with convenience or judgment sampling. Thus, great care should be used in interpreting the results based on nonprobability sampling methods.

Summary

In this chapter we presented the concepts of sampling and sampling distributions. We demonstrated how a simple random sample can be selected from a finite population and discussed how a random sample can be collected from an infinite population. The data collected from such samples can be used to develop point estimates of population parameters. Because different samples provide different values for the point estimators, point estimators such as $\bar{x}$ and $\bar{p}$ are random variables. The probability distribution of such a random variable is called a sampling distribution. In particular, we described the sampling distributions of the sample mean $\bar{x}$ and the sample proportion $\bar{p}$.

In considering the characteristics of the sampling distributions of $\bar{x}$ and $\bar{p}$, we stated that $E(\bar{x}) = \mu$ and $E(\bar{p}) = p$. After developing the standard deviation or standard error formulas for these estimators, we described the conditions necessary for the sampling distributions of $\bar{x}$ and $\bar{p}$ to follow a normal distribution. Other sampling methods, including stratified random sampling, cluster sampling, systematic sampling, convenience sampling, and judgment sampling, were discussed.

Glossary

Sampled population The population from which the sample is taken.

Frame A listing of the elements the sample will be selected from.

Parameter A numerical characteristic of a population, such as a population mean μ, a population standard deviation σ, a population proportion p, and so on.

Simple random sample A simple random sample of size n from a finite population of size N is a sample selected such that each possible sample of size n has the same probability of being selected.

Sampling without replacement Once an element has been included in the sample, it is removed from the population and cannot be selected a second time.

Sampling with replacement Once an element has been included in the sample, it is returned to the population. A previously selected element can be selected again and therefore may appear in the sample more than once.

Random sample A random sample from an infinite population is a sample selected such that the following conditions are satisfied: (1) Each element selected comes from the same population; (2) each element is selected independently.

Sample statistic A sample characteristic, such as a sample mean $\bar{x}$, a sample standard deviation s, a sample proportion $\bar{p}$, and so on. The value of the sample statistic is used to estimate the value of the corresponding population parameter.

Point estimator The sample statistic, such as $\bar{x}$, s, or $\bar{p}$, that provides the point estimate of the population parameter.

Point estimate The value of a point estimator used in a particular instance as an estimate of a population parameter.

Target population The population for which statistical inferences such as point estimates are made. It is important for the target population to correspond as closely as possible to the sampled population.

Sampling distribution A probability distribution consisting of all possible values of a sample statistic.

Unbiased A property of a point estimator that is present when the expected value of the point estimator is equal to the population parameter it estimates.

Finite population correction factor The term $\sqrt{(N-n)/(N-1)}$ that is used in the formulas for $\sigma_{\bar{x}}$ and $\sigma_{\bar{p}}$ whenever a finite population, rather than an infinite population, is being sampled. The generally accepted rule of thumb is to ignore the finite population correction factor whenever $n/N \leq .05$.

Standard error The standard deviation of a point estimator.

Central limit theorem A theorem that enables one to use the normal probability distribution to approximate the sampling distribution of $\bar{x}$ whenever the sample size is large.

Stratified random sampling A probability sampling method in which the population is first divided into strata and a simple random sample is then taken from each stratum.

Cluster sampling A probability sampling method in which the population is first divided into clusters and then a simple random sample of the clusters is taken.

Systematic sampling A probability sampling method in which we randomly select one of the first k elements and then select every kth element thereafter.

Convenience sampling A nonprobability method of sampling whereby elements are selected for the sample on the basis of convenience.

Judgment sampling A nonprobability method of sampling whereby elements are selected for the sample based on the judgment of the person doing the study.

Key Formulas

Expected Value of $\bar{x}$

$$E(\bar{x}) = \mu \qquad (7.1)$$

Standard Deviation of $\bar{x}$ (Standard Error)

Finite Population   *Infinite Population*

$$\sigma_{\bar{x}} = \sqrt{\frac{N-n}{N-1}}\left(\frac{\sigma}{\sqrt{n}}\right) \qquad \sigma_{\bar{x}} = \frac{\sigma}{\sqrt{n}} \qquad (7.2)$$

Expected Value of $\bar{p}$

$$E(\bar{p}) = p \qquad (7.4)$$

Standard Deviation of $\bar{p}$ (Standard Error)

Finite Population   *Infinite Population*

$$\sigma_{\bar{p}} = \sqrt{\frac{N-n}{N-1}}\sqrt{\frac{p(1-p)}{n}} \qquad \sigma_{\bar{p}} = \sqrt{\frac{p(1-p)}{n}} \qquad (7.5)$$

Supplementary Exercises

42. Lori Jeffrey is a successful sales representative for a major publisher of college textbooks. Historically, Lori obtains a book adoption on 25% of her sales calls. Viewing her sales calls for one month as a sample of all possible sales calls, assume that a statistical analysis of the data yields a standard error of the proportion of .0625.
    a. How large was the sample used in this analysis? That is, how many sales calls did Lori make during the month?
    b. Let $\bar{p}$ indicate the sample proportion of book adoptions obtained during the month. Show the sampling distribution of $\bar{p}$.
    c. Using the sampling distribution of $\bar{p}$, compute the probability that Lori will obtain book adoptions on 30% or more of her sales calls during a one-month period.

43. The proportion of individuals insured by the All-Driver Automobile Insurance Company who received at least one traffic ticket during a five-year period is .15.
    a. Show the sampling distribution of $\bar{p}$ if a random sample of 150 insured individuals is used to estimate the proportion having received at least one ticket.
    b. What is the probability that the sample proportion will be within ±.03 of the population proportion?

44. Advertisers contract with Internet service providers and search engines to place ads on websites. They pay a fee based on the number of potential customers who click on their ad. Unfortunately, click fraud—the practice of someone clicking on an ad solely for the purpose of driving up advertising revenue—has become a problem. Forty percent of advertisers claim they have been a victim of click fraud (*BusinessWeek,* March 13, 2006).

Suppose a simple random sample of 380 advertisers will be taken to learn more about how they are affected by this practice.
   a. What is the probability that the sample proportion will be within $\pm.04$ of the population proportion experiencing click fraud?
   b. What is the probability that the sample proportion will be greater than .45?

45. A market research firm conducts telephone surveys with a 40% historical response rate. What is the probability that in a new sample of 400 telephone numbers, at least 150 individuals will cooperate and respond to the questions? In other words, what is the probability that the sample proportion will be at least $150/400 = .375$?

46. About 28% of private companies are owned by women (*The Cincinnati Enquirer*, January 26, 2006). Answer the following questions based on a sample of 240 private companies.
   a. Show the sampling distribution of $\bar{p}$, the sample proportion of companies that are owned by women.
   b. What is the probability that the sample proportion will be within $\pm.04$ of the population proportion?
   c. What is the probability that the sample proportion will be within $\pm.02$ of the population proportion?

47. A researcher reports survey results by stating that the standard error of the mean is 20. The population standard deviation is 500.
   a. How large was the sample used in this survey?
   b. What is the probability that the point estimate was within ± 25 of the population mean?

48. A production process is checked periodically by a quality control inspector. The inspector selects simple random samples of 30 finished products and computes the sample mean product weights $\bar{x}$. If test results over a long period of time show that 5% of the $\bar{x}$ values are over 2.1 pounds and 5% are under 1.9 pounds, what are the mean and the standard deviation for the population of products produced with this process?

49. After deducting grants based on need, the average cost to attend the University of Southern California (USC) is $27,175 (*U.S. News & World Report, America's Best Colleges*, 2009 ed.). Assume the population standard deviation is $7400. Suppose that a random sample of 60 USC students will be taken from this population.
   a. What is the value of the standard error of the mean?
   b. What is the probability that the sample mean will be more than $27,175?
   c. What is the probability that the sample mean will be within $1000 of the population mean?
   d. How would the probability in part (c) change if the sample size were increased to 100?

50. Three firms carry inventories that differ in size. Firm A's inventory contains 2000 items, firm B's inventory contains 5000 items, and firm C's inventory contains 10,000 items. The population standard deviation for the cost of the items in each firm's inventory is $\sigma = 144$. A statistical consultant recommends that each firm take a sample of 50 items from its inventory to provide statistically valid estimates of the average cost per item. Managers of the small firm state that because it has the smallest population, it should be able to make the estimate from a much smaller sample than that required by the larger firms. However, the consultant states that to obtain the same standard error and thus the same precision in the sample results, all firms should use the same sample size regardless of population size.
   a. Using the finite population correction factor, compute the standard error for each of the three firms given a sample of size 50.
   b. What is the probability that for each firm the sample mean $\bar{x}$ will be within ± 25 of the population mean μ?

51. *BusinessWeek* surveyed MBA alumni 10 years after graduation (*BusinessWeek*, September 22, 2003). One finding was that alumni spend an average of $115.50 per week eating out socially. You have been asked to conduct a follow-up study by taking a sample of 40 of these MBA alumni. Assume the population standard deviation is $35.
   a. Show the sampling distribution of $\bar{x}$, the sample mean weekly expenditure for the 40 MBA alumni.

b. What is the probability that the sample mean will be within $10 of the population mean?
c. Suppose you find a sample mean of $100. What is the probability of finding a sample mean of $100 or less? Would you consider this sample to be an unusually low spending group of alumni? Why or why not?

52. The mean television viewing time for Americans is 15 hours per week (*Money*, November 2003). Suppose a sample of 60 Americans is taken to further investigate viewing habits. Assume the population standard deviation for weekly viewing time is $\sigma = 4$ hours.
    a. What is the probability that the sample mean will be within 1 hour of the population mean?
    b. What is the probability that the sample mean will be within 45 minutes of the population mean?

53. *U.S. News & World Report* publishes comprehensive information on America's best colleges (*America's Best Colleges*, 2009 ed.). Among other things, it provides a listing of the 133 best national universities. You would like to take a sample of these universities for a follow-up study on their students. Begin at the bottom of the third column of random digits in Table 7.1. Ignoring the first two digits in each five-number group and using the three-digit random numbers beginning with 959, read *up* the column to identify the number (from 1 to 133) of the first seven universities to be included in a simple random sample. Continue by starting at the bottom of the fourth and fifth columns and reading up if necessary.

54. Americans have become increasingly concerned about the rising cost of Medicare. In 1990, the average annual Medicare spending per enrollee was $3267; in 2003, the average annual Medicare spending per enrollee was $6883 (*Money*, Fall 2003). Suppose you hired a consulting firm to take a sample of fifty 2003 Medicare enrollees to further investigate the nature of expenditures. Assume the population standard deviation for 2003 was $2000.
    a. Show the sampling distribution of the mean amount of Medicare spending for a sample of fifty 2003 enrollees.
    b. What is the probability that the sample mean will be within ±$300 of the population mean?
    c. What is the probability that the sample mean will be greater than $7500? If the consulting firm tells you the sample mean for the Medicare enrollees it interviewed was $7500, would you question whether the firm followed correct simple random sampling procedures? Why or why not?

Appendix 7.1 Random Sampling Using Minitab

If a list of the elements in a population is available in a Minitab file, Minitab can be used to select a simple random sample. For example, a list of the top 100 metropolitan areas in the United States and Canada is provided in column 1 of the data set MetAreas (*Places Rated Almanac—The Millennium Edition 2000*). Column 2 contains the overall rating of each metropolitan area. The first 10 metropolitan areas in the data set and their corresponding ratings are shown in Table 7.6.

Suppose that you would like to select a simple random sample of 30 metropolitan areas in order to do an in-depth study of the cost of living in the United States and Canada. The following steps can be used to select the sample.

WEB file
MetAreas

Step 1. Select the **Calc** pull-down menu
Step 2. Choose **Random Data**
Step 3. Choose **Sample From Columns**
Step 4. When the Sample From Columns dialog box appears:
      Enter 30 in the **Number of rows to sample** box
      Enter C1 C2 in the **From columns** box below
      Enter C3 C4 in the **Store samples in** box
Step 5. Click **OK**

The random sample of 30 metropolitan areas appears in columns C3 and C4.

Appendix 7.2 Random Sampling Using Excel

If a list of the elements in a population is available in an Excel file, Excel can be used to select a simple random sample. For example, a list of the top 100 metropolitan areas in the United States and Canada is provided in column A of the data set MetAreas (*Places Rated Almanac—The Millennium Edition 2000*). Column B contains the overall rating of each metropolitan area. The first 10 metropolitan areas in the data set and their corresponding ratings are shown in Table 7.6. Assume that you would like to select a simple random sample of 30 metropolitan areas in order to do an in-depth study of the cost of living in the United States and Canada.

The rows of any Excel data set can be placed in a random order by adding an extra column to the data set and filling the column with random numbers using the =RAND() function. Then, using Excel's sort ascending capability on the random number column, the rows of the data set will be reordered randomly. The random sample of size *n* appears in the first *n* rows of the reordered data set.

In the MetAreas data set, labels are in row 1 and the 100 metropolitan areas are in rows 2 to 101. The following steps can be used to select a simple random sample of 30 metropolitan areas.

Step 1. Enter =RAND() in cell C2
Step 2. Copy cell C2 to cells C3:C101
Step 3. Select any cell in Column C
Step 4. Click the **Home** tab on the Ribbon
Step 5. In the **Editing** group, click **Sort & Filter**
Step 6. Click **Sort Smallest to Largest**

The random sample of 30 metropolitan areas appears in rows 2 to 31 of the reordered data set. The random numbers in column C are no longer necessary and can be deleted if desired.

Appendix 7.3 Random Sampling Using StatTools

If a list of the elements in a population is available in an Excel file, StatTools Random Sample Utility can be used to select a simple random sample. For example, a list of the top 100 metropolitan areas in the United States and Canada is provided in column A of the data set

TABLE 7.6 OVERALL RATING FOR THE FIRST 10 METROPOLITAN AREAS IN THE DATA SET METAREAS

Metropolitan Area	Rating
Albany, NY	64.18
Albuquerque, NM	66.16
Appleton, WI	60.56
Atlanta, GA	69.97
Austin, TX	71.48
Baltimore, MD	69.75
Birmingham, AL	69.59
Boise City, ID	68.36
Boston, MA	68.99
Buffalo, NY	66.10

Appendix 7.3 Random Sampling Using StatTools

MetAreas (*Places Rated Almanac—The Millennium Edition 2000*). Column B contains the overall rating of each metropolitan area. Assume that you would like to select a simple random sample of 30 metropolitan areas in order to do an in-depth study of the cost of living in the United States and Canada.

Begin by using the Data Set Manager to create a StatTools data set for these data using the procedure described in the appendix to Chapter 1. The following steps will generate a simple random sample of 30 metropolitan areas.

WEB file
MetAreas

Step 1. Click the **StatTools** tab on the Ribbon
Step 2. In the **Data Group** click **Data Utilities**
Step 3. Choose the **Random Sample** option
Step 4. When the StatTools—Random Sample Utility dialog box appears:
  In the **Variables** section:
    Select **Metropolitan Area**
    Select **Rating**
  In the **Options** section:
    Enter 1 in the **Number of Samples** box
    Enter 30 in the **Sample Size** box
  Click **OK**

The random sample of 30 metropolitan areas will appear in columns A and B of the worksheet entitled Random Sample.

CHAPTER 8

Interval Estimation

CONTENTS

STATISTICS IN PRACTICE: FOOD LION

8.1 POPULATION MEAN: σ KNOWN
Margin of Error and the Interval Estimate
Practical Advice

8.2 POPULATION MEAN: σ UNKNOWN
Margin of Error and the Interval Estimate
Practical Advice
Using a Small Sample
Summary of Interval Estimation Procedures

8.3 DETERMINING THE SAMPLE SIZE

8.4 POPULATION PROPORTION
Determining the Sample Size

STATISTICS in PRACTICE

FOOD LION*
SALISBURY, NORTH CAROLINA

Founded in 1957 as Food Town, Food Lion is one of the largest supermarket chains in the United States, with 1300 stores in 11 Southeastern and Mid-Atlantic states. The company sells more than 24,000 different products and offers nationally and regionally advertised brand-name merchandise, as well as a growing number of high-quality private label products manufactured especially for Food Lion. The company maintains its low price leadership and quality assurance through operating efficiencies such as standard store formats, innovative warehouse design, energy-efficient facilities, and data synchronization with suppliers. Food Lion looks to a future of continued innovation, growth, price leadership, and service to its customers.

Being in an inventory-intense business, Food Lion made the decision to adopt the LIFO (last-in, first-out) method of inventory valuation. This method matches current costs against current revenues, which minimizes the effect of radical price changes on profit and loss results. In addition, the LIFO method reduces net income thereby reducing income taxes during periods of inflation.

Food Lion establishes a LIFO index for each of seven inventory pools: Grocery, Paper/Household, Pet Supplies, Health & Beauty Aids, Dairy, Cigarette/Tobacco, and Beer/Wine. For example, a LIFO index of 1.008 for the Grocery pool would indicate that the company's grocery inventory value at current costs reflects a .8% increase due to inflation over the most recent one-year period.

A LIFO index for each inventory pool requires that the year-end inventory count for each product be valued at the current year-end cost and at the preceding year-end cost. To avoid excessive time and expense associated with counting the inventory in all 1200 store locations, Food Lion selects a random sample of 50 stores. Year-end physical inventories are taken in each of the sample stores. The current-year and preceding-year costs for each item are then used to construct the required LIFO indexes for each inventory pool.

For a recent year, the sample estimate of the LIFO index for the Health & Beauty Aids inventory pool was 1.015. Using a 95% confidence level, Food Lion computed a margin of error of .006 for the sample estimate. Thus, the interval from 1.009 to 1.021 provided a 95% confidence interval estimate of the population LIFO index. This level of precision was judged to be very good.

In this chapter you will learn how to compute the margin of error associated with sample estimates. You will also learn how to use this information to construct and interpret interval estimates of a population mean and a population proportion.

Fresh bread arriving at a Food Lion Store.

---

*The authors are indebted to Keith Cunningham, Tax Director, and Bobby Harkey, Staff Tax Accountant, at Food Lion for providing this Statistics in Practice.

In Chapter 7, we stated that a point estimator is a sample statistic used to estimate a population parameter. For instance, the sample mean $\bar{x}$ is a point estimator of the population mean μ and the sample proportion $\bar{p}$ is a point estimator of the population proportion p. Because a point estimator cannot be expected to provide the exact value of the population parameter, an **interval estimate** is often computed by adding and subtracting a value, called the **margin of error**, to the point estimate. The general form of an interval estimate is as follows:

$$\text{Point estimate} \pm \text{Margin of error}$$

The purpose of an interval estimate is to provide information about how close the point estimate, provided by the sample, is to the value of the population parameter.

In this chapter we show how to compute interval estimates of a population mean μ and a population proportion p. The general form of an interval estimate of a population mean is

$$\bar{x} \pm \text{Margin of error}$$

Similarly, the general form of an interval estimate of a population proportion is

$$\bar{p} \pm \text{Margin of error}$$

The sampling distributions of $\bar{x}$ and $\bar{p}$ play key roles in computing these interval estimates.

8.1 Population Mean: σ Known

In order to develop an interval estimate of a population mean, either the population standard deviation σ or the sample standard deviation s must be used to compute the margin of error. In most applications σ is not known, and s is used to compute the margin of error. In some applications, however, large amounts of relevant historical data are available and can be used to estimate the population standard deviation prior to sampling. Also, in quality control applications where a process is assumed to be operating correctly, or "in control," it is appropriate to treat the population standard deviation as known. We refer to such cases as the *σ known* case. In this section we introduce an example in which it is reasonable to treat σ as known and show how to construct an interval estimate for this case.

Each week Lloyd's Department Store selects a simple random sample of 100 customers in order to learn about the amount spent per shopping trip. With x representing the amount spent per shopping trip, the sample mean $\bar{x}$ provides a point estimate of μ, the mean amount spent per shopping trip for the population of all Lloyd's customers. Lloyd's has been using the weekly survey for several years. Based on the historical data, Lloyd's now assumes a known value of $\sigma = \$20$ for the population standard deviation. The historical data also indicate that the population follows a normal distribution.

During the most recent week, Lloyd's surveyed 100 customers ($n = 100$) and obtained a sample mean of $\bar{x} = \$82$. The sample mean amount spent provides a point estimate of the population mean amount spent per shopping trip, μ. In the discussion that follows, we show how to compute the margin of error for this estimate and develop an interval estimate of the population mean.

Margin of Error and the Interval Estimate

In Chapter 7 we showed that the sampling distribution of $\bar{x}$ can be used to compute the probability that $\bar{x}$ will be within a given distance of μ. In the Lloyd's example, the historical data show that the population of amounts spent is normally distributed with a standard deviation of $\sigma = 20$. So, using what we learned in Chapter 7, we can conclude that the sampling distribution of $\bar{x}$ follows a normal distribution with a standard error of $\sigma_{\bar{x}} = \sigma/\sqrt{n} = 20/\sqrt{100} = 2$. This sampling distribution is shown in Figure 8.1.[1] Because

---

[1] We use the fact that the population of amounts spent has a normal distribution to conclude that the sampling distribution of $\bar{x}$ has a normal distribution. If the population did not have a normal distribution, we could rely on the central limit theorem and the sample size of $n = 100$ to conclude that the sampling distribution of $\bar{x}$ is approximately normal. In either case, the sampling distribution of $\bar{x}$ would appear as shown in Figure 8.1.

8.1 Population Mean: σ Known

FIGURE 8.1 SAMPLING DISTRIBUTION OF THE SAMPLE MEAN AMOUNT SPENT FROM SIMPLE RANDOM SAMPLES OF 100 CUSTOMERS

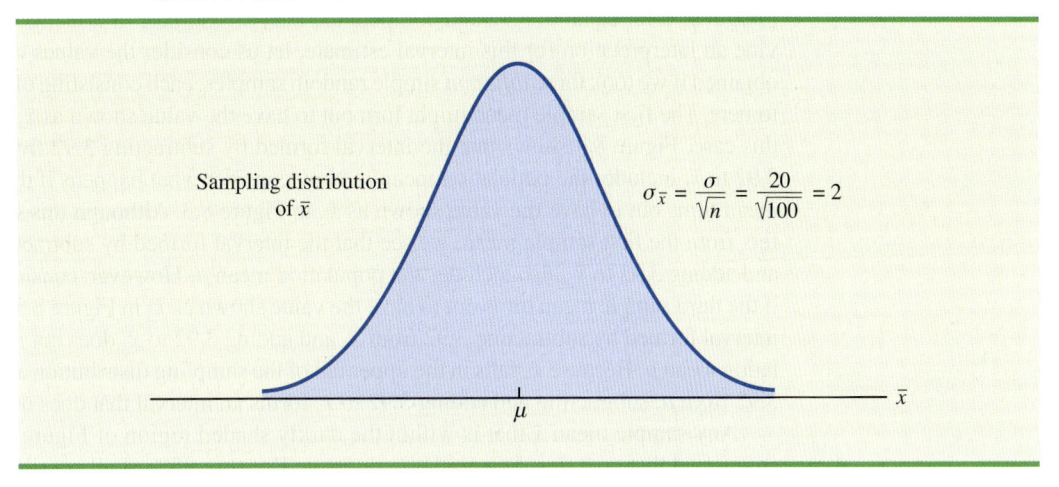

Sampling distribution of $\bar{x}$

$$\sigma_{\bar{x}} = \frac{\sigma}{\sqrt{n}} = \frac{20}{\sqrt{100}} = 2$$

the sampling distribution shows how values of $\bar{x}$ are distributed around the population mean μ, the sampling distribution of $\bar{x}$ provides information about the possible differences between $\bar{x}$ and μ.

Using the standard normal probability table, we find that 95% of the values of any normally distributed random variable are within ± 1.96 standard deviations of the mean. Thus, when the sampling distribution of $\bar{x}$ is normally distributed, 95% of the $\bar{x}$ values must be within $\pm 1.96\sigma_{\bar{x}}$ of the mean μ. In the Lloyd's example we know that the sampling distribution of $\bar{x}$ is normally distributed with a standard error of $\sigma_{\bar{x}} = 2$. Because $\pm 1.96\sigma_{\bar{x}} = 1.96(2) = 3.92$, we can conclude that 95% of all $\bar{x}$ values obtained using a sample size of $n = 100$ will be within ± 3.92 of the population mean μ. See Figure 8.2.

FIGURE 8.2 SAMPLING DISTRIBUTION OF $\bar{x}$ SHOWING THE LOCATION OF SAMPLE MEANS THAT ARE WITHIN 3.92 OF μ

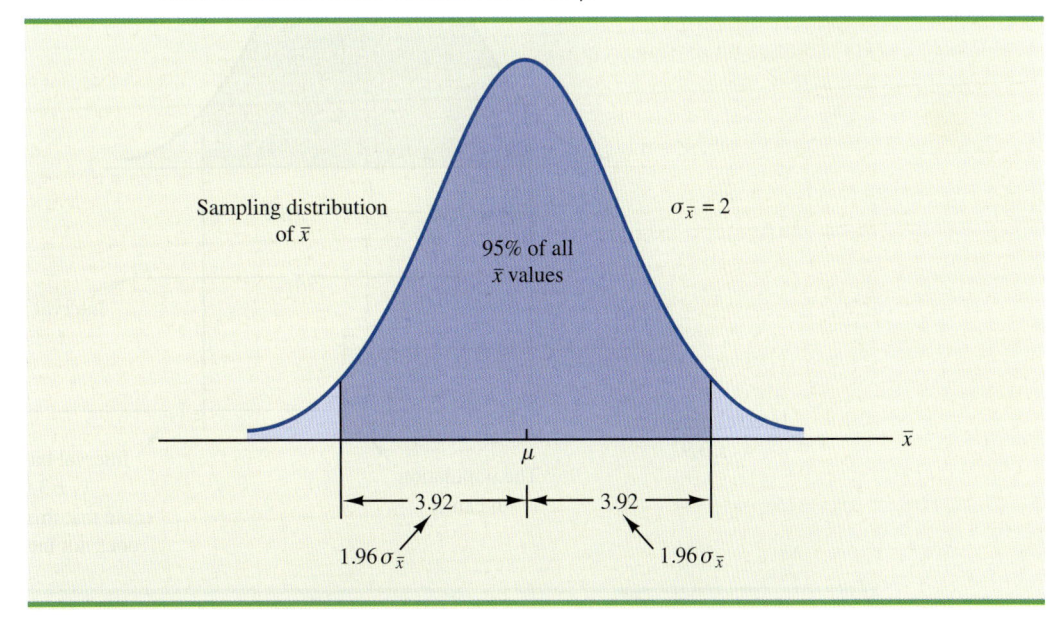

Sampling distribution of $\bar{x}$

$\sigma_{\bar{x}} = 2$

95% of all $\bar{x}$ values

3.92   3.92

$1.96\sigma_{\bar{x}}$ $1.96\sigma_{\bar{x}}$

In the introduction to this chapter we said that the general form of an interval estimate of the population mean μ is $\bar{x} \pm$ margin of error. For the Lloyd's example, suppose we set the margin of error equal to 3.92 and compute the interval estimate of μ using $\bar{x} \pm 3.92$. To provide an interpretation for this interval estimate, let us consider the values of $\bar{x}$ that could be obtained if we took three *different* simple random samples, each consisting of 100 Lloyd's customers. The first sample mean might turn out to have the value shown as $\bar{x}_1$ in Figure 8.3. In this case, Figure 8.3 shows that the interval formed by subtracting 3.92 from $\bar{x}_1$ and adding 3.92 to $\bar{x}_1$ includes the population mean μ. Now consider what happens if the second sample mean turns out to have the value shown as $\bar{x}_2$ in Figure 8.3. Although this sample mean differs from the first sample mean, we see that the interval formed by subtracting 3.92 from $\bar{x}_2$ and adding 3.92 to $\bar{x}_2$ also includes the population mean μ. However, consider what happens if the third sample mean turns out to have the value shown as $\bar{x}_3$ in Figure 8.3. In this case, the interval formed by subtracting 3.92 from $\bar{x}_3$ and adding 3.92 to $\bar{x}_3$ does not include the population mean μ. Because $\bar{x}_3$ falls in the upper tail of the sampling distribution and is farther than 3.92 from μ, subtracting and adding 3.92 to $\bar{x}_3$ forms an interval that does not include μ.

Any sample mean $\bar{x}$ that is within the darkly shaded region of Figure 8.3 will provide an interval that contains the population mean μ. Because 95% of all possible sample means are in the darkly shaded region, 95% of all intervals formed by subtracting 3.92 from $\bar{x}$ and adding 3.92 to $\bar{x}$ will include the population mean μ.

Recall that during the most recent week, the quality assurance team at Lloyd's surveyed 100 customers and obtained a sample mean amount spent of $\bar{x} = 82$. Using $\bar{x} \pm 3.92$ to

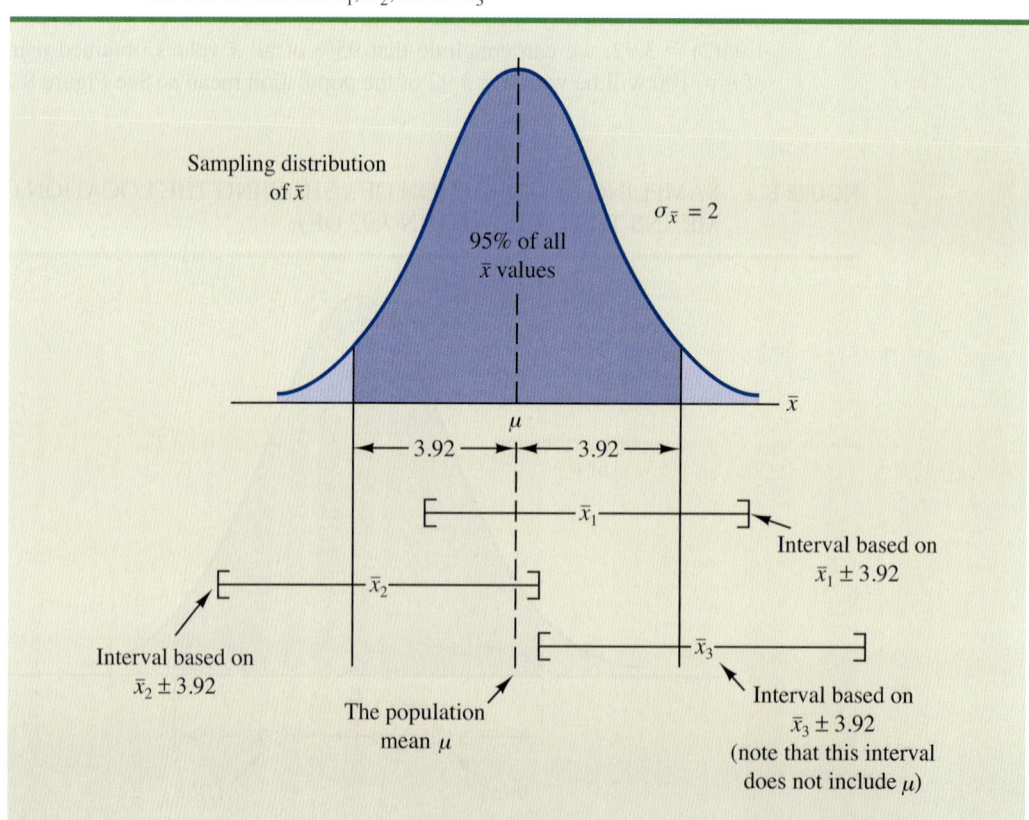

FIGURE 8.3 INTERVALS FORMED FROM SELECTED SAMPLE MEANS AT LOCATIONS $\bar{x}_1$, $\bar{x}_2$, AND $\bar{x}_3$

8.1 Population Mean: σ Known

construct the interval estimate, we obtain 82 ± 3.92. Thus, the specific interval estimate of μ based on the data from the most recent week is 82 − 3.92 = 78.08 to 82 + 3.92 = 85.92. Because 95% of all the intervals constructed using $\bar{x} \pm 3.92$ will contain the population mean, we say that we are 95% confident that the interval 78.08 to 85.92 includes the population mean μ. We say that this interval has been established at the 95% **confidence level**. The value .95 is referred to as the **confidence coefficient**, and the interval 78.08 to 85.92 is called the 95% **confidence interval**.

This discussion provides insight as to why the interval is called a 95% confidence interval.

With the margin of error given by $z_{\alpha/2}(\sigma/\sqrt{n})$, the general form of an interval estimate of a population mean for the σ known case follows.

INTERVAL ESTIMATE OF A POPULATION MEAN: σ KNOWN

$$\bar{x} \pm z_{\alpha/2} \frac{\sigma}{\sqrt{n}} \qquad (8.1)$$

where $(1 - \alpha)$ is the confidence coefficient and $z_{\alpha/2}$ is the z value providing an area of $\alpha/2$ in the upper tail of the standard normal probability distribution.

Let us use expression (8.1) to construct a 95% confidence interval for the Lloyd's example. For a 95% confidence interval, the confidence coefficient is $(1 - \alpha) = .95$ and thus, $\alpha = .05$. Using the standard normal probability table, an area of $\alpha/2 = .05/2 = .025$ in the upper tail provides $z_{.025} = 1.96$. With the Lloyd's sample mean $\bar{x} = 82$, $\sigma = 20$, and a sample size $n = 100$, we obtain

$$82 \pm 1.96 \frac{20}{\sqrt{100}}$$

$$82 \pm 3.92$$

Thus, using expression (8.1), the margin of error is 3.92 and the 95% confidence interval is 82 − 3.92 = 78.08 to 82 + 3.92 = 85.92.

Although a 95% confidence level is frequently used, other confidence levels such as 90% and 99% may be considered. Values of $z_{\alpha/2}$ for the most commonly used confidence levels are shown in Table 8.1. Using these values and expression (8.1), the 90% confidence interval for the Lloyd's example is

$$82 \pm 1.645 \frac{20}{\sqrt{100}}$$

$$82 \pm 3.29$$

TABLE 8.1 VALUES OF $z_{\alpha/2}$ FOR THE MOST COMMONLY USED CONFIDENCE LEVELS

Confidence Level	α	α/2	$z_{\alpha/2}$
90%	.10	.05	1.645
95%	.05	.025	1.960
99%	.01	.005	2.576

Thus, at 90% confidence, the margin of error is 3.29 and the confidence interval is $82 - 3.29 = 78.71$ to $82 + 3.29 = 85.29$. Similarly, the 99% confidence interval is

$$82 \pm 2.576 \frac{20}{\sqrt{100}}$$

$$82 \pm 5.15$$

Thus, at 99% confidence, the margin of error is 5.15 and the confidence interval is $82 - 5.15 = 76.85$ to $82 + 5.15 = 87.15$.

Comparing the results for the 90%, 95%, and 99% confidence levels, we see that in order to have a higher degree of confidence, the margin of error and thus the width of the confidence interval must be larger.

Practical Advice

If the population follows a normal distribution, the confidence interval provided by expression (8.1) is exact. In other words, if expression (8.1) were used repeatedly to generate 95% confidence intervals, exactly 95% of the intervals generated would contain the population mean. If the population does not follow a normal distribution, the confidence interval provided by expression (8.1) will be approximate. In this case, the quality of the approximation depends on both the distribution of the population and the sample size.

In most applications, a sample size of $n \geq 30$ is adequate when using expression (8.1) to develop an interval estimate of a population mean. If the population is not normally distributed but is roughly symmetric, sample sizes as small as 15 can be expected to provide good approximate confidence intervals. With smaller sample sizes, expression (8.1) should only be used if the analyst believes, or is willing to assume, that the population distribution is at least approximately normal.

NOTES AND COMMENTS

1. The interval estimation procedure discussed in this section is based on the assumption that the population standard deviation σ is known. By σ known we mean that historical data or other information are available that permit us to obtain a good estimate of the population standard deviation prior to taking the sample that will be used to develop an estimate of the population mean. So technically we don't mean that σ is actually known with certainty. We just mean that we obtained a good estimate of the standard deviation prior to sampling and thus we won't be using the same sample to estimate both the population mean and the population standard deviation.

2. The sample size n appears in the denominator of the interval estimation expression (8.1). Thus, if a particular sample size provides too wide an interval to be of any practical use, we may want to consider increasing the sample size. With n in the denominator, a larger sample size will provide a smaller margin of error, a narrower interval, and greater precision. The procedure for determining the size of a simple random sample necessary to obtain a desired precision is discussed in Section 8.3.

Exercises

Methods

1. A simple random sample of 40 items resulted in a sample mean of 25. The population standard deviation is $\sigma = 5$.
    a. What is the standard error of the mean, $\sigma_{\bar{x}}$?
    b. At 95% confidence, what is the margin of error?

8.1 Population Mean: σ Known

2. A simple random sample of 50 items from a population with $\sigma = 6$ resulted in a sample mean of 32.
   a. Provide a 90% confidence interval for the population mean.
   b. Provide a 95% confidence interval for the population mean.
   c. Provide a 99% confidence interval for the population mean.

3. A simple random sample of 60 items resulted in a sample mean of 80. The population standard deviation is $\sigma = 15$.
   a. Compute the 95% confidence interval for the population mean.
   b. Assume that the same sample mean was obtained from a sample of 120 items. Provide a 95% confidence interval for the population mean.
   c. What is the effect of a larger sample size on the interval estimate?

4. A 95% confidence interval for a population mean was reported to be 152 to 160. If $\sigma = 15$, what sample size was used in this study?

Applications

5. In an effort to estimate the mean amount spent per customer for dinner at a major Atlanta restaurant, data were collected for a sample of 49 customers. Assume a population standard deviation of $5.
   a. At 95% confidence, what is the margin of error?
   b. If the sample mean is $24.80, what is the 95% confidence interval for the population mean?

6. Nielsen Media Research conducted a study of household television viewing times during the 8 P.M. to 11 P.M. time period. The data contained in the file named Nielsen are consistent with the findings reported (*The World Almanac*, 2003). Based upon past studies, the population standard deviation is assumed known with $\sigma = 3.5$ hours. Develop a 95% confidence interval estimate of the mean television viewing time per week during the 8 P.M. to 11 P.M. time period.

7. *The Wall Street Journal* reported that automobile crashes cost the United States $162 billion annually (*The Wall Street Journal*, March 5, 2008). The average cost per person for crashes in the Tampa, Florida, area was reported to be $1599. Suppose this average cost was based on a sample of 50 persons who had been involved in car crashes and that the population standard deviation is $\sigma = \$600$. What is the margin of error for a 95% confidence interval? What would you recommend if the study required a margin of error of $150 or less?

8. The National Quality Research Center at the University of Michigan provides a quarterly measure of consumer opinions about products and services (*The Wall Street Journal*, February 18, 2003). A survey of 10 restaurants in the Fast Food/Pizza group showed a sample mean customer satisfaction index of 71. Past data indicate that the population standard deviation of the index has been relatively stable with $\sigma = 5$.
   a. What assumption should the researcher be willing to make if a margin of error is desired?
   b. Using 95% confidence, what is the margin of error?
   c. What is the margin of error if 99% confidence is desired?

9. AARP reported on a study conducted to learn how long it takes individuals to prepare their federal income tax return (*AARP Bulletin*, April 2008). The data contained in the file named TaxReturn are consistent with the study results. These data provide the time in hours required for 40 individuals to complete their federal income tax returns. Using past years' data, the population standard deviation can be assumed known with $\sigma = 9$ hours. What is the 95% confidence interval estimate of the mean time it takes an individual to complete a federal income tax return?

10. *Playbill* magazine reported that the mean annual household income of its readers is $119,155 (*Playbill*, January 2006). Assume this estimate of the mean annual household income is based on a sample of 80 households, and, based on past studies, the population standard deviation is known to be $\sigma = \$30{,}000$.

a. Develop a 90% confidence interval estimate of the population mean.
b. Develop a 95% confidence interval estimate of the population mean.
c. Develop a 99% confidence interval estimate of the population mean.
d. Discuss what happens to the width of the confidence interval as the confidence level is increased. Does this result seem reasonable? Explain.

8.2 Population Mean: σ Unknown

When developing an interval estimate of a population mean, we usually do not have a good estimate of the population standard deviation either. In these cases, we must use the same sample to estimate both μ and σ. This situation represents the **σ unknown** case. When s is used to estimate σ, the margin of error and the interval estimate for the population mean are based on a probability distribution known as the **t distribution**. Although the mathematical development of the t distribution is based on the assumption of a normal distribution for the population we are sampling from, research shows that the t distribution can be successfully applied in many situations where the population deviates significantly from normal. Later in this section we provide guidelines for using the t distribution if the population is not normally distributed.

William Sealy Gosset, writing under the name "Student," is the founder of the t distribution. Gosset, an Oxford graduate in mathematics, worked for the Guinness Brewery in Dublin, Ireland. He developed the t distribution while working on small-scale materials and temperature experiments.

The t distribution is a family of similar probability distributions, with a specific t distribution depending on a parameter known as the **degrees of freedom**. The t distribution with one degree of freedom is unique, as is the t distribution with two degrees of freedom, with three degrees of freedom, and so on. As the number of degrees of freedom increases, the difference between the t distribution and the standard normal distribution becomes smaller and smaller. Figure 8.4 shows t distributions with 10 and 20 degrees of freedom and their relationship to the standard normal probability distribution. Note that a t distribution with more degrees of freedom exhibits less variability and more

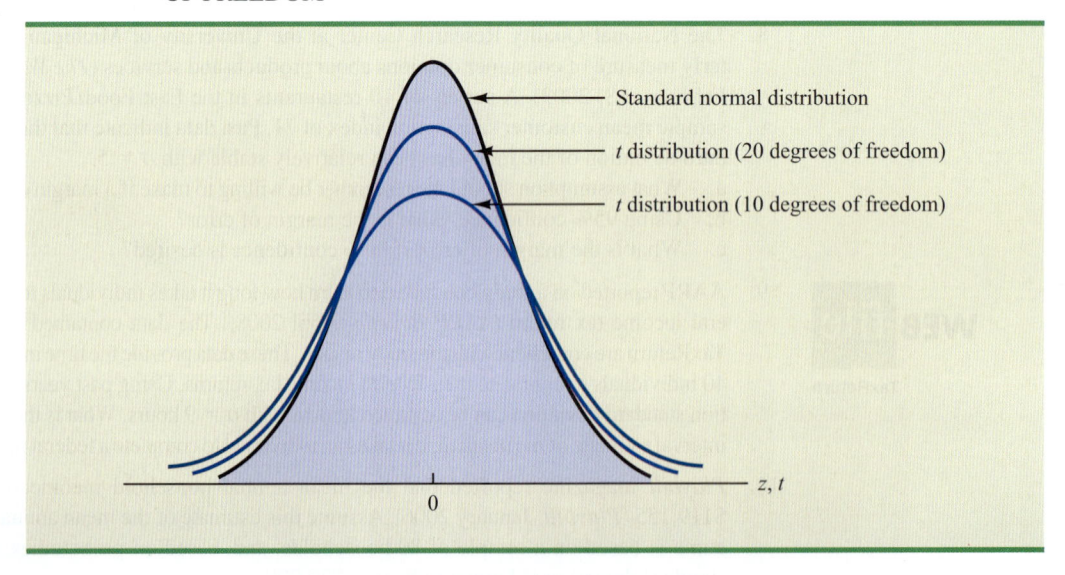

FIGURE 8.4 COMPARISON OF THE STANDARD NORMAL DISTRIBUTION WITH t DISTRIBUTIONS HAVING 10 AND 20 DEGREES OF FREEDOM

closely resembles the standard normal distribution. Note also that the mean of the t distribution is zero.

We place a subscript on t to indicate the area in the upper tail of the t distribution. For example, just as we used $z_{.025}$ to indicate the z value providing a .025 area in the upper tail of a standard normal distribution, we will use $t_{.025}$ to indicate a .025 area in the upper tail of a t distribution. In general, we will use the notation $t_{\alpha/2}$ to represent a t value with an area of $\alpha/2$ in the upper tail of the t distribution. See Figure 8.5.

Table 2 in Appendix B contains a table for the t distribution. A portion of this table is shown in Table 8.2. Each row in the table corresponds to a separate t distribution with the degrees of freedom shown. For example, for a t distribution with 9 degrees of freedom, $t_{.025} = 2.262$. Similarly, for a t distribution with 60 degrees of freedom, $t_{.025} = 2.000$. As the degrees of freedom continue to increase, $t_{.025}$ approaches $z_{.025} = 1.96$. In fact, the standard normal distribution z values can be found in the infinite degrees of freedom row (labeled ∞) of the t distribution table. If the degrees of freedom exceed 100, the infinite degrees of freedom row can be used to approximate the actual t value; in other words, for more than 100 degrees of freedom, the standard normal z value provides a good approximation to the t value.

As the degrees of freedom increase, the t distribution approaches the standard normal distribution.

Margin of Error and the Interval Estimate

In Section 8.1 we showed that an interval estimate of a population mean for the σ known case is

$$\bar{x} \pm z_{\alpha/2} \frac{\sigma}{\sqrt{n}}$$

To compute an interval estimate of μ for the σ unknown case, the sample standard deviation s is used to estimate σ, and $z_{\alpha/2}$ is replaced by the t distribution value $t_{\alpha/2}$. The margin

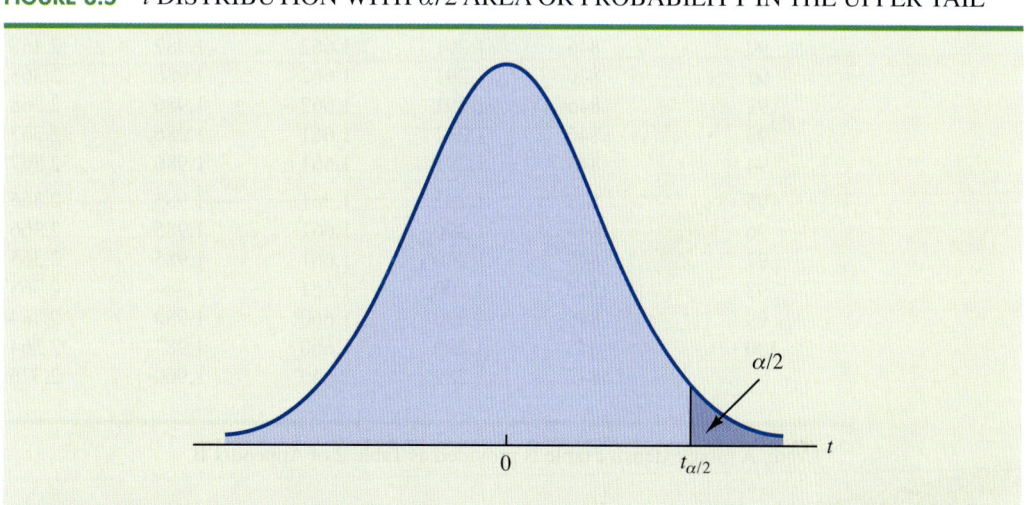

FIGURE 8.5 t DISTRIBUTION WITH $\alpha/2$ AREA OR PROBABILITY IN THE UPPER TAIL

TABLE 8.2 SELECTED VALUES FROM THE *t* DISTRIBUTION TABLE*

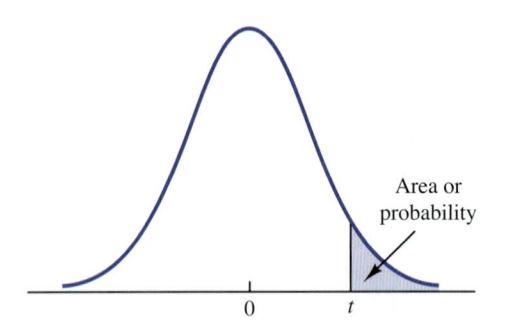

Degrees of Freedom	.20	.10	.05	.025	.01	.005
1	1.376	3.078	6.314	12.706	31.821	63.656
2	1.061	1.886	2.920	4.303	6.965	9.925
3	.978	1.638	2.353	3.182	4.541	5.841
4	.941	1.533	2.132	2.776	3.747	4.604
5	.920	1.476	2.015	2.571	3.365	4.032
6	.906	1.440	1.943	2.447	3.143	3.707
7	.896	1.415	1.895	2.365	2.998	3.499
8	.889	1.397	1.860	2.306	2.896	3.355
9	.883	1.383	1.833	2.262	2.821	3.250
⋮	⋮	⋮	⋮	⋮	⋮	⋮
60	.848	1.296	1.671	2.000	2.390	2.660
61	.848	1.296	1.670	2.000	2.389	2.659
62	.847	1.295	1.670	1.999	2.388	2.657
63	.847	1.295	1.669	1.998	2.387	2.656
64	.847	1.295	1.669	1.998	2.386	2.655
65	.847	1.295	1.669	1.997	2.385	2.654
66	.847	1.295	1.668	1.997	2.384	2.652
67	.847	1.294	1.668	1.996	2.383	2.651
68	.847	1.294	1.668	1.995	2.382	2.650
69	.847	1.294	1.667	1.995	2.382	2.649
⋮	⋮	⋮	⋮	⋮	⋮	⋮
90	.846	1.291	1.662	1.987	2.368	2.632
91	.846	1.291	1.662	1.986	2.368	2.631
92	.846	1.291	1.662	1.986	2.368	2.630
93	.846	1.291	1.661	1.986	2.367	2.630
94	.845	1.291	1.661	1.986	2.367	2.629
95	.845	1.291	1.661	1.985	2.366	2.629
96	.845	1.290	1.661	1.985	2.366	2.628
97	.845	1.290	1.661	1.985	2.365	2.627
98	.845	1.290	1.661	1.984	2.365	2.627
99	.845	1.290	1.660	1.984	2.364	2.626
100	.845	1.290	1.660	1.984	2.364	2.626
∞	.842	1.282	1.645	1.960	2.326	2.576

Note: A more extensive table is provided as Table 2 of Appendix B.

8.2 Population Mean: σ Unknown

of error is then given by $t_{\alpha/2} s/\sqrt{n}$. With this margin of error, the general expression for an interval estimate of a population mean when σ is unknown follows.

> **INTERVAL ESTIMATE OF A POPULATION MEAN: σ UNKNOWN**
>
> $$\bar{x} \pm t_{\alpha/2} \frac{s}{\sqrt{n}} \quad (8.2)$$
>
> where s is the sample standard deviation, $(1 - \alpha)$ is the confidence coefficient, and $t_{\alpha/2}$ is the t value providing an area of $\alpha/2$ in the upper tail of the t distribution with $n - 1$ degrees of freedom.

The reason the number of degrees of freedom associated with the t value in expression (8.2) is $n - 1$ concerns the use of s as an estimate of the population standard deviation σ. The expression for the sample standard deviation is

$$s = \sqrt{\frac{\Sigma(x_i - \bar{x})^2}{n - 1}}$$

Degrees of freedom refer to the number of independent pieces of information that go into the computation of $\Sigma(x_i - \bar{x})^2$. The n pieces of information involved in computing $\Sigma(x_i - \bar{x})^2$ are as follows: $x_1 - \bar{x}, x_2 - \bar{x}, \ldots, x_n - \bar{x}$. In Section 3.2 we indicated that $\Sigma(x_i - \bar{x}) = 0$ for any data set. Thus, only $n - 1$ of the $x_i - \bar{x}$ values are independent; that is, if we know $n - 1$ of the values, the remaining value can be determined exactly by using the condition that the sum of the $x_i - \bar{x}$ values must be 0. Thus, $n - 1$ is the number of degrees of freedom associated with $\Sigma(x_i - \bar{x})^2$ and hence the number of degrees of freedom for the t distribution in expression (8.2).

To illustrate the interval estimation procedure for the σ unknown case, we will consider a study designed to estimate the mean credit card debt for the population of U.S. households. A sample of $n = 70$ households provided the credit card balances shown in Table 8.3. For this situation, no previous estimate of the population standard deviation σ is available. Thus, the sample data must be used to estimate both the population mean and the population standard deviation. Using the data in Table 8.3, we compute the sample mean $\bar{x} = \$9312$ and the sample standard deviation $s = \$4007$. With 95% confidence and $n - 1 = 69$ degrees of

TABLE 8.3 CREDIT CARD BALANCES FOR A SAMPLE OF 70 HOUSEHOLDS ($)

WEB file
NewBalance

9430	14661	7159	9071	9691	11032
7535	12195	8137	3603	11448	6525
4078	10544	9467	16804	8279	5239
5604	13659	12595	13479	5649	6195
5179	7061	7917	14044	11298	12584
4416	6245	11346	6817	4353	15415
10676	13021	12806	6845	3467	15917
1627	9719	4972	10493	6191	12591
10112	2200	11356	615	12851	9743
6567	10746	7117	13627	5337	10324
13627	12744	9465	12557	8372	
18719	5742	19263	6232	7445	

freedom, Table 8.2 can be used to obtain the appropriate value for $t_{.025}$. We want the t value in the row with 69 degrees of freedom, and the column corresponding to .025 in the upper tail. The value shown is $t_{.025} = 1.995$.

We use expression (8.2) to compute an interval estimate of the population mean credit card balance.

$$9312 \pm 1.995 \frac{4007}{\sqrt{70}}$$

$$9312 \pm 955$$

The point estimate of the population mean is $9312, the margin of error is $955, and the 95% confidence interval is 9312 − 955 = $8357 to 9312 + 955 = $10,267. Thus, we are 95% confident that the mean credit card balance for the population of all households is between $8357 and $10,267.

The procedures used by Minitab, Excel, and StatTools to develop confidence intervals for a population mean are described in Appendixes 8.1, 8.2, and 8.3. For the household credit card balances study, the results of the Minitab interval estimation procedure are shown in Figure 8.6. The sample of 70 households provides a sample mean credit card balance of $9312, a sample standard deviation of $4007, a standard error of the mean of $479, and a 95% confidence interval of $8357 to $10,267.

Practical Advice

If the population follows a normal distribution, the confidence interval provided by expression (8.2) is exact and can be used for any sample size. If the population does not follow a normal distribution, the confidence interval provided by expression (8.2) will be approximate. In this case, the quality of the approximation depends on both the distribution of the population and the sample size.

Larger sample sizes are needed if the distribution of the population is highly skewed or includes outliers.

In most applications, a sample size of $n \geq 30$ is adequate when using expression (8.2) to develop an interval estimate of a population mean. However, if the population distribution is highly skewed or contains outliers, most statisticians would recommend increasing the sample size to 50 or more. If the population is not normally distributed but is roughly symmetric, sample sizes as small as 15 can be expected to provide good approximate confidence intervals. With smaller sample sizes, expression (8.2) should be used only if the analyst believes, or is willing to assume, that the population distribution is at least approximately normal.

Using a Small Sample

In the following example we develop an interval estimate for a population mean when the sample size is small. As we already noted, an understanding of the distribution of the population becomes a factor in deciding whether the interval estimation procedure provides acceptable results.

Scheer Industries is considering a new computer-assisted program to train maintenance employees to do machine repairs. In order to fully evaluate the program, the director of

FIGURE 8.6 MINITAB CONFIDENCE INTERVAL FOR THE CREDIT CARD BALANCE SURVEY

```
Variable      N     Mean     StDev    SE Mean        95% CI
NewBalance    70    9312     4007       479       (8357, 10267)
```

8.2 Population Mean: σ Unknown

TABLE 8.4 TRAINING TIME IN DAYS FOR A SAMPLE OF 20 SCHEER INDUSTRIES EMPLOYEES

52	59	54	42
44	50	42	48
55	54	60	55
44	62	62	57
45	46	43	56

manufacturing requested an estimate of the population mean time required for maintenance employees to complete the computer-assisted training.

A sample of 20 employees is selected, with each employee in the sample completing the training program. Data on the training time in days for the 20 employees are shown in Table 8.4. A histogram of the sample data appears in Figure 8.7. What can we say about the distribution of the population based on this histogram? First, the sample data do not support the conclusion that the distribution of the population is normal, yet we do not see any evidence of skewness or outliers. Therefore, using the guidelines in the previous subsection, we conclude that an interval estimate based on the t distribution appears acceptable for the sample of 20 employees.

We continue by computing the sample mean and sample standard deviation as follows.

$$\bar{x} = \frac{\Sigma x_i}{n} = \frac{1030}{20} = 51.5 \text{ days}$$

$$s = \sqrt{\frac{\Sigma(x_i - \bar{x})^2}{n-1}} = \sqrt{\frac{889}{20-1}} = 6.84 \text{ days}$$

FIGURE 8.7 HISTOGRAM OF TRAINING TIMES FOR THE SCHEER INDUSTRIES SAMPLE

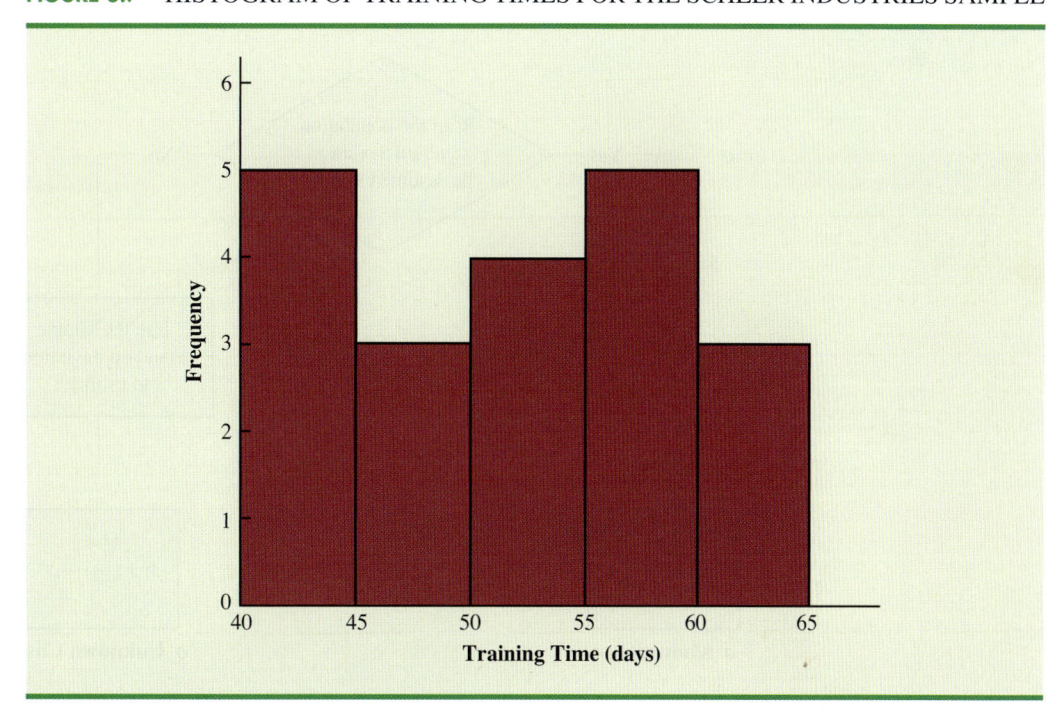

For a 95% confidence interval, we use Table 2 of Appendix B and $n - 1 = 19$ degrees of freedom to obtain $t_{.025} = 2.093$. Expression (8.2) provides the interval estimate of the population mean.

$$51.5 \pm 2.093\left(\frac{6.84}{\sqrt{20}}\right)$$

$$51.5 \pm 3.2$$

The point estimate of the population mean is 51.5 days. The margin of error is 3.2 days and the 95% confidence interval is $51.5 - 3.2 = 48.3$ days to $51.5 + 3.2 = 54.7$ days.

Using a histogram of the sample data to learn about the distribution of a population is not always conclusive, but in many cases it provides the only information available. The histogram, along with judgment on the part of the analyst, can often be used to decide whether expression (8.2) can be used to develop the interval estimate.

Summary of Interval Estimation Procedures

We provided two approaches to developing an interval estimate of a population mean. For the σ known case, σ and the standard normal distribution are used in expression (8.1) to compute the margin of error and to develop the interval estimate. For the σ unknown case, the sample standard deviation s and the t distribution are used in expression (8.2) to compute the margin of error and to develop the interval estimate.

A summary of the interval estimation procedures for the two cases is shown in Figure 8.8. In most applications, a sample size of $n \geq 30$ is adequate. If the population has a normal or approximately normal distribution, however, smaller sample sizes may be used.

FIGURE 8.8 SUMMARY OF INTERVAL ESTIMATION PROCEDURES FOR A POPULATION MEAN

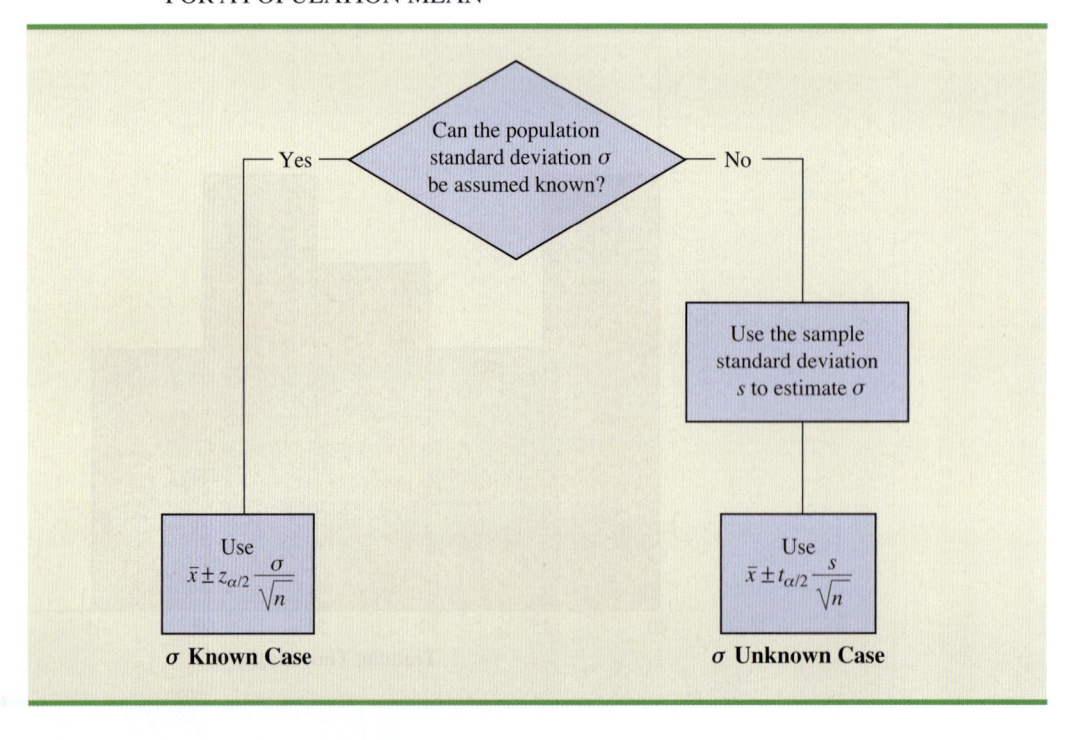

For the σ unknown case, a sample size of $n \geq 50$ is recommended if the population distribution is believed to be highly skewed or has outliers.

NOTES AND COMMENTS

1. When σ is known, the margin of error, $z_{\alpha/2}(\sigma/\sqrt{n})$, is fixed and is the same for all samples of size n. When σ is unknown, the margin of error, $t_{\alpha/2}(s/\sqrt{n})$, varies from sample to sample. This variation occurs because the sample standard deviation s varies depending upon the sample selected. A large value for s provides a larger margin of error, while a small value for s provides a smaller margin of error.

2. What happens to confidence interval estimates when the population is skewed? Consider a population that is skewed to the right with large data values stretching the distribution to the right. When such skewness exists, the sample mean $\bar{x}$ and the sample standard deviation s are positively correlated. Larger values of s tend to be associated with larger values of $\bar{x}$. Thus, when $\bar{x}$ is larger than the population mean, s tends to be larger than σ. This skewness causes the margin of error, $t_{\alpha/2}(s/\sqrt{n})$, to be larger than it would be with σ known. The confidence interval with the larger margin of error tends to include the population mean μ more often than it would if the true value of σ were used. But when $\bar{x}$ is smaller than the population mean, the correlation between $\bar{x}$ and s causes the margin of error to be small. In this case, the confidence interval with the smaller margin of error tends to miss the population mean more than it would if we knew σ and used it. For this reason, we recommend using larger sample sizes with highly skewed population distributions.

Exercises

Methods

11. For a t distribution with 16 degrees of freedom, find the area, or probability, in each region.
    a. To the right of 2.120
    b. To the left of 1.337
    c. To the left of -1.746
    d. To the right of 2.583
    e. Between -2.120 and 2.120
    f. Between -1.746 and 1.746

12. Find the t value(s) for each of the following cases.
    a. Upper tail area of .025 with 12 degrees of freedom
    b. Lower tail area of .05 with 50 degrees of freedom
    c. Upper tail area of .01 with 30 degrees of freedom
    d. Where 90% of the area falls between these two t values with 25 degrees of freedom
    e. Where 95% of the area falls between these two t values with 45 degrees of freedom

13. The following sample data are from a normal population: 10, 8, 12, 15, 13, 11, 6, 5.
    a. What is the point estimate of the population mean?
    b. What is the point estimate of the population standard deviation?
    c. With 95% confidence, what is the margin of error for the estimation of the population mean?
    d. What is the 95% confidence interval for the population mean?

14. A simple random sample with $n = 54$ provided a sample mean of 22.5 and a sample standard deviation of 4.4.
    a. Develop a 90% confidence interval for the population mean.
    b. Develop a 95% confidence interval for the population mean.

c. Develop a 99% confidence interval for the population mean.
d. What happens to the margin of error and the confidence interval as the confidence level is increased?

Applications

15. Sales personnel for Skillings Distributors submit weekly reports listing the customer contacts made during the week. A sample of 65 weekly reports showed a sample mean of 19.5 customer contacts per week. The sample standard deviation was 5.2. Provide 90% and 95% confidence intervals for the population mean number of weekly customer contacts for the sales personnel.

16. The mean number of hours of flying time for pilots at Continental Airlines is 49 hours per month (*The Wall Street Journal,* February 25, 2003). Assume that this mean was based on actual flying times for a sample of 100 Continental pilots and that the sample standard deviation was 8.5 hours.
    a. At 95% confidence, what is the margin of error?
    b. What is the 95% confidence interval estimate of the population mean flying time for the pilots?
    c. The mean number of hours of flying time for pilots at United Airlines is 36 hours per month. Use your results from part (b) to discuss differences between the flying times for the pilots at the two airlines. *The Wall Street Journal* reported United Airlines as having the highest labor cost among all airlines. Does the information in this exercise provide insight as to why United Airlines might expect higher labor costs?

17. The International Air Transport Association surveys business travelers to develop quality ratings for transatlantic gateway airports. The maximum possible rating is 10. Suppose a simple random sample of 50 business travelers is selected and each traveler is asked to provide a rating for the Miami International Airport. The ratings obtained from the sample of 50 business travelers follow.

6	4	6	8	7	7	6	3	3	8	10	4	8
7	8	7	5	9	5	8	4	3	8	5	5	4
4	4	8	4	5	6	2	5	9	9	8	4	8
9	9	5	9	7	8	3	10	8	9	6		

Develop a 95% confidence interval estimate of the population mean rating for Miami.

18. Older people often have a hard time finding work. AARP reported on the number of weeks it takes a worker aged 55 plus to find a job. The data on number of weeks spent searching for a job contained in the file JobSearch are consistent with the AARP findings (*AARP Bulletin,* April 2008).
    a. Provide a point estimate of the population mean number of weeks it takes a worker aged 55 plus to find a job.
    b. At 95% confidence, what is the margin of error?
    c. What is the 95% confidence interval estimate of the mean?
    d. Discuss the degree of skewness found in the sample data. What suggestion would you make for a repeat of this study?

19. The average cost per night of a hotel room in New York City is $273 (*SmartMoney,* March 2009). Assume this estimate is based on a sample of 45 hotels and that the sample standard deviation is $65.
    a. With 95% confidence, what is the margin of error?
    b. What is the 95% confidence interval estimate of the population mean?
    c. Two years ago the average cost of a hotel room in New York City was $229. Discuss the change in cost over the two-year period.

20. Is your favorite TV program often interrupted by advertising? CNBC presented statistics on the average number of programming minutes in a half-hour sitcom (CNBC, February 23, 2006). The following data (in minutes) are representative of its findings.

21.06	22.24	20.62
21.66	21.23	23.86
23.82	20.30	21.52
21.52	21.91	23.14
20.02	22.20	21.20
22.37	22.19	22.34
23.36	23.44	

Assume the population is approximately normal. Provide a point estimate and a 95% confidence interval for the mean number of programming minutes during a half-hour television sitcom.

21. Consumption of alcoholic beverages by young women of drinking age has been increasing in the United Kingdom, the United States, and Europe (*The Wall Street Journal*, February 15, 2006). Data (annual consumption in liters) consistent with the findings reported in *The Wall Street Journal* article are shown for a sample of 20 European young women.

266	82	199	174	97
170	222	115	130	169
164	102	113	171	0
93	0	93	110	130

Assuming the population is roughly symmetric, construct a 95% confidence interval for the mean annual consumption of alcoholic beverages by European young women.

22. Disney's *Hannah Montana: The Movie* opened on Easter weekend in April 2009. Over the three-day weekend, the movie became the number-one box office attraction (*The Wall Street Journal*, April 13, 2009). The ticket sales revenue in dollars for a sample of 25 theaters is as follows.

20,200	10,150	13,000	11,320	9,700
8,350	7,300	14,000	9,940	11,200
10,750	6,240	12,700	7,430	13,500
13,900	4,200	6,750	6,700	9,330
13,185	9,200	21,400	11,380	10,800

a. What is the 95% confidence interval estimate for the mean ticket sales revenue per theater? Interpret this result.
b. Using the movie ticket price of $7.16 per ticket, what is the estimate of the mean number of customers per theater?
c. The movie was shown in 3118 theaters. Estimate the total number of customers who saw *Hannah Montana: The Movie* and the total box office ticket sales for the three-day weekend.

8.3 Determining the Sample Size

If a desired margin of error is selected prior to sampling, the procedures in this section can be used to determine the sample size necessary to satisfy the margin of error requirement.

In providing practical advice in the two preceding sections, we commented on the role of the sample size in providing good approximate confidence intervals when the population is not normally distributed. In this section, we focus on another aspect of the sample size issue. We describe how to choose a sample size large enough to provide a desired margin of error. To understand how this process works, we return to the σ known case presented in Section 8.1. Using expression (8.1), the interval estimate is

$$\bar{x} \pm z_{\alpha/2} \frac{\sigma}{\sqrt{n}}$$

The quantity $z_{\alpha/2}(\sigma/\sqrt{n})$ is the margin of error. Thus, we see that $z_{\alpha/2}$, the population standard deviation σ, and the sample size n combine to determine the margin of error. Once we select a confidence coefficient, $1 - \alpha$, $z_{\alpha/2}$ can be determined. Then, if we have a value for σ, we can determine the sample size n needed to provide any desired margin of error. Development of the formula used to compute the required sample size n follows.

Let E = the desired margin of error:

$$E = z_{\alpha/2} \frac{\sigma}{\sqrt{n}}$$

Solving for $\sqrt{n}$, we have

$$\sqrt{n} = \frac{z_{\alpha/2}\sigma}{E}$$

Squaring both sides of this equation, we obtain the following expression for the sample size.

Equation (8.3) can be used to provide a good sample size recommendation. However, judgment on the part of the analyst should be used to determine whether the final sample size should be adjusted upward.

SAMPLE SIZE FOR AN INTERVAL ESTIMATE OF A POPULATION MEAN

$$n = \frac{(z_{\alpha/2})^2 \sigma^2}{E^2} \tag{8.3}$$

This sample size provides the desired margin of error at the chosen confidence level.

In equation (8.3), E is the margin of error that the user is willing to accept, and the value of $z_{\alpha/2}$ follows directly from the confidence level to be used in developing the interval estimate. Although user preference must be considered, 95% confidence is the most frequently chosen value ($z_{.025} = 1.96$).

Finally, use of equation (8.3) requires a value for the population standard deviation σ. However, even if σ is unknown, we can use equation (8.3) provided we have a preliminary or *planning value* for σ. In practice, one of the following procedures can be chosen.

A planning value for the population standard deviation σ must be specified before the sample size can be determined. Three methods of obtaining a planning value for σ are discussed here.

1. Use the estimate of the population standard deviation computed from data of previous studies as the planning value for σ.
2. Use a pilot study to select a preliminary sample. The sample standard deviation from the preliminary sample can be used as the planning value for σ.
3. Use judgment or a "best guess" for the value of σ. For example, we might begin by estimating the largest and smallest data values in the population. The difference between the largest and smallest values provides an estimate of the range for the data. Finally, the range divided by 4 is often suggested as a rough approximation of the standard deviation and thus an acceptable planning value for σ.

Let us demonstrate the use of equation (8.3) to determine the sample size by considering the following example. A previous study that investigated the cost of renting automobiles in the United States found a mean cost of approximately $55 per day for renting a midsize automobile. Suppose that the organization that conducted this study would like to conduct a new study in order to estimate the population mean daily rental cost for a midsize automobile in the United States. In designing the new study, the project director specifies that the population mean daily rental cost be estimated with a margin of error of $2 and a 95% level of confidence.

The project director specified a desired margin of error of $E = 2$, and the 95% level of confidence indicates $z_{.025} = 1.96$. Thus, we need only a planning value for the population standard deviation σ in order to compute the required sample size. At this point, an analyst reviewed the sample data from the previous study and found that the sample standard deviation for the daily rental cost was $9.65. Using 9.65 as the planning value for σ, we obtain

Equation (8.3) provides the minimum sample size needed to satisfy the desired margin of error requirement. If the computed sample size is not an integer, rounding up to the next integer value will provide a margin of error slightly smaller than required.

$$n = \frac{(z_{\alpha/2})^2 \sigma^2}{E^2} = \frac{(1.96)^2(9.65)^2}{2^2} = 89.43$$

Thus, the sample size for the new study needs to be at least 89.43 midsize automobile rentals in order to satisfy the project director's $2 margin-of-error requirement. In cases where the computed n is not an integer, we round up to the next integer value; hence, the recommended sample size is 90 midsize automobile rentals.

Exercises

Methods

23. How large a sample should be selected to provide a 95% confidence interval with a margin of error of 10? Assume that the population standard deviation is 40.

24. The range for a set of data is estimated to be 36.
    a. What is the planning value for the population standard deviation?
    b. At 95% confidence, how large a sample would provide a margin of error of 3?
    c. At 95% confidence, how large a sample would provide a margin of error of 2?

Applications

25. Refer to the Scheer Industries example in Section 8.2. Use 6.84 days as a planning value for the population standard deviation.
    a. Assuming 95% confidence, what sample size would be required to obtain a margin of error of 1.5 days?
    b. If the precision statement was made with 90% confidence, what sample size would be required to obtain a margin of error of 2 days?

26. The average cost of a gallon of unleaded gasoline in Greater Cincinnati was reported to be $2.41 (*The Cincinnati Enquirer*, February 3, 2006). During periods of rapidly changing prices, the newspaper samples service stations and prepares reports on gasoline prices frequently. Assume the standard deviation is $.15 for the price of a gallon of unleaded regular gasoline, and recommend the appropriate sample size for the newspaper to use if it wishes to report a margin of error at 95% confidence.
    a. Suppose the desired margin of error is $.07.
    b. Suppose the desired margin of error is $.05.
    c. Suppose the desired margin of error is $.03.

27. Annual starting salaries for college graduates with degrees in business administration are generally expected to be between $30,000 and $45,000. Assume that a 95% confidence interval estimate of the population mean annual starting salary is desired. What is the planning value for the population standard deviation? How large a sample should be taken if the desired margin of error is
    a. $500?
    b. $200?
    c. $100?
    d. Would you recommend trying to obtain the $100 margin of error? Explain.

28. An online survey by ShareBuilder, a retirement plan provider, and Harris Interactive reported that 60% of female business owners are not confident they are saving enough for retirement (*SmallBiz*, Winter 2006). Suppose we would like to do a follow-up study to determine how much female business owners are saving each year toward retirement and want to use $100 as the desired margin of error for an interval estimate of the population mean. Use $1100 as a planning value for the standard deviation and recommend a sample size for each of the following situations.
    a. A 90% confidence interval is desired for the mean amount saved.
    b. A 95% confidence interval is desired for the mean amount saved.

c. A 99% confidence interval is desired for the mean amount saved.
d. When the desired margin of error is set, what happens to the sample size as the confidence level is increased? Would you recommend using a 99% confidence interval in this case? Discuss.

29. The travel-to-work time for residents of the 15 largest cities in the United States is reported in the *2003 Information Please Almanac*. Suppose that a preliminary simple random sample of residents of San Francisco is used to develop a planning value of 6.25 minutes for the population standard deviation.
   a. If we want to estimate the population mean travel-to-work time for San Francisco residents with a margin of error of 2 minutes, what sample size should be used? Assume 95% confidence.
   b. If we want to estimate the population mean travel-to-work time for San Francisco residents with a margin of error of 1 minute, what sample size should be used? Assume 95% confidence.

30. During the first quarter of 2003, the price/earnings (P/E) ratio for stocks listed on the New York Stock Exchange generally ranged from 5 to 60 (*The Wall Street Journal*, March 7, 2003). Assume that we want to estimate the population mean P/E ratio for all stocks listed on the exchange. How many stocks should be included in the sample if we want a margin of error of 3? Use 95% confidence.

8.4 Population Proportion

In the introduction to this chapter we said that the general form of an interval estimate of a population proportion p is

$$\bar{p} \pm \text{Margin of error}$$

The sampling distribution of $\bar{p}$ plays a key role in computing the margin of error for this interval estimate.

In Chapter 7 we said that the sampling distribution of $\bar{p}$ can be approximated by a normal distribution whenever $np \geq 5$ and $n(1-p) \geq 5$. Figure 8.9 shows the normal approximation

FIGURE 8.9 NORMAL APPROXIMATION OF THE SAMPLING DISTRIBUTION OF $\bar{p}$

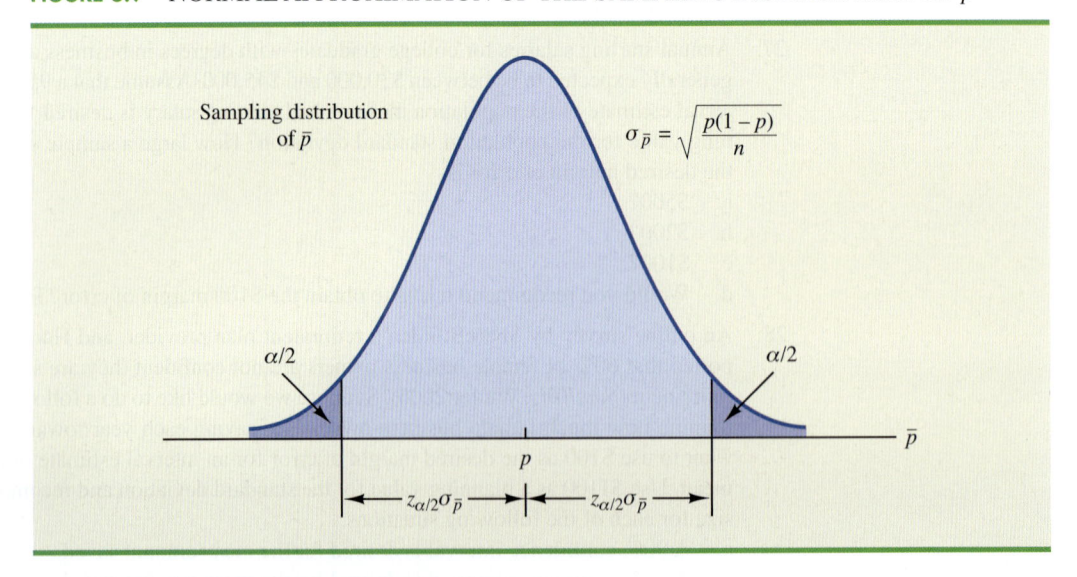

8.4 Population Proportion

of the sampling distribution of $\bar{p}$. The mean of the sampling distribution of $\bar{p}$ is the population proportion p, and the standard error of $\bar{p}$ is

$$\sigma_{\bar{p}} = \sqrt{\frac{p(1-p)}{n}} \quad (8.4)$$

Because the sampling distribution of $\bar{p}$ is normally distributed, if we choose $z_{\alpha/2}\sigma_{\bar{p}}$ as the margin of error in an interval estimate of a population proportion, we know that $100(1-\alpha)\%$ of the intervals generated will contain the true population proportion. But $\sigma_{\bar{p}}$ cannot be used directly in the computation of the margin of error because p will not be known; p is what we are trying to estimate. So $\bar{p}$ is substituted for p and the margin of error for an interval estimate of a population proportion is given by

$$\text{Margin of error} = z_{\alpha/2}\sqrt{\frac{\bar{p}(1-\bar{p})}{n}} \quad (8.5)$$

With this margin of error, the general expression for an interval estimate of a population proportion is as follows.

INTERVAL ESTIMATE OF A POPULATION PROPORTION

$$\bar{p} \pm z_{\alpha/2}\sqrt{\frac{\bar{p}(1-\bar{p})}{n}} \quad (8.6)$$

where $1 - \alpha$ is the confidence coefficient and $z_{\alpha/2}$ is the z value providing an area of $\alpha/2$ in the upper tail of the standard normal distribution.

When developing confidence intervals for proportions, the quantity $z_{\alpha/2}\sqrt{\bar{p}(1-\bar{p})/n}$ provides the margin of error.

WEB file

TeeTimes

The following example illustrates the computation of the margin of error and interval estimate for a population proportion. A national survey of 900 women golfers was conducted to learn how women golfers view their treatment at golf courses in the United States. The survey found that 396 of the women golfers were satisfied with the availability of tee times. Thus, the point estimate of the proportion of the population of women golfers who are satisfied with the availability of tee times is $396/900 = .44$. Using expression (8.6) and a 95% confidence level,

$$\bar{p} \pm z_{\alpha/2}\sqrt{\frac{\bar{p}(1-\bar{p})}{n}}$$

$$.44 \pm 1.96\sqrt{\frac{.44(1-.44)}{900}}$$

$$.44 \pm .0324$$

Thus, the margin of error is .0324 and the 95% confidence interval estimate of the population proportion is .4076 to .4724. Using percentages, the survey results enable us to state with 95% confidence that between 40.76% and 47.24% of all women golfers are satisfied with the availability of tee times.

Determining the Sample Size

Let us consider the question of how large the sample size should be to obtain an estimate of a population proportion at a specified level of precision. The rationale for the sample size determination in developing interval estimates of p is similar to the rationale used in Section 8.3 to determine the sample size for estimating a population mean.

Previously in this section we said that the margin of error associated with an interval estimate of a population proportion is $z_{\alpha/2}\sqrt{\bar{p}(1-\bar{p})/n}$. The margin of error is based on the value of $z_{\alpha/2}$, the sample proportion $\bar{p}$, and the sample size n. Larger sample sizes provide a smaller margin of error and better precision.

Let E denote the desired margin of error.

$$E = z_{\alpha/2}\sqrt{\frac{\bar{p}(1-\bar{p})}{n}}$$

Solving this equation for n provides a formula for the sample size that will provide a margin of error of size E.

$$n = \frac{(z_{\alpha/2})^2 \bar{p}(1-\bar{p})}{E^2}$$

Note, however, that we cannot use this formula to compute the sample size that will provide the desired margin of error because $\bar{p}$ will not be known until after we select the sample. What we need, then, is a planning value for $\bar{p}$ that can be used to make the computation. Using p^* to denote the planning value for $\bar{p}$, the following formula can be used to compute the sample size that will provide a margin of error of size E.

SAMPLE SIZE FOR AN INTERVAL ESTIMATE OF A POPULATION PROPORTION

$$n = \frac{(z_{\alpha/2})^2 p^*(1-p^*)}{E^2} \tag{8.7}$$

In practice, the planning value p^* can be chosen by one of the following procedures.

1. Use the sample proportion from a previous sample of the same or similar units.
2. Use a pilot study to select a preliminary sample. The sample proportion from this sample can be used as the planning value, p^*.
3. Use judgment or a "best guess" for the value of p^*.
4. If none of the preceding alternatives apply, use a planning value of $p^* = .50$.

Let us return to the survey of women golfers and assume that the company is interested in conducting a new survey to estimate the current proportion of the population of women golfers who are satisfied with the availability of tee times. How large should the sample be if the survey director wants to estimate the population proportion with a margin of error of .025 at 95% confidence? With $E = .025$ and $z_{\alpha/2} = 1.96$, we need a planning value p^* to answer the sample size question. Using the previous survey result of $\bar{p} = .44$ as the planning value p^*, equation (8.7) shows that

$$n = \frac{(z_{\alpha/2})^2 p^*(1-p^*)}{E^2} = \frac{(1.96)^2(.44)(1-.44)}{(.025)^2} = 1514.5$$

8.4 Population Proportion

TABLE 8.5 SOME POSSIBLE VALUES FOR $p^*(1 - p^*)$

p^*	$p^*(1 - p^*)$	
.10	(.10)(.90) = .09	
.30	(.30)(.70) = .21	
.40	(.40)(.60) = .24	
.50	(.50)(.50) = .25	← Largest value for $p^*(1 - p^*)$
.60	(.60)(.40) = .24	
.70	(.70)(.30) = .21	
.90	(.90)(.10) = .09	

Thus, the sample size must be at least 1514.5 women golfers to satisfy the margin of error requirement. Rounding up to the next integer value indicates that a sample of 1515 women golfers is recommended to satisfy the margin of error requirement.

The fourth alternative suggested for selecting a planning value p^* is to use $p^* = .50$. This value of p^* is frequently used when no other information is available. To understand why, note that the numerator of equation (8.7) shows that the sample size is proportional to the quantity $p^*(1 - p^*)$. A larger value for the quantity $p^*(1 - p^*)$ will result in a larger sample size. Table 8.5 gives some possible values of $p^*(1 - p^*)$. Note that the largest value of $p^*(1 - p^*)$ occurs when $p^* = .50$. Thus, in case of any uncertainty about an appropriate planning value, we know that $p^* = .50$ will provide the largest sample size recommendation. In effect, we play it safe by recommending the largest necessary sample size. If the sample proportion turns out to be different from the .50 planning value, the margin of error will be smaller than anticipated. Thus, in using $p^* = .50$, we guarantee that the sample size will be sufficient to obtain the desired margin of error.

In the survey of women golfers example, a planning value of $p^* = .50$ would have provided the sample size

$$n = \frac{(z_{\alpha/2})^2 p^*(1 - p^*)}{E^2} = \frac{(1.96)^2(.50)(1 - .50)}{(.025)^2} = 1536.6$$

Thus, a slightly larger sample size of 1537 women golfers would be recommended.

NOTES AND COMMENTS

The desired margin of error for estimating a population proportion is almost always .10 or less. In national public opinion polls conducted by organizations such as Gallup and Harris, a .03 or .04 margin of error is common. With such margins of error, equation (8.7) will almost always provide a sample size that is large enough to satisfy the requirements of $np \geq 5$ and $n(1 - p) \geq 5$ for using a normal distribution as an approximation for the sampling distribution of $\bar{x}$.

Exercises

Methods

31. A simple random sample of 400 individuals provides 100 Yes responses.
    a. What is the point estimate of the proportion of the population that would provide Yes responses?
    b. What is your estimate of the standard error of the proportion, $\sigma_{\bar{p}}$?
    c. Compute the 95% confidence interval for the population proportion.

32. A simple random sample of 800 elements generates a sample proportion $\bar{p} = .70$.
    a. Provide a 90% confidence interval for the population proportion.
    b. Provide a 95% confidence interval for the population proportion.

33. In a survey, the planning value for the population proportion is $p^* = .35$. How large a sample should be taken to provide a 95% confidence interval with a margin of error of .05?

34. At 95% confidence, how large a sample should be taken to obtain a margin of error of .03 for the estimation of a population proportion? Assume that past data are not available for developing a planning value for p^*.

Applications

35. The Consumer Reports National Research Center conducted a telephone survey of 2000 adults to learn about the major economic concerns for the future (*Consumer Reports,* January 2009). The survey results showed that 1760 of the respondents think the future health of Social Security is a major economic concern.
    a. What is the point estimate of the population proportion of adults who think the future health of Social Security is a major economic concern?
    b. At 90% confidence, what is the margin of error?
    c. Develop a 90% confidence interval for the population proportion of adults who think the future health of Social Security is a major economic concern.
    d. Develop a 95% confidence interval for this population proportion.

36. According to statistics reported on CNBC, a surprising number of motor vehicles are not covered by insurance (CNBC, February 23, 2006). Sample results, consistent with the CNBC report, showed 46 of 200 vehicles were not covered by insurance.
    a. What is the point estimate of the proportion of vehicles not covered by insurance?
    b. Develop a 95% confidence interval for the population proportion.

37. Towers Perrin, a New York human resources consulting firm, conducted a survey of 1100 employees at medium-sized and large companies to determine how dissatisfied employees were with their jobs (*The Wall Street Journal,* January 29, 2003). Representative data are shown in the file JobSatisfaction. A response of Yes indicates the employee strongly disliked the current work experience.
    a. What is the point estimate of the proportion of the population of employees who strongly dislike their current work experience?
    b. At 95% confidence, what is the margin of error?
    c. What is the 95% confidence interval for the proportion of the population of employees who strongly dislike their current work experience?
    d. Towers Perrin estimates that it costs employers one-third of an hourly employee's annual salary to find a successor and as much as 1.5 times the annual salary to find a successor for a highly compensated employee. What message did this survey send to employers?

38. According to Thomson Financial, through January 25, 2006, the majority of companies reporting profits had beaten estimates (*BusinessWeek,* February 6, 2006). A sample of 162 companies showed that 104 beat estimates, 29 matched estimates, and 29 fell short.
    a. What is the point estimate of the proportion that fell short of estimates?
    b. Determine the margin of error and provide a 95% confidence interval for the proportion that beat estimates.
    c. How large a sample is needed if the desired margin of error is .05?

39. The percentage of people not covered by health care insurance in 2003 was 15.6% (*Statistical Abstract of the United States,* 2006). A congressional committee has been charged with conducting a sample survey to obtain more current information.
    a. What sample size would you recommend if the committee's goal is to estimate the current proportion of individuals without health care insurance with a margin of error of .03? Use a 95% confidence level.
    b. Repeat part (a) using a 99% confidence level.

40. For many years businesses have struggled with the rising cost of health care. But recently, the increases have slowed due to less inflation in health care prices and employees paying for a larger portion of health care benefits. A recent Mercer survey showed that 52% of U.S. employers were likely to require higher employee contributions for health care coverage in 2009 (*BusinessWeek*, February 16, 2009). Suppose the survey was based on a sample of 800 companies. Compute the margin of error and a 95% confidence interval for the proportion of companies likely to require higher employee contributions for health care coverage in 2009.

41. America's young people are heavy Internet users; 87% of Americans ages 12 to 17 are Internet users (*The Cincinnati Enquirer*, February 7, 2006). MySpace was voted the most popular website by 9% in a sample survey of Internet users in this age group. Suppose 1400 youths participated in the survey. What is the margin of error, and what is the interval estimate of the population proportion for which MySpace is the most popular website? Use a 95% confidence level.

42. A poll for the presidential campaign sampled 491 potential voters in June. A primary purpose of the poll was to obtain an estimate of the proportion of potential voters who favored each candidate. Assume a planning value of $p^* = .50$ and a 95% confidence level.
    a. For $p^* = .50$, what was the planned margin of error for the June poll?
    b. Closer to the November election, better precision and smaller margins of error are desired. Assume the following margins of error are requested for surveys to be conducted during the presidential campaign. Compute the recommended sample size for each survey.

Survey	Margin of Error
September	.04
October	.03
Early November	.02
Pre-Election Day	.01

43. A Phoenix Wealth Management/Harris Interactive survey of 1500 individuals with net worth of $1 million or more provided a variety of statistics on wealthy people (*BusinessWeek*, September 22, 2003). The previous three-year period had been bad for the stock market, which motivated some of the questions asked.
    a. The survey reported that 53% of the respondents lost 25% or more of their portfolio value over the past three years. Develop a 95% confidence interval for the proportion of wealthy people who lost 25% or more of their portfolio value over the past three years.
    b. The survey reported that 31% of the respondents feel they have to save more for retirement to make up for what they lost. Develop a 95% confidence interval for the population proportion.
    c. Five percent of the respondents gave $25,000 or more to charity over the previous year. Develop a 95% confidence interval for the proportion who gave $25,000 or more to charity.
    d. Compare the margin of error for the interval estimates in parts (a), (b), and (c). How is the margin of error related to $\bar{p}$? When the same sample is being used to estimate a variety of proportions, which of the proportions should be used to choose the planning value p^*? Why do you think $p^* = .50$ is often used in these cases?

Summary

In this chapter we presented methods for developing interval estimates of a population mean and a population proportion. A point estimator may or may not provide a good estimate of a population parameter. The use of an interval estimate provides a measure of the precision of an estimate. Both the interval estimate of the population mean and the population proportion are of the form: point estimate ± margin of error.

We presented interval estimates for a population mean for two cases. In the σ known case, historical data or other information is used to develop an estimate of σ prior to taking a sample. Analysis of new sample data then proceeds based on the assumption that σ is known. In the σ unknown case, the sample data are used to estimate both the population mean and the population standard deviation. The final choice of which interval estimation procedure to use depends upon the analyst's understanding of which method provides the best estimate of σ.

In the σ known case, the interval estimation procedure is based on the assumed value of σ and the use of the standard normal distribution. In the σ unknown case, the interval estimation procedure uses the sample standard deviation s and the t distribution. In both cases the quality of the interval estimates obtained depends on the distribution of the population and the sample size. If the population is normally distributed, the interval estimates will be exact in both cases, even for small sample sizes. If the population is not normally distributed, the interval estimates obtained will be approximate. Larger sample sizes will provide better approximations, but the more highly skewed the population is, the larger the sample size needs to be to obtain a good approximation. Practical advice about the sample size necessary to obtain good approximations was included in Sections 8.1 and 8.2. In most cases a sample of size 30 or more will provide good approximate confidence intervals.

The general form of the interval estimate for a population proportion is $\bar{p} \pm$ margin of error. In practice the sample sizes used for interval estimates of a population proportion are generally large. Thus, the interval estimation procedure is based on the standard normal distribution.

Often a desired margin of error is specified prior to developing a sampling plan. We showed how to choose a sample size large enough to provide the desired precision.

Glossary

Interval estimate An estimate of a population parameter that provides an interval believed to contain the value of the parameter. For the interval estimates in this chapter, it has the form: point estimate $\pm$ margin of error.

Margin of error The $\pm$ value added to and subtracted from a point estimate in order to develop an interval estimate of a population parameter.

σ known The case when historical data or other information provides a good value for the population standard deviation prior to taking a sample. The interval estimation procedure uses this known value of σ in computing the margin of error.

Confidence level The confidence associated with an interval estimate. For example, if an interval estimation procedure provides intervals such that 95% of the intervals formed using the procedure will include the population parameter, the interval estimate is said to be constructed at the 95% confidence level.

Confidence coefficient The confidence level expressed as a decimal value. For example, .95 is the confidence coefficient for a 95% confidence level.

Confidence interval Another name for an interval estimate.

σ unknown The more common case when no good basis exists for estimating the population standard deviation prior to taking the sample. The interval estimation procedure uses the sample standard deviation s in computing the margin of error.

t distribution A family of probability distributions that can be used to develop an interval estimate of a population mean whenever the population standard deviation σ is unknown and is estimated by the sample standard deviation s.

Degrees of freedom A parameter of the t distribution. When the t distribution is used in the computation of an interval estimate of a population mean, the appropriate t distribution has $n - 1$ degrees of freedom, where n is the size of the simple random sample.

Key Formulas

Interval Estimate of a Population Mean: σ Known

$$\bar{x} \pm z_{\alpha/2} \frac{\sigma}{\sqrt{n}} \tag{8.1}$$

Interval Estimate of a Population Mean: σ Unknown

$$\bar{x} \pm t_{\alpha/2} \frac{s}{\sqrt{n}} \tag{8.2}$$

Sample Size for an Interval Estimate of a Population Mean

$$n = \frac{(z_{\alpha/2})^2 \sigma^2}{E^2} \tag{8.3}$$

Interval Estimate of a Population Proportion

$$\bar{p} \pm z_{\alpha/2} \sqrt{\frac{\bar{p}(1-\bar{p})}{n}} \tag{8.6}$$

Sample Size for an Interval Estimate of a Population Proportion

$$n = \frac{(z_{\alpha/2})^2 p^*(1-p^*)}{E^2} \tag{8.7}$$

Supplementary Exercises

44. A survey conducted by the American Automobile Association showed that a family of four spends an average of $215.60 per day while on vacation. Suppose a sample of 64 families of four vacationing at Niagara Falls resulted in a sample mean of $252.45 per day and a sample standard deviation of $74.50.
    a. Develop a 95% confidence interval estimate of the mean amount spent per day by a family of four visiting Niagara Falls.
    b. Based on the confidence interval from part (a), does it appear that the population mean amount spent per day by families visiting Niagara Falls differs from the mean reported by the American Automobile Association? Explain.

45. The 92 million Americans of age 50 and over control 50% of all discretionary income (*AARP Bulletin,* March 2008). AARP estimated that the average annual expenditure on restaurants and carryout food was $1873 for individuals in this age group. Suppose this estimate is based on a sample of 80 persons and that the sample standard deviation is $550.
    a. At 95% confidence, what is the margin of error?
    b. What is the 95% confidence interval for the population mean amount spent on restaurants and carryout food?
    c. What is your estimate of the total amount spent by Americans of age 50 and over on restaurants and carryout food?
    d. If the amount spent on restaurants and carryout food is skewed to the right, would you expect the median amount spent to be greater or less than $1873?

46. A sample survey of 54 discount brokers showed that the mean price charged for a trade of 100 shares at $50 per share was $33.77 (*AAII Journal,* February 2006). The survey is conducted annually. With the historical data available, assume a known population standard deviation of $15.
    a. Using the sample data, what is the margin of error associated with a 95% confidence interval?
    b. Develop a 95% confidence interval for the mean price charged by discount brokers for a trade of 100 shares at $50 per share.

47. Many stock market observers say that when the P/E ratio for stocks gets over 20, the market is overvalued. The P/E ratio is the stock price divided by the most recent 12 months of earnings. Suppose you are interested in seeing whether the current market is overvalued and would also like to know what proportion of companies pay dividends. A random sample of 30 companies listed on the New York Stock Exchange (NYSE) is provided (*Barron's,* January 19, 2004).

WEBfile NYSEStocks

Company	Dividend	P/E Ratio	Company	Dividend	P/E Ratio
Albertsons	Yes	14	NY Times A	Yes	25
BRE Prop	Yes	18	Omnicare	Yes	25
CityNtl	Yes	16	PallCp	Yes	23
DelMonte	No	21	PubSvcEnt	Yes	11
EnrgzHldg	No	20	SensientTch	Yes	11
Ford Motor	Yes	22	SmtProp	Yes	12
Gildan A	No	12	TJX Cos	Yes	21
HudsnUtdBcp	Yes	13	Thomson	Yes	30
IBM	Yes	22	USB Hldg	Yes	12
JeffPilot	Yes	16	US Restr	Yes	26
KingswayFin	No	6	Varian Med	No	41
Libbey	Yes	13	Visx	No	72
MasoniteIntl	No	15	Waste Mgt	No	23
Motorola	Yes	68	Wiley A	Yes	21
Ntl City	Yes	10	Yum Brands	No	18

a. What is a point estimate of the P/E ratio for the population of stocks listed on the New York Stock Exchange? Develop a 95% confidence interval.
b. Based on your answer to part (a), do you believe that the market is overvalued?
c. What is a point estimate of the proportion of companies on the NYSE that pay dividends? Is the sample size large enough to justify using the normal distribution to construct a confidence interval for this proportion? Why or why not?

48. US Airways conducted a number of studies that indicated a substantial savings could be obtained by encouraging Dividend Miles frequent flyer customers to redeem miles and schedule award flights online (*US Airways Attaché,* February 2003). One study collected data on the amount of time required to redeem miles and schedule an award flight over the telephone. A sample showing the time in minutes required for each of 150 award flights scheduled by telephone is contained in the data set Flights. Use Minitab or Excel to help answer the following questions.

WEBfile Flights

a. What is the sample mean number of minutes required to schedule an award flight by telephone?
b. What is the 95% confidence interval for the population mean time to schedule an award flight by telephone?
c. Assume a telephone ticket agent works 7.5 hours per day. How many award flights can one ticket agent be expected to handle a day?
d. Discuss why this information supported US Airways' plans to use an online system to reduce costs.

49. A recent article reported that there are approximately 11 minutes of actual playing time in a typical National Football League (NFL) game (*The Wall Street Journal,* January 15, 2010). The article included information about the amount of time devoted to replays, the amount of time devoted to commercials, and the amount of time the players spend standing around between plays. Data consistent with the findings published in *The Wall Street Journal* are in the file named Standing. These data provide the amount of time players spend standing around between plays for a sample of 60 NFL games.

WEBfile Standing

a. Use the Standing data set to develop a point estimate of the number of minutes during an NFL game that players are standing around between plays. Compare this to the actual playing time reported in the article. Are you surprised?
b. What is the sample standard deviation?
c. Develop a 95% confidence interval for the number of minutes players spend standing around between plays.

Supplementary Exercises

50. Although airline schedules and cost are important factors for business travelers when choosing an airline carrier, a *USA Today* survey found that business travelers list an airline's frequent flyer program as the most important factor. From a sample of $n = 1993$ business travelers who responded to the survey, 618 listed a frequent flyer program as the most important factor.
    a. What is the point estimate of the proportion of the population of business travelers who believe a frequent flyer program is the most important factor when choosing an airline carrier?
    b. Develop a 95% confidence interval estimate of the population proportion.
    c. How large a sample would be required to report the margin of error of .01 at 95% confidence? Would you recommend that *USA Today* attempt to provide this degree of precision? Why or why not?

51. A well-known bank credit card firm wishes to estimate the proportion of credit card holders who carry a nonzero balance at the end of the month and incur an interest charge. Assume that the desired margin of error is .03 at 98% confidence.
    a. How large a sample should be selected if it is anticipated that roughly 70% of the firm's card holders carry a nonzero balance at the end of the month?
    b. How large a sample should be selected if no planning value for the proportion could be specified?

52. Workers in several industries were surveyed to determine the proportion of workers who feel their industry is understaffed. In the government sector, 37% of the respondents said they were understaffed, in the health care sector 33% said they were understaffed, and in the education sector 28% said they were understaffed (*USA Today*, January 11, 2010). Suppose that 200 workers were surveyed in each industry.
    a. Construct a 95% confidence interval for the proportion of workers in each of these industries who feel their industry is understaffed.
    b. Assuming the same sample size will be used in each industry, how large would the sample need to be to ensure that the margin of error is .05 or less for each of the three confidence intervals?

53. Which would be hardest for you to give up: Your computer or your television? In a survey of 1677 U.S. Internet users, 74% of the young tech elite (average age of 22) say their computer would be very hard to give up (*PC Magazine*, February 3, 2004). Only 48% say their television would be very hard to give up.
    a. Develop a 95% confidence interval for the proportion of the young tech elite that would find it very hard to give up their computer.
    b. Develop a 99% confidence interval for the proportion of the young tech elite that would find it very hard to give up their television.
    c. In which case, part (a) or part (b), is the margin of error larger? Explain why.

54. The *2003 Statistical Abstract of the United States* reported the percentage of people 18 years of age and older who smoke. Suppose that a study designed to collect new data on smokers and nonsmokers uses a preliminary estimate of the proportion who smoke of .30.
    a. How large a sample should be taken to estimate the proportion of smokers in the population with a margin of error of .02? Use 95% confidence.
    b. Assume that the study uses your sample size recommendation in part (a) and finds 520 smokers. What is the point estimate of the proportion of smokers in the population?
    c. What is the 95% confidence interval for the proportion of smokers in the population?

55. Cincinnati/Northern Kentucky International Airport had the second highest on-time arrival rate for 2005 among the nation's busiest airports (*The Cincinnati Enquirer*, February 3, 2006). Assume the findings were based on 455 on-time arrivals out of a sample of 550 flights.
    a. Develop a point estimate of the on-time arrival rate (proportion of flights arriving on time) for the airport.
    b. Construct a 95% confidence interval for the on-time arrival rate of the population of all flights at the airport during 2005.

56. Annual salary plus bonus data for chief executive officers are presented in the *BusinessWeek* Annual Pay Survey. A preliminary sample showed that the standard deviation is $675 with data provided in thousands of dollars. How many chief executive officers should be in a sample if we want to estimate the population mean annual salary plus bonus with a margin of error of $100,000? (*Note:* The desired margin of error would be $E = 100$ if the data are in thousands of dollars.) Use 95% confidence.

57. A *USA Today*/CNN/Gallup survey of 369 working parents found 200 who said they spend too little time with their children because of work commitments.
    a. What is the point estimate of the proportion of the population of working parents who feel they spend too little time with their children because of work commitments?
    b. At 95% confidence, what is the margin of error?
    c. What is the 95% confidence interval estimate of the population proportion of working parents who feel they spend too little time with their children because of work commitments?

58. The National Center for Education Statistics reported that 47% of college students work to pay for tuition and living expenses. Assume that a sample of 450 college students was used in the study.
    a. Provide a 95% confidence interval for the population proportion of college students who work to pay for tuition and living expenses.
    b. Provide a 99% confidence interval for the population proportion of college students who work to pay for tuition and living expenses.
    c. What happens to the margin of error as the confidence is increased from 95% to 99%?

59. Mileage tests are conducted for a particular model of automobile. If a 98% confidence interval with a margin of error of 1 mile per gallon is desired, how many automobiles should be used in the test? Assume that preliminary mileage tests indicate the standard deviation is 2.6 miles per gallon.

60. In developing patient appointment schedules, a medical center wants to estimate the mean time that a staff member spends with each patient. How large a sample should be taken if the desired margin of error is two minutes at a 95% level of confidence? How large a sample should be taken for a 99% level of confidence? Use a planning value for the population standard deviation of eight minutes.

Case Problem 1 *Young Professional* Magazine

Young Professional magazine was developed for a target audience of recent college graduates who are in their first 10 years in a business/professional career. In its two years of publication, the magazine has been fairly successful. Now the publisher is interested in expanding the magazine's advertising base. Potential advertisers continually ask about the demographics and interests of subscribers to *Young Professional*. To collect this information, the magazine commissioned a survey to develop a profile of its subscribers. The survey results will be used to help the magazine choose articles of interest and provide advertisers with a profile of subscribers. As a new employee of the magazine, you have been asked to help analyze the survey results.

Some of the survey questions follow:

1. What is your age?_____
2. Are you: Male_____   Female_____
3. Do you plan to make any real estate purchases in the next two years? Yes_____ No_____
4. What is the approximate total value of financial investments, exclusive of your home, owned by you or members of your household?_____

TABLE 8.6 PARTIAL SURVEY RESULTS FOR *YOUNG PROFESSIONAL* MAGAZINE

Age	Gender	Real Estate Purchases	Value of Investments($)	Number of Transactions	Broadband Access	Household Income($)	Children
38	Female	No	12,200	4	Yes	75,200	Yes
30	Male	No	12,400	4	Yes	70,300	Yes
41	Female	No	26,800	5	Yes	48,200	No
28	Female	Yes	19,600	6	No	95,300	No
31	Female	Yes	15,100	5	No	73,300	Yes
⋮	⋮	⋮	⋮	⋮	⋮	⋮	⋮

5. How many stock/bond/mutual fund transactions have you made in the past year?_____
6. Do you have broadband access to the Internet at home? Yes_____ No_____
7. Please indicate your total household income last year._____
8. Do you have children? Yes_____ No_____

The file entitled Professional contains the responses to these questions. Table 8.6 shows the portion of the file pertaining to the first five survey respondents.

Managerial Report

Prepare a managerial report summarizing the results of the survey. In addition to statistical summaries, discuss how the magazine might use these results to attract advertisers. You might also comment on how the survey results could be used by the magazine's editors to identify topics that would be of interest to readers. Your report should address the following issues, but do not limit your analysis to just these areas.

1. Develop appropriate descriptive statistics to summarize the data.
2. Develop 95% confidence intervals for the mean age and household income of subscribers.
3. Develop 95% confidence intervals for the proportion of subscribers who have broadband access at home and the proportion of subscribers who have children.
4. Would *Young Professional* be a good advertising outlet for online brokers? Justify your conclusion with statistical data.
5. Would this magazine be a good place to advertise for companies selling educational software and computer games for young children?
6. Comment on the types of articles you believe would be of interest to readers of *Young Professional*.

Case Problem 2 Gulf Real Estate Properties

Gulf Real Estate Properties, Inc., is a real estate firm located in southwest Florida. The company, which advertises itself as "expert in the real estate market," monitors condominium sales by collecting data on location, list price, sale price, and number of days it takes to sell each unit. Each condominium is classified as Gulf View if it is located directly on the Gulf of Mexico or No Gulf View if it is located on the bay or a golf course, near but not on the Gulf. Sample data from the Multiple Listing Service in Naples, Florida, provided sales data for 40 Gulf View condominiums and 18 No Gulf View condominiums.* Prices are in thousands of dollars. The data are shown in Table 8.7.

---

*Data based on condominium sales reported in the Naples MLS (Coldwell Banker, June 2000).

TABLE 8.7 SALES DATA FOR GULF REAL ESTATE PROPERTIES

Gulf View Condominiums			No Gulf View Condominiums		
List Price ($1000s)	Sale Price ($1000s)	Days to Sell	List Price ($1000s)	Sale Price ($1000s)	Days to Sell
495.0	475.0	130	217.0	217.0	182
379.0	350.0	71	148.0	135.5	338
529.0	519.0	85	186.5	179.0	122
552.5	534.5	95	239.0	230.0	150
334.9	334.9	119	279.0	267.5	169
550.0	505.0	92	215.0	214.0	58
169.9	165.0	197	279.0	259.0	110
210.0	210.0	56	179.9	176.5	130
975.0	945.0	73	149.9	144.9	149
314.0	314.0	126	235.0	230.0	114
315.0	305.0	88	199.8	192.0	120
885.0	800.0	282	210.0	195.0	61
975.0	975.0	100	226.0	212.0	146
469.0	445.0	56	149.9	146.5	137
329.0	305.0	49	160.0	160.0	281
365.0	330.0	48	322.0	292.5	63
332.0	312.0	88	187.5	179.0	48
520.0	495.0	161	247.0	227.0	52
425.0	405.0	149			
675.0	669.0	142			
409.0	400.0	28			
649.0	649.0	29			
319.0	305.0	140			
425.0	410.0	85			
359.0	340.0	107			
469.0	449.0	72			
895.0	875.0	129			
439.0	430.0	160			
435.0	400.0	206			
235.0	227.0	91			
638.0	618.0	100			
629.0	600.0	97			
329.0	309.0	114			
595.0	555.0	45			
339.0	315.0	150			
215.0	200.0	48			
395.0	375.0	135			
449.0	425.0	53			
499.0	465.0	86			
439.0	428.5	158			

WEB file
GulfProp

Managerial Report

1. Use appropriate descriptive statistics to summarize each of the three variables for the 40 Gulf View condominiums.
2. Use appropriate descriptive statistics to summarize each of the three variables for the 18 No Gulf View condominiums.
3. Compare your summary results. Discuss any specific statistical results that would help a real estate agent understand the condominium market.

4. Develop a 95% confidence interval estimate of the population mean sales price and population mean number of days to sell for Gulf View condominiums. Interpret your results.
5. Develop a 95% confidence interval estimate of the population mean sales price and population mean number of days to sell for No Gulf View condominiums. Interpret your results.
6. Assume the branch manager requested estimates of the mean selling price of Gulf View condominiums with a margin of error of $40,000 and the mean selling price of No Gulf View condominiums with a margin of error of $15,000. Using 95% confidence, how large should the sample sizes be?
7. Gulf Real Estate Properties just signed contracts for two new listings: a Gulf View condominium with a list price of $589,000 and a No Gulf View condominium with a list price of $285,000. What is your estimate of the final selling price and number of days required to sell each of these units?

Case Problem 3 Metropolitan Research, Inc.

Metropolitan Research, Inc., a consumer research organization, conducts surveys designed to evaluate a wide variety of products and services available to consumers. In one particular study, Metropolitan looked at consumer satisfaction with the performance of automobiles produced by a major Detroit manufacturer. A questionnaire sent to owners of one of the manufacturer's full-sized cars revealed several complaints about early transmission problems. To learn more about the transmission failures, Metropolitan used a sample of actual transmission repairs provided by a transmission repair firm in the Detroit area. The following data show the actual number of miles driven for 50 vehicles at the time of transmission failure.

85,092	32,609	59,465	77,437	32,534	64,090	32,464	59,902
39,323	89,641	94,219	116,803	92,857	63,436	65,605	85,861
64,342	61,978	67,998	59,817	101,769	95,774	121,352	69,568
74,276	66,998	40,001	72,069	25,066	77,098	69,922	35,662
74,425	67,202	118,444	53,500	79,294	64,544	86,813	116,269
37,831	89,341	73,341	85,288	138,114	53,402	85,586	82,256
77,539	88,798						

Managerial Report

1. Use appropriate descriptive statistics to summarize the transmission failure data.
2. Develop a 95% confidence interval for the mean number of miles driven until transmission failure for the population of automobiles with transmission failure. Provide a managerial interpretation of the interval estimate.
3. Discuss the implication of your statistical findings in terms of the belief that some owners of the automobiles experienced early transmission failures.
4. How many repair records should be sampled if the research firm wants the population mean number of miles driven until transmission failure to be estimated with a margin of error of 5000 miles? Use 95% confidence.
5. What other information would you like to gather to evaluate the transmission failure problem more fully?

Appendix 8.1 Interval Estimation Using Minitab

We describe the use of Minitab in constructing confidence intervals for a population mean and a population proportion.

Population Mean: σ Known

We illustrate interval estimation using the Lloyd's example in Section 8.1. The amounts spent per shopping trip for the sample of 100 customers are in column C1 of a Minitab worksheet. The population standard deviation $\sigma = 20$ is assumed known. The following steps can be used to compute a 95% confidence interval estimate of the population mean.

Step 1. Select the **Stat** menu
Step 2. Choose **Basic Statistics**
Step 3. Choose **1-Sample Z**
Step 4. When the 1-Sample Z (Test and Confidence Interval) dialog box appears:
    Enter C1 in the **Samples in columns** box
    Enter 20 in the **Standard deviation** box
Step 5. Click **OK**

The Minitab default is a 95% confidence level. In order to specify a different confidence level such as 90%, add the following to step 4.

Select **Options**
When the 1-Sample Z-Options dialog box appears:
    Enter 90 in the **Confidence level** box
Click **OK**

Population Mean: σ Unknown

We illustrate interval estimation using the data in Table 8.3 showing the credit card balances for a sample of 70 households. The data are in column C1 of a Minitab worksheet. In this case the population standard deviation σ will be estimated by the sample standard deviation s. The following steps can be used to compute a 95% confidence interval estimate of the population mean.

Step 1. Select the **Stat** menu
Step 2. Choose **Basic Statistics**
Step 3. Choose **1-Sample t**
Step 4. When the 1-Sample t (Test and Confidence Interval) dialog box appears:
    Enter C1 in the **Samples in columns** box
Step 5. Click **OK**

The Minitab default is a 95% confidence level. In order to specify a different confidence level such as 90%, add the following to step 4.

Select **Options**
When the 1-Sample t-Options dialog box appears:
    Enter 90 in the **Confidence level** box
Click **OK**

Population Proportion

We illustrate interval estimation using the survey data for women golfers presented in Section 8.4. The data are in column C1 of a Minitab worksheet. Individual responses are recorded as Yes if the golfer is satisfied with the availability of tee times and No otherwise. The following steps can be used to compute a 95% confidence interval estimate of the proportion of women golfers who are satisfied with the availability of tee times.

Step 1. Select the **Stat** menu
Step 2. Choose **Basic Statistics**
Step 3. Choose **1 Proportion**
Step 4. When the 1 Proportion (Test and Confidence Interval) dialog box appears:
   Enter C1 in the **Samples in columns** box
Step 5. Select **Options**
Step 6. When the 1 Proportion-Options dialog box appears:
   Select **Use test and interval based on normal distribution**
   Click **OK**
Step 7. Click **OK**

The Minitab default is a 95% confidence level. In order to specify a different confidence level such as 90%, enter 90 in the **Confidence Level** box when the 1 Proportion-Options dialog box appears in step 6.

Note: Minitab's 1 Proportion routine uses an alphabetical ordering of the responses and selects the *second response* for the population proportion of interest. In the women golfers example, Minitab used the alphabetical ordering No-Yes and then provided the confidence interval for the proportion of Yes responses. Because Yes was the response of interest, the Minitab output was fine. However, if Minitab's alphabetical ordering does not provide the response of interest, select any cell in the column and use the sequence: Editor > Column > Value Order. It will provide you with the option of entering a user-specified order, but you must list the response of interest second in the define-an-order box.

Appendix 8.2 Interval Estimation Using Excel

We describe the use of Excel in constructing confidence intervals for a population mean and a population proportion.

Population Mean: σ Known

We illustrate interval estimation using the Lloyd's example in Section 8.1. The population standard deviation $\sigma = 20$ is assumed known. The amounts spent for the sample of 100 customers are in column A of an Excel worksheet. The following steps can be used to compute the margin of error for an estimate of the population mean. We begin by using Excel's Descriptive Statistics Tool described in Chapter 3.

Step 1. Click the **Data** tab on the Ribbon
Step 2. In the **Analysis** group, click **Data Analysis**
Step 3. Choose **Descriptive Statistics** from the list of Analysis Tools
Step 4. When the Descriptive Statistics dialog box appears:
   Enter A1:A101 in the **Input Range** box
   Select **Grouped by Columns**
   Select **Labels in First Row**
   Select **Output Range**
   Enter C1 in the **Output Range** box
   Select **Summary Statistics**
   Click **OK**

The summary statistics will appear in columns C and D. Continue by computing the margin of error using Excel's Confidence function as follows:

Step 5. Select cell C16 and enter the label Margin of Error
Step 6. Select cell D16 and enter the Excel formula =CONFIDENCE(.05,20,100)

The three parameters of the Confidence function are

Alpha = 1 − confidence coefficient = 1 − .95 = .05
The population standard deviation = 20
The sample size = 100 (*Note:* This parameter appears as Count in cell D15.)

The point estimate of the population mean is in cell D3 and the margin of error is in cell D16. The point estimate (82) and the margin of error (3.92) allow the confidence interval for the population mean to be easily computed.

Population Mean: σ Unknown

We illustrate interval estimation using the data in Table 8.2, which show the credit card balances for a sample of 70 households. The data are in column A of an Excel worksheet. The following steps can be used to compute the point estimate and the margin of error for an interval estimate of a population mean. We will use Excel's Descriptive Statistics Tool described in Chapter 3.

Step 1. Click the **Data** tab on the Ribbon
Step 2. In the **Analysis** group, click **Data Analysis**
Step 3. Choose **Descriptive Statistics** from the list of Analysis Tools
Step 4. When the Descriptive Statistics dialog box appears:
  Enter A1:A71 in the **Input Range** box
  Select **Grouped by Columns**
  Select **Labels in First Row**
  Select **Output Range**
   Enter C1 in the Output Range box
  Select **Summary Statistics**
  Select **Confidence Level for Mean**
   Enter 95 in the Confidence Level for Mean box
  Click **OK**

The summary statistics will appear in columns C and D. The point estimate of the population mean appears in cell D3. The margin of error, labeled "Confidence Level(95.0%)," appears in cell D16. The point estimate ($9312) and the margin of error ($955) allow the confidence interval for the population mean to be easily computed. The output from this Excel procedure is shown in Figure 8.10.

Population Proportion

We illustrate interval estimation using the survey data for women golfers presented in Section 8.4. The data are in column A of an Excel worksheet. Individual responses are recorded as Yes if the golfer is satisfied with the availability of tee times and No otherwise. Excel does not offer a built-in routine to handle the estimation of a population proportion; however, it is relatively easy to develop an Excel template that can be used for this purpose. The template shown in Figure 8.11 provides the 95% confidence interval estimate of the proportion of women golfers who are satisfied with the availability of tee times. Note that the background worksheet in Figure 8.11 shows the cell formulas that provide the interval estimation results shown in the foreground worksheet. The following steps are necessary to use the template for this data set.

Step 1. Enter the data range A2:A901 into the =COUNTA cell formula in cell D3
Step 2. Enter Yes as the response of interest in cell D4
Step 3. Enter the data range A2:A901 into the =COUNTIF cell formula in cell D5
Step 4. Enter .95 as the confidence coefficient in cell D8

FIGURE 8.10 INTERVAL ESTIMATION OF THE POPULATION MEAN CREDIT CARD BALANCE USING EXCEL

	A	B	C	D	E	F
1	NewBalance		NewBalance			
2	9430					
3	7535		Mean	9312		Point Estimate
4	4078		Standard Error	478.9281		
5	5604		Median	9466		
6	5179		Mode	13627		
7	4416		Standard Deviation	4007		
8	10676		Sample Variance	16056048		
9	1627		Kurtosis	−0.296		
10	10112		Skewness	0.18792		
11	6567		Range	18648		
12	13627		Minimum	615		
13	18719		Maximum	19263		
14	14661		Sum	651840		
15	12195		Count	70		Margin of Error
16	10544		Confidence Level(95.0%)	955.4354		
17	13659					
70	9743					
71	10324					
71						

Note: Rows 18 to 69 are hidden.

The template automatically provides the confidence interval in cells D15 and D16.

This template can be used to compute the confidence interval for a population proportion for other applications. For instance, to compute the interval estimate for a new data set, enter the new sample data into column A of the worksheet and then make the changes to the four cells as shown. If the new sample data have already been summarized, the sample data do not have to be entered into the worksheet. In this case, enter the sample size into cell D3 and the sample proportion into cell D6; the worksheet template will then provide the confidence interval for the population proportion. The worksheet in Figure 8.11 is available in the file Interval p on the website that accompanies this book.

Appendix 8.3 Interval Estimation Using StatTools

In this appendix we show how StatTools can be used to develop an interval estimate of a population mean for the σ unknown case, to select a sample size for the σ unknown case, and to develop an interval estimate of a population proportion.

Population Mean: σ Unknown Case

In this case the population standard deviation σ will be estimated by the sample standard deviation s. We use the credit card balance data in Table 8.3 to illustrate. Begin by using the Data Set Manager to create a StatTools data set for these data using the procedure described in the appendix to Chapter 1. The following steps can be used to compute a 95% confidence interval estimate of the population mean.

Step 1. Click the **StatTools** tab on the Ribbon
Step 2. In the **Analyses** group, click **Statistical Inference**

342 Chapter 8 Interval Estimation

FIGURE 8.11 EXCEL TEMPLATE FOR INTERVAL ESTIMATION OF A POPULATION PROPORTION

	A	B	C	D	E
1	Response		Interval Estimate of a Population Proportion		
2	Yes				
3	No		Sample Size	=COUNTA(A2:A901)	
4	Yes		Response of Interest	Yes	
5	Yes		Count for Response	=COUNTIF(A2:A901,D4)	
6	No		Sample Proportion	=D5/D3	
7	No				
8	No		Confidence Coefficient	0.95	
9	Yes		z Value	=NORMSINV(0.5+D8/2)	
10	Yes				
11	Yes		Standard Error	=SQRT(D6*(1-D6)/D3)	
12	No		Margin of Error	=D9*D11	
13	No				
14	Yes		Point Estimate	=D6	
15	No		Lower Limit	=D14-D12	
16	No		Upper Limit	=D14+D12	
17	Yes				
18	No				
901	Yes				
902					

	A	B	C	D	E	F	G
1	Response		Interval Estimate of a Population Proportion				
2	Yes						
3	No		Sample Size	900		Enter the response of interest	
4	Yes		Response of Interest	Yes			
5	Yes		Count for Response	396			
6	No		Sample Proportion	0.4400			
7	No						
8	No		Confidence Coefficient	0.95		Enter the confidence coefficient	
9	Yes		z Value	1.960			
10	Yes						
11	Yes		Standard Error	0.0165			
12	No		Margin of Error	0.0324			
13	No						
14	Yes		Point Estimate	0.4400			
15	No		Lower Limit	0.4076			
16	No		Upper Limit	0.4724			
17	Yes						
18	No						
901	Yes						
902							

Note: Rows 19 to 900 are hidden.

WEBfile
NewBalance

Step 3. Choose **Confidence Interval**
Step 4. Choose Mean/Std. Deviation
Step 5. When the StatTools—Confidence Interval for Mean/Std. Deviation dialog box appears:
　　For **Analysis Type** choose **One-Sample Analysis**
　　In the **Variables** section, select **NewBalance**
　　In the **Confidence Intervals to Calculate** section:
　　　Select the **For the Mean** option
　　　Select 95% for the **Confidence Level**
　　Click **OK**

Some descriptive statistics and the confidence interval will appear.

Determining the Sample Size

In Section 8.3 we showed how to determine the sample size needed to provide a desired margin of error. The example used involved a study designed to estimate the population mean daily rental cost for a midsize automobile in the United States. The project director specified that the population mean daily rental cost be estimated with a margin of error of $2 and a 95% level of confidence. Sample data from a previous study provided a sample standard deviation of $9.65; this value was used as the planning value for the population standard deviation. The following steps can be used to compute the recommended sample size required to provide a 95% confidence interval estimate of the population mean with a margin of error of $2.

The half-length of interval is the margin of error.

Step 1. Click the **StatTools** tab on the Ribbon
Step 2. In the **Analyses** group, click **Statistical Inference**
Step 3. Choose the **Sample Size Selection** option
Step 4. When the StatTools—Sample Size Selection dialog box appears:
   In the **Parameter to Estimate** section, select **Mean**
   In the **Confidence Interval Specification** section:
      Select **95%** for the **Confidence Level**
      Enter **2** in the **Half-Length of Interval** box
      Enter **9.65** in the **Estimated Std Dev** box
   Click **OK**

The output showing a recommended sample size of 90 will appear.

Population Proportion

We illustrate using the survey data for women golfers presented in Section 8.4. Begin by using the Data Set Manager to create a StatTools data set for these data using the procedure described in the appendix to Chapter 1. The following steps can be used to compute a 95% confidence interval estimate of the population mean.

WEB file
TeeTimes

Step 1. Click the **StatTools** tab on the Ribbon
Step 2. In the **Analyses** group, click **Statistical Inference**
Step 3. Choose **Confidence Interval**
Step 4. Choose **Proportion**
Step 5. When the StatTools—Confidence Interval for Proportion dialog box appears:
   For **Analysis Type** choose **One-Sample Analysis**
   In the **Variables** section, select **Response**
   In the **Categories to Analyze** section, select **Yes**
   In the **Options** section, enter 95% in the **Confidence Level** box Click **OK**

Some descriptive statistics and the confidence interval will appear.

StatTools also provides a capability for determining the appropriate sample size to provide a desired margin of error. The steps are similar to those for determining the sample size in the previous subsection.

CHAPTER 9

Hypothesis Tests

CONTENTS

**STATISTICS IN PRACTICE:
JOHN MORRELL & COMPANY**

9.1 DEVELOPING NULL AND ALTERNATIVE HYPOTHESES
The Alternative Hypothesis as a Research Hypothesis
The Null Hypothesis as an Assumption to Be Challenged
Summary of Forms for Null and Alternative Hypotheses

9.2 TYPE I AND TYPE II ERRORS

9.3 POPULATION MEAN: σ KNOWN
One-Tailed Test
Two-Tailed Test
Summary and Practical Advice
Relationship Between Interval Estimation and Hypothesis Testing

9.4 POPULATION MEAN: σ UNKNOWN
One-Tailed Test
Two-Tailed Test
Summary and Practical Advice

9.5 POPULATION PROPORTION
Summary

STATISTICS *in* PRACTICE

JOHN MORRELL & COMPANY*
CINCINNATI, OHIO

John Morrell & Company, which began in England in 1827, is considered the oldest continuously operating meat manufacturer in the United States. It is a wholly owned and independently managed subsidiary of Smithfield Foods, Smithfield, Virginia. John Morrell & Company offers an extensive product line of processed meats and fresh pork to consumers under 13 regional brands including John Morrell, E-Z-Cut, Tobin's First Prize, Dinner Bell, Hunter, Kretschmar, Rath, Rodeo, Shenson, Farmers Hickory Brand, Iowa Quality, and Peyton's. Each regional brand enjoys high brand recognition and loyalty among consumers.

Market research at Morrell provides management with up-to-date information on the company's various products and how the products compare with competing brands of similar products. A recent study compared a Beef Pot Roast made by Morrell to similar beef products from two major competitors. In the three-product comparison test, a sample of consumers was used to indicate how the products rated in terms of taste, appearance, aroma, and overall preference.

One research question concerned whether the Beef Pot Roast made by Morrell was the preferred choice of more than 50% of the consumer population. Letting p indicate the population proportion preferring Morrell's product, the hypothesis test for the research question is as follows:

$$H_0: p \leq .50$$
$$H_a: p > .50$$

The null hypothesis H_0 indicates the preference for Morrell's product is less than or equal to 50%. If the sample data support rejecting H_0 in favor of the alternative

---
*The authors are indebted to Marty Butler, Vice President of Marketing, John Morrell, for providing this Statistics in Practice.

Hypothesis testing helps John Morrell & Company analyze market research about their products.

hypothesis H_a, Morrell will draw the research conclusion that in a three-product comparison, its Beef Pot Roast is preferred by more than 50% of the consumer population.

In an independent taste test study using a sample of 224 consumers in Cincinnati, Milwaukee, and Los Angeles, 150 consumers selected the Beef Pot Roast made by Morrell as the preferred product. Using statistical hypothesis testing procedures, the null hypothesis H_0 was rejected. The study provided statistical evidence supporting H_a and the conclusion that the Morrell product is preferred by more than 50% of the consumer population.

The point estimate of the population proportion was $\bar{p} = 150/224 = .67$. Thus, the sample data provided support for a food magazine advertisement showing that in a three-product taste comparison, Beef Pot Roast made by Morrell was "preferred 2 to 1 over the competition."

In this chapter we will discuss how to formulate hypotheses and how to conduct tests like the one used by Morrell. Through the analysis of sample data, we will be able to determine whether a hypothesis should or should not be rejected.

In Chapters 7 and 8 we showed how a sample could be used to develop point and interval estimates of population parameters. In this chapter we continue the discussion of statistical inference by showing how hypothesis testing can be used to determine whether a statement about the value of a population parameter should or should not be rejected.

In hypothesis testing we begin by making a tentative assumption about a population parameter. This tentative assumption is called the **null hypothesis** and is denoted by H_0. We then define another hypothesis, called the **alternative hypothesis**, which is the opposite of what is stated in the null hypothesis. The alternative hypothesis is denoted by H_a.

The hypothesis testing procedure uses data from a sample to test the two competing statements indicated by H_0 and H_a.

This chapter shows how hypothesis tests can be conducted about a population mean and a population proportion. We begin by providing examples that illustrate approaches to developing null and alternative hypotheses.

9.1 Developing Null and Alternative Hypotheses

It is not always obvious how the null and alternative hypotheses should be formulated. Care must be taken to structure the hypotheses appropriately so that the hypothesis testing conclusion provides the information the researcher or decision maker wants. The context of the situation is very important in determining how the hypotheses should be stated. All hypothesis testing applications involve collecting a sample and using the sample results to provide evidence for drawing a conclusion. Good questions to consider when formulating the null and alternative hypotheses are, What is the purpose of collecting the sample? What conclusions are we hoping to make?

Learning to correctly formulate hypotheses will take some practice. Expect some initial confusion over the proper choice of the null and alternative hypotheses. The examples in this section are intended to provide guidelines.

In the chapter introduction, we stated that the null hypothesis H_0 is a tentative assumption about a population parameter such as a population mean or a population proportion. The alternative hypothesis H_a is a statement that is the opposite of what is stated in the null hypothesis. In some situations it is easier to identify the alternative hypothesis first and then develop the null hypothesis. In other situations it is easier to identify the null hypothesis first and then develop the alternative hypothesis. We will illustrate these situations in the following examples.

The Alternative Hypothesis as a Research Hypothesis

Many applications of hypothesis testing involve an attempt to gather evidence in support of a research hypothesis. In these situations, it is often best to begin with the alternative hypothesis and make it the conclusion that the researcher hopes to support. Consider a particular automobile that currently attains a fuel efficiency of 24 miles per gallon in city driving. A product research group has developed a new fuel injection system designed to increase the miles-per-gallon rating. The group will run controlled tests with the new fuel injection system looking for statistical support for the conclusion that the new fuel injection system provides more miles per gallon than the current system.

Several new fuel injection units will be manufactured, installed in test automobiles, and subjected to research-controlled driving conditions. The sample mean miles per gallon for these automobiles will be computed and used in a hypothesis test to determine if it can be concluded that the new system provides more than 24 miles per gallon. In terms of the population mean miles per gallon μ, the research hypothesis $\mu > 24$ becomes the alternative hypothesis. Since the current system provides an average or mean of 24 miles per gallon, we will make the tentative assumption that the new system is not any better than the current system and choose $\mu \leq 24$ as the null hypothesis. The null and alternative hypotheses are:

$$H_0: \mu \leq 24$$
$$H_a: \mu > 24$$

If the sample results lead to the conclusion to reject H_0, the inference can be made that $H_a: \mu > 24$ is true. The researchers have the statistical support to state that the new fuel injection system increases the mean number of miles per gallon. The production of automobiles with the new fuel injection system should be considered. However, if the sample results lead to the conclusion that H_0 cannot be rejected, the researchers cannot conclude

The conclusion that the research hypothesis is true is made if the sample data provide sufficient evidence to show that the null hypothesis can be rejected.

that the new fuel injection system is better than the current system. Production of automobiles with the new fuel injection system on the basis of better gas mileage cannot be justified. Perhaps more research and further testing can be conducted.

Successful companies stay competitive by developing new products, new methods, new systems, and the like that are better than what is currently available. Before adopting something new, it is desirable to conduct research to determine if there is statistical support for the conclusion that the new approach is indeed better. In such cases, the research hypothesis is stated as the alternative hypothesis. For example, a new teaching method is developed that is believed to be better than the current method. The alternative hypothesis is that the new method is better. The null hypothesis is that the new method is no better than the old method. A new sales force bonus plan is developed in an attempt to increase sales. The alternative hypothesis is that the new bonus plan increases sales. The null hypothesis is that the new bonus plan does not increase sales. A new drug is developed with the goal of lowering blood pressure more than an existing drug. The alternative hypothesis is that the new drug lowers blood pressure more than the existing drug. The null hypothesis is that the new drug does not provide lower blood pressure than the existing drug. In each case, rejection of the null hypothesis H_0 provides statistical support for the research hypothesis. We will see many examples of hypothesis tests in research situations such as these throughout this chapter and in the remainder of the text.

The Null Hypothesis as an Assumption to Be Challenged

Of course, not all hypothesis tests involve research hypotheses. In the following discussion we consider applications of hypothesis testing where we begin with a belief or an assumption that a statement about the value of a population parameter is true. We will then use a hypothesis test to challenge the assumption and determine if there is statistical evidence to conclude that the assumption is incorrect. In these situations, it is helpful to develop the null hypothesis first. The null hypothesis H_0 expresses the belief or assumption about the value of the population parameter. The alternative hypothesis H_a is that the belief or assumption is incorrect.

As an example, consider the situation of a manufacturer of soft drink products. The label on a soft drink bottle states that it contains 67.6 fluid ounces. We consider the label correct provided the population mean filling weight for the bottles is *at least* 67.6 fluid ounces. Without any reason to believe otherwise, we would give the manufacturer the benefit of the doubt and assume that the statement provided on the label is correct. Thus, in a hypothesis test about the population mean fluid weight per bottle, we would begin with the assumption that the label is correct and state the null hypothesis as $\mu \geq 67.6$. The challenge to this assumption would imply that the label is incorrect and the bottles are being underfilled. This challenge would be stated as the alternative hypothesis $\mu < 67.6$. Thus, the null and alternative hypotheses are

$$H_0: \mu \geq 67.6$$
$$H_a: \mu < 67.6$$

A manufacturer's product information is usually assumed to be true and stated as the null hypothesis. The conclusion that the information is incorrect can be made if the null hypothesis is rejected.

A government agency with the responsibility for validating manufacturing labels could select a sample of soft drink bottles, compute the sample mean filling weight, and use the sample results to test the preceding hypotheses. If the sample results lead to the conclusion to reject H_0, the inference that $H_a: \mu < 67.6$ is true can be made. With this statistical support, the agency is justified in concluding that the label is incorrect and underfilling of the bottles is occurring. Appropriate action to force the manufacturer to comply with labeling standards would be considered. However, if the sample results indicate H_0 cannot be rejected, the assumption that the manufacturer's labeling is correct cannot be rejected. With this conclusion, no action would be taken.

Let us now consider a variation of the soft drink bottle filling example by viewing the same situation from the manufacturer's point of view. The bottle-filling operation has been designed to fill soft drink bottles with 67.6 fluid ounces as stated on the label. The company does not want to underfill the containers because that could result in an underfilling complaint from customers or, perhaps, a government agency. However, the company does not want to overfill containers either because putting more soft drink than necessary into the containers would be an unnecessary cost. The company's goal would be to adjust the bottle-filling operation so that the population mean filling weight per bottle is 67.6 fluid ounces as specified on the label.

Although this is the company's goal, from time to time any production process can get out of adjustment. If this occurs in our example, underfilling or overfilling of the soft drink bottles will occur. In either case, the company would like to know about it in order to correct the situation by readjusting the bottle-filling operation to the designed 67.6 fluid ounces. In a hypothesis testing application, we would again begin with the assumption that the production process is operating correctly and state the null hypothesis as $\mu = 67.6$ fluid ounces. The alternative hypothesis that challenges this assumption is that $\mu \neq 67.6$, which indicates either overfilling or underfilling is occurring. The null and alternative hypotheses for the manufacturer's hypothesis test are

$$H_0: \mu = 67.6$$
$$H_a: \mu \neq 67.6$$

Suppose that the soft drink manufacturer uses a quality control procedure to periodically select a sample of bottles from the filling operation and computes the sample mean filling weight per bottle. If the sample results lead to the conclusion to reject H_0, the inference is made that $H_a: \mu \neq 67.6$ is true. We conclude that the bottles are not being filled properly and the production process should be adjusted to restore the population mean to 67.6 fluid ounces per bottle. However, if the sample results indicate H_0 cannot be rejected, the assumption that the manufacturer's bottle filling operation is functioning properly cannot be rejected. In this case, no further action would be taken and the production operation would continue to run.

The two preceding forms of the soft drink manufacturing hypothesis test show that the null and alternative hypotheses may vary depending upon the point of view of the researcher or decision maker. To correctly formulate hypotheses it is important to understand the context of the situation and structure the hypotheses to provide the information the researcher or decision maker wants.

Summary of Forms for Null and Alternative Hypotheses

The hypothesis tests in this chapter involve two population parameters: the population mean and the population proportion. Depending on the situation, hypothesis tests about a population parameter may take one of three forms: Two use inequalities in the null hypothesis; the third uses an equality in the null hypothesis. For hypothesis tests involving a population mean, we let μ_0 denote the hypothesized value and we must choose one of the following three forms for the hypothesis test.

The three possible forms of hypotheses H_0 and H_a are shown here. Note that the equality always appears in the null hypothesis H_0.

$$H_0: \mu \geq \mu_0 \qquad H_0: \mu \leq \mu_0 \qquad H_0: \mu = \mu_0$$
$$H_a: \mu < \mu_0 \qquad H_a: \mu > \mu_0 \qquad H_a: \mu \neq \mu_0$$

For reasons that will be clear later, the first two forms are called one-tailed tests. The third form is called a two-tailed test.

In many situations, the choice of H_0 and H_a is not obvious and judgment is necessary to select the proper form. However, as the preceding forms show, the equality part of the

Exercises

1. The manager of the Danvers-Hilton Resort Hotel stated that the mean guest bill for a weekend is $600 or less. A member of the hotel's accounting staff noticed that the total charges for guest bills have been increasing in recent months. The accountant will use a sample of future weekend guest bills to test the manager's claim.
   a. Which form of the hypotheses should be used to test the manager's claim? Explain.

$$H_0: \mu \geq 600 \qquad H_0: \mu \leq 600 \qquad H_0: \mu = 600$$
$$H_a: \mu < 600 \qquad H_a: \mu > 600 \qquad H_a: \mu \neq 600$$

   b. What conclusion is appropriate when H_0 cannot be rejected?
   c. What conclusion is appropriate when H_0 can be rejected?

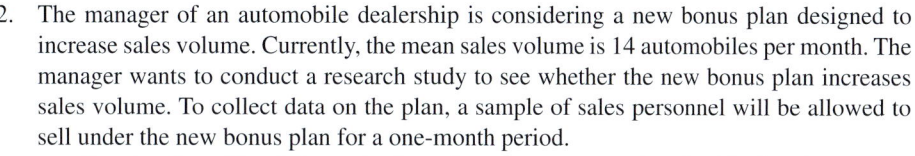

2. The manager of an automobile dealership is considering a new bonus plan designed to increase sales volume. Currently, the mean sales volume is 14 automobiles per month. The manager wants to conduct a research study to see whether the new bonus plan increases sales volume. To collect data on the plan, a sample of sales personnel will be allowed to sell under the new bonus plan for a one-month period.
   a. Develop the null and alternative hypotheses most appropriate for this situation.
   b. Comment on the conclusion when H_0 cannot be rejected.
   c. Comment on the conclusion when H_0 can be rejected.

3. A production line operation is designed to fill cartons with laundry detergent to a mean weight of 32 ounces. A sample of cartons is periodically selected and weighed to determine whether underfilling or overfilling is occurring. If the sample data lead to a conclusion of underfilling or overfilling, the production line will be shut down and adjusted to obtain proper filling.
   a. Formulate the null and alternative hypotheses that will help in deciding whether to shut down and adjust the production line.
   b. Comment on the conclusion and the decision when H_0 cannot be rejected.
   c. Comment on the conclusion and the decision when H_0 can be rejected.

4. Because of high production-changeover time and costs, a director of manufacturing must convince management that a proposed manufacturing method reduces costs before the new method can be implemented. The current production method operates with a mean cost of $220 per hour. A research study will measure the cost of the new method over a sample production period.
   a. Develop the null and alternative hypotheses most appropriate for this study.
   b. Comment on the conclusion when H_0 cannot be rejected.
   c. Comment on the conclusion when H_0 can be rejected.

9.2 Type I and Type II Errors

The null and alternative hypotheses are competing statements about the population. Either the null hypothesis H_0 is true or the alternative hypothesis H_a is true, but not both. Ideally the hypothesis testing procedure should lead to the acceptance of H_0 when H_0 is true and the

TABLE 9.1 ERRORS AND CORRECT CONCLUSIONS IN HYPOTHESIS TESTING

	Population Condition	
Conclusion	H_0 True	H_a True
Accept H_0	Correct Conclusion	Type II Error
Reject H_0	Type I Error	Correct Conclusion

rejection of H_0 when H_a is true. Unfortunately, the correct conclusions are not always possible. Because hypothesis tests are based on sample information, we must allow for the possibility of errors. Table 9.1 illustrates the two kinds of errors that can be made in hypothesis testing.

The first row of Table 9.1 shows what can happen if the conclusion is to accept H_0. If H_0 is true, this conclusion is correct. However, if H_a is true, we make a **Type II error**; that is, we accept H_0 when it is false. The second row of Table 9.1 shows what can happen if the conclusion is to reject H_0. If H_0 is true, we make a **Type I error**; that is, we reject H_0 when it is true. However, if H_a is true, rejecting H_0 is correct.

Recall the hypothesis testing illustration discussed in Section 9.1 in which an automobile product research group developed a new fuel injection system designed to increase the miles-per-gallon rating of a particular automobile. With the current model obtaining an average of 24 miles per gallon, the hypothesis test was formulated as follows.

$$H_0: \mu \leq 24$$
$$H_a: \mu > 24$$

The alternative hypothesis, $H_a: \mu > 24$, indicates that the researchers are looking for sample evidence to support the conclusion that the population mean miles per gallon with the new fuel injection system is greater than 24.

In this application, the Type I error of rejecting H_0 when it is true corresponds to the researchers claiming that the new system improves the miles-per-gallon rating ($\mu > 24$) when in fact the new system is not any better than the current system. In contrast, the Type II error of accepting H_0 when it is false corresponds to the researchers concluding that the new system is not any better than the current system ($\mu \leq 24$) when in fact the new system improves miles-per-gallon performance.

For the miles-per-gallon rating hypothesis test, the null hypothesis is $H_0: \mu \leq 24$. Suppose the null hypothesis is true as an equality; that is, $\mu = 24$. The probability of making a Type I error when the null hypothesis is true as an equality is called the **level of significance**. Thus, for the miles-per-gallon rating hypothesis test, the level of significance is the probability of rejecting $H_0: \mu \leq 24$ when $\mu = 24$. Because of the importance of this concept, we now restate the definition of level of significance.

LEVEL OF SIGNIFICANCE

The level of significance is the probability of making a Type I error when the null hypothesis is true as an equality.

The Greek symbol α (alpha) is used to denote the level of significance, and common choices for α are .05 and .01.

9.2 Type I and Type II Errors

In practice, the person responsible for the hypothesis test specifies the level of significance. By selecting α, that person is controlling the probability of making a Type I error. If the cost of making a Type I error is high, small values of α are preferred. If the cost of making a Type I error is not too high, larger values of α are typically used. Applications of hypothesis testing that only control for the Type I error are called *significance tests*. Many applications of hypothesis testing are of this type.

Although most applications of hypothesis testing control for the probability of making a Type I error, they do not always control for the probability of making a Type II error. Hence, if we decide to accept H_0, we cannot determine how confident we can be with that decision. Because of the uncertainty associated with making a Type II error when conducting significance tests, statisticians usually recommend that we use the statement "do not reject H_0" instead of "accept H_0." Using the statement "do not reject H_0" carries the recommendation to withhold both judgment and action. In effect, by not directly accepting H_0, the statistician avoids the risk of making a Type II error. Whenever the probability of making a Type II error has not been determined and controlled, we will not make the statement "accept H_0." In such cases, only two conclusions are possible: *do not reject H_0 or reject H_0*.

Although controlling for a Type II error in hypothesis testing is not common, it can be done. More advanced texts describe procedures for determining and controlling the probability of making a Type II error.* If proper controls have been established for this error, action based on the "accept H_0" conclusion can be appropriate.

If the sample data are consistent with the null hypothesis H_0, we will follow the practice of concluding "do not reject H_0." This conclusion is preferred over "accept H_0," because the conclusion to accept H_0 puts us at risk of making a Type II error.

NOTES AND COMMENTS

Walter Williams, syndicated columnist and professor of economics at George Mason University, points out that the possibility of making a Type I or a Type II error is always present in decision making (*The Cincinnati Enquirer*, August 14, 2005). He notes that the Food and Drug Administration (FDA) runs the risk of making these errors in its drug approval process. The FDA must either approve a new drug or not approve it. Thus the FDA runs the risk of approving a new drug that is not safe and effective, or failing to approve a new drug that is safe and effective. Regardless of the decision made, the possibility of making a costly error cannot be eliminated.

Exercises

SELF test

5. Nielsen reported that young men in the United States watch 56.2 minutes of prime-time TV daily (*The Wall Street Journal Europe*, November 18, 2003). A researcher believes that young men in Germany spend more time watching prime-time TV. A sample of German young men will be selected by the researcher and the time they spend watching TV in one day will be recorded. The sample results will be used to test the following null and alternative hypotheses.

$$H_0: \mu \leq 56.2$$
$$H_a: \mu > 56.2$$

   a. What is the Type I error in this situation? What are the consequences of making this error?
   b. What is the Type II error in this situation? What are the consequences of making this error?

6. The label on a 3-quart container of orange juice states that the orange juice contains an average of 1 gram of fat or less. Answer the following questions for a hypothesis test that could be used to test the claim on the label.
   a. Develop the appropriate null and alternative hypotheses.

---

*See, for example, D. R. Anderson, D. J. Sweeney, and T. A. Williams, *Statistics for Business and Economics*, 11th edition (Cincinnati, OH: South-Western/Cengage Learning, 2011).

b. What is the Type I error in this situation? What are the consequences of making this error?
c. What is the Type II error in this situation? What are the consequences of making this error?

7. Carpetland salespersons average $8000 per week in sales. Steve Contois, the firm's vice president, proposes a compensation plan with new selling incentives. Steve hopes that the results of a trial selling period will enable him to conclude that the compensation plan increases the average sales per salesperson.
   a. Develop the appropriate null and alternative hypotheses.
   b. What is the Type I error in this situation? What are the consequences of making this error?
   c. What is the Type II error in this situation? What are the consequences of making this error?

8. Suppose a new production method will be implemented if a hypothesis test supports the conclusion that the new method reduces the mean operating cost per hour.
   a. State the appropriate null and alternative hypotheses if the mean cost for the current production method is $220 per hour.
   b. What is the Type I error in this situation? What are the consequences of making this error?
   c. What is the Type II error in this situation? What are the consequences of making this error?

9.3 Population Mean: σ Known

In Chapter 8 we said that the σ known case corresponds to applications in which historical data and/or other information are available that enable us to obtain a good estimate of the population standard deviation prior to sampling. In such cases the population standard deviation can, for all practical purposes, be considered known. In this section we show how to conduct a hypothesis test about a population mean for the σ known case.

The methods presented in this section are exact if the sample is selected from a population that is normally distributed. In cases where it is not reasonable to assume the population is normally distributed, these methods are still applicable if the sample size is large enough. We provide some practical advice concerning the population distribution and the sample size at the end of this section.

One-Tailed Test

One-tailed tests about a population mean take one of the following two forms.

Lower Tail Test	Upper Tail Test
$H_0: \mu \geq \mu_0$	$H_0: \mu \leq \mu_0$
$H_a: \mu < \mu_0$	$H_a: \mu > \mu_0$

Let us consider an example involving a lower tail test.

The Federal Trade Commission (FTC) periodically conducts statistical studies designed to test the claims that manufacturers make about their products. For example, the label on a large can of Hilltop Coffee states that the can contains 3 pounds of coffee. The FTC knows that Hilltop's production process cannot place exactly 3 pounds of coffee in each can, even if the mean filling weight for the population of all cans filled is 3 pounds per can. However, as long as the population mean filling weight is at least 3 pounds per can, the rights of consumers will be protected. Thus, the FTC interprets the label information on a large can of coffee as a claim by Hilltop that the population mean filling weight is at least 3 pounds per can. We will show how the FTC can check Hilltop's claim by conducting a lower tail hypothesis test.

The first step is to develop the null and alternative hypotheses for the test. If the population mean filling weight is at least 3 pounds per can, Hilltop's claim is correct. This establishes the null hypothesis for the test. However, if the population mean weight is less than 3 pounds per can, Hilltop's claim is incorrect. This establishes the alternative hypothesis. With μ denoting the population mean filling weight, the null and alternative hypotheses are as follows:

$$H_0: \mu \geq 3$$
$$H_a: \mu < 3$$

Note that the hypothesized value of the population mean is $\mu_0 = 3$.

If the sample data indicate that H_0 cannot be rejected, the statistical evidence does not support the conclusion that a label violation has occurred. Hence, no action should be taken against Hilltop. However, if the sample data indicate H_0 can be rejected, we will conclude that the alternative hypothesis, $H_a: \mu < 3$, is true. In this case a conclusion of underfilling and a charge of a label violation against Hilltop would be justified.

Suppose a sample of 36 cans of coffee is selected and the sample mean $\bar{x}$ is computed as an estimate of the population mean μ. If the value of the sample mean $\bar{x}$ is less than 3 pounds, the sample results will cast doubt on the null hypothesis. What we want to know is how much less than 3 pounds must $\bar{x}$ be before we would be willing to declare the difference significant and risk making a Type I error by falsely accusing Hilltop of a label violation. A key factor in addressing this issue is the value the decision maker selects for the level of significance.

As noted in the preceding section, the level of significance, denoted by α, is the probability of making a Type I error by rejecting H_0 when the null hypothesis is true as an equality. The decision maker must specify the level of significance. If the cost of making a Type I error is high, a small value should be chosen for the level of significance. If the cost is not high, a larger value is more appropriate. In the Hilltop Coffee study, the director of the FTC's testing program made the following statement: "If the company is meeting its weight specifications at $\mu = 3$, I do not want to take action against them. But, I am willing to risk a 1% chance of making such an error." From the director's statement, we set the level of significance for the hypothesis test at $\alpha = .01$. Thus, we must design the hypothesis test so that the probability of making a Type I error when $\mu = 3$ is .01.

For the Hilltop Coffee study, by developing the null and alternative hypotheses and specifying the level of significance for the test, we carry out the first two steps required in conducting every hypothesis test. We are now ready to perform the third step of hypothesis testing: collect the sample data and compute the value of what is called a test statistic.

Test statistic For the Hilltop Coffee study, previous FTC tests show that the population standard deviation can be assumed known with a value of $\sigma = .18$. In addition, these tests also show that the population of filling weights can be assumed to have a normal distribution. From the study of sampling distributions in Chapter 7 we know that if the population from which we are sampling is normally distributed, the sampling distribution of $\bar{x}$ will also be normally distributed. Thus, for the Hilltop Coffee study, the sampling distribution of $\bar{x}$ is normally distributed. With a known value of $\sigma = .18$ and a sample size of $n = 36$, Figure 9.1 shows the sampling distribution of $\bar{x}$ when the null hypothesis is true as an equality; that is, when $\mu = \mu_0 = 3$.[1] Note that the standard error of $\bar{x}$ is given by $\sigma_{\bar{x}} = \sigma/\sqrt{n} = .18/\sqrt{36} = .03$.

The standard error of $\bar{x}$ is the standard deviation of the sampling distribution of $\bar{x}$.

Because the sampling distribution of $\bar{x}$ is normally distributed, the sampling distribution of

$$z = \frac{\bar{x} - \mu_0}{\sigma_{\bar{x}}} = \frac{\bar{x} - 3}{.03}$$

---

[1] In constructing sampling distributions for hypothesis tests, it is assumed that H_0 is satisfied as an equality.

FIGURE 9.1 SAMPLING DISTRIBUTION OF $\bar{x}$ FOR THE HILLTOP COFFEE STUDY WHEN THE NULL HYPOTHESIS IS TRUE AS AN EQUALITY ($\mu = 3$)

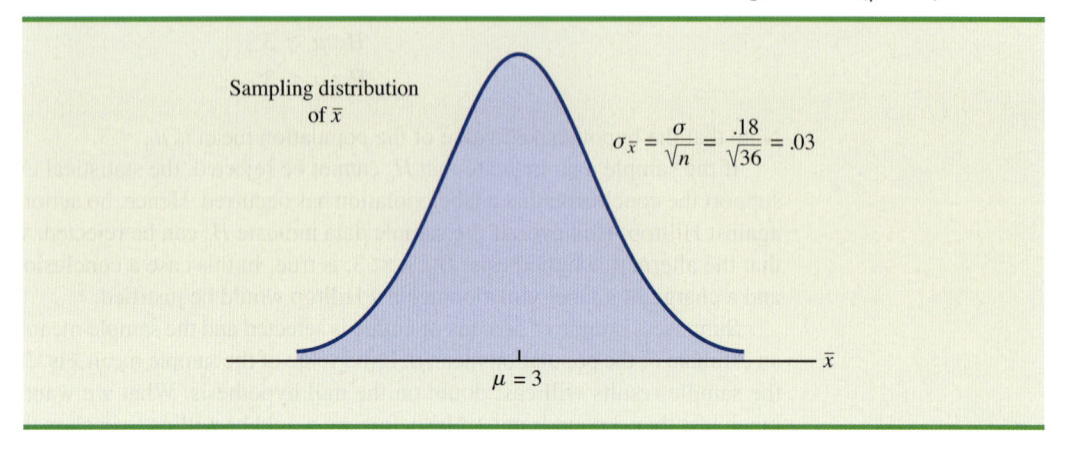

Sampling distribution of $\bar{x}$

$\sigma_{\bar{x}} = \dfrac{\sigma}{\sqrt{n}} = \dfrac{.18}{\sqrt{36}} = .03$

$\mu = 3$

is a standard normal distribution. A value of $z = -1$ means that the value of $\bar{x}$ is one standard error below the hypothesized value of the mean, a value of $z = -2$ means that the value of $\bar{x}$ is two standard errors below the hypothesized value of the mean, and so on. We can use the standard normal probability table to find the lower tail probability corresponding to any z value. For instance, the lower tail area at $z = -3.00$ is .0013. Hence, the probability of obtaining a value of z that is three or more standard errors below the mean is .0013. As a result, the probability of obtaining a value of $\bar{x}$ that is three or more standard errors below the hypothesized population mean $\mu_0 = 3$ is also .0013. Such a result is unlikely if the null hypothesis is true.

For hypothesis tests about a population mean in the σ known case, we use the standard normal random variable z as a **test statistic** to determine whether $\bar{x}$ deviates from the hypothesized value of μ enough to justify rejecting the null hypothesis. With $\sigma_{\bar{x}} = \sigma/\sqrt{n}$, the test statistic is as follows.

TEST STATISTIC FOR HYPOTHESIS TESTS ABOUT A POPULATION MEAN: σ KNOWN

$$z = \dfrac{\bar{x} - \mu_0}{\sigma/\sqrt{n}} \tag{9.1}$$

The key question for a lower tail test is, How small must the test statistic z be before we choose to reject the null hypothesis? Two approaches can be used to answer this question: the *p*-value approach and the critical value approach.

p-value approach The *p*-value approach uses the value of the test statistic z to compute a probability called a *p-value*.

A small p-value indicates that the value of the test statistic is unusual given the assumption that H_0 is true.

---

p-VALUE

A *p*-value is a probability that provides a measure of the evidence against the null hypothesis provided by the sample. Smaller *p*-values indicate more evidence against H_0.

The *p*-value is used to determine whether the null hypothesis should be rejected.

9.3 Population Mean: σ Known

Let us see how the *p*-value is computed and used. The value of the test statistic is used to compute the *p*-value. The method used depends on whether the test is a lower tail, an upper tail, or a two-tailed test. For a lower tail test, the *p*-value is the probability of obtaining a value for the test statistic as small as or smaller than that provided by the sample. Thus, to compute the *p*-value for the lower tail test in the σ known case, we must find using the standard normal distribution, the probability that *z* is less than or equal to the value of the test statistic. After computing the *p*-value, we must then decide whether it is small enough to reject the null hypothesis; as we will show, this decision involves comparing the *p*-value to the level of significance.

Let us now compute the *p*-value for the Hilltop Coffee lower tail test. Suppose the sample of 36 Hilltop coffee cans provides a sample mean of $\bar{x} = 2.92$ pounds. Is $\bar{x} = 2.92$ small enough to cause us to reject H_0? Because this is a lower tail test, the *p*-value is the area under the standard normal curve for values of $z \leq$ the value of the test statistic. Using $\bar{x} = 2.92$, $\sigma = .18$, and $n = 36$, we compute the value of the test statistic *z*.

$$z = \frac{\bar{x} - \mu_0}{\sigma/\sqrt{n}} = \frac{2.92 - 3}{.18/\sqrt{36}} = -2.67$$

Thus, the *p*-value is the probability that *z* is less than or equal to -2.67 (the lower tail area corresponding to the value of the test statistic).

Using the standard normal probability table, we find that the lower tail area at $z = -2.67$ is .0038. Figure 9.2 shows that $\bar{x} = 2.92$ corresponds to $z = -2.67$ and a *p*-value = .0038. This *p*-value indicates a small probability of obtaining a sample mean of $\bar{x} = 2.92$ (and a test statistic of -2.67) or smaller when sampling from a population with $\mu = 3$. This

FIGURE 9.2 *p*-VALUE FOR THE HILLTOP COFFEE STUDY WHEN $\bar{x} = 2.92$ AND $z = -2.67$

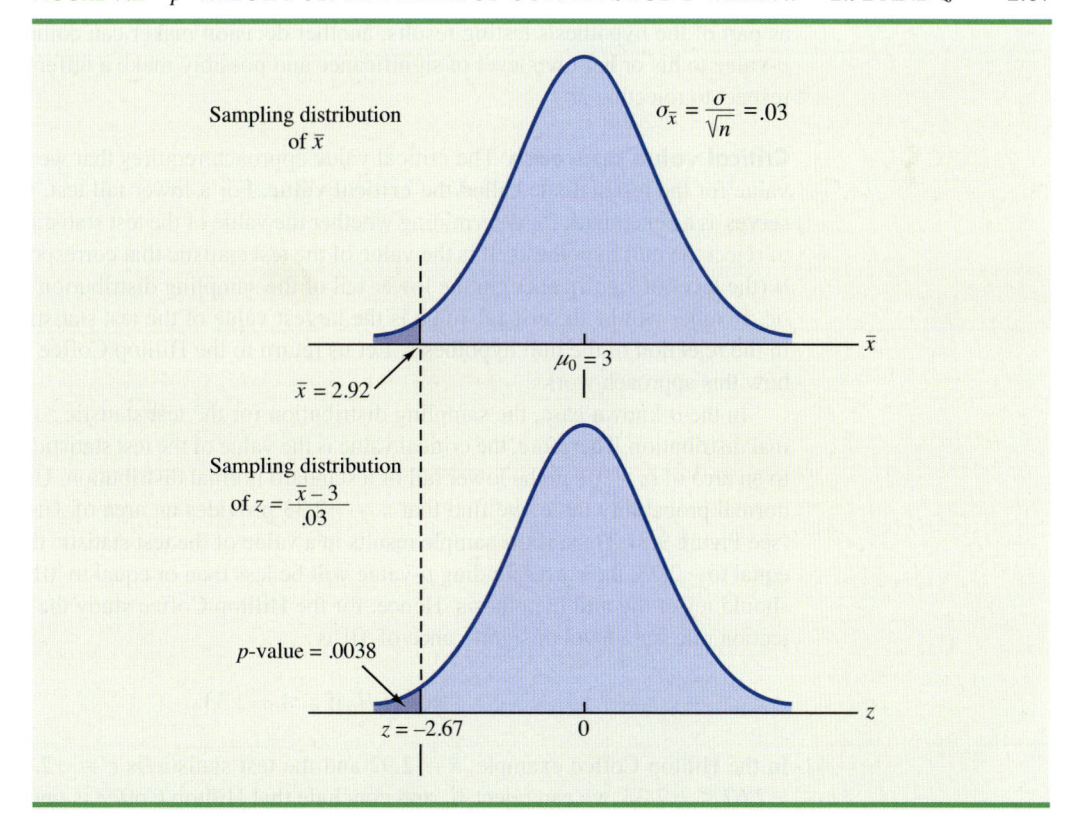

p-value does not provide much support for the null hypothesis, but is it small enough to cause us to reject H_0? The answer depends upon the level of significance for the test.

As noted previously, the director of the FTC's testing program selected a value of .01 for the level of significance. The selection of $\alpha = .01$ means that the director is willing to tolerate a probability of .01 of rejecting the null hypothesis when it is true as an equality ($\mu_0 = 3$). The sample of 36 coffee cans in the Hilltop Coffee study resulted in a p-value = .0038, which means that the probability of obtaining a value of $\bar{x} = 2.92$ or less when the null hypothesis is true as an equality is .0038. Because .0038 is less than or equal to $\alpha = .01$, we reject H_0. Therefore, we find sufficient statistical evidence to reject the null hypothesis at the .01 level of significance.

We can now state the general rule for determining whether the null hypothesis can be rejected when using the p-value approach. For a level of significance α, the rejection rule using the p-value approach is as follows.

> **REJECTION RULE USING p-VALUE**
>
> Reject H_0 if p-value $\leq \alpha$

In the Hilltop Coffee test, the p-value of .0038 resulted in the rejection of the null hypothesis. Although the basis for making the rejection decision involves a comparison of the p-value to the level of significance specified by the FTC director, the observed p-value of .0038 means that we would reject H_0 for any value of $\alpha \geq .0038$. For this reason, the p-value is also called the *observed level of significance*.

Different decision makers may express different opinions concerning the cost of making a Type I error and may choose a different level of significance. By providing the p-value as part of the hypothesis testing results, another decision maker can compare the reported p-value to his or her own level of significance and possibly make a different decision with respect to rejecting H_0.

Critical value approach The critical value approach requires that we first determine a value for the test statistic called the **critical value**. For a lower tail test, the critical value serves as a benchmark for determining whether the value of the test statistic is small enough to reject the null hypothesis. It is the value of the test statistic that corresponds to an area of α (the level of significance) in the lower tail of the sampling distribution of the test statistic. In other words, the critical value is the largest value of the test statistic that will result in the rejection of the null hypothesis. Let us return to the Hilltop Coffee example and see how this approach works.

In the σ known case, the sampling distribution for the test statistic z is a standard normal distribution. Therefore, the critical value is the value of the test statistic that corresponds to an area of $\alpha = .01$ in the lower tail of a standard normal distribution. Using the standard normal probability table, we find that $z = -2.33$ provides an area of .01 in the lower tail (see Figure 9.3). Thus, if the sample results in a value of the test statistic that is less than or equal to -2.33, the corresponding p-value will be less than or equal to .01; in this case, we should reject the null hypothesis. Hence, for the Hilltop Coffee study the critical value rejection rule for a level of significance of .01 is

$$\text{Reject } H_0 \text{ if } z \leq -2.33$$

In the Hilltop Coffee example, $\bar{x} = 2.92$ and the test statistic is $z = -2.67$. Because $z = -2.67 < -2.33$, we can reject H_0 and conclude that Hilltop Coffee is underfilling cans.

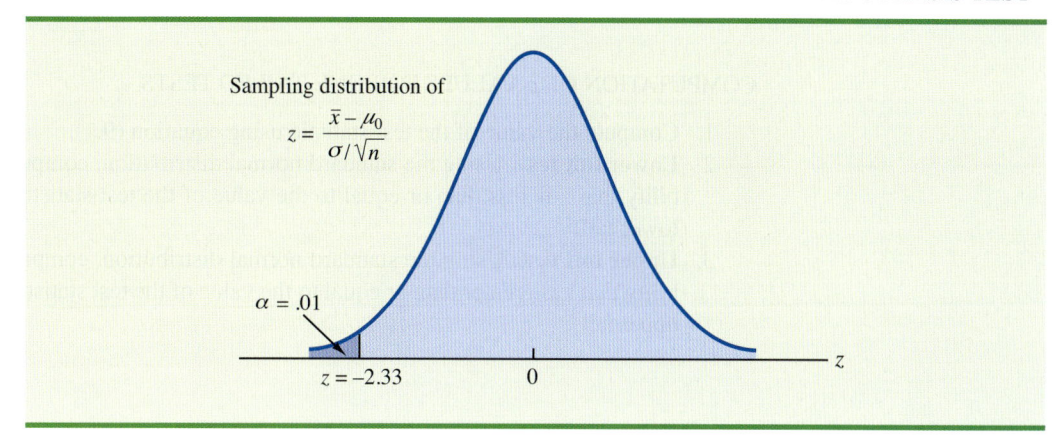

FIGURE 9.3 CRITICAL VALUE = −2.33 FOR THE HILLTOP COFFEE HYPOTHESIS TEST

Sampling distribution of
$$z = \frac{\bar{x} - \mu_0}{\sigma/\sqrt{n}}$$

$\alpha = .01$

$z = -2.33 \qquad 0$

We can generalize the rejection rule for the critical value approach to handle any level of significance. The rejection rule for a lower tail test follows.

REJECTION RULE FOR A LOWER TAIL TEST: CRITICAL VALUE APPROACH

$$\text{Reject } H_0 \text{ if } z \leq -z_\alpha$$

where $-z_\alpha$ is the critical value; that is, the z value that provides an area of α in the lower tail of the standard normal distribution.

Summary The p-value approach to hypothesis testing and the critical value approach will always lead to the same rejection decision; that is, whenever the p-value is less than or equal to α, the value of the test statistic will be less than or equal to the critical value. The advantage of the p-value approach is that the p-value tells us *how* significant the results are (the observed level of significance). If we use the critical value approach, we only know that the results are significant at the stated level of significance.

At the beginning of this section, we said that one-tailed tests about a population mean take one of the following two forms:

Lower Tail Test	Upper Tail Test
$H_0: \mu \geq \mu_0$	$H_0: \mu \leq \mu_0$
$H_a: \mu < \mu_0$	$H_a: \mu > \mu_0$

We used the Hilltop Coffee study to illustrate how to conduct a lower tail test. We can use the same general approach to conduct an upper tail test. The test statistic z is still computed using equation (9.1). But, for an upper tail test, the p-value is the probability of obtaining a value for the test statistic as large as or larger than that provided by the sample. Thus, to compute the p-value for the upper tail test in the σ known case, we must find using the standard normal distribution, the probability that z is greater than or equal to the value of the test statistic. Using the critical value approach causes us to reject the null hypothesis if the value of the test statistic is greater than or equal to the critical value z_α; in other words, we reject H_0 if $z \geq z_\alpha$.

Let us summarize the steps involved in computing *p*-values for one-tailed hypothesis tests.

> COMPUTATION OF *p*-VALUES FOR ONE-TAILED TESTS
>
> 1. Compute the value of the test statistic using equation (9.1).
> 2. **Lower tail test:** Using the standard normal distribution, compute the probability that z is less than or equal to the value of the test statistic (area in the lower tail).
> 3. **Upper tail test:** Using the standard normal distribution, compute the probability that z is greater than or equal to the value of the test statistic (area in the upper tail).

Two-Tailed Test

In hypothesis testing, the general form for a **two-tailed test** about a population mean is as follows:

$$H_0: \mu = \mu_0$$
$$H_a: \mu \neq \mu_0$$

In this subsection we show how to conduct a two-tailed test about a population mean for the σ known case. As an illustration, we consider the hypothesis testing situation facing MaxFlight, Inc.

The U.S. Golf Association (USGA) establishes rules that manufacturers of golf equipment must meet if their products are to be acceptable for use in USGA events. MaxFlight uses a high-technology manufacturing process to produce golf balls with a mean driving distance of 295 yards. Sometimes, however, the process gets out of adjustment and produces golf balls with a mean driving distance different from 295 yards. When the mean distance falls below 295 yards, the company worries about losing sales because the golf balls do not provide as much distance as advertised. When the mean distance passes 295 yards, MaxFlight's golf balls may be rejected by the USGA for exceeding the overall distance standard concerning carry and roll.

MaxFlight's quality control program involves taking periodic samples of 50 golf balls to monitor the manufacturing process. For each sample, a hypothesis test is conducted to determine whether the process has fallen out of adjustment. Let us develop the null and alternative hypotheses. We begin by assuming that the process is functioning correctly; that is, the golf balls being produced have a mean distance of 295 yards. This assumption establishes the null hypothesis. The alternative hypothesis is that the mean distance is not equal to 295 yards. With a hypothesized value of $\mu_0 = 295$, the null and alternative hypotheses for the MaxFlight hypothesis test are as follows:

$$H_0: \mu = 295$$
$$H_a: \mu \neq 295$$

If the sample mean $\bar{x}$ is significantly less than 295 yards or significantly greater than 295 yards, we will reject H_0. In this case, corrective action will be taken to adjust the manufacturing process. On the other hand, if $\bar{x}$ does not deviate from the hypothesized mean $\mu_0 = 295$ by a significant amount, H_0 will not be rejected and no action will be taken to adjust the manufacturing process.

The quality control team selected $\alpha = .05$ as the level of significance for the test. Data from previous tests conducted when the process was known to be in adjustment show that

9.3 Population Mean: σ Known

FIGURE 9.4 SAMPLING DISTRIBUTION OF $\bar{x}$ FOR THE MAXFLIGHT HYPOTHESIS TEST

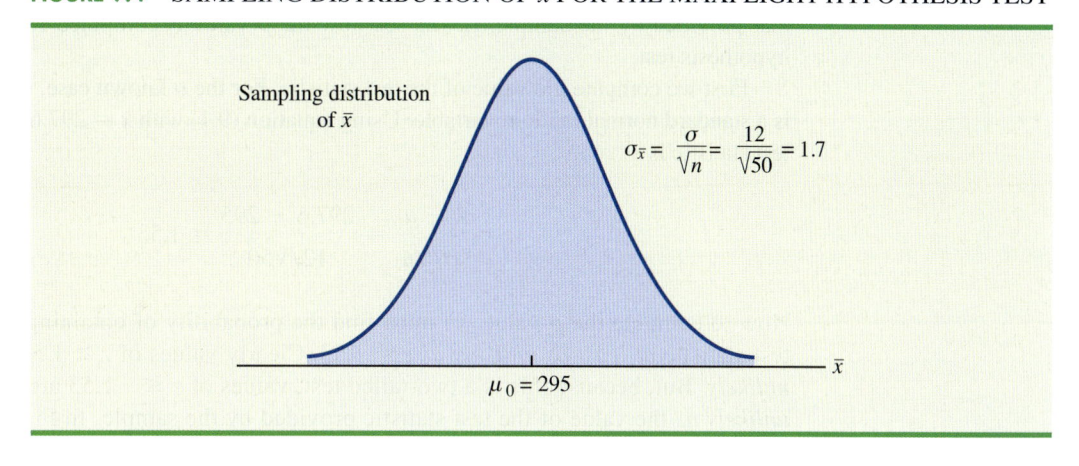

Sampling distribution of $\bar{x}$

$\sigma_{\bar{x}} = \dfrac{\sigma}{\sqrt{n}} = \dfrac{12}{\sqrt{50}} = 1.7$

$\mu_0 = 295$

the population standard deviation can be assumed known with a value of $\sigma = 12$. Thus, with a sample size of $n = 50$, the standard error of $\bar{x}$ is

$$\sigma_{\bar{x}} = \dfrac{\sigma}{\sqrt{n}} = \dfrac{12}{\sqrt{50}} = 1.7$$

Because the sample size is large, the central limit theorem (see Chapter 7) allows us to conclude that the sampling distribution of $\bar{x}$ can be approximated by a normal distribution. Figure 9.4 shows the sampling distribution of $\bar{x}$ for the MaxFlight hypothesis test with a hypothesized population mean of $\mu_0 = 295$.

Suppose that a sample of 50 golf balls is selected and that the sample mean is $\bar{x} = 297.6$ yards. This sample mean provides support for the conclusion that the population mean is larger than 295 yards. Is this value of $\bar{x}$ enough larger than 295 to cause us to reject H_0 at the .05 level of significance? In the previous section we described two approaches that can be used to answer this question: the *p*-value approach and the critical value approach.

WEB file
GolfTest

***p*-value approach** Recall that the *p*-value is a probability used to determine whether the null hypothesis should be rejected. For a two-tailed test, values of the test statistic in *either* tail provide evidence against the null hypothesis. For a two-tailed test, the *p*-value is

FIGURE 9.5 *p*-VALUE FOR THE MAXFLIGHT HYPOTHESIS TEST

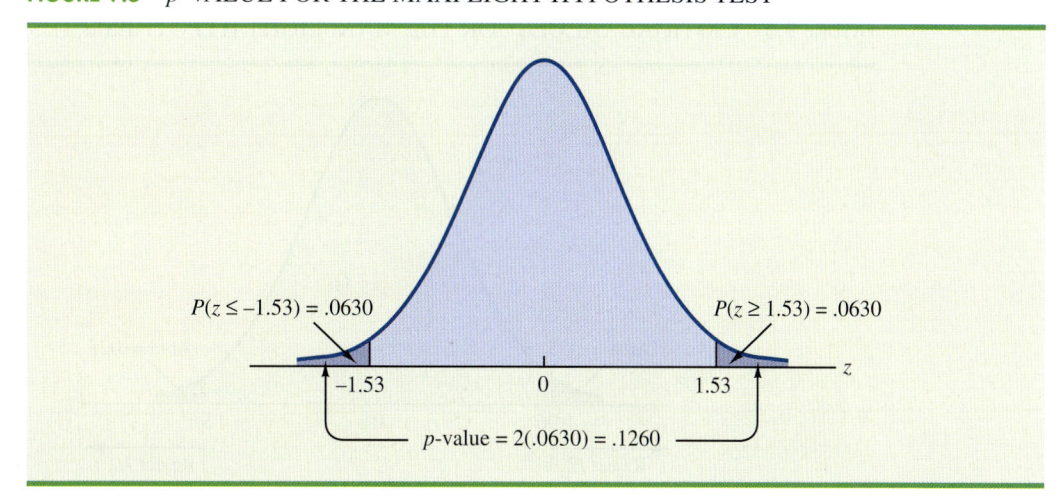

$P(z \leq -1.53) = .0630$ $P(z \geq 1.53) = .0630$

-1.53 0 1.53

p-value = 2(.0630) = .1260

the probability of obtaining a value for the test statistic *as unlikely as or more unlikely than* that provided by the sample. Let us see how the *p*-value is computed for the MaxFlight hypothesis test.

First we compute the value of the test statistic. For the σ known case, the test statistic z is a standard normal random variable. Using equation (9.1) with $\bar{x} = 297.6$, the value of the test statistic is

$$z = \frac{\bar{x} - \mu_0}{\sigma/\sqrt{n}} = \frac{297.6 - 295}{12/\sqrt{50}} = 1.53$$

Now to compute the *p*-value we must find the probability of obtaining a value for the test statistic *at least as unlikely as* $z = 1.53$. Clearly values of $z \geq 1.53$ are *at least as unlikely*. But, because this is a two-tailed test, values of $z \leq -1.53$ are also *at least as unlikely as* the value of the test statistic provided by the sample. In Figure 9.5, we see that the two-tailed *p*-value in this case is given by $P(z \leq -1.53) + P(z \geq 1.53)$. Because the normal curve is symmetric, we can compute this probability by finding the upper tail area at $z = 1.53$ and doubling it. The table for the standard normal distribution shows that $p(z < 1.53)$ is .9370. Thus, $p(z \geq 1.53)$ is $1.0000 - .9370 = .0630$. Doubling this, we find the *p*-value for the MaxFlight two-tailed hypothesis test is *p*-value = 2(.0630) = .1260.

Next we compare the *p*-value to the level of significance to see whether the null hypothesis should be rejected. With a level of significance of $\alpha = .05$, we do not reject H_0 because the *p*-value = .1260 > .05. Because the null hypothesis is not rejected, no action will be taken to adjust the MaxFlight manufacturing process.

Let us summarize the steps involved in computing *p*-values for two-tailed hypothesis tests.

COMPUTATION OF *p*-VALUES FOR TWO-TAILED TESTS

1. Compute the value of the test statistic using equation (9.1).
2. If the value of the test statistic is in the upper tail, compute the probability that *z* is greater than or equal to the value of the test statistic (the upper tail area). If the value of the test statistic is in the lower tail, compute the probability that *z* is less than or equal to the value of the test statistic (the lower tail area).
3. Double the probability (or tail area) from step 2 to obtain the *p*-value.

FIGURE 9.6   CRITICAL VALUES FOR THE MAXFLIGHT HYPOTHESIS TEST

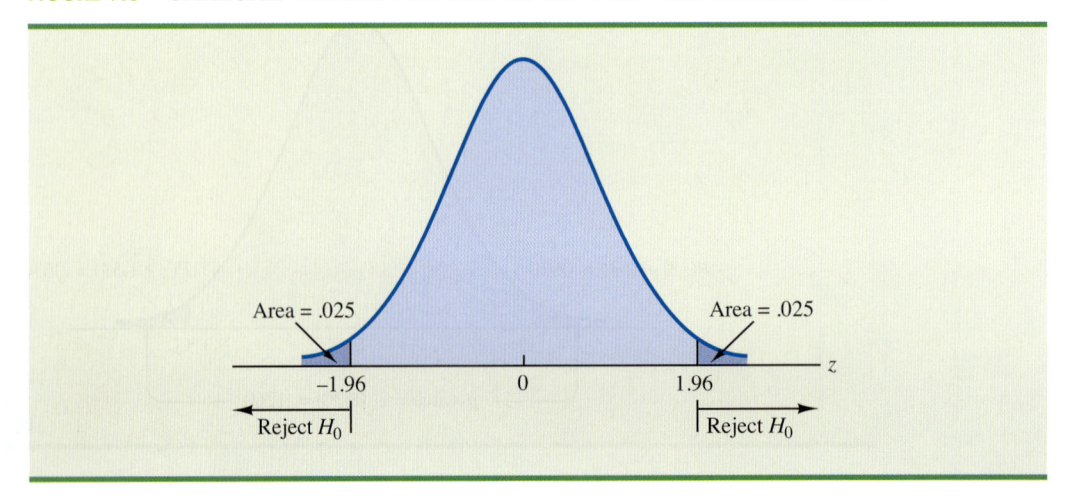

Critical value approach Before leaving this section, let us see how the test statistic z can be compared to a critical value to make the hypothesis testing decision for a two-tailed test. Figure 9.6 shows that the critical values for the test will occur in both the lower and upper tails of the standard normal distribution. With a level of significance of $\alpha = .05$, the area in each tail corresponding to the critical values is $\alpha/2 = .05/2 = .025$. Using the standard normal probability table, we find the critical values for the test statistic are $-z_{.025} = -1.96$ and $z_{.025} = 1.96$. Thus, using the critical value approach, the two-tailed rejection rule is

$$\text{Reject } H_0 \text{ if } z \leq -1.96 \text{ or if } z \geq 1.96$$

Because the value of the test statistic for the MaxFlight study is $z = 1.53$, the statistical evidence will not permit us to reject the null hypothesis at the .05 level of significance.

Summary and Practical Advice

We presented examples of a lower tail test and a two-tailed test about a population mean. Based upon these examples, we can now summarize the hypothesis testing procedures about a population mean for the σ known case as shown in Table 9.2. Note that μ_0 is the hypothesized value of the population mean.

The hypothesis testing steps followed in the two examples presented in this section are common to every hypothesis test.

STEPS OF HYPOTHESIS TESTING

Step 1. Develop the null and alternative hypotheses.
Step 2. Specify the level of significance.
Step 3. Collect the sample data and compute the value of the test statistic.

p-Value Approach

Step 4. Use the value of the test statistic to compute the *p*-value.
Step 5. Reject H_0 if the *p*-value $\leq \alpha$.

Critical Value Approach

Step 4. Use the level of significance to determine the critical value and the rejection rule.
Step 5. Use the value of the test statistic and the rejection rule to determine whether to reject H_0.

TABLE 9.2 SUMMARY OF HYPOTHESIS TESTS ABOUT A POPULATION MEAN: σ KNOWN CASE

	Lower Tail Test	Upper Tail Test	Two-Tailed Test
Hypotheses	$H_0: \mu \geq \mu_0$ $H_a: \mu < \mu_0$	$H_0: \mu \leq \mu_0$ $H_a: \mu > \mu_0$	$H_0: \mu = \mu_0$ $H_a: \mu \neq \mu_0$
Test Statistic	$z = \dfrac{\bar{x} - \mu_0}{\sigma/\sqrt{n}}$	$z = \dfrac{\bar{x} - \mu_0}{\sigma/\sqrt{n}}$	$z = \dfrac{\bar{x} - \mu_0}{\sigma/\sqrt{n}}$
Rejection Rule: *p*-Value Approach	Reject H_0 if *p*-value $\leq \alpha$	Reject H_0 if *p*-value $\leq \alpha$	Reject H_0 if *p*-value $\leq \alpha$
Rejection Rule: Critical Value Approach	Reject H_0 if $z \leq -z_\alpha$	Reject H_0 if $z \geq z_\alpha$	Reject H_0 if $z \leq -z_{\alpha/2}$ or if $z \geq z_{\alpha/2}$

Practical advice about the sample size for hypothesis tests is similar to the advice we provided about the sample size for interval estimation in Chapter 8. In most applications, a sample size of $n \geq 30$ is adequate when using the hypothesis testing procedure described in this section. In cases where the sample size is less than 30, the distribution of the population from which we are sampling becomes an important consideration. If the population is normally distributed, the hypothesis testing procedure that we described is exact and can be used for any sample size. If the population is not normally distributed but is at least roughly symmetric, sample sizes as small as 15 can be expected to provide acceptable results.

Relationship Between Interval Estimation and Hypothesis Testing

In Chapter 8 we showed how to develop a confidence interval estimate of a population mean. For the σ known case, the $(1 - \alpha)\%$ confidence interval estimate of a population mean is given by

$$\bar{x} \pm z_{\alpha/2} \frac{\sigma}{\sqrt{n}}$$

In this section, we showed that a two-tailed hypothesis test about a population mean takes the following form:

$$H_0: \mu = \mu_0$$
$$H_a: \mu \neq \mu_0$$

where μ_0 is the hypothesized value for the population mean.

Suppose that we follow the procedure described in Chapter 8 for constructing a $100(1 - \alpha)\%$ confidence interval for the population mean. We know that $100(1 - \alpha)\%$ of the confidence intervals generated will contain the population mean and $100\alpha\%$ of the confidence intervals generated will not contain the population mean. Thus, if we reject H_0 whenever the confidence interval does not contain μ_0, we will be rejecting the null hypothesis when it is true ($\mu = \mu_0$) with probability α. Recall that the level of significance is the probability of rejecting the null hypothesis when it is true. So constructing a $100(1 - \alpha)\%$ confidence interval and rejecting H_0 whenever the interval does not contain μ_0 is equivalent to conducting a two-tailed hypothesis test with α as the level of significance. The procedure for using a confidence interval to conduct a two-tailed hypothesis test can now be summarized.

A CONFIDENCE INTERVAL APPROACH TO TESTING A HYPOTHESIS OF THE FORM

$$H_0: \mu = \mu_0$$
$$H_a: \mu \neq \mu_0$$

1. Select a simple random sample from the population and use the value of the sample mean $\bar{x}$ to develop the confidence interval for the population mean μ.

$$\bar{x} \pm z_{\alpha/2} \frac{\sigma}{\sqrt{n}}$$

2. If the confidence interval contains the hypothesized value μ_0, do not reject H_0. Otherwise, reject[2] H_0.

For a two-tailed hypothesis test, the null hypothesis can be rejected if the confidence interval does not include μ_0.

---

[2]To be consistent with the rule for rejecting H_0 when the p-value $\leq \alpha$, we would also reject H_0 using the confidence interval approach if μ_0 happens to be equal to one of the end points of the $100(1 - \alpha)\%$ confidence interval.

Let us illustrate by conducting the MaxFlight hypothesis test using the confidence interval approach. The MaxFlight hypothesis test takes the following form:

$$H_0: \mu = 295$$
$$H_a: \mu \neq 295$$

To test these hypotheses with a level of significance of $\alpha = .05$, we sampled 50 golf balls and found a sample mean distance of $\bar{x} = 297.6$ yards. Recall that the population standard deviation is $\sigma = 12$. Using these results with $z_{.025} = 1.96$, we find that the 95% confidence interval estimate of the population mean is

$$\bar{x} \pm z_{.025} \frac{\sigma}{\sqrt{n}}$$

$$297.6 \pm 1.96 \frac{12}{\sqrt{50}}$$

$$297.6 \pm 3.3$$

or

$$294.3 \text{ to } 300.9$$

This finding enables the quality control manager to conclude with 95% confidence that the mean distance for the population of golf balls is between 294.3 and 300.9 yards. Because the hypothesized value for the population mean, $\mu_0 = 295$, is in this interval, the hypothesis testing conclusion is that the null hypothesis, $H_0: \mu = 295$, cannot be rejected.

Note that this discussion and example pertain to two-tailed hypothesis tests about a population mean. However, the same confidence interval and two-tailed hypothesis testing relationship exists for other population parameters. The relationship can also be extended to one-tailed tests about population parameters. Doing so, however, requires the development of one-sided confidence intervals, which are rarely used in practice.

NOTES AND COMMENTS

We have shown how to use *p*-values. The smaller the *p*-value the greater the evidence against H_0 and the more the evidence in favor of H_a. Here are some guidelines statisticians suggest for interpreting small *p*-values.

- Less than .01—Overwhelming evidence to conclude H_a is true.
- Between .01 and .05—Strong evidence to conclude H_a is true.
- Between .05 and .10—Weak evidence to conclude H_a is true.
- Greater than .10—Insufficient evidence to conclude H_a is true.

Exercises

Note to Student: Some of the exercises that follow ask you to use the *p*-value approach and others ask you to use the critical value approach. Both methods will provide the same hypothesis testing conclusion. We provide exercises with both methods to give you practice using both. In later sections and in following chapters, we will generally emphasize the *p*-value approach as the preferred method, but you may select either based on personal preference.

Methods

9. Consider the following hypothesis test:

$$H_0: \mu \geq 20$$
$$H_a: \mu < 20$$

A sample of 50 provided a sample mean of 19.4. The population standard deviation is 2.
   a. Compute the value of the test statistic.
   b. What is the *p*-value?
   c. Using $\alpha = .05$, what is your conclusion?
   d. What is the rejection rule using the critical value? What is your conclusion?

10. **SELF test** Consider the following hypothesis test:

$$H_0: \mu \leq 25$$
$$H_a: \mu > 25$$

A sample of 40 provided a sample mean of 26.4. The population standard deviation is 6.
   a. Compute the value of the test statistic.
   b. What is the *p*-value?
   c. At $\alpha = .01$, what is your conclusion?
   d. What is the rejection rule using the critical value? What is your conclusion?

11. **SELF test** Consider the following hypothesis test:

$$H_0: \mu = 15$$
$$H_a: \mu \neq 15$$

A sample of 50 provided a sample mean of 14.15. The population standard deviation is 3.
   a. Compute the value of the test statistic.
   b. What is the *p*-value?
   c. At $\alpha = .05$, what is your conclusion?
   d. What is the rejection rule using the critical value? What is your conclusion?

12. Consider the following hypothesis test:

$$H_0: \mu \geq 80$$
$$H_a: \mu < 80$$

A sample of 100 is used and the population standard deviation is 12. Compute the *p*-value and state your conclusion for each of the following sample results. Use $\alpha = .01$.
   a. $\bar{x} = 78.5$
   b. $\bar{x} = 77$
   c. $\bar{x} = 75.5$
   d. $\bar{x} = 81$

13. Consider the following hypothesis test:

$$H_0: \mu \leq 50$$
$$H_a: \mu > 50$$

A sample of 60 is used and the population standard deviation is 8. Use the critical value approach to state your conclusion for each of the following sample results. Use $\alpha = .05$.
   a. $\bar{x} = 52.5$
   b. $\bar{x} = 51$
   c. $\bar{x} = 51.8$

14. Consider the following hypothesis test:

$$H_0: \mu = 22$$
$$H_a: \mu \neq 22$$

A sample of 75 is used and the population standard deviation is 10. Compute the *p*-value and state your conclusion for each of the following sample results. Use $\alpha = .01$.
   a. $\bar{x} = 23$
   b. $\bar{x} = 25.1$
   c. $\bar{x} = 20$

Applications

15. Individuals filing federal income tax returns prior to March 31 received an average refund of $1056. Consider the population of "last-minute" filers who mail their tax return during the last five days of the income tax period (typically April 10 to April 15).
    a. A researcher suggests that a reason individuals wait until the last five days is that on average these individuals receive lower refunds than do early filers. Develop appropriate hypotheses such that rejection of H_0 will support the researcher's contention.
    b. For a sample of 400 individuals who filed a tax return between April 10 and 15, the sample mean refund was $910. Based on prior experience a population standard deviation of $\sigma = \$1600$ may be assumed. What is the *p*-value?
    c. At $\alpha = .05$, what is your conclusion?
    d. Repeat the preceding hypothesis test using the critical value approach.

16. In a study entitled How Undergraduate Students Use Credit Cards, it was reported that undergraduate students have a mean credit card balance of $3173 (*Sallie Mae*, April 2009). This figure was an all-time high and had increased 44% over the previous five years. Assume that a current study is being conducted to determine if it can be concluded that the mean credit card balance for undergraduate students has continued to increase compared to the April 2009 report. Based on previous studies, use a population standard deviation $\sigma = \$1000$.
    a. State the null and alternative hypotheses.
    b. What is the *p*-value for a sample of 180 undergraduate students with a sample mean credit card balance of $3325?
    c. Using a .05 level of significance, what is your conclusion?

17. Wall Street securities firms paid out record year-end bonuses of $125,500 per employee for 2005 (*Fortune*, February 6, 2006). Suppose we would like to take a sample of employees at the Jones & Ryan securities firm to see whether the mean year-end bonus is different from the reported mean of $125,500 for the population.
    a. State the null and alternative hypotheses you would use to test whether the year-end bonuses paid by Jones & Ryan were different from the population mean.
    b. Suppose a sample of 40 Jones & Ryan employees showed a sample mean year-end bonus of $118,000. Assume a population standard deviation of $\sigma = \$30,000$ and compute the *p*-value.
    c. With $\alpha = .05$ as the level of significance, what is your conclusion?
    d. Repeat the preceding hypothesis test using the critical value approach.

18. The average annual total return for U.S. Diversified Equity mutual funds from 1999 to 2003 was 4.1% (*BusinessWeek*, January 26, 2004). A researcher would like to conduct a hypothesis test to see whether the returns for mid-cap growth funds over the same period are significantly different from the average for U.S. Diversified Equity funds.

a. Formulate the hypotheses that can be used to determine whether the mean annual return for mid-cap growth funds differ from the mean for U.S. Diversified Equity funds.
b. A sample of 40 mid-cap growth funds provides a mean return of $\bar{x} = 3.4\%$. Assume the population standard deviation for mid-cap growth funds is known from previous studies to be $\sigma = 2\%$. Use the sample results to compute the test statistic and p-value for the hypothesis test.
c. At $\alpha = .05$, what is your conclusion?

19. The Internal Revenue Service (IRS) provides a toll-free help line for taxpayers to call in and get answers to questions as they prepare their tax returns. In recent years, the IRS has been inundated with taxpayer calls and has redesigned its phone service as well as posted answers to frequently asked questions on its website (*The Cincinnati Enquirer*, January 7, 2010). According to a report by a taxpayer advocate, callers using the new system can expect to wait on hold for an unreasonably long time of 12 minutes before being able to talk to an IRS employee. Suppose you select a sample of 50 callers after the new phone service has been implemented; the sample results show a mean waiting time of 10 minutes before an IRS employee comes on the line. Based upon data from past years, you decide it is reasonable to assume that the standard deviation of waiting time is 8 minutes. Using your sample results, can you conclude that the actual mean waiting time turned out to be significantly less than the 12-minute claim made by the taxpayer advocate? Use $\alpha = .05$.

20. For the United States, the mean monthly Internet bill is $32.79 per household (CNBC, January 18, 2006). A sample of 50 households in a southern state showed a sample mean of $30.63. Use a population standard deviation of $\sigma = \$5.60$.
    a. Formulate hypotheses for a test to determine whether the sample data support the conclusion that the mean monthly Internet bill in the southern state is less than the national mean of $32.79.
    b. What is the value of the test statistic?
    c. What is the p-value?
    d. At $\alpha = .01$, what is your conclusion?

21. Fowle Marketing Research, Inc., bases charges to a client on the assumption that telephone surveys can be completed in a mean time of 15 minutes or less. If a longer mean survey time is necessary, a premium rate is charged. A sample of 35 surveys provided the survey times shown in the file named Fowle. Based upon past studies, the population standard deviation is assumed known with $\sigma = 4$ minutes. Is the premium rate justified?
    a. Formulate the null and alternative hypotheses for this application.
    b. Compute the value of the test statistic.
    c. What is the p-value?
    d. At $\alpha = .01$, what is your conclusion?

22. CCN and ActMedia provided a television channel targeted to individuals waiting in supermarket checkout lines. The channel showed news, short features, and advertisements. The length of the program was based on the assumption that the population mean time a shopper stands in a supermarket checkout line is 8 minutes. A sample of actual waiting times will be used to test this assumption and determine whether actual mean waiting time differs from this standard.
    a. Formulate the hypotheses for this application.
    b. A sample of 120 shoppers showed a sample mean waiting time of 8.4 minutes. Assume a population standard deviation of $\sigma = 3.2$ minutes. What is the p-value?
    c. At $\alpha = .05$, what is your conclusion?
    d. Compute a 95% confidence interval for the population mean. Does it support your conclusion?

9.4 Population Mean: σ Unknown

In this section we describe how to conduct hypothesis tests about a population mean for the σ unknown case. Because the σ unknown case corresponds to situations in which an estimate of the population standard deviation cannot be developed prior to sampling, the sample must be used to develop an estimate of both μ and σ. Thus, to conduct a hypothesis test about a population mean for the σ unknown case, the sample mean $\bar{x}$ is used as an estimate of μ and the sample standard deviation s is used as an estimate of σ.

The steps of the hypothesis testing procedure for the σ unknown case are the same as those for the σ known case described in Section 9.3. But, with σ unknown, the computation of the test statistic and p-value is a bit different. Recall that for the σ known case, the sampling distribution of the test statistic has a standard normal distribution. For the σ unknown case, however, the sampling distribution of the test statistic follows the t distribution; it has slightly more variability because the sample is used to develop estimates of both μ and σ.

In Section 8.2 we showed that an interval estimate of a population mean for the σ unknown case is based on a probability distribution known as the t distribution. Hypothesis tests about a population mean for the σ unknown case are also based on the t distribution. For the σ unknown case, the test statistic has a t distribution with $n - 1$ degrees of freedom.

TEST STATISTIC FOR HYPOTHESIS TESTS ABOUT A POPULATION MEAN: σ UNKNOWN

$$t = \frac{\bar{x} - \mu_0}{s/\sqrt{n}} \tag{9.2}$$

In Chapter 8 we said that the t distribution is based on an assumption that the population from which we are sampling has a normal distribution. However, research shows that this assumption can be relaxed considerably when the sample size is large enough. We provide some practical advice concerning the population distribution and sample size at the end of the section.

One-Tailed Test

Let us consider an example of a one-tailed test about a population mean for the σ unknown case. A business travel magazine wants to classify transatlantic gateway airports according to the mean rating for the population of business travelers. A rating scale with a low score of 0 and a high score of 10 will be used, and airports with a population mean rating greater than 7 will be designated as superior service airports. The magazine staff surveyed a sample of 60 business travelers at each airport to obtain the ratings data. The sample for London's Heathrow Airport provided a sample mean rating of $\bar{x} = 7.25$ and a sample standard deviation of $s = 1.052$. Do the data indicate that Heathrow should be designated as a superior service airport?

We want to develop a hypothesis test for which the decision to reject H_0 will lead to the conclusion that the population mean rating for the Heathrow Airport is *greater* than 7. Thus, an upper tail test with $H_a: \mu > 7$ is required. The null and alternative hypotheses for this upper tail test are as follows:

$$H_0: \mu \leq 7$$
$$H_a: \mu > 7$$

We will use $\alpha = .05$ as the level of significance for the test.

Using equation (9.2) with $\bar{x} = 7.25$, $\mu_0 = 7$, $s = 1.052$, and $n = 60$, the value of the test statistic is

$$t = \frac{\bar{x} - \mu_0}{s/\sqrt{n}} = \frac{7.25 - 7}{1.052/\sqrt{60}} = 1.84$$

The sampling distribution of t has $n - 1 = 60 - 1 = 59$ degrees of freedom. Because the test is an upper tail test, the p-value is $P(t \geq 1.84)$; that is, the upper tail area corresponding to the value of the test statistic.

The t distribution table provided in most textbooks will not contain sufficient detail to determine the exact p-value, such as the p-value corresponding to $t = 1.84$. For instance, using Table 2 in Appendix B, the t distribution with 59 degrees of freedom provides the following information.

Area in Upper Tail	.20	.10	.05	.025	.01	.005
t Value (59 df)	.848	1.296	1.671	2.001	2.391	2.662

$t = 1.84$

We see that $t = 1.84$ is between 1.671 and 2.001. Although the table does not provide the exact p-value, the values in the "Area in Upper Tail" row show that the p-value must be less than .05 and greater than .025. With a level of significance of $\alpha = .05$, this placement is all we need to know to make the decision to reject the null hypothesis and conclude that Heathrow should be classified as a superior service airport.

Appendix F shows how to compute p-values using Excel or Minitab.

Because it is cumbersome to use a t table to compute p-values, and only approximate values are obtained, we show how to compute the exact p-value using Excel or Minitab. The directions can be found in Appendix F at the end of this text. Using Excel or Minitab with $t = 1.84$ provides the upper tail p-value of .0354 for the Heathrow Airport hypothesis test. With $.0354 < .05$, we reject the null hypothesis and conclude that Heathrow should be classified as a superior service airport.

Two-Tailed Test

To illustrate how to conduct a two-tailed test about a population mean for the σ unknown case, let us consider the hypothesis testing situation facing Holiday Toys. The company manufactures and distributes its products through more than 1000 retail outlets. In planning production levels for the coming winter season, Holiday must decide how many units of each product to produce prior to knowing the actual demand at the retail level. For this year's most important new toy, Holiday's marketing director is expecting demand to average 40 units per retail outlet. Prior to making the final production decision based upon this estimate, Holiday decided to survey a sample of 25 retailers in order to develop more information about the demand for the new product. Each retailer was provided with information about the features of the new toy along with the cost and the suggested selling price. Then each retailer was asked to specify an anticipated order quantity.

With μ denoting the population mean order quantity per retail outlet, the sample data will be used to conduct the following two-tailed hypothesis test:

$$H_0: \mu = 40$$
$$H_a: \mu \neq 40$$

9.4 Population Mean: σ Unknown

If H_0 cannot be rejected, Holiday will continue its production planning based on the marketing director's estimate that the population mean order quantity per retail outlet will be $\mu = 40$ units. However, if H_0 is rejected, Holiday will immediately reevaluate its production plan for the product. A two-tailed hypothesis test is used because Holiday wants to reevaluate the production plan if the population mean quantity per retail outlet is less than anticipated or greater than anticipated. Because no historical data are available (it's a new product), the population mean μ and the population standard deviation must both be estimated using $\bar{x}$ and s from the sample data.

The sample of 25 retailers provided a mean of $\bar{x} = 37.4$ and a standard deviation of $s = 11.79$ units. Before going ahead with the use of the t distribution, the analyst constructed a histogram of the sample data in order to check on the form of the population distribution. The histogram of the sample data showed no evidence of skewness or any extreme outliers, so the analyst concluded that the use of the t distribution with $n - 1 = 24$ degrees of freedom was appropriate. Using equation (9.2) with $\bar{x} = 37.4$, $\mu_0 = 40$, $s = 11.79$, and $n = 25$, the value of the test statistic is

$$t = \frac{\bar{x} - \mu_0}{s/\sqrt{n}} = \frac{37.4 - 40}{11.79/\sqrt{25}} = -1.10$$

Because we have a two-tailed test, the p-value is two times the area under the curve of the t distribution for $t \leq -1.10$. Using Table 2 in Appendix B, we see that the t distribution table for 24 degrees of freedom provides the following information.

Area in Upper Tail	.20	.10	.05	.025	.01	.005
t-Value (24 df)	.857	1.318	1.711	2.064	2.492	2.797

$t = 1.10$

The t distribution table contains only positive t values. Because the t distribution is symmetric, however, the upper tail area at $t = 1.10$ is the same as the lower tail area at $t = -1.10$. We see that $t = 1.10$ is between 0.857 and 1.318. From the "Area in Upper Tail" row, we see that the area in the tail to the right of $t = 1.10$ is between .20 and .10. When we double these amounts, we see that the p-value must be between .40 and .20. With a level of significance of $\alpha = .05$, we now know that the p-value is greater than α. Therefore, H_0 cannot be rejected. Sufficient evidence is not available to conclude that Holiday should change its production plan for the coming season.

Appendix F shows how the p-value for this test can be computed using Excel or Minitab. The p-value obtained is .2822. With a level of significance of $\alpha = .05$, we cannot reject H_0 because $.2822 > .05$.

The test statistic can also be compared to the critical value to make the two-tailed hypothesis testing decision. With $\alpha = .05$ and the t distribution with 24 degrees of freedom, $-t_{.025} = -2.064$ and $t_{.025} = 2.064$ are the critical values for the two-tailed test. The rejection rule using the test statistic is

$$\text{Reject } H_0 \text{ if } t \leq -2.064 \text{ or if } t \geq 2.064$$

Based on the test statistic $t = -1.10$, H_0 cannot be rejected. This result indicates that Holiday should continue its production planning for the coming season based on the expectation that $\mu = 40$.

TABLE 9.3 SUMMARY OF HYPOTHESIS TESTS ABOUT A POPULATION MEAN: σ UNKNOWN CASE

	Lower Tail Test	Upper Tail Test	Two-Tailed Test
Hypotheses	$H_0: \mu \geq \mu_0$ $H_a: \mu < \mu_0$	$H_0: \mu \leq \mu_0$ $H_a: \mu > \mu_0$	$H_0: \mu = \mu_0$ $H_a: \mu \neq \mu_0$
Test Statistic	$t = \dfrac{\bar{x} - \mu_0}{s/\sqrt{n}}$	$t = \dfrac{\bar{x} - \mu_0}{s/\sqrt{n}}$	$t = \dfrac{\bar{x} - \mu_0}{s/\sqrt{n}}$
Rejection Rule: p-Value Approach	Reject H_0 if p-value $\leq \alpha$	Reject H_0 if p-value $\leq \alpha$	Reject H_0 if p-value $\leq \alpha$
Rejection Rule: Critical Value Approach	Reject H_0 if $t \leq -t_\alpha$	Reject H_0 if $t \geq t_\alpha$	Reject H_0 if $t \leq -t_{\alpha/2}$ or if $t \geq t_{\alpha/2}$

Summary and Practical Advice

Table 9.3 provides a summary of the hypothesis testing procedures about a population mean for the σ unknown case. The key difference between these procedures and the ones for the σ known case is that s is used, instead of σ, in the computation of the test statistic. For this reason, the test statistic follows the t distribution.

The applicability of the hypothesis testing procedures of this section is dependent on the distribution of the population being sampled from and the sample size. When the population is normally distributed, the hypothesis tests described in this section provide exact results for any sample size. When the population is not normally distributed, the procedures are approximations. Nonetheless, we find that sample sizes of 30 or greater will provide good results in most cases. If the population is approximately normal, small sample sizes (e.g., $n < 15$) can provide acceptable results. If the population is highly skewed or contains outliers, sample sizes approaching 50 are recommended.

Exercises

Methods

23. Consider the following hypothesis test:

    $$H_0: \mu \leq 12$$
    $$H_a: \mu > 12$$

    A sample of 25 provided a sample mean $\bar{x} = 14$ and a sample standard deviation $s = 4.32$.
    a. Compute the value of the test statistic.
    b. Use the t distribution table (Table 2 in Appendix B) to compute a range for the p-value.
    c. At $\alpha = .05$, what is your conclusion?
    d. What is the rejection rule using the critical value? What is your conclusion?

24. Consider the following hypothesis test:

    $$H_0: \mu = 18$$
    $$H_a: \mu \neq 18$$

9.4 Population Mean: σ Unknown

A sample of 48 provided a sample mean $\bar{x} = 17$ and a sample standard deviation $s = 4.5$.
a. Compute the value of the test statistic.
b. Use the t distribution table (Table 2 in Appendix B) to compute a range for the p-value.
c. At $\alpha = .05$, what is your conclusion?
d. What is the rejection rule using the critical value? What is your conclusion?

25. Consider the following hypothesis test:

$$H_0: \mu \geq 45$$
$$H_a: \mu < 45$$

A sample of 36 is used. Identify the p-value and state your conclusion for each of the following sample results. Use $\alpha = .01$.
a. $\bar{x} = 44$ and $s = 5.2$
b. $\bar{x} = 43$ and $s = 4.6$
c. $\bar{x} = 46$ and $s = 5.0$

26. Consider the following hypothesis test:

$$H_0: \mu = 100$$
$$H_a: \mu \neq 100$$

A sample of 65 is used. Identify the p-value and state your conclusion for each of the following sample results. Use $\alpha = .05$.
a. $\bar{x} = 103$ and $s = 11.5$
b. $\bar{x} = 96.5$ and $s = 11.0$
c. $\bar{x} = 102$ and $s = 10.5$

Applications

27. The Employment and Training Administration reported that the U.S. mean unemployment insurance benefit was $238 per week (*The World Almanac*, 2003). A researcher in the state of Virginia anticipated that sample data would show evidence that the mean weekly unemployment insurance benefit in Virginia was below the national average.
    a. Develop appropriate hypotheses such that rejection of H_0 will support the researcher's contention.
    b. For a sample of 100 individuals, the sample mean weekly unemployment insurance benefit was $231 with a sample standard deviation of $80. What is the p-value?
    c. At $\alpha = .05$, what is your conclusion?
    d. Repeat the preceding hypothesis test using the critical value approach.

28. A shareholders' group, in lodging a protest, claimed that the mean tenure for a chief executive office (CEO) was at least nine years. A survey of companies reported in *The Wall Street Journal* found a sample mean tenure of $\bar{x} = 7.27$ years for CEOs with a standard deviation of $s = 6.38$ years (*The Wall Street Journal*, January 2, 2007).
    a. Formulate hypotheses that can be used to challenge the validity of the claim made by the shareholders' group.
    b. Assume 85 companies were included in the sample. What is the p-value for your hypothesis test?
    c. At $\alpha = .01$, what is your conclusion?

29. The cost of a one-carat VS2 clarity, H color diamond from Diamond Source USA is $5600 (Diamond Source website, March 2003). A midwestern jeweler makes calls to contacts in the diamond district of New York City to see whether the mean price of diamonds there differs from $5600.
    a. Formulate hypotheses that can be used to determine whether the mean price in New York City differs from $5600.
    b. A sample of 25 New York City contacts provided the prices shown in the file named Diamonds. What is the p-value?

c. At $\alpha = .05$, can the null hypothesis be rejected? What is your conclusion?
d. Repeat the preceding hypothesis test using the critical value approach.

30. AOL Time Warner Inc.'s CNN has been the longtime ratings leader of cable television news. Nielsen Media Research indicated that the mean CNN viewing audience was 600,000 viewers per day during 2002 (*The Wall Street Journal,* March 10, 2003). Assume that for a sample of 40 days during the first half of 2003, the daily audience was 612,000 viewers with a sample standard deviation of 65,000 viewers.
   a. What are the hypotheses if CNN management would like information on any change in the CNN viewing audience?
   b. What is the *p*-value?
   c. Select your own level of significance. What is your conclusion?
   d. What recommendation would you make to CNN management in this application?

31. The Coca-Cola Company reported that the mean per capita annual sales of its beverages in the United States was 423 eight-ounce servings (Coca-Cola Company website, February 3, 2009). Suppose you are curious whether the consumption of Coca-Cola beverages is higher in Atlanta, Georgia, the location of Coca-Cola's corporate headquarters. A sample of 36 individuals from the Atlanta area showed a sample mean annual consumption of 460.4 eight-ounce servings with a standard deviation of $s = 101.9$ ounces. Using $\alpha = .05$, do the sample results support the conclusion that mean annual consumption of Coca-Cola beverage products is higher in Atlanta?

32. According to the National Automobile Dealers Association, the mean price for used cars is $10,192. A manager of a Kansas City used car dealership reviewed a sample of 50 recent used car sales at the dealership in an attempt to determine whether the population mean price for used cars at this particular dealership differed from the national mean. The prices for the sample of 50 cars are shown in the file named UsedCars.
   a. Formulate the hypotheses that can be used to determine whether a difference exists in the mean price for used cars at the dealership.
   b. What is the *p*-value?
   c. At $\alpha = .05$, what is your conclusion?

33. Annual per capita consumption of milk is 21.6 gallons (*Statistical Abstract of the United States: 2006*). Being from the Midwest, you believe milk consumption is higher there and wish to support your opinion. A sample of 16 individuals from the midwestern town of Webster City showed a sample mean annual consumption of 24.1 gallons with a standard deviation of $s = 4.8$.
   a. Develop a hypothesis test that can be used to determine whether the mean annual consumption in Webster City is higher than the national mean.
   b. What is a point estimate of the difference between mean annual consumption in Webster City and the national mean?
   c. At $\alpha = .05$, test for a significant difference. What is your conclusion?

34. Joan's Nursery specializes in custom-designed landscaping for residential areas. The estimated labor cost associated with a particular landscaping proposal is based on the number of plantings of trees, shrubs, and so on to be used for the project. For cost-estimating purposes, managers use two hours of labor time for the planting of a medium-sized tree. Actual times from a sample of 10 plantings during the past month follow (times in hours).

   1.7   1.5   2.6   2.2   2.4   2.3   2.6   3.0   1.4   2.3

   With a .05 level of significance, test to see whether the mean tree-planting time differs from two hours.
   a. State the null and alternative hypotheses.
   b. Compute the sample mean.
   c. Compute the sample standard deviation.
   d. What is the *p*-value?
   e. What is your conclusion?

9.5 Population Proportion

In this section we show how to conduct a hypothesis test about a population proportion p. Using p_0 to denote the hypothesized value for the population proportion, the three forms for a hypothesis test about a population proportion are as follows.

$$H_0: p \geq p_0 \qquad H_0: p \leq p_0 \qquad H_0: p = p_0$$
$$H_a: p < p_0 \qquad H_a: p > p_0 \qquad H_a: p \neq p_0$$

The first form is called a lower tail test, the second form is called an upper tail test, and the third form is called a two-tailed test.

Hypothesis tests about a population proportion are based on the difference between the sample proportion $\bar{p}$ and the hypothesized population proportion p_0. The methods used to conduct the hypothesis test are similar to those used for hypothesis tests about a population mean. The only difference is that we use the sample proportion and its standard error to compute the test statistic. The p-value approach or the critical value approach is then used to determine whether the null hypothesis should be rejected.

Let us consider an example involving a situation faced by Pine Creek golf course. Over the past year, 20% of the players at Pine Creek were women. In an effort to increase the proportion of women players, Pine Creek implemented a special promotion designed to attract women golfers. One month after the promotion was implemented, the course manager requested a statistical study to determine whether the proportion of women players at Pine Creek had increased. Because the objective of the study is to determine whether the proportion of women golfers increased, an upper tail test with $H_a: p > .20$ is appropriate. The null and alternative hypotheses for the Pine Creek hypothesis test are as follows:

$$H_0: p \leq .20$$
$$H_a: p > .20$$

If H_0 can be rejected, the test results will give statistical support for the conclusion that the proportion of women golfers increased and the promotion was beneficial. The course manager specified that a level of significance of $\alpha = .05$ be used in carrying out this hypothesis test.

The next step of the hypothesis testing procedure is to select a sample and compute the value of an appropriate test statistic. To show how this step is done for the Pine Creek upper tail test, we begin with a general discussion of how to compute the value of the test statistic for any form of a hypothesis test about a population proportion. The sampling distribution of $\bar{p}$, the point estimator of the population parameter p, is the basis for developing the test statistic.

When the null hypothesis is true as an equality, the expected value of $\bar{p}$ equals the hypothesized value p_0; that is, $E(\bar{p}) = p_0$. The standard error of $\bar{p}$ is given by

$$\sigma_{\bar{p}} = \sqrt{\frac{p_0(1 - p_0)}{n}}$$

In Chapter 7 we said that if $np \geq 5$ and $n(1 - p) \geq 5$, the sampling distribution of $\bar{p}$ can be approximated by a normal distribution.[3] Under these conditions, which usually apply in practice, the quantity

---

[3] In most applications involving hypothesis tests of a population proportion, sample sizes are large enough to use the normal approximation. The exact sampling distribution of $\bar{p}$ is discrete, with the probability for each value of $\bar{p}$ given by the binomial distribution. So hypothesis testing is a bit more complicated for small samples when the normal approximation cannot be used.

$$z = \frac{\bar{p} - p_0}{\sigma_{\bar{p}}} \qquad (9.3)$$

has a standard normal probability distribution. With $\sigma_{\bar{p}} = \sqrt{p_0(1-p_0)/n}$, the standard normal random variable z is the test statistic used to conduct hypothesis tests about a population proportion.

> **TEST STATISTIC FOR HYPOTHESIS TESTS ABOUT A POPULATION PROPORTION**
>
> $$z = \frac{\bar{p} - p_0}{\sqrt{\frac{p_0(1-p_0)}{n}}} \qquad (9.4)$$

We can now compute the test statistic for the Pine Creek hypothesis test. Suppose a random sample of 400 players was selected, and that 100 of the players were women. The proportion of women golfers in the sample is

$$\bar{p} = \frac{100}{400} = .25$$

Using equation (9.4), the value of the test statistic is

$$z = \frac{\bar{p} - p_0}{\sqrt{\frac{p_0(1-p_0)}{n}}} = \frac{.25 - .20}{\sqrt{\frac{.20(1-.20)}{400}}} = \frac{.05}{.02} = 2.50$$

Because the Pine Creek hypothesis test is an upper tail test, the p-value is the probability that z is greater than or equal to $z = 2.50$; that is, it is the upper tail area corresponding to $z \geq 2.50$. Using the standard normal probability table, we find that the lower tail area for $z = 2.50$ is .9938. Thus, the p-value for the Pine Creek test is $1.0000 - .9938 = .0062$. Figure 9.7 shows this p-value calculation.

Recall that the course manager specified a level of significance of $\alpha = .05$. A p-value $= .0062 < .05$ gives sufficient statistical evidence to reject H_0 at the .05 level of significance.

FIGURE 9.7 CALCULATION OF THE p-VALUE FOR THE PINE CREEK HYPOTHESIS TEST

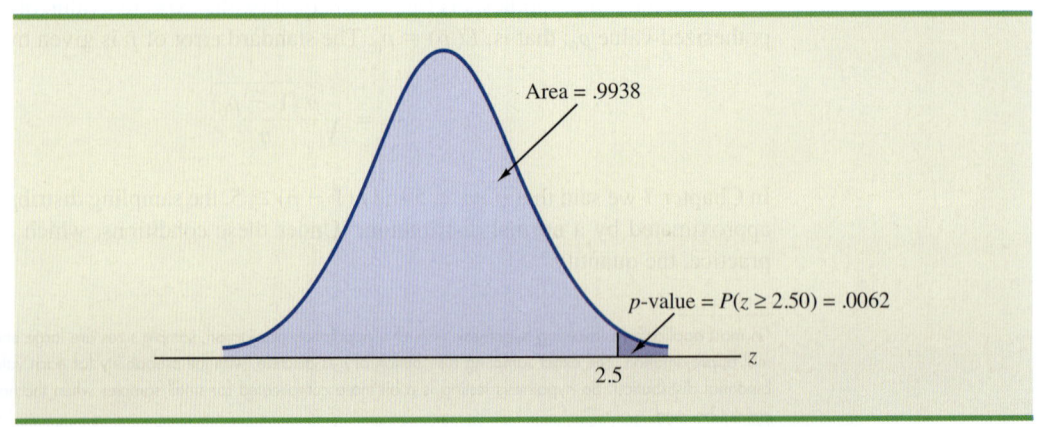

9.5 Population Proportion

TABLE 9.4 SUMMARY OF HYPOTHESIS TESTS ABOUT A POPULATION PROPORTION

	Lower Tail Test	Upper Tail Test	Two-Tailed Test
Hypotheses	$H_0: p \geq p_0$ $H_a: p < p_0$	$H_0: p \leq p_0$ $H_a: p > p_0$	$H_0: p = p_0$ $H_a: p \neq p_0$
Test Statistic	$z = \dfrac{\bar{p} - p_0}{\sqrt{\dfrac{p_0(1-p_0)}{n}}}$	$z = \dfrac{\bar{p} - p_0}{\sqrt{\dfrac{p_0(1-p_0)}{n}}}$	$z = \dfrac{\bar{p} - p_0}{\sqrt{\dfrac{p_0(1-p_0)}{n}}}$
Rejection Rule: *p*-Value Approach	Reject H_0 if *p*-value $\leq \alpha$	Reject H_0 if *p*-value $\leq \alpha$	Reject H_0 if *p*-value $\leq \alpha$
Rejection Rule: Critical Value Approach	Reject H_0 if $z \leq -z_\alpha$	Reject H_0 if $z \geq z_\alpha$	Reject H_0 if $z \leq -z_{\alpha/2}$ or if $z \geq z_{\alpha/2}$

Thus, the test provides statistical support for the conclusion that the special promotion increased the proportion of women players at the Pine Creek golf course.

The decision whether to reject the null hypothesis can also be made using the critical value approach. The critical value corresponding to an area of .05 in the upper tail of a normal probability distribution is $z_{.05} = 1.645$. Thus, the rejection rule using the critical value approach is to reject H_0 if $z \geq 1.645$. Because $z = 2.50 > 1.645$, H_0 is rejected.

Again, we see that the *p*-value approach and the critical value approach lead to the same hypothesis testing conclusion, but the *p*-value approach provides more information. With a *p*-value = .0062, the null hypothesis would be rejected for any level of significance greater than or equal to .0062.

Summary

The procedure used to conduct a hypothesis test about a population proportion is similar to the procedure used to conduct a hypothesis test about a population mean. Although we only illustrated how to conduct a hypothesis test about a population proportion for an upper tail test, similar procedures can be used for lower tail and two-tailed tests. Table 9.4 provides a summary of the hypothesis tests about a population proportion. We assume that $np \geq 5$ and $n(1 - p) \geq 5$; thus the normal probability distribution can be used to approximate the sampling distribution of $\bar{p}$.

Exercises

Methods

35. Consider the following hypothesis test:

$$H_0: p = .20$$
$$H_a: p \neq .20$$

A sample of 400 provided a sample proportion $\bar{p} = .175$.
   a. Compute the value of the test statistic.
   b. What is the *p*-value?
   c. At $\alpha = .05$, what is your conclusion?
   d. What is the rejection rule using the critical value? What is your conclusion?

36. Consider the following hypothesis test:

$$H_0: p \geq .75$$
$$H_a: p < .75$$

A sample of 300 items was selected. Compute the *p*-value and state your conclusion for each of the following sample results. Use $\alpha = .05$.
   a. $\bar{p} = .68$
   b. $\bar{p} = .72$
   c. $\bar{p} = .70$
   d. $\bar{p} = .77$

Applications

37. A study found that, in 2005, 12.5% of U.S. workers belonged to unions (*The Wall Street Journal*, January 21, 2006). Suppose a sample of 400 U.S. workers is collected in 2006 to determine whether union efforts to organize have increased union membership.
   a. Formulate the hypotheses that can be used to determine whether union membership increased in 2006.
   b. If the sample results show that 52 of the workers belonged to unions, what is the *p*-value for your hypothesis test?
   c. At $\alpha = .05$, what is your conclusion?

38. A study by *Consumer Reports* showed that 64% of supermarket shoppers believe supermarket brands to be as good as national name brands. To investigate whether this result applies to its own product, the manufacturer of a national name-brand ketchup asked a sample of shoppers whether they believed that supermarket ketchup was as good as the national brand ketchup.
   a. Formulate the hypotheses that could be used to determine whether the percentage of supermarket shoppers who believe that the supermarket ketchup was as good as the national brand ketchup differed from 64%.
   b. If a sample of 100 shoppers showed 52 stating that the supermarket brand was as good as the national brand, what is the *p*-value?
   c. At $\alpha = .05$, what is your conclusion?
   d. Should the national brand ketchup manufacturer be pleased with this conclusion? Explain.

39. According to the Pew Internet & American Life Project, 75% of American adults use the Internet (Pew Internet website, April 19, 2008). The Pew project authors also reported on the percentage of Americans who use the Internet by age group. The data in the file AgeGroup are consistent with their findings. These data were obtained from a sample of 100 Internet users in the 30–49 age group and 200 Internet users in the 50–64 age group. A Yes indicates the survey repondent had used the Internet; a No indicates the survey repondent had not.
   a. Formulate hypotheses that could be used to determine whether the percentage of Internet users in the two age groups differs from the overall average of 75%.
   b. Estimate the proportion of Internet users in the 30–49 age group. Does this proportion differ significantly from the overall proportion of .75? Use $\alpha = .05$.
   c. Estimate the proportion of Internet users in the 50–64 age group. Does this proportion differ significantly from the overall proportion of .75? Use $\alpha = .05$.
   d. Would you expect the proportion of users in the 18–29 age group to be larger or smaller than the proportion for the 30–49 age group? Support you conclusion with the results obtained in parts (b) and (c).

40. In 2008, 46% of business owners gave a holiday gift to their employees. A 2009 survey of business owners indicated that 35% planned to provide a holiday gift to their employees (Radio WEZV, Myrtle Beach, SC, November 11, 2009). Suppose the survey results are based on a sample of 60 business owners.
   a. How many business owners in the survey planned to provide a holiday gift to their employees in 2009?

b. Suppose the business owners in the sample did as they planned. Compute the *p*-value for a hypothesis test that can be used to determine if the proportion of business owners providing holiday gifts had decreased from the 2008 level.

c. Using a .05 level of significance, would you conclude that the proportion of business owners providing gifts decreased? What is the smallest level of significance for which you could draw such a conclusion?

41. Speaking to a group of analysts in January 2006, a brokerage firm executive claimed that at least 70% of investors are currently confident of meeting their investment objectives. A UBS Investor Optimism Survey, conducted over the period January 2 to January 15, found that 67% of investors were confident of meeting their investment objectives (CNBC, January 20, 2006).

   a. Formulate the hypotheses that can be used to test the validity of the brokerage firm executive's claim.
   b. Assume the UBS Investor Optimism Survey collected information from 300 investors. What is the *p*-value for the hypothesis test?
   c. At $\alpha = .05$, should the executive's claim be rejected?

42. According to the University of Nevada Center for Logistics Management, 6% of all merchandise sold in the United States is returned (*BusinessWeek*, January 15, 2007). A Houston department store sampled 80 items sold in January and found that 12 of the items were returned.

   a. Construct a point estimate of the proportion of items returned for the population of sales transactions at the Houston store.
   b. Construct a 95% confidence interval for the porportion of returns at the Houston store.
   c. Is the proportion of returns at the Houston store significantly different from the returns for the nation as a whole? Provide statistical support for your answer.

43. Eagle Outfitters is a chain of stores specializing in outdoor apparel and camping gear. It is considering a promotion that involves mailing discount coupons to all its credit card customers. This promotion will be considered a success if more than 10% of those receiving the coupons use them. Before going national with the promotion, coupons were sent to a sample of 100 credit card customers.

   a. Develop hypotheses that can be used to test whether the population proportion of those who will use the coupons is sufficient to go national.
   b. The file Eagle contains the sample data. Develop a point estimate of the population proportion.
   c. Use $\alpha = .05$ to conduct your hypothesis test. Should Eagle go national with the promotion?

44. In a cover story, *BusinessWeek* published information about sleep habits of Americans (*BusinessWeek*, January 26, 2004). The article noted that sleep deprivation causes a number of problems, including highway deaths. Fifty-one percent of adult drivers admit to driving while drowsy. A researcher hypothesized that this issue was an even bigger problem for night shift workers.

   a. Formulate the hypotheses that can be used to help determine whether more than 51% of the population of night shift workers admit to driving while drowsy.
   b. A sample of 400 night shift workers identified those who admitted to driving while drowsy. See the Drowsy file. What is the sample proportion? What is the *p*-value?
   c. At $\alpha = .01$, what is your conclusion?

45. Many investors and financial analysts believe the Dow Jones Industrial Average (DJIA) provides a good barometer of the overall stock market. On January 31, 2006, 9 of the 30 stocks making up the DJIA increased in price (*The Wall Street Journal*, February 1, 2006). On the basis of this fact, a financial analyst claims we can assume that 30% of the stocks traded on the New York Stock Exchange (NYSE) went up the same day.

   a. Formulate null and alternative hypotheses to test the analyst's claim.
   b. A sample of 50 stocks traded on the NYSE that day showed that 24 went up. What is your point estimate of the population proportion of stocks that went up?
   c. Conduct your hypothesis test using $\alpha = .01$ as the level of significance. What is your conclusion?

Summary

Hypothesis testing is a statistical procedure that uses sample data to determine whether a statement about the value of a population parameter should or should not be rejected. The hypotheses are two competing statements about a population parameter. One statement is called the null hypothesis (H_0), and the other statement is called the alternative hypothesis (H_a). In Section 9.1 we provided guidelines for developing hypotheses for situations frequently encountered in practice.

Whenever historical data or other information provide a basis for assuming that the population standard deviation is known, the hypothesis testing procedure for the population mean is based on the standard normal distribution. Whenever σ is unknown, the sample standard deviation s is used to estimate σ and the hypothesis testing procedure is based on the t distribution. In both cases, the quality of results depends on both the form of the population distribution and the sample size. If the population has a normal distribution, both hypothesis testing procedures are applicable, even with small sample sizes. If the population is not normally distributed, larger sample sizes are needed. General guidelines about the sample size were provided in Sections 9.3 and 9.4. In the case of hypothesis tests about a population proportion, the hypothesis testing procedure uses a test statistic based on the standard normal distribution.

In all cases, the value of the test statistic can be used to compute a p-value for the test. A p-value is a probability used to determine whether the null hypothesis should be rejected. If the p-value is less than or equal to the level of significance α, the null hypothesis can be rejected.

Hypothesis testing conclusions can also be made by comparing the value of the test statistic to a critical value. For lower tail tests, the null hypothesis is rejected if the value of the test statistic is less than or equal to the critical value. For upper tail tests, the null hypothesis is rejected if the value of the test statistic is greater than or equal to the critical value. Two-tailed tests consist of two critical values: one in the lower tail of the sampling distribution and one in the upper tail. In this case, the null hypothesis is rejected if the value of the test statistic is less than or equal to the critical value in the lower tail or greater than or equal to the critical value in the upper tail.

Glossary

Null hypothesis The hypothesis tentatively assumed true in the hypothesis testing procedure.
Alternative hypothesis The hypothesis concluded to be true if the null hypothesis is rejected.
Type I error The error of rejecting H_0 when it is true.
Type II error The error of accepting H_0 when it is false.
Level of significance The probability of making a Type I error when the null hypothesis is true as an equality.
One-tailed test A hypothesis test in which rejection of the null hypothesis occurs for values of the test statistic in one tail of its sampling distribution.
Test statistic A statistic whose value helps determine whether a null hypothesis should be rejected.
***p*-value** A probability that provides a measure of the evidence against the null hypothesis given by the sample. Smaller p-values indicate more evidence against H_0. For a lower tail test, the p-value is the probability of obtaining a value for the test statistic as small as or smaller than that provided by the sample. For an upper tail test, the p-value is the probability of obtaining a value for the test statistic as large as or larger than that provided by the sample. For a two-tailed test, the p-value is the probability of obtaining a value for the test statistic at least as unlikely as or more unlikely than that provided by the sample.

Critical value A value that is compared with the test statistic to determine whether H_0 should be rejected.

Two-tailed test A hypothesis test in which rejection of the null hypothesis occurs for values of the test statistic in either tail of its sampling distribution.

Key Formulas

Test Statistic for Hypothesis Tests About a Population Mean: σ Known

$$z = \frac{\bar{x} - \mu_0}{\sigma/\sqrt{n}} \tag{9.1}$$

Test Statistic for Hypothesis Tests About a Population Mean: σ Unknown

$$t = \frac{\bar{x} - \mu_0}{s/\sqrt{n}} \tag{9.2}$$

Test Statistic for Hypothesis Tests About a Population Proportion

$$z = \frac{\bar{p} - p_0}{\sqrt{\dfrac{p_0(1 - p_0)}{n}}} \tag{9.4}$$

Supplementary Exercises

46. On Friday, Wall Street traders were anxiously awaiting the federal government's release of numbers on the January increase in nonfarm payrolls. The early consensus estimate among economists was for a growth of 250,000 new jobs (CNBC, February 3, 2006). However, a sample of 20 economists taken Thursday afternoon provided a sample mean of 266,000 with a sample standard deviation of 24,000. Financial analysts often call such a sample mean, based on late-breaking news, the *whisper number*. Treat the "consensus estimate" as the population mean. Conduct a hypothesis test to determine whether the whisper number justifies a conclusion of a statistically significant increase in the consensus estimate of economists. Use $\alpha = .01$ as the level of significance.

47. *Playbill* is a magazine distributed around the country to people attending musicals and other theatrical productions. The mean annual household income for the population of *Playbill* readers is $119,155 (*Playbill,* January 2006). Assume the standard deviation is $\sigma = \$20,700$. A San Francisco civic group has asserted that the mean for theatergoers in the Bay Area is higher. A sample of 60 theater attendees in the Bay Area showed a sample mean household income of $126,100.
    a. Develop hypotheses that can be used to determine whether the sample data support the conclusion that theater attendees in the Bay Area have a higher mean household income than that for all *Playbill* readers.
    b. What is the *p*-value based on the sample of 60 theater attendees in the Bay Area?
    c. Use $\alpha = .01$ as the level of significance. What is your conclusion?

48. At Western University the historical mean of scholarship examination scores for freshman applications is 900. A historical population standard deviation $\sigma = 180$ is assumed known. Each year, the assistant dean uses a sample of applications to determine whether the mean examination score for the new freshman applications has changed.
    a. State the hypotheses.
    b. What is the 95% confidence interval estimate of the population mean examination score if a sample of 200 applications provided a sample mean $\bar{x} = 935$?

c. Use the confidence interval to conduct a hypothesis test. Using $\alpha = .05$, what is your conclusion?
d. What is the *p*-value?

49. The chamber of commerce of a Florida Gulf Coast community advertises that area residential property is available at a mean cost of $125,000 or less per lot. Suppose a sample of 32 properties provided a sample mean of $130,000 per lot and a sample standard deviation of $12,500. Use a .05 level of significance to test the validity of the advertising claim.

50. Data released by the National Center for Health Statistics showed that the mean age at which women had their first child was 25.0 in 2006 (*The Wall Street Journal*, February 4, 2009). The reporter, Sue Shellenbarger, noted that this was the first decrease in the average age at which women had their first child in several years. A recent sample of 42 women provided the data in the website file named FirstBirth concerning the age at which these women had their first child. Do the data indicate a change from 2006 in the mean age at which women had their first child? Use $\alpha = .05$.

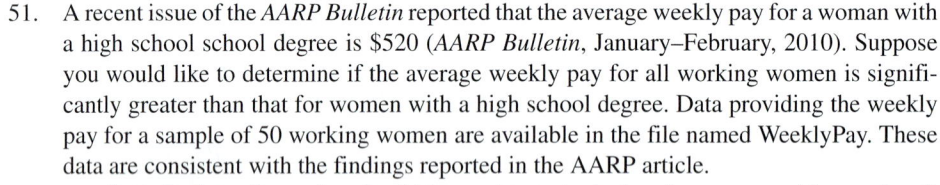

51. A recent issue of the *AARP Bulletin* reported that the average weekly pay for a woman with a high school school degree is $520 (*AARP Bulletin*, January–February, 2010). Suppose you would like to determine if the average weekly pay for all working women is significantly greater than that for women with a high school degree. Data providing the weekly pay for a sample of 50 working women are available in the file named WeeklyPay. These data are consistent with the findings reported in the AARP article.
    a. State the hypotheses that should be used to test whether the mean weekly pay for all women is significantly greater than the mean weekly pay for women with a high school degree.
    b. Use the data in the file named WeeklyPay to compute the sample mean, the test statistic, and the *p*-value.
    c. Use $\alpha = .05$. What is your conclusion?
    d. Repeat the hypothesis test using the critical value approach.

52. A production line operates with a mean filling weight of 16 ounces per container. Overfilling or underfilling presents a serious problem and when detected requires the operator to shut down the production line to readjust the filling mechanism. From past data, a population standard deviation $\sigma = .8$ ounces is assumed. A quality control inspector selects a sample of 30 items every hour and at that time makes the decision of whether to shut down the line for readjustment. The level of significance is $\alpha = .05$.
    a. State the hypothesis test for this quality control application.
    b. If a sample mean of $\bar{x} = 16.32$ ounces were found, what is the *p*-value? What action would you recommend?
    c. If a sample mean of $\bar{x} = 15.82$ ounces were found, what is the *p*-value? What action would you recommend?
    d. Use the critical value approach. What is the rejection rule for the preceding hypothesis testing procedure? Repeat parts (b) and (c). Do you reach the same conclusion?

53. The U.S. Energy Administration reported that the mean price for a gallon of regular gasoline in the United States was $2.357 (U.S. Energy Administration, January 30, 2006). Data for a sample of regular gasoline prices at 50 service stations in the Lower Atlantic states are contained in the data file named Gasoline. Conduct a hypothesis test to determine whether the mean price for a gallon of gasoline in the Lower Atlantic states is different from the national mean. Use $\alpha = .05$ for the level of significance, and state your conclusion.

54. According to the federal government, 24% of workers covered by their company's health care plan were not required to contribute to the premium (*Statistical Abstract of the United States: 2006*). A recent study found that 81 out of 400 workers sampled were not required to contribute to their company's health care plan.
    a. Develop hypotheses that can be used to test whether the percent of workers not required to contribute to their company's health care plan has declined.

b. What is a point estimate of the proportion receiving free company-sponsored health care insurance?
c. Has a statistically significant decline occurred in the proportion of workers receiving free company-sponsored health care insurance? Use $\alpha = .05$.

55. A radio station in Myrtle Beach announced that at least 90% of the hotels and motels would be full for the Memorial Day weekend. The station advised listeners to make reservations in advance if they planned to be in the resort over the weekend. On Saturday night a sample of 58 hotels and motels showed 49 with a no-vacancy sign and 9 with vacancies. What is your reaction to the radio station's claim after seeing the sample evidence? Use $\alpha = .05$ in making the statistical test. What is the *p*-value?

56. During the 2004 election year, new polling results were reported daily. In an IBD/TIPP poll of 910 adults, 503 respondents reported that they were optimistic about the national outlook, and President Bush's leadership index jumped 4.7 points to 55.3 (*Investor's Business Daily,* January 14, 2004).
   a. What is the sample proportion of respondents who are optimistic about the national outlook?
   b. A campaign manager wants to claim that this poll indicates that the majority of adults are optimistic about the national outlook. Construct a hypothesis test so that rejection of the null hypothesis will permit the conclusion that the proportion optimistic is greater than 50%.
   c. Use the polling data to compute the *p*-value for the hypothesis test in part (b). Explain to the manager what this *p*-value means about the level of significance of the results.

57. Virtual call centers are staffed by individuals working out of their homes. Most home agents earn $10 to $15 per hour without benefits versus $7 to $9 per hour with benefits at a traditional call center (*BusinessWeek,* January 23, 2006). Regional Airways is considering employing home agents, but only if a level of customer satisfaction greater than 80% can be maintained. A test was conducted with home service agents. In a sample of 300 customers, 252 reported that they were satisfied with service.
   a. Develop hypotheses for a test to determine whether the sample data support the conclusion that customer service with home agents meets the Regional Airways criterion.
   b. What is your point estimate of the percentage of satisfied customers?
   c. What is the *p*-value provided by the sample data?
   d. What is your hypothesis testing conclusion? Use $\alpha = .05$ as the level of significance.

58. A recent article concerning bullish and bearish sentiment about the stock market reported that 41% of investors responding to an American Institute of Individual Investors (AAII) poll were bullish on the market and 26% were bearish (*USA Today,* January 11, 2010). The article also reported that the long-term average measure of bullishness is .39 or 39%. Suppose the AAII poll used a sample size of 450. Using .39 (the long-term average) as the population proportion of investors who are bullish, conduct a hypothesis test to determine if the current proportion of investors who are bullish is significantly greater than the long-term average proportion.
   a. State the appropriate hypotheses for your significance test.
   b. Use the sample results to compute the test statistic and the *p*-value.
   c. Using $\alpha = .10$, what is your conclusion?

59. On December 25, 2009, an airline passenger was subdued while attempting to blow up a Northwest Airlines flight headed for Detroit, Michigan. The passenger had smuggled explosives hidden in his underwear past a metal detector at an airport screening facility. As a result, the Transportation Security Administration (TSA) proposed installing full-body scanners to replace the metal detectors at the nation's largest airports. This proposal resulted in strong objections from privacy advocates who considered the scanners an invasion of privacy. On January 5–6, 2010, *USA Today* conducted a poll of 542 adults to learn what proportion of airline travelers approved of using full-body scanners (*USA Today,* January 11, 2010).

The poll results showed that 455 of the respondents felt that full-body scanners would improve airline security and 423 indicated that they approved of using the devices.

a. Conduct a hypothesis test to determine if the results of the poll justify concluding that over 80% of airline travelers feel that the use of full-body scanners will improve airline security. Use $\alpha = .05$.

b. Suppose the TSA will go forward with the installation and mandatory use of full-body scanners if over 75% of airline travelers approve of using the devices. You have been told to conduct a statistical analysis using the poll results to determine if the TSA should require mandatory use of the full-body scanners. Because this is viewed as a very sensitive decision, use $\alpha = .01$. What is your recommendation?

Case Problem 1 Quality Associates, Inc.

Quality Associates, Inc., a consulting firm, advises its clients about sampling and statistical procedures that can be used to control their manufacturing processes. In one particular application, a client gave Quality Associates a sample of 800 observations taken during a time in which that client's process was operating satisfactorily. The sample standard deviation for these data was .21; hence, with so much data, the population standard deviation was assumed to be .21. Quality Associates then suggested that random samples of size 30 be taken periodically to monitor the process on an ongoing basis. By analyzing the new samples, the client could quickly learn whether the process was operating satisfactorily. When the process was not operating satisfactorily, corrective action could be taken to eliminate the problem. The design specification indicated the mean for the process should be 12. The hypothesis test suggested by Quality Associates follows.

$$H_0: \mu = 12$$
$$H_a: \mu \neq 12$$

Corrective action will be taken anytime H_0 is rejected.

The following samples were collected at hourly intervals during the first day of operation of the new statistical process control procedure. These data are available in the data set Quality.

Sample 1	Sample 2	Sample 3	Sample 4
11.55	11.62	11.91	12.02
11.62	11.69	11.36	12.02
11.52	11.59	11.75	12.05
11.75	11.82	11.95	12.18
11.90	11.97	12.14	12.11
11.64	11.71	11.72	12.07
11.80	11.87	11.61	12.05
12.03	12.10	11.85	11.64
11.94	12.01	12.16	12.39
11.92	11.99	11.91	11.65
12.13	12.20	12.12	12.11
12.09	12.16	11.61	11.90
11.93	12.00	12.21	12.22
12.21	12.28	11.56	11.88
12.32	12.39	11.95	12.03
11.93	12.00	12.01	12.35
11.85	11.92	12.06	12.09
11.76	11.83	11.76	11.77

Sample 1	Sample 2	Sample 3	Sample 4
12.16	12.23	11.82	12.20
11.77	11.84	12.12	11.79
12.00	12.07	11.60	12.30
12.04	12.11	11.95	12.27
11.98	12.05	11.96	12.29
12.30	12.37	12.22	12.47
12.18	12.25	11.75	12.03
11.97	12.04	11.96	12.17
12.17	12.24	11.95	11.94
11.85	11.92	11.89	11.97
12.30	12.37	11.88	12.23
12.15	12.22	11.93	12.25

Managerial Report

1. Conduct a hypothesis test for each sample at the .01 level of significance and determine what action, if any, should be taken. Provide the test statistic and p-value for each test.
2. Compute the standard deviation for each of the four samples. Does the assumption of .21 for the population standard deviation appear reasonable?
3. Compute limits for the sample mean $\bar{x}$ around $\mu = 12$ such that, as long as a new sample mean is within those limits, the process will be considered to be operating satisfactorily. If $\bar{x}$ exceeds the upper limit or if $\bar{x}$ is below the lower limit, corrective action will be taken. These limits are referred to as upper and lower control limits for quality control purposes.
4. Discuss the implications of changing the level of significance to a larger value. What mistake or error could increase if the level of significance is increased?

Case Problem 2 Ethical Behavior of Business Students at Bayview Universtiy

During the global recession of 2008 and 2009, there were many accusations of unethical behavior by Wall Street executives, financial managers, and other corporate officers. At that time, an article appeared that suggested that part of the reason for such unethical business behavior may stem from the fact that cheating has become more prevalent among business students (*Chronicle of Higher Education*, February 10, 2009). The article reported that 56% of business students admitted to cheating at some time during their academic career as compared to 47% of nonbusiness students.

Cheating has been a concern of the dean of the College of Business at Bayview University for several years. Some faculty members in the college believe that cheating is more widespread at Bayview than at other universities, while other faculty members think that cheating is not a major problem in the college. To resolve some of these issues, the dean commissioned a study to assess the current ethical behavior of business students at Bayview. As part of this study, an anonymous exit survey was administered to a sample of 90 business students from this year's graduating class. Responses to the following questions were used to obtain data regarding three types of cheating.

During your time at Bayview, did you ever present work copied off the Internet as your own?

    Yes _____   No _____

During your time at Bayview, did you ever copy answers off another student's exam?

    Yes _____   No _____

During your time at Bayview, did you ever collaborate with other students on projects that were supposed to be completed individually?

    Yes _____   No _____

Any student who answered Yes to one or more of these questions was considered to have been involved in some type of cheating. A portion of the data collected follows. The complete data set is in the file named Bayview.

Student	Copied from Internet	Copied on Exam	Collaborated on Individual Project	Gender
1	No	No	No	Female
2	No	No	No	Male
3	Yes	No	Yes	Male
4	Yes	Yes	No	Male
5	No	No	Yes	Male
6	Yes	No	No	Female
.	.	.	.	.
.	.	.	.	.
.	.	.	.	.
88	No	No	No	Male
89	No	Yes	Yes	Male
90	No	No	No	Female

Managerial Report

Prepare a report for the dean of the college that summarizes your assessment of the nature of cheating by business students at Bayview University. Be sure to include the following items in your report.

1. Use descriptive statistics to summarize the data and comment on your findings.
2. Develop 95% confidence intervals for the proportion of all students, the proportion of male students, and the proportion of female students who were involved in some type of cheating.
3. Conduct a hypothesis test to determine if the proportion of business students at Bayview University who were involved in some type of cheating is less than that of business students at other institutions as reported by the *Chronicle of Higher Education*.
4. Conduct a hypothesis test to determine if the proportion of business students at Bayview University who were involved in some form of cheating is less than that of nonbusiness students at other institutions as reported by the *Chronicle of Higher Education*.
5. What advice would you give to the dean based upon your analysis of the data?

Appendix 9.1 Hypothesis Testing Using Minitab

We describe the use of Minitab to conduct hypothesis tests about a population mean and a population proportion.

Population Mean: σ Known

We illustrate using the MaxFlight golf ball distance example in Section 9.3. The data are in column C1 of a Minitab worksheet. The population standard deviation $\sigma = 12$ is assumed known and the level of significance is $\alpha = .05$. The following steps can be used to test the hypothesis $H_0: \mu = 295$ versus $H_a: \mu \neq 295$.

Step 1. Select the **Stat** menu
Step 2. Choose **Basic Statistics**
Step 3. Choose **1-Sample Z**
Step 4. When the 1-Sample Z (Test and Confidence Interval) dialog box appears:
    Enter C1 in the **Samples in columns** box
    Enter 12 in the **Standard deviation** box
    Select **Perform Hypothesis Test**
    Enter 295 in the **Hypothesized mean** box
    Select **Options**
Step 5. When the 1-Sample Z-Options dialog box appears:
    Enter 95 in the **Confidence level** box*
    Select **not equal** in the **Alternative** box
    Click **OK**
Step 6. Click **OK**

In addition to the hypothesis testing results, Minitab provides a 95% confidence interval for the population mean.

The procedure can be easily modified for a one-tailed hypothesis test by selecting the less than or greater than option in the **Alternative** box in step 5.

Population Mean: σ Unknown

The ratings that 60 business travelers gave for Heathrow Airport are entered in column C1 of a Minitab worksheet. The level of significance for the test is $\alpha = .05$, and the population standard deviation σ will be estimated by the sample standard deviation s. The following steps can be used to test the hypothesis $H_0: \mu \leq 7$ against $H_a: \mu > 7$.

Step 1. Select the **Stat** menu
Step 2. Choose **Basic Statistics**
Step 3. Choose **1-Sample t**
Step 4. When the 1-Sample t (Test and Confidence Interval) dialog box appears:
    Enter C1 in the **Samples in columns** box
    Select **Perform Hypothesis Test**
    Enter 7 in the **Hypothesized mean** box
    Select **Options**
Step 5. When the 1-Sample t-options dialog box appears:
    Enter 95 in the **Confidence level** box
    Select **greater than** in the **Alternative** box
    Click **OK**
Step 6. Click **OK**

---

*Minitab provides both hypothesis testing and interval estimation results simultaneously. The user may select any confidence level for the interval estimate of the population mean: 95% confidence is suggested here.

The Heathrow Airport rating study involved a greater than alternative hypothesis. The preceding steps can be easily modified for other hypothesis tests by selecting the less than or not equal options in the **Alternative** box in step 5.

Population Proportion

We illustrate using the Pine Creek golf course example in Section 9.5. The data with responses Female and Male are in column C1 of a Minitab worksheet. Minitab uses an alphabetical ordering of the responses and selects the *second response* for the population proportion of interest. In this example, Minitab uses the alphabetical ordering Female-Male to provide results for the population proportion of Male responses. Because Female is the response of interest, we change Minitab's ordering as follows: Select any cell in the column and use the sequence: Editor > Column > Value Order. Then choose the option of entering a user-specified order. Enter Male-Female in the **Define-an-order** box and click OK. Minitab's 1 Proportion routine will then provide the hypothesis test results for the population proportion of female golfers. We proceed as follows:

Step 1. Select the **Stat** menu
Step 2. Choose **Basic Statistics**
Step 3. Choose **1 Proportion**
Step 4. When the 1 Proportion (Test and Confidence Interval) dialog box appears:
　　　　Enter C1 in the **Samples in Columns** box
　　　　Select **Perform Hypothesis Test**
　　　　Enter .20 in the **Hypothesized proportion** box
　　　　Select **Options**
Step 5. When the 1 Proportion-Options dialog box appears:
　　　　Enter 95 in the **Confidence level** box
　　　　Select greater than in the **Alternative** box
　　　　Select **Use test and interval based on normal distribution**
　　　　Click **OK**
Step 6. Click **OK**

Appendix 9.2 Hypothesis Testing Using Excel

Excel does not provide built-in routines for the hypothesis tests presented in this chapter. To handle these situations, we present Excel worksheets that we designed to use as templates for testing hypotheses about a population mean and a population proportion. The worksheets are easy to use and can be modified to handle any sample data. The worksheets are available on the website that accompanies this book.

Population Mean: σ Known

We illustrate using the MaxFlight golf ball distance example in Section 9.3. The data are in column A of an Excel worksheet. The population standard deviation $\sigma = 12$ is assumed known and the level of significance is $\alpha = .05$. The following steps can be used to test the hypothesis $H_0: \mu = 295$ versus $H_a: \mu \neq 295$.

Refer to Figure 9.8 as we describe the procedure. The worksheet in the background shows the cell formulas used to compute the results shown in the foreground worksheet. The data are entered into cells A2:A51. The following steps are necessary to use the template for this data set.

Appendix 9.2 Hypothesis Testing Using Excel

Step 1. Enter the data range A2:A51 into the =COUNT cell formula in cell D4
Step 2. Enter the data range A2:A51 into the =AVERAGE cell formula in cell D5
Step 3. Enter the population standard deviation $\sigma = 12$ into cell D6
Step 4. Enter the hypothesized value for the population mean 295 into cell D8

The remaining cell formulas automatically provide the standard error, the value of the test statistic z, and three p-values. Because the alternative hypothesis ($\mu_0 \neq 295$) indicates a

FIGURE 9.8 EXCEL WORKSHEET FOR HYPOTHESIS TESTS ABOUT A POPULATION MEAN WITH σ KNOWN

	A	B	C	D	E
1	Yards		Hypothesis Test About a Population Mean		
2	303		With σ Known		
3	282				
4	289		Sample Size	=COUNT(A2:A51)	
5	298		Sample Mean	=AVERAGE(A2:A51)	
6	283		Population Std. Deviation	12	
7	317				
8	297		Hypothesized Value	295	
9	308				
10	317		Standard Error	=D6/SQRT(D4)	
11	293		Test Statistic z	=(D5-D8)/D10	
12	284				
13	290		p-value (Lower Tail)	=NORMSDIST(D11)	
14	304		p-value (Upper Tail)	=1-D13	
15	290		p-value (Two Tail)	=2*MIN(D13,D14)	
16	311				
17	305				
49	303				
50	301				
51	292				
52					

Note: Rows 18 to 48 are hidden.

	A	B	C	D	E
1	Yards		Hypothesis Test About a Population Mean		
2	303		With σ Known		
3	282				
4	289		Sample Size	50	
5	298		Sample Mean	297.6	
6	283		Population Std. Deviation	12	
7	317				
8	297		Hypothesized Value	295	
9	308				
10	317		Standard Error	1.70	
11	293		Test Statistic z	1.53	
12	284				
13	290		p-value (Lower Tail)	0.9372	
14	304		p-value (Upper Tail)	0.0628	
15	290		p-value (Two Tail)	0.1255	
16	311				
17	305				
49	303				
50	301				
51	292				
52					

two-tailed test, the *p*-value (Two Tail) in cell D15 is used to make the rejection decision. With *p*-value = .1255 > α = .05, the null hypothesis cannot be rejected. The *p*-values in cells D13 or D14 would be used if the hypotheses involved a one-tailed test.

This template can be used to make hypothesis testing computations for other applications. For instance, to conduct a hypothesis test for a new data set, enter the new sample data into column A of the worksheet. Modify the formulas in cells D4 and D5 to correspond to the new data range. Enter the population standard deviation into cell D6 and the hypothesized value for the population mean into cell D8 to obtain the results. If the new sample data have already been summarized, the new sample data do not have to be entered into the worksheet. In this case, enter the sample size into cell D4, the sample mean into cell D5, the population standard deviation into cell D6, and the hypothesized value for the population mean into cell D8 to obtain the results. The worksheet in Figure 9.8 is available in the file Hyp Sigma Known on the website that accompanies this book.

Population Mean: σ Unknown

We illustrate using the Heathrow Airport rating example in Section 9.4. The data are in column A of an Excel worksheet. The population standard deviation σ is unknown and will be estimated by the sample standard deviation s. The level of significance is α = .05. The following steps can be used to test the hypothesis $H_0: \mu \leq 7$ versus $H_a: \mu > 7$.

Refer to Figure 9.9 as we describe the procedure. The background worksheet shows the cell formulas used to compute the results shown in the foreground version of the worksheet. The data are entered into cells A2:A61. The following steps are necessary to use the template for this data set.

Step 1. Enter the data range A2:A61 into the =COUNT cell formula in cell D4
Step 2. Enter the data range A2:A61 into the =AVERAGE cell formula in cell D5
Step 3. Enter the data range A2:A61 into the =STDEV cell formula in cell D6
Step 4. Enter the hypothesized value for the population mean 7 into cell D8

The remaining cell formulas automatically provide the standard error, the value of the test statistic *t*, the number of degrees of freedom, and three *p*-values. Because the alternative hypothesis ($\mu > 7$) indicates an upper tail test, the *p*-value (Upper Tail) in cell D15 is used to make the decision. With *p*-value = .0353 < α = .05, the null hypothesis is rejected. The *p*-values in cells D14 or D16 would be used if the hypotheses involved a lower tail test or a two-tailed test.

This template can be used to make hypothesis testing computations for other applications. For instance, to conduct a hypothesis test for a new data set, enter the new sample data into column A of the worksheet and modify the formulas in cells D4, D5, and D6 to correspond to the new data range. Enter the hypothesized value for the population mean into cell D8 to obtain the results. If the new sample data have already been summarized, the new sample data do not have to be entered into the worksheet. In this case, enter the sample size into cell D4, the sample mean into cell D5, the sample standard deviation into cell D6, and the hypothesized value for the population mean into cell D8 to obtain the results. The worksheet in Figure 9.9 is available in the file Hyp Sigma Unknown on the website that accompanies this book.

Population Proportion

We illustrate using the Pine Creek golf course survey data presented in Section 9.5. The data of Male or Female golfer are in column A of an Excel worksheet. Refer to Figure 9.10 as we describe the procedure. The background worksheet shows the cell formulas used to compute the results shown in the foreground worksheet. The data are entered into cells A2:A401. The following steps can be used to test the hypothesis H_0: $p \leq .20$ versus H_a: $p > .20$.

Appendix 9.2 Hypothesis Testing Using Excel   **389**

FIGURE 9.9 EXCEL WORKSHEET FOR HYPOTHESIS TESTS ABOUT A POPULATION MEAN WITH σ UNKNOWN

	A	B	C	D	E
1	Rating		Hypothesis Test About a Population Mean		
2	5		With σ Unknown		
3	7				
4	8		Sample Size	=COUNT(A2:A61)	
5	7		Sample Mean	=AVERAGE(A2:A61)	
6	8		Sample Std. Deviation	=STDEV(A2:A61)	
7	8				
8	8		Hypothesized Value	7	
9	7				
10	8		Standard Error	=D6/SQRT(D4)	
11	10		Test Statistic t	=(D5-D8)/D10	
12	6		Degrees of Freedom	=D4-1	
13	7				
14	8		p-value (Lower Tail)	=IF(D11<0,TDIST(-D11,D12,1),1-TDIST(D11,D12,1))	
15	8		p-value (Upper Tail)	=1-D14	
16	9		p-value (Two Tail)	=2*MIN(D14,D15)	
17	7				
59	7				
60	7				
61	8				
62					

	A	B	C	D	E
1	Rating		Hypothesis Test About a Population Mean		
2	5		With σ Unknown		
3	7				
4	8		Sample Size	60	
5	7		Sample Mean	7.25	
6	8		Sample Std. Deviation	1.05	
7	8				
8	8		Hypothesized Value	7	
9	7				
10	8		Standard Error	0.136	
11	10		Test Statistic t	1.841	
12	6		Degrees of Freedom	59	
13	7				
14	8		p-value (Lower Tail)	0.9647	
15	8		p-value (Upper Tail)	0.0353	
16	9		p-value (Two Tail)	0.0706	
17	7				
59	7				
60	7				
61	8				
62					

Note: Rows 18 to 58 are hidden.

Step 1. Enter the data range A2:A401 into the =COUNTA cell formula in cell D3
Step 2. Enter Female as the response of interest in cell D4
Step 3. Enter the data range A2:A401 into the =COUNTIF cell formula in cell D5
Step 4. Enter the hypothesized value for the population proportion .20 into cell D8

The remaining cell formulas automatically provide the standard error, the value of the test statistic z, and three p-values. Because the alternative hypothesis ($p > .20$) indicates an upper tail test, the p-value (Upper Tail) in cell D14 is used to make the decision. With p-value $= .0062 < \alpha = .05$, the null hypothesis is rejected. The p-values in cells D13 or D15 would be used if the hypothesis involved a lower tail test or a two-tailed test.

FIGURE 9.10 EXCEL WORKSHEET FOR HYPOTHESIS TESTS ABOUT A POPULATION PROPORTION

	A	B	C	D	E
1	Golfer		Hypothesis Test About a Population Proportion		
2	Female				
3	Male		Sample Size	=COUNTA(A2:A401)	
4	Female		Response of Interest	Female	
5	Male		Count for Response	=COUNTIF(A2:A401,D4)	
6	Male		Sample Proportion	=D5/D3	
7	Female				
8	Male		Hypothesized Value	0.20	
9	Male				
10	Female		Standard Error	=SQRT(D8*(1-D8)/D3)	
11	Male		Test Statistic z	=(D6-D8)/D10	
12	Male				
13	Male		p-value (Lower Tail)	=NORMSDIST(D11)	
14	Male		p-value (Upper Tail)	=1-D13	
15	Male		p-value (Two Tail)	=2*MIN(D13,D14)	
16	Female				
400	Male				
401	Male				
402					

Note: Rows 17 to 399 are hidden.

	A	B	C	D	E
1	Golfer		Hypothesis Test About a Population Proportion		
2	Female				
3	Male		Sample Size	400	
4	Female		Response of Interest	Female	
5	Male		Count for Response	100	
6	Male		Sample Proportion	0.2500	
7	Female				
8	Male		Hypothesized Value	0.20	
9	Male				
10	Female		Standard Error	0.0200	
11	Male		Test Statistic z	2.50	
12	Male				
13	Male		p-value (Lower Tail)	0.9938	
14	Male		p-value (Upper Tail)	0.0062	
15	Male		p-value (Two Tail)	0.0124	
16	Female				
400	Male				
401	Male				
402					

This template can be used to make hypothesis testing computations for other applications. For instance, to conduct a hypothesis test for a new data set, enter the new sample data into column A of the worksheet. Modify the formulas in cells D3 and D5 to correspond to the new data range. Enter the response of interest into cell D4 and the hypothesized value for the population proportion into cell D8 to obtain the results. If the new sample data have already been summarized, the new sample data do not have to be entered into the worksheet. In this case, enter the sample size into cell D3, the sample proportion into cell D6, and the hypothesized value for the population proportion into cell D8 to obtain the results. The worksheet in Figure 9.10 is available in the file Hypothesis p on the website that accompanies this book.

Appendix 9.3 Hypothesis Testing Using StatTools

In this appendix we show how StatTools can be used to conduct hypothesis tests about a population mean for the σ unknown case and about a population proportion.

Population Mean: σ Unknown Case

In this case the population standard deviation σ will be estimated by the sample standard deviation s. We use the example discussed in Section 9.4 involving ratings that 60 business travelers gave for Heathrow Airport.

Begin by using the Data Set Manager to create a StatTools data set for these data using the procedure described in the appendix in Chapter 1. The following steps can be used to test the hypothesis $H_0: \mu \leq 7$ against $H_a: \mu > 7$.

Step 1. Click the **StatTools** tab on the Ribbon
Step 2. In the **Analyses** group, click **Statistical Inference**
Step 3. Choose the **Hypothesis Test** option
Step 4. Choose Mean/Std. Deviation
Step 5. When the StatTools—Hypothesis Test for Mean/Std. Deviation dialog box appears:
    For **Analysis Type,** choose **One-Sample Analysis**
    In the **Variables** section, select **Rating**
    In the **Hypothesis Tests to Perform** section:
      Select the **Mean** option
      Enter 7 in the **Null Hypothesis Value** box
      Select **Greater Than Null Value (One-Tailed Test)** in the **Alternative Hypothesis** box
    If selected, remove the check in the **Standard Deviation** box
    Click **OK**

The results from the hypothesis test will appear. They include the p-value and the value of the test statistic.

Population Proportion

We illustrate using the Pine Creek golf course example in Section 9.5. Begin by using the Data Set Manager to create a StatTools data set for these data using the procedure described in the appendix to Chapter 1. The following steps can be used to conduct a hypothesis test of the population proportion.

Step 1. Click the **StatTools** tab on the Ribbon
Step 2. In the **Analyses** group, click **Statistical Inference**
Step 3. Choose **Hypothesis Test**
Step 4. Choose **Proportion**
Step 5. When the StatTools—Hypothesis Test for Proportion dialog box appears:
    For **Analysis Type** choose **One-Sample Analysis**
    In the **Variables** section, select **Golfer**
    In the **Categories to Analyze** section, select **Female**
    In the **Hypotheses About Proportion** section:
      Enter .20 in the **Null Hypothesis Value** box
      In the **Alternaive Hypotheses** box, choose **Greater Than Null Value (One-Tailed Test)**
    Click **OK**

The results from the hypothesis test will appear. They include the p-value and the value of the test statistic.

CHAPTER 10

Comparisons Involving Means, Experimental Design, and Analysis of Variance

CONTENTS

STATISTICS IN PRACTICE: U.S. FOOD AND DRUG ADMINISTRATION

10.1 INFERENCES ABOUT THE DIFFERENCE BETWEEN TWO POPULATION MEANS: σ_1 AND σ_2 KNOWN
Interval Estimation of $\mu_1 - \mu_2$
Hypothesis Tests About $\mu_1 - \mu_2$
Practical Advice

10.2 INFERENCES ABOUT THE DIFFERENCE BETWEEN TWO POPULATION MEANS: σ_1 AND σ_2 UNKNOWN
Interval Estimation of $\mu_1 - \mu_2$
Hypothesis Tests About $\mu_1 - \mu_2$
Practical Advice

10.3 INFERENCES ABOUT THE DIFFERENCE BETWEEN TWO POPULATION MEANS: MATCHED SAMPLES

10.4 AN INTRODUCTION TO EXPERIMENTAL DESIGN AND ANALYSIS OF VARIANCE
Data Collection
Assumptions for Analysis of Variance
Analysis of Variance: A Conceptual Overview

10.5 ANALYSIS OF VARIANCE AND THE COMPLETELY RANDOMIZED DESIGN
Between-Treatments Estimate of Population Variance
Within-Treatments Estimate of Population Variance
Comparing the Variance Estimates: The F Test
ANOVA Table
Computer Results for Analysis of Variance
Testing for the Equality of k Population Means: An Observational Study

STATISTICS *in* PRACTICE

U.S. FOOD AND DRUG ADMINISTRATION
WASHINGTON, D.C.

It is the responsibility of the U.S. Food and Drug Administration (FDA), through its Center for Drug Evaluation and Research (CDER), to ensure that drugs are safe and effective. But CDER does not do the actual testing of new drugs itself. It is the responsibility of the company seeking to market a new drug to test it and submit evidence that it is safe and effective. CDER statisticians and scientists then review the evidence submitted.

Companies seeking approval of a new drug conduct extensive statistical studies to support their application. The testing process in the pharmaceutical industry usually consists of three stages: (1) preclinical testing, (2) testing for long-term usage and safety, and (3) clinical efficacy testing. At each successive stage, the chance that a drug will pass the rigorous tests decreases; however, the cost of further testing increases dramatically. Industry surveys indicate that on average the research and development for one new drug costs $250 million and takes 12 years. Hence, it is important to eliminate unsuccessful new drugs in the early stages of the testing process, as well as to identify promising ones for further testing.

Statistics plays a major role in pharmaceutical research, where government regulations are stringent and rigorously enforced. In preclinical testing, a two- or three-population statistical study typically is used to determine whether a new drug should continue to be studied in the long-term usage and safety program. The populations may consist of the new drug, a control, and a standard drug. The preclinical testing process begins when a new drug is sent to the pharmacology group for evaluation of efficacy—the capacity of the drug to produce the desired effects. As part of the process, a statistician is asked to design an experiment that can be used to test the new drug. The design must specify the sample size and the statistical methods of analysis. In a two-population study, one sample is used to obtain data on the efficacy of the new drug (population 1) and a second sample is used to obtain data on the efficacy of a standard drug (population 2). Depending on the intended use, the new and standard drugs are tested in such disciplines as

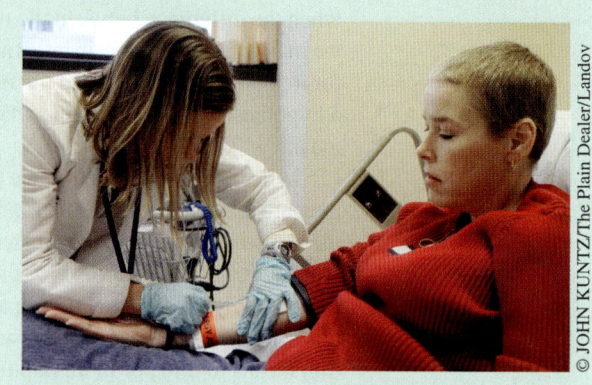

Extensive statistical studies are conducted before a new drug is approved.

neurology, cardiology, and immunology. In most studies, the statistical method involves hypothesis testing for the difference between the means of the new drug population and the standard drug population. If a new drug lacks efficacy or produces undesirable effects in comparison with the standard drug, the new drug is rejected and withdrawn from further testing. Only new drugs that show promising comparisons with the standard drugs are forwarded to the long-term usage and safety testing program.

Further data collection and multipopulation studies are conducted in the long-term usage and safety testing program and in the clinical testing programs. The FDA requires that statistical methods be defined prior to such testing to avoid data-related biases. In addition, to avoid human biases, some of the clinical trials are double or triple blind. That is, neither the subject nor the investigator knows what drug is administered to whom. If the new drug meets all requirements in relation to the standard drug, a new drug application (NDA) is filed with the FDA. The application is rigorously scrutinized by statisticians and scientists at the agency.

In this chapter you will learn how to construct interval estimates and make hypothesis tests about means and proportions with two populations. Techniques will be presented for analyzing independent random samples as well as matched samples.

In Chapters 8 and 9 we showed how to develop interval estimates and conduct hypothesis tests for situations involving a single population mean and a single population proportion. In Sections 10.1 to 10.3 we continue our discussion of statistical inference by showing how interval estimates and hypothesis tests can be developed for situations involving two populations when the difference between the two population means is of prime importance. For example, we may want to develop an interval estimate of the difference between the mean starting salary for a population of men and the mean starting salary for a population of women, or conduct a hypothesis test to determine whether any difference is present between the two population means.

In Section 10.4, we introduce the basic principles of an experimental study and show how they are used in a completely randomized design. We also provide a conceptual overview of the statistical procedure called analysis of variance (ANOVA). In Section 10.5 we show how ANOVA can be used to test for the equality of k population means using data obtained from a completely randomized design as well as data obtained from an observational study. So, in this sense, ANOVA extends the statistical material in Sections 10.1 to 10.3 from two population means to three or more population means.

We begin our discussion of statistical inference about two populations by showing how to develop interval estimates and conduct hypothesis tests about the difference between the means of two populations when the standard deviations of the two populations are assumed known.

10.1 Inferences About the Difference Between Two Population Means: σ_1 and σ_2 Known

Letting μ_1 denote the mean of population 1 and μ_2 denote the mean of population 2, we will focus on inferences about the difference between the means: $\mu_1 - \mu_2$. To make an inference about this difference, we select a random sample of n_1 elements from population 1 and a second random sample of n_2 elements from population 2. The two samples, taken separately and independently, are referred to as **independent random samples**. In this section, we assume that information is available such that the two population standard deviations, σ_1 and σ_2, can be assumed known prior to collecting the samples. We refer to this situation as the σ_1 and σ_2 known case. In the following example we show how to compute a margin of error and develop an interval estimate of the difference between the two population means when σ_1 and σ_2 are known.

Interval Estimation of $\mu_1 - \mu_2$

Greystone Department Stores, Inc., operates two stores in Buffalo, New York: One is in the inner city and the other is in a suburban shopping center. The regional manager noticed that products that sell well in one store do not always sell well in the other. The manager believes this situation may be attributable to differences in customer demographics at the two locations. Customers may differ in age, education, income, and so on. Suppose the manager asks us to investigate the difference between the mean ages of the customers who shop at the two stores.

Let us define population 1 as all customers who shop at the inner-city store and population 2 as all customers who shop at the suburban store.

μ_1 = mean of population 1 (i.e., the mean age of all customers who shop at the inner-city store)

μ_2 = mean of population 2 (i.e., the mean age of all customers who shop at the suburban store)

The difference between the two population means is $\mu_1 - \mu_2$.

10.1 Inferences About the Difference Between Two Population Means: σ_1 and σ_2 Known

To estimate $\mu_1 - \mu_2$, we will select a random sample of n_1 customers from population 1 and a sample of n_2 customers from population 2. We then compute the two sample means.

$\bar{x}_1$ = sample mean age for the random sample of n_1 inner-city customers
$\bar{x}_2$ = sample mean age for the random sample of n_2 suburban customers

The point estimator of the difference between the two population means is the difference between the two sample means.

POINT ESTIMATOR OF THE DIFFERENCE BETWEEN TWO POPULATION MEANS

$$\bar{x}_1 - \bar{x}_2 \tag{10.1}$$

Figure 10.1 provides an overview of the process used to estimate the difference between two population means based on two independent random samples.

As with other point estimators, the point estimator $\bar{x}_1 - \bar{x}_2$ has a standard error that describes the variation in the sampling distribution of the estimator. With two independent random samples, the standard error of $\bar{x}_1 - \bar{x}_2$ is as follows.

The standard error of $\bar{x}_1 - \bar{x}_2$ is the standard deviation of the sampling distribution of $\bar{x}_1 - \bar{x}_2$.

STANDARD ERROR OF $\bar{x}_1 - \bar{x}_2$

$$\sigma_{\bar{x}_1 - \bar{x}_2} = \sqrt{\frac{\sigma_1^2}{n_1} + \frac{\sigma_2^2}{n_2}} \tag{10.2}$$

If both populations have a normal distribution, or if the sample sizes are large enough that the central limit theorem enables us to conclude that the sampling distributions of $\bar{x}_1$ and $\bar{x}_2$ can be approximated by a normal distribution, the sampling distribution of $\bar{x}_1 - \bar{x}_2$ will have a normal distribution with mean given by $\mu_1 - \mu_2$.

FIGURE 10.1 ESTIMATING THE DIFFERENCE BETWEEN TWO POPULATION MEANS

Population 1
Inner-City Store Customers
μ_1 = mean age of inner-city store customers

Population 2
Suburban Store Customers
μ_2 = mean age of suburban store customers

$\mu_1 - \mu_2$ = difference between the mean ages

Two Independent Simple Random Samples

Random sample of n_1 inner-city customers
$\bar{x}_1$ = sample mean age for the inner-city store customers

Random sample of n_2 suburban customers
$\bar{x}_2$ = sample mean age for the suburban store customers

$\bar{x}_1 - \bar{x}_2$ = Point estimator of $\mu_1 - \mu_2$

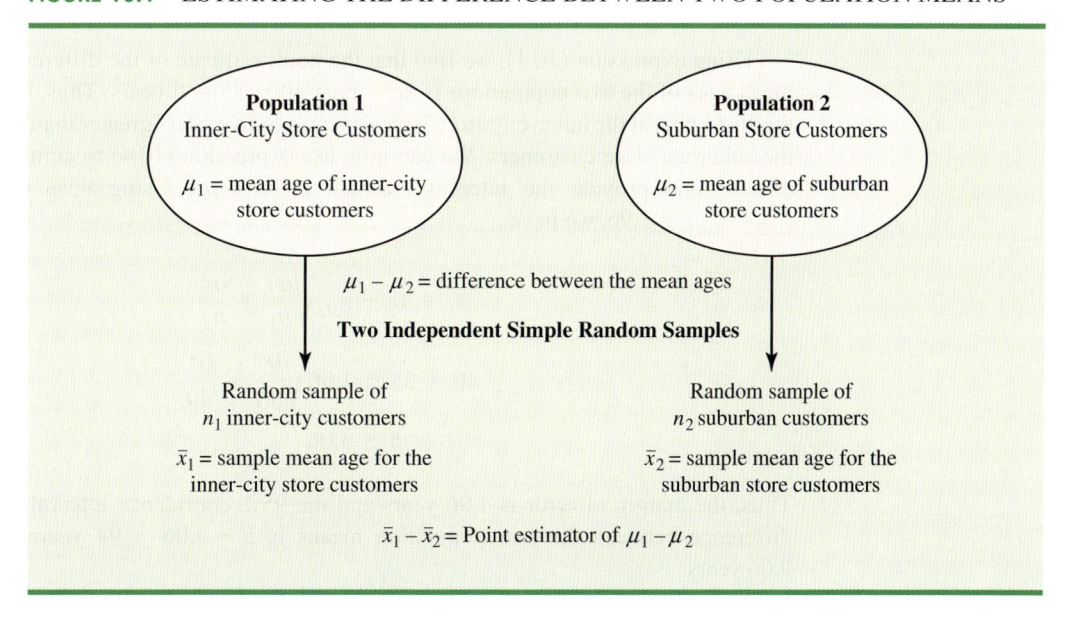

As we showed in Chapter 8, an interval estimate is given by a point estimate ± a margin of error. In the case of estimation of the difference between two population means, an interval estimate will take the following form:

$$\bar{x}_1 - \bar{x}_2 \pm \text{Margin of error}$$

With the sampling distribution of $\bar{x}_1 - \bar{x}_2$ having a normal distribution, we can write the margin of error as follows:

The margin of error is given by multiplying the standard error by $z_{\alpha/2}$.

$$\text{Margin of error} = z_{\alpha/2}\sigma_{\bar{x}_1-\bar{x}_2} = z_{\alpha/2}\sqrt{\frac{\sigma_1^2}{n_1} + \frac{\sigma_2^2}{n_2}} \quad (10.3)$$

Thus the interval estimate of the difference between two population means is as follows.

INTERVAL ESTIMATE OF THE DIFFERENCE BETWEEN TWO POPULATION MEANS: σ_1 AND σ_2 KNOWN

$$\bar{x}_1 - \bar{x}_2 \pm z_{\alpha/2}\sqrt{\frac{\sigma_1^2}{n_1} + \frac{\sigma_2^2}{n_2}} \quad (10.4)$$

where $1 - \alpha$ is the confidence coefficient.

Let us return to the Greystone example. Based on data from previous customer demographic studies, the two population standard deviations are known with $\sigma_1 = 9$ years and $\sigma_2 = 10$ years. The data collected from the two independent random samples of Greystone customers provided the following results:

	Inner-City Store	Suburban Store
Sample Size	$n_1 = 36$	$n_2 = 49$
Sample Mean	$\bar{x}_1 = 40$ years	$\bar{x}_2 = 35$ years

Using expression (10.1), we find that the point estimate of the difference between the mean ages of the two populations is $\bar{x}_1 - \bar{x}_2 = 40 - 35 = 5$ years. Thus, we estimate that the customers at the inner-city store have a mean age five years greater than the mean age of the suburban store customers. We can now use expression (10.4) to compute the margin of error and provide the interval estimate of $\mu_1 - \mu_2$. Using 95% confidence and $z_{\alpha/2} = z_{.025} = 1.96$, we have

$$\bar{x}_1 - \bar{x}_2 \pm z_{\alpha/2}\sqrt{\frac{\sigma_1^2}{n_1} + \frac{\sigma_2^2}{n_2}}$$

$$40 - 35 \pm 1.96\sqrt{\frac{9^2}{36} + \frac{10^2}{49}}$$

$$5 \pm 4.06$$

Thus, the margin of error is 4.06 years and the 95% confidence interval estimate of the difference between the two population means is $5 - 4.06 = .94$ years to $5 + 4.06 = 9.06$ years.

10.1 Inferences About the Difference Between Two Population Means: σ_1 and σ_2 Known

Hypothesis Tests About $\mu_1 - \mu_2$

Let us consider hypothesis tests about the difference between two population means. Using D_0 to denote the hypothesized difference between μ_1 and μ_2, the three forms for a hypothesis test are as follows:

$$H_0: \mu_1 - \mu_2 \geq D_0 \qquad H_0: \mu_1 - \mu_2 \leq D_0 \qquad H_0: \mu_1 - \mu_2 = D_0$$
$$H_a: \mu_1 - \mu_2 < D_0 \qquad H_a: \mu_1 - \mu_2 > D_0 \qquad H_a: \mu_1 - \mu_2 \neq D_0$$

In many applications, $D_0 = 0$. Using the two-tailed test as an example, when $D_0 = 0$ the null hypothesis is $H_0: \mu_1 - \mu_2 = 0$. In this case, the null hypothesis is that μ_1 and μ_2 are equal. Rejection of H_0 leads to the conclusion that $H_a: \mu_1 - \mu_2 \neq 0$ is true; that is, μ_1 and μ_2 are not equal.

The steps for conducting hypothesis tests presented in Chapter 9 are applicable here. We must choose a level of significance, compute the value of the test statistic, and find the p-value to determine whether the null hypothesis should be rejected. With two independent random samples, we showed that the point estimator $\bar{x}_1 - \bar{x}_2$ has a standard error $\sigma_{\bar{x}_1 - \bar{x}_2}$ given by expression (10.2) and, when the sample sizes are large enough, the distribution of $\bar{x}_1 - \bar{x}_2$ can be described by a normal distribution. In this case, the test statistic for the difference between two population means when σ_1 and σ_2 are known is as follows.

TEST STATISTIC FOR HYPOTHESIS TESTS ABOUT $\mu_1 - \mu_2$: σ_1 AND σ_2 KNOWN

$$z = \frac{(\bar{x}_1 - \bar{x}_2) - D_0}{\sqrt{\dfrac{\sigma_1^2}{n_1} + \dfrac{\sigma_2^2}{n_2}}} \tag{10.5}$$

Let us demonstrate the use of this test statistic in the following hypothesis testing example.

As part of a study to evaluate differences in education quality between two training centers, a standardized examination is given to individuals who are trained at the centers. The difference between the mean examination scores is used to assess quality differences between the centers. The population means for the two centers are as follows.

$\mu_1 =$ the mean examination score for the population of individuals trained at center A

$\mu_2 =$ the mean examination score for the population of individuals trained at center B

We begin with the tentative assumption that no difference exists between the training quality provided at the two centers. Hence, in terms of the mean examination scores, the null hypothesis is that $\mu_1 - \mu_2 = 0$. If sample evidence leads to the rejection of this hypothesis, we will conclude that the mean examination scores differ for the two populations. This conclusion indicates a quality differential between the two centers and suggests that a follow-up study investigating the reason for the differential may be warranted. The null and alternative hypotheses for this two-tailed test are written as follows:

$$H_0: \mu_1 - \mu_2 = 0$$
$$H_a: \mu_1 - \mu_2 \neq 0$$

The standardized examination given previously in a variety of settings always resulted in an examination score standard deviation near 10 points. Thus, we will use this information to assume that the population standard deviations are known with $\sigma_1 = 10$ and $\sigma_2 = 10$. An $\alpha = .05$ level of significance is specified for the study.

Independent random samples of $n_1 = 30$ individuals from training center A and $n_2 = 40$ individuals from training center B are taken. The respective sample means are $\bar{x}_1 = 82$ and $\bar{x}_2 = 78$. Do these data suggest a significant difference between the population means at the two training centers? To help answer this question, we compute the test statistic using equation (10.5).

$$z = \frac{(\bar{x}_1 - \bar{x}_2) - D_0}{\sqrt{\frac{\sigma_1^2}{n_1} + \frac{\sigma_2^2}{n_2}}} = \frac{(82 - 78) - 0}{\sqrt{\frac{10^2}{30} + \frac{10^2}{40}}} = 1.66$$

Next let us compute the *p*-value for this two-tailed test. Because the test statistic z is in the upper tail, we first compute $P(z \geq 1.66)$. Using the standard normal distribution table, the area to the left of $z = 1.66$ is .9515. Thus, the area in the upper tail of the distribution is $1.0000 - .9515 = .0485$. Because this test is a two-tailed test, we must double the tail area: *p*-value = 2(.0485) = .0970. Following the usual rule to reject H_0 if *p*-value $\leq \alpha$, we see that the *p*-value of .0970 does not allow us to reject H_0 at the .05 level of significance. The sample results do not provide sufficient evidence to conclude the training centers differ in quality.

In this chapter we will use the *p*-value approach to hypothesis testing as described in Chapter 9. However, if you prefer, the test statistic and the critical value rejection rule may be used. With $\alpha = .05$ and $z_{\alpha/2} = z_{.025} = 1.96$, the rejection rule employing the critical value approach would be "reject H_0 if $z \leq -1.96$ or if $z \geq 1.96$." With $z = 1.66$, we reach the same "do not reject H_0" conclusion.

In the preceding example, we demonstrated a two-tailed hypothesis test about the difference between two population means. Lower tail and upper tail tests can also be considered. These tests use the same test statistic as given in equation (10.5). The procedure for computing the *p*-value and the rejection rules for these one-tailed tests is the same as those presented in Chapter 9.

Practical Advice

In most applications of the interval estimation and hypothesis testing procedures presented in this section, random samples with $n_1 \geq 30$ and $n_2 \geq 30$ are adequate. In cases where either or both sample sizes are less than 30, the distributions of the populations become important considerations. In general, with smaller sample sizes, it is more important for the analyst to be satisfied that it is reasonable to assume that the distributions of the two populations are at least approximately normal.

Exercises

Methods

1. The following results come from two independent random samples taken of two populations.

Sample 1	Sample 2
$n_1 = 50$	$n_2 = 35$
$\bar{x}_1 = 13.6$	$\bar{x}_2 = 11.6$
$\sigma_1 = 2.2$	$\sigma_2 = 3.0$

a. What is the point estimate of the difference between the two population means?
b. Provide a 90% confidence interval for the difference between the two population means.
c. Provide a 95% confidence interval for the difference between the two population means.

2. Consider the following hypothesis test.

$$H_0: \mu_1 - \mu_2 \leq 0$$
$$H_a: \mu_1 - \mu_2 > 0$$

The following results are for two independent random samples taken from the two populations.

Sample 1	Sample 2
$n_1 = 40$	$n_2 = 50$
$\bar{x}_1 = 25.2$	$\bar{x}_2 = 22.8$
$\sigma_1 = 5.2$	$\sigma_2 = 6.0$

a. What is the value of the test statistic?
b. What is the *p*-value?
c. With $\alpha = .05$, what is your hypothesis testing conclusion?

3. Consider the following hypothesis test.

$$H_0: \mu_1 - \mu_2 = 0$$
$$H_a: \mu_1 - \mu_2 \neq 0$$

The following results are for two independent samples taken from the two populations.

Sample 1	Sample 2
$n_1 = 80$	$n_2 = 70$
$\bar{x}_1 = 104$	$\bar{x}_2 = 106$
$\sigma_1 = 8.4$	$\sigma_2 = 7.6$

a. What is the value of the test statistic?
b. What is the *p*-value?
c. With $\alpha = .05$, what is your hypothesis testing conclusion?

Applications

4. *Condé Nast Traveler* conducts an annual survey in which readers rate their favorite cruise ship. All ships are rated on a 100-point scale, with higher values indicating better service. A sample of 37 ships that carry fewer than 500 passengers resulted in an average rating of 85.36, and a sample of 44 ships that carry 500 or more passengers provided an average rating of 81.40 (*Condé Nast Traveler*, February 2008). Assume that the population standard deviation is 4.55 for ships that carry fewer than 500 passengers and 3.97 for ships that carry 500 or more passengers.
   a. What is the point estimate of the difference between the population mean rating for ships that carry fewer than 500 passengers and the population mean rating for ships that carry 500 or more passengers?
   b. At 95% confidence, what is the margin of error?
   c. What is a 95% confidence interval estimate of the difference between the population mean ratings for the two sizes of ships?

5. The average expenditure on Valentine's Day was expected to be $100.89 (*USA Today*, February 13, 2006). Do male and female consumers differ in the amounts they spend? The average expenditure in a sample survey of 40 male consumers was $135.67, and the average expenditure in a sample survey of 30 female consumers was $68.64. Based on past

surveys, the standard deviation for male consumers is assumed to be $35, and the standard deviation for female consumers is assumed to be $20.
   a. What is the point estimate of the difference between the population mean expenditure for males and the population mean expenditure for females?
   b. At 99% confidence, what is the margin of error?
   c. Develop a 99% confidence interval for the difference between the two population means.

6. Suppose that you are responsible for making arrangements for a business convention. Because of budget cuts due to the recent recession, you have been charged with choosing a city for the convention that has the least expensive hotel rooms. You have narrowed your choices to Atlanta and Houston. The file named Hotel contains samples of prices for rooms in Atlanta and Houston that are consistent with the results reported by Smith Travel Research (*SmartMoney*, March 2009). Because considerable historical data on the prices of rooms in both cities are available, the population standard deviations for the prices can be assumed to be $20 in Atlanta and $25 in Houston. Based on the sample data, can you conclude that the mean price of a hotel room in Atlanta is lower than one in Houston?

7. During the 2003 season, Major League Baseball took steps to speed up the play of baseball games in order to maintain fan interest (CNN Headline News, September 30, 2003). The following results come from a sample of 60 games played during the summer of 2002 and a sample of 50 games played during the summer of 2003. The sample mean shows the mean duration of the games included in each sample.

2002 Season	2003 Season
$n_1 = 60$	$n_2 = 50$
$\bar{x}_1 = 2$ hours, 52 minutes	$\bar{x}_2 = 2$ hours, 46 minutes

   a. A research hypothesis was that the steps taken during the 2003 season would reduce the population mean duration of baseball games. Formulate the null and alternative hypotheses.
   b. What is the point estimate of the reduction in the mean duration of games during the 2003 season?
   c. Historical data indicate a population standard deviation of 12 minutes is a reasonable assumption for both years. Conduct the hypothesis test and report the *p*-value. At a .05 level of significance, what is your conclusion?
   d. Provide a 95% confidence interval estimate of the reduction in the mean duration of games during the 2003 season.
   e. What was the percentage reduction in the mean time of baseball games during the 2003 season? Should management be pleased with the results of the statistical analysis? Discuss. Should the length of baseball games continue to be an issue in future years? Explain.

8. Will improving customer service result in higher stock prices for the companies providing the better service? "When a company's satisfaction score has improved over the prior year's results and is above the national average (currently 75.7), studies show its shares have a good chance of outperforming the broad stock market in the long run" (*BusinessWeek*, March 2, 2009). The following satisfaction scores of three companies for the 4th quarters of 2007 and 2008 were obtained from the American Customer Satisfaction Index. Assume that the scores are based on a poll of 60 customers from each company. Because the polling has been done for several years, the standard deviation can be assumed to equal 6 points in each case.

Company	2007 Score	2008 Score
Rite Aid	73	76
Expedia	75	77
J.C. Penney	77	78

a. For Rite Aid, is the increase in the satisfaction score from 2007 to 2008 statistically significant? Use $\alpha = .05$. What can you conclude?
b. Can you conclude that the 2008 score for Rite Aid is above the national average of 75.7? Use $\alpha = .05$.
c. For Expedia, is the increase from 2007 to 2008 statistically significant? Use $\alpha = .05$.
d. When conducting a hypothesis test with the values given for the standard deviation, sample size, and α, how large must the increase from 2007 to 2008 be for it to be statistically significant?
e. Use the result of part (d) to state whether the increase for J.C. Penney from 2007 to 2008 is statistically significant.

10.2 Inferences About the Difference Between Two Population Means: σ_1 and σ_2 Unknown

In this section we extend the discussion of inferences about the difference between two population means to the case when the two population standard deviations, σ_1 and σ_2, are unknown. In this case, we will use the sample standard deviations, s_1 and s_2, to estimate the unknown population standard deviations. When we use the sample standard deviations, the interval estimation and hypothesis testing procedures will be based on the t distribution rather than the standard normal distribution.

Interval Estimation of $\mu_1 - \mu_2$

In the following example we show how to compute a margin of error and develop an interval estimate of the difference between two population means when σ_1 and σ_2 are unknown. Clearwater National Bank is conducting a study designed to identify differences between checking account practices by customers at two of its branch banks. A sample of 28 checking accounts is selected from the Cherry Grove Branch and an independent sample of 22 checking accounts is selected from the Beechmont Branch. The current checking account balance is recorded for each of the checking accounts. A summary of the account balances follows:

	Cherry Grove	Beechmont
Sample Size	$n_1 = 28$	$n_2 = 22$
Sample Mean	$\bar{x}_1 = \$1025$	$\bar{x}_2 = \$910$
Sample Standard Deviation	$s_1 = \$150$	$s_2 = \$125$

Clearwater National Bank would like to estimate the difference between the mean checking account balance maintained by the population of Cherry Grove customers and the population of Beechmont customers. Let us develop the margin of error and an interval estimate of the difference between these two population means.

In Section 10.1, we provided the following interval estimate for the case when the population standard deviations, σ_1 and σ_2, are known.

$$\bar{x}_1 - \bar{x}_2 \pm z_{\alpha/2} \sqrt{\frac{\sigma_1^2}{n_1} + \frac{\sigma_2^2}{n_2}}$$

When σ_1 and σ_2 are estimated by s_1 and s_2, the t distribution is used to make inferences about the difference between two population means.

With σ_1 and σ_2 unknown, we will use the sample standard deviations s_1 and s_2 to estimate σ_1 and σ_2 and replace $z_{\alpha/2}$ with $t_{\alpha/2}$. As a result, the interval estimate of the difference between two population means is given by the following expression.

INTERVAL ESTIMATE OF THE DIFFERENCE BETWEEN TWO POPULATION MEANS: σ_1 AND σ_2 UNKNOWN

$$\bar{x}_1 - \bar{x}_2 \pm t_{\alpha/2} \sqrt{\frac{s_1^2}{n_1} + \frac{s_2^2}{n_2}} \qquad (10.6)$$

where $1 - \alpha$ is the confidence coefficient.

In this expression, the use of the t distribution is an approximation, but it provides excellent results and is relatively easy to use. The only difficulty that we encounter in using expression (10.6) is determining the appropriate degrees of freedom for $t_{\alpha/2}$. Statistical software packages compute the appropriate degrees of freedom automatically. The formula used is as follows.

DEGREES OF FREEDOM: t DISTRIBUTION WITH TWO INDEPENDENT RANDOM SAMPLES

$$df = \frac{\left(\frac{s_1^2}{n_1} + \frac{s_2^2}{n_2}\right)^2}{\frac{1}{n_1 - 1}\left(\frac{s_1^2}{n_1}\right)^2 + \frac{1}{n_2 - 1}\left(\frac{s_2^2}{n_2}\right)^2} \qquad (10.7)$$

Let us return to the Clearwater National Bank example and show how to use expression (10.6) to provide a 95% confidence interval estimate of the difference between the population mean checking account balances at the two branch banks. The sample data show $n_1 = 28$, $\bar{x}_1 = \$1025$, and $s_1 = \$150$ for the Cherry Grove branch, and $n_2 = 22$, $\bar{x}_2 = \$910$, and $s_2 = \$125$ for the Beechmont branch. The calculation for degrees of freedom for $t_{\alpha/2}$ is as follows:

$$df = \frac{\left(\frac{s_1^2}{n_1} + \frac{s_2^2}{n_2}\right)^2}{\frac{1}{n_1 - 1}\left(\frac{s_1^2}{n_1}\right)^2 + \frac{1}{n_2 - 1}\left(\frac{s_2^2}{n_2}\right)^2} = \frac{\left(\frac{150^2}{28} + \frac{125^2}{22}\right)^2}{\frac{1}{28 - 1}\left(\frac{150^2}{28}\right)^2 + \frac{1}{22 - 1}\left(\frac{125^2}{22}\right)^2} = 47.8$$

We round the noninteger degrees of freedom *down* to 47 to provide a larger t value and a more conservative interval estimate. Using the t distribution table with 47 degrees of freedom, we find $t_{.025} = 2.012$. Using expression (10.6), we develop the 95% confidence interval estimate of the difference between the two population means as follows:

$$\bar{x}_1 - \bar{x}_2 \pm t_{.025} \sqrt{\frac{s_1^2}{n_1} + \frac{s_2^2}{n_2}}$$

$$1025 - 910 \pm 2.012 \sqrt{\frac{150^2}{28} + \frac{125^2}{22}}$$

$$115 \pm 78$$

The point estimate of the difference between the population mean checking account balances at the two branches is $115. The margin of error is $78, and the 95% confidence interval estimate of the difference between the two population means is $115 - 78 = \$37$ to $115 + 78 = \$193$.

This suggestion should help if you are using equation (10.7) to calculate the degrees of freedom by hand.

The computation of the degrees of freedom (equation 10.7) is cumbersome if you are doing the calculation by hand, but it is easily implemented with a computer software package. However, note that the expressions s_1^2/n_1 and s_2^2/n_2 appear in both expression (10.6) and equation (10.7). These values only need to be computed once in order to evaluate both (10.6) and (10.7).

Hypothesis Tests About $\mu_1 - \mu_2$

Let us now consider hypothesis tests about the difference between the means of two populations when the population standard deviations σ_1 and σ_2 are unknown. Letting D_0 denote the hypothesized difference between μ_1 and μ_2, Section 10.1 showed that the test statistic used for the case where σ_1 and σ_2 are known is as follows.

$$z = \frac{(\bar{x}_1 - \bar{x}_2) - D_0}{\sqrt{\dfrac{\sigma_1^2}{n_1} + \dfrac{\sigma_2^2}{n_2}}}$$

The test statistic, z, follows the standard normal distribution.

When σ_1 and σ_2 are unknown, we use s_1 as an estimator of σ_1 and s_2 as an estimator of σ_2. Substituting these sample standard deviations for σ_1 and σ_2 provides the following test statistic when σ_1 and σ_2 are unknown.

TEST STATISTIC FOR HYPOTHESIS TESTS ABOUT $\mu_1 - \mu_2$: σ_1 AND σ_2 UNKNOWN

$$t = \frac{(\bar{x}_1 - \bar{x}_2) - D_0}{\sqrt{\dfrac{s_1^2}{n_1} + \dfrac{s_2^2}{n_2}}} \tag{10.8}$$

The degrees of freedom for t are given by equation (10.7).

Let us demonstrate the use of this test statistic in the following hypothesis testing example.

Consider a new computer software package developed to help systems analysts reduce the time required to design, develop, and implement an information system. To evaluate the benefits of the new software package, a random sample of 24 systems analysts is selected. Each analyst is given specifications for a hypothetical information system. Then 12 of the analysts are instructed to produce the information system by using current technology. The other 12 analysts are trained in the use of the new software package and then instructed to use it to produce the information system.

This study involves two populations: a population of systems analysts using the current technology and a population of systems analysts using the new software package. In terms of the time required to complete the information system design project, the population means are as follow.

μ_1 = the mean project completion time for systems analysts using the current technology

μ_2 = the mean project completion time for systems analysts using the new software package

The researcher in charge of the new software evaluation project hopes to show that the new software package will provide a shorter mean project completion time. Thus, the researcher is looking for evidence to conclude that μ_2 is less than μ_1; in this case, the

TABLE 10.1 COMPLETION TIME DATA AND SUMMARY STATISTICS FOR THE SOFTWARE TESTING STUDY

	Current Technology	New Software
	300	274
	280	220
	344	308
	385	336
	372	198
	360	300
	288	315
	321	258
	376	318
	290	310
	301	332
	283	263
Summary Statistics		
Sample size	$n_1 = 12$	$n_2 = 12$
Sample mean	$\bar{x}_1 = 325$ hours	$\bar{x}_2 = 286$ hours
Sample standard deviation	$s_1 = 40$	$s_2 = 44$

WEB file
SoftwareTest

difference between the two population means, $\mu_1 - \mu_2$, will be greater than zero. The research hypothesis $\mu_1 - \mu_2 > 0$ is stated as the alternative hypothesis. Thus, the hypothesis test becomes

$$H_0: \mu_1 - \mu_2 \leq 0$$
$$H_a: \mu_1 - \mu_2 > 0$$

We will use $\alpha = .05$ as the level of significance.

Suppose that the 24 analysts complete the study with the results shown in Table 10.1. Using the test statistic in equation (10.8), we have

$$t = \frac{(\bar{x}_1 - \bar{x}_2) - D_0}{\sqrt{\dfrac{s_1^2}{n_1} + \dfrac{s_2^2}{n_2}}} = \frac{(325 - 286) - 0}{\sqrt{\dfrac{40^2}{12} + \dfrac{44^2}{12}}} = 2.27$$

Computing the degrees of freedom using equation (10.7), we have

$$df = \frac{\left(\dfrac{s_1^2}{n_1} + \dfrac{s_2^2}{n_2}\right)^2}{\dfrac{1}{n_1 - 1}\left(\dfrac{s_1^2}{n_1}\right)^2 + \dfrac{1}{n_2 - 1}\left(\dfrac{s_2^2}{n_2}\right)^2} = \frac{\left(\dfrac{40^2}{12} + \dfrac{44^2}{12}\right)^2}{\dfrac{1}{12 - 1}\left(\dfrac{40^2}{12}\right)^2 + \dfrac{1}{12 - 1}\left(\dfrac{44^2}{12}\right)^2} = 21.8$$

Rounding down, we will use a t distribution with 21 degrees of freedom. This row of the t distribution table is as follows:

Area in Upper Tail	.20	.10	.05	.025	.01	.005
t-Value (21 df)	0.859	1.323	1.721	2.080	2.518	2.831

$t = 2.27$

FIGURE 10.2 MINITAB OUTPUT FOR THE HYPOTHESIS TEST OF THE CURRENT AND NEW SOFTWARE TECHNOLOGY

```
Two-sample T for Current vs New
              N      Mean     StDev    SE Mean
Current      12     325.0      40.0       12
New          12     286.0      44.0       13

Difference = mu Current - mu New
Estimate for difference:  39.0000
95% lower bound for difference = 9.5
T-Test of difference = 0 (vs >):  T-Value = 2.27   P-Value = 0.017   DF = 21
```

Using the t distribution table, we can only determine a range for the p-value. Use of Excel or Minitab shows the exact p-value = .017.

With an upper tail test, the *p*-value is the area in the upper tail corresponding to $t = 2.27$. From the preceding results, we see that the *p*-value is between .025 and .01. Thus, the *p*-value is less than $\alpha = .05$ and H_0 is rejected. The sample results enable the researcher to conclude that $\mu_1 - \mu_2 > 0$, or $\mu_1 > \mu_2$. Thus, the research study supports the conclusion that the new software package provides a smaller population mean completion time.

Minitab or Excel can be used to analyze data for testing hypotheses about the difference between two population means. The Minitab output comparing the current and new software technology is shown in Figure 10.2. The last line of the output shows $t = 2.27$ and *p*-value $= .017$. Note that Minitab used equation (10.7) to compute 21 degrees of freedom for this analysis.

Practical Advice

Whenever possible, equal sample sizes, $n_1 = n_2$, are recommended.

The interval estimation and hypothesis testing procedures presented in this section are robust and can be used with relatively small sample sizes. In most applications, equal or nearly equal sample sizes such that the total sample size $n_1 + n_2$ is at least 20 can be expected to provide very good results even if the populations are not normal. Larger sample sizes are recommended if the distributions of the populations are highly skewed or contain outliers. Smaller sample sizes should only be used if the analyst is satisfied that the distributions of the populations are at least approximately normal.

NOTES AND COMMENTS

Another approach used to make inferences about the difference between two population means when σ_1 and σ_2 are unknown is based on the assumption that the two population standard deviations are equal ($\sigma_1 = \sigma_2 = \sigma$). Under this assumption, the two sample standard deviations are combined to provide the following *pooled sample variance:*

$$s_p^2 = \frac{(n_1 - 1)s_1^2 + (n_2 - 1)s_2^2}{n_1 + n_2 - 2}$$

The *t* test statistic becomes

$$t = \frac{(\bar{x}_1 - \bar{x}_2) - D_0}{s_p \sqrt{\frac{1}{n_1} + \frac{1}{n_2}}}$$

and has $n_1 + n_2 - 2$ degrees of freedom. At this point, the computation of the *p*-value and the interpretation of the sample results are identical to the procedures discussed earlier in this section.

A difficulty with this procedure is that the assumption that the two population standard deviations are equal is usually difficult to verify. Unequal population standard deviations are frequently encountered. Using the pooled procedure may not provide satisfactory results, especially if the sample sizes n_1 and n_2 are quite different.

The *t* procedure that we presented in this section does not require the assumption of equal population standard deviations and can be applied whether the population standard deviations are equal or not. It is a more general procedure and is recommended for most applications.

Exercises

Methods

9. The following results are for independent random samples taken from two populations.

Sample 1	Sample 2
$n_1 = 20$	$n_2 = 30$
$\bar{x}_1 = 22.5$	$\bar{x}_2 = 20.1$
$s_1 = 2.5$	$s_2 = 4.8$

   a. What is the point estimate of the difference between the two population means?
   b. What is the degrees of freedom for the *t* distribution?
   c. At 95% confidence, what is the margin of error?
   d. What is the 95% confidence interval for the difference between the two population means?

10. Consider the following hypothesis test.

$$H_0: \mu_1 - \mu_2 = 0$$
$$H_a: \mu_1 - \mu_2 \neq 0$$

The following results are from independent random samples taken from two populations.

Sample 1	Sample 2
$n_1 = 35$	$n_2 = 40$
$\bar{x}_1 = 13.6$	$\bar{x}_2 = 10.1$
$s_1 = 5.2$	$s_2 = 8.5$

   a. What is the value of the test statistic?
   b. What is the degrees of freedom for the *t* distribution?
   c. What is the *p*-value?
   d. At $\alpha = .05$, what is your conclusion?

11. Consider the following data for two independent random samples taken from two normal populations.

Sample 1	10	7	13	7	9	8
Sample 2	8	7	8	4	6	9

   a. Compute the two sample means.
   b. Compute the two sample standard deviations.
   c. What is the point estimate of the difference between the two population means?
   d. What is the 90% confidence interval estimate of the difference between the two population means?

Applications

12. The U.S. Department of Transportation provides the number of miles that residents of the 75 largest metropolitan areas travel per day in a car. Suppose that for a random sample of 50 Buffalo residents the mean is 22.5 miles a day and the standard deviation is 8.4 miles a

day, and for an independent random sample of 40 Boston residents the mean is 18.6 miles a day and the standard deviation is 7.4 miles a day.

a. What is the point estimate of the difference between the mean number of miles that Buffalo residents travel per day and the mean number of miles that Boston residents travel per day?

b. What is the 95% confidence interval for the difference between the two population means?

13. FedEx and United Parcel Service (UPS) are the world's two leading cargo carriers by volume and revenue (*The Wall Street Journal*, January 27, 2004). According to the Airports Council International, the Memphis International Airport (FedEx) and the Louisville International Airport (UPS) are 2 of the 10 largest cargo airports in the world. The following random samples show the tons of cargo per day handled by these airports. Data are in thousands of tons.

Memphis

9.1	15.1	8.8	10.0	7.5	10.5
8.3	9.1	6.0	5.8	12.1	9.3

Louisville

4.7	5.0	4.2	3.3	5.5
2.2	4.1	2.6	3.4	7.0

a. Compute the sample mean and sample standard deviation for each airport.

b. What is the point estimate of the difference between the two population means? Interpret this value in terms of the higher-volume airport and a comparison of the volume difference between the two airports.

c. Develop a 95% confidence interval of the difference between the daily population means for the two airports.

14. Are nursing salaries in Tampa, Florida, lower than those in Dallas, Texas? Salary data show staff nurses in Tampa earn less than staff nurses in Dallas (*The Tampa Tribune*, January 15, 2007). Suppose that in a follow-up study of 40 staff nurses in Tampa and 50 staff nurses in Dallas you obtain the following results.

Tampa	Dallas
$n_1 = 40$	$n_2 = 50$
$\bar{x}_1 = \$56,100$	$\bar{x}_2 = \$59,400$
$s_1 = \$6,000$	$s_2 = \$7,000$

a. Formulate a hypothesis so that, if the null hypothesis is rejected, we can conclude that salaries for staff nurses in Tampa are significantly lower than for those in Dallas. Use $\alpha = .05$.

b. What is the value of the test statistic?

c. What is the *p*-value?

d. What is your conclusion?

15. Commercial real estate prices and rental rates suffered substantial declines in the past year (*Newsweek*, July 27, 2009). These declines were particularly severe in Asia; annual lease rates in Tokyo, Hong Kong, and Singapore declined by 40% or more. Even with such large declines, annual lease rates in Asia were still higher than those in many cities in Europe and the United States. Annual lease rates for a sample of 30 commercial properties in Hong Kong showed an average of $1,114 per square meter with a standard deviation of $230.

Annual lease rates for a sample of 40 commercial properties in Paris showed an average lease rate of $989 per square meter with a standard deviation of $195.

a. On the basis of the sample results, can we conclude that the mean annual lease rate is higher in Hong Kong than in Paris? Develop appropriate null and alternative hypotheses.
b. Use $\alpha = .01$. What is your conclusion? Are rental rates higher in Hong Kong?

16. The College Board provided comparisons of SAT scores based on the highest level of education attained by the test taker's parents. A research hypothesis was that students whose parents had attained a higher level of education would on average score higher on the SAT. The overall mean SAT math score is 515 (*College Board website*, November 15, 2009). SAT math scores for independent samples of students follow. The first sample shows the SAT math test scores for students whose parents are college graduates with a bachelor's degree. The second sample shows the SAT math test scores for students whose parents are high school graduates but do not have a college degree.

	Student's Parents		
College Grads		**High School Grads**	
485	487	442	492
534	533	580	478
650	526	479	425
554	410	486	485
550	515	528	390
572	578	524	535
497	448		
592	469		

a. Formulate the hypotheses that can be used to determine whether the sample data support the hypothesis that students show a higher population mean math score on the SAT if their parents attained a higher level of education.
b. What is the point estimate of the difference between the means for the two populations?
c. Compute the p-value for the hypothesis test.
d. At $\alpha = .05$, what is your conclusion?

17. Periodically, Merrill Lynch customers are asked to evaluate Merrill Lynch financial consultants and services. Higher ratings on the client satisfaction survey indicate better service, with 7 the maximum service rating. Independent samples of service ratings for two financial consultants are summarized here. Consultant A has 10 years of experience, whereas consultant B has 1 year of experience. Use $\alpha = .05$ and test to see whether the consultant with more experience has the higher population mean service rating.

Consultant A	Consultant B
$n_1 = 16$	$n_2 = 10$
$\bar{x}_1 = 6.82$	$\bar{x}_2 = 6.25$
$s_1 = .64$	$s_2 = .75$

a. State the null and alternative hypotheses.
b. Compute the value of the test statistic.
c. What is the p-value?
d. What is your conclusion?

WEBfile

SAT

18. Educational testing companies provide tutoring, classroom learning, and practice tests in an effort to help students perform better on tests such as the SAT. The test preparation companies claim that their courses will improve SAT score performances by an average of 120 points (*The Wall Street Journal,* January 23, 2003). A researcher is uncertain of this claim and believes that 120 points may be an overstatement in an effort to encourage students to take the test preparation course. In an evaluation study of one test preparation service, the researcher collects SAT score data for 35 students who took the test preparation course and 48 students who did not take the course. The file named SAT contains the scores for this study.
   a. Formulate the hypotheses that can be used to test the researcher's belief that the improvement in SAT scores may be less than the stated average of 120 points.
   b. Using $\alpha = .05$, what is your conclusion?
   c. What is the point estimate of the improvement in the average SAT scores provided by the test preparation course? Provide a 95% confidence interval estimate of the improvement.
   d. What advice would you have for the researcher after seeing the confidence interval?

10.3 Inferences About the Difference Between Two Population Means: Matched Samples

Suppose employees at a manufacturing company can use two different methods to perform a production task. To maximize production output, the company wants to identify the method with the smaller population mean completion time. Let μ_1 denote the population mean completion time for production method 1 and μ_2 denote the population mean completion time for production method 2. With no preliminary indication of the preferred production method, we begin by tentatively assuming that the two production methods have the same population mean completion time. Thus, the null hypothesis is $H_0: \mu_1 - \mu_2 = 0$. If this hypothesis is rejected, we can conclude that the population mean completion times differ. In this case, the method providing the smaller mean completion time would be recommended. The null and alternative hypotheses are written as follows.

$$H_0: \mu_1 - \mu_2 = 0$$
$$H_a: \mu_1 - \mu_2 \neq 0$$

In choosing the sampling procedure that will be used to collect production time data and test the hypotheses, we consider two alternative designs. One is based on independent samples and the other is based on **matched samples**.

1. *Independent sample design:* A random sample of workers is selected and each worker in the sample uses method 1. A second independent random sample of workers is selected and each worker in this sample uses method 2. The test of the difference between population means is based on the procedures in Section 10.2.
2. *Matched sample design:* One random sample of workers is selected. Each worker first uses one method and then uses the other method. The order of the two methods is assigned randomly to the workers, with some workers performing method 1 first and others performing method 2 first. Each worker provides a pair of data values, one value for method 1 and another value for method 2.

In the matched sample design the two production methods are tested under similar conditions (i.e., with the same workers); hence this design often leads to a smaller sampling error than the independent sample design. The primary reason is that in a matched sample design, variation among workers is eliminated because the same workers are used for both production methods.

TABLE 10.2 TASK COMPLETION TIMES FOR A MATCHED SAMPLE DESIGN

Worker	Completion Time for Method 1 (minutes)	Completion Time for Method 2 (minutes)	Difference in Completion Times (d_i)
1	6.0	5.4	.6
2	5.0	5.2	−.2
3	7.0	6.5	.5
4	6.2	5.9	.3
5	6.0	6.0	.0
6	6.4	5.8	.6

Let us demonstrate the analysis of a matched sample design by assuming it is the method used to test the difference between population means for the two production methods. A random sample of six workers is used. The data on completion times for the six workers are given in Table 10.2. Note that each worker provides a pair of data values, one for each production method. Also note that the last column contains the difference in completion times d_i for each worker in the sample.

The key to the analysis of the matched sample design is to realize that we consider only the column of differences. Therefore, we have six data values (.6, −.2, .5, .3, .0, and .6) that will be used to analyze the difference between population means of the two production methods.

Let μ_d = the mean of the *difference* in values for the population of workers. With this notation, the null and alternative hypotheses are rewritten as follows:

$$H_0: \mu_d = 0$$
$$H_a: \mu_d \neq 0$$

If H_0 is rejected, we can conclude that the population mean completion times differ.

The d notation is a reminder that the matched sample provides *difference* data. The sample mean and sample standard deviation for the six difference values in Table 10.2 follow.

Other than the use of the d notation, the formulas for the sample mean and sample standard deviation are the same ones used previously in the text.

$$\bar{d} = \frac{\Sigma d_i}{n} = \frac{1.8}{6} = .30$$

$$s_d = \sqrt{\frac{\Sigma(d_i - \bar{d})^2}{n-1}} = \sqrt{\frac{.56}{5}} = .335$$

It is not necessary to make the assumption that the population has a normal distribution if the sample size is large. Sample size guidelines for using the t distribution were presented in Chapters 8 and 9.

With the small sample of $n = 6$ workers, we need to make the assumption that the population of differences has a normal distribution. This assumption is necessary so that we may use the t distribution for hypothesis testing and interval estimation procedures. Based on this assumption, the following test statistic has a t distribution with $n - 1$ degrees of freedom.

TEST STATISTIC FOR HYPOTHESIS TESTS INVOLVING MATCHED SAMPLES

$$t = \frac{\bar{d} - \mu_d}{s_d/\sqrt{n}} \quad (10.9)$$

10.3 Inferences About the Difference Between Two Population Means: Matched Samples

Once the difference data are computed, the t distribution procedure for matched samples is the same as the one-population estimation and hypothesis testing procedures described in Chapters 8 and 9.

Let us use equation (10.9) to test the hypotheses $H_0: \mu_d = 0$ and $H_a: \mu_d \neq 0$, using $\alpha = .05$. Substituting the sample results $\bar{d} = .30$, $s_d = .335$, and $n = 6$ into equation (10.9), we compute the value of the test statistic.

$$t = \frac{\bar{d} - \mu_d}{s_d/\sqrt{n}} = \frac{.30 - 0}{.335/\sqrt{6}} = 2.20$$

Now let us compute the *p*-value for this two-tailed test. Because $t = 2.20 > 0$, the test statistic is in the upper tail of the *t* distribution. With $t = 2.20$, the area in the upper tail can be found by using the *t* distribution table with degrees of freedom $= n - 1 = 6 - 1 = 5$. Information from the 5 degrees of freedom row of the *t* distribution table is as follows:

Area in Upper Tail	.20	.10	.05	.025	.01	.005
t-Value (5 *df*)	0.920	1.476	2.015	2.571	3.365	4.032

$$\uparrow$$
$$t = 2.20$$

Thus, we see that the area in the upper tail is between .05 and .025. Because this test is a two-tailed test, we double these values to conclude that the *p*-value is between .10 and .05. This *p*-value is greater than $\alpha = .05$. Thus, the null hypothesis $H_0: \mu_d = 0$ is not rejected. Using Excel or Minitab and the data in Table 10.2, we find the exact *p*-value $= .080$.

In addition we can obtain an interval estimate of the difference between the two population means by using the single population methodology of Chapter 8. At 95% confidence, the calculation follows.

$$\bar{d} \pm t_{.025}\frac{s_d}{\sqrt{n}}$$

$$.3 \pm 2.571\left(\frac{.335}{\sqrt{6}}\right)$$

$$.3 \pm .35$$

Thus, the margin of error is .35 and the 95% confidence interval for the difference between the population means of the two production methods is $-.05$ minute to .65 minute.

NOTES AND COMMENTS

1. In the example presented in this section, workers performed the production task with first one method and then the other method. This example illustrates a matched sample design in which each sampled element (worker) provides a pair of data values. It is also possible to use different but "similar" elements to provide the pair of data values. For example, a worker at one location could be matched with a similar worker at another location (similarity based on age, education, gender, experience, etc.). The pairs of workers would provide the difference data that could be used in the matched sample analysis.

2. A matched sample procedure for inferences about two population means generally provides better precision than the independent sample approach; therefore it is the recommended design. However, in some applications the matching cannot be achieved, or perhaps the time and cost associated with matching are excessive. In such cases, the independent sample design should be used.

Exercises

Methods

19. Consider the following hypothesis test.

$$H_0: \mu_d \leq 0$$
$$H_a: \mu_d > 0$$

The following data are from matched samples taken from two populations.

	Population	
Element	1	2
1	21	20
2	28	26
3	18	18
4	20	20
5	26	24

a. Compute the difference value for each element.
b. Compute $\bar{d}$.
c. Compute the standard deviation s_d.
d. Conduct a hypothesis test using $\alpha = .05$. What is your conclusion?

20. The following data are from matched samples taken from two populations.

	Population	
Element	1	2
1	11	8
2	7	8
3	9	6
4	12	7
5	13	10
6	15	15
7	15	14

a. Compute the difference value for each element.
b. Compute $\bar{d}$.
c. Compute the standard deviation s_d.
d. What is the point estimate of the difference between the two population means?
e. Provide a 95% confidence interval for the difference between the two population means.

Applications

21. A market research firm used a sample of individuals to rate the purchase potential of a particular product before and after the individuals saw a new television commercial about the product. The purchase potential ratings were based on a 0 to 10 scale, with higher values indicating a higher purchase potential. The null hypothesis stated that the mean rating "after" would be less than or equal to the mean rating "before." Rejection of this hypothesis would show that the commercial improved the mean purchase potential rating. Use $\alpha = .05$ and the following data to test the hypothesis and comment on the value of the commercial.

	Purchase Rating			Purchase Rating	
Individual	After	Before	Individual	After	Before
1	6	5	5	3	5
2	6	4	6	9	8
3	7	7	7	7	5
4	4	3	8	6	6

22. Per share earnings data comparing the current quarter's earnings with the previous quarter are in the file entitled Earnings2005 (*The Wall Street Journal,* January 27, 2006). Provide a 95% confidence interval estimate of the difference between the population mean for the current quarter versus the previous quarter. Have earnings increased?

23. Bank of America's Consumer Spending Survey collected data on annual credit card charges in seven different categories of expenditures: transportation, groceries, dining out, household expenses, home furnishings, apparel, and entertainment (*US Airways Attaché,* December 2003). Using data from a sample of 42 credit card accounts, assume that each account was used to identify the annual credit card charges for groceries (population 1) and the annual credit card charges for dining out (population 2). Using the difference data, the sample mean difference was $\bar{d} = \$850$, and the sample standard deviation was $s_d = \$1123$.
    a. Formulate the null and alternative hypotheses to test for no difference between the population mean credit card charges for groceries and the population mean credit card charges for dining out.
    b. Use a .05 level of significance. Can you conclude that the population means differ? What is the *p*-value?
    c. Which category, groceries or dining out, has a higher population mean annual credit card charge? What is the point estimate of the difference between the population means? What is the 95% confidence interval estimate of the difference between the population means?

24. Airline travelers often choose which airport to fly from based on flight cost. Cost data (in dollars) for a sample of flights to eight cities from Dayton, Ohio, and Louisville, Kentucky, were collected to help determine which of the two airports was more costly to fly from (*The Cincinnati Enquirer,* February 19, 2006). A researcher argued that it is significantly more costly to fly out of Dayton than Louisville. Use the sample data to see whether they support the researcher's argument. Use $\alpha = .05$ as the level of significance.

Destination	Dayton	Louisville
Chicago O'Hare	$319	$142
Grand Rapids, Michigan	192	213
Portland, Oregon	503	317
Atlanta	256	387
Seattle	339	317
South Bend, Indiana	379	167
Miami	268	273
Dallas–Ft. Worth	288	274

25. In recent years, a growing array of entertainment options competes for consumer time. By 2004, cable television and radio surpassed broadcast television, recorded music, and the daily newspaper to become the two entertainment media with the greatest usage (*The Wall Street Journal,* January 26, 2004). Researchers used a sample of 15 individuals and collected data on the hours per week spent watching cable television and hours per week spent listening to the radio.

Individual	Television	Radio	Individual	Television	Radio
1	22	25	9	21	21
2	8	10	10	23	23
3	25	29	11	14	15
4	22	19	12	14	18
5	12	13	13	14	17
6	26	28	14	16	15
7	22	23	15	24	23
8	19	21			

a. Use a .05 level of significance and test for a difference between the population mean usage for cable television and radio. What is the *p*-value?

b. What is the sample mean number of hours per week spent watching cable television? What is the sample mean number of hours per week spent listening to radio? Which medium has the greater usage?

26. Scores in the first and fourth (final) rounds for a sample of 20 golfers who competed in PGA tournaments are shown in the following table (*Golfweek*, February 14, 2009, and February 28, 2009). Suppose you would like to determine if the mean score for the first round of a PGA Tour event is significantly different from the mean score for the fourth and final round. Does the pressure of playing in the final round cause scores to go up? Or does the increased player concentration cause scores to come down?

Player	First Round	Final Round	Player	First Round	Final Round
Michael Letzig	70	72	Aron Price	72	72
Scott Verplank	71	72	Charles Howell	72	70
D. A. Points	70	75	Jason Dufner	70	73
Jerry Kelly	72	71	Mike Weir	70	77
Soren Hansen	70	69	Carl Pettersson	68	70
D. J. Trahan	67	67	Bo Van Pelt	68	65
Bubba Watson	71	67	Ernie Els	71	70
Reteif Goosen	68	75	Cameron Beckman	70	68
Jeff Klauk	67	73	Nick Watney	69	68
Kenny Perry	70	69	Tommy Armour III	67	71

a. Use $\alpha = .10$ to test for a statistically significantly difference between the population means for first-and fourth-round scores. What is the *p*-value? What is your conclusion?

b. What is the point estimate of the difference between the two population means? For which round is the population mean score lower?

c. What is the margin of error for a 90% confidence interval estimate for the difference between the population means? Could this confidence interval have been used to test the hypothesis in part (a)? Explain.

10.4 An Introduction to Experimental Design and Analysis of Variance

In Chapter 1 we stated that statistical studies can be classified as either experimental or observational. In an experimental statistical study, an experiment is conducted to generate

10.4 An Introduction to Experimental Design and Analysis of Variance

the data. An experiment begins with identifying a variable of interest. Then one or more other variables, thought to be related, are identified and controlled, and data are collected about how those variables influence the variable of interest.

In an observational study, data are usually obtained through sample surveys and not a controlled experiment. Good sample designs are employed, but the rigorous controls associated with an experimental statistical study are often not possible. For instance, in a study of the relationship between smoking and lung cancer the researcher cannot assign a smoking habit to subjects. The researcher is restricted to simply observing the effects of smoking on people who already smoke and the effects of not smoking on people who already do not smoke.

In this section we introduce the basic principles of an experimental study and show how they are used in a completely randomized design. We also provide a conceptual overview of the statistical procedure called analysis of variance (ANOVA). In the following section we show how ANOVA can be used to test for the equality of k population means using data obtained from a completely randomized design as well as data obtained from an observational study. So, in this sense, ANOVA extends the statistical material in the preceding sections from two population means to three or more population means. In later chapters, we will see that ANOVA plays a key role in analyzing the results of regression studies involving both experimental and observational data.

As an example of an experimental statistical study, let us consider the problem facing Chemitech, Inc. Chemitech developed a new filtration system for municipal water supplies. The components for the new filtration system will be purchased from several suppliers, and Chemitech will assemble the components at its plant in Columbia, South Carolina. The industrial engineering group is responsible for determining the best assembly method for the new filtration system. After considering a variety of possible approaches, the group narrows the alternatives to three: method A, method B, and method C. These methods differ in the sequence of steps used to assemble the system. Managers at Chemitech want to determine which assembly method can produce the greatest number of filtration systems per week.

In the Chemitech experiment, assembly method is the independent variable or **factor**. Because three assembly methods correspond to this factor, we say that three treatments are associated with this factor; each **treatment** corresponds to one of the three assembly methods. The Chemitech problem is an example of a **single-factor experiment**; it involves one qualitative factor (method of assembly). More complex experiments may consist of multiple factors; some factors may be qualitative and others may be quantitative.

The three assembly methods or treatments define the three populations of interest for the Chemitech experiment. One population is all Chemitech employees who use assembly method A, another is those who use method B, and the third is those who use method C. Note that for each population the dependent or **response variable** is the number of filtration systems assembled per week, and the primary statistical objective of the experiment is to determine whether the mean number of units produced per week is the same for all three populations (methods).

Suppose a random sample of three employees is selected from all assembly workers at the Chemitech production facility. In experimental design terminology, the three randomly selected workers are the **experimental units**. The experimental design that we will use for the Chemitech problem is called a **completely randomized design**. This type of design requires that each of the three assembly methods or treatments be assigned randomly to one of the experimental units or workers. For example, method A might be randomly assigned to the second worker, method B to the first worker, and method C to the third worker. The concept of *randomization,* as illustrated in this example, is an important principle of all experimental designs.

Note that this experiment would result in only one measurement or number of units assembled for each treatment. To obtain additional data for each assembly method, we

---

Sir Ronald Alymer Fisher (1890–1962) invented the branch of statistics known as experimental design. In addition to being accomplished in statistics, he was a noted scientist in the field of genetics.

Cause-and-effect relationships can be difficult to establish in observational studies; such relationships are easier to establish in experimental studies.

Randomization is the process of assigning the treatments to the experimental units at random. Prior to the work of Sir R. A. Fisher, treatments were assigned on a systematic or subjective basis.

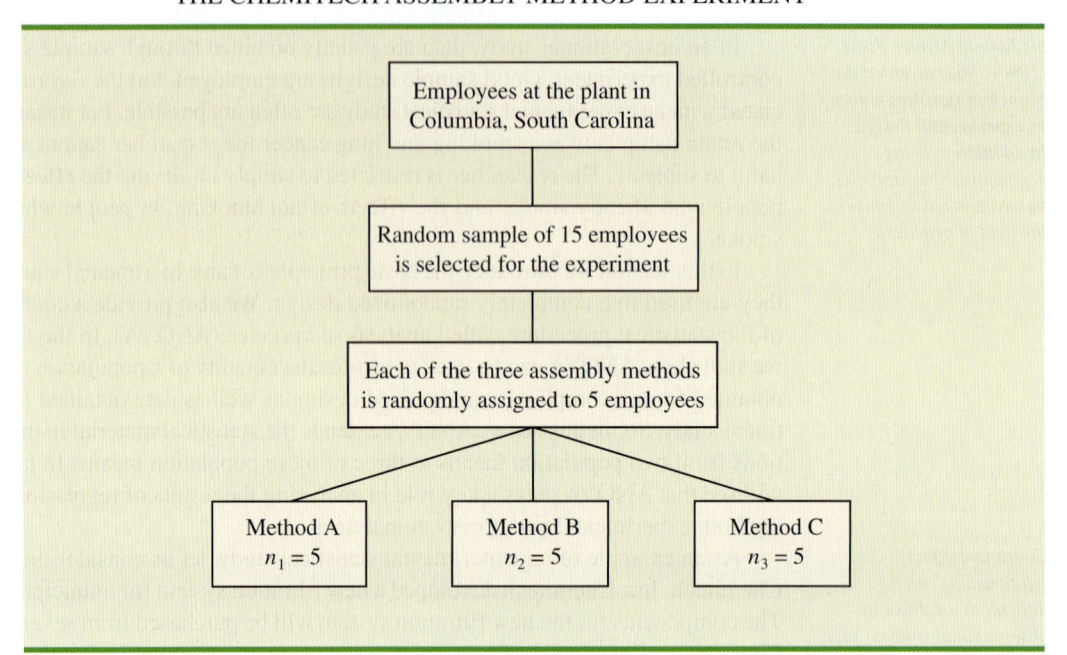

FIGURE 10.3 COMPLETELY RANDOMIZED DESIGN FOR EVALUATING THE CHEMITECH ASSEMBLY METHOD EXPERIMENT

```
        Employees at the plant in
        Columbia, South Carolina
                   |
        Random sample of 15 employees
        is selected for the experiment
                   |
        Each of the three assembly methods
        is randomly assigned to 5 employees
          /        |        \
   Method A    Method B    Method C
   n₁ = 5      n₂ = 5      n₃ = 5
```

must repeat or replicate the basic experimental process. Suppose, for example, that instead of selecting just three workers at random we selected 15 workers and then randomly assigned each of the three treatments to 5 of the workers. Because each method of assembly is assigned to 5 workers, we say that five replicates have been obtained. The process of *replication* is another important principle of experimental design. Figure 10.3 shows the completely randomized design for the Chemitech experiment.

Data Collection

Once we are satisfied with the experimental design, we proceed by collecting and analyzing the data. In the Chemitech case, the employees would be instructed in how to perform the assembly method assigned to them and then would begin assembling the new filtration systems using that method. After this assignment and training, the number of units assembled by each employee during one week is as shown in Table 10.3. The sample means, sample variances, and sample standard deviations for each assembly method are also provided. Thus, the sample mean number of units produced using method A is 62; the sample mean using method B is 66; and the sample mean using method C is 52. From these data, method B appears to result in higher production rates than either of the other methods.

The real issue is whether the three sample means observed are different enough for us to conclude that the means of the populations corresponding to the three methods of assembly are different. To write this question in statistical terms, we introduce the following notation.

μ_1 = mean number of units produced per week using method A
μ_2 = mean number of units produced per week using method B
μ_3 = mean number of units produced per week using method C

TABLE 10.3 NUMBER OF UNITS PRODUCED BY 15 WORKERS

	Method A	Method B	Method C
	58	58	48
	64	69	57
	55	71	59
	66	64	47
	67	68	49
Sample mean	62	66	52
Sample variance	27.5	26.5	31.0
Sample standard deviation	5.244	5.148	5.568

Although we will never know the actual values of μ_1, μ_2, and μ_3, we want to use the sample means to test the following hypotheses.

If H_0 is rejected, we cannot conclude that all population means are different. Rejecting H_0 means that at least two population means have different values.

$$H_0: \mu_1 = \mu_2 = \mu_3$$
$$H_a: \text{Not all population means are equal}$$

As we will demonstrate shortly, analysis of variance (ANOVA) is the statistical procedure used to determine whether the observed differences in the three sample means are large enough to reject H_0.

Assumptions for Analysis of Variance

Three assumptions are required to use analysis of variance.

If the sample sizes are equal, analysis of variance is not sensitive to departures from the assumption of normally distributed populations.

1. **For each population, the response variable is normally distributed.** Implication: In the Chemitech experiment the number of units produced per week (response variable) must be normally distributed for each assembly method.
2. **The variance of the response variable, denoted σ^2, is the same for all of the populations.** Implication: In the Chemitech experiment, the variance of the number of units produced per week must be the same for each assembly method.
3. **The observations must be independent.** Implication: In the Chemitech experiment, the number of units produced per week for each employee must be independent of the number of units produced per week for any other employee.

Analysis of Variance: A Conceptual Overview

If the means for the three populations are equal, we would expect the three sample means to be close together. In fact, the closer the three sample means are to one another, the more evidence we have for the conclusion that the population means are equal. Alternatively, the more the sample means differ, the more evidence we have for the conclusion that the population means are not equal. In other words, if the variability among the sample means is "small," it supports H_0; if the variability among the sample means is "large," it supports H_a.

If the null hypothesis, $H_0: \mu_1 = \mu_2 = \mu_3$, is true, we can use the variability among the sample means to develop an estimate of σ^2. First, note that if the assumptions for analysis of variance are satisfied, each sample will have come from the same normal distribution with mean μ and variance σ^2. Recall from Chapter 7 that the sampling distribution of the

FIGURE 10.4 SAMPLING DISTRIBUTION OF $\bar{x}$ GIVEN H_0 IS TRUE

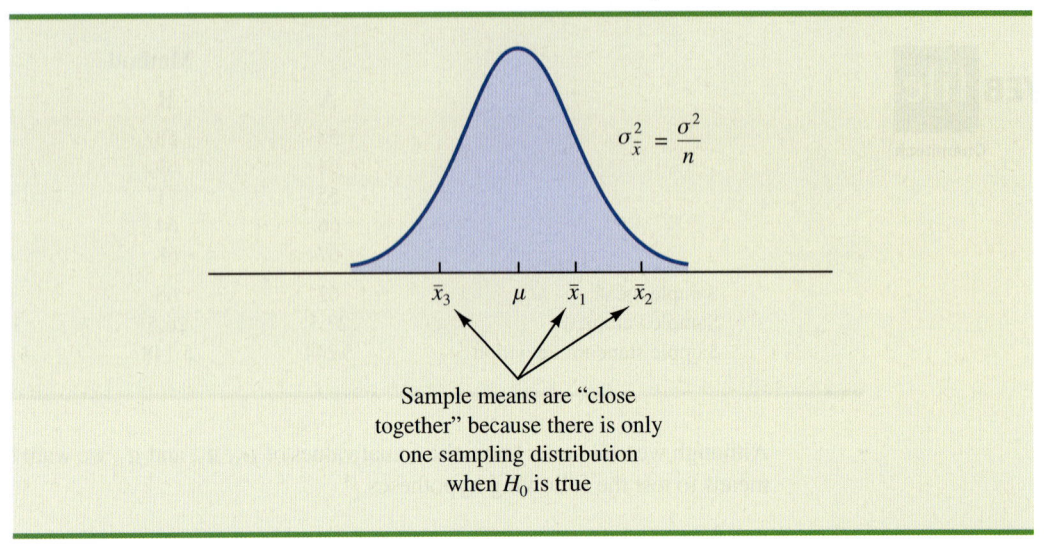

$$\sigma_{\bar{x}}^2 = \frac{\sigma^2}{n}$$

$\bar{x}_3 \quad \mu \quad \bar{x}_1 \quad \bar{x}_2$

Sample means are "close together" because there is only one sampling distribution when H_0 is true

sample mean $\bar{x}$ for a simple random sample of size n from a normal population will be normally distributed with mean μ and variance σ^2/n. Figure 10.4 illustrates such a sampling distribution.

Thus, if the null hypothesis is true, we can think of each of the three sample means, $\bar{x}_1 = 62$, $\bar{x}_2 = 66$, and $\bar{x}_3 = 52$ from Table 10.3, as values drawn at random from the sampling distribution shown in Figure 10.4. In this case, the mean and variance of the three $\bar{x}$ values can be used to estimate the mean and variance of the sampling distribution. When the sample sizes are equal, as in the Chemitech experiment, the best estimate of the mean of the sampling distribution of $\bar{x}$ is the mean or average of the sample means. Thus, in the Chemitech experiment, an estimate of the mean of the sampling distribution of $\bar{x}$ is $(62 + 66 + 52)/3 = 60$. We refer to this estimate as the *overall sample mean*. An estimate of the variance of the sampling distribution of $\bar{x}$, $\sigma_{\bar{x}}^2$, is provided by the variance of the three sample means.

$$s_{\bar{x}}^2 = \frac{(62-60)^2 + (66-60)^2 + (52-60)^2}{3-1} = \frac{104}{2} = 52$$

Because $\sigma_{\bar{x}}^2 = \sigma^2/n$, solving for σ^2 gives

$$\sigma^2 = n\sigma_{\bar{x}}^2$$

Hence,

$$\text{Estimate of } \sigma^2 = n \text{ (Estimate of } \sigma_{\bar{x}}^2) = ns_{\bar{x}}^2 = 5(52) = 260$$

The result, $ns_{\bar{x}}^2 = 260$, is referred to as the *between-treatments* estimate of σ^2.

The between-treatments estimate of σ^2 is based on the assumption that the null hypothesis is true. In this case, each sample comes from the same population, and there is only one sampling distribution of $\bar{x}$. To illustrate what happens when H_0 is false, suppose the population means all differ. Note that because the three samples are from normal populations with different means, they will result in three different sampling distributions. Figure 10.5 shows that in this case, the sample means are not as close together as they were

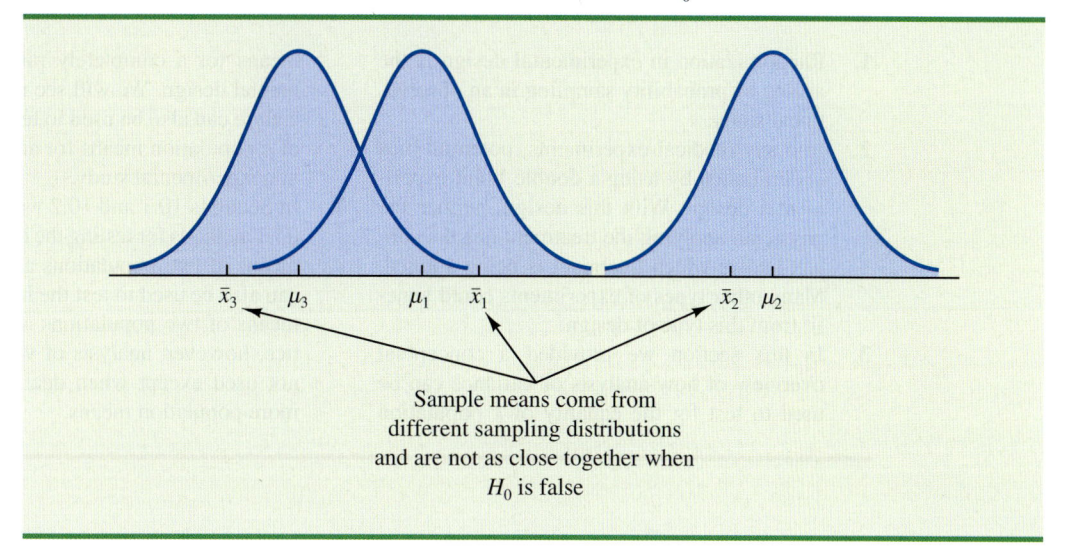

FIGURE 10.5 SAMPLING DISTRIBUTIONS OF $\bar{x}$ GIVEN H_0 IS FALSE

Sample means come from different sampling distributions and are not as close together when H_0 is false

when H_0 was true. Thus, $s_{\bar{x}}^2$ will be larger, causing the between-treatments estimate of σ^2 to be larger. In general, when the population means are not equal, the between-treatments estimate will overestimate the population variance σ^2.

The variation within each of the samples also has an effect on the conclusion we reach in analysis of variance. When a random sample is selected from each population, each of the sample variances provides an unbiased estimate of σ^2. Hence, we can combine or pool the individual estimates of σ^2 into one overall estimate. The estimate of σ^2 obtained in this way is called the *pooled* or *within-treatments* estimate of σ^2. Because each sample variance provides an estimate of σ^2 based only on the variation within each sample, the within-treatments estimate of σ^2 is not affected by whether the population means are equal. When the sample sizes are equal, the within-treatments estimate of σ^2 can be obtained by computing the average of the individual sample variances. For the Chemitech experiment we obtain

$$\text{Within-treatments estimate of } \sigma^2 = \frac{27.5 + 26.5 + 31.0}{3} = \frac{85}{3} = 28.33$$

In the Chemitech experiment, the between-treatments estimate of σ^2 (260) is much larger than the within-treatments estimate of σ^2 (28.33). In fact, the ratio of these two estimates is $260/28.33 = 9.18$. Recall, however, that the between-treatments approach provides a good estimate of σ^2 only if the null hypothesis is true; if the null hypothesis is false, the between-treatments approach overestimates σ^2. The within-treatments approach provides a good estimate of σ^2 in either case. Thus, if the null hypothesis is true, the two estimates will be similar and their ratio will be close to 1. If the null hypothesis is false, the between-treatments estimate will be larger than the within-treatments estimate, and their ratio will be large. In the next section we will show how large this ratio must be to reject H_0.

In summary, the logic behind ANOVA is based on the development of two independent estimates of the common population variance σ^2. One estimate of σ^2 is based on the variability among the sample means themselves, and the other estimate of σ^2 is based on the variability of the data within each sample. By comparing these two estimates of σ^2, we will be able to determine whether the population means are equal.

NOTES AND COMMENTS

1. Randomization in experimental design is the analog of probability sampling in an observational study.
2. In many medical experiments, potential bias is eliminated by using a double-blind experimental design. With this design, neither the physician applying the treatment nor the subject knows which treatment is being applied. Many other types of experiments could benefit from this type of design.
3. In this section we provided a conceptual overview of how analysis of variance can be used to test for the equality of k population means for a completely randomized experimental design. We will see that the same procedure can also be used to test for the equality of k population means for an observational or nonexperimental study.
4. In Sections 10.1 and 10.2 we presented statistical methods for testing the hypothesis that the means of two populations are equal. ANOVA can also be used to test the hypothesis that the means of two populations are equal. In practice, however, analysis of variance is usually not used except when dealing with three or more population means.

10.5 Analysis of Variance and the Completely Randomized Design

In this section we show how analysis of variance can be used to test for the equality of k population means for a completely randomized design. The general form of the hypotheses tested is

$$H_0: \mu_1 = \mu_2 = \cdots = \mu_k$$
$$H_a: \text{Not all population means are equal}$$

where

$$\mu_j = \text{mean of the } j\text{th population}$$

We assume that a simple random sample of size n_j has been selected from each of the k populations or treatments. For the resulting sample data, let

x_{ij} = value of observation i for treatment j
n_j = number of observations for treatment j
$\bar{x}_j$ = sample mean for treatment j
s_j^2 = sample variance for treatment j
s_j = sample standard deviation for treatment j

The formulas for the sample mean and sample variance for treatment j are as follow.

$$\bar{x}_j = \frac{\sum_{i=1}^{n_j} x_{ij}}{n_j} \tag{10.10}$$

10.5 Analysis of Variance and the Completely Randomized Design

$$s_j^2 = \frac{\sum_{i=1}^{n_j}(x_{ij} - \bar{x}_j)^2}{n_j - 1} \tag{10.11}$$

The overall sample mean, denoted $\bar{\bar{x}}$, is the sum of all the observations divided by the total number of observations. That is,

$$\bar{\bar{x}} = \frac{\sum_{j=1}^{k}\sum_{i=1}^{n_j} x_{ij}}{n_T} \tag{10.12}$$

where

$$n_T = n_1 + n_2 + \cdots + n_k \tag{10.13}$$

If the size of each sample is n, $n_T = kn$; in this case equation (10.12) reduces to

$$\bar{\bar{x}} = \frac{\sum_{j=1}^{k}\sum_{i=1}^{n_j} x_{ij}}{kn} = \frac{\sum_{j=1}^{k}\sum_{i=1}^{n_j} x_{ij}/n}{k} = \frac{\sum_{j=1}^{k} \bar{x}_j}{k} \tag{10.14}$$

In other words, whenever the sample sizes are the same, the overall sample mean is just the average of the k sample means.

Because each sample in the Chemitech experiment consists of $n = 5$ observations, the overall sample mean can be computed by using equation (10.14). For the data in Table 10.3 we obtained the following result.

$$\bar{\bar{x}} = \frac{62 + 66 + 52}{3} = 60$$

If the null hypothesis is true ($\mu_1 = \mu_2 = \mu_3 = \mu$), the overall sample mean of 60 is the best estimate of the population mean μ.

Between-Treatments Estimate of Population Variance

In the preceding section, we introduced the concept of a between-treatments estimate of σ^2 and showed how to compute it when the sample sizes were equal. This estimate of σ^2 is called the *mean square due to treatments* and is denoted MSTR. The general formula for computing MSTR is

$$\text{MSTR} = \frac{\sum_{j=1}^{k} n_j(\bar{x}_j - \bar{\bar{x}})^2}{k - 1} \tag{10.15}$$

The numerator in equation (10.15) is called the *sum of squares due to treatments* and is denoted SSTR. The denominator, $k - 1$, represents the degrees of freedom associated with SSTR. Hence, the mean square due to treatments can be computed using the following formula.

MEAN SQUARE DUE TO TREATMENTS

$$\text{MSTR} = \frac{\text{SSTR}}{k-1} \qquad (10.16)$$

where

$$\text{SSTR} = \sum_{j=1}^{k} n_j(\bar{x}_j - \bar{\bar{x}})^2 \qquad (10.17)$$

If H_0 is true, MSTR provides an unbiased estimate of σ^2. However, if the means of the k populations are not equal, MSTR is not an unbiased estimate of σ^2; in fact, in that case, MSTR should overestimate σ^2.

For the Chemitech data in Table 10.3, we obtain the following results.

$$\text{SSTR} = \sum_{j=1}^{k} n_j(\bar{x}_j - \bar{\bar{x}})^2 = 5(62 - 60)^2 + 5(66 - 60)^2 + 5(52 - 60)^2 = 520$$

$$\text{MSTR} = \frac{\text{SSTR}}{k-1} = \frac{520}{2} = 260$$

Within-Treatments Estimate of Population Variance

Earlier, we introduced the concept of a within-treatments estimate of σ^2 and showed how to compute it when the sample sizes were equal. This estimate of σ^2 is called the *mean square due to error* and is denoted MSE. The general formula for computing MSE is

$$\text{MSE} = \frac{\sum_{j=1}^{k}(n_j - 1)s_j^2}{n_T - k} \qquad (10.18)$$

The numerator in equation (10.18) is called the *sum of squares due to error* and is denoted SSE. The denominator of MSE is referred to as the degrees of freedom associated with SSE. Hence, the formula for MSE can also be stated as follows.

MEAN SQUARE DUE TO ERROR

$$\text{MSE} = \frac{\text{SSE}}{n_T - k} \qquad (10.19)$$

where

$$\text{SSE} = \sum_{j=1}^{k}(n_j - 1)s_j^2 \qquad (10.20)$$

Note that MSE is based on the variation within each of the treatments; it is not influenced by whether the null hypothesis is true. Thus, MSE always provides an unbiased estimate of σ^2.

For the Chemitech data in Table 10.3 we obtain the following results.

$$\text{SSE} = \sum_{j=1}^{k}(n_j - 1)s_j^2 = (5-1)27.5 + (5-1)26.5 + (5-1)31 = 340$$

$$\text{MSE} = \frac{\text{SSE}}{n_T - k} = \frac{340}{15 - 3} = \frac{340}{12} = 28.33$$

10.5 Analysis of Variance and the Completely Randomized Design

Comparing the Variance Estimates: The *F* Test

If the null hypothesis is true, MSTR and MSE provide two independent, unbiased estimates of σ^2. If the ANOVA assumptions are also valid, the sampling distribution of MSTR/MSE is an **F distribution** with numerator degrees of freedom equal to $k - 1$ and denominator degrees of freedom equal to $n_T - k$. The general shape of the *F* distribution is shown in Figure 10.6. If the null hypothesis is true, the value of MSTR/MSE should appear to have been selected from this *F* distribution.

However, if the null hypothesis is false, the value of MSTR/MSE will be inflated because MSTR overestimates σ^2. Hence, we will reject H_0 if the resulting value of MSTR/MSE appears to be too large to have been selected from an *F* distribution with $k - 1$ numerator degrees of freedom and $n_T - k$ denominator degrees of freedom. Because the decision to reject H_0 is based on the value of MSTR/MSE, the test statistic used to test for the equality of *k* population means is as follows.

TEST STATISTIC FOR THE EQUALITY OF *k* POPULATION MEANS

$$F = \frac{\text{MSTR}}{\text{MSE}} \qquad (10.21)$$

The test statistic follows an *F* distribution with $k - 1$ degrees of freedom in the numerator and $n_T - k$ degrees of freedom in the denominator.

Let us return to the Chemitech experiment and use a level of significance $\alpha = .05$ to conduct the hypothesis test. The value of the test statistic is

$$F = \frac{\text{MSTR}}{\text{MSE}} = \frac{260}{28.33} = 9.18$$

The numerator degrees of freedom is $k - 1 = 3 - 1 = 2$ and the denominator degrees of freedom is $n_T - k = 15 - 3 = 12$. Because we will only reject the null hypothesis for large values of the test statistic, the *p*-value is the upper tail area of the *F* distribution corresponding to the value of the test statistic $F = 9.18$. Figure 10.6 shows the sampling distribution of $F = \text{MSTR}/\text{MSE}$, the value of the test statistic, and the upper tail area that is the *p*-value for the hypothesis test.

FIGURE 10.6 COMPUTATION OF *p*-VALUE USING THE SAMPLING DISTRIBUTION OF MSTR/MSE

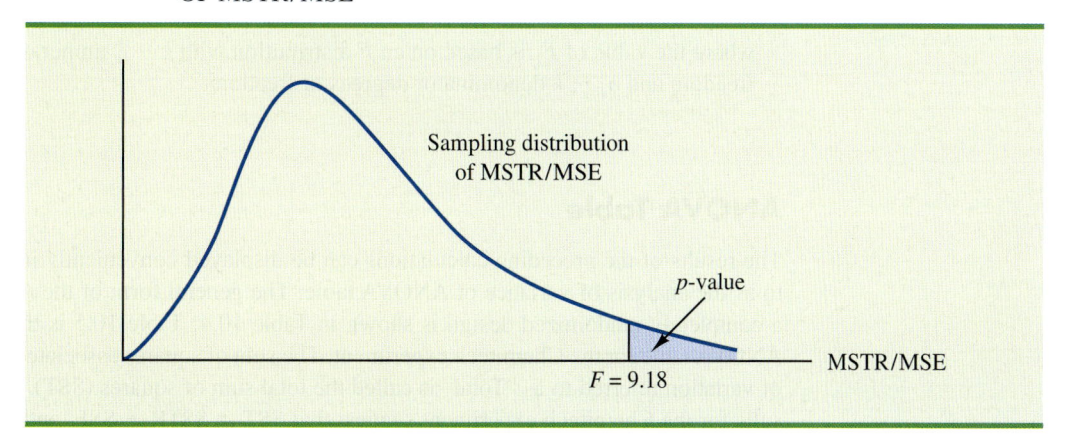

From Table 4 of Appendix B we find the following areas in the upper tail of an F distribution with 2 numerator degrees of freedom and 12 denominator degrees of freedom.

Area in Upper Tail	.10	.05	.025	.01
F Value ($df_1 = 2, df_2 = 12$)	2.81	3.89	5.10	6.93

$F = 9.18$

Appendix F shows how to compute p-values using Minitab or Excel.

Because $F = 9.18$ is greater than 6.93, the area in the upper tail at $F = 9.18$ is less than .01. Thus, the p-value is less than .01. Minitab or Excel can be used to show that the exact p-value is .004. With p-value $\leq \alpha = .05$, H_0 is rejected. The test provides sufficient evidence to conclude that the means of the three populations are not equal. In other words, analysis of variance supports the conclusion that the population mean number of units produced per week for the three assembly methods are not equal.

As with other hypothesis testing procedures, the critical value approach may also be used. With $\alpha = .05$, the critical F value occurs with an area of .05 in the upper tail of an F distribution with 2 and 12 degrees of freedom. From the F distribution table, we find $F_{.05} = 3.89$. Hence, the appropriate upper tail rejection rule for the Chemitech experiment is

$$\text{Reject } H_0 \text{ if } F \geq 3.89$$

With $F = 9.18$, we reject H_0 and conclude that the means of the three populations are not equal. A summary of the overall procedure for testing for the equality of k population means follows.

TEST FOR THE EQUALITY OF k POPULATION MEANS

$$H_0: \mu_1 = \mu_2 = \cdots = \mu_k$$
$$H_a: \text{Not all population means are equal}$$

TEST STATISTIC

$$F = \frac{\text{MSTR}}{\text{MSE}}$$

REJECTION RULE

p-value approach: Reject H_0 if p-value $\leq \alpha$
Critical value approach: Reject H_0 if $F \geq F_\alpha$

where the value of F_α is based on an F distribution with $k - 1$ numerator degrees of freedom and $n_T - k$ denominator degrees of freedom.

ANOVA Table

The results of the preceding calculations can be displayed conveniently in a table referred to as the analysis of variance or ANOVA table. The general form of the ANOVA table for a completely randomized design is shown in Table 10.4; Table 10.5 is the corresponding ANOVA table for the Chemitech experiment. The sum of squares associated with the source of variation referred to as "Total" is called the total sum of squares (SST). Note that the results for the Chemitech experiment suggest that SST = SSTR + SSE, and that the degrees

10.5 Analysis of Variance and the Completely Randomized Design

TABLE 10.4 ANOVA TABLE FOR A COMPLETELY RANDOMIZED DESIGN

Source of Variation	Sum of Squares	Degrees of Freedom	Mean Square	F	p-value
Treatments	SSTR	$k - 1$	$\text{MSTR} = \dfrac{\text{SSTR}}{k - 1}$	$\dfrac{\text{MSTR}}{\text{MSE}}$	
Error	SSE	$n_T - k$	$\text{MSE} = \dfrac{\text{SSE}}{n_T - k}$		
Total	SST	$n_T - 1$			

TABLE 10.5 ANALYSIS OF VARIANCE TABLE FOR THE CHEMITECH EXPERIMENT

Source of Variation	Sum of Squares	Degrees of Freedom	Mean Square	F	p-value
Treatments	520	2	260.00	9.18	.004
Error	340	12	28.33		
Total	860	14			

of freedom associated with this total sum of squares is the sum of the degrees of freedom associated with the sum of squares due to treatments and the sum of squares due to error.

We point out that SST divided by its degrees of freedom $n_T - 1$ is nothing more than the overall sample variance that would be obtained if we treated the entire set of 15 observations as one data set. With the entire data set as one sample, the formula for computing the total sum of squares, SST, is

$$\text{SST} = \sum_{j=1}^{k}\sum_{i=1}^{n_j}(x_{ij} - \bar{\bar{x}})^2 \tag{10.22}$$

It can be shown that the results we observed for the analysis of variance table for the Chemitech experiment also apply to other problems. That is,

$$\text{SST} = \text{SSTR} + \text{SSE} \tag{10.23}$$

Analysis of variance can be thought of as a statistical procedure for partitioning the total sum of squares into separate components.

In other words, SST can be partitioned into two sums of squares: the sum of squares due to treatments and the sum of squares due to error. Note also that the degrees of freedom corresponding to SST, $n_T - 1$, can be partitioned into the degrees of freedom corresponding to SSTR, $k - 1$, and the degrees of freedom corresponding to SSE, $n_T - k$. The analysis of variance can be viewed as the process of **partitioning** the total sum of squares and the degrees of freedom into their corresponding sources: treatments and error. Dividing the sum of squares by the appropriate degrees of freedom provides the variance estimates, the F value, and the p-value used to test the hypothesis of equal population means.

Computer Results for Analysis of Variance

Using statistical computer packages, analysis of variance computations with large sample sizes or a large number of populations can be performed easily. Appendixes 10.2, 10.4, and 10.6 show the steps required to use Minitab, Excel, and StatTools to perform the analysis of variance computations. In Figure 10.7 we show output for the Chemitech experiment

FIGURE 10.7 MINITAB OUTPUT FOR THE CHEMITECH EXPERIMENT ANALYSIS OF VARIANCE

```
Source    DF        SS          MS         F        P
Factor     2      520.0       260.0      9.18    0.004
Error     12      340.0        28.3
Total     14      860.0

S = 5.323      R-Sq = 60.47%     R-Sq(adj) = 53.88%

                                   Individual 95% CIs For Mean Based on
                                   Pooled StDev
Level    N     Mean     StDev   ---+---------+---------+---------+------
A        5    62.000    5.244                          (--------*--------)
B        5    66.000    5.148                                 (-------*-------)
C        5    52.000    5.568   (-------*-------)
                                ---+---------+---------+---------+------
Pooled StDev = 5.323              49.0      56.0      63.0      70.0
```

obtained using Minitab. The first part of the computer output contains the familiar ANOVA table format. Comparing Figure 10.7 with Table 10.5, we see that the same information is available, although some of the headings are slightly different. The heading Source is used for the source of variation column, Factor identifies the treatments row, and the sum of squares and degrees of freedom columns are interchanged.

Note that following the ANOVA table the computer output contains the respective sample sizes, the sample means, and the standard deviations. In addition, Minitab provides a figure that shows individual 95% confidence interval estimates of each population mean. In developing these confidence interval estimates, Minitab uses MSE as the estimate of σ^2. Thus, the square root of MSE provides the best estimate of the population standard deviation σ. This estimate of σ on the computer output is Pooled StDev; it is equal to 5.323. To provide an illustration of how these interval estimates are developed, we will compute a 95% confidence interval estimate of the population mean for method A.

From our study of interval estimation in Chapter 8, we know that the general form of an interval estimate of a population mean is

$$\bar{x} \pm t_{\alpha/2} \frac{s}{\sqrt{n}} \tag{10.24}$$

where s is the estimate of the population standard deviation σ. Because the best estimate of σ is provided by the Pooled StDev, we use a value of 5.323 for s in expression (10.24). The degrees of freedom for the t value is 12, the degrees of freedom associated with the error sum of squares. Hence, with $t_{.025} = 2.179$ we obtain

$$62 \pm 2.179 \frac{5.323}{\sqrt{5}} = 62 \pm 5.19$$

Thus, the individual 95% confidence interval for method A goes from $62 - 5.19 = 56.81$ to $62 + 5.19 = 67.19$. Because the sample sizes are equal for the Chemitech experiment, the individual confidence intervals for methods B and C are also constructed by adding and subtracting 5.19 from each sample mean. Thus, in the figure provided by Minitab we see that the widths of the confidence intervals are the same.

Testing for the Equality of k Population Means: An Observational Study

We have shown how analysis of variance can be used to test for the equality of k population means for a completely randomized experimental design. It is important to understand that ANOVA can also be used to test for the equality of three or more population means using data obtained from an observational study. As an example, let us consider the situation at National Computer Products, Inc. (NCP).

NCP manufactures printers and fax machines at plants located in Atlanta, Dallas, and Seattle. To measure how much employees at these plants know about quality management, a random sample of six employees was selected from each plant and the employees selected were given a quality awareness examination. The examination scores for these 18 employees are shown in Table 10.6. The sample means, sample variances, and sample standard deviations for each group are also provided. Managers want to use these data to test the hypothesis that the mean examination score is the same for all three plants.

We define population 1 as all employees at the Atlanta plant, population 2 as all employees at the Dallas plant, and population 3 as all employees at the Seattle plant. Let

μ_1 = mean examination score for population 1
μ_2 = mean examination score for population 2
μ_3 = mean examination score for population 3

Although we will never know the actual values of μ_1, μ_2, and μ_3, we want to use the sample results to test the following hypotheses.

$H_0: \mu_1 = \mu_2 = \mu_3$
$H_a:$ Not all population means are equal

Exercise 34 will ask you to analyze the NCP data using the analysis of variance procedure.

Note that the hypothesis test for the NCP observational study is exactly the same as the hypothesis test for the Chemitech experiment. Indeed, the same analysis of variance methodology we used to analyze the Chemitech experiment can also be used to analyze the data from the NCP observational study.

Even though the same ANOVA methodology is used for the analysis, it is worth noting how the NCP observational statistical study differs from the Chemitech experimental statistical study. The individuals who conducted the NCP study had no control over how the plants were

TABLE 10.6 EXAMINATION SCORES FOR 18 EMPLOYEES

	Plant 1 Atlanta	Plant 2 Dallas	Plant 3 Seattle
	85	71	59
	75	75	64
	82	73	62
	76	74	69
	71	69	75
	85	82	67
Sample mean	79	74	66
Sample variance	34	20	32
Sample standard deviation	5.83	4.47	5.66

assigned to individual employees. That is, the plants were already in operation and a particular employee worked at one of the three plants. All that NCP could do was to select a random sample of 6 employees from each plant and administer the quality awareness examination. To be classified as an experimental study, NCP would have had to be able to randomly select 18 employees and then assign the plants to each employee in a random fashion.

Exercises

Methods

27. The following data are from a completely randomized design.

	Treatment A	Treatment B	Treatment C
	162	142	126
	142	156	122
	165	124	138
	145	142	140
	148	136	150
	174	152	128
Sample mean	156	142	134
Sample variance	164.4	131.2	110.4

a. Compute the sum of squares between treatments.
b. Compute the mean square between treatments.
c. Compute the sum of squares due to error.
d. Compute the mean square due to error.
e. Set up the ANOVA table for this problem.
f. At the $\alpha = .05$ level of significance, test whether the means for the three treatments are equal.

28. In a completely randomized design, seven experimental units were used for each of the five levels of the factor. Complete the following ANOVA table.

Source of Variation	Sum of Squares	Degrees of Freedom	Mean Square	F	p-value
Treatments	300				
Error					
Total	460				

29. Refer to exercise 28.
a. What hypotheses are implied in this problem?
b. At the $\alpha = .05$ level of significance, can we reject the null hypothesis in part (a)? Explain.

30. In an experiment designed to test the output levels of three different treatments, the following results were obtained: SST = 400, SSTR = 150, n_T = 19. Set up the ANOVA table and test for any significant difference between the mean output levels of the three treatments. Use $\alpha = .05$.

31. In a completely randomized design, 12 experimental units were used for the first treatment, 15 for the second treatment, and 20 for the third treatment. Complete the following analysis of variance. At a .05 level of significance, is there a significant difference between the treatments?

10.5 Analysis of Variance and the Completely Randomized Design

Source of Variation	Sum of Squares	Degrees of Freedom	Mean Square	F	p-value
Treatments	1200				
Error					
Total	1800				

32. Develop the analysis of variance computations for the following completely randomized design. At $\alpha = .05$, is there a significant difference between the treatment means?

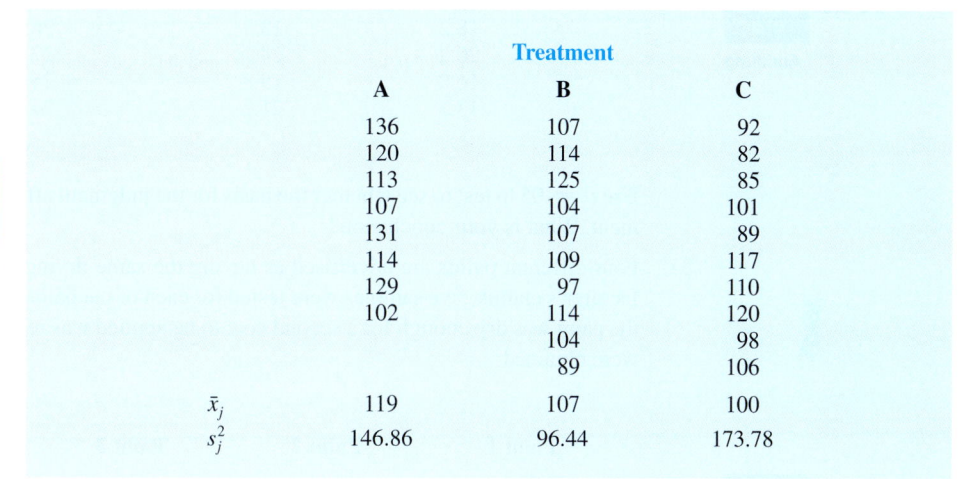

WEB file
RandomDesign

	Treatment		
	A	B	C
	136	107	92
	120	114	82
	113	125	85
	107	104	101
	131	107	89
	114	109	117
	129	97	110
	102	114	120
		104	98
		89	106
$\bar{x}_j$	119	107	100
s_j^2	146.86	96.44	173.78

Applications

33. Three different methods for assembling a product were proposed by an industrial engineer. To investigate the number of units assembled correctly with each method, 30 employees were randomly selected and randomly assigned to the three proposed methods in such a way that each method was used by 10 workers. The number of units assembled correctly was recorded, and the analysis of variance procedure was applied to the resulting data set. The following results were obtained: SST = 10,800; SSTR = 4560.
    a. Set up the ANOVA table for this problem.
    b. Use $\alpha = .05$ to test for any significant difference in the means for the three assembly methods.

34. Refer to the NCP data in Table 10.6. Set up the ANOVA table and test for any significant difference in the mean examination score for the three plants. Use $\alpha = .05$.

35. To study the effect of temperature on yield in a chemical process, five batches were produced at each of three temperature levels. The results follow. Construct an analysis of variance table. Use a .05 level of significance to test whether the temperature level has an effect on the mean yield of the process.

	Temperature		
	50° C	60° C	70° C
	34	30	23
	24	31	28
	36	34	28
	39	23	30
	32	27	31

36. Auditors must make judgments about various aspects of an audit on the basis of their own direct experience, indirect experience, or a combination of the two. In a study, auditors were asked to make judgments about the frequency of errors to be found in an audit. The judgments by the auditors were then compared to the actual results. Suppose the following data were obtained from a similar study; lower scores indicate better judgments.

Direct	Indirect	Combination
17.0	16.6	25.2
18.5	22.2	24.0
15.8	20.5	21.5
18.2	18.3	26.8
20.2	24.2	27.5
16.0	19.8	25.8
13.3	21.2	24.2

Use $\alpha = .05$ to test to see whether the basis for the judgment affects the quality of the judgment. What is your conclusion?

37. Four different paints are advertised as having the same drying time. To check the manufacturer's claims, five samples were tested for each of the paints. The time in minutes until the paint was dry enough for a second coat to be applied was recorded. The following data were obtained.

Paint 1	Paint 2	Paint 3	Paint 4
128	144	133	150
137	133	143	142
135	142	137	135
124	146	136	140
141	130	131	153

At the $\alpha = .05$ level of significance, test to see whether the mean drying time is the same for each type of paint.

38. The *Consumer Reports* Restaurant Customer Satisfaction Survey is based upon 148,599 visits to full-service restaurant chains (Consumer Reports website). One of the variables in the study is meal price, the average amount paid per person for dinner and drinks, minus the tip. Suppose a reporter for the *Sun Coast Times* thought that it would be of interest to her readers to conduct a similar study for restaurants located on the Grand Strand section in Myrtle Beach, South Carolina. The reporter selected a sample of eight seafood restaurants, eight Italian restaurants, and eight steakhouses. The following data show the meal prices ($) obtained for the 24 restaurants sampled. Use $\alpha = .05$ to test whether there is a significant difference among the mean meal price for the three types of restaurants.

Italian	Seafood	Steakhouse
$12	$16	$24
13	18	19
15	17	23
17	26	25
18	23	21
20	15	22
17	19	27
24	18	31

Summary

In this chapter we discussed procedures for developing interval estimates and conducting hypothesis tests involving two populations. First, we showed how to make inferences about the difference between two population means when independent random samples are selected. We first considered the case where the population standard deviations, σ_1 and σ_2, could be assumed known. The standard normal distribution z was used to develop the interval estimate and served as the test statistic for hypothesis tests. We then considered the case where the population standard deviations were unknown and estimated by the sample standard deviations s_1 and s_2. In this case, the t distribution was used to develop the interval estimate and served as the test statistic for hypothesis tests.

Inferences about the difference between two population means were then discussed for the matched sample design. In the matched sample design each element provides a pair of data values, one from each population. The difference between the paired data values is then used in the statistical analysis. The matched sample design is generally preferred to the independent sample design because the matched-sample procedure often improves the precision of the estimate.

In the final two sections we provided an introduction to experimental design and the analysis of variance (ANOVA). Experimental studies differ from observational studies in the sense that an experiment is conducted to generate the data. The completely randomized design was described and the analysis of variance was used to test for a treatment effect. The same analysis of variance procedure can be used to test for the difference among k population means in an observational study.

Glossary

Independent random samples Samples selected from two populations in such a way that the elements making up one sample are chosen independently of the elements making up the other sample.

Matched samples Samples in which each data value of one sample is matched with a corresponding data value of the other sample.

Factor Another word for the independent variable of interest.

Treatments Different levels of a factor.

Single-factor experiment An experiment involving only one factor with k populations or treatments.

Response variable Another word for the dependent variable of interest.

Experimental units The elements of interest in the experiment.

Completely randomized design An experimental design in which the treatments are randomly assigned to the experimental units.

F distribution A probability distribution based on the ratio of two independent estimates of the variance of a normal population. The F distribution is used in hypothesis tests about the equality of k population means.

Partitioning The process of allocating the total sum of squares and degrees of freedom to the various components.

Key Formulas

Point Estimator of the Difference Between Two Population Means

$$\bar{x}_1 - \bar{x}_2 \qquad (10.1)$$

Standard Error of $\bar{x}_1 - \bar{x}_2$

$$\sigma_{\bar{x}_1 - \bar{x}_2} = \sqrt{\frac{\sigma_1^2}{n_1} + \frac{\sigma_2^2}{n_2}} \quad (10.2)$$

Interval Estimate of the Difference Between Two Population Means: σ_1 and σ_2 Known

$$\bar{x}_1 - \bar{x}_2 \pm z_{\alpha/2} \sqrt{\frac{\sigma_1^2}{n_1} + \frac{\sigma_2^2}{n_2}} \quad (10.4)$$

Test Statistic for Hypothesis Tests About $\mu_1 - \mu_2$: σ_1 and σ_2 Known

$$z = \frac{(\bar{x}_1 - \bar{x}_2) - D_0}{\sqrt{\frac{\sigma_1^2}{n_1} + \frac{\sigma_2^2}{n_2}}} \quad (10.5)$$

Interval Estimate of the Difference Between Two Population Means: σ_1 and σ_2 Unknown

$$\bar{x}_1 - \bar{x}_2 \pm t_{\alpha/2} \sqrt{\frac{s_1^2}{n_1} + \frac{s_2^2}{n_2}} \quad (10.6)$$

Degrees of Freedom: t Distribution with Two Independent Random Samples

$$df = \frac{\left(\frac{s_1^2}{n_1} + \frac{s_2^2}{n_2}\right)^2}{\frac{1}{n_1 - 1}\left(\frac{s_1^2}{n_1}\right)^2 + \frac{1}{n_2 - 1}\left(\frac{s_2^2}{n_2}\right)^2} \quad (10.7)$$

Test Statistic for Hypothesis Tests About $\mu_1 - \mu_2$: σ_1 and σ_2 Unknown

$$t = \frac{(\bar{x}_1 - \bar{x}_2) - D_0}{\sqrt{\frac{s_1^2}{n_1} + \frac{s_2^2}{n_2}}} \quad (10.8)$$

Test Statistic for Hypothesis Tests Involving Matched Samples

$$t = \frac{\bar{d} - \mu_d}{s_d / \sqrt{n}} \quad (10.9)$$

Sample Mean for Treatment j

$$\bar{x}_j = \frac{\sum_{i=1}^{n_j} x_{ij}}{n_j} \quad (10.10)$$

Sample Variance for Treatment j

$$s_j^2 = \frac{\sum_{i=1}^{n_j} (x_{ij} - \bar{x}_j)^2}{n_j - 1} \quad (10.11)$$

Overall Sample Mean

$$\bar{\bar{x}} = \frac{\sum_{j=1}^{k}\sum_{i=1}^{n_j} x_{ij}}{n_T} \qquad (10.12)$$

where

$$n_T = n_1 + n_2 + \cdots + n_k \qquad (10.13)$$

Mean Square Due to Treatments

$$\text{MSTR} = \frac{\text{SSTR}}{k-1} \qquad (10.16)$$

Sum of Squares Due to Treatments

$$\text{SSTR} = \sum_{j=1}^{k} n_j(\bar{x}_j - \bar{\bar{x}})^2 \qquad (10.17)$$

Mean Square Due to Error

$$\text{MSE} = \frac{\text{SSE}}{n_T - k} \qquad (10.19)$$

Sum of Squares Due to Error

$$\text{SSE} = \sum_{j=1}^{k} (n_j - 1)s_j^2 \qquad (10.20)$$

Test Statistic for the Equality of k Population Means

$$F = \frac{\text{MSTR}}{\text{MSE}} \qquad (10.21)$$

Total Sum of Squares

$$\text{SST} = \sum_{j=1}^{k}\sum_{i=1}^{n_j} (x_{ij} - \bar{\bar{x}})^2 \qquad (10.22)$$

Partitioning of Sum of Squares

$$\text{SST} = \text{SSTR} + \text{SSE} \qquad (10.23)$$

Supplementary Exercises

39. Safegate Foods, Inc., is redesigning the checkout lanes in its supermarkets throughout the country and is considering two designs. Tests on customer checkout times conducted at two stores where the two new systems have been installed result in the following summary of the data.

System A	System B
$n_1 = 120$	$n_2 = 100$
$\bar{x}_1 = 4.1$ minutes	$\bar{x}_2 = 3.4$ minutes
$\sigma_1 = 2.2$ minutes	$\sigma_2 = 1.5$ minutes

Test at the .05 level of significance to determine whether the population mean checkout times of the two systems differ. Which system is preferred?

40. How much is the cost of a hospital stay increasing? The mean cost of one day in a semiprivate room was reported to be $4848 in 2005 and $5260 in 2006 (*The Wall Street Journal*, January 2, 2007). Assume the estimate for 2005 is a sample mean based on a sample size of 80 and the estimate for 2006 is a sample mean based on a sample size of 60.
    a. Develop a point estimate of the increase in the cost of a semiprivate hospital room from 2005 to 2006.
    b. Historical data indicate that a population standard deviation of $800 is a reasonable assumption for both years. Compute the margin of error for your estimate in part (a). Use 95% confidence.
    c. Develop a 95% confidence interval estimate of the increase in cost for a semiprivate room.

41. Home values tend to increase over time under normal conditions, but the recession of 2008 and 2009 has reportedly caused the sales price of existing homes to fall nationwide (*BusinessWeek*, March 9, 2009). You would like to see if the data support this conclusion. The file HomePrices contains data on 30 existing home sales in 2006 and 40 existing home sales in 2009.
    a. Provide a point estimate of the difference between the population mean prices for the two years.
    b. Develop a 99% confidence interval estimate of the difference between the resale prices of houses in 2006 and 2009.
    c. Would you feel justified in concluding that resale prices of existing homes have declined from 2006 to 2009? Why or why not?

42. Mutual funds are classified as *load* or *no-load* funds. Load funds require an investor to pay an initial fee based on a percentage of the amount invested in the fund. The no-load funds do not require this initial fee. Some financial advisors argue that the load mutual funds may be worth the extra fee because these funds provide a higher mean rate of return than the no-load mutual funds. A sample of 30 load mutual funds and a sample of 30 no-load mutual funds were selected. Data were collected on the annual return for the funds over a five-year period. The data are contained in the data set Mutual. The data for the first five load and first five no-load mutual funds are as follows.

Mutual Funds—Load	Return	Mutual Funds—No Load	Return
American National Growth	15.51	Amana Income Fund	13.24
Arch Small Cap Equity	14.57	Berger One Hundred	12.13
Bartlett Cap Basic	17.73	Columbia International Stock	12.17
Calvert World International	10.31	Dodge & Cox Balanced	16.06
Colonial Fund A	16.23	Evergreen Fund	17.61

a. Formulate H_0 and H_a such that rejection of H_0 leads to the conclusion that the load mutual funds have a higher mean annual return over the five-year period.
b. Use the 60 mutual funds in the data set Mutual to conduct the hypothesis test. What is the *p*-value? At $\alpha = .05$, what is your conclusion?

Supplementary Exercises

43. The National Association of Home Builders provided data on the cost of the most popular home remodeling projects. Sample data on cost in thousands of dollars for two types of remodeling projects are as follows.

Kitchen	Master Bedroom	Kitchen	Master Bedroom
25.2	18.0	23.0	17.8
17.4	22.9	19.7	24.6
22.8	26.4	16.9	21.0
21.9	24.8	21.8	
19.7	26.9	23.6	

   a. Develop a point estimate of the difference between the population mean remodeling costs for the two types of projects.
   b. Develop a 90% confidence interval for the difference between the two population means.

44. In early 2009, the economy was experiencing a recession. But how was the recession affecting the stock market? Shown are data from a sample of 15 companies. Shown for each company is the price per share of stock on January 1 and April 30 (*The Wall Street Journal*, May 1, 2009).

Company	January 1 ($)	April 30 ($)
Applied Materials	10.13	12.21
Bank of New York	28.33	25.48
Chevron	73.97	66.10
Cisco Systems	16.30	19.32
Coca-Cola	45.27	43.05
Comcast	16.88	15.46
Ford Motors	2.29	5.98
General Electric	16.20	12.65
Johnson & Johnson	59.83	52.36
JP Morgan Chase	31.53	33.00
Microsoft	19.44	20.26
Oracle	17.73	19.34
Pfizer	17.71	13.36
Philip Morris	43.51	36.18
Procter & Gamble	61.82	49.44

   a. What is the change in the mean price per share of stock over the four-month period?
   b. Provide a 90% confident interval estimate of the change in the mean price per share of stock. Interpret the results.
   c. What was the percentage change in the mean price per share of stock over the four-month period?
   d. If this same percentage change were to occur for the next four months and again for the four months after that, what would be the mean price per share of stock at the end of the year 2009?

45. A study reported in the *Journal of Small Business Management* concluded that self-employed individuals do not experience higher job satisfaction than individuals who are not self-employed. In this study, job satisfaction is measured using 18 items, each of which is rated using a Likert-type scale with 1–5 response options ranging from strong agreement to strong disagreement. A higher score on this scale indicates a higher degree of job satisfaction. The sum of the ratings for the 18 items, ranging from 18–90, is used as the measure of job satisfaction. Suppose that this approach was used to measure the job satisfaction for lawyers, physical therapists, cabinetmakers, and systems analysts. The results obtained for a sample of 10 individuals from each profession follow.

Lawyer	Physical Therapist	Cabinetmaker	Systems Analyst
44	55	54	44
42	78	65	73
74	80	79	71
42	86	69	60
53	60	79	64
50	59	64	66
45	62	59	41
48	52	78	55
64	55	84	76
38	50	60	62

At the $\alpha = .05$ level of significance, test for any difference in the job satisfaction among the four professions.

46. *Money* magazine reports percentage returns and expense ratios for stock and bond funds. The following data are the expense ratios for 10 midcap stock funds, 10 small-cap stock funds, 10 hybrid stock funds, and 10 specialty stock funds (*Money*, March 2003).

Midcap	Small-Cap	Hybrid	Specialty
1.2	2.0	2.0	1.6
1.1	1.2	2.7	2.7
1.0	1.7	1.8	2.6
1.2	1.8	1.5	2.5
1.3	1.5	2.5	1.9
1.8	2.3	1.0	1.5
1.4	1.9	0.9	1.6
1.4	1.3	1.9	2.7
1.0	1.2	1.4	2.2
1.4	1.3	0.3	0.7

Use $\alpha = .05$ to test for any significant difference in the mean expense ratio among the four types of stock funds.

47. The U.S. Census Bureau computes quarterly vacancy and homeownership rates by state and metropolitan statistical area. Each metropolitan statistical area (MSA) has at least one urbanized area of 50,000 or more inhabitants. The following data are the rental vacancy rates (%) for MSAs in four geographic regions of the United States for the first quarter of 2008 (U.S. Census Bureau website, January 2009).

Midwest	Northeast	South	West
16.2	2.7	16.6	7.9
10.1	11.5	8.5	6.6
8.6	6.6	12.1	6.9
12.3	7.9	9.8	5.6
10.0	5.3	9.3	4.3
16.9	10.7	9.1	15.2
16.9	8.6	5.6	5.7
5.4	5.5	9.4	4.0
18.1	12.7	11.6	12.3
11.9	8.3	15.6	3.6
11.0	6.7	18.3	11.0
9.6	14.2	13.4	12.1

Midwest	Northeast	South	West
7.6	1.7	6.5	8.7
12.9	3.6	11.4	5.0
12.2	11.5	13.1	4.7
13.6	16.3	4.4	3.3
		8.2	3.4
		24.0	5.5
		12.2	
		22.6	
		12.0	
		14.5	
		12.6	
		9.5	
		10.1	

Use $\alpha = .05$ to test whether there the mean vacancy rate is the same for each geographic region.

48. Three different assembly methods have been proposed for a new product. A completely randomized experimental design was chosen to determine which assembly method results in the greatest number of parts produced per hour, and 30 workers were randomly selected and assigned to use one of the proposed methods. The number of units produced by each worker follows.

Method

A	B	C
97	93	99
73	100	94
93	93	87
100	55	66
73	77	59
91	91	75
100	85	84
86	73	72
92	90	88
95	83	86

Use these data and test to see whether the mean number of parts produced is the same with each method. Use $\alpha = .05$.

49. In a study conducted to investigate browsing activity by shoppers, each shopper was initially classified as a nonbrowser, light browser, or heavy browser. For each shopper, the study obtained a measure to determine how comfortable the shopper was in a store. Higher scores indicated greater comfort. Suppose the following data were collected. Use $\alpha = .05$ to test for differences among comfort levels for the three types of browsers.

Nonbrowser	Light Browser	Heavy Browser
4	5	5
5	6	7
6	5	5
3	4	7
3	7	4
4	4	6
5	6	5
4	5	7

Case Problem 1 Par, Inc.

Par, Inc., is a major manufacturer of golf equipment. Management believes that Par's market share could be increased with the introduction of a cut-resistant, longer-lasting golf ball. Therefore, the research group at Par has been investigating a new golf ball coating designed to resist cuts and provide a more durable ball. The tests with the coating have been promising.

One of the researchers voiced concern about the effect of the new coating on driving distances. Par would like the new cut-resistant ball to offer driving distances comparable to those of the current-model golf ball. To compare the driving distances for the two balls, 40 balls of both the new and current models were subjected to distance tests. The testing was performed with a mechanical hitting machine so that any difference between the mean distances for the two models could be attributed to a difference in the two models. The results of the tests, with distances measured to the nearest yard, follow. These data are available on the website that accompanies the text.

WEB file
Golf

Model		Model		Model		Model	
Current	New	Current	New	Current	New	Current	New
264	277	270	272	263	274	281	283
261	269	287	259	264	266	274	250
267	263	289	264	284	262	273	253
272	266	280	280	263	271	263	260
258	262	272	274	260	260	275	270
283	251	275	281	283	281	267	263
258	262	265	276	255	250	279	261
266	289	260	269	272	263	274	255
259	286	278	268	266	278	276	263
270	264	275	262	268	264	262	279

Managerial Report

1. Formulate and present the rationale for a hypothesis test that Par could use to compare the driving distances of the current and new golf balls.
2. Analyze the data to provide the hypothesis testing conclusion. What is the *p*-value for your test? What is your recommendation for Par, Inc.?
3. Provide descriptive statistical summaries of the data for each model.
4. What is the 95% confidence interval for the population mean of each model, and what is the 95% confidence interval for the difference between the means of the two populations?
5. Do you see a need for larger sample sizes and more testing with the golf balls? Discuss.

Case Problem 2 Wentworth Medical Center

As part of a long-term study of individuals 65 years of age or older, sociologists and physicians at the Wentworth Medical Center in upstate New York investigated the relationship between geographic location and depression. A sample of 60 individuals, all in reasonably good health, was selected; 20 individuals were residents of Florida, 20 were residents of New York, and 20 were residents of North Carolina. Each of the individuals sampled was given a standardized test to measure depression. The data collected follow; higher test scores indicate higher levels of depression. These data are available on the data disk in the file Medical1.

A second part of the study considered the relationship between geographic location and depression for individuals 65 years of age or older who had a chronic health condition such as arthritis, hypertension, and/or heart ailment. A sample of 60 individuals with such conditions was identified. Again, 20 were residents of Florida, 20 were residents of New York, and 20 were residents of North Carolina. The levels of depression recorded for this study follow. These data are available on the website accompanying the text in the file named Medical2.

Data from Medical1			Data from Medical2		
Florida	New York	North Carolina	Florida	New York	North Carolina
3	8	10	13	14	10
7	11	7	12	9	12
7	9	3	17	15	15
3	7	5	17	12	18
8	8	11	20	16	12
8	7	8	21	24	14
8	8	4	16	18	17
5	4	3	14	14	8
5	13	7	13	15	14
2	10	8	17	17	16
6	6	8	12	20	18
2	8	7	9	11	17
6	12	3	12	23	19
6	8	9	15	19	15
9	6	8	16	17	13
7	8	12	15	14	14
5	5	6	13	9	11
4	7	3	10	14	12
7	7	8	11	13	13
3	8	11	17	11	11

Managerial Report

1. Use descriptive statistics to summarize the data from the two studies. What are your preliminary observations about the depression scores?
2. Use analysis of variance on both data sets. State the hypotheses being tested in each case. What are your conclusions?
3. Use inferences about individual treatment means where appropriate. What are your conclusions?

Case Problem 3 Compensation for Sales Professionals

Suppose that a local chapter of sales professionals in the greater San Francisco area conducted a survey of its membership to study the relationship, if any, between the years of experience and salary for individuals employed in inside and outside sales positions. On the survey, respondents were asked to specify one of three levels of years of experience: low (1–10 years), medium (11–20 years), and high (21 or more years). A portion of the data obtained follow. The complete data set, consisting of 120 observations, is available on the website accompanying the text in the file named SalesSalary.

WEBfile
SalesSalary

Observation	Salary $	Position	Experience
1	53938	Inside	Medium
2	52694	Inside	Medium
3	70515	Outside	Low
4	52031	Inside	Medium
5	62283	Outside	Low
6	57718	Inside	Low
7	79081	Outside	High
8	48621	Inside	Low
9	72835	Outside	High
10	54768	Inside	Medium
⋮	⋮	⋮	⋮
115	58080	Inside	High
116	78702	Outside	Medium
117	83131	Outside	Medium
118	57788	Inside	High
119	53070	Inside	Medium
120	60259	Outside	Low

Managerial Report

1. Use descriptive statistics to summarize the data.
2. Develop a 95% confidence interval estimate of the mean annual salary for all salespersons, regardless of years of experience and type of position.
3. Develop a 95% confidence interval estimate of the mean salary for inside salespersons.
4. Develop a 95% confidence interval estimate of the mean salary for outside salespersons.
5. Use analysis of variance to test for any significant differences due to position. Use a .05 level of significance, and for now, ignore the effect of years of experience.
6. Use analysis of variance to test for any significant differences due to years of experience. Use a .05 level of significance, and for now, ignore the effect of position.

Appendix 10.1 Inferences About Two Populations Using Minitab

We describe the use of Minitab to develop interval estimates and conduct hypothesis tests about the difference between two population means and the difference between two population proportions. Minitab provides both interval estimation and hypothesis testing results within the same module. Thus, the Minitab procedure is the same for both types of inferences. In the examples that follow, we will demonstrate interval estimation and hypothesis testing for the same two samples. We note that Minitab does not provide a routine for inferences about the difference between two population means when the population standard deviations σ_1 and σ_2 are known.

Difference Between Two Population Means: σ_1 and σ_2 Unknown

WEBfile
CheckAcct

We will use the data for the checking account balances example presented in Section 10.2. The checking account balances at the Cherry Grove branch are in column C1, and the checking account balances at the Beechmont branch are in column C2. In this example, we will use the Minitab 2-Sample *t* procedure to provide a 95% confidence interval estimate of the difference between population means for the checking account balances at the two branch banks. The output of the procedure also provides the *p*-value for the hypothesis test: $H_0: \mu_1 - \mu_2 = 0$ versus $H_a: \mu_1 - \mu_2 \neq 0$. The following steps are necessary to execute the procedure:

Step 1. Select the **Stat** menu
Step 2. Choose **Basic Statistics**
Step 3. Choose **2-Sample t**
Step 4. When the 2-Sample t (Test and Confidence Interval) dialog box appears:
Select **Samples in different columns**
Enter C1 in the **First** box
Enter C2 in the **Second** box
Select **Options**
Step 5. When the 2-Sample t—Options dialog box appears:
Enter 95 in the **Confidence level** box
Enter 0 in the **Test difference** box
Enter not equal in the **Alternative** box
Click **OK**
Step 6. Click **OK**

The 95% confidence interval estimate is $37 to $193, as described in Section 10.2. The *p*-value = .005 shows the null hypothesis of equal population means can be rejected at the $\alpha = .01$ level of significance. In other applications, step 5 may be modified to provide different confidence levels, different hypothesized values, and different forms of the hypotheses.

Difference Between Two Population Means with Matched Samples

We use the data on production times in Table 10.2 to illustrate the matched-sample procedure. The completion times for method 1 are entered into column C1 and the completion times for method 2 are entered into column C2. The Minitab steps for a matched sample are as follows:

Step 1. Select the **Stat** menu
Step 2. Choose **Basic Statistics**
Step 3. Choose **Paired t**
Step 4. When the Paired t (Test and Confidence Interval) dialog box appears:
Select **Samples in columns**
Enter C1 in the **First sample** box
Enter C2 in the **Second sample** box
Select **Options**
Step 5. When the Paired t—Options dialog box appears:
Enter 95 in the **Confidence level**
Enter 0 in the **Test mean** box
Enter not equal in the **Alternative** box
Click **OK**
Step 6. Click **OK**

The 95% confidence interval estimate is $-.05$ to .65, as described in Section 10.3. The *p*-value = .08 shows that the null hypothesis of no difference in completion times cannot be rejected at $\alpha = .05$. Step 5 may be modified to provide different confidence levels, different hypothesized values, and different forms of the hypothesis.

Appendix 10.2 Analysis of Variance Using Minitab

Completely Randomized Design

In Section 10.5 we showed how analysis of variance could be used to test for the equality of k population means using data from a completely randomized design. To illustrate how Minitab can be used for this type of experimental design, we show how to test whether the mean number of units produced per week is the same for each assembly method in the Chemitech experiment introduced in Section 10.4. The sample data are entered into the first three columns of a Minitab worksheet; column 1 is labeled A, column 2 is labeled B, and column 3 is labeled C. The following steps produce the Minitab output in Figure 10.7.

Step 1. Select the **Stat** menu
Step 2. Choose **ANOVA**
Step 3. Choose **One-way (Unstacked)**
Step 4. When the One-way Analysis of Variance dialog box appears:
  Enter C1-C3 in the **Responses (in separate columns)** box
  Click **OK**

Appendix 10.3 Inferences About Two Populations Using Excel

We describe the use of Excel to conduct hypothesis tests about the difference between two population means.* We begin with inferences about the difference between the means of two populations when the population standard deviations σ_1 and σ_2 are known.

Difference Between Two Population Means: σ_1 and σ_2 Known

We will use the examination scores for the two training centers discussed in Section 10.1. The label Center A is in cell A1 and the label Center B is in cell B1. The examination scores for Center A are in cells A2:A31 and examination scores for Center B are in cells B2:B41. The population standard deviations are assumed known with $\sigma_1 = 10$ and $\sigma_2 = 10$. The Excel routine will request the input of variances which are $\sigma_1^2 = 100$ and $\sigma_2^2 = 100$. The following steps can be used to conduct a hypothesis test about the difference between the two population means.

Step 1. Click the **Data** tab on the Ribbon
Step 2. In the **Analysis** group, click **Data Analysis**
Step 3. When the Data Analysis dialog box appears:
  Choose **z-Test: Two Sample for Means**
  Click **OK**
Step 4. When the z-Test: Two Sample for Means dialog box appears:
  Enter A1:A31 in the **Variable 1 Range** box
  Enter B1:B41 in the **Variable 2 Range** box
  Enter 0 in the **Hypothesized Mean Difference** box
  Enter 100 in the **Variable 1 Variance (known)** box
  Enter 100 in the **Variable 2 Variance (known)** box
  Select **Labels**
  Enter .05 in the **Alpha** box

---

*Excel's data analysis tools provide hypothesis testing procedures for the difference between two population means. No routines are available for interval estimation of the difference between two population means nor for inferences about the difference between two population proportions.

Select **Output Range** and enter C1 in the box
Click **OK**

The two-tailed *p*-value is denoted P(Z<=z) two-tail. Its value of .0977 does not allow us to reject the null hypothesis at $\alpha = .05$.

Difference Between Two Population Means: σ_1 and σ_2 Unknown

WEB file
SoftwareTest

We use the data for the software testing study in Table 10.1. The data are already entered into an Excel worksheet with the label Current in cell A1 and the label New in cell B1. The completion times for the current technology are in cells A2:A13, and the completion times for the new software are in cells B2:B13. The following steps can be used to conduct a hypothesis test about the difference between two population means with σ_1 and σ_2 unknown.

Step 1. Click the **Data** tab on the Ribbon
Step 2. In the **Analysis** group, click **Data Analysis**
Step 3. When the Data Analysis dialog box appears:
 Choose **t-Test: Two Sample Assuming Unequal Variances**
 Click **OK**
Step 4. When the t-Test: Two Sample Assuming Unequal Variances dialog box appears:
 Enter A1:A13 in the **Variable 1 Range** box
 Enter B1:B13 in the **Variable 2 Range** box
 Enter 0 in the **Hypothesized Mean Difference** box
 Select **Labels**
 Enter .05 in the **Alpha** box
 Select **Output Range** and enter C1 in the box
 Click **OK**

The appropriate *p*-value is denoted P(T<=t) one-tail. Its value of .017 allows us to reject the null hypothesis at $\alpha = .05$.

Difference Between Two Population Means with Matched Samples

WEB file
Matched

We use the matched-sample completion times in Table 10.2 to illustrate. The data are entered into a worksheet with the label Method 1 in cell A1 and the label Method 2 in cell B2. The completion times for method 1 are in cells A2:A7 and the completion times for method 2 are in cells B2:B7. The Excel procedure uses the steps previously described for the *t*-Test except the user chooses the ***t*-Test: Paired Two Sample for Means** data analysis tool in step 3. The variable 1 range is A1:A7 and the variable 2 range is B1:B7.

The appropriate *p*-value is denoted P(T<=t) two-tail. Its value of .08 does not allow us to reject the null hypothesis at $\alpha = .05$.

Appendix 10.4 Analysis of Variance Using Excel

Completely Randomized Design

WEB file
Chemitech

In Section 10.5 we showed how analysis of variance could be used to test for the equality of *k* population means using data from a completely randomized design. To illustrate how Excel can be used to test for the equality of *k* population means for this type of experimental design, we show how to test whether the mean number of units produced per week is the same for each assembly method in the Chemitech experiment introduced in Section 10.4. The sample data are entered into worksheet rows 2 to 6 of columns A, B, and C

FIGURE 10.8 EXCEL SOLUTION FOR THE CHEMITECH EXPERIMENT

	A	B	C	D	E	F	G	H
1	Method A	Method B	Method C					
2	58	58	48					
3	64	69	57					
4	55	71	59					
5	66	64	47					
6	67	68	49					
7								
8	Anova: Single Factor							
9								
10	SUMMARY							
11	*Groups*	*Count*	*Sum*	*Average*	*Variance*			
12	Method A	5	310	62	27.5			
13	Method B	5	330	66	26.5			
14	Method C	5	260	52	31			
15								
16								
17	ANOVA							
18	*Source of Variation*	*SS*	*df*	*MS*	*F*	*P-value*	*F crit*	
19	Between Groups	520	2	260	9.1765	0.0038	3.8853	
20	Within Groups	340	12	28.3333				
21								
22	Total	860	14					
23								
24								

as shown in Figure 10.8. The following steps are used to obtain the output shown in cells A8:G22; the ANOVA portion of this output corresponds to the ANOVA table shown in Table 10.5.

    Step 1. Click the **Data** tab on the Ribbon
    Step 2. In the **Analysis** group, click **Data Analysis**
    Step 3. Choose **Anova: Single Factor** from the list of Analysis Tools
          Click **OK**
    Step 4. When the Anova: Single Factor dialog box appears:
          Enter A1:C6 in **Input Range** box
          Select **Columns**
          Select **Labels in First Row**
          Select **Output Range** and enter A8 in the box
          Click **OK**

Appendix 10.5 Inferences About Two Populations Using StatTools

In this appendix we show how StatTools can be used to develop interval estimates and conduct hypothesis tests about the difference between two population means for the σ_1 and σ_2 unknown case. We also show how StatTools can be used to do the same for the matched samples case.

Interval Estimation of μ_1 and μ_2

WEB file
CheckAcct

We will use the data for the checking account balances example presented in Section 10.2. Begin by using the Data Set Manager to create a StatTools data set for these data using the procedure described in the appendix in Chapter 1. The following steps can be used to compute a 95% confidence interval estimate of the difference between the two population means.

Step 1. Click the **StatTools** tab on the Ribbon
Step 2. In the **Analyses** group, click **Statistical Inference**
Step 3. Select the **Confidence Interval** option
Step 4. Choose Mean/Std. Deviation
Step 5. When the StatTools—Confidence Interval for Mean/Std. Deviation dialog box appears:
  For **Analysis Type**, choose **Two-Sample Analysis**
  In the **Variables** section,
    Select **Cherry Grove**
    Select **Beechmont**
  In the **Confidence Intervals to Calculate** section,
    Select the **For the Difference of Means** option
    Select 95% for the **Confidence Level**
  Click **OK**

Because the sample size for Cherry Grove ($n_1 = 28$) differs from the sample size for Beechmont ($n_2 = 22$), StatTools will inform you of this difference after you click OK in step 5. A dialog box will appear saying "The variable Beechmont contains missing data, which this analysis will ignore." Click OK. A Choose Variable Ordering dialog box then appears, indicating that the analysis will compare the difference between the Cherry Grove data set and the Beechmont data set. Click OK and the StatTools interval estimation output will appear.

Hypothesis Tests About μ_1 and μ_2

WEB file
SoftwareTest

We will use the software evaluation example and the completion time data presented in Table 10.1. Begin by using the Data Set Manager to create a StatTools data set for these data using the procedure described in the appendix in Chapter 1. The following steps can be used to test the hypothesis: $H_0: \mu_1 - \mu_2 \leq 0$ against $H_a: \mu_1 - \mu_2 > 0$.

Step 1. Click the **StatTools** tab on the Ribbon
Step 2. In the **Analyses** group, click **Statistical Inference**
Step 3. Select the **Hypothesis Test** option
Step 4. Choose Mean/Std. Deviation
Step 5. When the StatTools—Hypothesis Test for Mean/Std. Deviation dialog box appears:
  For **Analysis Type**, choose **Two-Sample Analysis**
  In the **Variables** section,
    Select **Current**
    Select **New**
  In the **Hypothesis Test to Perform** section,
    Select **Difference of Means**
    Enter 0 in the **Null Hypothesis Value** box
    Select **Greater Than Null Value (One-Tailed Test)** in the **Alternative Hypothesis** box
  Click **OK**
  When the Choose Variable Ordering dialog box appears, click **OK**

The results of the hypothesis test will then appear.

Inferences About the Difference Between Two Population Means: Matched Samples

StatTools can be used to develop interval estimates and conduct hypothesis tests for the difference between population means for the matched samples case. We will use the matched-sample completion times in Table 10.2 to illustrate.

Begin by using the Data Set Manager to create a StatTools data set for these data using the procedure described in the appendix in Chapter 1. The following steps can be used to compute a 95% confidence interval estimate of the difference between the population mean completion times.

Step 1. Click the **StatTools** tab on the Ribbon
Step 2. In the **Analyses** group, click **Statistical Inference**
Step 3. Select the **Confidence Interval** option
Step 4. Choose Mean/Std. Deviation
Step 5. When the StatTools—Confidence Interval for Mean/Std. Deviation dialog box appears:
　For **Analysis Type**, choose **Paired-Sample Analysis**
　In the **Variables** section,
　　Select **Method 1**
　　Select **Method 2**
　In the **Confidence Intervals to Calculate** section,
　　Select the **For the Difference of Means** option
　　Select 95% for the **Confidence Level**
　　If selected, remove the check in the **For the Standard Deviation box**
　Click **OK**
　When the Choose Variable Ordering dialog box appears, click **OK**

The confidence interval will appear.

Conducting hypothesis tests for the matched samples case is very similar to conducting hypothesis tests for the difference in two means shown previously. Choose the Hypothesis Test option in step 3. Then when the Hypothesis Test for Mean/Std. Deviation dialog box appears, describe the type of hypothesis test desired.

Appendix 10.6　Analysis of Variance Using StatTools

In this appendix we show how StatTools can be used to test for the equality of *k* population means for a completely randomized design. We use the Chemitech data in Table 10.3 to illustrate. Begin by using the Data Set Manager to create a StatTools data set for these data using the procedure described in the appendix in Chapter 1. The following steps can be used to test for the equality of the three population means.

Step 1. Click the **StatTools** tab on the Ribbon
Step 2. In the **Analyses** group, click **Statistical Inference**
Step 3. Select the **One-Way ANOVA** option
Step 4. When the StatTools-One-Way ANOVA dialog box appears:
　In the **Variables** section:
　　Click the **Format button** and select **Unstacked**
　　Select **Method A**
　　Select **Method B**
　　Select **Method C**
　Select 95% in the **Confidence Level** box
　Click **OK**

Appendix 10.6 Analysis of Variance Using StatTools

Note that in step 4 we selected the Unstacked option after clicking the Format button. The Unstacked option means that the data for the three treatments appear in separate columns of the worksheet. In a stacked format, only two columns would be used. For example, the data could have been organized as follows:

	A	B	C
1	Method	Units Produced	
2	Method A	58	
3	Method A	64	
4	Method A	55	
5	Method A	66	
6	Method A	67	
7	Method B	58	
8	Method B	69	
9	Method B	71	
10	Method B	64	
11	Method B	68	
12	Method C	48	
13	Method C	57	
14	Method C	59	
15	Method C	47	
16	Method C	49	
17			

Data are frequently recorded in a stacked format. For stacked data, simply select the Stacked option after clicking the Format button.

CHAPTER 11

Comparisons Involving Proportions and a Test of Independence

CONTENTS

STATISTICS IN PRACTICE:
UNITED WAY

11.1 INFERENCES ABOUT THE DIFFERENCE BETWEEN TWO POPULATION PROPORTIONS
Interval Estimation of $p_1 - p_2$
Hypothesis Tests About $p_1 - p_2$

11.2 HYPOTHESIS TEST FOR PROPORTIONS OF A MULTINOMIAL POPULATION

11.3 TEST OF INDEPENDENCE

STATISTICS in PRACTICE

UNITED WAY*
ROCHESTER, NEW YORK

United Way of Greater Rochester is a nonprofit organization dedicated to improving the quality of life for all people in the seven counties it serves by meeting the community's most important human care needs.

The annual United Way/Red Cross fund-raising campaign, conducted each spring, funds hundreds of programs offered by more than 200 service providers. These providers meet a wide variety of human needs—physical, mental, and social—and serve people of all ages, backgrounds, and economic means.

Because of enormous volunteer involvement, United Way of Greater Rochester is able to hold its operating costs at just eight cents of every dollar raised.

The United Way of Greater Rochester decided to conduct a survey to learn more about community perceptions of charities. Focus-group interviews were held with professional, service, and general worker groups to get preliminary information on perceptions. The information obtained was then used to help develop the questionnaire for the survey. The questionnaire was pretested, modified, and distributed to 440 individuals; 323 completed questionnaires were obtained.

A variety of descriptive statistics, including frequency distributions and crosstabulations, were provided from the data collected. An important part of the analysis involved the use of contingency tables and chi-square tests of independence. One use of such statistical tests was to determine whether perceptions of administrative expenses were independent of occupation.

The hypotheses for the test of independence were

H_0: Perception of United Way administrative expenses is independent of the occupation of the respondent.

United Way contributions provide support for over 200 community partners.

H_a: Perception of United Way administrative expenses is not independent of the occupation of the respondent.

Two questions in the survey provided the data for the statistical test. One question obtained data on perceptions of the percentage of funds going to administrative expenses (up to 10%, 11–20%, and 21% or more). The other question asked for the occupation of the respondent.

The chi-square test at a .05 level of significance led to rejection of the null hypothesis of independence and to the conclusion that perceptions of United Way's administrative expenses did vary by occupation. Actual administrative expenses were less than 9%, but 35% of the respondents perceived that administrative expenses were 21% or more. Hence, many had inaccurate perceptions of administrative costs. In this group, production-line, clerical, sales, and professional-technical employees had more inaccurate perceptions than other groups.

The community perceptions study helped United Way of Rochester to develop adjustments to its programs and fund-raising activities. In this chapter, you will learn how a statistical test of independence, such as that described here, is conducted.

---

*The authors are indebted to Dr. Philip R. Tyler, Marketing Consultant to the United Way, for providing this Statistics in Practice.

Many statistical applications call for a comparison of population proportions. In Section 11.1, we describe statistical inferences concerning differences in the proportions for two populations. Two samples are required, one from each population, and the statistical inference is based on the two sample proportions. The second section looks at a hypothesis test comparing the proportions of a single multinomial population with the proportions stated in a null hypothesis. One sample from the multinomial population is used, and the hypothesis test is based on comparing the sample proportions with those stated in the null hypothesis. In the last section of the chapter, we show how contingency tables can be used to test for the independence of two variables. One sample is used for the test of independence, but measures on two variables are required for each sampled element. Both Sections 11.2 and 11.3 rely on the use of a chi-square statistical test.

11.1 Inferences About the Difference Between Two Population Proportions

Letting p_1 denote the proportion for population 1 and p_2 denote the proportion for population 2, we next consider inferences about the difference between the two population proportions: $p_1 - p_2$. To make an inference about this difference, we will select two independent random samples consisting of n_1 units from population 1 and n_2 units from population 2.

Interval Estimation of $p_1 - p_2$

In the following example, we show how to compute a margin of error and develop an interval estimate of the difference between two population proportions.

A tax preparation firm is interested in comparing the quality of work at two of its regional offices. By randomly selecting samples of tax returns prepared at each office and verifying the sample returns' accuracy, the firm will be able to estimate the proportion of erroneous returns prepared at each office. Of particular interest is the difference between these proportions.

p_1 = proportion of erroneous returns for population 1 (office 1)
p_2 = proportion of erroneous returns for population 2 (office 2)
$\bar{p}_1$ = sample proportion for a simple random sample from population 1
$\bar{p}_2$ = sample proportion for a simple random sample from population 2

The difference between the two population proportions is given by $p_1 - p_2$. The point estimator of $p_1 - p_2$ is as follows.

> **POINT ESTIMATOR OF THE DIFFERENCE BETWEEN TWO POPULATION PROPORTIONS**
>
> $$\bar{p}_1 - \bar{p}_2 \tag{11.1}$$

Thus, the point estimator of the difference between two population proportions is the difference between the sample proportions of two independent simple random samples.

As with other point estimators, the point estimator $\bar{p}_1 - \bar{p}_2$ has a sampling distribution that reflects the possible values of $\bar{p}_1 - \bar{p}_2$ if we repeatedly took two independent random samples. The mean of this sampling distribution is $p_1 - p_2$ and the standard error of $\bar{p}_1 - \bar{p}_2$ is as follows:

> **STANDARD ERROR OF $\bar{p}_1 - \bar{p}_2$**
>
> $$\sigma_{\bar{p}_1 - \bar{p}_2} = \sqrt{\frac{p_1(1 - p_1)}{n_1} + \frac{p_2(1 - p_2)}{n_2}} \tag{11.2}$$

11.1 Inferences About the Difference Between Two Population Proportions

If the sample sizes are large enough that n_1p_1, $n_1(1-p_1)$, n_2p_2, and $n_2(1-p_2)$ are all greater than or equal to 5, the sampling distribution of $\bar{p}_1 - \bar{p}_2$ can be approximated by a normal distribution.

As we showed previously, an interval estimate is given by a point estimate ± a margin of error. In the estimation of the difference between two population proportions, an interval estimate will take the following form:

$$\bar{p}_1 - \bar{p}_2 \pm \text{Margin of error}$$

With the sampling distribution of $\bar{p}_1 - \bar{p}_2$ approximated by a normal distribution, we would like to use $z_{\alpha/2}\sigma_{\bar{p}_1-\bar{p}_2}$ as the margin of error. However, $\sigma_{\bar{p}_1-\bar{p}_2}$ given by equation (11.2) cannot be used directly because the two population proportions, p_1 and p_2, are unknown. Using the sample proportion $\bar{p}_1$ to estimate p_1 and the sample proportion $\bar{p}_2$ to estimate p_2, the margin of error is as follows.

$$\text{Margin of error} = z_{\alpha/2}\sqrt{\frac{\bar{p}_1(1-\bar{p}_1)}{n_1} + \frac{\bar{p}_2(1-\bar{p}_2)}{n_2}} \quad (11.3)$$

The general form of an interval estimate of the difference between two population proportions is as follows.

INTERVAL ESTIMATE OF THE DIFFERENCE BETWEEN TWO POPULATION PROPORTIONS

$$\bar{p}_1 - \bar{p}_2 \pm z_{\alpha/2}\sqrt{\frac{\bar{p}_1(1-\bar{p}_1)}{n_1} + \frac{\bar{p}_2(1-\bar{p}_2)}{n_2}} \quad (11.4)$$

where $1 - \alpha$ is the confidence coefficient.

Returning to the tax preparation example, we find that independent simple random samples from the two offices provide the following information.

Office 1	Office 2
$n_1 = 250$	$n_2 = 300$
Number of returns with errors = 35	Number of returns with errors = 27

WEB file

TaxPrep

The sample proportions for the two offices follow.

$$\bar{p}_1 = \frac{35}{250} = .14$$

$$\bar{p}_2 = \frac{27}{300} = .09$$

The point estimate of the difference between the proportions of erroneous tax returns for the two populations is $\bar{p}_1 - \bar{p}_2 = .14 - .09 = .05$. Thus, we estimate that office 1 has a .05, or 5%, greater error rate than office 2.

Expression (11.4) can now be used to provide a margin of error and interval estimate of the difference between the two population proportions. Using a 90% confidence interval with $z_{\alpha/2} = z_{.05} = 1.645$, we have

$$\bar{p}_1 - \bar{p}_2 \pm z_{\alpha/2}\sqrt{\frac{\bar{p}_1(1-\bar{p}_1)}{n_1} + \frac{\bar{p}_2(1-\bar{p}_2)}{n_2}}$$

$$.14 - .09 \pm 1.645\sqrt{\frac{.14(1-.14)}{250} + \frac{.09(1-.09)}{300}}$$

$$.05 \pm .045$$

Thus, the margin of error is .045, and the 90% confidence interval is .005 to .095.

Hypothesis Tests About $p_1 - p_2$

Let us now consider hypothesis tests about the difference between the proportions of two populations. We focus on tests involving no difference between the two population proportions. In this case, the three forms for a hypothesis test are as follows:

All hypotheses considered use 0 as the difference of interest.

$$H_0: p_1 - p_2 \geq 0 \qquad H_0: p_1 - p_2 \leq 0 \qquad H_0: p_1 - p_2 = 0$$
$$H_a: p_1 - p_2 < 0 \qquad H_a: p_1 - p_2 > 0 \qquad H_a: p_1 - p_2 \neq 0$$

When we assume H_0 is true as an equality, we have $p_1 - p_2 = 0$, which is the same as saying that the population proportions are equal, $p_1 = p_2$.

We will base the test statistic on the sampling distribution of the point estimator $\bar{p}_1 - \bar{p}_2$. In equation (11.2), we showed that the standard error of $\bar{p}_1 - \bar{p}_2$ is given by

$$\sigma_{\bar{p}_1 - \bar{p}_2} = \sqrt{\frac{p_1(1-p_1)}{n_1} + \frac{p_2(1-p_2)}{n_2}}$$

Under the assumption H_0 is true as an equality, the population proportions are equal and $p_1 = p_2 = p$. In this case, $\sigma_{\bar{p}_1 - \bar{p}_2}$ becomes

---

STANDARD ERROR OF $\bar{p}_1 - \bar{p}_2$ WHEN $p_1 = p_2 = p$

$$\sigma_{\bar{p}_1 - \bar{p}_2} = \sqrt{\frac{p(1-p)}{n_1} + \frac{p(1-p)}{n_2}} = \sqrt{p(1-p)\left(\frac{1}{n_1} + \frac{1}{n_2}\right)} \qquad (11.5)$$

With p unknown, we pool, or combine, the point estimators from the two samples ($\bar{p}_1$ and $\bar{p}_2$) to obtain a single point estimator of p as follows.

POOLED ESTIMATOR OF p WHEN $p_1 = p_2 = p$

$$\bar{p} = \frac{n_1 \bar{p}_1 + n_2 \bar{p}_2}{n_1 + n_2} \qquad (11.6)$$

This **pooled estimator of p** is a weighted average of $\bar{p}_1$ and $\bar{p}_2$.

Substituting $\bar{p}$ for p in equation (11.5), we obtain an estimate of the standard error of $\bar{p}_1 - \bar{p}_2$. This estimate of the standard error is used in the test statistic. The general form of the test statistic for hypothesis tests about the difference between two population proportions is the point estimator divided by the estimate of $\sigma_{\bar{p}_1 - \bar{p}_2}$.

TEST STATISTIC FOR HYPOTHESIS TESTS ABOUT $p_1 - p_2$

$$z = \frac{(\bar{p}_1 - \bar{p}_2)}{\sqrt{\bar{p}(1-\bar{p})\left(\frac{1}{n_1} + \frac{1}{n_2}\right)}} \qquad (11.7)$$

11.1 Inferences About the Difference Between Two Population Proportions

This test statistic applies to large sample situations where $n_1 p_1$, $n_1(1 - p_1)$, $n_2 p_2$, and $n_2(1 - p_2)$ are all greater than or equal to 5.

Let us return to the tax preparation firm example and assume that the firm wants to use a hypothesis test to determine whether the error proportions differ between the two offices. A two-tailed test is required. The null and alternative hypotheses are as follows:

$$H_0: p_1 - p_2 = 0$$
$$H_a: p_1 - p_2 \neq 0$$

If H_0 is rejected, the firm can conclude that the error rates at the two offices differ. We will use $\alpha = .10$ as the level of significance.

The sample data previously collected showed $\bar{p}_1 = .14$ for the $n_1 = 250$ returns sampled at office 1 and $\bar{p}_2 = .09$ for the $n_2 = 300$ returns sampled at office 2. We continue by computing the pooled estimate of p.

$$\bar{p} = \frac{n_1 \bar{p}_1 + n_2 \bar{p}_2}{n_1 + n_2} = \frac{250(.14) + 300(.09)}{250 + 300} = .1127$$

Using this pooled estimate and the difference between the sample proportions, the value of the test statistic is as follows.

$$z = \frac{(\bar{p}_1 - \bar{p}_2)}{\sqrt{\bar{p}(1 - \bar{p})\left(\frac{1}{n_1} + \frac{1}{n_2}\right)}} = \frac{(.14 - .09)}{\sqrt{.1127(1 - .1127)\left(\frac{1}{250} + \frac{1}{300}\right)}} = 1.85$$

In computing the p-value for this two-tailed test, we first note that $z = 1.85$ is in the upper tail of the standard normal distribution. Using $z = 1.85$ and the standard normal distribution table, we find the area in the upper tail is $1.0000 - .9678 = .0322$. Doubling this area for a two-tailed test, we find the p-value $= 2(.0322) = .0644$. With the p-value less than $\alpha = .10$, H_0 is rejected at the .10 level of significance. The firm can conclude that the error rates differ between the two offices. This hypothesis testing conclusion is consistent with the earlier interval estimation results that showed the interval estimate of the difference between the population error rates at the two offices to be .005 to .095, with Office 1 having the higher error rate.

Exercises

Methods

SELF test

1. Consider the following results for independent samples taken from two populations.

Sample 1	Sample 2
$n_1 = 400$	$n_2 = 300$
$\bar{p}_1 = .48$	$\bar{p}_2 = .36$

    a. What is the point estimate of the difference between the two population proportions?
    b. Develop a 90% confidence interval for the difference between the two population proportions.
    c. Develop a 95% confidence interval for the difference between the two population proportions.

2. Consider the following hypothesis test.

$$H_0: p_1 - p_2 = 0$$
$$H_a: p_1 - p_2 \neq 0$$

The following results are for independent samples taken from the two populations.

Sample 1	Sample 2
$n_1 = 100$	$n_2 = 140$
$\bar{p}_1 = .28$	$\bar{p}_2 = .20$

a. What is the pooled estimate of p?
b. What is the p-value?
c. What is your conclusion?

3. Consider the hypothesis test

$$H_0: p_1 - p_2 \leq 0$$
$$H_a: p_1 - p_2 > 0$$

The following results are for independent samples taken from the two populations.

Sample 1	Sample 2
$n_1 = 200$	$n_2 = 300$
$\bar{p}_1 = .22$	$\bar{p}_2 = .16$

a. What is the p-value?
b. With $\alpha = .05$, what is your hypothesis testing conclusion?

Applications

4. The Professional Golf Association (PGA) measured the putting accuracy of professional golfers playing on the PGA Tour and the best amateur golfers playing in the World Amateur Championship (*Golf Magazine*, January 2007). A sample of 1075 six-foot putts by professional golfers found 688 made putts. A sample of 1200 six-foot putts by amateur golfers found 696 made putts.
   a. Estimate the proportion of made 6-foot putts by professional golfers. Estimate the proportion of made 6-foot putts by amateur golfers. Which group had a better putting accuracy?
   b. What is the point estimate of the difference between the proportions of the two populations? What does this estimate tell you about the percentage of putts made by the two groups of golfers?
   c. What is the 95% confidence interval for the difference between the two population proportions? Interpret this confidence interval in terms of the percentage of putts made by the two groups of golfers.

5. Slot machines are the favorite game at casinos throughout the United States (*Harrah's Survey 2002: Profile of the American Gambler*). The following sample data show the number of women and number of men who selected slot machines as their favorite game.

	Women	Men
Sample Size	320	250
Favorite Game—Slots	256	165

   a. What is the point estimate of the proportion of women who say slots is their favorite game?

b. What is the point estimate of the proportion of men who say slots is their favorite game?
c. Provide a 95% confidence interval estimate of the difference between the proportion of women and proportion of men who say slots is their favorite game.

6. *BusinessWeek* reported that there seems to be a difference by age group in how well people like life in Russia (*BusinessWeek*, March 10, 2008). The following sample data are consistent with the *BusinessWeek* findings and show the responses by age group to the question: "Do you like life in Russia?"

	Russian Age Group	
	17–26	40 and over
Sample	300	260
Responded Yes	192	117

a. What is the point estimate of the proportion of Russians aged 17 to 26 who like life in Russia?
b. What is the point estimate of the proportion of Russians aged 40 and over who like life in Russia?
c. Provide a 95% confidence interval estimate of the difference between the proportion of young Russians aged 17 to 26 and older Russians aged 40 and over who like life in Russia.

7. An American Automobile Association (AAA) study investigated the question of whether a man or a woman was more likely to stop and ask for directions (AAA, January 2006). The situation referred to in the study stated the following: "If you and your spouse are driving together and become lost, would you stop and ask for directions?" A sample representative of the data used by AAA showed 300 of 811 women said that they would stop and ask for directions, while 255 of 750 men said that they would stop and ask for directions.
a. The AAA research hypothesis was that women would be more likely to say that they would stop and ask for directions. Formulate the null and alternative hypotheses for this study.
b. What is the percentage of women who indicated that they would stop and ask for directions?
c. What is the percentage of men who indicated that they would stop and ask for directions?
d. At $\alpha = .05$, test the hypothesis. What is the *p*-value, and what conclusion would you expect AAA to draw from this study?

8. Chicago O'Hare and Atlanta Hartsfield-Jackson are the two busiest airports in the United States. The congestion often leads to delayed flight arrivals as well as delayed flight departures. The Bureau of Transportation tracks the on-time and delayed performance at major airports (*Travel & Leisure*, November 2006). A flight is considered delayed if it is more than 15 minutes behind schedule. The following sample data show the delayed departures at Chicago O'Hare and Atlanta Hartsfield-Jackson airports.

	Chicago O'Hare	Atlanta Hartsfield-Jackson
Flights	900	1200
Delayed Departures	252	312

a. State the hypotheses that can be used to determine whether the population proportions of delayed departures differ at these two airports.

b. What is the point estimate of the proportion of flights that have delayed departures at Chicago O'Hare?
c. What is the point estimate of the proportion of flights that have delayed departures at Atlanta Hartsfield-Jackson?
d. What is the *p*-value for the hypothesis test? What is your conclusion?

9. A 2003 *New York Times*/CBS News poll sampled 523 adults who were planning a vacation during the next six months and found that 141 were expecting to travel by airplane (New York Times News Service, March 2, 2003). A similar survey question in a May 1993 *New York Times*/CBS News poll found that of 477 adults who were planning a vacation in the next six months, 81 were expecting to travel by airplane.
   a. State the hypotheses that can be used to determine whether a significant change occurred in the population proportion planning to travel by airplane over the 10-year period.
   b. What is the sample proportion expecting to travel by airplane in 2003? In 1993?
   c. Use $\alpha = .01$ and test for a significant difference. What is your conclusion?
   d. Discuss reasons that might provide an explanation for this conclusion.

10. During the 2003 Super Bowl, Miller Lite Beer's commercial referred to as "The Miller Lite Girls" ranked among the top three most effective advertisements aired during the Super Bowl (*USA Today*, December 29, 2003). The survey of advertising effectiveness, conducted by *USA Today*'s Ad Track poll, reported separate samples by respondent age group to learn about how the Super Bowl advertisement appealed to different age groups. The following sample data apply to the "The Miller Lite Girls" commercial.

Age Group	Sample Size	Liked the Ad a Lot
Under 30	100	49
30 to 49	150	54

   a. Formulate a hypothesis test that can be used to determine whether the population proportions for the two age groups differ.
   b. What is the point estimate of the difference between the two population proportions?
   c. Conduct the hypothesis test and report the *p*-value. At $\alpha = .05$, what is your conclusion?
   d. Discuss the appeal of the advertisements to the younger and the older age groups. Would the Miller Lite organization find the results of the *USA Today* Ad Track poll encouraging? Explain.

11.2 Hypothesis Test for Proportions of a Multinomial Population

In this section, we consider hypothesis tests concerning the proportion of elements in a population belonging to each of several classes or categories. In contrast to the preceding section, we deal with a single population: a **multinomial population**. The parameters of the multinomial population are the proportion of elements belonging to each category; the hypothesis tests we describe concern the value of those parameters. The multinomial probability distribution can be thought of as an extension of the binomial distribution to the case of three or more categories of outcomes. On each trial of a multinomial experiment, one and only one of the outcomes occurs. Each trial of the experiment is assumed independent, and the probabilities of the outcomes stay the same for each trial.

As an example, consider the market share study being conducted by Scott Marketing Research. Over the past year market shares stabilized at 30% for company A, 50% for

11.2 Hypothesis Test for Proportions of a Multinomial Population

The assumptions for the multinomial experiment parallel those for the binomial experiment with the exception that the multinomial has three or more outcomes per trial.

company B, and 20% for company C. Recently company C developed a "new and improved" product to replace its current entry in the market. Company C retained Scott Marketing Research to determine whether the new product will alter market shares.

In this case, the population of interest is a multinomial population; each customer is classified as buying from company A, company B, or company C. Thus, we have a multinomial population with three outcomes. Let us use the following notation for the proportions.

$$p_A = \text{market share for company A}$$
$$p_B = \text{market share for company B}$$
$$p_C = \text{market share for company C}$$

Scott Marketing Research will conduct a sample survey and compute the proportion preferring each company's product. A hypothesis test will then be conducted to see whether the new product caused a change in market shares. Assuming that company C's new product will not alter the market shares, the null and alternative hypotheses are stated as follows.

$$H_0: p_A = .30, p_B = .50, \text{ and } p_C = .20$$
H_a: The population proportions are not
$$p_A = .30, p_B = .50, \text{ and } p_C = .20$$

If the sample results lead to the rejection of H_0, Scott Marketing Research will have evidence that the introduction of the new product affects market shares.

Let us assume that the market research firm has used a consumer panel of 200 customers for the study. Each individual was asked to specify a purchase preference among the three alternatives: company A's product, company B's product, and company C's new product. The 200 responses are summarized here.

The consumer panel of 200 customers in which each individual is asked to select one of three alternatives is equivalent to a multinomial experiment consisting of 200 trials.

	Observed Frequency	
Company A's Product	Company B's Product	Company C's New Product
48	98	54

We now can perform a **goodness of fit test** that will determine whether the sample of 200 customer purchase preferences is consistent with the null hypothesis. The goodness of fit test is based on a comparison of the sample of *observed* results with the *expected* results under the assumption that the null hypothesis is true. Hence, the next step is to compute expected purchase preferences for the 200 customers under the assumption that $p_A = .30, p_B = .50,$ and $p_C = .20$. Doing so provides the expected results.

	Expected Frequency	
Company A's Product	Company B's Product	Company C's New Product
200(.30) = 60	200(.50) = 100	200(.20) = 40

Thus, we see that the expected frequency for each category is found by multiplying the sample size of 200 by the hypothesized proportion for the category.

The goodness of fit test now focuses on the differences between the observed frequencies and the expected frequencies. Large differences between observed and expected

458 Chapter 11 Comparisons Involving Proportions and a Test of Independence

frequencies cast doubt on the assumption that the hypothesized proportions or market shares are correct. Whether the differences between the observed and expected frequencies are "large" or "small" is a question answered with the aid of the following test statistic.

TEST STATISTIC FOR GOODNESS OF FIT

$$\chi^2 = \sum_{i=1}^{k} \frac{(f_i - e_i)^2}{e_i} \qquad (11.8)$$

where

f_i = observed frequency for category i
e_i = expected frequency for category i
k = the number of categories

Note: The test statistic has a chi-square distribution with $k - 1$ degrees of freedom provided that the expected frequencies are 5 *or more* for all categories.

Let us continue with the Scott Market Research example and use the sample data to test the hypothesis that the multinomial population retains the proportions $p_A = .30$, $p_B = .50$, and $p_C = .20$. We will use an $\alpha = .05$ level of significance. We proceed by using the observed and expected frequencies to compute the value of the test statistic. With the expected frequencies all 5 or more, the computation of the chi-square test statistic is shown in Table 11.1. Thus, we have $\chi^2 = 7.34$.

The test for goodness of fit is always a one-tailed test with the rejection occurring in the upper tail of the chi-square distribution.

We will reject the null hypothesis if the differences between the observed and expected frequencies are *large*. Large differences between the observed and expected frequencies will result in a large value for the test statistic. Thus the test of goodness of fit will always be an upper tail test. We can use the upper tail area for the test statistic and the *p*-value approach to determine whether the null hypothesis can be rejected. With $k - 1 = 3 - 1 = 2$ degrees of freedom, Table 11.2 shows the following areas in the upper tail and their corresponding chi-square (χ^2) values:

Area in Upper Tail	.10	.05	.025	.01
χ^2 Value (2 df)	4.605	5.991	7.378	9.210

$\chi^2 = 7.34$

TABLE 11.1 COMPUTATION OF THE CHI-SQUARE TEST STATISTIC FOR THE SCOTT MARKETING RESEARCH MARKET SHARE STUDY

Category	Hypothesized Proportion	Observed Frequency (f_i)	Expected Frequency (e_i)	Difference ($f_i - e_i$)	Squared Difference ($f_i - e_i)^2$	Squared Difference Divided by Expected Frequency $(f_i - e_i)^2/e_i$
Company A	.30	48	60	−12	144	2.40
Company B	.50	98	100	−2	4	0.04
Company C	.20	54	40	14	196	4.90
Total		200				$\chi^2 = 7.34$

11.2 Hypothesis Test for Proportions of a Multinomial Population

TABLE 11.2 SELECTED VALUES FROM THE CHI-SQUARE DISTRIBUTION TABLE*

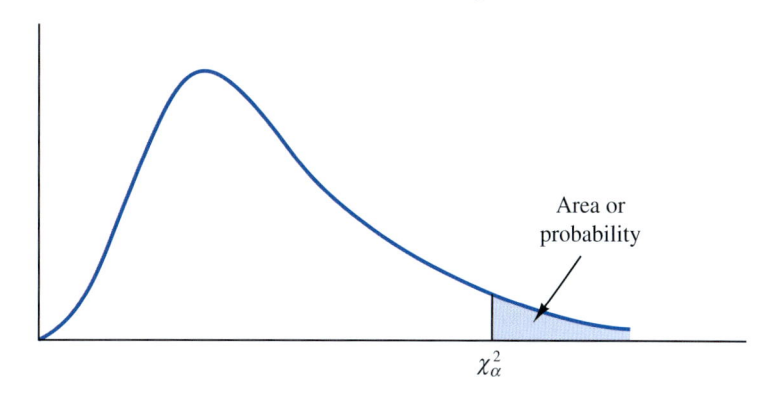

Degrees of Freedom	.99	.975	.95	.90	.10	.05	.025	.01
1	.000	.001	.004	.016	2.706	3.841	5.024	6.635
2	.020	.051	.103	.211	4.605	5.991	7.378	9.210
3	.115	.216	.352	.584	6.251	7.815	9.348	11.345
4	.297	.484	.711	1.064	7.779	9.488	11.143	13.277
5	.554	.831	1.145	1.610	9.236	11.070	12.832	15.086
6	.872	1.237	1.635	2.204	10.645	12.592	14.449	16.812
7	1.239	1.690	2.167	2.833	12.017	14.067	16.013	18.475
8	1.647	2.180	2.733	3.490	13.362	15.507	17.535	20.090
9	2.088	2.700	3.325	4.168	14.684	16.919	19.023	21.666
10	2.558	3.247	3.940	4.865	15.987	18.307	20.483	23.209
11	3.053	3.816	4.575	5.578	17.275	19.675	21.920	24.725
12	3.571	4.404	5.226	6.304	18.549	21.026	23.337	26.217
13	4.107	5.009	5.892	7.041	19.812	22.362	24.736	27.688
14	4.660	5.629	6.571	7.790	21.064	23.685	26.119	29.141
15	5.229	6.262	7.261	8.547	22.307	24.996	27.488	30.578
16	5.812	6.908	7.962	9.312	23.542	26.296	28.845	32.000
17	6.408	7.564	8.672	10.085	24.769	27.587	30.191	33.409
18	7.015	8.231	9.390	10.865	25.989	28.869	31.526	34.805
19	7.633	8.907	10.117	11.651	27.204	30.144	32.852	36.191
20	8.260	9.591	10.851	12.443	28.412	31.410	34.170	37.566
21	8.897	10.283	11.591	13.240	29.615	32.671	35.479	38.932
22	9.542	10.982	12.338	14.041	30.813	33.924	36.781	40.289
23	10.196	11.689	13.091	14.848	32.007	35.172	38.076	41.638
24	10.856	12.401	13.848	15.659	33.196	36.415	39.364	42.980
25	11.524	13.120	14.611	16.473	34.382	37.652	40.646	44.314
26	12.198	13.844	15.379	17.292	35.563	38.885	41.923	45.642
27	12.878	14.573	16.151	18.114	36.741	40.113	43.195	46.963
28	13.565	15.308	16.928	18.939	37.916	41.337	44.461	48.278
29	14.256	16.047	17.708	19.768	39.087	42.557	45.722	49.588
30	14.953	16.791	18.493	20.599	40.256	43.773	46.979	50.892
40	22.164	24.433	26.509	29.051	51.805	55.758	59.342	63.691
60	37.485	40.482	43.188	46.459	74.397	79.082	83.298	88.379
80	53.540	57.153	60.391	64.278	96.578	101.879	106.629	112.329
100	70.065	74.222	77.929	82.358	118.498	124.342	129.561	135.807

*Note: A more extensive table is provided as Table 3 of Appendix B.

The test statistic $\chi^2 = 7.34$ is between 5.991 and 7.378. Thus, the corresponding upper tail area or p-value must be between .05 and .025. With p-value $\leq \alpha = .05$, we reject H_0 and conclude that the introduction of the new product by company C will alter the current market share structure. Minitab or Excel procedures provided in Appendix F at the back of the book can be used to show that $\chi^2 = 7.34$ provides a p-value $= .0255$.

Instead of using the p-value, we could use the critical value approach to draw the same conclusion. With $\alpha = .05$ and 2 degrees of freedom, the critical value for the test statistic is $\chi^2_{.05} = 5.991$. The upper tail rejection rule becomes

$$\text{Reject } H_0 \text{ if } \chi^2 \geq 5.991$$

With $7.34 > 5.991$, we reject H_0. The p-value approach and critical value approach provide the same hypothesis testing conclusion.

Although no further conclusions can be made as a result of the test, we can compare the observed and expected frequencies informally to obtain an idea of how the market share structure may change. Considering company C, we find that the observed frequency of 54 is larger than the expected frequency of 40. Because the expected frequency was based on current market shares, the larger observed frequency suggests that the new product will have a positive effect on company C's market share. Comparisons of the observed and expected frequencies for the other two companies indicate that company C's gain in market share will hurt company A more than company B.

Let us summarize the general steps that can be used to conduct a goodness of fit test for a hypothesized multinomial population distribution.

MULTINOMIAL DISTRIBUTION GOODNESS OF FIT TEST: A SUMMARY

1. State the null and alternative hypotheses.

   H_0: The population follows a multinomial distribution with specified probabilities for each of the k categories

   H_a: The population does not follow a multinomial distribution with the specified probabilities for each of the k categories

2. Select a random sample and record the observed frequencies f_i for each category.
3. Assume the null hypothesis is true and determine the expected frequency e_i in each category by multiplying the category probability by the sample size.
4. Compute the value of the test statistic.

$$\chi^2 = \sum_{i=1}^{k} \frac{(f_i - e_i)^2}{e_i}$$

5. Rejection rule:

   p-value approach: Reject H_0 if p-value $\leq \alpha$

   Critical value approach: Reject H_0 if $\chi^2 \geq \chi^2_\alpha$

   where α is the level of significance for the test and there are $k - 1$ degrees of freedom.

Exercises

Methods

11. Test the following hypotheses by using the χ^2 goodness of fit test.

    $$H_0: p_A = .40, p_B = .40, \text{ and } p_C = .20$$
    $$H_a: \text{The population proportions are not}$$
    $$p_A = .40, p_B = .40, \text{ and } p_C = .20$$

    A sample of size 200 yielded 60 in category A, 120 in category B, and 20 in category C. Use $\alpha = .01$ and test to see whether the proportions are as stated in H_0.
    a. Use the *p*-value approach.
    b. Repeat the test using the critical value approach.

12. Suppose we have a multinomial population with four categories: A, B, C, and D. The null hypothesis is that the proportion of items is the same in every category. The null hypothesis is

    $$H_0: p_A = p_B = p_C = p_D = .25$$

    A sample of size 300 yielded the following results.

    A: 85   B: 95   C: 50   D: 70

    Use $\alpha = .05$ to determine whether H_0 should be rejected. What is the *p*-value?

Applications

13. During the first 13 weeks of the network television season, the Saturday evening 8:00 P.M. to 9:00 P.M. audience proportions were recorded as ABC 29%, CBS 28%, NBC 25%, and independents 18%. A sample of 300 homes two weeks after a Saturday night schedule revision yielded the following viewing audience data: ABC 95 homes, CBS 70 homes, NBC 89 homes, and independents 46 homes. Test with $\alpha = .05$ to determine whether the viewing audience proportions changed.

14. M&M/MARS, makers of M&M's® Chocolate Candies, conducted a national poll in which more than 10 million people indicated their preference for a new color. The tally of this poll resulted in the replacement of tan-colored M&Ms with a new blue color. In the brochure "Colors," made available by M&M/MARS Consumer Affairs, the distribution of colors for the plain candies is as follows:

Brown	Yellow	Red	Orange	Green	Blue
30%	20%	20%	10%	10%	10%

    In a follow-up study, samples of 1-pound bags were used to determine whether the reported percentages were indeed valid. The following results were obtained for one sample of 506 plain candies.

Brown	Yellow	Red	Orange	Green	Blue
177	135	79	41	36	38

    Use $\alpha = .05$ to determine whether these data support the percentages reported by the company.

15. Where do women most often buy casual clothing? Data from the U.S. Shopper Database provided the following percentages for women shopping at each of the various outlets (*The Wall Street Journal*, January 28, 2004).

Outlet	Percentage	Outlet	Percentage
Walmart	24	Kohl's	8
Traditional department stores	11	Mail order	12
JC Penney	8	Other	37

The other category included outlets such as Target, Kmart, and Sears as well as numerous smaller specialty outlets. No individual outlet in this group accounted for more than 5% of the women shoppers. A recent survey using a sample of 140 women shoppers in Atlanta, Georgia, found 42 Walmart, 20 traditional department store, 8 JC Penney, 10 Kohl's, 21 mail order, and 39 other outlet shoppers. Does this sample suggest that women shoppers in Atlanta differ from the shopping preferences expressed in the U.S. Shopper Database? What is the *p*-value? Use $\alpha = .05$. What is your conclusion?

16. The American Bankers Association collects data on the use of credit cards, debit cards, personal checks, and cash when consumers pay for in-store purchases (*The Wall Street Journal*, December 16, 2003). In 1999, the following usages were reported.

In-Store Purchase	Percentage
Credit card	22
Debit card	21
Personal check	18
Cash	39

A sample taken in 2003 found that for 220 in-stores purchases, 46 used a credit card, 67 used a debit card, 33 used a personal check, and 74 used cash.

  a. At $\alpha = .01$, can we conclude that a change occurred in how customers paid for in-store purchases over the four-year period from 1999 to 2003? What is the *p*-value?
  b. Compute the percentage of use for each method of payment using the 2003 sample data. What appears to have been the major change or changes over the four-year period?
  c. In 2003, what percentage of payments was made using plastic (credit card or debit card)?

17. *The Wall Street Journal*'s Shareholder Scoreboard tracks the performance of 1000 major U.S. companies (*The Wall Street Journal*, March 10, 2003). The performance of each company is rated based on the annual total return, including stock price changes and the reinvestment of dividends. Ratings are assigned by dividing all 1000 companies into five groups from A (top 20%), B (next 20%), to E (bottom 20%). Shown here are the one-year ratings for a sample of 60 of the largest companies. Do the largest companies differ in performance from the performance of the 1000 companies in the Shareholder Scoreboard? Use $\alpha = .05$.

A	B	C	D	E
5	8	15	20	12

18. How well do airline companies serve their customers? A study showed the following customer ratings: 3% excellent, 28% good, 45% fair, and 24% poor. In a similar study of service by telephone companies, assume that a sample of 400 adults found the following

customer ratings: 24 excellent, 124 good, 172 fair, and 80 poor. Is the distribution of the customer ratings for telephone companies different from the distribution of customer ratings for airline companies? Test with $\alpha = .01$. What is your conclusion?

11.3 Test of Independence

Another important application of the chi-square distribution involves using sample data to test for the independence of two variables. Let us illustrate the test of independence by considering the study conducted by the Alber's Brewery of Tucson, Arizona. Alber's manufactures and distributes three types of beer: light, regular, and dark. In an analysis of the market segments for the three beers, the firm's market research group raised the question of whether preferences for the three beers differ among male and female beer drinkers. If beer preference is independent of the gender of the beer drinker, one advertising campaign will be initiated for all of Alber's beers. However, if beer preference depends on the gender of the beer drinker, the firm will tailor its promotions to different target markets.

A test of independence addresses the question of whether the beer preference (light, regular, or dark) is independent of the gender of the beer drinker (male, female). The hypotheses for this test of independence are

H_0: Beer preference is independent of the gender of the beer drinker
H_a: Beer preference is not independent of the gender of the beer drinker

To test whether two variables are independent, one sample is selected and crosstabulation is used to summarize the data for the two variables simultaneously.

Table 11.3 can be used to describe the situation being studied. After identification of the population as all male and female beer drinkers, a sample can be selected and each individual asked to state his or her preference for the three Alber's beers. Every individual in the sample will be classified in one of the six cells in the table. For example, an individual may be a male preferring regular beer [cell (1,2)], a female preferring light beer [cell (2,1)], a female preferring dark beer [cell (2,3)], and so on. Because we have listed all possible combinations of beer preference and gender or, in other words, listed all possible contingencies, Table 11.3 is called a **contingency table**. The test of independence uses the contingency table format and for that reason is sometimes referred to as a *contingency table test*.

Suppose a simple random sample of 150 beer drinkers is selected. After tasting each beer, the individuals in the sample are asked to state their preference or first choice. The crosstabulation in Table 11.4 summarizes the responses for the study. As we see, the data for the test of independence are collected in terms of counts or frequencies for each cell or category. Of the 150 individuals in the sample, 20 were men who favored light beer, 40 were men who favored regular beer, 20 were men who favored dark beer, and so on.

The data in Table 11.4 are the observed frequencies for the six classes or categories. If we can determine the expected frequencies under the assumption of independence between

TABLE 11.3 CONTINGENCY TABLE FOR BEER PREFERENCE AND GENDER OF BEER DRINKER

		Beer Preference		
		Light	Regular	Dark
Gender	Male	cell(1,1)	cell(1,2)	cell(1,3)
	Female	cell(2,1)	cell(2,2)	cell(2,3)

TABLE 11.4 SAMPLE RESULTS FOR BEER PREFERENCES OF MALE AND FEMALE BEER DRINKERS (OBSERVED FREQUENCIES)

		Light	Regular	Dark	Total
Gender	Male	20	40	20	80
	Female	30	30	10	70
	Total	50	70	30	150

Beer Preference

beer preference and gender of the beer drinker, we can use the chi-square distribution to determine whether there is a significant difference between observed and expected frequencies.

Expected frequencies for the cells of the contingency table are based on the following rationale. First we assume that the null hypothesis of independence between beer preference and gender of the beer drinker is true. Then we note that in the entire sample of 150 beer drinkers, a total of 50 prefer light beer, 70 prefer regular beer, and 30 prefer dark beer. In terms of fractions we conclude that $50/150 = 1/3$ of the beer drinkers prefer light beer, $70/150 = 7/15$ prefer regular beer, and $30/150 = 1/5$ prefer dark beer. If the *independence* assumption is valid, we argue that these fractions must be applicable to both male and female beer drinkers. Thus, under the assumption of independence, we would expect the sample of 80 male beer drinkers to show that $(1/3)80 = 26.67$ prefer light beer, $(7/15)80 = 37.33$ prefer regular beer, and $(1/5)80 = 16$ prefer dark beer. Application of the same fractions to the 70 female beer drinkers provides the expected frequencies shown in Table 11.5.

Let e_{ij} denote the expected frequency for the contingency table category in row i and column j. With this notation, let us reconsider the expected frequency calculation for males (row $i = 1$) who prefer regular beer (column $j = 2$); that is, expected frequency e_{12}. Following the preceding argument for the computation of expected frequencies, we can show that

$$e_{12} = (7/15)80 = 37.33$$

This expression can be written slightly differently as

$$e_{12} = (7/15)80 = (70/150)80 = \frac{(80)(70)}{150} = 37.33$$

Note that 80 in the expression is the total number of males (row 1 total), 70 is the total number of individuals preferring regular beer (column 2 total), and 150 is the total sample size. Hence, we see that

$$e_{12} = \frac{\text{(Row 1 Total)(Column 2 Total)}}{\text{Sample Size}}$$

TABLE 11.5 EXPECTED FREQUENCIES IF BEER PREFERENCE IS INDEPENDENT OF THE GENDER OF THE BEER DRINKER

		Light	Regular	Dark	Total
Gender	Male	26.67	37.33	16.00	80
	Female	23.33	32.67	14.00	70
	Total	50.00	70.00	30.00	150

Beer Preference

11.3 Test of Independence

Generalization of the expression shows that the following formula provides the expected frequencies for a contingency table in the test of independence.

EXPECTED FREQUENCIES FOR CONTINGENCY TABLES UNDER THE ASSUMPTION OF INDEPENDENCE

$$e_{ij} = \frac{(\text{Row } i \text{ Total})(\text{Column } j \text{ Total})}{\text{Sample Size}} \quad (11.9)$$

Using the formula for male beer drinkers who prefer dark beer, we find an expected frequency of $e_{13} = (80)(30)/150 = 16.00$, as shown in Table 11.5. Use equation (11.9) to verify the other expected frequencies shown in Table 11.5.

The test procedure for comparing the observed frequencies of Table 11.4 with the expected frequencies of Table 11.5 is similar to the goodness of fit calculations made in Section 11.2. Specifically, the χ^2 value based on the observed and expected frequencies is computed as follows.

TEST STATISTIC FOR INDEPENDENCE

$$\chi^2 = \sum_i \sum_j \frac{(f_{ij} - e_{ij})^2}{e_{ij}} \quad (11.10)$$

where

f_{ij} = observed frequency for contingency table category in row i and column j

e_{ij} = expected frequency for contingency table category in row i and column j based on the assumption of independence

Note: With n rows and m columns in the contingency table, the test statistic has a chi-square distribution with $(n - 1)(m - 1)$ degrees of freedom provided that the expected frequencies are five or more for all categories.

The double summation in equation (11.10) is used to indicate that the calculation must be made for all the cells in the contingency table.

By reviewing the expected frequencies in Table 11.5, we see that the expected frequencies are five or more for each category. We therefore proceed with the computation of the chi-square test statistic. The calculations necessary to compute the chi-square test statistic for determining whether beer preference is independent of the gender of the beer drinker are shown in Table 11.6. We see that the value of the test statistic is $\chi^2 = 6.12$.

The number of degrees of freedom for the appropriate chi-square distribution is computed by multiplying the number of rows minus 1 by the number of columns minus 1. With two rows and three columns, we have $(2 - 1)(3 - 1) = 2$ degrees of freedom. Just like the test for goodness of fit, the test for independence rejects H_0 if the differences between observed and expected frequencies provide a large value for the test statistic. Thus the test for

The test for independence is always a one-tailed test with the rejection region in the upper tail of the chi-square distribution.

TABLE 11.6 COMPUTATION OF THE CHI-SQUARE TEST STATISTIC FOR DETERMINING WHETHER BEER PREFERENCE IS INDEPENDENT OF THE GENDER OF THE BEER DRINKER

Gender	Beer Preference	Observed Frequency (f_{ij})	Expected Frequency (e_{ij})	Difference ($f_{ij} - e_{ij}$)	Squared Difference ($f_{ij} - e_{ij})^2$	Squared Difference Divided by Expected Frequency ($f_{ij} - e_{ij})^2/e_{ij}$
Male	Light	20	26.67	−6.67	44.44	1.67
Male	Regular	40	37.33	2.67	7.11	0.19
Male	Dark	20	16.00	4.00	16.00	1.00
Female	Light	30	23.33	6.67	44.44	1.90
Female	Regular	30	32.67	−2.67	7.11	0.22
Female	Dark	10	14.00	−4.00	16.00	1.14
	Total	150				$\chi^2 = 6.12$

independence is also an upper tail test. Using the chi-square table, we find the following information for 2 degrees of freedom.

Area in Upper Tail	.10	.05	.025	.01
χ^2 Value (2 df)	4.605	5.991	7.378	9.210

$\chi^2 = 6.12$

The test statistic $\chi^2 = 6.12$ is between 5.991 and 7.378. Thus, the corresponding upper tail area or *p*-value is between .05 and .025. The Minitab or Excel procedures in Appendix F can be used to show that *p*-value = .0469. With *p*-value ≤ α = .05, we reject the null hypothesis and conclude that beer preference is not independent of the gender of the beer drinker.

Computer software packages such as Minitab and Excel can be used to simplify the computations required for tests of independence. The input to these computer procedures is the contingency table of observed frequencies shown in Table 11.4. The software then computes the expected frequencies, the value of the χ^2 test statistic, and the *p*-value automatically. The Minitab and Excel procedures that can be used to conduct these tests of independence are presented in Appendixes 11.1 and 11.2. The Minitab output for the Alber's Brewery test of independence is shown in Figure 11.1.

FIGURE 11.1 MINITAB OUTPUT FOR THE ALBER'S BREWERY TEST OF INDEPENDENCE

```
Expected counts are printed below observed counts

            Light    Regular     Dark     Total
     1         20         40       20        80
             26.67      37.33    16.00

     2         30         30       10        70
             23.33      32.67    14.00

Total          50         70       30       150

Chi-Sq = 6.122, DF = 2, P-Value = 0.047
```

11.3 Test of Independence

Although no further conclusions can be made as a result of the test, we can compare the observed and expected frequencies informally to obtain an idea about the dependence between beer preference and gender. Refer to Tables 11.4 and 11.5. We see that male beer drinkers have higher observed than expected frequencies for both regular and dark beers, whereas female beer drinkers have a higher observed than expected frequency only for light beer. These observations give us insight about the beer preference differences between male and female beer drinkers.

Let us summarize the steps in a contingency table test of independence.

TEST OF INDEPENDENCE: A SUMMARY

1. State the null and alternative hypotheses.

   H_0: The column variable is independent of the row variable
   H_a: The column variable is not independent of the row variable

2. Select a random sample and record the observed frequencies for each cell of the contingency table.
3. Use equation (12.2) to compute the expected frequency for each cell.
4. Use equation (12.3) to compute the value of the test statistic.
5. Rejection rule:

   p-value approach: Reject H_0 if p-value $\leq \alpha$
   Critical value approach: Reject H_0 if $\chi^2 \geq \chi^2_\alpha$

   where α is the level of significance, with n rows and m columns providing $(n-1)(m-1)$ degrees of freedom.

NOTES AND COMMENTS

The test statistic for the chi-square tests in this chapter requires an expected frequency of five for each category. When a category has fewer than five, it is often appropriate to combine two adjacent categories to obtain an expected frequency of five or more in each category.

Exercises

Methods

19. The following 2 × 3 contingency table contains observed frequencies for a sample of 200. Test for independence of the row and column variables using the χ^2 test with $\alpha = .05$.

	Column Variable		
Row Variable	**A**	**B**	**C**
P	20	44	50
Q	30	26	30

20. The following 3 × 3 contingency table contains observed frequencies for a sample of 240. Test for independence of the row and column variables using the χ^2 test with $\alpha = .05$.

	Column Variable		
Row Variable	A	B	C
P	20	30	20
Q	30	60	25
R	10	15	30

Applications

21. One of the questions on the *BusinessWeek* Subscriber Study was, "In the past 12 months, when traveling for business, what type of airline ticket did you purchase most often?" The data obtained are shown in the following contingency table.

	Type of Flight	
Type of Ticket	Domestic Flights	International Flights
First class	29	22
Business/executive class	95	121
Full fare economy/coach class	518	135

Use $\alpha = .05$ and test for the independence of type of flight and type of ticket. What is your conclusion?

22. Visa Card USA studied how frequently consumers of various age groups use plastic cards (debit and credit cards) when making purchases (Associated Press, January 16, 2006). Sample data for 300 customers show the use of plastic cards by four age groups.

	Age Group			
Payment	18–24	25–34	35–44	45 and over
Plastic	21	27	27	36
Cash or check	21	36	42	90

a. Test for the independence between method of payment and age group. What is the *p*-value? Using $\alpha = .05$, what is your conclusion?
b. If method of payment and age group are not independent, what observation can you make about how different age groups use plastic to make purchases?
c. What implications does this study have for companies such as Visa, MasterCard, and Discover?

23. With double-digit annual percentage increases in the cost of health insurance, more and more workers are likely to lack health insurance coverage (*USA Today*, January 23, 2004). The following sample data provide a comparison of workers with and without health insurance coverage for small, medium, and large companies. For the purposes of this study, small companies are companies that have fewer than 100 employees. Medium companies have 100 to 999 employees, and large companies have 1000 or more employees. Sample data are reported for 50 employees of small companies, 75 employees of medium companies, and 100 employees of large companies.

11.3 Test of Independence

	Health Insurance		
Size of Company	Yes	No	Total
Small	36	14	50
Medium	65	10	75
Large	88	12	100

a. Conduct a test of independence to determine whether employee health insurance coverage is independent of the size of the company. Use $\alpha = .05$. What is the *p*-value, and what is your conclusion?

b. The *USA Today* article indicated employees of small companies are more likely to lack health insurance coverage. Use percentages based on the preceding data to support this conclusion.

24. *Consumer Reports* measures owner satisfaction of various automobiles by asking the survey question, "Considering factors such as price, performance, reliability, comfort and enjoyment, would you purchase this automobile if you had it to do all over again?" (Consumer Reports website, January 2009). Sample data for 300 owners of four popular midsize sedans are as follows.

	Automobile				
Purchase Again	Chevrolet Impala	Ford Taurus	Honda Accord	Toyota Camry	Total
Yes	49	44	60	46	199
No	37	27	18	19	101

a. Conduct a test of independence to determine if the owner's intent to purchase again is independent of the automobile. Use a .05 level of significance. What is your conclusion?

b. *Consumer Reports* provides an owner satisfaction score for each automobile by reporting the percentage of owners who would purchase the same automobile if they could do it all over again. What are the *Consumer Reports* owner satisfaction scores for the Chevrolet Impala, Ford Taurus, Honda Accord, and Toyota Camry? Rank the four automobiles in terms of owner satisfaction.

c. Twenty-three different automobiles were reviewed in the *Consumer Reports* midsize sedan class. The overall owner satisfaction score for all automobiles in this class was 69. How do the United States manufactured automobiles (Impala and Taurus) compare to the Japanese manufactured automobiles (Accord and Camry) in terms of owner satisfaction? What is the implication of these findings on the future market share for these automobiles?

25. FlightStats, Inc., collects data on the number of flights scheduled and the number of flights flown at major airports throughout the United States. FlightStats data showed that 56% of flights scheduled at Newark, La Guardia, and Kennedy airports were flown during a three-day snowstorm (*The Wall Street Journal*, February 21, 2006). All airlines say they always operate within set safety parameters—if conditions are too poor, they don't fly. The following data show a sample of 400 scheduled flights during the snowstorm.

	Airline				
Did It Fly?	American	Continental	Delta	United	Total
Yes	48	69	68	25	210
No	52	41	62	35	190

Use the chi-square test of independence with a .05 level of significance to analyze the data. What is your conclusion? Do you have a preference for which airline you would choose to fly during similar snowstorm conditions? Explain.

26. As the price of oil rises, there is increased worldwide interest in alternate sources of energy. A *Financial Times*/Harris Poll surveyed people in six countries to assess attitudes toward a variety of alternate forms of energy (Harris Interactive website, February 27, 2008). The data in the following table are a portion of the poll's findings concerning whether people favor or oppose the building of new nuclear power plants.

	\multicolumn{6}{c}{Country}
Response	Great Britain	France	Italy	Spain	Germany	United States
Strongly favor	141	161	298	133	128	204
Favor more than oppose	348	366	309	222	272	326
Oppose more than favor	381	334	219	311	322	316
Strongly oppose	217	215	219	443	389	174

   a. How large was the sample in this poll?
   b. Conduct a hypothesis test to determine whether people's attitude toward building new nuclear power plants is independent of country. What is your conclusion?
   c. Using the percentage of respondents who "strongly favor" and "favor more than oppose," which country has the most favorable attitude toward building new nuclear power plants? Which country has the least favorable attitude?

27. The National Sleep Foundation used a survey to determine whether hours of sleeping per night are independent of age (*Newsweek,* January 19, 2004). The following show the hours of sleep on weeknights for a sample of individuals age 49 and younger and for a sample of individuals age 50 and older.

	\multicolumn{4}{c}{Hours of Sleep}	
Age	Fewer than 6	6 to 6.9	7 to 7.9	8 or more	Total
49 or younger	38	60	77	65	240
50 or older	36	57	75	92	260

   a. Conduct a test of independence to determine whether the hours of sleep on weeknights are independent of age. Use $\alpha = .05$. What is the *p*-value, and what is your conclusion?
   b. What is your estimate of the percentage of people who sleep fewer than 6 hours, 6 to 6.9 hours, 7 to 7.9 hours, and 8 or more hours on weeknights?

28. Samples taken in three cities, Anchorage, Atlanta, and Minneapolis, were used to learn about the percentage of married couples with both the husband and the wife in the workforce (*USA Today,* January 15, 2006). Analyze the following data to see whether both the husband and wife being in the workforce is independent of location. Use a .05 level of significance. What is your conclusion? What is the overall estimate of the percentage of married couples with both the husband and the wife in the workforce?

	\multicolumn{3}{c}{Location}	
In Workforce	Anchorage	Atlanta	Minneapolis
Both	57	70	63
Only one	33	50	27

Summary

In this chapter, we described statistical procedures for comparisons involving proportions and the contingency table test for independence of two variables. In the first section, we compared a proportion for one population with the same proportion from another population. We described how to construct an interval estimate for the difference between the proportions and how to conduct a hypothesis test to learn whether the difference between the proportions was statistically significant.

In the second section, we focused on a single multinomial population. There we saw how to conduct hypothesis tests to determine whether the sample proportions for the categories of the multinomial population were significantly different from the hypothesized values. The chi-square goodness of fit test was used to make the comparison.

The final section was concerned with tests of independence for two variables. A test of independence for two variables is an extension of the methodology employed in the goodness of fit test for a multinomial population. A contingency table is used to determine the observed and expected frequencies. Then a chi-square value is computed. Large chi-square values, caused by large differences between observed and expected frequencies, lead to the rejection of the null hypothesis of independence.

Glossary

Pooled estimator of p An estimator of a population proportion obtained by computing a weighted average of the sample proportions obtained from two independent samples.
Multinomial population A population in which each element is assigned to one and only one of several categories. The multinomial distribution extends the binomial distribution from two to three or more outcomes.
Goodness of fit test A statistical test conducted to determine whether to reject a hypothesized probability distribution for a population.
Contingency table A table used to summarize observed and expected frequencies for a test of independence.

Key Formulas

Point Estimator of the Difference Between Two Population Proportions

$$\bar{p}_1 - \bar{p}_2 \qquad (11.1)$$

Standard Error of $\bar{p}_1 - \bar{p}_2$

$$\sigma_{\bar{p}_1 - \bar{p}_2} = \sqrt{\frac{p_1(1-p_1)}{n_1} + \frac{p_2(1-p_2)}{n_2}} \qquad (11.2)$$

Interval Estimate of the Difference Between Two Population Proportions

$$\bar{p}_1 - \bar{p}_2 \pm z_{\alpha/2} \sqrt{\frac{\bar{p}_1(1-\bar{p}_1)}{n_1} + \frac{\bar{p}_2(1-\bar{p}_2)}{n_2}} \qquad (11.4)$$

Standard Error of $\bar{p}_1 - \bar{p}_2$ when $p_1 = p_2 = p$

$$\sigma_{\bar{p}_1 - \bar{p}_2} = \sqrt{p(1-p)\left(\frac{1}{n_1} + \frac{1}{n_2}\right)} \qquad (11.5)$$

Pooled Estimator of p when $p_1 = p_2 = p$

$$\bar{p} = \frac{n_1\bar{p}_1 + n_2\bar{p}_2}{n_1 + n_2} \quad (11.6)$$

Test Statistic for Hypothesis Tests About $p_1 - p_2$

$$z = \frac{(\bar{p}_1 - \bar{p}_2)}{\sqrt{\bar{p}(1 - \bar{p})\left(\frac{1}{n_1} + \frac{1}{n_2}\right)}} \quad (11.7)$$

Test Statistic for Goodness of Fit

$$\chi^2 = \sum_{i=1}^{k} \frac{(f_i - e_i)^2}{e_i} \quad (11.8)$$

Expected Frequencies for Contingency Tables Under the Assumption of Independence

$$e_{ij} = \frac{(\text{Row } i \text{ Total})(\text{Column } j \text{ Total})}{\text{Sample Size}} \quad (11.9)$$

Test Statistic for Independence

$$\chi^2 = \sum_i \sum_j \frac{(f_{ij} - e_{ij})^2}{e_{ij}} \quad (11.10)$$

Supplementary Exercises

29. A regional transit authority is concerned about the number of riders on one of its bus routes. In setting up the route, the assumption is that the number of riders is the same on every day from Monday through Friday. Using the following data, test with $\alpha = .05$ to determine whether the transit authority's assumption is correct.

Day	Number of Riders
Monday	13
Tuesday	16
Wednesday	28
Thursday	17
Friday	16

30. The results of *Computerworld*'s Annual Job Satisfaction Survey showed that 28% of information systems (IS) managers are very satisfied with their job, 46% are somewhat satisfied, 12% are neither satisfied nor dissatisfied, 10% are somewhat dissatisfied, and 4% are very dissatisfied. Suppose that a sample of 500 computer programmers yielded the following results.

Category	Number of Respondents
Very satisfied	105
Somewhat satisfied	235
Neither	55
Somewhat dissatisfied	90
Very dissatisfied	15

Use $\alpha = .05$ and test to determine whether the job satisfaction for computer programmers is different from the job satisfaction for IS managers.

31. A sample of parts provided the following contingency table data on part quality by production shift.

Shift	Number Good	Number Defective
First	368	32
Second	285	15
Third	176	24

Use $\alpha = .05$ and test the hypothesis that part quality is independent of the production shift. What is your conclusion?

32. Vacation occupancy rates were expected to be up during March 2008 in Myrtle Beach, South Carolina (*The Sun News*, February 29, 2008). Data in the file Occupancy will allow you to replicate the findings presented in the newspaper. The data show units rented and not rented for a random sample of vacation properties during the first week of March 2007 and March 2008.
   a. Estimate the proportion of units rented during the first week of March 2007 and the first week of March 2008.
   b. Provide a 95% confidence interval for the difference in proportions.
   c. On the basis of your findings, does it appear March rental rates for 2008 will be up from those a year earlier?

33. Seven percent of mutual fund investors rate corporate stocks "very safe," 58% rate them "somewhat safe," 24% rate them "not very safe," 4% rate them "not at all safe," and 7% are "not sure." A *BusinessWeek*/Harris poll asked 529 mutual fund investors how they would rate corporate bonds on safety. The responses are as follows.

Safety Rating	Frequency
Very safe	48
Somewhat safe	323
Not very safe	79
Not at all safe	16
Not sure	63
Total	529

Do mutual fund investors' attitudes toward corporate bonds differ from their attitudes toward corporate stocks? Support your conclusion with a statistical test. Use $\alpha = .01$.

34. *The Wall Street Journal* Subscriber Study showed data on the employment status of subscribers. Sample results corresponding to subscribers of the eastern and western editions are shown here.

	Region	
Employment Status	Eastern Edition	Western Edition
Full-time	1105	574
Part-time	31	15
Self-employed/consultant	229	186
Not employed	485	344

Use $\alpha = .05$ and test the hypothesis that employment status is independent of the region. What is your conclusion?

35. Since 2000, the Toyota Camry, Honda Accord, and Ford Taurus have been the three best-selling passenger cars in the United States. Sales data for 2003 indicated market shares among the top three as follows: Toyota Camry 37%, Honda Accord 34%, and Ford Taurus 29% (*The World Almanac*, 2004). Assume a sample of 1200 sales of passenger cars during the first quarter of 2004 shows the following.

Passenger Car	**Units Sold**
Toyota Camry	480
Honda Accord	390
Ford Taurus	330

Can these data be used to conclude that the market shares among the top three passenger cars have changed during the first quarter of 2004? What is the *p*-value? Use a .05 level of significance. What is your conclusion?

36. A lending institution supplied the following data on loan approvals by four loan officers. Use $\alpha = .05$ and test to determine whether the loan approval decision is independent of the loan officer reviewing the loan application.

	Loan Approval Decision	
Loan Officer	**Approved**	**Rejected**
Miller	24	16
McMahon	17	13
Games	35	15
Runk	11	9

37. In setting sales quotas, the marketing manager makes the assumption that order potentials are the same for each of four sales territories. A sample of 200 sales follows. Should the manager's assumption be rejected? Use $\alpha = .05$.

Sales Territories			
I	II	III	IV
60	45	59	36

38. A Pew Research Center survey asked respondents if they would rather live in a place with a slower pace of life or a place with a faster pace of life (*USA Today*, February 13, 2009). Consider the following data showing a sample of preferences expressed by 150 men and 150 women.

	Preferred Pace of Life		
Respondent	Slower	No Preference	Faster
Men	102	9	39
Women	111	12	27

a. Combine the samples of men and women. What is the overall percentage of respondents who prefer to live in a place with a slower pace of life? What is the overall percentage of respondents who prefer to live in a place with a faster pace of life? What is your conclusion?

b. Is the preferred pace of life independent of the respondent? Use $\alpha = .05$. What is your conclusion? What is your recommendation?

39. The bullish sentiment of individual investors was 27.6% (*AAII Journal*, February 2009). The bullish sentiment was reported to be 48.7% one week earlier and 39.7% one month earlier. The sentiment measures were based on a poll conducted by the American Association of Individual Investors. Assume that each bullish sentiment measure was based on a sample of 240 investors.

    a. Develop a 95% confidence interval for the difference between the bullish sentiment measures for the most recent two weeks.
    b. Develop hypotheses so that rejection of the null hypothesis will allow us to conclude that the most recent bullish sentiment is weaker than that of one month earlier.
    c. Conduct a hypotheses test of part (b) using $\alpha = .01$. What is your conclusion?

40. Barna Research Group collected data showing church attendance by age group (*USA Today*, November 20, 2003). Use the sample data to determine whether attending church is independent of age. Use a .05 level of significance. What is your conclusion? What conclusion can you draw about church attendance as individuals grow older?

	Church Attendance		
Age	Yes	No	Total
20 to 29	31	69	100
30 to 39	63	87	150
40 to 49	94	106	200
50 to 59	72	78	150

41. Medical tests were conducted to learn about drug-resistant tuberculosis. Of 142 cases tested in New Jersey, 9 were found to be drug-resistant. Of 268 cases tested in Texas, 5 were found to be drug-resistant. Do these data suggest a statistically significant difference between the proportions of drug-resistant cases in the two states? Use a .02 level of significance. What is the *p*-value, and what is your conclusion?

42. The office occupancy rates were reported for four California metropolitan areas. Do the following data suggest that the office vacancies were independent of metropolitan area? Use a .05 level of significance. What is your conclusion?

Occupancy Status	Los Angeles	San Diego	San Francisco	San Jose
Occupied	160	116	192	174
Vacant	40	34	33	26

43. A large automobile insurance company selected samples of single and married male policyholders and recorded the number who made an insurance claim over the preceding three-year period.

Single Policyholders	Married Policyholders
$n_1 = 400$	$n_2 = 900$
Number making claims = 76	Number making claims = 90

a. Use $\alpha = .05$. Test to determine whether the claim rates differ between single and married male policyholders.
b. Provide a 95% confidence interval for the difference between the proportions for the two populations.

44. A salesperson makes four calls per day. A sample of 100 days gives the following frequencies of sales volumes.

Number of Sales	Observed Frequency (days)
0	30
1	32
2	25
3	10
4	3
Total	100

Records show that sales are made to 30% of all sales calls. Assuming independent sales calls, the number of sales per day should follow a binomial distribution. The binomial probability function presented in Chapter 5 is

$$f(x) = \frac{n!}{x!(n-x)!} p^x (1-p)^{n-x}$$

For this exercise, assume that the population has a binomial distribution with $n = 4$, $p = .30$, and $x = 0, 1, 2, 3$, and 4.

a. Compute the expected frequencies for $x = 0, 1, 2, 3$, and 4 by using the binomial probability function. Combine categories if necessary to satisfy the requirement that the expected frequency is five or more for all categories.
b. Use the goodness of fit test to determine whether the assumption of a binomial distribution should be rejected. Use $\alpha = .05$. Because no parameters of the binomial distribution were estimated from the sample data, the degrees of freedom are $k - 1$ when k is the number of categories.

45. Jupiter Media used a survey to determine how people use their free time. Watching television was the most popular activity selected by both men and women (*The Wall Street Journal,* January 26, 2004). The proportion of men and the proportion of women who selected watching television as their most popular leisure time activity can be estimated from the following sample data.

Gender	Sample Size	Watching Television
Men	800	248
Women	600	156

a. State the hypotheses that can be used to test for a difference between the proportion for the population of men and the proportion for the population of women who selected watching television as their most popular leisure time activity.
b. What is the sample proportion of men who selected watching television as their most popular leisure time activity? What is the sample proportion of women?

c. Conduct the hypothesis test and compute the *p*-value. At a .05 level of significance, what is your conclusion?
d. What is the margin of error and 95% confidence interval estimate of the difference between the population proportions?

Case Problem A Bipartisan Agenda for Change

In a study conducted by Zogby International for the *Democrat and Chronicle,* more than 700 New Yorkers were polled to determine whether the New York state government works. Respondents surveyed were asked questions involving pay cuts for state legislators, restrictions on lobbyists, terms limits for legislators, and whether state citizens should be able to put matters directly on the state ballot for a vote. The results regarding several proposed reforms had broad support, crossing all demographic and political lines.

Suppose that a follow-up survey of 100 individuals who live in the western region of New York was conducted. The party affiliation (Democrat, Independent, Republican) of each individual surveyed was recorded, as well as the responses to the following three questions.

1. Should legislative pay be cut for every day the state budget is late?
   Yes ____ No ____
2. Should there be more restrictions on lobbyists?
   Yes ____ No ____
3. Should there be term limits requiring that legislators serve a fixed number of years?
   Yes ____ No ____

The responses were coded using 1 for a Yes response and 2 for a No response. The complete data set is available on the data disk in the data set named NYReform.

Managerial Report

1. Use descriptive statistics to summarize the data from this study. What are your preliminary conclusions about the independence of the response (Yes or No) and party affiliation for each of the three questions in the survey?
2. With regard to question 1, test for the independence of the response (Yes and No) and party affiliation. Use $\alpha = .05$.
3. With regard to question 2, test for the independence of the response (Yes and No) and party affiliation. Use $\alpha = .05$.
4. With regard to question 3, test for the independence of the response (Yes and No) and party affiliation. Use $\alpha = .05$.
5. Does it appear that there is broad support for change across all political lines? Explain.

Appendix 11.1 Inferences About Two Population Proportions Using Minitab

Confidence Intervals and Hypothesis Tests

We describe the use of Minitab to develop confidence intervals and conduct hypothesis tests about the difference between two population proportions. We will use the data on tax preparation errors presented in Section 11.1. The sample results for 250 tax returns prepared at office 1 are in column C1 and the sample results for 300 tax returns prepared at office 2 are in column C2. Yes denotes an error was found in the tax return and No indicates no error was found. The procedure we describe provides both a 90% confidence interval estimate of the difference between the two population proportions and hypothesis testing results for $H_0: p_1 - p_2 = 0$ vs. $H_a: p_1 - p_2 \neq 0$.

Step 1. Select the **Stat** menu
Step 2. Choose **Basic Statistics**
Step 3. Choose **2 Proportions**
Step 4. When the 2 Proportions (Test and Confidence Interval) dialog box appears:
  Select **Samples in different columns**
  Enter C1 in the **First** box
  Enter C2 in the **Second** box
  Select **Options**
Step 5. When the 2 Proportions-Options dialog box appears:
  Enter 90 in the **Confidence level** box
  Enter 0 in the **Test difference** box
  Enter not equal in the **Alternative** box
  Select **Use pooled estimate of p for test**
  Click **OK**
Step 6. Click **OK**

Step 5 may be modified to provide different confidence levels, different hypothesized values, and different forms of the hypotheses.

In the tax preparation example, the data are qualitative. Yes and No are used to indicate whether an error is present. In modules involving proportions, Minitab calculates proportions for the response coming second in alphabetic order. Thus, in the tax preparation example, Minitab computes the proportion of Yes responses, which is the proportion we wanted.

If Minitab's alphabetical ordering does not compute the proportion for the response of interest, we can fix it. Select any cell in the data column, go to the Minitab menu bar, and select Editor > Column > Value Order. This sequence will provide the option of entering a user-specified order. Simply make sure that the response of interest is listed second in the define-an-order box. Minitab's 2 Proportion routine will then provide the confidence interval and hypothesis testing results for the population proportion of interest.

Finally, we note that Minitab's 2 Proportion routine uses a computational procedure different from the procedure described in the text. Thus, the Minitab output may provide slightly different interval estimates and slightly different *p*-values. However, results from the two methods should be close and are expected to provide the same interpretation and conclusion.

Appendix 11.2 Tests of Goodness of Fit and Independence Using Minitab

Goodness of Fit Test

This Minitab procedure can be used for a goodness of fit test of a multinomial population. The user must obtain the observed frequency and the hypothesized proportion for each of the *k* categories. The observed frequencies are entered in Column C1 and the hypothesized proportions are entered in Column C2. Using the Scott Marketing Research example presented in Section 11.2. Column C1 is labeled Observed and Column C2 is labeled Proportion. Enter the observed frequencies 48, 98, and 54 in Column C1 and enter the hypothesized proportions .30, .50, and .20 in Column C2. The Minitab steps for the goodness of fit test follow.

Step 1. Select the **Stat** menu
Step 2. Select **Tables**
Step 3. Choose **Chi-Square Goodness of Fit Test (One Variable)**
Step 4. When the Chi-Square Goodness of Fit Test dialog box appears:
    Select **Observed counts**
    Enter C1 in the **Observed counts** box
    Select **Specific proportions**
    Enter C2 in the **Specific proportions** box
    Click **OK**

Test of Independence

We begin with a new Minitab worksheet and enter the observed frequency data for the Alber's Brewery example from Section 11.3 into columns 1, 2, and 3, respectively. Thus, we entered the observed frequencies corresponding to a light beer preference (20 and 30) in C1, the observed frequencies corresponding to a regular beer preference (40 and 30) in C2, and the observed frequencies corresponding to a dark beer preference (20 and 10) in C3. The Minitab steps for the test of independence are as follows.

Step 1. Select the **Stat** menu
Step 2. Select **Tables**
Step 3. Choose **Chi-Square Test (Two-Way Table in Worksheet)**
Step 4. When the Chi-Square Test dialog box appears:
    Enter C1-C3 in the **Columns containing the table** box
    Click **OK**

Appendix 11.3 Tests of Goodness of Fit and Independence Using Excel

Goodness of Fit Test

This Excel procedure can be used for a goodness of fit test for a multinomial distribution. The user must obtain the observed frequencies, calculate the expected frequencies, and enter both the observed and expected frequencies in an Excel worksheet.

The observed frequencies and expected frequencies for the Scott Market Research example presented in Section 11.2 are entered in columns A and B as shown in Figure 11.2. The test statistic $\chi^2 = 7.34$ is calculated in column D. With $k = 3$ categories, the user enters the degrees of freedom $k - 1 = 3 - 1 = 2$ in cell D11. The CHIDIST function provides the p-value in cell D13. The background worksheet shows the cell formulas.

Test of Independence

The Excel procedure for the test of independence requires the user to obtain the observed frequencies and enter them in the worksheet. The Alber's Brewery example from Section 11.3 provides the observed frequencies, which are entered in cells B7 to D8 as shown in the worksheet in Figure 11.3. The cell formulas in the background worksheet show the procedure used

480   Chapter 11   Comparisons Involving Proportions and a Test of Independence

FIGURE 11.2   EXCEL WORKSHEET FOR THE SCOTT MARKETING RESEARCH GOODNESS OF FIT TEST

	A	B	C	D	E
1	Goodness of Fit Test				
2					
3	Observed	Expected			
4	Frequency	Frequency		Calculations	
5	48	60		=(A5-B5)^2/B5	
6	98	100		=(A6-B6)^2/B6	
7	54	40		=(A7-B7)^2/B7	
8					
9			Test Statistic	=SUM(D5:D7)	
10					
11			Degrees of Freedom	2	
12					
13			p-Value	=CHIDIST(D9,D11)	
14					

	A	B	C	D	E
1	Goodness of Fit Test				
2					
3	Observed	Expected			
4	Frequency	Frequency		Calculations	
5	48	60		2.40	
6	98	100		0.04	
7	54	40		4.90	
8					
9			Test Statistic	7.34	
10					
11			Degrees of Freedom	2	
12					
13			p-Value	0.0255	
14					

to compute the expected frequencies. With two rows and three columns, the user enters the degrees of freedom $(2 - 1)(3 - 1) = 2$ in cell E22. The CHITEST function provides the *p*-value in cell E24.

Appendix 11.4   Inferences About Two Population Proportions Using StatTools

Confidence Intervals

WEB file
TaxPrep

We use the data on tax preparation errors presented in Section 11.1. The sample results for 250 tax returns prepared at office 1 are in column C1 and the sample results for 300 tax returns prepared at office 2 are in column C2. Yes denotes an error was found in the tax return and No indicates no error was found. Begin by using the Data Set Manager to create a StatTools data set using the procedure described in the appendix to Chapter 1. The following steps will provide a 90% confidence interval estimate of the difference between the two population proportions.

Appendix 11.4 Inferences About Two Population Proportions Using StatTools

FIGURE 11.3 EXCEL WORKSHEET FOR THE ALBER'S BREWERY TEST OF INDEPENDENCE

	A	B	C	D	E	F
1	Test of Independence					
2						
3	Observed Frequencies					
4						
5			Beer Peference			
6	Gender	Light	Regular	Dark	Total	
7	Male	20	40	20	=SUM(B7:D7)	
8	Female	30	30	10	=SUM(B8:D8)	
9	Total	=SUM(B7:B8)	=SUM(C7:C8)	=SUM(D7:D8)	=SUM(E7:E8)	
10						
11						
12	Expected Frequencies					
13						
14			Beer Peference			
15	Gender	Light	Regular	Dark	Total	
16	Male	=E7*B$9/$E$9	=E7*C$9/$E$9	=E7*D$9/$E$9	=SUM(B16:D16)	
17	Female	=E8*B$9/$E$9	=E8*C$9/$E$9	=E8*D$9/$E$9	=SUM(B17:D17)	
18	Total	=SUM(B16:B17)	=SUM(C16:C17)	=SUM(D16:D17)	=SUM(E16:E17)	
19						
20				Test Statistic	=CHIINV(E24,E22)	
21						
22				Degrees of Freedom	2	
23						
24				p-value	=CHITEST(B7:D8,B16:D17)	
25						

	A	B	C	D	E	F
1	Test of Independence					
2						
3	Observed Frequencies					
4						
5			Beer Peference			
6	Gender	Light	Regular	Dark	Total	
7	Male	20	40	20	80	
8	Female	30	30	10	70	
9	Total	50	70	30	150	
10						
11						
12	Expected Frequencies					
13						
14			Beer Peference			
15	Gender	Light	Regular	Dark	Total	
16	Male	26.67	37.33	16	80	
17	Female	23.33	32.67	14	70	
18	Total	50	70	30	150	
19						
20				Test Statistic	6.12	
21						
22				Degrees of Freedom	2	
23						
24				p-value	0.0468	
25						

Step 1. Click the **StatTools** tab on the Ribbon
Step 2. In the **Analyses** group, click **Statistical Inference**
Step 3. Choose **Confidence Interval**
Step 4. Choose **Proportion**
Step 5. When the StatTools—Confidence Interval for Proportion dialog box appears:
    In the **Analysis Type** box, select **Two-Sample Analysis**
    In the **Variables** section, select both **Office 1** and **Office 2**
    In the **Categories to Analyze** section, select **Yes**
    In the **Options** section, enter 90% in the **Confidence Level** box
    Click **OK**

Hypothesis Tests

TaxPrep

We use the data on tax preparation errors presented in Section 11.1. Begin by using the Data Set Manager to create a StatTools data set using the procedure described in the appendix to Chapter 1. The follow steps will test the hypothesis that there is no difference between the two population proportions.

Step 1. Click the **StatTools** tab on the Ribbon
Step 2. In the **Analyses** group, click **Statistical Inference**
Step 3. Choose **Hypothesis Test**
Step 4. Choose **Proportion**
Step 5. When the StatTools—Hypothesis Test for Proportion dialog box appears:
    In the **Analysis Type** box, select **Two-Sample Analysis**
    In the **Variables** section, select both **Office 1** and **Office 2**
    In the **Categories to Analyze** section, select **Yes**
    In the **Hypothesis About Difference Between Proportions** section:
        Enter 0 in the **Null Hypothesis Value** box
        Select **Not Equal to Null Value (Two-Tailed Test)**
    Click **OK**
Step 6. When the StatTools dialog box appears:
    Click **OK**
Step 7. When the Choose Variable Ordering dialog box appears:
    Click **OK**

Appendix 11.5 Test of Independence Using StatTools

Albers

We begin with a new Excel worksheet and enter the observed frequency data for the Alber's Brewery example from Section 11.3. The row labels Male and Female are entered in cells A2 and A3. The column labels Light, Regular, and Dark are entered in cells B1, C1, and D1. The corresponding observed frequencies 20, 40, 20, 30, 30, and 10 are entered in cells B2 to D3 providing the Alber's Brewery contingency table. The StatTools steps for the test of independence are as follows.

Step 1. Click the **StatTools** tab on the Ribbon
Step 2. In the **Analysis Group**, click **Statistical Inference**
Step 3. Choose **Chi-Square Independence Test**
Step 4. When the StatTools—Chi-Square Test for Independence dialog box appears:
    Enter A1:D3 in the **Contingency Table Range** box
    Select **Table Contains Row and Column Headers**
    Click **OK**

CHAPTER 12

Simple Linear Regression

CONTENTS

STATISTICS IN PRACTICE:
ALLIANCE DATA SYSTEMS

12.1 SIMPLE LINEAR
REGRESSION MODEL
Regression Model
and Regression
Equation
Estimated Regression
Equation

12.2 LEAST SQUARES METHOD

12.3 COEFFICIENT OF
DETERMINATION
Correlation Coefficient

12.4 MODEL ASSUMPTIONS

12.5 TESTING FOR
SIGNIFICANCE
Estimate of σ^2
t Test
Confidence Interval for β_1
F Test
Some Cautions About
the Interpretation of
Significance Tests

12.6 USING THE ESTIMATED
REGRESSION EQUATION
FOR ESTIMATION AND
PREDICTION
Point Estimation
Interval Estimation
Confidence Interval for the Mean
Value of y
Prediction Interval for an
Individual Value of y

12.7 COMPUTER SOLUTION

12.8 RESIDUAL ANALYSIS:
VALIDATING MODEL
ASSUMPTIONS
Residual Plot Against x
Residual Plot Against $\hat{y}$

STATISTICS *in* PRACTICE

ALLIANCE DATA SYSTEMS*
DALLAS, TEXAS

Alliance Data Systems (ADS) provides transaction processing, credit services, and marketing services for clients in the rapidly growing customer relationship management (CRM) industry. ADS clients are concentrated in four industries: retail, petroleum/convenience stores, utilities, and transportation. In 1983, Alliance began offering end-to-end credit processing services to the retail, petroleum, and casual dining industries; today it employs more than 6500 employees who provide services to clients around the world. Operating more than 140,000 point-of-sale terminals in the United States alone, ADS processes in excess of 2.5 billion transactions annually. The company ranks second in the United States in private label credit services by representing 49 private label programs with nearly 72 million cardholders. In 2001, ADS made an initial public offering and is now listed on the New York Stock Exchange.

As one of its marketing services, ADS designs direct mail campaigns and promotions. With its database containing information on the spending habits of more than 100 million consumers, ADS can target those consumers most likely to benefit from a direct mail promotion. The Analytical Development Group uses regression analysis to build models that measure and predict the responsiveness of consumers to direct market campaigns. Some regression models predict the probability of purchase for individuals receiving a promotion, and others predict the amount spent by those consumers making a purchase.

For one particular campaign, a retail store chain wanted to attract new customers. To predict the effect of the campaign, ADS analysts selected a sample from the consumer database, sent the sampled individuals promotional materials, and then collected transaction data on the consumers' response. Sample data were collected on the amount of purchase made by the consumers responding to the campaign, as well as a variety of consumer-specific variables thought to be useful in predicting sales. The consumer-specific variable that contributed most to predicting the amount purchased was the total amount of credit purchases at related stores over the past 39 months. ADS analysts developed an estimated regression equation relating the amount of purchase to the amount spent at related stores:

$$\hat{y} = 26.7 + 0.00205x$$

where

$\hat{y}$ = amount of purchase
x = amount spent at related stores

Alliance Data Systems analysts discuss use of a regression model to predict sales for a direct marketing campaign.

Using this equation, we could predict that someone spending $10,000 over the past 39 months at related stores would spend $47.20 when responding to the direct mail promotion. In this chapter, you will learn how to develop this type of estimated regression equation.

The final model developed by ADS analysts also included several other variables that increased the predictive power of the preceding equation. Some of these variables included the absence/presence of a bank credit card, estimated income, and the average amount spent per trip at a selected store. In Chapter 13 we will learn how such additional variables can be incorporated into a multiple regression model.

---

*The authors are indebted to Philip Clemance, Director of Analytical Development at Alliance Data Systems, for providing this Statistics in Practice.

Managerial decisions often are based on the relationship between two or more variables. For example, after considering the relationship between advertising expenditures and sales, a marketing manager might attempt to predict sales for a given level of advertising expenditures. In another case, a public utility might use the relationship between the daily high temperature and the demand for electricity to predict electricity usage on the basis of next month's anticipated daily high temperatures. Sometimes a manager will rely on intuition to judge how two variables are related. However, if data can be obtained, a statistical procedure called *regression analysis* can be used to develop an equation showing how the variables are related.

In regression terminology, the variable being predicted is called the **dependent variable**. The variable or variables being used to predict the value of the dependent variable are called the **independent variables**. For example, in analyzing the effect of advertising expenditures on sales, a marketing manager's desire to predict sales would suggest making sales the dependent variable. Advertising expenditure would be the independent variable used to help predict sales. In statistical notation, y denotes the dependent variable and x denotes the independent variable.

In this chapter we consider the simplest type of regression analysis involving one independent variable and one dependent variable in which the relationship between the variables is approximated by a straight line. It is called **simple linear regression**. Regression analysis involving two or more independent variables is called multiple regression analysis; multiple regression is covered in Chapter 13.

The statistical methods used in studying the relationship between two variables were first employed by Sir Francis Galton (1822–1911). Galton was interested in studying the relationship between a father's height and the son's height. Galton's disciple, Karl Pearson (1857–1936), analyzed the relationship between the father's height and the son's height for 1078 pairs of subjects.

12.1 Simple Linear Regression Model

Armand's Pizza Parlors is a chain of Italian-food restaurants located in a five-state area. Armand's most successful locations are near college campuses. The managers believe that quarterly sales for these restaurants (denoted by y) are related positively to the size of the student population (denoted by x); that is, restaurants near campuses with a large student population tend to generate more sales than those located near campuses with a small student population. Using regression analysis, we can develop an equation showing how the dependent variable y is related to the independent variable x.

Regression Model and Regression Equation

In the Armand's Pizza Parlors example, the population consists of all the Armand's restaurants. For every restaurant in the population, there is a value of x (student population) and a corresponding value of y (quarterly sales). The equation that describes how y is related to x and an error term is called the **regression model**. The regression model used in simple linear regression follows.

SIMPLE LINEAR REGRESSION MODEL

$$y = \beta_0 + \beta_1 x + \epsilon \tag{12.1}$$

β_0 and β_1 are referred to as the parameters of the model, and ϵ (the Greek letter epsilon) is a random variable referred to as the error term. The error term accounts for the variability in y that cannot be explained by the linear relationship between x and y.

The population of all Armand's restaurants can also be viewed as a collection of subpopulations, one for each distinct value of x. For example, one subpopulation consists of all Armand's restaurants located near college campuses with 8000 students; another subpopulation consists of all Armand's restaurants located near college campuses with 9000 students; and so on. Each subpopulation has a corresponding distribution of y values. Thus, a distribution of y values is associated with restaurants located near campuses with 8000 students; a distribution of y values is associated with restaurants located near campuses with 9000 students; and so on. Each distribution of y values has its own mean or expected value. The equation that describes how the expected value of y, denoted $E(y)$, is related to x is called the **regression equation**. The regression equation for simple linear regression follows.

> **SIMPLE LINEAR REGRESSION EQUATION**
>
> $$E(y) = \beta_0 + \beta_1 x \tag{12.2}$$

The graph of the simple linear regression equation is a straight line; β_0 is the y-intercept of the regression line, β_1 is the slope, and $E(y)$ is the mean or expected value of y for a given value of x.

Examples of possible regression lines are shown in Figure 12.1. The regression line in Panel A shows that the mean value of y is related positively to x, with larger values of $E(y)$ associated with larger values of x. The regression line in Panel B shows the mean value of y is related negatively to x, with smaller values of $E(y)$ associated with larger values of x. The regression line in Panel C shows the case in which the mean value of y is not related to x; that is, the mean value of y is the same for every value of x.

Estimated Regression Equation

If the values of the population parameters β_0 and β_1 were known, we could use equation (12.2) to compute the mean value of y for a given value of x. In practice, the parameter values are not known and must be estimated using sample data. Sample statistics (denoted b_0 and b_1) are computed as estimates of the population parameters β_0 and β_1. Substituting the values of the sample statistics b_0 and b_1 for β_0 and β_1 in the regression equation, we obtain the

FIGURE 12.1 POSSIBLE REGRESSION LINES IN SIMPLE LINEAR REGRESSION

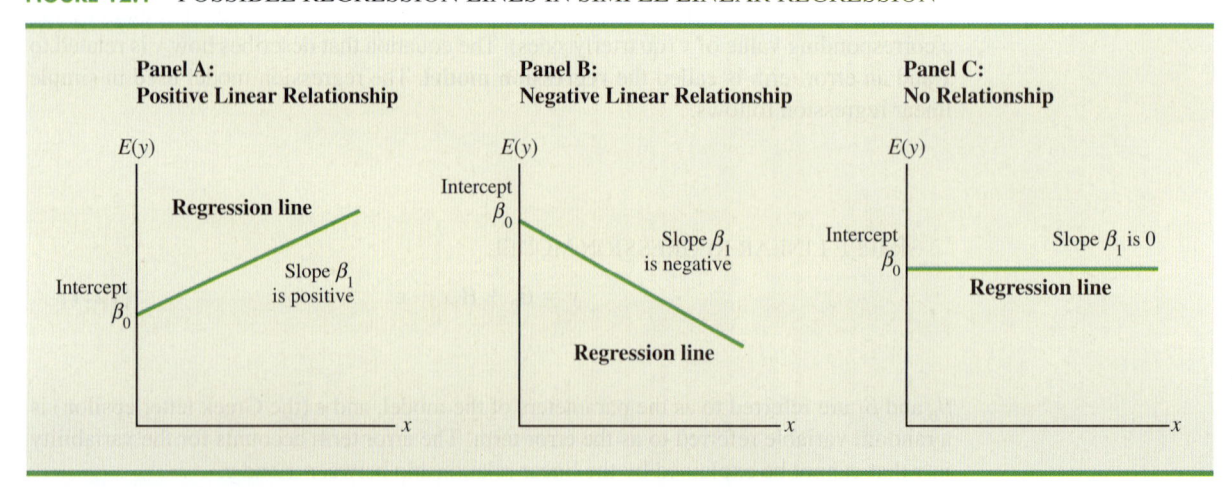

Panel A: Positive Linear Relationship — Regression line, Intercept β_0, Slope β_1 is positive

Panel B: Negative Linear Relationship — Intercept β_0, Slope β_1 is negative, Regression line

Panel C: No Relationship — Intercept β_0, Slope β_1 is 0, Regression line

estimated regression equation. The estimated regression equation for simple linear regression follows.

ESTIMATED SIMPLE LINEAR REGRESSION EQUATION

$$\hat{y} = b_0 + b_1 x \qquad (12.3)$$

The graph of the estimated simple linear regression equation is called the *estimated regression line*; b_0 is the y-intercept and b_1 is the slope. In the next section, we show how the least squares method can be used to compute the values of b_0 and b_1 in the estimated regression equation.

In general, $\hat{y}$ is the point estimator of $E(y)$, the mean value of y for a given value of x. Thus, to estimate the mean or expected value of quarterly sales for all restaurants located near campuses with 10,000 students, Armand's would substitute the value of 10,000 for x in equation (12.3). In some cases, however, Armand's may be more interested in predicting sales for one particular restaurant. For example, suppose Armand's would like to predict quarterly sales for the restaurant located near Talbot College, a school with 10,000 students. As it turns out, the best predictor of y for a given value of x is also provided by $\hat{y}$. Thus, to predict quarterly sales for the restaurant located near Talbot College, Armand's would also substitute the value of 10,000 for x in equation (12.3).

Because the value of $\hat{y}$ provides both a point estimate of $E(y)$ for a given value of x and a prediction of the value of y for a given value of x, we will refer to $\hat{y}$ simply as the *estimated value of y*. Figure 12.2 provides a summary of the estimation process for simple linear regression.

FIGURE 12.2 THE ESTIMATION PROCESS IN SIMPLE LINEAR REGRESSION

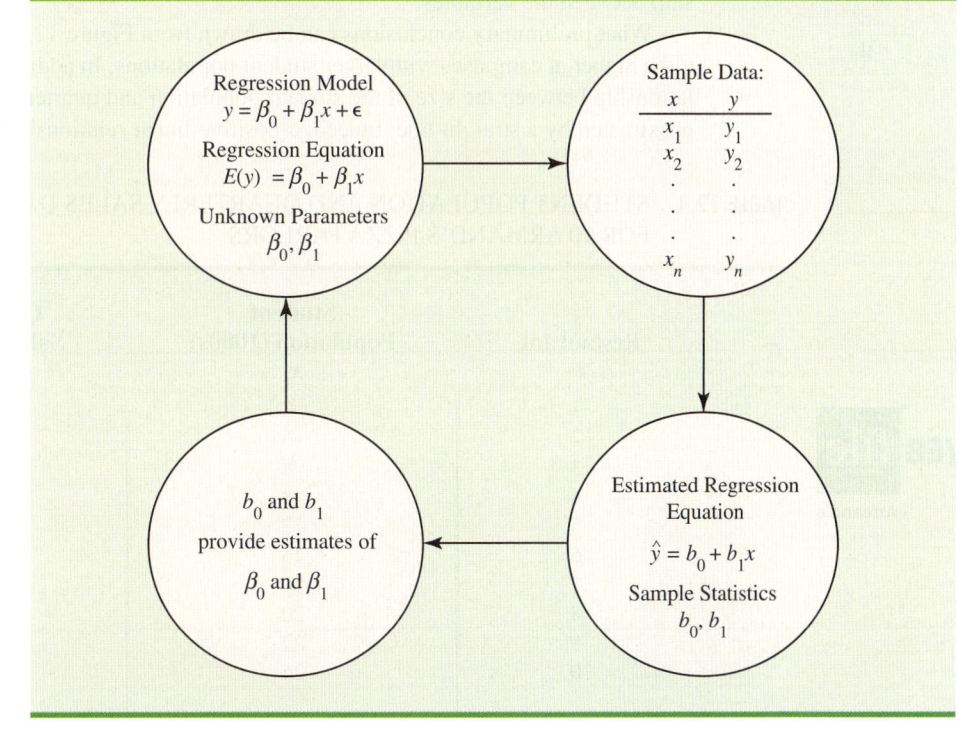

The estimation of β_0 and β_1 is a statistical process much like the estimation of μ discussed in Chapter 7. β_0 and β_1 are the unknown parameters of interest, and b_0 and b_1 are the sample statistics used to estimate the parameters.

Regression Model
$y = \beta_0 + \beta_1 x + \epsilon$
Regression Equation
$E(y) = \beta_0 + \beta_1 x$
Unknown Parameters
β_0, β_1

Sample Data:
x	y
x_1	y_1
x_2	y_2
.	.
.	.
x_n	y_n

Estimated Regression Equation
$\hat{y} = b_0 + b_1 x$
Sample Statistics
b_0, b_1

b_0 and b_1 provide estimates of β_0 and β_1

488 Chapter 12 Simple Linear Regression

NOTES AND COMMENTS

1. Regression analysis cannot be interpreted as a procedure for establishing a cause-and-effect relationship between variables. It can only indicate how or to what extent variables are associated with each other. Any conclusions about cause and effect must be based upon the judgment of those individuals most knowledgeable about the application.

2. The regression equation in simple linear regression is $E(y) = \beta_0 + \beta_1 x$. More advanced texts in regression analysis often write the regression equation as $E(y|x) = \beta_0 + \beta_1 x$ to emphasize that the regression equation provides the mean value of y for a given value of x.

12.2 Least Squares Method

In simple linear regression, each observation consists of two values: one for the independent variable and one for the dependent variable.

The **least squares method** is a procedure for using sample data to find the estimated regression equation. To illustrate the least squares method, suppose data were collected from a sample of 10 Armand's Pizza Parlor restaurants located near college campuses. For the ith observation or restaurant in the sample, x_i is the size of the student population (in thousands) and y_i is the quarterly sales (in thousands of dollars). The values of x_i and y_i for the 10 restaurants in the sample are summarized in Table 12.1. We see that restaurant 1, with $x_1 = 2$ and $y_1 = 58$, is near a campus with 2000 students and has quarterly sales of $58,000. Restaurant 2, with $x_2 = 6$ and $y_2 = 105$, is near a campus with 6000 students and has quarterly sales of $105,000. The largest sales value is for restaurant 10, which is near a campus with 26,000 students and has quarterly sales of $202,000.

Figure 12.3 is a scatter diagram of the data in Table 12.1. Student population is shown on the horizontal axis and quarterly sales is shown on the vertical axis. **Scatter diagrams** for regression analysis are constructed with the independent variable x on the horizontal axis and the dependent variable y on the vertical axis. The scatter diagram enables us to observe the data graphically and to draw preliminary conclusions about the possible relationship between the variables.

What preliminary conclusions can be drawn from Figure 12.3? Quarterly sales appear to be higher at campuses with larger student populations. In addition, for these data the relationship between the size of the student population and quarterly sales appears to be approximated by a straight line; indeed, a positive linear relationship is indicated between x

TABLE 12.1 STUDENT POPULATION AND QUARTERLY SALES DATA FOR 10 ARMAND'S PIZZA PARLORS

Restaurant i	Student Population (1000s) x_i	Quarterly Sales ($1000s) y_i
1	2	58
2	6	105
3	8	88
4	8	118
5	12	117
6	16	137
7	20	157
8	20	169
9	22	149
10	26	202

WEB file
Armand's

12.2 Least Squares Method

FIGURE 12.3 SCATTER DIAGRAM OF STUDENT POPULATION AND QUARTERLY SALES FOR ARMAND'S PIZZA PARLORS

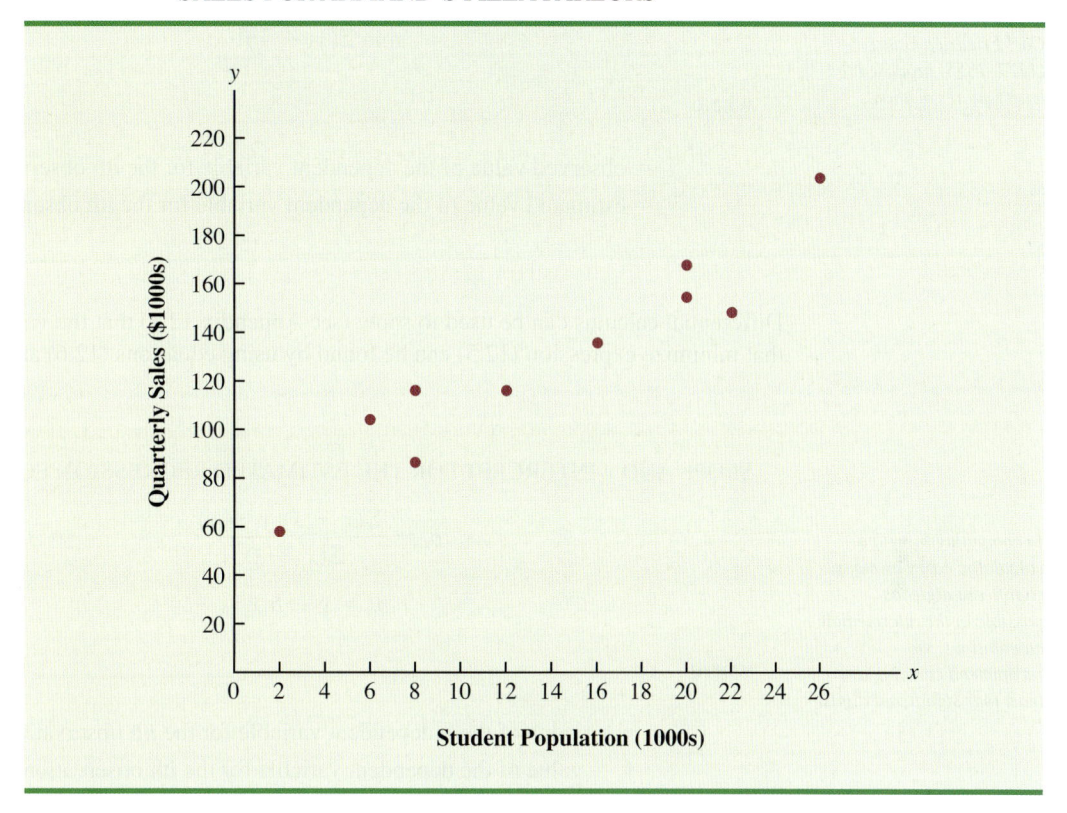

and y. We therefore choose the simple linear regression model to represent the relationship between quarterly sales and student population. Given that choice, our next task is to use the sample data in Table 12.1 to determine the values of b_0 and b_1 in the estimated simple linear regression equation. For the ith restaurant, the estimated regression equation provides

$$\hat{y}_i = b_0 + b_1 x_i \qquad (12.4)$$

where

$\hat{y}_i$ = estimated value of quarterly sales ($1000s) for the ith restaurant
b_0 = the y-intercept of the estimated regression line
b_1 = the slope of the estimated regression line
x_i = size of the student population (1000s) for the ith restaurant

With y_i denoting the observed (actual) sales for restaurant i and $\hat{y}_i$ in equation (12.4) representing the estimated value of sales for restaurant i, every restaurant in the sample will have an observed value of sales y_i and an estimated value of sales $\hat{y}_i$. For the estimated regression line to provide a good fit to the data, we want the differences between the observed sales values and the estimated sales values to be small.

The least squares method uses the sample data to provide the values of b_0 and b_1 that minimize the *sum of the squares of the deviations* between the observed values of the dependent variable y_i and the estimated values of the dependent variable $\hat{y}_i$. The criterion for the least squares method is given by expression (12.5).

Carl Friedrich Gauss (1777–1855) proposed the least squares method.

LEAST SQUARES CRITERION

$$\min \Sigma(y_i - \hat{y}_i)^2 \qquad (12.5)$$

where

y_i = observed value of the dependent variable for the ith observation
$\hat{y}_i$ = estimated value of the dependent variable for the ith observation

Differential calculus can be used to show (see Appendix 12.1) that the values of b_0 and b_1 that minimize expression (12.5) can be found by using equations (12.6) and (12.7).

SLOPE AND y-INTERCEPT FOR THE ESTIMATED REGRESSION EQUATION[1]

$$b_1 = \frac{\Sigma(x_i - \bar{x})(y_i - \bar{y})}{\Sigma(x_i - \bar{x})^2} \qquad (12.6)$$

$$b_0 = \bar{y} - b_1 \bar{x} \qquad (12.7)$$

In computing b_1 with a calculator, carry as many significant digits as possible in the intermediate calculations. We recommend carrying at least four significant digits.

where

x_i = value of the independent variable for the ith observation
y_i = value of the dependent variable for the ith observation
$\bar{x}$ = mean value for the independent variable
$\bar{y}$ = mean value for the dependent variable
n = total number of observations

Some of the calculations necessary to develop the least squares estimated regression equation for Armand's Pizza Parlors are shown in Table 12.2. With the sample of 10 restaurants, we have $n = 10$ observations. Because equations (12.6) and (12.7) require $\bar{x}$ and $\bar{y}$, we begin the calculations by computing $\bar{x}$ and $\bar{y}$.

$$\bar{x} = \frac{\Sigma x_i}{n} = \frac{140}{10} = 14$$

$$\bar{y} = \frac{\Sigma y_i}{n} = \frac{1300}{10} = 130$$

Using equations (12.6) and (12.7) and the information in Table 12.2, we can compute the slope and intercept of the estimated regression equation for Armand's Pizza Parlors. The calculation of the slope (b_1) proceeds as follows.

---

[1] An alternate formula for b_1 is

$$b_1 = \frac{\Sigma x_i y_i - (\Sigma x_i \Sigma y_i)/n}{\Sigma x_i^2 - (\Sigma x_i)^2/n}$$

This form of equation (12.6) is often recommended when using a calculator to compute b_1.

12.2 Least Squares Method

TABLE 12.2 CALCULATIONS FOR THE LEAST SQUARES ESTIMATED REGRESSION EQUATION FOR ARMAND'S PIZZA PARLORS

Restaurant i	x_i	y_i	$x_i - \bar{x}$	$y_i - \bar{y}$	$(x_i - \bar{x})(y_i - \bar{y})$	$(x_i - \bar{x})^2$
1	2	58	−12	−72	864	144
2	6	105	−8	−25	200	64
3	8	88	−6	−42	252	36
4	8	118	−6	−12	72	36
5	12	117	−2	−13	26	4
6	16	137	2	7	14	4
7	20	157	6	27	162	36
8	20	169	6	39	234	36
9	22	149	8	19	152	64
10	26	202	12	72	864	144
Totals	140	1300			2840	568
	Σx_i	Σy_i			$\Sigma(x_i - \bar{x})(y_i - \bar{y})$	$\Sigma(x_i - \bar{x})^2$

$$b_1 = \frac{\Sigma(x_i - \bar{x})(y_i - \bar{y})}{\Sigma(x_i - \bar{x})^2}$$

$$= \frac{2840}{568}$$

$$= 5$$

The calculation of the y-intercept (b_0) follows.

$$b_0 = \bar{y} - b_1\bar{x}$$
$$= 130 - 5(14)$$
$$= 60$$

Thus, the estimated regression equation is

$$\hat{y} = 60 + 5x$$

Figure 12.4 shows the graph of this equation on the scatter diagram.

The slope of the estimated regression equation ($b_1 = 5$) is positive, implying that as student population increases, sales increase. In fact, we can conclude (based on sales measured in $1000s and student population in 1000s) that an increase in the student population of 1000 is associated with an increase of $5000 in expected sales; that is, quarterly sales are expected to increase by $5 per student.

Using the estimated regression equation to make predictions outside the range of the values of the independent variable should be done with caution because outside that range we cannot be sure that the same relationship is valid.

If we believe the least squares estimated regression equation adequately describes the relationship between x and y, it would seem reasonable to use the estimated regression equation to predict the value of y for a given value of x. For example, if we wanted to predict quarterly sales for a restaurant to be located near a campus with 16,000 students, we would compute

$$\hat{y} = 60 + 5(16) = 140$$

Hence, we would predict quarterly sales of $140,000 for this restaurant. In the following sections we will discuss methods for assessing the appropriateness of using the estimated regression equation for estimation and prediction.

FIGURE 12.4 GRAPH OF THE ESTIMATED REGRESSION EQUATION FOR ARMAND'S PIZZA PARLORS: $\hat{y} = 60 + 5x$

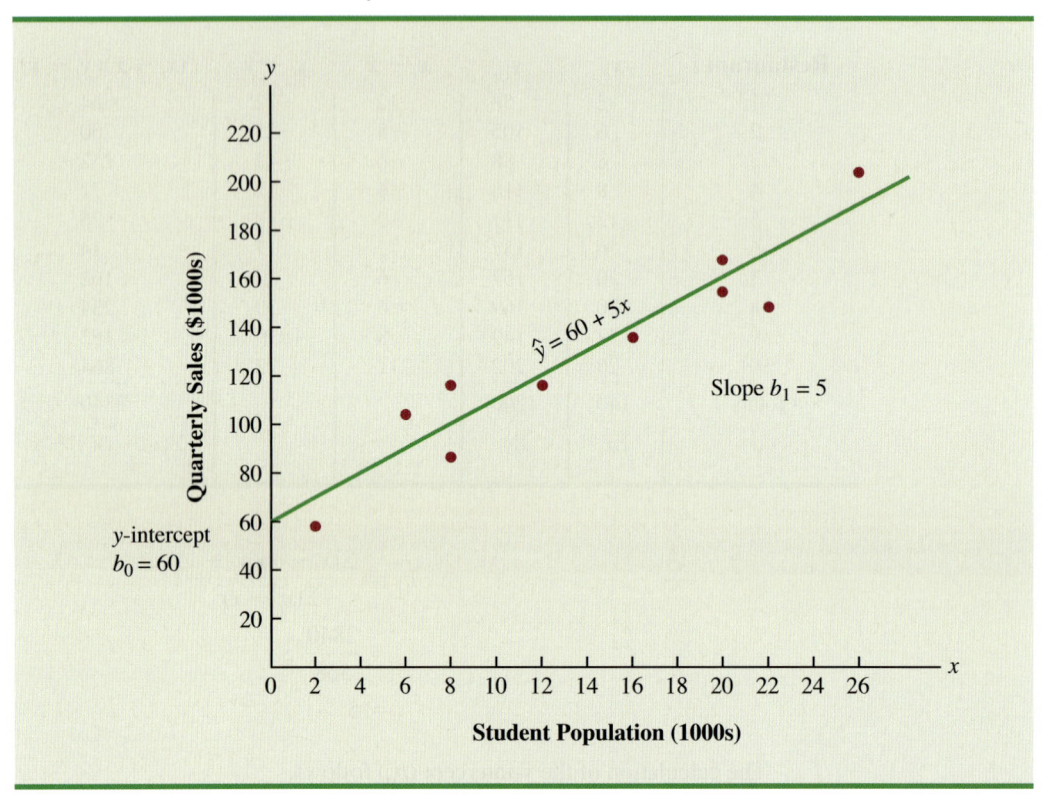

NOTES AND COMMENTS

The least squares method provides an estimated regression equation that minimizes the sum of squared deviations between the observed values of the dependent variable y_i and the estimated values of the dependent variable $\hat{y}_i$. This least squares criterion is used to choose the equation that provides the best fit. If some other criterion were used, such as minimizing the sum of the absolute deviations between y_i and $\hat{y}_i$, a different equation would be obtained. In practice, the least squares method is the most widely used.

Exercises

Methods

1. Given are five observations for two variables, x and y.

x_i	1	2	3	4	5
y_i	3	7	5	11	14

    a. Develop a scatter diagram for these data.
    b. What does the scatter diagram developed in part (a) indicate about the relationship between the two variables?

12.2 Least Squares Method

c. Try to approximate the relationship between x and y by drawing a straight line through the data.
d. Develop the estimated regression equation by computing the values of b_0 and b_1 using equations (12.6) and (12.7).
e. Use the estimated regression equation to predict the value of y when $x = 4$.

2. Given are five observations for two variables, x and y.

x_i	3	12	6	20	14
y_i	55	40	55	10	15

a. Develop a scatter diagram for these data.
b. What does the scatter diagram developed in part (a) indicate about the relationship between the two variables?
c. Try to approximate the relationship between x and y by drawing a straight line through the data.
d. Develop the estimated regression equation by computing the values of b_0 and b_1 using equations (12.6) and (12.7).
e. Use the estimated regression equation to predict the value of y when $x = 10$.

3. Given are five observations collected in a regression study on two variables.

x_i	2	6	9	13	20
y_i	7	18	9	26	23

a. Develop a scatter diagram for these data.
b. Develop the estimated regression equation for these data.
c. Use the estimated regression equation to predict the value of y when $x = 6$.

Applications

SELF test

4. The following data give the percentage of women working in five companies in the retail and trade industry. The percentage of management jobs held by women in each company is also shown.

% Working	67	45	73	54	61
% Management	49	21	65	47	33

a. Develop a scatter diagram for these data with the percentage of women working in the company as the independent variable.
b. What does the scatter diagram developed in part (a) indicate about the relationship between the two variables?
c. Try to approximate the relationship between the percentage of women working in the company and the percentage of management jobs held by women in that company.
d. Develop the estimated regression equation by computing the values of b_0 and b_1.
e. Predict the percentage of management jobs held by women in a company that has 60% women employees.

5. Elliptical trainers are becoming one of the more popular exercise machines. Their smooth and steady low-impact motion makes them a preferred choice for individuals with knee and ankle problems. But selecting the right trainer can be a difficult process. Price and quality are two important factors in any purchase decision. Are higher prices generally

associated with higher quality elliptical trainers? *Consumer Reports* conducted extensive tests to develop an overall rating based on ease of use, ergonomics, construction, and exercise range. The following data show the price and rating for eight elliptical trainers tested (*Consumer Reports,* February 2008).

Brand and Model	Price ($)	Rating
Precor 5.31	3700	87
Keys Fitness CG2	2500	84
Octane Fitness Q37e	2800	82
LifeFitness X1 Basic	1900	74
NordicTrack AudioStrider 990	1000	73
Schwinn 430	800	69
Vision Fitness X6100	1700	68
ProForm XP 520 Razor	600	55

a. Develop a scatter diagram with price as the independent variable.
b. An exercise equipment store that sells primarily higher priced equipment has a sign over the display area that says "Quality: You Get What You Pay For." Based upon your analysis of the data for ellipical trainers, do you think this sign fairly reflects the price-quality relationship for elliptical trainers?
c. Use the least squares method to develop the estimated regression equation.
d. Use the estimated regression equation to predict the rating for an ellipitical trainer with a price of $1500.

6. The cost of a previously owned car depends upon factors such as make and model, model year, mileage, condition, and whether the car is purchased from a dealer or from a private seller. To investigate the relationship between the car's mileage and the sales price, data were collected on the mileage and the sale price for 10 private sales of model year 2000 Honda Accords (PriceHub website, October 2008).

Miles (1000s)	Price ($1000s)
90	7.0
59	7.5
66	6.6
87	7.2
90	7.0
106	5.4
94	6.4
57	7.0
138	5.1
87	7.2

a. Develop a scatter diagram with miles as the independent variable.
b. What does the scatter diagram developed in part (a) indicate about the relationship between the two variables?
c. Use the least squares method to develop the estimated regression equation.
d. Provide an interpretation for the slope of the estimated regression equation.
e. Predict the sales price for a 2000 Honda Accord with 100,000 miles.

12.2 Least Squares Method

7. A sales manager collected the following data on annual sales and years of experience.

Salesperson	Years of Experience	Annual Sales ($1000s)
1	1	80
2	3	97
3	4	92
4	4	102
5	6	103
6	8	111
7	10	119
8	10	123
9	11	117
10	13	136

   a. Develop a scatter diagram for these data with years of experience as the independent variable.
   b. Develop an estimated regression equation that can be used to predict annual sales given the years of experience.
   c. Use the estimated regression equation to predict annual sales for a salesperson with 9 years of experience.

8. Bergans of Norway has been making outdoor gear since 1908. The following data show the temperature rating (F°) and the price ($) for 11 models of sleeping bags produced by Bergans (*Backpacker* 2006 Gear Guide).

Model	Temperature Rating (F°)	Price ($)
Ranger 3-Seasons	12	319
Ranger Spring	24	289
Ranger Winter	3	389
Rondane 3-Seasons	13	239
Rondane Summer	38	149
Rondane Winter	4	289
Senja Ice	5	359
Senja Snow	15	259
Senja Zero	25	229
Super Light	45	129
Tight & Light	25	199

   a. Develop a scatter diagram for these data with temperature rating (F°) as the independent variable.
   b. What does the scatter diagram developed in part (a) indicate about the relationship between temperature rating (F°) and price?
   c. Use the least squares method to develop the estimated regression equation.
   d. Predict the price for a sleeping bag with a temperature rating (F°) of 20.

9. To avoid extra checked-bag fees, airline travelers often pack as much as they can into their suitcase. Finding a rolling suitcase that is durable, has good capacity, and is easy to pull can be difficult. The following table shows the results of tests conducted by *Consumer Reports* for 10 rolling suitcases; higher scores indicate better overall test results (Consumer Reports website, October 2008).

Brand	Price ($)	Score
Briggs & Riley	325	72
Hartman	350	74
Heys	67	54
Kenneth Cole Reaction	120	54
Liz Claiborne	85	64
Samsonite	180	57
Titan	360	66
TravelPro	156	67
Tumi	595	87
Victorinox	400	77

a. Develop a scatter diagram with price as the independent variable.
b. What does the scatter diagram developed in part (a) indicate about the relationship between the two variables?
c. Use the least squares method to develop the estimated regression equation.
d. Provide an interpretation for the slope of the estimated regression equation.
e. The Eagle Creek Hovercraft suitcase has a price of $225. Predict the score for this suitcase using the estimated regression equation developed in part (c).

10. According to *Advertising Age*'s annual salary review, Mark Hurd, the 49-year-old chairman, president, and CEO of Hewlett-Packard Co., received an annual salary of $817,000, a bonus of more than $5 million, and other compensation exceeding $17 million. His total compensation was slightly better than the average CEO total pay of $12.4 million. The following table shows the age and annual salary (in thousands of dollars) for Mark Hurd and 14 other executives who led publicly held companies (*Advertising Age*, December 5, 2006).

Executive	Title	Company	Age	Salary ($1000s)
Charles Prince	Chmn/CEO	Citigroup	56	1000
Harold McGraw III	Chmn/Pres/CEO	McGraw-Hill Cos.	57	1172
James Dimon	Pres/CEO	JP Morgan Chase & Co.	50	1000
K. Rupert Murdoch	Chmn/CEO	News Corp.	75	4509
Kenneth D. Lewis	Chmn/Pres/CEO	Bank of America	58	1500
Kenneth I. Chenault	Chmn/CEO	American Express Co.	54	1092
Louis C. Camilleri	Chmn/CEO	Altria Group	51	1663
Mark V. Hurd	Chmn/Pres/CEO	Hewlett-Packard Co.	49	817
Martin S. Sorrell	CEO	WPP Group	61	1562
Robert L. Nardelli	Chmn/Pres/CEO	Home Depot	57	2164
Samuel J. Palmisano	Chmn/Pres/CEO	IBM Corp.	55	1680
David C. Novak	Chmn/Pres/CEO	Yum Brands	53	1173
Henry R. Silverman	Chmn/CEO	Cendant Corp.	65	3300
Robert C. Wright	Chmn/CEO	NBC Universal	62	2500
Sumner Redstone	Exec Chmn/Founder	Viacom	82	5807

a. Develop a scatter diagram for these data with the age of the executive as the independent variable.
b. What does the scatter diagram developed in part (a) indicate about the relationship between the two variables?
c. Develop the least squares estimated regression equation.
d. Suppose Bill Gustin is the 72-year-old chairman, president, and CEO of a major electronics company. Predict the annual salary for Bill Gustin.

11. Sporty cars are designed to provide better handling, acceleration, and a more responsive driving experience than a typical sedan. But, even within this select group of cars, performance as well as price can vary. *Consumer Reports* provided road-test scores and prices for the following 12 sporty cars (Consumer Reports website, October 2008). Prices are in thousands of dollars and road-test scores are based on a 0–100 rating scale, with higher values indicating better performance.

Car	Price ($1000s)	Road-Test Score
Chevrolet Cobalt SS	24.5	78
Dodge Caliber SRT4	24.9	56
Ford Mustang GT (V8)	29.0	73
Honda Civic Si	21.7	78
Mazda RX-8	31.3	86
Mini Cooper S	26.4	74
Mitsubishi Lancer Evolution GSR	38.1	83
Nissan Sentra SE-R Spec V	23.3	66
Suburu Impreza WRX	25.2	81
Suburu Impreza WRX Sti	37.6	89
Volkswagen GTI	24.0	83
Volkswagen R32	33.6	83

a. Develop a scatter diagram with price as the independent variable.
b. What does the scatter diagram developed in part (a) indicate about the relationship between the two variables?
c. Use the least squares method to develop the estimated regression equation.
d. Provide an interpretation for the slope of the estimated regression equation.
e. Another sporty car that *Consumer Reports* tested is the BMW 135i; the price for this car was $36,700. Predict the road-test score for the BMW 135i using the estimated regression equation developed in part (c).

12. A personal watercraft (PWC) is a vessel propelled by water jets, designed to be operated by a person sitting, standing, or kneeling on the vessel. In the early 1970s, Kawasaki Motors Corp. U.S.A. introduced the JET SKI® watercraft, the first commercially successful PWC. Today, *jetski* is commonly used as a generic term for personal watercraft. The following data show the weight (rounded to the nearest 10 lbs.) and the price (rounded to the nearest $50) for 10 three-seater personal watercraft (Jetski News website, 2006).

Make and Model	Weight (lbs.)	Price ($)
Honda AquaTrax F-12	750	9500
Honda AquaTrax F-12X	790	10500
Honda AquaTrax F-12X GPScape	800	11200
Kawasaki STX-12F Jetski	740	8500
Yamaha FX Cruiser Waverunner	830	10000
Yamaha FX High Output Waverunner	770	10000
Yamaha FX Waverunner	830	9300
Yamaha VX110 Deluxe Waverunner	720	7700
Yamaha VX110 Sport Waverunner	720	7000
Yamaha XLT1200 Waverunner	780	8500

a. Develop a scatter diagram for these data with weight as the independent variable.
b. What does the scatter diagram developed in part (a) indicate about the relationship between weight and price?
c. Use the least squares method to develop the estimated regression equation.
d. Predict the price for a three-seater PWC with a weight of 750 pounds.

e. The Honda AquaTrax F-12 weighs 750 pounds and has a price of $9500. Shouldn't the predicted price you developed in part (d) for a PWC with a weight of 750 pounds also be $9500?
f. The Kawasaki SX-R 800 Jetski has a seating capacity of one and weighs 350 pounds. Do you think the estimated regression equation developed in part (c) should be used to predict the price for this model?

13. To the Internal Revenue Service, the reasonableness of total itemized deductions depends on the taxpayer's adjusted gross income. Large deductions, which include charity and medical deductions, are more reasonable for taxpayers with large adjusted gross incomes. If a taxpayer claims larger than average itemized deductions for a given level of income, the chances of an IRS audit are increased. Data (in thousands of dollars) on adjusted gross income and the average or reasonable amount of itemized deductions follow.

Adjusted Gross Income ($1000s)	Reasonable Amount of Itemized Deductions ($1000s)
22	9.6
27	9.6
32	10.1
48	11.1
65	13.5
85	17.7
120	25.5

a. Develop a scatter diagram for these data with adjusted gross income as the independent variable.
b. Use the least squares method to develop the estimated regression equation.
c. Estimate a reasonable level of total itemized deductions for a taxpayer with an adjusted gross income of $52,500. If this taxpayer claimed itemized deductions of $20,400, would the IRS agent's request for an audit appear justified? Explain.

14. *PCWorld* rated four component characteristics for 10 ultraportable laptop computers: features, performance, design, and price. Each characteristic was rated using a 0–100 point scale. An overall rating, referred to as the *PCW World* Rating, was then developed for each laptop. The following table shows the features rating and the *PCW World* Rating for the 10 laptop computers (PC World website, February 5, 2009).

Model	Features Rating	PCW World Rating
Thinkpad X200	87	83
VGN-Z598U	85	82
U6V	80	81
Elitebook 2530P	75	78
X360	80	78
Thinkpad X300	76	78
Ideapad U110	81	77
Micro Express JFT2500	73	75
Toughbook W7	79	73
HP Voodoo Envy133	68	72

a. Develop a scatter diagram with the features rating as the independent variable.
b. What does the scatter diagram developed in part (a) indicate about the relationship between the two variables?
c. Use the least squares method to develop the estimated regression equation.
d. Estimate the *PCW World* Rating for a new laptop computer that has a features rating of 70.

12.3 Coefficient of Determination

For the Armand's Pizza Parlors example, we developed the estimated regression equation $\hat{y} = 60 + 5x$ to approximate the linear relationship between the size of the student population x and quarterly sales y. A question now is, How well does the estimated regression equation fit the data? In this section, we show that the **coefficient of determination** provides a measure of the goodness of fit for the estimated regression equation.

For the ith observation, the difference between the observed value of the dependent variable, y_i, and the estimated value of the dependent variable, $\hat{y}_i$, is called the ***i*th residual**. The ith residual represents the error in using $\hat{y}_i$ to estimate y_i. Thus, for the ith observation, the residual is $y_i - \hat{y}_i$. The sum of squares of these residuals or errors is the quantity that is minimized by the least squares method. This quantity, also known as the *sum of squares due to error*, is denoted by SSE.

SUM OF SQUARES DUE TO ERROR

$$SSE = \Sigma(y_i - \hat{y}_i)^2 \tag{12.8}$$

The value of SSE is a measure of the error in using the estimated regression equation to estimate the values of the dependent variable in the sample.

In Table 12.3 we show the calculations required to compute the sum of squares due to error for the Armand's Pizza Parlors example. For instance, for restaurant 1 the values of the independent and dependent variables are $x_1 = 2$ and $y_1 = 58$. Using the estimated regression equation, we find that the estimated value of quarterly sales for restaurant 1 is $\hat{y}_1 = 60 + 5(2) = 70$. Thus, the error in using $\hat{y}_1$ to estimate y_1 for restaurant 1 is $y_1 - \hat{y}_1 = 58 - 70 = -12$. The squared error, $(-12)^2 = 144$, is shown in the last column of Table 12.3. After computing and squaring the residuals for each restaurant in the sample, we sum them to obtain SSE = 1530. Thus, SSE = 1530 measures the error in using the estimated regression equation $\hat{y} = 60 + 5x$ to predict sales.

Now suppose we are asked to develop an estimate of quarterly sales without knowledge of the size of the student population. Without knowledge of any related variables, we would

TABLE 12.3 CALCULATION OF SSE FOR ARMAND'S PIZZA PARLORS

Restaurant i	x_i = Student Population (1000s)	y_i = Quarterly Sales ($1000s)	Predicted Sales $\hat{y}_i = 60 + 5x_i$	Error $y_i - \hat{y}_i$	Squared Error $(y_i - \hat{y}_i)^2$
1	2	58	70	−12	144
2	6	105	90	15	225
3	8	88	100	−12	144
4	8	118	100	18	324
5	12	117	120	−3	9
6	16	137	140	−3	9
7	20	157	160	−3	9
8	20	169	160	9	81
9	22	149	170	−21	441
10	26	202	190	12	144
				SSE =	1530

TABLE 12.4 COMPUTATION OF THE TOTAL SUM OF SQUARES FOR ARMAND'S PIZZA PARLORS

Restaurant i	x_i = Student Population (1000s)	y_i = Quarterly Sales ($1000s)	Deviation $y_i - \bar{y}$	Squared Deviation $(y_i - \bar{y})^2$
1	2	58	−72	5,184
2	6	105	−25	625
3	8	88	−42	1,764
4	8	118	−12	144
5	12	117	−13	169
6	16	137	7	49
7	20	157	27	729
8	20	169	39	1,521
9	22	149	19	361
10	26	202	72	5,184
				SST = 15,730

use the sample mean as an estimate of quarterly sales at any given restaurant. Table 12.2 showed that for the sales data, $\Sigma y_i = 1300$. Hence, the mean value of quarterly sales for the sample of 10 Armand's restaurants is $\bar{y} = \Sigma y_i/n = 1300/10 = 130$. In Table 12.4 we show the sum of squared deviations obtained by using the sample mean $\bar{y} = 130$ to estimate the value of quarterly sales for each restaurant in the sample. For the ith restaurant in the sample, the difference $y_i - \bar{y}$ provides a measure of the error involved in using $\bar{y}$ to estimate sales. The corresponding sum of squares, called the *total sum of squares,* is denoted SST.

TOTAL SUM OF SQUARES

$$\text{SST} = \Sigma(y_i - \bar{y})^2 \tag{12.9}$$

The sum at the bottom of the last column in Table 12.4 is the total sum of squares for Armand's Pizza Parlors; it is SST = 15,730.

With SST = 15,730 and SSE = 1530, the estimated regression line provides a much better fit to the data than the line y = ȳ.

In Figure 12.5 we show the estimated regression line $\hat{y} = 60 + 5x$ and the line corresponding to $\bar{y} = 130$. Note that the points cluster more closely around the estimated regression line than they do about the line $\bar{y} = 130$. For example, for the 10th restaurant in the sample we see that the error is much larger when $\bar{y} = 130$ is used as an estimate of y_{10} than when $\hat{y}_{10} = 60 + 5(26) = 190$ is used. We can think of SST as a measure of how well the observations cluster about the $\bar{y}$ line and SSE as a measure of how well the observations cluster about the $\hat{y}$ line.

To measure how much the $\hat{y}$ values on the estimated regression line deviate from $\bar{y}$, another sum of squares is computed. This sum of squares, called the *sum of squares due to regression,* is denoted SSR.

SUM OF SQUARES DUE TO REGRESSION

$$\text{SSR} = \Sigma(\hat{y}_i - \bar{y})^2 \tag{12.10}$$

12.3 Coefficient of Determination

FIGURE 12.5 DEVIATIONS ABOUT THE ESTIMATED REGRESSION LINE AND THE LINE $y = \bar{y}$ FOR ARMAND'S PIZZA PARLORS

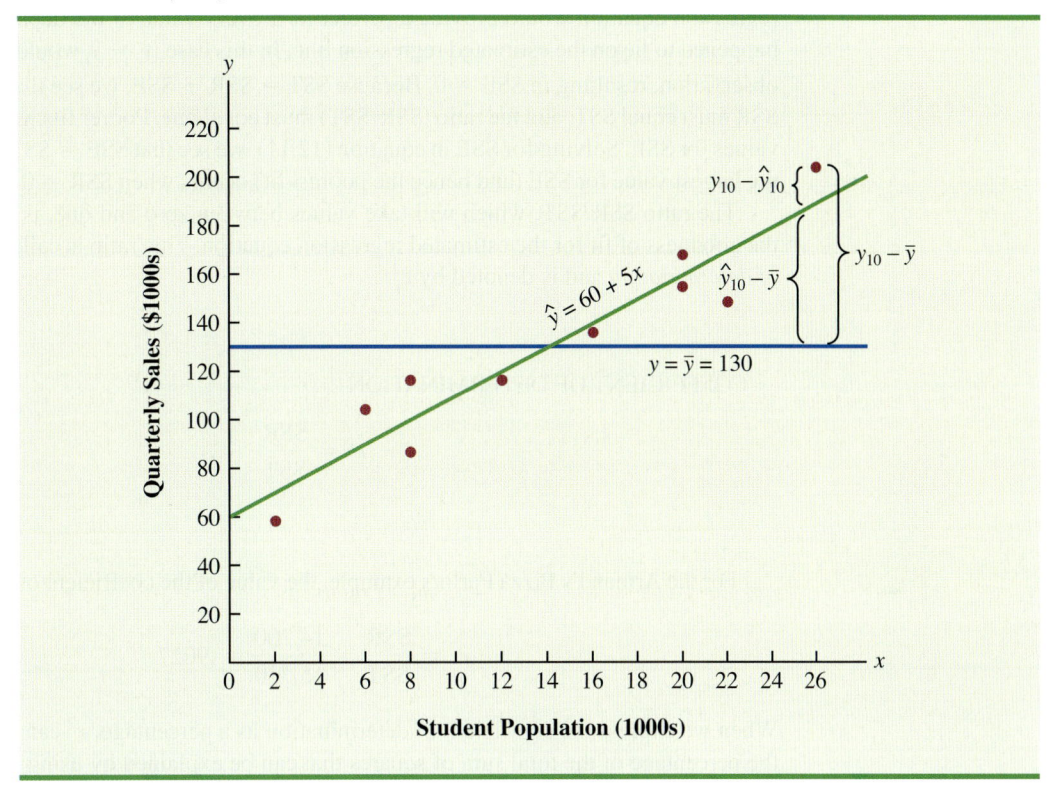

From the preceding discussion, we should expect that SST, SSR, and SSE are related. Indeed, the relationship among these three sums of squares provides one of the most important results in statistics.

SSR can be thought of as the explained portion of SST, and SSE can be thought of as the unexplained portion of SST.

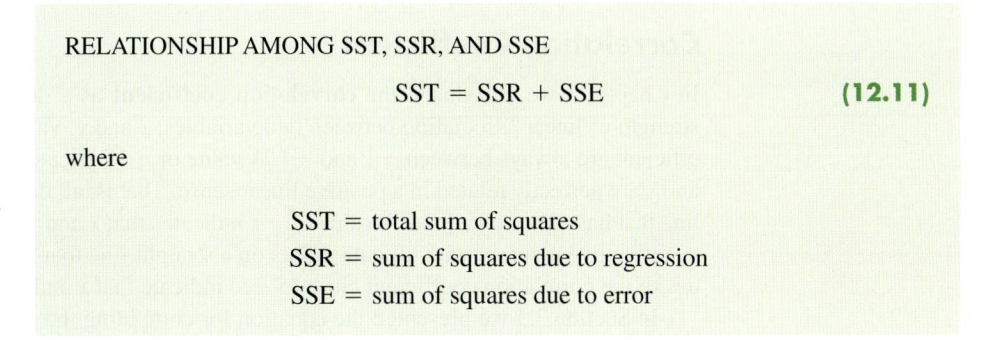

RELATIONSHIP AMONG SST, SSR, AND SSE

$$\text{SST} = \text{SSR} + \text{SSE} \quad (12.11)$$

where

SST = total sum of squares
SSR = sum of squares due to regression
SSE = sum of squares due to error

Equation (12.11) shows that the total sum of squares can be partitioned into two components, the sum of squares due to regression and the sum of squares due to error. Hence, if the values of any two of these sum of squares are known, the third sum of squares can be computed easily. For instance, in the Armand's Pizza Parlors example, we already know that SSE = 1530 and SST = 15,730; therefore, solving for SSR in equation (12.11), we find that the sum of squares due to regression is

$$\text{SSR} = \text{SST} - \text{SSE} = 15{,}730 - 1530 = 14{,}200$$

Now let us see how the three sums of squares, SST, SSR, and SSE, can be used to provide a measure of the goodness of fit for the estimated regression equation. The estimated regression equation would provide a perfect fit if every value of the dependent variable y_i happened to lie on the estimated regression line. In this case, $y_i - \hat{y}_i$ would be zero for each observation, resulting in SSE = 0. Because SST = SSR + SSE, we see that for a perfect fit SSR must equal SST, and the ratio (SSR/SST) must equal one. Poorer fits will result in larger values for SSE. Solving for SSE in equation (12.11), we see that SSE = SST − SSR. Hence, the largest value for SSE (and hence the poorest fit) occurs when SSR = 0 and SSE = SST.

The ratio SSR/SST, which will take values between zero and one, is used to evaluate the goodness of fit for the estimated regression equation. This ratio is called the *coefficient of determination* and is denoted by r^2.

> **COEFFICIENT OF DETERMINATION**
>
> $$r^2 = \frac{SSR}{SST} \qquad (12.12)$$

For the Armand's Pizza Parlors example, the value of the coefficient of determination is

$$r^2 = \frac{SSR}{SST} = \frac{14{,}200}{15{,}730} = .9027$$

When we express the coefficient of determination as a percentage, r^2 can be interpreted as the percentage of the total sum of squares that can be explained by using the estimated regression equation. For Armand's Pizza Parlors, we can conclude that 90.27% of the total sum of squares can be explained by using the estimated regression equation $\hat{y} = 60 + 5x$ to predict quarterly sales. In other words, 90.27% of the variability in sales can be explained by the linear relationship between the size of the student population and sales. We should be pleased to find such a good fit for the estimated regression equation.

Correlation Coefficient

In Chapter 3 we introduced the **correlation coefficient** as a descriptive measure of the strength of linear association between two variables, x and y. Values of the correlation coefficient are always between -1 and $+1$. A value of $+1$ indicates that the two variables x and y are perfectly related in a positive linear sense. That is, all data points are on a straight line that has a positive slope. A value of -1 indicates that x and y are perfectly related in a negative linear sense, with all data points on a straight line that has a negative slope. Values of the correlation coefficient close to zero indicate that x and y are not linearly related.

In Section 3.5 we presented the equation for computing the sample correlation coefficient. If a regression analysis has already been performed and the coefficient of determination r^2 computed, the sample correlation coefficient can be computed as follows.

> **SAMPLE CORRELATION COEFFICIENT**
>
> $$r_{xy} = (\text{sign of } b_1)\sqrt{\text{Coefficient of determination}}$$
> $$= (\text{sign of } b_1)\sqrt{r^2} \qquad (12.13)$$

where

$$b_1 = \text{the slope of the estimated regression equation } \hat{y} = b_0 + b_1x$$

The sign for the sample correlation coefficient is positive if the estimated regression equation has a positive slope ($b_1 > 0$) and negative if the estimated regression equation has a negative slope ($b_1 < 0$).

For the Armand's Pizza Parlor example, the value of the coefficient of determination corresponding to the estimated regression equation $\hat{y} = 60 + 5x$ is .9027. Because the slope of the estimated regression equation is positive, equation (12.13) shows that the sample correlation coefficient is $+\sqrt{.9027} = +.9501$. With a sample correlation coefficient of $r_{xy} = +.9501$, we would conclude that a strong positive linear association exists between x and y.

In the case of a linear relationship between two variables, both the coefficient of determination and the sample correlation coefficient provide measures of the strength of the relationship. The coefficient of determination provides a measure between zero and one, whereas the sample correlation coefficient provides a measure between -1 and $+1$. Although the sample correlation coefficient is restricted to a linear relationship between two variables, the coefficient of determination can be used for nonlinear relationships and for relationships that have two or more independent variables. Thus, the coefficient of determination provides a wider range of applicability.

NOTES AND COMMENTS

1. In developing the least squares estimated regression equation and computing the coefficient of determination, we made no probabilistic assumptions about the error term ϵ, and no statistical tests for significance of the relationship between x and y were conducted. Larger values of r^2 imply that the least squares line provides a better fit to the data; that is, the observations are more closely grouped about the least squares line. But, using only r^2, we can draw no conclusion about whether the relationship between x and y is statistically significant. Such a conclusion must be based on considerations that involve the sample size and the properties of the appropriate sampling distributions of the least squares estimators.

2. As a practical matter, for typical data found in the social sciences, values of r^2 as low as .25 are often considered useful. For data in the physical and life sciences, r^2 values of .60 or greater are often found; in fact, in some cases, r^2 values greater than .90 can be found. In business applications, r^2 values vary greatly, depending on the unique characteristics of each application.

Exercises

Methods

15. The data from exercise 1 follow.

x_i	1	2	3	4	5
y_i	3	7	5	11	14

    The estimated regression equation for these data is $\hat{y} = .20 + 2.60x$.
    a. Compute SSE, SST, and SSR using equations (12.8), (12.9), and (12.10).
    b. Compute the coefficient of determination r^2. Comment on the goodness of fit.
    c. Compute the sample correlation coefficient.

16. The data from exercise 2 follow.

x_i	3	12	6	20	14
y_i	55	40	55	10	15

The estimated regression equation for these data is $\hat{y} = 68 - 3x$.
   a. Compute SSE, SST, and SSR.
   b. Compute the coefficient of determination r^2. Comment on the goodness of fit.
   c. Compute the sample correlation coefficient.

17. The data from exercise 3 follow.

x_i	2	6	9	13	20
y_i	7	18	9	26	23

The estimated regression equation for these data is $\hat{y} = 7.6 + .9x$. What percentage of the total sum of squares can be accounted for by the estimated regression equation? What is the value of the sample correlation coefficient?

Applications

18. The following data are the monthly salaries y and the grade point averages x for students who obtained a bachelor's degree in business administration with a major in information systems. The estimated regression equation for these data is $\hat{y} = 1790.5 + 581.1x$.

GPA	Monthly Salary ($)
2.6	3300
3.4	3600
3.6	4000
3.2	3500
3.5	3900
2.9	3600

   a. Compute SST, SSR, and SSE.
   b. Compute the coefficient of determination r^2. Comment on the goodness of fit.
   c. What is the value of the sample correlation coefficient?

19. In exercise 7 a sales manager collected the following data on x = annual sales and y = years of experience. The estimated regression equation for these data is $\hat{y} = 80 + 4x$.

Salesperson	Years of Experience	Annual Sales ($1000s)
1	1	80
2	3	97
3	4	92
4	4	102
5	6	103
6	8	111
7	10	119
8	10	123
9	11	117
10	13	136

12.3 Coefficient of Determination

a. Compute SST, SSR, and SSE.
b. Compute the coefficient of determination r^2. Comment on the goodness of fit.
c. What is the value of the sample correlation coefficient?

20. *Consumer Reports* provided extensive testing and ratings for more than 100 HDTVs. An overall score, based primarily on picture quality, was developed for each model. In general, a higher overall score indicates better performance. The following data show the price and overall score for the ten 42-inch plasma televisions (*Consumer Reports*, March 2006).

Brand	Price	Score
Dell	2800	62
Hisense	2800	53
Hitachi	2700	44
JVC	3500	50
LG	3300	54
Maxent	2000	39
Panasonic	4000	66
Phillips	3000	55
Proview	2500	34
Samsung	3000	39

a. Use these data to develop an estimated regression equation that could be used to estimate the overall score for a 42-inch plasma television given the price.
b. Compute r^2. Did the estimated regression equation provide a good fit?
c. Estimate the overall score for a 42-inch plasma television with a price of $3200.

21. An important application of regression analysis in accounting is in the estimation of cost. By collecting data on volume and cost and using the least squares method to develop an estimated regression equation relating volume and cost, an accountant can estimate the cost associated with a particular manufacturing volume. Consider the following sample of production volumes and total cost data for a manufacturing operation.

Production Volume (units)	Total Cost ($)
400	4000
450	5000
550	5400
600	5900
700	6400
750	7000

a. Use these data to develop an estimated regression equation that could be used to predict the total cost for a given production volume.
b. What is the variable cost per unit produced?
c. Compute the coefficient of determination. What percentage of the variation in total cost can be explained by production volume?
d. The company's production schedule shows 500 units must be produced next month. What is the estimated total cost for this operation?

22. Refer to exercise 5, where the following data were used to investigate whether higher prices are generally associated with higher ratings for elliptical trainers (*Consumer Reports*, February 2008).

WEB file

Ellipticals

Brand and Model	Price ($)	Rating
Precor 5.31	3700	87
Keys Fitness CG2	2500	84
Octane Fitness Q37e	2800	82
LifeFitness X1 Basic	1900	74
NordicTrack AudioStrider 990	1000	73
Schwinn 430	800	69
Vision Fitness X6100	1700	68
ProForm XP 520 Razor	600	55

With $x =$ price ($) and $y =$ rating, the estimated regression equation is $\hat{y} = 58.158 + .008449x$. For these data, SSE = 173.88.

a. Compute the coefficient of determination r^2.
b. Did the estimated regression equation provide a good fit? Explain.
c. What is the value of the sample correlation coefficient? Does it reflect a strong or weak relationship between price and rating?

12.4 Model Assumptions

In conducting a regression analysis, we begin by making an assumption about the appropriate model for the relationship between the dependent and independent variable(s). For the case of simple linear regression, the assumed regression model is

$$y = \beta_0 + \beta_1 x + \epsilon$$

Then the least squares method is used to develop values for b_0 and b_1, the estimates of the model parameters β_0 and β_1, respectively. The resulting estimated regression equation is

$$\hat{y} = b_0 + b_1 x$$

We saw that the value of the coefficient of determination (r^2) is a measure of the goodness of fit of the estimated regression equation. However, even with a large value of r^2, the estimated regression equation should not be used until further analysis of the appropriateness of the assumed model has been conducted. An important step in determining whether the assumed model is appropriate involves testing for the significance of the relationship. The tests of significance in regression analysis are based on the following assumptions about the error term ϵ.

ASSUMPTIONS ABOUT THE ERROR TERM ϵ IN THE REGRESSION MODEL

$$y = \beta_0 + \beta_1 x + \epsilon$$

1. The error term ϵ is a random variable with a mean or expected value of zero; that is, $E(\epsilon) = 0$.
   Implication: β_0 and β_1 are constants, therefore $E(\beta_0) = \beta_0$ and $E(\beta_1) = \beta_1$; thus, for a given value of x, the expected value of y is

$$E(y) = \beta_0 + \beta_1 x \quad (12.14)$$

(continued)

As we indicated previously, equation (12.14) is referred to as the regression equation.
2. The variance of ϵ, denoted by σ^2, is the same for all values of x.
   Implication: The variance of y about the regression line equals σ^2 and is the same for all values of x.
3. The values of ϵ are independent.
   Implication: The value of ϵ for a particular value of x is not related to the value of ϵ for any other value of x; thus, the value of y for a particular value of x is not related to the value of y for any other value of x.
4. The error term ϵ is a normally distributed random variable.
   Implication: Because y is a linear function of ϵ, y is also a normally distributed random variable.

Figure 12.6 illustrates the model assumptions and their implications; note that in this graphical interpretation, the value of $E(y)$ changes according to the specific value of x considered. However, regardless of the x value, the probability distribution of ϵ and hence the probability distributions of y are normally distributed, each with the same variance. The specific value of the error ϵ at any particular point depends on whether the actual value of y is greater than or less than $E(y)$.

At this point, we must keep in mind that we are also making an assumption or hypothesis about the form of the relationship between x and y. That is, we assume that a straight

FIGURE 12.6 ASSUMPTIONS FOR THE REGRESSION MODEL

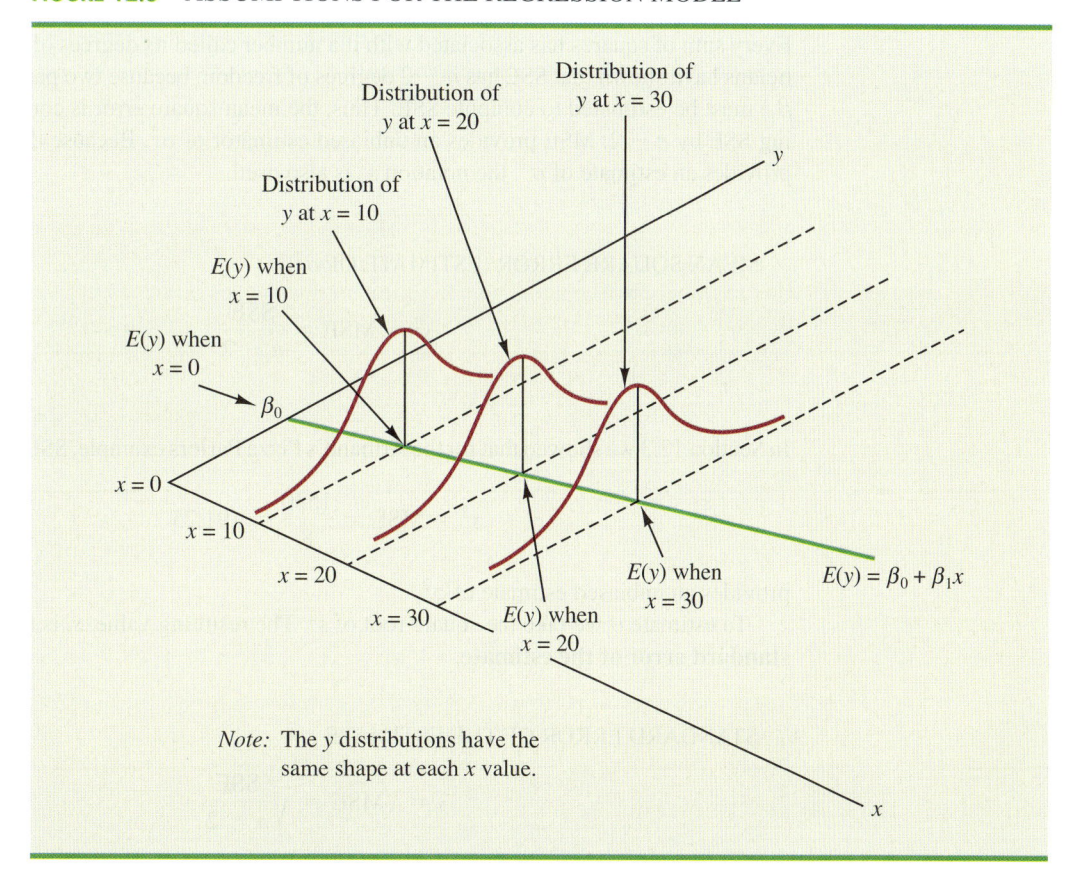

line represented by $\beta_0 + \beta_1 x$ is the basis for the relationship between the variables. We must not lose sight of the fact that some other model, for instance $y = \beta_0 + \beta_1 x^2 + \epsilon$, may turn out to be a better model for the underlying relationship.

12.5 Testing for Significance

In a simple linear regression equation, the mean or expected value of y is a linear function of x: $E(y) = \beta_0 + \beta_1 x$. If the value of β_1 is zero, $E(y) = \beta_0 + (0)x = \beta_0$. In this case, the mean value of y does not depend on the value of x and hence we would conclude that x and y are not linearly related. Alternatively, if the value of β_1 is not equal to zero, we would conclude that the two variables are related. Thus, to test for a significant regression relationship, we must conduct a hypothesis test to determine whether the value of β_1 is zero. Two tests are commonly used. Both require an estimate of σ^2, the variance of ϵ in the regression model.

Estimate of σ^2

From the regression model and its assumptions we can conclude that σ^2, the variance of ϵ, also represents the variance of the y values about the regression line. Recall that the deviations of the y values about the estimated regression line are called residuals. Thus, SSE, the sum of squared residuals, is a measure of the variability of the actual observations about the estimated regression line. The **mean square error** (MSE) provides the estimate of σ^2; it is SSE divided by its degrees of freedom.

With $\hat{y}_i = b_0 + b_1 x_i$, SSE can be written as

$$SSE = \Sigma(y_i - \hat{y}_i)^2 = \Sigma(y_i - b_0 - b_1 x_i)^2$$

Every sum of squares has associated with it a number called its degrees of freedom. Statisticians have shown that SSE has $n - 2$ degrees of freedom because two parameters (β_0 and β_1) must be estimated to compute SSE. Thus, the mean square error is computed by dividing SSE by $n - 2$. MSE provides an unbiased estimator of σ^2. Because the value of MSE provides an estimate of σ^2, the notation s^2 is also used.

MEAN SQUARE ERROR (ESTIMATE OF σ^2)

$$s^2 = MSE = \frac{SSE}{n - 2} \tag{12.15}$$

In Section 12.3 we showed that for the Armand's Pizza Parlors example, SSE = 1530; hence,

$$s^2 = MSE = \frac{1530}{8} = 191.25$$

provides an unbiased estimate of σ^2.

To estimate σ we take the square root of s^2. The resulting value, s, is referred to as the **standard error of the estimate**.

STANDARD ERROR OF THE ESTIMATE

$$s = \sqrt{MSE} = \sqrt{\frac{SSE}{n - 2}} \tag{12.16}$$

For the Armand's Pizza Parlors example, $s = \sqrt{MSE} = \sqrt{191.25} = 13.829$. In the following discussion, we use the standard error of the estimate in the tests for a significant relationship between x and y.

t Test

The simple linear regression model is $y = \beta_0 + \beta_1 x + \epsilon$. If x and y are linearly related, we must have $\beta_1 \neq 0$. The purpose of the *t* test is to see whether we can conclude that $\beta_1 \neq 0$. We will use the sample data to test the following hypotheses about the parameter β_1.

$$H_0: \beta_1 = 0$$
$$H_a: \beta_1 \neq 0$$

If H_0 is rejected, we will conclude that $\beta_1 \neq 0$ and that a statistically significant relationship exists between the two variables. However, if H_0 cannot be rejected, we will have insufficient evidence to conclude that a significant relationship exists. The properties of the sampling distribution of b_1, the least squares estimator of β_1, provide the basis for the hypothesis test.

First, let us consider what would happen if we used a different random sample for the same regression study. For example, suppose that Armand's Pizza Parlors used the sales records of a different sample of 10 restaurants. A regression analysis of this new sample might result in an estimated regression equation similar to our previous estimated regression equation $\hat{y} = 60 + 5x$. However, it is doubtful that we would obtain exactly the same equation (with an intercept of exactly 60 and a slope of exactly 5). Indeed, b_0 and b_1, the least squares estimators, are sample statistics with their own sampling distributions. The properties of the sampling distribution of b_1 follow.

SAMPLING DISTRIBUTION OF b_1

Expected Value
$$E(b_1) = \beta_1$$

Standard Deviation
$$\sigma_{b_1} = \frac{\sigma}{\sqrt{\Sigma(x_i - \bar{x})^2}} \quad (12.17)$$

Distribution Form
Normal

Note that the expected value of b_1 is equal to β_1, so b_1 is an unbiased estimator of β_1.

Because we do not know the value of σ, we develop an estimate of σ_{b_1}, denoted s_{b_1}, by estimating σ with s in equation (12.17). Thus, we obtain the following estimate of σ_{b_1}.

The standard deviation of b_1 is also referred to as the standard error of b_1. Thus, s_{b_1} provides an estimate of the standard error of b_1.

ESTIMATED STANDARD DEVIATION OF b_1

$$s_{b_1} = \frac{s}{\sqrt{\Sigma(x_i - \bar{x})^2}} \quad (12.18)$$

For Armand's Pizza Parlors, $s = 13.829$. Hence, using $\Sigma(x_i - \bar{x})^2 = 568$ as shown in Table 12.2, we have

$$s_{b_1} = \frac{13.829}{\sqrt{568}} = .5803$$

as the estimated standard deviation of b_1.

The t test for a significant relationship is based on the fact that the test statistic

$$\frac{b_1 - \beta_1}{s_{b_1}}$$

follows a t distribution with $n - 2$ degrees of freedom. If the null hypothesis is true, then $\beta_1 = 0$ and $t = b_1/s_{b_1}$.

Let us conduct this test of significance for Armand's Pizza Parlors at the $\alpha = .01$ level of significance. The test statistic is

$$t = \frac{b_1}{s_{b_1}} = \frac{5}{.5803} = 8.62$$

Appendixes 12.3 and 12.4 show how Minitab and Excel can be used to compute the p-value.

The t distribution table shows that with $n - 2 = 10 - 2 = 8$ degrees of freedom, $t = 3.355$ provides an area of .005 in the upper tail. Thus, the area in the upper tail of the t distribution corresponding to the test statistic $t = 8.62$ must be less than .005. Because this test is a two-tailed test, we double this value to conclude that the p-value associated with $t = 8.62$ must be less than $2(.005) = .01$. Excel or Minitab show the p-value $= .000$. Because the p-value is less than $\alpha = .01$, we reject H_0 and conclude that β_1 is not equal to zero. This evidence is sufficient to conclude that a significant relationship exists between student population and quarterly sales. A summary of the t test for significance in simple linear regression follows.

---

t TEST FOR SIGNIFICANCE IN SIMPLE LINEAR REGRESSION

$$H_0: \beta_1 = 0$$
$$H_a: \beta_1 \neq 0$$

TEST STATISTIC

$$t = \frac{b_1}{s_{b_1}} \quad (12.19)$$

REJECTION RULE

p-value approach: Reject H_0 if p-value $\leq \alpha$
Critical value approach: Reject H_0 if $t \leq -t_{\alpha/2}$ or if $t \geq t_{\alpha/2}$

where $t_{\alpha/2}$ is based on a t distribution with $n - 2$ degrees of freedom.

Confidence Interval for β_1

The form of a confidence interval for β_1 is as follows:

$$b_1 \pm t_{\alpha/2} s_{b_1}$$

12.5 Testing for Significance

The point estimator is b_1 and the margin of error is $t_{\alpha/2} s_{b_1}$. The confidence coefficient associated with this interval is $1 - \alpha$, and $t_{\alpha/2}$ is the t value providing an area of $\alpha/2$ in the upper tail of a t distribution with $n - 2$ degrees of freedom. For example, suppose that we wanted to develop a 99% confidence interval estimate of β_1 for Armand's Pizza Parlors. From Table 2 of Appendix B we find that the t value corresponding to $\alpha = .01$ and $n - 2 = 10 - 2 = 8$ degrees of freedom is $t_{.005} = 3.355$. Thus, the 99% confidence interval estimate of β_1 is

$$b_1 \pm t_{\alpha/2} s_{b_1} = 5 \pm 3.355(.5803) = 5 \pm 1.95$$

or 3.05 to 6.95.

In using the t test for significance, the hypotheses tested were

$$H_0: \beta_1 = 0$$
$$H_a: \beta_1 \neq 0$$

At the $\alpha = .01$ level of significance, we can use the 99% confidence interval as an alternative for drawing the hypothesis testing conclusion for the Armand's data. Because 0, the hypothesized value of β_1, is not included in the confidence interval (3.05 to 6.95), we can reject H_0 and conclude that a significant statistical relationship exists between the size of the student population and quarterly sales. In general, a confidence interval can be used to test any two-sided hypothesis about β_1. If the hypothesized value of β_1 is contained in the confidence interval, do not reject H_0. Otherwise, reject H_0.

F Test

An F test, based on the F probability distribution, can also be used to test for significance in regression. With only one independent variable, the F test will provide the same conclusion as the t test; that is, if the t test indicates $\beta_1 \neq 0$ and hence a significant relationship, the F test will also indicate a significant relationship. But with more than one independent variable, only the F test can be used to test for an overall significant relationship.

The logic behind the use of the F test for determining whether the regression relationship is statistically significant is based on the development of two independent estimates of σ^2. We explained how MSE provides an estimate of σ^2. If the null hypothesis $H_0: \beta_1 = 0$ is true, the sum of squares due to regression, SSR, divided by its degrees of freedom provides another independent estimate of σ^2. This estimate is called the *mean square due to regression,* or simply the *mean square regression,* and is denoted MSR. In general,

$$\text{MSR} = \frac{\text{SSR}}{\text{Regression degrees of freedom}}$$

For the models we consider in this text, the regression degrees of freedom is always equal to the number of independent variables in the model:

$$\text{MSR} = \frac{\text{SSR}}{\text{Number of independent variables}} \quad \textbf{(12.20)}$$

Because we consider only regression models with one independent variable in this chapter, we have MSR = SSR/1 = SSR. Hence, for Armand's Pizza Parlors, MSR = SSR = 14,200.

If the null hypothesis ($H_0: \beta_1 = 0$) is true, MSR and MSE are two independent estimates of σ^2 and the sampling distribution of MSR/MSE follows an F distribution with numerator degrees of freedom equal to one and denominator degrees of freedom equal to $n - 2$. Therefore, when $\beta_1 = 0$, the value of MSR/MSE should be close to 1. However, if the null hypothesis is false ($\beta_1 \neq 0$), MSR will overestimate σ^2 and the value of MSR/MSE will be inflated; thus, large values of MSR/MSE lead to the rejection of H_0 and the conclusion that the relationship between x and y is statistically significant.

Let us conduct the F test for the Armand's Pizza Parlors example. The test statistic is

$$F = \frac{\text{MSR}}{\text{MSE}} = \frac{14{,}200}{191.25} = 74.25$$

The F test and the t test provide identical results for simple linear regression.

The F distribution table (Table 4 of Appendix B) shows that with one degree of freedom in the numerator and $n - 2 = 10 - 2 = 8$ degrees of freedom in the denominator, $F = 11.26$ provides an area of .01 in the upper tail. Thus, the area in the upper tail of the F distribution corresponding to the test statistic $F = 74.25$ must be less than .01. Thus, we conclude that the p-value must be less than .01. Excel or Minitab show the p-value $= .000$. Because the p-value is less than $\alpha = .01$, we reject H_0 and conclude that a significant relationship exists between the size of the student population and quarterly sales. A summary of the F test for significance in simple linear regression follows.

If H_0 is false, MSE still provides an unbiased estimate of σ^2 and MSR overestimates σ^2. If H_0 is true, both MSE and MSR provide unbiased estimates of σ^2; in this case the value of MSR/MSE should be close to 1.

F TEST FOR SIGNIFICANCE IN SIMPLE LINEAR REGRESSION

$$H_0: \beta_1 = 0$$
$$H_a: \beta_1 \neq 0$$

TEST STATISTIC

$$F = \frac{\text{MSR}}{\text{MSE}} \tag{12.21}$$

REJECTION RULE

p-value approach: Reject H_0 if p-value $\leq \alpha$

Critical value approach: Reject H_0 if $F \geq F_\alpha$

where F_α is based on an F distribution with 1 degree of freedom in the numerator and $n - 2$ degrees of freedom in the denominator.

In Chapter 10 we covered analysis of variance (ANOVA) and showed how an **ANOVA table** could be used to provide a convenient summary of the computational aspects of analysis of variance. A similar ANOVA table can be used to summarize the results of the F test for significance in regression. Table 12.5 is the general form of the ANOVA table for simple linear regression. Table 12.6 is the ANOVA table with the F test computations performed for Armand's Pizza Parlors. Regression, Error, and Total are the labels for the three sources of variation, with SSR, SSE, and SST appearing as the corresponding sum of squares in column 2. The degrees of freedom, 1 for SSR, $n - 2$ for SSE, and $n - 1$ for SST, are shown in column 3. Column 4 contains the values of MSR and MSE, column 5 contains the value of $F = \text{MSR}/\text{MSE}$, and column 6 contains the p-value corresponding to the F value in column 5. Almost all computer printouts of regression analysis include an ANOVA table summary of the F test for significance.

12.5 Testing for Significance

TABLE 12.5 GENERAL FORM OF THE ANOVA TABLE FOR SIMPLE LINEAR REGRESSION

In every analysis of variance table the total sum of squares is the sum of the regression sum of squares and the error sum of squares; in addition, the total degrees of freedom is the sum of the regression degrees of freedom and the error degrees of freedom.

Source of Variation	Sum of Squares	Degrees of Freedom	Mean Square	F	p-value
Regression	SSR	1	$\text{MSR} = \dfrac{\text{SSR}}{1}$	$F = \dfrac{\text{MSR}}{\text{MSE}}$	
Error	SSE	$n - 2$	$\text{MSE} = \dfrac{\text{SSE}}{n - 2}$		
Total	SST	$n - 1$			

Some Cautions About the Interpretation of Significance Tests

Regression analysis, which can be used to identify how variables are associated with one another, cannot be used as evidence of a cause-and-effect relationship.

Rejecting the null hypothesis $H_0: \beta_1 = 0$ and concluding that the relationship between x and y is significant do not enable us to conclude that a cause-and-effect relationship is present between x and y. Concluding a cause-and-effect relationship is warranted only if the analyst can provide some type of theoretical justification that the relationship is in fact causal. In the Armand's Pizza Parlors example, we can conclude that there is a significant relationship between the size of the student population x and quarterly sales y; moreover, the estimated regression equation $\hat{y} = 60 + 5x$ provides the least squares estimate of the relationship. We cannot, however, conclude that changes in student population x *cause* changes in quarterly sales y just because we identified a statistically significant relationship. The appropriateness of such a cause-and-effect conclusion is left to supporting theoretical justification and to good judgment on the part of the analyst. Armand's managers felt that increases in the student population were a likely cause of increased quarterly sales. Thus, the result of the significance test enabled them to conclude that a cause-and-effect relationship was present.

In addition, just because we are able to reject $H_0: \beta_1 = 0$ and demonstrate statistical significance does not enable us to conclude that the relationship between x and y is linear. We can state only that x and y are related and that a linear relationship explains a significant portion of the variability in y over the range of values for x observed in the sample. Figure 12.7 illustrates this situation. The test for significance calls for the rejection of the null hypothesis $H_0: \beta_1 = 0$ and leads to the conclusion that x and y are significantly related, but the figure shows that the actual relationship between x and y is not linear. Although the

TABLE 12.6 ANOVA TABLE FOR THE ARMAND'S PIZZA PARLORS PROBLEM

Source of Variation	Sum of Squares	Degrees of Freedom	Mean Square	F	p-value
Regression	14,200	1	$\dfrac{14{,}200}{1} = 14{,}200$	$\dfrac{14{,}200}{191.25} = 74.25$	.000
Error	1,530	8	$\dfrac{1530}{8} = 191.25$		
Total	15,730	9			

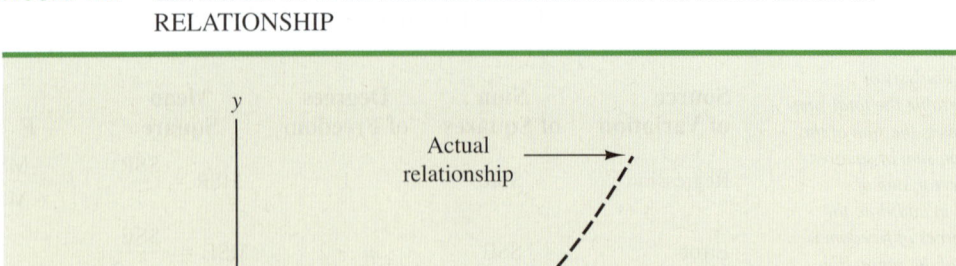

FIGURE 12.7 EXAMPLE OF A LINEAR APPROXIMATION OF A NONLINEAR RELATIONSHIP

linear approximation provided by $\hat{y} = b_0 + b_1 x$ is good over the range of x values observed in the sample, it becomes poor for x values outside that range.

Given a significant relationship, we should feel confident in using the estimated regression equation for predictions corresponding to x values within the range of the x values observed in the sample. For Armand's Pizza Parlors, this range corresponds to values of x between 2 and 26. Unless other reasons indicate that the model is valid beyond this range, predictions outside the range of the independent variable should be made with caution. For Armand's Pizza Parlors, because the regression relationship has been found significant at the .01 level, we should feel confident using it to predict sales for restaurants where the associated student population is between 2000 and 26,000.

NOTES AND COMMENTS

1. The assumptions made about the error term (Section 12.4) are what allow the tests of statistical significance in this section. The properties of the sampling distribution of b_1 and the subsequent t and F tests follow directly from these assumptions.
2. Do not confuse statistical significance with practical significance. With very large sample sizes, statistically significant results can be obtained for small values of b_1; in such cases, one must exercise care in concluding that the relationship has practical significance.
3. A test of significance for a linear relationship between x and y can also be performed by using the sample correlation coefficient r_{xy}. With ρ_{xy} denoting the population correlation coefficient, the hypotheses are as follows.

$$H_0: \rho_{xy} = 0$$
$$H_a: \rho_{xy} \neq 0$$

A significant relationship can be concluded if H_0 is rejected. The details of this test are provided in more advanced texts. However, the t and F tests presented previously in this section give the same result as the test for significance using the correlation coefficient. Conducting a test for significance using the correlation coefficient therefore is not necessary if a t or F test has already been conducted.

12.5 Testing for Significance

Exercises

Methods

23. The data from exercise 1 follow.

x_i	1	2	3	4	5
y_i	3	7	5	11	14

    a. Compute the mean square error using equation (12.15).
    b. Compute the standard error of the estimate using equation (12.16).
    c. Compute the estimated standard deviation of b_1 using equation (12.18).
    d. Use the t test to test the following hypotheses ($\alpha = .05$):

    $$H_0: \beta_1 = 0$$
    $$H_a: \beta_1 \neq 0$$

    e. Use the F test to test the hypotheses in part (d) at a .05 level of significance. Present the results in the analysis of variance table format.

24. The data from exercise 2 follow.

x_i	3	12	6	20	14
y_i	55	40	55	10	15

    a. Compute the mean square error using equation (12.15).
    b. Compute the standard error of the estimate using equation (12.16).
    c. Compute the estimated standard deviation of b_1 using equation (12.18).
    d. Use the t test to test the following hypotheses ($\alpha = .05$):

    $$H_0: \beta_1 = 0$$
    $$H_a: \beta_1 \neq 0$$

    e. Use the F test to test the hypotheses in part (d) at a .05 level of significance. Present the results in the analysis of variance table format.

25. The data from exercise 3 follow.

x_i	2	6	9	13	20
y_i	7	18	9	26	23

    a. What is the value of the standard error of the estimate?
    b. Test for a significant relationship by using the t test. Use $\alpha = .05$.
    c. Use the F test to test for a significant relationship. Use $\alpha = .05$. What is your conclusion?

Applications

26. In exercise 18 the data on grade point average and monthly salary were as follows.

GPA	Monthly Salary ($)	GPA	Monthly Salary ($)
2.6	3300	3.2	3500
3.4	3600	3.5	3900
3.6	4000	2.9	3600

a. Does the t test indicate a significant relationship between grade point average and monthly salary? What is your conclusion? Use $\alpha = .05$.
b. Test for a significant relationship using the F test. What is your conclusion? Use $\alpha = .05$.
c. Show the ANOVA table.

27. The number of megapixels in a digital camera is one of the most important factors in determining picture quality. But, do digital cameras with more megapixels cost more? The following data show the number of megapixels and the price ($) for 10 digital cameras (*Consumer Reports*, March 2009).

Brand and Model	Megapixels	Price ($)
Canon PowerShot SD1100 IS	8	180
Casio Exilim Card EX-S10	10	200
Sony Cyber-shot DSC-T70	7	230
Pentax Optio M50	8	120
Canon PowerShot G10	15	470
Canon PowerShot A590 IS	8	140
Canon PowerShot E1	10	180
Fujifilm FinePix F00FD	12	310
Sony Cyber-shot DSC-W170	10	250
Canon PowerShot A470	7	110

a. Use these data to develop an estimated regression equation that can be used to predict the price of a digital camera given the number of megapixels.
b. At the .05 level of significance, are the number of megapixels and the price related? Explain.
c. Would you feel comfortable using the estimated regression equation developed in part (a) to predict the price of a digital camera given the number of megapixels? Explain.
d. The Kodak EasyShare Z1012 IS digital camera has 10 megapixels. Predict the price of this camera using the estimated regression equation developed in part (a).

28. In exercise 8, data on x = temperature rating (F°) and y = price ($) for 11 sleeping bags manufactured by Bergans of Norway provided the estimated regression equation $\hat{y} = 359.2668 - 5.2772x$. At the .05 level of significance, test whether temperature rating and price are related. Show the ANOVA table. What is your conclusion?

29. Refer to exercise 21, where data on production volume and cost were used to develop an estimated regression equation relating production volume and cost for a particular manufacturing operation. Use $\alpha = .05$ to test whether the production volume is significantly related to the total cost. Show the ANOVA table. What is your conclusion?

30. Refer to excercise 5, where the following data were used to investigate whether higher prices are generally associated with higher ratings for elliptical trainers (*Consumer Reports*, February 2008).

Brand and Model	Price ($)	Rating
Precor 5.31	3700	87
Keys Fitness CG2	2500	84
Octane Fitness Q37e	2800	82
LifeFitness X1 Basic	1900	74
NordicTrack AudioStrider 990	1000	73
Schwinn 430	800	69
Vision Fitness X6100	1700	68
ProForm XP 520 Razor	600	55

With x = price ($) and y = rating, the estimated regression equation is $\hat{y} = 58.158 + .008449x$. For these data, SSE = 173.88 and SST = 756. Does the evidence indicate a significant relationship between price and rating?

31. In exercise 20, data on x = price ($) and y = overall score for ten 42-inch plasma televisions tested by *Consumer Reports* provided the estimated regression equation $\hat{y} = 12.0169 + .0127x$. For these data SSE = 540.04 and SST = 982.40. Use the F test to determine whether the price for a 42-inch plasma television and the overall score are related at the .05 level of significance.

12.6 Using the Estimated Regression Equation for Estimation and Prediction

When using the simple linear regression model we are making an assumption about the relationship between x and y. We then use the least squares method to obtain the estimated simple linear regression equation. If a significant relationship exists between x and y, and the coefficient of determination shows that the fit is good, the estimated regression equation should be useful for estimation and prediction.

Point Estimation

In the Armand's Pizza Parlors example, the estimated regression equation $\hat{y} = 60 + 5x$ provides an estimate of the relationship between the size of the student population x and quarterly sales y. We can use the estimated regression equation to develop a point estimate of the mean value of y for a particular value of x or to predict an individual value of y corresponding to a given value of x. For instance, suppose Armand's managers want a point estimate of the mean quarterly sales for all restaurants located near college campuses with 10,000 students. Using the estimated regression equation $\hat{y} = 60 + 5x$, we see that for $x = 10$ (or 10,000 students), $\hat{y} = 60 + 5(10) = 110$. Thus, a point estimate of the mean quarterly sales for all restaurants located near campuses with 10,000 students is $110,000.

Now suppose Armand's managers want to predict sales for an individual restaurant located near Talbot College, a school with 10,000 students. In this case we are not interested in the mean value for all restaurants located near campuses with 10,000 students; we are just interested in predicting quarterly sales for one individual restaurant. As it turns out, the point estimate for an individual value of y is the same as the point estimate for the mean value of y. Hence, we would predict quarterly sales of $\hat{y} = 60 + 5(10) = 110$ or $110,000 for this one restaurant.

Interval Estimation

Confidence intervals and prediction intervals show the precision of the regression results. Narrower intervals provide a higher degree of precision.

Point estimates do not provide any information about the precision associated with an estimate. For that we must develop interval estimates much like those in Chapters 8, 10, and 11. The first type of interval estimate, a **confidence interval**, is an interval estimate of the *mean value of y* for a given value of x. The second type of interval estimate, a **prediction interval**, is used whenever we want an interval estimate of an *individual value of y* for a given value of x. The point estimate of the mean value of y is the same as the prediction of an individual value of y. But the interval estimates we obtain for the two cases are different. The margin of error is larger for a prediction interval.

Confidence Interval for the Mean Value of y

The estimated regression equation provides a point estimate of the mean value of y for a given value of x. In developing the confidence interval, we will use the following notation.

x_p = the particular or given value of the independent variable x
y_p = the value of the dependent variable y corresponding to the given x_p
$E(y_p)$ = the mean or expected value of the dependent variable y corresponding to the given x_p
$\hat{y}_p = b_0 + b_1 x_p$ = the point estimate of $E(y_p)$ when $x = x_p$

Using this notation to estimate the mean sales for all Armand's restaurants located near a campus with 10,000 students, we have $x_p = 10$, and $E(y_p)$ denotes the unknown mean value of sales for all restaurants where $x_p = 10$. The point estimate of $E(y_p)$ is provided by $\hat{y}_p = 60 + 5(10) = 110$.

In general, we cannot expect $\hat{y}_p$ to equal $E(y_p)$ exactly. If we want to make an inference about how close $\hat{y}_p$ is to the true mean value $E(y_p)$, we will have to estimate the variance of $\hat{y}_p$. The formula for estimating the variance of $\hat{y}_p$ given x_p, denoted by $s_{\hat{y}_p}^2$, is

$$s_{\hat{y}_p}^2 = s^2 \left[\frac{1}{n} + \frac{(x_p - \bar{x})^2}{\Sigma(x_i - \bar{x})^2} \right] \tag{12.22}$$

The estimate of the standard deviation of $\hat{y}_p$ is given by the square root of equation (12.22).

$$s_{\hat{y}_p} = s \sqrt{\frac{1}{n} + \frac{(x_p - \bar{x})^2}{\Sigma(x_i - \bar{x})^2}} \tag{12.23}$$

The computational results for Armand's Pizza Parlors in Section 12.5 provided $s = 13.829$. With $x_p = 10$, $\bar{x} = 14$, and $\Sigma(x_i - \bar{x})^2 = 568$, we can use equation (12.23) to obtain

$$s_{\hat{y}_p} = 13.829 \sqrt{\frac{1}{10} + \frac{(10-14)^2}{568}}$$
$$= 13.829 \sqrt{.1282} = 4.95$$

The general expression for a confidence interval follows.

> **CONFIDENCE INTERVAL FOR $E(y_p)$**
>
> $$\hat{y}_p \pm t_{\alpha/2} s_{\hat{y}_p} \tag{12.24}$$
>
> where the confidence coefficient is $1 - \alpha$ and $t_{\alpha/2}$ is based on a t distribution with $n - 2$ degrees of freedom.

The margin of error associated with this confidence interval is $t_{\alpha/2} s_{\hat{y}_p}$.

Using expression (12.24) to develop a 95% confidence interval of the mean quarterly sales for all Armand's restaurants located near campuses with 10,000 students, we need the value of t for $\alpha/2 = .025$ and $n - 2 = 10 - 2 = 8$ degrees of freedom. Using Table 2 of Appendix B, we have $t_{.025} = 2.306$. Thus, with $\hat{y}_p = 110$ and a margin of error of $t_{\alpha/2} s_{\hat{y}_p} = 2.306(4.95) = 11.415$, the 95% confidence interval estimate is

$$110 \pm 11.415$$

FIGURE 12.8 CONFIDENCE INTERVALS FOR THE MEAN SALES y AT GIVEN VALUES OF STUDENT POPULATION x

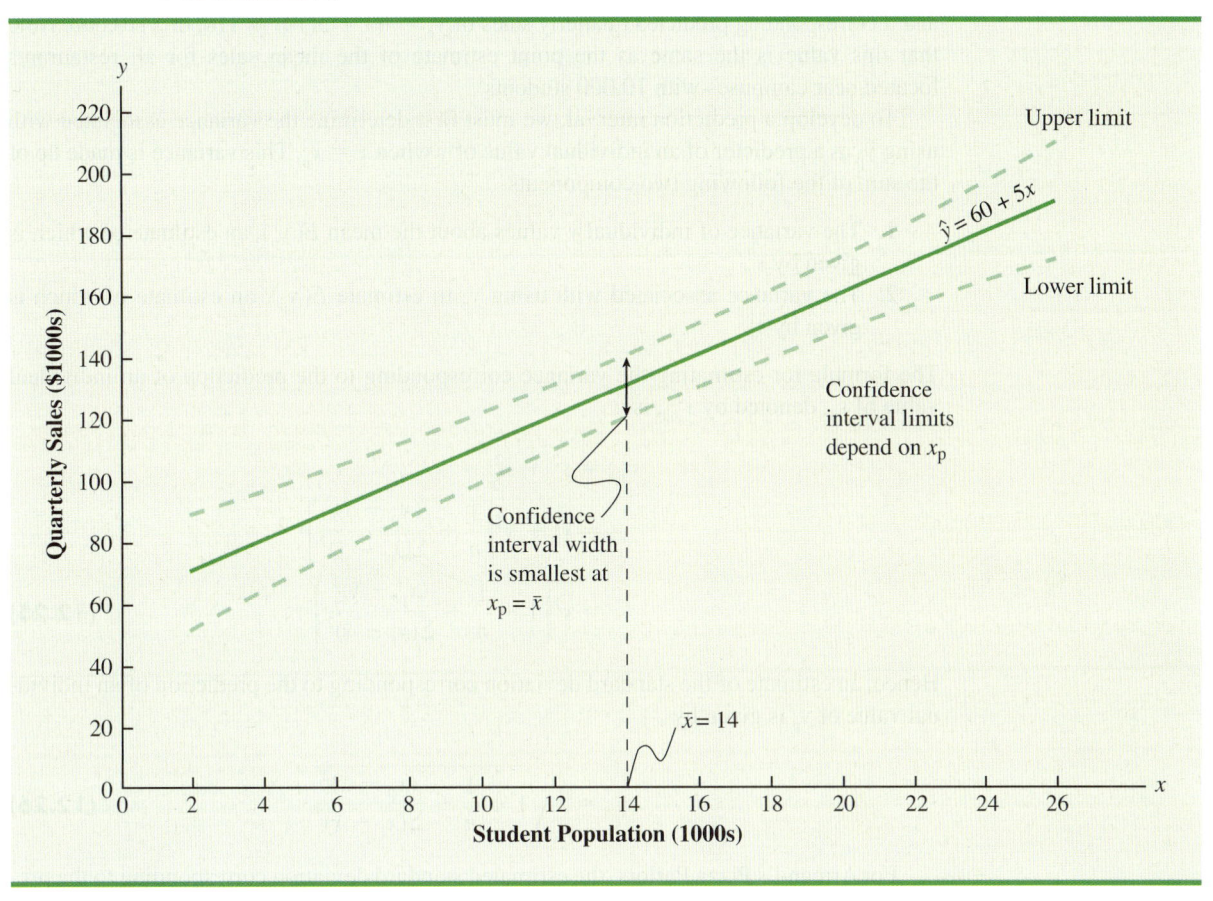

In dollars, the 95% confidence interval for the mean quarterly sales of all restaurants near campuses with 10,000 students is $110,000 ± $11,415. Therefore, the 95% confidence interval for the mean quarterly sales when the student population is 10,000 is $98,585 to $121,415.

Note that the estimated standard deviation of $\hat{y}_p$ given by equation (12.23) is smallest when $x_p = \bar{x}$ and the quantity $x_p − \bar{x} = 0$. In this case, the estimated standard deviation of $\hat{y}_p$ becomes

$$s_{\hat{y}_p} = s\sqrt{\frac{1}{n} + \frac{(\bar{x} − \bar{x})^2}{\Sigma(x_i − \bar{x})^2}} = s\sqrt{\frac{1}{n}}$$

This result implies that we can make the best or most precise estimate of the mean value of y whenever $x_p = \bar{x}$. In fact, the further x_p is from $\bar{x}$ the larger $x_p − \bar{x}$ becomes. As a result, confidence intervals for the mean value of y will become wider as x_p deviates more from $\bar{x}$. This pattern is shown graphically in Figure 12.8.

Prediction Interval for an Individual Value of y

Suppose that instead of estimating the mean value of sales for all Armand's restaurants located near campuses with 10,000 students, we want to predict the sales for an individual restaurant located near Talbot College, a school with 10,000 students. As noted previously, the predictor of y_p, the value of y corresponding to the given x_p, is provided by the estimated regression equation $\hat{y}_p = b_0 + b_1 x_p$. For the restaurant at Talbot College, we have $x_p = 10$ and a corresponding predicted quarterly sales of $\hat{y}_p = 60 + 5(10) = 110$, or $110,000. Note that this value is the same as the point estimate of the mean sales for all restaurants located near campuses with 10,000 students.

To develop a prediction interval, we must first determine the variance associated with using $\hat{y}_p$ as a predictor of an individual value of y when $x = x_p$. This variance is made up of the sum of the following two components.

1. The variance of individual y values about the mean $E(y_p)$, an estimate of which is given by s^2
2. The variance associated with using $\hat{y}_p$ to estimate $E(y_p)$, an estimate of which is given by $s_{\hat{y}_p}^2$

The formula for estimating the variance corresponding to the prediction of an individual value of y_p, denoted by s_{ind}^2, is

$$\begin{aligned} s_{ind}^2 &= s^2 + s_{\hat{y}_p}^2 \\ &= s^2 + s^2 \left[\frac{1}{n} + \frac{(x_p - \bar{x})^2}{\Sigma(x_i - \bar{x})^2} \right] \\ &= s^2 \left[1 + \frac{1}{n} + \frac{(x_p - \bar{x})^2}{\Sigma(x_i - \bar{x})^2} \right] \end{aligned} \quad (12.25)$$

Hence, an estimate of the standard deviation corresponding to the prediction of an individual value of y_p is given by

$$s_{ind} = s \sqrt{1 + \frac{1}{n} + \frac{(x_p - \bar{x})^2}{\Sigma(x_i - \bar{x})^2}} \quad (12.26)$$

For Armand's Pizza Parlors, the estimated standard deviation corresponding to the prediction of sales for one specific restaurant located near a campus with 10,000 students is computed as follows.

$$\begin{aligned} s_{ind} &= 13.829 \sqrt{1 + \frac{1}{10} + \frac{(10 - 14)^2}{568}} \\ &= 13.829 \sqrt{1.1282} \\ &= 14.69 \end{aligned}$$

The general expression for a prediction interval follows.

> **PREDICTION INTERVAL FOR y_p**
>
> $$\hat{y}_p \pm t_{\alpha/2} s_{ind} \quad (12.27)$$
>
> where the confidence coefficient is $1 - \alpha$ and $t_{\alpha/2}$ is based on a t distribution with $n - 2$ degrees of freedom.

The margin of error associated with this prediction interval is $t_{\alpha/2} s_{ind}$.

The 95% prediction interval for quarterly sales at Armand's Talbot College restaurant can be found by using $t_{.025} = 2.306$ and $s_{ind} = 14.69$. Thus, with $\hat{y}_p = 110$ and a margin of error of $t_{\alpha/2} s_{ind} = 2.306(14.69) = 33.875$, the 95% prediction interval is

$$110 \pm 33.875$$

12.6 Using the Estimated Regression Equation for Estimation and Prediction   521

FIGURE 12.9 CONFIDENCE AND PREDICTION INTERVALS FOR SALES y AT GIVEN VALUES OF STUDENT POPULATION x

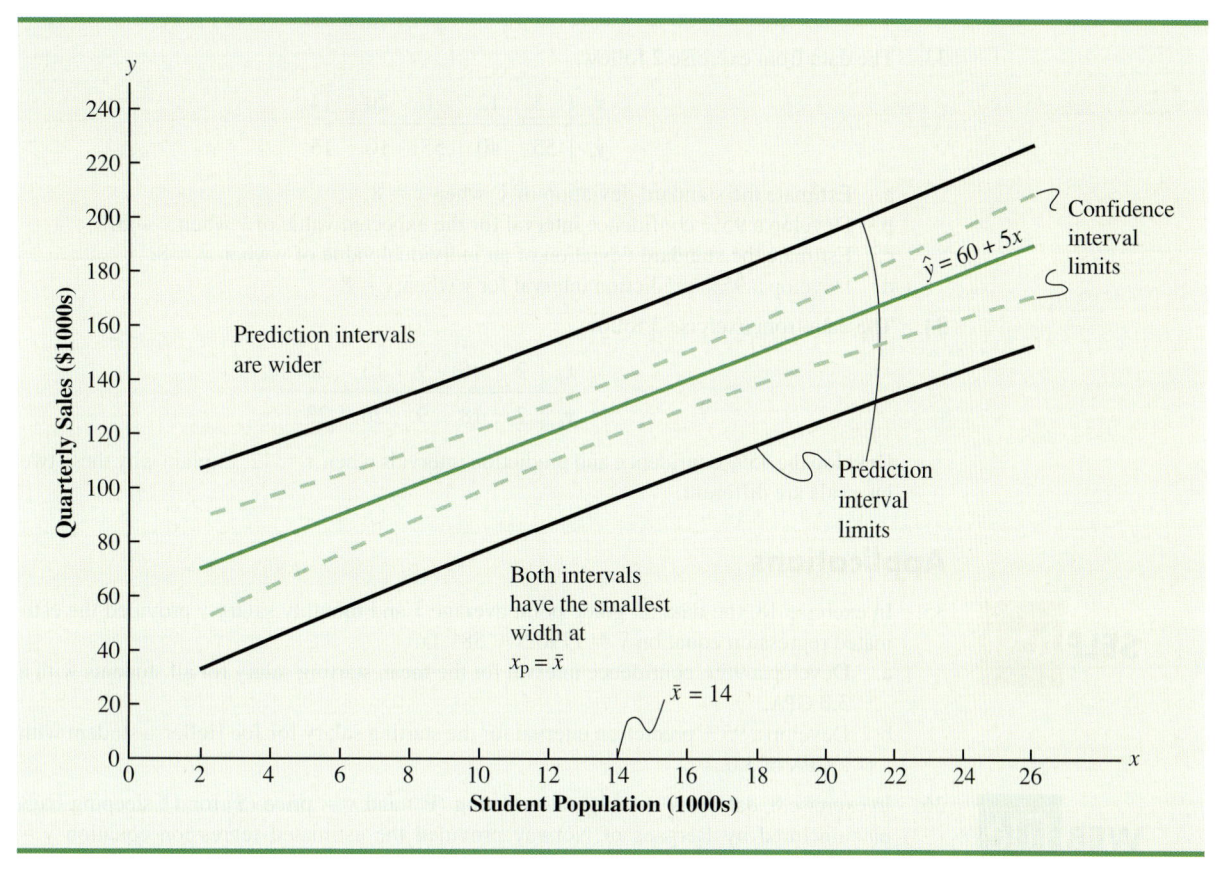

In dollars, this prediction interval is $110,000 ± $33,875 or $76,125 to $143,875. Note that the prediction interval for an individual restaurant located near a campus with 10,000 students is wider than the confidence interval for the mean sales of all restaurants located near campuses with 10,000 students. The difference reflects the fact that we are able to estimate the mean value of y more precisely than we can predict an individual value of y.

Both confidence interval estimates and prediction interval estimates are most precise when the value of the independent variable is $x_p = \bar{x}$. The general shapes of confidence intervals and the wider prediction intervals are shown together in Figure 12.9.

In general, the lines for the confidence interval limits and the prediction interval limits both have curvature.

Exercises

Methods

32. The data from exercise 1 follow.

x_i	1	2	3	4	5
y_i	3	7	5	11	14

a. Use equation (12.23) to estimate the standard deviation of $\hat{y}_p$ when $x = 4$.
b. Use expression (12.24) to develop a 95% confidence interval for the expected value of y when $x = 4$.

c. Use equation (12.26) to estimate the standard deviation of an individual value of y when $x = 4$.
d. Use expression (12.27) to develop a 95% prediction interval for y when $x = 4$.

33. The data from exercise 2 follow.

x_i	3	12	6	20	14
y_i	55	40	55	10	15

a. Estimate the standard deviation of $\hat{y}_p$ when $x = 8$.
b. Develop a 95% confidence interval for the expected value of y when $x = 8$.
c. Estimate the standard deviation of an individual value of y when $x = 8$.
d. Develop a 95% prediction interval for y when $x = 8$.

34. The data from exercise 3 follow.

x_i	2	6	9	13	20
y_i	7	18	9	26	23

Develop the 95% confidence and prediction intervals when $x = 12$. Explain why these two intervals are different.

Applications

35. In exercise 18, the data on grade point average x and monthly salary y provided the estimated regression equation $\hat{y} = 1790.5 + 581.1x$.
   a. Develop a 95% confidence interval for the mean starting salary for all students with a 3.0 GPA.
   b. Develop a 95% prediction interval for the starting salary for Joe Heller, a student with a GPA of 3.0.

36. In exercise 8, data on $x =$ temperature rating (F°) and $y =$ price ($) for 11 sleeping bags manufactured by Bergans of Norway provided the estimated regression equation $\hat{y} = 359.2668 - 5.2772x$. For these data $s = 37.9372$.
   a. Develop a point estimate of the price for a sleeping bag with a temperature rating of 30.
   b. Develop a 95% confidence interval for the mean overall temperature rating for all sleeping bags with a temperature rating of 30.
   c. Suppose that Bergans developed a new model with a temperature rating of 30. Develop a 95% prediction interval for the price of this new model.
   d. Discuss the differences in your answers to parts (b) and (c).

37. In exercise 13, data were given on the adjusted gross income x and the amount of itemized deductions taken by taxpayers. Data were reported in thousands of dollars. With the estimated regression equation $\hat{y} = 4.68 + .16x$, the point estimate of a reasonable level of total itemized deductions for a taxpayer with an adjusted gross income of $52,500 is $13,080.
   a. Develop a 95% confidence interval for the mean amount of total itemized deductions for all taxpayers with an adjusted gross income of $52,500.
   b. Develop a 95% prediction interval estimate for the amount of total itemized deductions for a particular taxpayer with an adjusted gross income of $52,500.
   c. If the particular taxpayer referred to in part (b) claimed total itemized deductions of $20,400, would the IRS agent's request for an audit appear to be justified?
   d. Use your answer to part (b) to give the IRS agent a guideline as to the amount of total itemized deductions a taxpayer with an adjusted gross income of $52,500 should claim before an audit is recommended.

38. Refer to Exercise 21, where data on the production volume x and total cost y for a particular manufacturing operation were used to develop the estimated regression equation $\hat{y} = 1246.67 + 7.6x$.
   a. The company's production schedule shows that 500 units must be produced next month. What is the point estimate of the total cost for next month?

b. Develop a 99% prediction interval for the total cost for next month.
c. If an accounting cost report at the end of next month shows that the actual production cost during the month was $6000, should managers be concerned about incurring such a high total cost for the month? Discuss.

39. Almost all U.S. light-rail systems use electric cars that run on tracks built at street level. The Federal Transit Administration claims light-rail is one of the safest modes of travel, with an accident rate of .99 accidents per million passenger miles as compared to 2.29 for buses. The following data show the miles of track and the weekday ridership in thousands of passengers for six light-rail systems (*USA Today*, January 7, 2003).

City	Miles of Track	Ridership (1000s)
Cleveland	15	15
Denver	17	35
Portland	38	81
Sacramento	21	31
San Diego	47	75
San Jose	31	30
St. Louis	34	42

a. Use these data to develop an estimated regression equation that could be used to predict the ridership given the miles of track.
b. Did the estimated regression equation provide a good fit? Explain.
c. Develop a 95% confidence interval for the mean weekday ridership for all light-rail systems with 30 miles of track.
d. Suppose that Charlotte is considering construction of a light-rail system with 30 miles of track. Develop a 95% prediction interval for the weekday ridership for the Charlotte system. Do you think that the prediction interval you developed would be of value to Charlotte planners in anticipating the number of weekday riders for their new light-rail system? Explain.

12.7 Computer Solution

Performing the regression analysis computations without the help of a computer can be quite time consuming. In this section we discuss how the computational burden can be minimized by using a computer software package such as Minitab.

We entered Armand's student population and sales data into a Minitab worksheet. The independent variable was named Pop and the dependent variable was named Sales to assist with interpretation of the computer output. Using Minitab, we obtained the printout for Armand's Pizza Parlors shown in Figure 12.10.[2] The interpretation of this printout follows.

1. Minitab prints the estimated regression equation as Sales = 60.0 + 5.00 Pop.
2. A table is printed that shows the values of the coefficients b_0 and b_1, the standard deviation of each coefficient, the t value obtained by dividing each coefficient value by its standard deviation, and the p-value associated with the t test. Because the p-value is zero (to three decimal places), the sample results indicate that the null hypothesis (H_0: $\beta_1 = 0$) should be rejected. Alternatively, we could compare 8.62 (located in the t-ratio column) to the appropriate critical value. This procedure for the t test was described in Section 12.5.

---
[2]The Minitab steps necessary to generate the output are given in Appendix 12.3.

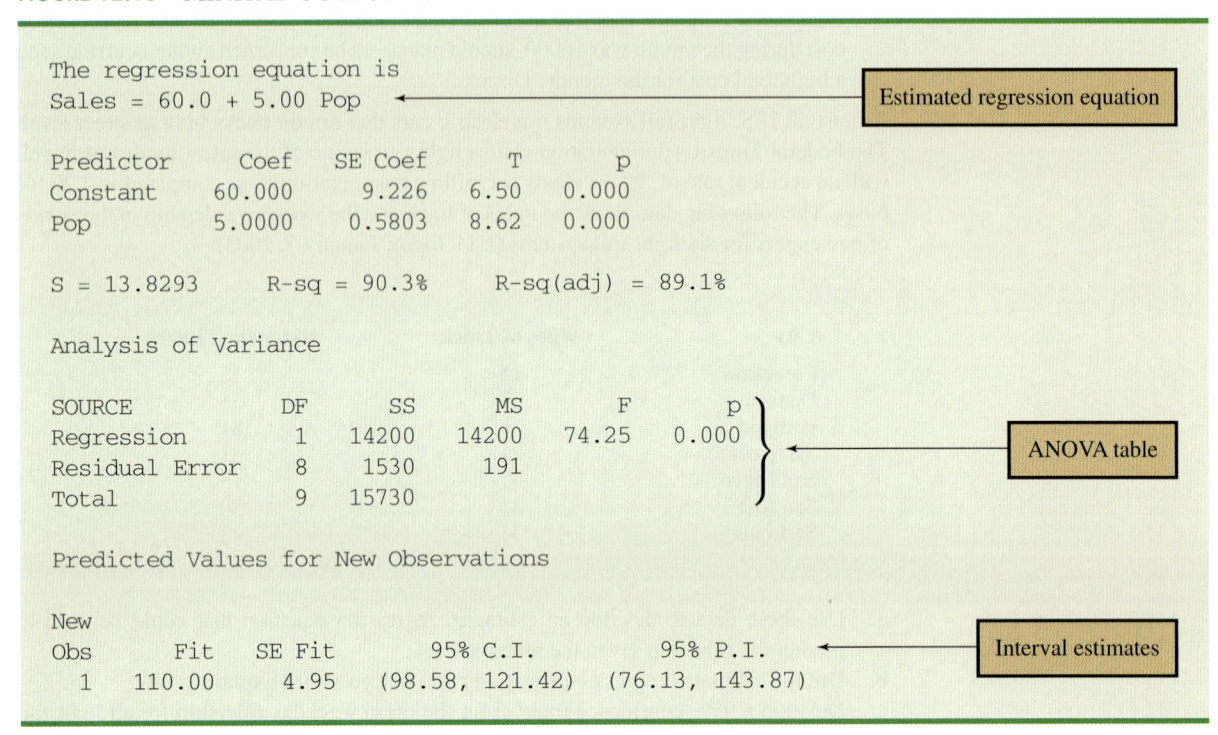

FIGURE 12.10 MINITAB OUTPUT FOR THE ARMAND'S PIZZA PARLORS PROBLEM

```
The regression equation is
Sales = 60.0 + 5.00 Pop          ◄──────  Estimated regression equation

Predictor      Coef     SE Coef      T        p
Constant     60.000      9.226    6.50    0.000
Pop           5.0000     0.5803   8.62    0.000

S = 13.8293     R-sq = 90.3%     R-sq(adj) = 89.1%

Analysis of Variance

SOURCE            DF       SS       MS       F        p  ⎫
Regression         1    14200    14200    74.25    0.000 ⎬  ◄── ANOVA table
Residual Error     8     1530      191                   ⎪
Total              9    15730                            ⎭

Predicted Values for New Observations

New
Obs     Fit    SE Fit       95% C.I.           95% P.I.       ◄── Interval estimates
 1    110.00    4.95    (98.58, 121.42)    (76.13, 143.87)
```

3. Minitab prints the standard error of the estimate, $s = 13.8293$, as well as information about the goodness of fit. Note that "R-sq = 90.3%" is the coefficient of determination expressed as a percentage. The value "R-Sq(adj) = 89.1%" is discussed in Chapter 13.
4. The ANOVA table is printed below the heading Analysis of Variance. Minitab uses the label Residual Error for the error source of variation. Note that DF is an abbreviation for degrees of freedom and that MSR is given as 14,200 and MSE as 191. The ratio of these two values provides the F value of 74.25 and the corresponding p-value of 0.000. Because the p-value is zero (to three decimal places), the relationship between Sales and Pop is judged statistically significant.
5. The 95% confidence interval estimate of the expected sales and the 95% prediction interval estimate of sales for an individual restaurant located near a campus with 10,000 students are printed below the ANOVA table. The confidence interval is (98.58, 121.42) and the prediction interval is (76.13, 143.87) as we showed in Section 12.6.

Exercises

Applications

40. The commercial division of a real estate firm is conducting a regression analysis of the relationship between x, annual gross rents (in thousands of dollars), and y, selling price (in thousands of dollars) for apartment buildings. Data were collected on several properties recently sold and the following computer output was obtained.

```
The regression equation is
Y = 20.0 + 7.21 X

Predictor      Coef      SE Coef        T
Constant     20.000      3.2213       6.21
X             7.210      1.3626       5.29

Analysis of Variance

SOURCE            DF        SS
Regression         1    41587.3
Residual Error     7
Total              8    51984.1
```

a. How many apartment buildings were in the sample?
b. Write the estimated regression equation.
c. What is the value of s_{b_1}?
d. Use the F statistic to test the significance of the relationship at a .05 level of significance.
e. Estimate the selling price of an apartment building with gross annual rents of $50,000.

41. Following is a portion of the computer output for a regression analysis relating y = maintenance expense (dollars per month) to x = usage (hours per week) of a particular brand of computer terminal.

```
The regression equation is
Y = 6.1092 + .8951 X

Predictor      Coef      SE Coef
Constant     6.1092      0.9361
X            0.8951      0.1490

Analysis of Variance

SOURCE            DF        SS         MS
Regression         1    1575.76    1575.76
Residual Error     8     349.14      43.64
Total              9    1924.90
```

a. Write the estimated regression equation.
b. Use a t test to determine whether monthly maintenance expense is related to usage at the .05 level of significance.
c. Use the estimated regression equation to predict monthly maintenance expense for any terminal that is used 25 hours per week.

42. A regression model relating x, number of salespersons at a branch office, to y, annual sales at the office (in thousands of dollars), provided the following computer output from a regression analysis of the data.

```
The regression equation is
Y = 80.0 + 50.00 X

Predictor         Coef      SE Coef        T
Constant          80.0       11.333      7.06
X                 50.0        5.482      9.12

Analysis of Variance

SOURCE            DF          SS          MS
Regression         1        6828.6      6828.6
Residual Error    28        2298.8        82.1
Total             29        9127.4
```

a. Write the estimated regression equation.
b. How many branch offices were involved in the study?
c. Compute the F statistic and test the significance of the relationship at a .05 level of significance.
d. Predict the annual sales at the Memphis branch office. This branch employs 12 salespersons.

43. Out-of-state tuition and fees at the top graduate schools of business can be very expensive, but the starting salary and bonus paid to graduates from many of these schools can be substantial. The following data show the out-of-state tuition and fees (rounded to the nearest $1000) and the average starting salary and bonus paid to recent graduates (rounded to the nearest $1000) for a sample of 20 graduate schools of business (*U.S. News & World Report 2009 Edition America's Best Graduate Schools*).

School	Tuition & Fees ($1000s)	Salary & Bonus ($1000s)
Arizona State University	28	98
Babson College	35	94
Cornell University	44	119
Georgetown University	40	109
Georgia Institute of Technology	30	88
Indiana University—Bloomington	35	105
Michigan State University	26	99
Northwestern University	44	123
Ohio State University	35	97
Purdue University—West Lafayette	33	96
Rice University	36	102
Stanford University	46	135
University of California—Davis	35	89
University of Florida	23	71
University of Iowa	25	78
University of Minnesota—Twin Cities	37	100
University of Notre Dame	36	95
University of Rochester	38	99
University of Washington	30	94
University of Wisconsin—Madison	27	93

a. Develop a scatter diagram with salary and bonus as the dependent variable.
b. Does there appear to be any relationship between these variables? Explain.

12.8 Residual Analysis: Validating Model Assumptions

c. Develop an estimated regression equation that can be used to predict the starting salary and bonus paid to graduates given the cost of out-of-state tuition and fees at the school.
d. Test for a significant relationship at the .05 level of significance. What is your conclusion?
e. Did the estimated regression equation provide a good fit? Explain.
f. Suppose that we randomly select a recent graduate of the University of Virginia graduate school of business. The school has an out-of-state tuition and fees of $43,000. Estimate the starting salary and bonus for this graduate.

44. Automobile racing, high-performance driving schools, and driver education programs run by automobile clubs continue to grow in popularity. All these activities require the participant to wear a helmet that is certified by the Snell Memorial Foundation, a not-for-profit organization dedicated to research, education, testing, and development of helmet safety standards. Snell "SA" (Sports Application) rated professional helmets are designed for auto racing and provide extreme impact resistance and high fire protection. One of the key factors in selecting a helmet is weight, since lower weight helmets tend to place less stress on the neck. The following data show the weight and price for 18 SA helmets (SoloRacer website, April 20, 2008).

Helmet	Weight (oz)	Price ($)
Pyrotect Pro Airflow	64	248
Pyrotect Pro Airflow Graphics	64	278
RCi Full Face	64	200
RaceQuip RidgeLine	64	200
HJC AR-10	58	300
HJC Si-12	47	700
HJC HX-10	49	900
Impact Racing Super Sport	59	340
Zamp FSA-1	66	199
Zamp RZ-2	58	299
Zamp RZ-2 Ferrari	58	299
Zamp RZ-3 Sport	52	479
Zamp RZ-3 Sport Painted	52	479
Bell M2	63	369
Bell M4	62	369
Bell M4 Pro	54	559
G Force Pro Force 1	63	250
G Force Pro Force 1 Grafx	63	280

a. Develop a scatter diagram with weight as the independent variable.
b. Does there appear to be any relationship between these two variables?
c. Develop the estimated regression equation that could be used to predict the price given the weight.
d. Test for the significance of the relationship at the .05 level of significance.
e. Did the estimated regression equation provide a good fit? Explain.

12.8 Residual Analysis: Validating Model Assumptions

As we noted previously, the *residual* for observation i is the difference between the observed value of the dependent variable (y_i) and the estimated value of the dependent variable ($\hat{y}_i$).

RESIDUAL FOR OBSERVATION i

$$y_i - \hat{y}_i \tag{12.28}$$

where

y_i is the observed value of the dependent variable
$\hat{y}_i$ is the estimated value of the dependent variable

Residual analysis *is the primary tool for determining whether the assumed regression model is appropriate.*

In other words, the *i*th residual is the error resulting from using the estimated regression equation to predict the value of the dependent variable. The residuals for the Armand's Pizza Parlors example are computed in Table 12.7. The observed values of the dependent variable are in the second column and the estimated values of the dependent variable, obtained using the estimated regression equation $\hat{y} = 60 + 5x$, are in the third column. An analysis of the corresponding residuals in the fourth column will help determine whether the assumptions made about the regression model are appropriate.

Let us now review the regression assumptions for the Armand's Pizza Parlors example. A simple linear regression model was assumed.

$$y = \beta_0 + \beta_1 x + \epsilon \tag{12.29}$$

This model indicates that we assumed quarterly sales (y) to be a linear function of the size of the student population (x) plus an error term ϵ. In Section 12.4 we made the following assumptions about the error term ϵ.

1. $E(\epsilon) = 0$.
2. The variance of ϵ, denoted by σ^2, is the same for all values of x.
3. The values of ϵ are independent.
4. The error term ϵ has a normal distribution.

These assumptions provide the theoretical basis for the *t* test and the *F* test used to determine whether the relationship between x and y is significant, and for the confidence and prediction interval estimates presented in Section 12.6. If the assumptions about the error term ϵ appear questionable, the hypothesis tests about the significance of the regression relationship and the interval estimation results may not be valid.

TABLE 12.7 RESIDUALS FOR ARMAND'S PIZZA PARLORS

Student Population x_i	Sales y_i	Estimated Sales $\hat{y}_i = 60 + 5x_i$	Residuals $y_i - \hat{y}_i$
2	58	70	−12
6	105	90	15
8	88	100	−12
8	118	100	18
12	117	120	−3
16	137	140	−3
20	157	160	−3
20	169	160	9
22	149	170	−21
26	202	190	12

12.8 Residual Analysis: Validating Model Assumptions

The residuals provide the best information about ϵ; hence an analysis of the residuals is an important step in determining whether the assumptions for ϵ are appropriate. Much of residual analysis is based on an examination of graphical plots. In this section, we discuss the following residual plots.

1. A plot of the residuals against values of the independent variable x
2. A plot of residuals against the predicted values of the dependent variable $\hat{y}$

Residual Plot Against x

A **residual plot** against the independent variable x is a graph in which the values of the independent variable are represented by the horizontal axis and the corresponding residual values are represented by the vertical axis. A point is plotted for each residual. The first coordinate for each point is given by the value of x_i and the second coordinate is given by the corresponding value of the residual $y_i - \hat{y}_i$. For a residual plot against x with the Armand's Pizza Parlors data from Table 12.7, the coordinates of the first point are $(2, -12)$, corresponding to $x_1 = 2$ and $y_1 - \hat{y}_1 = -12$; the coordinates of the second point are $(6, 15)$, corresponding to $x_2 = 6$ and $y_2 - \hat{y}_2 = 15$; and so on. Figure 12.11 shows the resulting residual plot.

Before interpreting the results for this residual plot, let us consider some general patterns that might be observed in any residual plot. Three examples appear in Figure 12.12. If the assumption that the variance of ϵ is the same for all values of x and the assumed regression model is an adequate representation of the relationship between the variables, the residual plot should give an overall impression of a horizontal band of points such as the one in Panel A of Figure 12.12. However, if the variance of ϵ is not the same for all values of x—for example, if variability about the regression line is greater for larger values of x—a pattern such as the one in Panel B of Figure 12.12 could be observed. In this case, the assumption of a constant variance of ϵ is violated. Another possible residual plot is shown in Panel C. In this case, we would conclude that the assumed regression model is not an

FIGURE 12.11 PLOT OF THE RESIDUALS AGAINST THE INDEPENDENT VARIABLE x FOR ARMAND'S PIZZA PARLORS

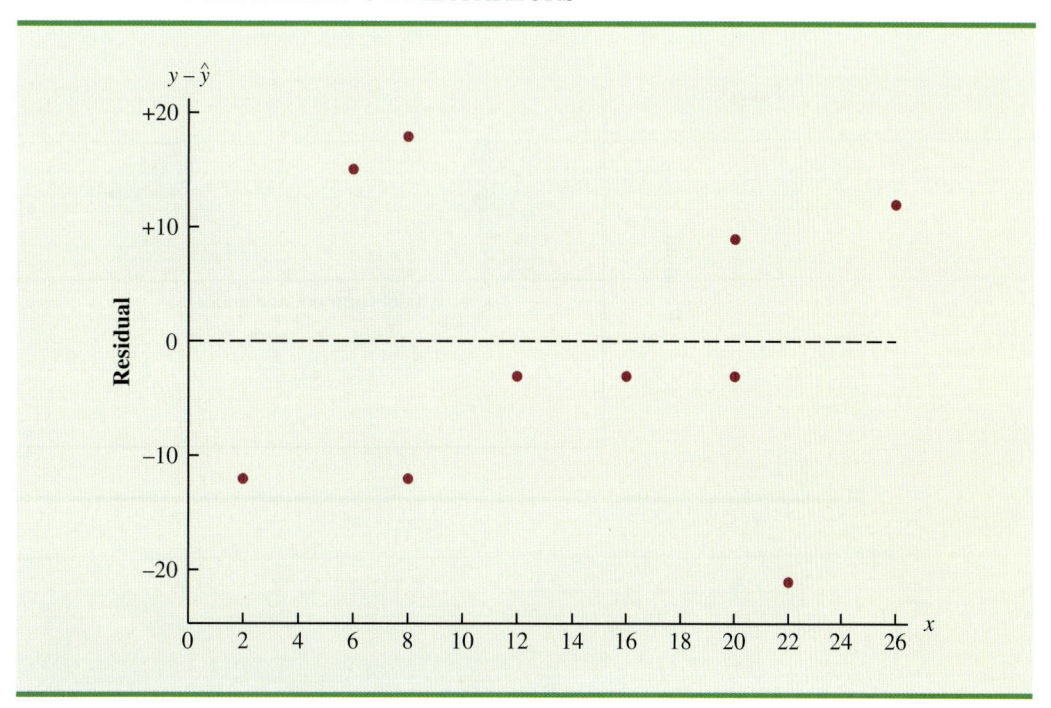

530 Chapter 12 Simple Linear Regression

FIGURE 12.12 RESIDUAL PLOTS FROM THREE REGRESSION STUDIES

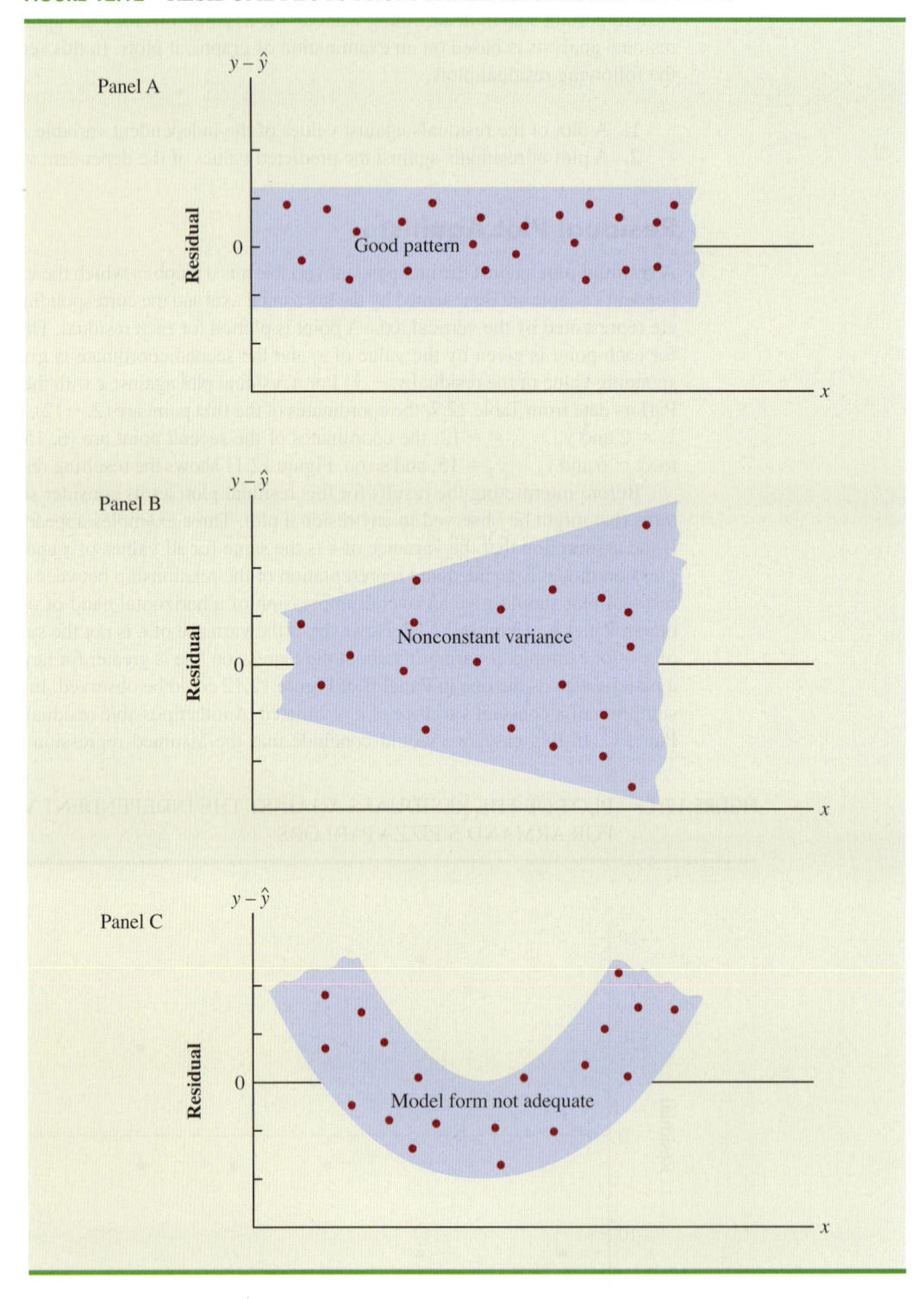

12.8 Residual Analysis: Validating Model Assumptions

adequate representation of the relationship between the variables. A curvilinear regression model or multiple regression model should be considered.

Now let us return to the residual plot for Armand's Pizza Parlors shown in Figure 12.11. The residuals appear to approximate the horizontal pattern in Panel A of Figure 12.12. Hence, we conclude that the residual plot does not provide evidence that the assumptions made for Armand's regression model should be challenged. At this point, we are confident in the conclusion that Armand's simple linear regression model is valid.

Experience and good judgment are always factors in the effective interpretation of residual plots. Seldom does a residual plot conform precisely to one of the patterns in Figure 12.12. Yet analysts who frequently conduct regression studies and frequently review residual plots become adept at understanding the differences between patterns that are reasonable and patterns that indicate the assumptions of the model should be questioned. A residual plot provides one technique to assess the validity of the assumptions for a regression model.

Residual Plot Against $\hat{y}$

Another residual plot represents the predicted value of the dependent variable $\hat{y}$ on the horizontal axis and the residual values on the vertical axis. A point is plotted for each residual. The first coordinate for each point is given by $\hat{y}_i$ and the second coordinate is given by the corresponding value of the ith residual $y_i - \hat{y}_i$. With the Armand's data from Table 12.7, the coordinates of the first point are $(70, -12)$, corresponding to $\hat{y}_1 = 70$ and $y_1 - \hat{y}_1 = -12$; the coordinates of the second point are $(90, 15)$; and so on. Figure 12.13 provides the residual plot. Note that the pattern of this residual plot is the same as the pattern of the residual plot against the independent variable x. It is not a pattern that would lead us to question the model assumptions. For simple linear regression, both the residual plot against x and the residual plot against $\hat{y}$ provide the same pattern. For multiple regression analysis, the residual plot against $\hat{y}$ is more widely used because of the presence of more than one independent variable.

FIGURE 12.13 PLOT OF THE RESIDUALS AGAINST THE PREDICTED VALUES $\hat{y}$ FOR ARMAND'S PIZZA PARLORS

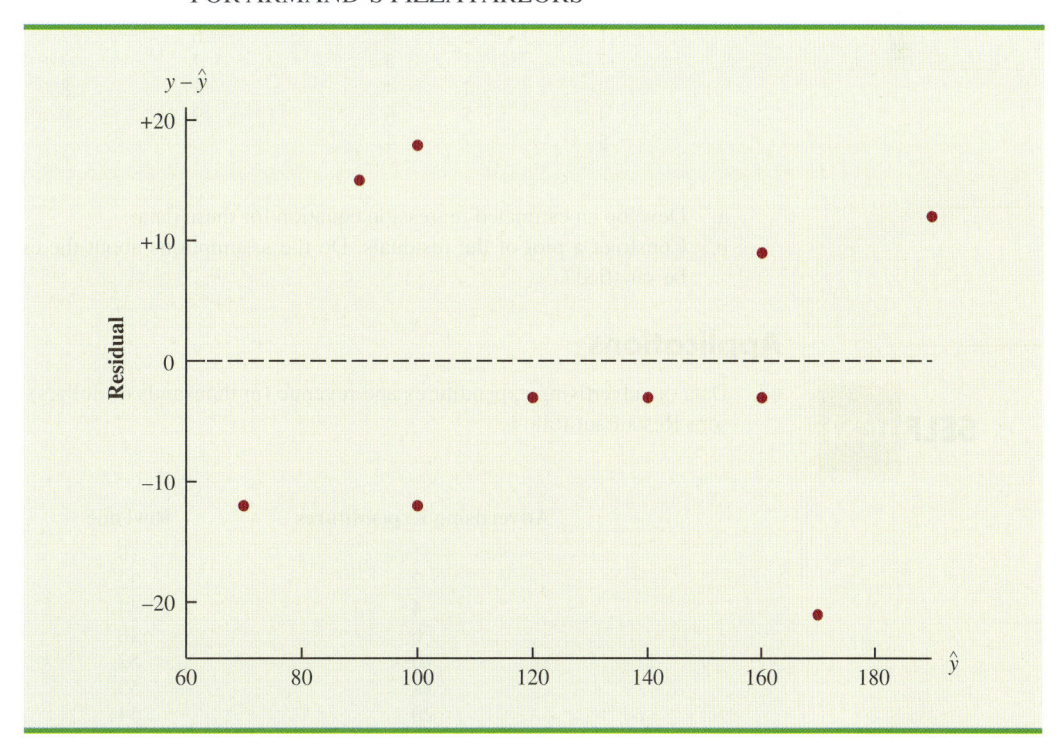

NOTES AND COMMENTS

1. We use residual plots to validate the assumptions of a regression model. If our review indicates that one or more assumptions are questionable, a different regression model or a transformation of the data should be considered. The appropriate corrective action when the assumptions are violated must be based on good judgment; recommendations from an experienced statistician can be valuable.

2. Analysis of residuals is the primary method statisticians use to verify that the assumptions associated with a regression model are valid. Even if no violations are found, it does not necessarily follow that the model will yield good predictions. However, if additional statistical tests support the conclusion of significance and the coefficient of determination is large, we should be able to develop good estimates and predictions using the estimated regression equation.

Exercises

Methods

45. Given are data for two variables, x and y.

x_i	6	11	15	18	20
y_i	6	8	12	20	30

    a. Develop an estimated regression equation for these data.
    b. Compute the residuals.
    c. Develop a plot of the residuals against the independent variable x. Do the assumptions about the error terms seem to be satisfied?

46. The following data were used in a regression study.

Observation	x_i	y_i	Observation	x_i	y_i
1	2	4	6	7	6
2	3	5	7	7	9
3	4	4	8	8	5
4	5	6	9	9	11
5	7	4			

    a. Develop an estimated regression equation for these data.
    b. Construct a plot of the residuals. Do the assumptions about the error term seem to be satisfied?

Applications

47. Data on advertising expenditures and revenue (in thousands of dollars) for the Four Seasons Restaurant follow.

Advertising Expenditures	Revenue
1	19
2	32
4	44
6	40
10	52
14	53
20	54

a. Let x equal advertising expenditures and y equal revenue. Use the method of least squares to develop a straight line approximation of the relationship between the two variables.
b. Test whether revenue and advertising expenditures are related at a .05 level of significance.
c. Prepare a residual plot of $y - \hat{y}$ versus $\hat{y}$. Use the result from part (a) to obtain the values of $\hat{y}$.
d. What conclusions can you draw from residual analysis? Should this model be used, or should we look for a better one?

48. Refer to exercise 7, where an estimated regression equation relating years of experience and annual sales was developed.
    a. Compute the residuals and construct a residual plot for this problem.
    b. Do the assumptions about the error terms seem reasonable in light of the residual plot?

49. Recent family home sales in San Antonio provided the following data (San Antonio Realty Watch website, November 2008).

Square Footage	Price ($)
1580	142,500
1572	145,000
1352	115,000
2224	155,900
1556	95,000
1435	128,000
1438	100,000
1089	55,000
1941	142,000
1698	115,000
1539	115,000
1364	105,000
1979	155,000
2183	132,000
2096	140,000
1400	85,000
2372	145,000
1752	155,000
1386	80,000
1163	100,000

a. Develop the estimated regression equation that can be used to predict the sales prices given the square footage.
b. Construct a residual plot against the independent variable.
c. Do the assumptions about the error term and model form seem reasonable in light of the residual plot?

Summary

In this chapter we showed how regression analysis can be used to determine how a dependent variable y is related to an independent variable x. In simple linear regression, the regression model is $y = \beta_0 + \beta_1 x + \epsilon$. The simple linear regression equation $E(y) = \beta_0 + \beta_1 x$ describes how the mean or expected value of y is related to x. We used sample data and the least squares method to develop the estimated regression equation $\hat{y} = b_0 + b_1 x$. In effect, b_0 and b_1 are the sample statistics used to estimate the unknown model parameters β_0 and β_1.

The coefficient of determination was presented as a measure of the goodness of fit for the estimated regression equation; it can be interpreted as the proportion of the variation in the dependent variable y that can be explained by the estimated regression equation. We reviewed correlation as a descriptive measure of the strength of a linear relationship between two variables.

The assumptions about the regression model and its associated error term ϵ were discussed, and t and F tests, based on those assumptions, were presented as a means for determining whether the relationship between two variables is statistically significant. We showed how to use the estimated regression equation to develop confidence interval estimates of the mean value of y and prediction interval estimates of individual values of y.

The chapter concluded with a section on the computer solution of regression problems and a section on the use of residual analysis to validate the model assumptions.

Glossary

Dependent variable The variable that is being predicted or explained. It is denoted by y.

Independent variable The variable that is doing the predicting or explaining. It is denoted by x.

Simple linear regression Regression analysis involving one independent variable and one dependent variable in which the relationship between the variables is approximated by a straight line.

Regression model The equation that describes how y is related to x and an error term; in simple linear regression, the regression model is $y = \beta_0 + \beta_1 x + \epsilon$.

Regression equation The equation that describes how the mean or expected value of the dependent variable is related to the independent variable; in simple linear regression, $E(y) = \beta_0 + \beta_1 x$.

Estimated regression equation The estimate of the regression equation developed from sample data by using the least squares method. For simple linear regression, the estimated regression equation is $\hat{y} = b_0 + b_1 x$.

Least squares method A procedure used to develop the estimated regression equation. The objective is to minimize $\Sigma(y_i - \hat{y}_i)^2$.

Scatter diagram A graph of bivariate data in which the independent variable is on the horizontal axis and the dependent variable is on the vertical axis.

Coefficient of determination A measure of the goodness of fit of the estimated regression equation. It can be interpreted as the proportion of the variability in the dependent variable y that is explained by the estimated regression equation.

ith residual The difference between the observed value of the dependent variable and the value predicted using the estimated regression equation; for the ith observation the ith residual is $y_i - \hat{y}_i$.

Correlation coefficient A measure of the strength of the linear relationship between two variables (previously discussed in Chapter 3).

Mean square error The unbiased estimate of the variance of the error term σ^2. It is denoted by MSE or s^2.

Standard error of the estimate The square root of the mean square error, denoted by s. It is the estimate of σ, the standard deviation of the error term ϵ.

ANOVA table The analysis of variance table used to summarize the computations associated with the F test for significance.

Confidence interval The interval estimate of the mean value of y for a given value of x.

Prediction interval The interval estimate of an individual value of y for a given value of x.

Residual analysis The analysis of the residuals used to determine whether the assumptions made about the regression model appear to be valid. Residual analysis is also used to identify outliers and influential observations.

Residual plot Graphical representation of the residuals that can be used to determine whether the assumptions made about the regression model appear to be valid.

Key Formulas

Simple Linear Regression Model

$$y = \beta_0 + \beta_1 x + \epsilon \tag{12.1}$$

Simple Linear Regression Equation

$$E(y) = \beta_0 + \beta_1 x \tag{12.2}$$

Estimated Simple Linear Regression Equation

$$\hat{y} = b_0 + b_1 x \tag{12.3}$$

Least Squares Criterion

$$\min \Sigma(y_i - \hat{y}_i)^2 \tag{12.5}$$

Slope and y-Intercept for the Estimated Regression Equation

$$b_1 = \frac{\Sigma(x_i - \bar{x})(y_i - \bar{y})}{\Sigma(x_i - \bar{x})^2} \tag{12.6}$$

$$b_0 = \bar{y} - b_1 \bar{x} \tag{12.7}$$

Sum of Squares Due to Error

$$SSE = \Sigma(y_i - \hat{y}_i)^2 \tag{12.8}$$

Total Sum of Squares

$$SST = \Sigma(y_i - \bar{y})^2 \tag{12.9}$$

Sum of Squares Due to Regression

$$SSR = \Sigma(\hat{y}_i - \bar{y})^2 \tag{12.10}$$

Relationship Among SST, SSR, and SSE

$$SST = SSR + SSE \tag{12.11}$$

Coefficient of Determination

$$r^2 = \frac{SSR}{SST} \tag{12.12}$$

Sample Correlation Coefficient

$$\begin{aligned} r_{xy} &= (\text{sign of } b_1)\sqrt{\text{Coefficient of determination}} \\ &= (\text{sign of } b_1)\sqrt{r^2} \end{aligned} \tag{12.13}$$

Mean Square Error (Estimate of σ^2)

$$s^2 = MSE = \frac{SSE}{n - 2} \tag{12.15}$$

Standard Error of the Estimate

$$s = \sqrt{\text{MSE}} = \sqrt{\frac{\text{SSE}}{n-2}} \tag{12.16}$$

Standard Deviation of b_1

$$\sigma_{b_1} = \frac{\sigma}{\sqrt{\Sigma(x_i - \bar{x})^2}} \tag{12.17}$$

Estimated Standard Deviation of b_1

$$s_{b_1} = \frac{s}{\sqrt{\Sigma(x_i - \bar{x})^2}} \tag{12.18}$$

t Test Statistic

$$t = \frac{b_1}{s_{b_1}} \tag{12.19}$$

Mean Square Regression

$$\text{MSR} = \frac{\text{SSR}}{\text{Number of independent variables}} \tag{12.20}$$

F Test Statistic

$$F = \frac{\text{MSR}}{\text{MSE}} \tag{12.21}$$

Estimated Standard Deviation of $\hat{y}_p$

$$s_{\hat{y}_p} = s\sqrt{\frac{1}{n} + \frac{(x_p - \bar{x})^2}{\Sigma(x_i - \bar{x})^2}} \tag{12.23}$$

Confidence Interval for $E(y_p)$

$$\hat{y}_p \pm t_{\alpha/2} s_{\hat{y}_p} \tag{12.24}$$

Estimated Standard Deviation of an Individual Value

$$s_{\text{ind}} = s\sqrt{1 + \frac{1}{n} + \frac{(x_p - \bar{x})^2}{\Sigma(x_i - \bar{x})^2}} \tag{12.26}$$

Prediction Interval for y_p

$$\hat{y}_p \pm t_{\alpha/2} s_{\text{ind}} \tag{12.27}$$

Residual for Observation i

$$y_i - \hat{y}_i \tag{12.28}$$

Supplementary Exercises

50. The data in the following table show the number of shares selling (millions) and the expected price (average of projected low price and projected high price) for 10 selected initial public stock offerings.

Company	Shares Selling (millions)	Expected Price ($)
American Physician	5.0	15
Apex Silver Mines	9.0	14
Dan River	6.7	15
Franchise Mortgage	8.75	17
Gene Logic	3.0	11
International Home Foods	13.6	19
PRT Group	4.6	13
Rayovac	6.7	14
RealNetworks	3.0	10
Software AG Systems	7.7	13

a. Develop an estimated regression equation with the number of shares selling as the independent variable and the expected price as the dependent variable.
b. At the .05 level of significance, is there a significant relationship between the two variables?
c. Did the estimated regression equation provide a good fit? Explain.
d. Use the estimated regression equation to estimate the expected price for a firm considering an initial public offering of 6 million shares.

51. The following data show Morningstar's Fair Value estimate and the Share Price for 28 companies. Fair Value is an estimate of a company's value per share that takes into account estimates of the company's growth, profitability, riskiness, and other factors over the next five years (*Morningstar Stocks 500*, 2008 edition).

Company	Fair Value ($)	Share Price ($)
Air Products and Chemicals	80	98.63
Allied Waste Industries	17	11.02
America Mobile	83	61.39
AT&T	35	41.56
Bank of America	70	41.26
Barclays PLC	68	40.37
Citigroup	53	29.44
Costco Wholesale Corp.	75	69.76
Covidien, Ltd.	58	44.29
Darden Restaurants	52	27.71
Dun & Bradstreet	87	88.63
Equifax	42	36.36
Gannett Co.	38	39.00
Genuine Parts	48	46.30
GlaxoSmithKline PLC	57	50.39
Iron Mountain	33	37.02
ITT Corporation	83	66.04
Johnson & Johnson	80	66.70
Las Vegas Sands	98	103.05
Macrovision	23	18.33
Marriott International	39	34.18
Nalco Holding Company	29	24.18
National Interstate	25	33.10
Portugal Telecom	15	13.02
Qualcomm	48	39.35
Royal Dutch Shell Ltd.	87	84.20
SanDisk	60	33.17
Time Warner	42	27.60

a. Develop the estimated regression equation that could be used to estimate the Share Price given the Fair Value.
b. At the .05 level of significance, is there a significant relationship between the two variables?
c. Use the estimated regression equation to estimate the Share Price for a company that has a Fair Value of $50.
d. Do you believe the estimated regression equation would provide a good prediction of the share price? Use r^2 to support your answer.

52. One of the biggest changes in higher education in recent years has been the growth of online universities. The Online Education Database is an independent organization whose mission is to build a comprehensive list of the top accredited online colleges. The following table shows the retention rate (%) and the graduation rate (%) for 29 online colleges (Online Education Database website, January 2009).

College	Retention Rate (%)	Graduation Rate (%)
Western International University	7	25
South University	51	25
University of Phoenix	4	28
American InterContinental University	29	32
Franklin University	33	33
Devry University	47	33
Tiffin University	63	34
Post University	45	36
Peirce College	60	36
Everest University	62	36
Upper Iowa University	67	36
Dickinson State University	65	37
Western Governors University	78	37
Kaplan University	75	38
Salem International University	54	39
Ashford University	45	41
ITT Technical Institute	38	44
Berkeley College	51	45
Grand Canyon University	69	46
Nova Southeastern University	60	47
Westwood College	37	48
Everglades University	63	50
Liberty University	73	51
LeTourneau University	78	52
Rasmussen College	48	53
Keiser University	95	55
Herzing College	68	56
National University	100	57
Florida National College	100	61

a. Develop a scatter diagram with retention rate as the independent variable. What does the scatter diagram indicate about the relationship between the two variables?
b. Develop the estimated regression equation.
c. Test for a significant relationship. Use $\alpha = .05$.

d. Did the estimated regression equation provide a good fit?
e. Suppose you were the president of South University. After reviewing the results, would you have any concerns about the performance of your university as compared to other online universities?
f. Suppose you were the president of the University of Phoenix. After reviewing the results, would you have any concerns about the performance of your university as compared to other online universities?

53. Jensen Tire & Auto is in the process of deciding whether to purchase a maintenance contract for its new computer wheel alignment and balancing machine. Managers feel that maintenance expense should be related to usage, and they collected the following information on weekly usage (hours) and annual maintenance expense (in hundreds of dollars).

Weekly Usage (hours)	Annual Maintenance Expense
13	17.0
10	22.0
20	30.0
28	37.0
32	47.0
17	30.5
24	32.5
31	39.0
40	51.5
38	40.0

a. Develop the estimated regression equation that relates annual maintenance expense to weekly usage.
b. Test the significance of the relationship in part (a) at a .05 level of significance.
c. Jensen expects to use the new machine 30 hours per week. Develop a 95% prediction interval for the company's annual maintenance expense.
d. If the maintenance contract costs $3000 per year, would you recommend purchasing it? Why or why not?

54. In a manufacturing process the assembly line speed (feet per minute) was thought to affect the number of defective parts found during the inspection process. To test this theory, managers devised a situation in which the same batch of parts was inspected visually at a variety of line speeds. They collected the following data.

Line Speed	Number of Defective Parts Found
20	21
20	19
40	15
30	16
60	14
40	17

a. Develop the estimated regression equation that relates line speed to the number of defective parts found.
b. At a .05 level of significance, determine whether line speed and number of defective parts found are related.

540 Chapter 12 Simple Linear Regression

c. Did the estimated regression equation provide a good fit to the data?
d. Develop a 95% confidence interval to predict the mean number of defective parts for a line speed of 50 feet per minute.

55. A sociologist was hired by a large city hospital to investigate the relationship between the number of unauthorized days that employees are absent per year and the distance (miles) between home and work for the employees. A sample of 10 employees was chosen, and the following data were collected.

Distance to Work (miles)	Number of Days Absent
1	8
3	5
4	8
6	7
8	6
10	3
12	5
14	2
14	4
18	2

a. Develop a scatter diagram for these data. Does a linear relationship appear reasonable? Explain.
b. Develop the least squares estimated regression equation.
c. Is there a significant relationship between the two variables? Use $\alpha = .05$.
d. Did the estimated regression equation provide a good fit? Explain.
e. Use the estimated regression equation developed in part (b) to develop a 95% confidence interval for the expected number of days absent for employees living 5 miles from the company.

56. The regional transit authority for a major metropolitan area wants to determine whether there is any relationship between the age of a bus and the annual maintenance cost. A sample of 10 buses resulted in the following data.

Age of Bus (years)	Maintenance Cost ($)
1	350
2	370
2	480
2	520
2	590
3	550
4	750
4	800
5	790
5	950

a. Develop the least squares estimated regression equation.
b. Test to see whether the two variables are significantly related with $\alpha = .05$.
c. Did the least squares line provide a good fit to the observed data? Explain.
d. Develop a 95% prediction interval for the maintenance cost for a specific bus that is 4 years old.

57. A marketing professor at Givens College is interested in the relationship between hours spent studying and total points earned in a course. Data collected on 10 students who took the course last quarter follow.

Hours Spent Studying	Total Points Earned
45	40
30	35
90	75
60	65
105	90
65	50
90	90
80	80
55	45
75	65

a. Develop an estimated regression equation showing how total points earned is related to hours spent studying.
b. Test the significance of the model with $\alpha = .05$.
c. Predict the total points earned by Mark Sweeney. He spent 95 hours studying.
d. Develop a 95% prediction interval for the total points earned by Mark Sweeney.

58. Reuters reported the market beta for Xerox was 1.22 (Reuters website, January 30, 2009). Market betas for individual stocks are determined by simple linear regression. For each stock, the dependent variable is its quarterly percentage return (capital appreciation plus dividends) minus the percentage return that could be obtained from a risk-free investment (the Treasury Bill rate is used as the risk-free rate). The independent variable is the quarterly percentage return (capital appreciation plus dividends) for the stock market (S&P 500) minus the percentage return from a risk-free investment. An estimated regression equation is developed with quarterly data; the market beta for the stock is the slope of the estimated regression equation (b_1). The value of the market beta is often interpreted as a measure of the risk associated with the stock. Market betas greater than 1 indicate that the stock is more volatile than the market average; market betas less than 1 indicate that the stock is less volatile than the market average. Suppose that the following figures are the differences between the percentage return and the risk-free return for 10 quarters for the S&P 500 and Horizon Technology.

S&P 500	Horizon
1.2	−0.7
−2.5	−2.0
−3.0	−5.5
2.0	4.7
5.0	1.8
1.2	4.1
3.0	2.6
−1.0	2.0
.5	−1.3
2.5	5.5

a. Develop an estimated regression equation that can be used to determine the market beta for Horizon Technology. What is Horizon Technology's market beta?
b. Test for a significant relationship at the .05 level of significance.
c. Did the estimated regression equation provide a good fit? Explain.
d. Use the market betas of Xerox and Horizon Technology to compare the risk associated with the two stocks.

59. The Transactional Records Access Clearinghouse at Syracuse University reported data showing the odds of an Internal Revenue Service audit. The following table shows the average adjusted gross income reported and the percent of the returns that were audited for 20 selected IRS districts.

District	Adjusted Gross Income ($)	Percent Audited
Los Angeles	36,664	1.3
Sacramento	38,845	1.1
Atlanta	34,886	1.1
Boise	32,512	1.1
Dallas	34,531	1.0
Providence	35,995	1.0
San Jose	37,799	0.9
Cheyenne	33,876	0.9
Fargo	30,513	0.9
New Orleans	30,174	0.9
Oklahoma City	30,060	0.8
Houston	37,153	0.8
Portland	34,918	0.7
Phoenix	33,291	0.7
Augusta	31,504	0.7
Albuquerque	29,199	0.6
Greensboro	33,072	0.6
Columbia	30,859	0.5
Nashville	32,566	0.5
Buffalo	34,296	0.5

a. Develop the estimated regression equation that could be used to predict the percent audited given the average adjusted gross income reported.
b. At the .05 level of significance, determine whether the adjusted gross income and the percent audited are related.
c. Did the estimated regression equation provide a good fit? Explain.
d. Use the estimated regression equation developed in part (a) to calculate a 95% confidence interval for the expected percent audited for districts with an average adjusted gross income of $35,000.

60. The Australian Public Service Commission's State of the Service Report 2002–2003 reported job satisfaction ratings for employees. One of the survey questions asked employees to choose the five most important workplace factors (from a list of factors) that most affected how satisfied they were with their job. Respondents were then asked to indicate their level of satisfaction with their top five factors. The following data show the percentage of employees who nominated the factor in their top five, and a corresponding satisfaction rating measured using the percentage of employees who nominated the factor in the top five and who were "very satisfied" or "satisfied" with the factor in their current workplace (Australian Public Service Commission website).

Case Problem 1 Measuring Stock Market Risk

WEB file
JobSat

Workplace Factor	Top Five (%)	Satisfaction Rating (%)
Appropriate workload	30	49
Chance to be creative/innovative	38	64
Chance to make a useful contribution to society	40	67
Duties/expectations made clear	40	69
Flexible working arrangements	55	86
Good working relationships	60	85
Interesting work provided	48	74
Opportunities for career development	33	43
Opportunities to develop my skills	46	66
Opportunities to utilize my skills	50	70
Regular feedback/recognition for effort	42	53
Salary	47	62
Seeing tangible results from my work	42	69

a. Develop a scatter diagram with Top Five (%) on the horizontal axis and Satisfaction Rating (%) on the vertical axis.
b. What does the scatter diagram developed in part (a) indicate about the relationship between the two variables?
c. Develop the estimated regression equation that could be used to predict the Satisfaction Rating (%) given the Top Five (%).
d. Test for a significant relationship at the .05 level of significance.
e. Did the estimated regression equation provide a good fit? Explain.
f. What is the value of the sample correlation coefficient?

Case Problem 1 Measuring Stock Market Risk

One measure of the risk or volatility of an individual stock is the standard deviation of the total return (capital appreciation plus dividends) over several periods of time. Although the standard deviation is easy to compute, it does not take into account the extent to which the price of a given stock varies as a function of a standard market index, such as the S&P 500. As a result, many financial analysts prefer to use another measure of risk referred to as *beta*.

Betas for individual stocks are determined by simple linear regression. The dependent variable is the total return for the stock and the independent variable is the total return for the stock market.* For this case problem we will use the S&P 500 index as the measure of the total return for the stock market, and an estimated regression equation will be developed using monthly data. The beta for the stock is the slope of the estimated regression equation (b_1). The data contained in the file named Beta provides the total return (capital appreciation plus dividends) over 36 months for eight widely traded common stocks and the S&P 500.

WEB file
Beta

The value of beta for the stock market will always be 1; thus, stocks that tend to rise and fall with the stock market will also have a beta close to 1. Betas greater than 1 indicate that the stock is more volatile than the market, and betas less than 1 indicate that the stock is less volatile than the market. For instance, if a stock has a beta of 1.4, it is 40% *more* volatile than the market, and if a stock has a beta of .4, it is 60% *less* volatile than the market.

---

*Various sources use different approaches for computing betas. For instance, some sources subtract the return that could be obtained from a risk-free investment (e.g., T-bills) from the dependent variable and the independent variable before computing the estimated regression equation. Some also use different indexes for the total return of the stock market; for instance, *Value Line* computes betas using the New York Stock Exchange composite index.

Managerial Report

You have been assigned to analyze the risk characteristics of these stocks. Prepare a report that includes but is not limited to the following items.

a. Compute descriptive statistics for each stock and the S&P 500. Comment on your results. Which stocks are the most volatile?
b. Compute the value of beta for each stock. Which of these stocks would you expect to perform best in an up market? Which would you expect to hold their value best in a down market?
c. Comment on how much of the return for the individual stocks is explained by the market.

Case Problem 2 U.S. Department of Transportation

As part of a study on transportation safety, the U.S. Department of Transportation collected data on the number of fatal accidents per 1000 licenses and the percentage of licensed drivers under the age of 21 in a sample of 42 cities. Data collected over a one-year period follow. These data are contained in the file named Safety.

Percent Under 21	Fatal Accidents per 1000 Licenses	Percent Under 21	Fatal Accidents per 1000 Licenses
13	2.962	17	4.100
12	0.708	8	2.190
8	0.885	16	3.623
12	1.652	15	2.623
11	2.091	9	0.835
17	2.627	8	0.820
18	3.830	14	2.890
8	0.368	8	1.267
13	1.142	15	3.224
8	0.645	10	1.014
9	1.028	10	0.493
16	2.801	14	1.443
12	1.405	18	3.614
9	1.433	10	1.926
10	0.039	14	1.643
9	0.338	16	2.943
11	1.849	12	1.913
12	2.246	15	2.814
14	2.855	13	2.634
14	2.352	9	0.926
11	1.294	17	3.256

Managerial Report

1. Develop numerical and graphical summaries of the data.
2. Use regression analysis to investigate the relationship between the number of fatal accidents and the percentage of drivers under the age of 21. Discuss your findings.
3. What conclusion and recommendations can you derive from your analysis?

Case Problem 3 Alumni Giving

Alumni donations are an important source of revenue for colleges and universities. If administrators could determine the factors that influence increases in the percentage of alumni who make a donation, they might be able to implement policies that could lead to increased revenues. Research shows that students who are more satisfied with their contact with teachers are more likely to graduate. As a result, one might suspect that smaller class sizes and lower student-faculty ratios might lead to a higher percentage of satisfied graduates, which in turn might lead to increases in the percentage of alumni who make a donation. Table 12.8 shows data for 48 national universities (*America's Best Colleges,* Year 2000 ed.). The column labeled % of Classes Under 20 shows the percentage of classes offered with fewer than 20 students. The column labeled Student/Faculty Ratio is the number of students enrolled divided by the total number of faculty. Finally, the column labeled Alumni Giving Rate is the percentage of alumni that made a donation to the university.

Managerial Report

1. Develop numerical and graphical summaries of the data.
2. Use regression analysis to develop an estimated regression equation that could be used to predict the alumni giving rate given the percentage of classes with fewer than 20 students.
3. Use regression analysis to develop an estimated regression equation that could be used to predict the alumni giving rate given the student-faculty ratio.
4. Which of the two estimated regression equations provides the best fit? For this estimated regression equation, perform an analysis of the residuals and discuss your findings and conclusions.
5. What conclusions and recommendations can you derive from your analysis?

Case Problem 4 PGA Tour Statistics

The Professional Golfers Association (PGA) maintains data on performance and earnings for members of the PGA Tour. The top 125 players based on total earnings in PGA Tour events are exempt for the following season. Making the top 125 money list is important because a player who is "exempt" has qualified to be a full-time member of the PGA tour for the following season.

During recent years on the PGA Tour there have been significant advances in the technology of golf balls and golf clubs, and this technology has been one of the major reasons for the increase in the average driving distance of PGA Tour players. In 1992, the average driving distance was 260 yards, but in 2003 this increased to 286 yards. PGA Tour pros are hitting the ball farther than ever before, but how important is driving distance in terms of a player's performance? And what effect has this increased distance had on the players' accuracy? To investigate these issues, year-end performance data for the 125 players who had the highest total earnings in PGA Tour events for 2008 are contained in the file named PGATour (PGA Tour website, 2009). Each row of the data set corresponds to a PGA Tour player, and the data have been sorted based upon total earnings. Descriptions for the data follow.

Money: Total earnings in PGA Tour events.

Scoring Average: The average number of strokes per completed round.

TABLE 12.8 DATA FOR 48 NATIONAL UNIVERSITIES

	% of Classes Under 20	Student/Faculty Ratio	Alumni Giving Rate
Boston College	39	13	25
Brandeis University	68	8	33
Brown University	60	8	40
California Institute of Technology	65	3	46
Carnegie Mellon University	67	10	28
Case Western Reserve Univ.	52	8	31
College of William and Mary	45	12	27
Columbia University	69	7	31
Cornell University	72	13	35
Dartmouth College	61	10	53
Duke University	68	8	45
Emory University	65	7	37
Georgetown University	54	10	29
Harvard University	73	8	46
Johns Hopkins University	64	9	27
Lehigh University	55	11	40
Massachusetts Inst. of Technology	65	6	44
New York University	63	13	13
Northwestern University	66	8	30
Pennsylvania State Univ.	32	19	21
Princeton University	68	5	67
Rice University	62	8	40
Stanford University	69	7	34
Tufts University	67	9	29
Tulane University	56	12	17
U. of California—Berkeley	58	17	18
U. of California—Davis	32	19	7
U. of California—Irvine	42	20	9
U. of California—Los Angeles	41	18	13
U. of California—San Diego	48	19	8
U. of California—Santa Barbara	45	20	12
U. of Chicago	65	4	36
U. of Florida	31	23	19
U. of Illinois—Urbana Champaign	29	15	23
U. of Michigan—Ann Arbor	51	15	13
U. of North Carolina—Chapel Hill	40	16	26
U. of Notre Dame	53	13	49
U. of Pennsylvania	65	7	41
U. of Rochester	63	10	23
U. of Southern California	53	13	22
U. of Texas—Austin	39	21	13
U. of Virginia	44	13	28
U. of Washington	37	12	12
U. of Wisconsin—Madison	37	13	13
Vanderbilt University	68	9	31
Wake Forest University	59	11	38
Washington University—St. Louis	73	7	33
Yale University	77	7	50

WEB file
Alumni

DrDist (Driving Distance): DrDist is the average number of yards per measured drive. On the PGA Tour, driving distance is measured on two holes per round. Care is taken to select two holes which face in opposite directions to counteract the effect of wind. Drives are measured to the point at which they come to rest regardless of whether they are in the fairway or not.

DrAccu (Driving Accuracy): The percentage of time a tee shot comes to rest in the fairway (regardless of club). Driving accuracy is measured on every hole, excluding par 3's.

GIR (Greens in Regulation): The percentage of time a player was able to hit the green in regulation. A green is considered hit in regulation if any portion of the ball is touching the putting surface after the GIR stroke has been taken. The GIR stroke is determined by subtracting 2 from par (first stroke on a par 3, second on a par 4, third on a par 5). In other words, a green is considered hit in regulation if the player has reached the putting surface in par minus two strokes.

Managerial Report

1. Develop numerical and graphical summaries of the data.
2. Use regression analysis to investigate the relationship between Scoring Average and DrDist. Does it appear that players who drive the ball farther have lower average scores?
3. Use regression analysis to investigate the relationship between Scoring Average and DrAccu. Does it appear that players who are more accurate in hitting the fairway have lower average scores?
4. Use regression analysis to investigate the relationship between Scoring Average and GIR. Does it appear that players who are more accurate in hitting greens in regulation have lower average scores?
5. Which of the three variables (DrDist, DrAccu, and GIR) appears to be the most significant factor in terms of a player's average score?
6. Treating DrDist as the independent variable and DrAccu as the dependent variable, investigate the relationship between driving distance and driving accuracy.

Appendix 12.1 Regression Analysis Using Minitab

In Section 12.7 we discussed the computer solution of regression problems by showing Minitab's output for the Armand's Pizza Parlors problem. In this appendix, we describe the steps required to generate the Minitab computer solution. First, the data must be entered in a Minitab worksheet. Student population data are entered in column C1 and quarterly sales data are entered in column C2. The variable names Pop and Sales are entered as the column headings on the worksheet. In subsequent steps, we refer to the data by using the variable names Pop and Sales or the column indicators C1 and C2. The following steps describe how to use Minitab to produce the regression results shown in Figure 12.10.

Step 1. Select the **Stat** menu
Step 2. Select the **Regression** menu
Step 3. Choose **Regression**
Step 4. When the Regression dialog box appears:
    Enter Sales in the **Response** box
    Enter Pop in the **Predictors** box
    Click the **Options** button
When the Regression - Options dialog box appears:
    Enter 10 in the **Prediction intervals for new observations** box
    Click **OK**
When the Regression dialog box reappears:
    Click **OK**

The Minitab regression dialog box provides additional capabilities that can be obtained by selecting the desired options. For instance, to obtain a residual plot that shows the predicted value of the dependent variable $\hat{y}$ on the horizontal axis and the residual values on the vertical axis, step 4 would be as follows:

Step 4. When the Regression dialog box appears:
  Enter Sales in the **Response** box
  Enter Pop in the **Predictors** box
  Click the **Graphs** button
When the Regression - Graphs dialog box appears:
  Select **Regular** under Residuals for Plots
  Select **Residuals versus fits** under Residual Plots
  Click **OK**
When the Regression dialog box reappears:
  Click **OK**

Appendix 12.2 Regression Analysis Using Excel

In this appendix we will illustrate how Excel's Regression tool can be used to perform the regression analysis computations for the Armand's Pizza Parlors problem. Refer to Figure 12.14 as we describe the steps involved. The labels Restaurant, Population, and Sales are entered into cells A1:C1 of the worksheet. To identify each of the 10 observations, we entered the numbers 1 through 10 into cells A2:A11. The sample data are entered into cells B2:C11. The following steps describe how to use Excel to produce the regression results.

Step 1. Click the **Data** tab on the Ribbon
Step 2. In the **Analysis** group, click **Data Analysis**
Step 3. Choose **Regression** from the list of Analysis Tools
Step 4. Click **OK**
Step 5. When the Regression dialog box appears:
  Enter C1:C11 in the **Input Y Range** box
  Enter B1:B11 in the **Input X Range** box
  Select **Labels**
  Select **Confidence Level**
  Enter 99 in the **Confidence Level** box
  Select **Output Range**
  Enter A13 in the **Output Range** box
    (Any upper-left-hand corner cell indicating where the output is to begin may be entered here.)
  Click **OK**

The first section of the output, titled *Regression Statistics,* contains summary statistics such as the coefficient of determination (R Square). The second section of the output, entitled ANOVA, contains the analysis of variance table. The last section of the output, which is not titled, contains the estimated regression coefficients and related information. We will begin our discussion of the interpretation of the regression output with the information contained in cells A28:I30.

Interpretation of Estimated Regression Equation Output

The y intercept of the estimated regression line, $b_0 = 60$, is shown in cell B29, and the slope of the estimated regression line, $b_1 = 5$, is shown in cell B30. The label Intercept in cell A29 and the label Population in cell A30 are used to identify these two values.

Appendix 12.2 Regression Analysis Using Excel

FIGURE 12.14 EXCEL SOLUTION TO THE ARMAND'S PIZZA PARLORS PROBLEM

	A	B	C	D	E	F	G	H	I	J
1	Restaurant	Population	Sales							
2	1	2	58							
3	2	6	105							
4	3	8	88							
5	4	8	118							
6	5	12	117							
7	6	16	137							
8	7	20	157							
9	8	20	169							
10	9	22	149							
11	10	26	202							
12										
13	SUMMARY OUTPUT									
14										
15	*Regression Statistics*									
16	Multiple R	0.9501								
17	R Square	0.9027								
18	Adjusted R Square	0.8906								
19	Standard Error	13.8293								
20	Observations	10								
21										
22	ANOVA									
23		df	SS	MS	F	Significance F				
24	Regression	1	14200	14200	74.2484	2.55E-05				
25	Residual	8	1530	191.25						
26	Total	9	15730							
27										
28		Coefficients	Standard Error	t Stat	P-value	Lower 95%	Upper 95%	Lower 99.0%	Upper 99.0%	
29	Intercept	60	9.2260	6.5033	0.0002	38.7247	81.2753	29.0431	90.9569	
30	Population	5	0.5803	8.6167	2.55E-05	3.6619	6.3381	3.0530	6.9470	

In Section 12.5 we showed that the estimated standard deviation of b_1 is $s_{b_1} = .5803$. Note that the value in cell C30 is .5803. The label Standard Error in cell C28 is Excel's way of indicating that the value in cell C30 is the standard error, or standard deviation, of b_1. Recall that the *t* test for a significant relationship required the computation of the *t* statistic, $t = b_1/s_{b_1}$. For the Armand's data, the value of *t* that we computed was $t = 5/.5803 = 8.62$. The label in cell D28, *t Stat*, reminds us that cell D30 contains the value of the *t* test statistic.

The value in cell E30 is the *p*-value associated with the *t* test for significance. Excel has displayed the *p*-value in cell E30 using scientific notation. To obtain the decimal value, we move the decimal point 5 places to the left, obtaining a value of .0000255. Because the *p*-value = .0000255 < α = .01, we can reject H_0 and conclude that we have a significant relationship between student population and quarterly sales.

The information in cells F28:I30 can be used to develop confidence interval estimates of the *y*-intercept and slope of the estimated regression equation. Excel always provides the lower and upper limits for a 95% confidence interval. Recall that in step 4 we selected Confidence Level and entered 99 in the Confidence Level box. As a result, Excel's Regression tool also provides the lower and upper limits for a 99% confidence interval. The value in cell H30 is the lower limit for the 99% confidence interval estimate of β_1 and the value in cell I30 is the upper limit. Thus, after rounding, the 99% confidence interval estimate of β_1 is 3.05 to 6.95. The values in cells F30 and G30 provide the lower and upper limits for the 95% confidence interval. Thus, the 95% confidence interval is 3.66 to 6.34.

Interpretation of ANOVA Output

The information in cells A22:F26 is a summary of the analysis of variance computations. The three sources of variation are labeled Regression, Residual, and Total. The label *df* in cell B23 stands for degrees of freedom, the label *SS* in cell C23 stands for sum of squares, and the label *MS* in cell D23 stands for mean square.

In Section 12.5 we stated that the mean square error, obtained by dividing the error or residual sum of squares by its degrees of freedom, provides an estimate of σ^2. The value in cell D25, 191.25, is the mean square error for the Armand's regression output. In Section 12.5 we showed that an *F* test could also be used to test for significance in regression. The value in cell F24, .0000255, is the *p*-value associated with the *F* test for significance. Because the *p*-value = .0000255 < α = .01, we can reject H_0 and conclude that we have a significant relationship between student population and quarterly sales. The label Excel uses to identify the *p*-value for the *F* test for significance, shown in cell F23, is *Significance F*.

The label Significance F may be more meaningful if you think of the value in cell F24 as the observed level of significance for the F test.

Interpretation of Regression Statistics Output

The coefficient of determination, .9027, appears in cell B17; the corresponding label, R Square, is shown in cell A17. The square root of the coefficient of determination provides the sample correlation coefficient of .9501 shown in cell B16. Note that Excel uses the label Multiple R (cell A16) to identify this value. In cell A19, the label Standard Error is used to identify the value of the standard error of the estimate shown in cell B19. Thus, the standard error of the estimate is 13.8293. We caution the reader to keep in mind that in the Excel output, the label Standard Error appears in two different places. In the Regression Statistics section of the output, the label Standard Error refers to the estimate of σ. In the Estimated Regression Equation section of the output, the label *Standard Error* refers to s_{b_1}, the standard deviation of the sampling distribution of b_1.

Appendix 12.3 Regression Analysis Using StatTools

In this appendix we show how StatTools can be used to perform the regression analysis computations for the Armand's Pizza Parlors problem. Begin by using the Data Set Manager to create a StatTools data set for these data using the procedure described in the appendix in Chapter 1. The following steps describe how StatTools can be used to provide the regression results.

Step 1. Click the **StatTools** tab on the Ribbon
Step 2. In the **Analyses** group, click **Regression and Classification**
Step 3. Choose the **Regression** option
Step 4. When the StatTools - Regression dialog box appears:
Select **Multiple** in the **Regression Type** box
In the **Variables** section,
Click the **Format button** and select **Unstacked**
In the column labeled **I** select **Population**
In the column labeled **D** select **Sales**
Click **OK**

The regression analysis output will appear in a new worksheet.

Note that in step 4 we selected Multiple in the Regression Type box. In StatTools, the Multiple option is used for both simple linear regression and multiple regression. The StatTools - Regression dialog box contains a number of more advanced options for developing prediction interval estimates and producing residual plots. The StatTools Help facility provides information on using all of these options.

CHAPTER 13

Multiple Regression

CONTENTS

STATISTICS IN PRACTICE:
INTERNATIONAL PAPER

13.1 MULTIPLE REGRESSION MODEL
Regression Model and Regression Equation
Estimated Multiple Regression Equation

13.2 LEAST SQUARES METHOD
An Example: Butler Trucking Company
Note on Interpretation of Coefficients

13.3 MULTIPLE COEFFICIENT OF DETERMINATION

13.4 MODEL ASSUMPTIONS

13.5 TESTING FOR SIGNIFICANCE
F Test
t Test
Multicollinearity

13.6 USING THE ESTIMATED REGRESSION EQUATION FOR ESTIMATION AND PREDICTION

13.7 CATEGORICAL INDEPENDENT VARIABLES
An Example: Johnson Filtration, Inc.
Interpreting the Parameters
More Complex Categorical Variables

STATISTICS in PRACTICE

INTERNATIONAL PAPER*
PURCHASE, NEW YORK

International Paper is the world's largest paper and forest products company. The company employs more than 117,000 people in its operations in nearly 50 countries, and exports its products to more than 130 nations. International Paper produces building materials such as lumber and plywood; consumer packaging materials such as disposable cups and containers; industrial packaging materials such as corrugated boxes and shipping containers; and a variety of papers for use in photocopiers, printers, books, and advertising materials.

To make paper products, pulp mills process wood chips and chemicals to produce wood pulp. The wood pulp is then used at a paper mill to produce paper products. In the production of white paper products, the pulp must be bleached to remove any discoloration. A key bleaching agent used in the process is chlorine dioxide, which, because of its combustible nature, is usually produced at a pulp mill facility and then piped in solution form into the bleaching tower of the pulp mill. To improve one of the processes used to produce chlorine dioxide, researchers studied the process's control and efficiency. One aspect of the study looked at the chemical-feed rate for chlorine dioxide production.

To produce the chlorine dioxide, four chemicals flow at metered rates into the chlorine dioxide generator. The chlorine dioxide produced in the generator flows to an absorber where chilled water absorbs the chlorine dioxide gas to form a chlorine dioxide solution. The solution is then piped into the paper mill. A key part of controlling the process involves the chemical-feed rates. Historically, experienced operators set the chemical-feed rates, but this approach led to overcontrol by the operators. Consequently, chemical engineers at the mill requested that a set of control equations, one for each

Multiple regression analysis assisted in the development of a better bleaching process for making white paper products.

chemical feed, be developed to aid the operators in setting the rates.

Using multiple regression analysis, statistical analysts developed an estimated multiple regression equation for each of the four chemicals used in the process. Each equation related the production of chlorine dioxide to the amount of chemical used and the concentration level of the chlorine dioxide solution. The resulting set of four equations was programmed into a microcomputer at each mill. In the new system, operators enter the concentration of the chlorine dioxide solution and the desired production rate; the computer software then calculates the chemical feed needed to achieve the desired production rate. After the operators began using the control equations, the chlorine dioxide generator efficiency increased, and the number of times the concentrations fell within acceptable ranges increased significantly.

This example shows how multiple regression analysis can be used to develop a better bleaching process for producing white paper products. In this chapter we will discuss how computer software packages are used for such purposes. Most of the concepts introduced in Chapter 12 for simple linear regression can be directly extended to the multiple regression case.

---

*The authors are indebted to Marian Williams and Bill Griggs for providing this Statistics in Practice. This application was originally developed at Champion International Corporation, which became part of International Paper in 2000.

In Chapter 12 we presented simple linear regression and demonstrated its use in developing an estimated regression equation that describes the relationship between two variables. Recall that the variable being predicted or explained is called the dependent variable and the variable being used to predict or explain the dependent variable is called the independent variable. In this chapter we continue our study of regression analysis by considering situations involving two or more independent variables. This subject area, called **multiple regression analysis**, enables us to consider more factors and thus obtain better estimates than are possible with simple linear regression.

13.1 Multiple Regression Model

Multiple regression analysis is the study of how a dependent variable y is related to two or more independent variables. In the general case, we will use p to denote the number of independent variables.

Regression Model and Regression Equation

The concepts of a regression model and a regression equation introduced in the preceding chapter are applicable in the multiple regression case. The equation that describes how the dependent variable y is related to the independent variables $x_1, x_2, \ldots, x_p$ and an error term is called the **multiple regression model**. We begin with the assumption that the multiple regression model takes the following form.

> **MULTIPLE REGRESSION MODEL**
> $$y = \beta_0 + \beta_1 x_1 + \beta_2 x_2 + \cdots + \beta_p x_p + \epsilon \tag{13.1}$$

In the multiple regression model, $\beta_0, \beta_1, \beta_2, \ldots, \beta_p$ are the parameters and the error term ϵ (the Greek letter epsilon) is a random variable. A close examination of this model reveals that y is a linear function of $x_1, x_2, \ldots, x_p$ (the $\beta_0 + \beta_1 x_1 + \beta_2 x_2 + \cdots + \beta_p x_p$ part) plus the error term ϵ. The error term accounts for the variability in y that cannot be explained by the linear effect of the p independent variables.

In Section 13.4 we will discuss the assumptions for the multiple regression model and ϵ. One of the assumptions is that the mean or expected value of ϵ is zero. A consequence of this assumption is that the mean or expected value of y, denoted $E(y)$, is equal to $\beta_0 + \beta_1 x_1 + \beta_2 x_2 + \cdots + \beta_p x_p$. The equation that describes how the mean value of y is related to $x_1, x_2, \ldots, x_p$ is called the **multiple regression equation**.

> **MULTIPLE REGRESSION EQUATION**
> $$E(y) = \beta_0 + \beta_1 x_1 + \beta_2 x_2 + \cdots + \beta_p x_p \tag{13.2}$$

Estimated Multiple Regression Equation

If the values of $\beta_0, \beta_1, \beta_2, \ldots, \beta_p$ were known, equation (13.2) could be used to compute the mean value of y at given values of $x_1, x_2, \ldots, x_p$. Unfortunately, these parameter values will not, in general, be known and must be estimated from sample data. A simple random sample is used to compute sample statistics $b_0, b_1, b_2, \ldots, b_p$ that are used as the point

FIGURE 13.1 THE ESTIMATION PROCESS FOR MULTIPLE REGRESSION

In simple linear regression, b_0 and b_1 were the sample statistics used to estimate the parameters β_0 and β_1. Multiple regression parallels this statistical inference process, with b_0, $b_1, b_2, \ldots, b_p$ denoting the sample statistics used to estimate the parameters $\beta_0, \beta_1, \beta_2, \ldots, \beta_p$.

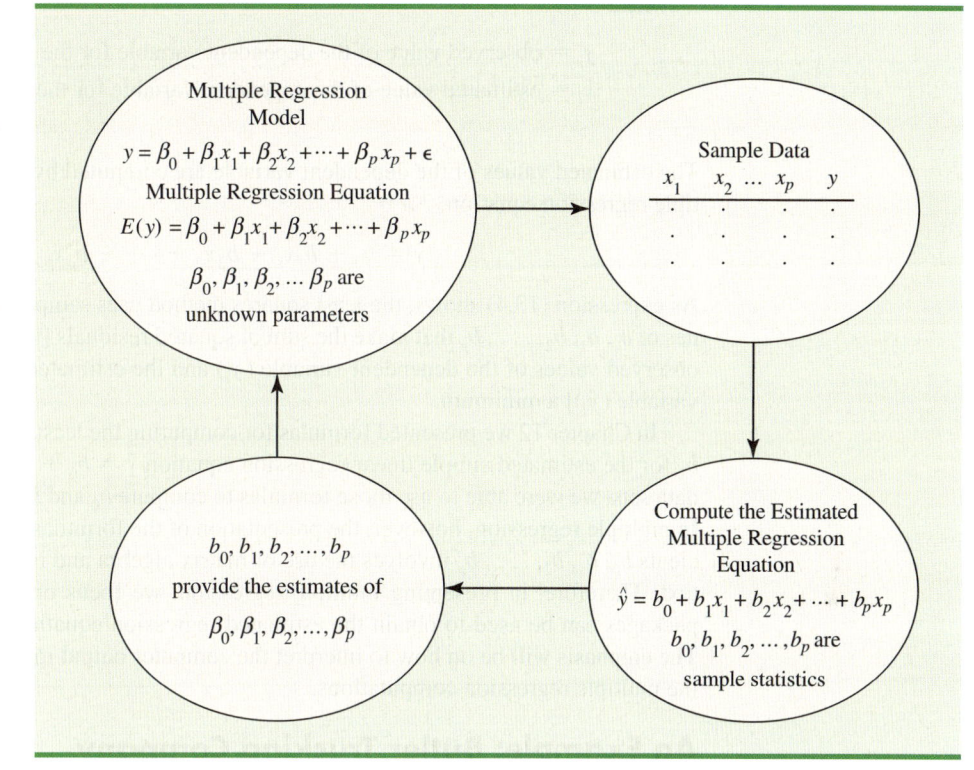

Multiple Regression Model
$y = \beta_0 + \beta_1 x_1 + \beta_2 x_2 + \cdots + \beta_p x_p + \epsilon$
Multiple Regression Equation
$E(y) = \beta_0 + \beta_1 x_1 + \beta_2 x_2 + \cdots + \beta_p x_p$
$\beta_0, \beta_1, \beta_2, \ldots \beta_p$ are unknown parameters

Sample Data:
$x_1 \quad x_2 \quad \ldots \quad x_p \quad y$

Compute the Estimated Multiple Regression Equation
$\hat{y} = b_0 + b_1 x_1 + b_2 x_2 + \cdots + b_p x_p$
$b_0, b_1, b_2, \ldots, b_p$ are sample statistics

$b_0, b_1, b_2, \ldots, b_p$ provide the estimates of $\beta_0, \beta_1, \beta_2, \ldots, \beta_p$

estimators of the parameters $\beta_0, \beta_1, \beta_2, \ldots, \beta_p$. These sample statistics provide the following **estimated multiple regression equation**.

ESTIMATED MULTIPLE REGRESSION EQUATION

$$\hat{y} = b_0 + b_1 x_1 + b_2 x_2 + \cdots + b_p x_p \tag{13.3}$$

where

$b_0, b_1, b_2, \ldots, b_p$ are the estimates of $\beta_0, \beta_1, \beta_2, \ldots, \beta_p$
$\hat{y}$ = estimated value of the dependent variable

The estimation process for multiple regression is shown in Figure 13.1.

13.2 Least Squares Method

In Chapter 12, we used the **least squares method** to develop the estimated regression equation that best approximated the straight line relationship between the dependent and independent variables. This same approach is used to develop the estimated multiple regression equation. The least squares criterion is restated as follows.

LEAST SQUARES CRITERION

$$\min \Sigma (y_i - \hat{y}_i)^2 \tag{13.4}$$

where

y_i = observed value of the dependent variable for the ith observation
$\hat{y}_i$ = estimated value of the dependent variable for the ith observation

The estimated values of the dependent variable are computed by using the estimated multiple regression equation,

$$\hat{y} = b_0 + b_1x_1 + b_2x_2 + \cdots + b_px_p$$

As expression (13.4) shows, the least squares method uses sample data to provide the values of $b_0, b_1, b_2, \ldots, b_p$ that make the sum of squared residuals [the deviations between the observed values of the dependent variable (y_i) and the estimated values of the dependent variable ($\hat{y}_i$)] a minimum.

In Chapter 12 we presented formulas for computing the least squares estimators b_0 and b_1 for the estimated simple linear regression equation $\hat{y} = b_0 + b_1x$. With relatively small data sets, we were able to use those formulas to compute b_0 and b_1 by manual calculations. In multiple regression, however, the presentation of the formulas for the regression coefficients $b_0, b_1, b_2, \ldots, b_p$ involves the use of matrix algebra and is beyond the scope of this text. Therefore, in presenting multiple regression, we focus on how computer software packages can be used to obtain the estimated regression equation and other information. The emphasis will be on how to interpret the computer output rather than on how to make the multiple regression computations.

An Example: Butler Trucking Company

As an illustration of multiple regression analysis, we will consider a problem faced by the Butler Trucking Company, an independent trucking company in southern California. A major portion of Butler's business involves deliveries throughout its local area. To develop better work schedules, the managers want to estimate the total daily travel time for their drivers.

Initially the managers believed that the total daily travel time would be closely related to the number of miles traveled in making the daily deliveries. A simple random sample of 10 driving assignments provided the data shown in Table 13.1 and the scatter diagram shown in Figure 13.2. After reviewing this scatter diagram, the managers hypothesized that the simple linear regression model $y = \beta_0 + \beta_1x_1 + \epsilon$ could be used to describe the relationship between the total travel time (y) and the number of miles traveled (x_1). To estimate

TABLE 13.1 PRELIMINARY DATA FOR BUTLER TRUCKING

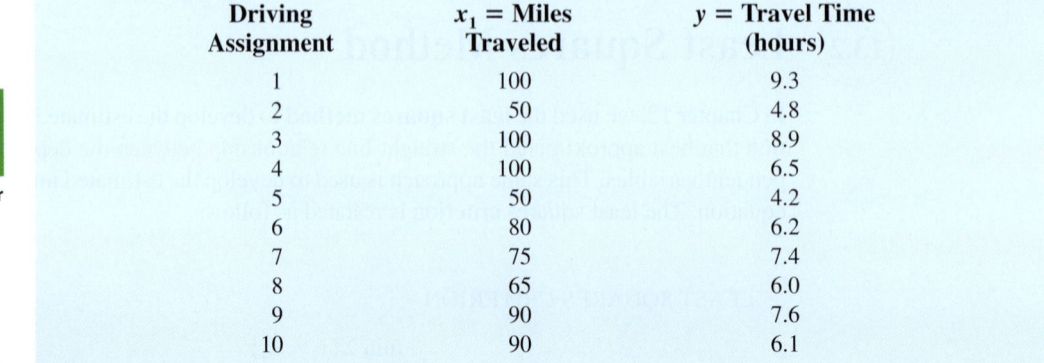

Driving Assignment	x_1 = Miles Traveled	y = Travel Time (hours)
1	100	9.3
2	50	4.8
3	100	8.9
4	100	6.5
5	50	4.2
6	80	6.2
7	75	7.4
8	65	6.0
9	90	7.6
10	90	6.1

13.2 Least Squares Method

FIGURE 13.2 SCATTER DIAGRAM OF PRELIMINARY DATA FOR BUTLER TRUCKING

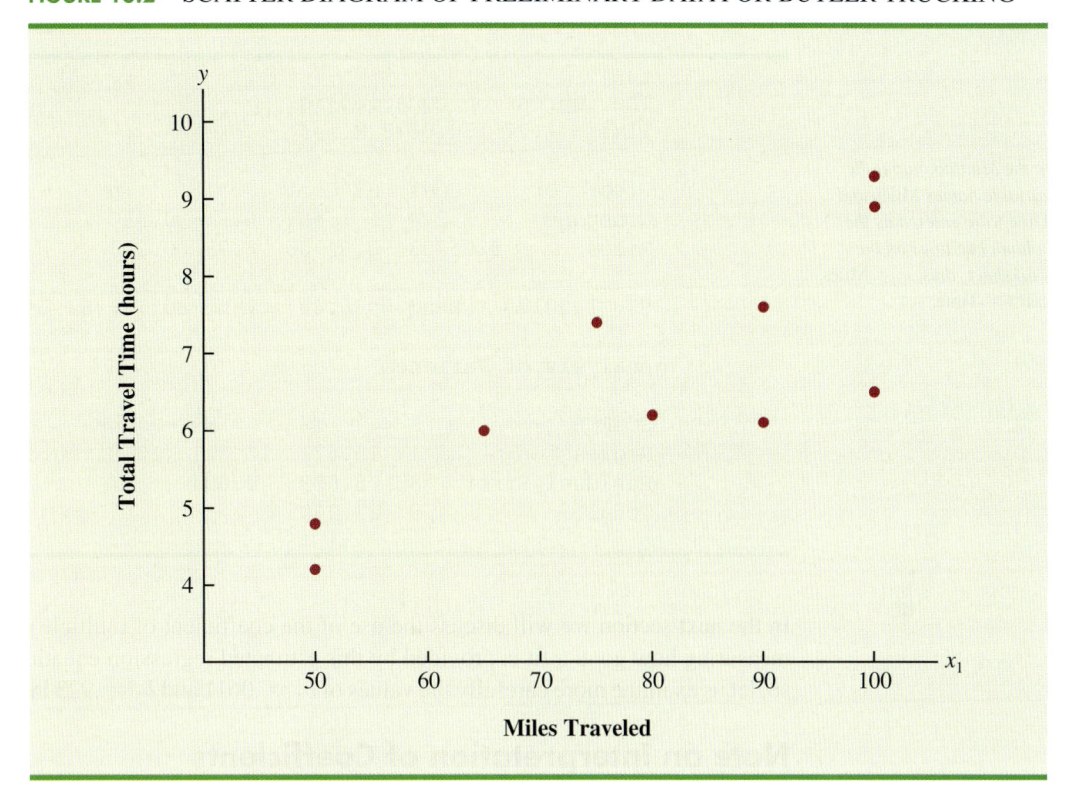

the parameters β_0 and β_1, the least squares method was used to develop the estimated regression equation.

$$\hat{y} = b_0 + b_1 x_1 \tag{13.5}$$

In Figure 13.3, we show the Minitab computer output from applying simple linear regression to the data in Table 13.1. The estimated regression equation is

$$\hat{y} = 1.27 + .0678 x_1$$

At the .05 level of significance, the F value of 15.81 and its corresponding p-value of .004 indicate that the relationship is significant; that is, we can reject H_0: $\beta_1 = 0$ because the p-value is less than $\alpha = .05$. Note that the same conclusion is obtained from the t value of 3.98 and its associated p-value of .004. Thus, we can conclude that the relationship between the total travel time and the number of miles traveled is significant; longer travel times are associated with more miles traveled. With a coefficient of determination (expressed as a percentage) of R-sq = 66.4%, we see that 66.4% of the variability in travel time can be explained by the linear effect of the number of miles traveled. This finding is fairly good, but the managers might want to consider adding a second independent variable to explain some of the remaining variability in the dependent variable.

In attempting to identify another independent variable, the managers felt that the number of deliveries could also contribute to the total travel time. The Butler Trucking data, with the number of deliveries added, are shown in Table 13.2. The Minitab computer solution with both miles traveled (x_1) and number of deliveries (x_2) as independent variables is shown in Figure 13.4. The estimated regression equation is

$$\hat{y} = -.869 + .0611 x_1 + .923 x_2 \tag{13.6}$$

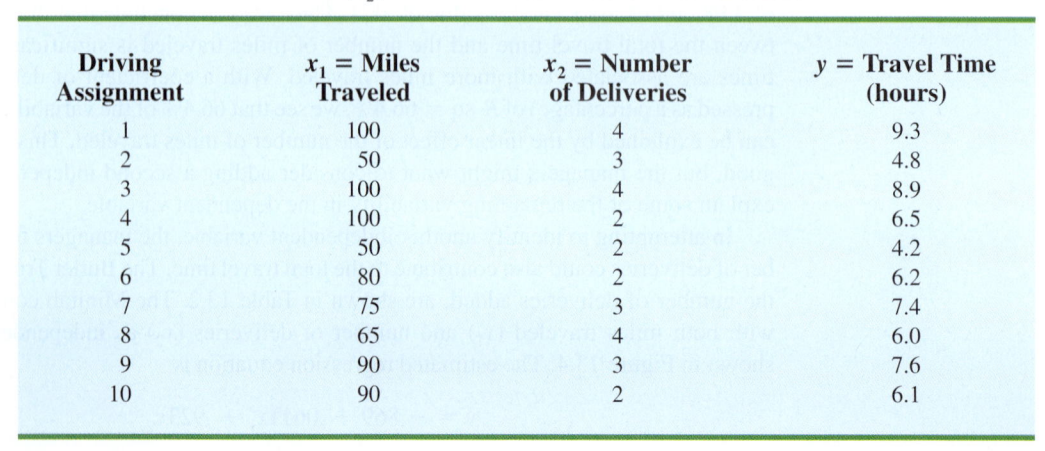

FIGURE 13.3 MINITAB OUTPUT FOR BUTLER TRUCKING WITH ONE INDEPENDENT VARIABLE

```
The regression equation is
Time = 1.27 + 0.0678 Miles

Predictor        Coef     SE Coef       T       p
Constant        1.274       1.401    0.91   0.390
Miles         0.06783     0.01706    3.98   0.004

S = 1.00179    R-sq = 66.4%    R-sq(adj) = 62.2%

Analysis of Variance

SOURCE            DF         SS         MS        F       p
Regression         1     15.871     15.871    15.81   0.004
Residual Error     8      8.029      1.004
Total              9     23.900
```

In the Minitab output the variable names Miles and Time were entered as the column headings on the worksheet; thus, x_1 = Miles and y = Time.

In the next section we will discuss the use of the coefficient of multiple determination in measuring how good a fit is provided by this estimated regression equation. Before doing so, let us examine more carefully the values of b_1 = .0611 and b_2 = .923 in equation (13.6).

Note on Interpretation of Coefficients

One observation can be made at this point about the relationship between the estimated regression equation with only the miles traveled as an independent variable and the equation that includes the number of deliveries as a second independent variable. The value of b_1 is not the same in both cases. In simple linear regression, we interpret b_1 as an estimate of the change in y for a one-unit change in the independent variable. In multiple regression analysis, this interpretation must be modified somewhat. That is, in multiple regression analysis, we interpret each regression coefficient as follows: b_i represents an estimate of the change in y corresponding to a one-unit change in x_i when all other independent variables are held constant. In the Butler Trucking example involving two independent variables, b_1 = .0611. Thus,

TABLE 13.2 DATA FOR BUTLER TRUCKING WITH MILES TRAVELED (x_1) AND NUMBER OF DELIVERIES (x_2) AS THE INDEPENDENT VARIABLES

Driving Assignment	x_1 = Miles Traveled	x_2 = Number of Deliveries	y = Travel Time (hours)
1	100	4	9.3
2	50	3	4.8
3	100	4	8.9
4	100	2	6.5
5	50	2	4.2
6	80	2	6.2
7	75	3	7.4
8	65	4	6.0
9	90	3	7.6
10	90	2	6.1

13.2 Least Squares Method

FIGURE 13.4 MINITAB OUTPUT FOR BUTLER TRUCKING WITH TWO INDEPENDENT VARIABLES

In the Minitab output the variable names Miles, Deliveries, *and* Time *were entered as the column headings on the worksheet; thus, x_1 =* Miles, x_2 = Deliveries, *and y =* Time.

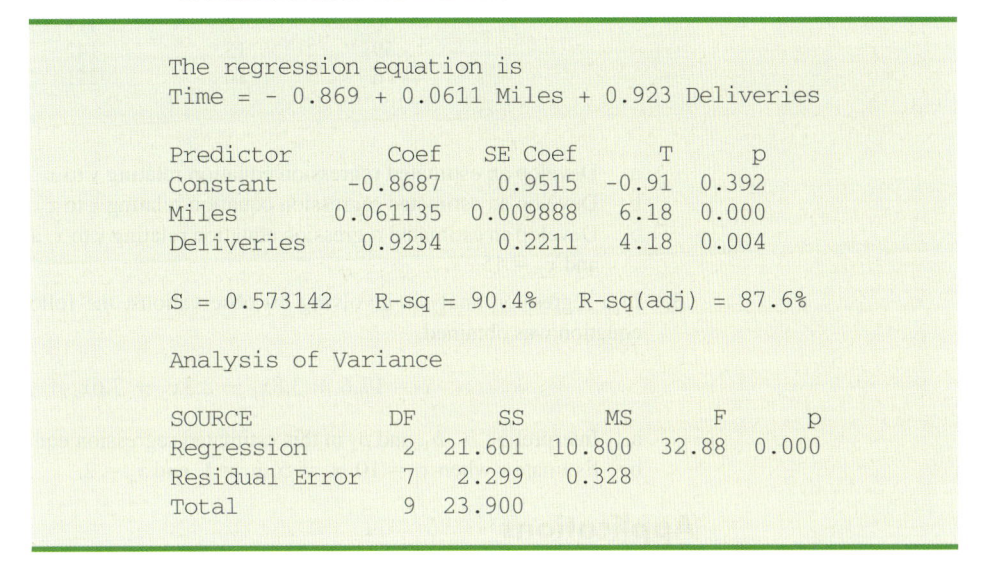

```
The regression equation is
Time = - 0.869 + 0.0611 Miles + 0.923 Deliveries

Predictor         Coef     SE Coef        T       p
Constant       -0.8687      0.9515    -0.91   0.392
Miles         0.061135    0.009888     6.18   0.000
Deliveries      0.9234      0.2211     4.18   0.004

S = 0.573142    R-sq = 90.4%    R-sq(adj) = 87.6%

Analysis of Variance

SOURCE          DF       SS       MS        F        p
Regression       2   21.601   10.800    32.88    0.000
Residual Error   7    2.299    0.328
Total            9   23.900
```

.0611 hours is an estimate of the expected increase in travel time corresponding to an increase of one mile in the distance traveled when the number of deliveries is held constant. Similarly, because b_2 = .923, an estimate of the expected increase in travel time corresponding to an increase of one delivery when the number of miles traveled is held constant is .923 hours.

Exercises

Note to student: The exercises involving data in this and subsequent sections were designed to be solved using a computer software package.

Methods

1. The estimated regression equation for a model involving two independent variables and 10 observations follows.

$$\hat{y} = 29.1270 + .5906x_1 + .4980x_2$$

   a. Interpret b_1 and b_2 in this estimated regression equation.
   b. Estimate y when x_1 = 180 and x_2 = 310.

2. Consider the following data for a dependent variable y and two independent variables, x_1 and x_2.

x_1	x_2	y
30	12	94
47	10	108
25	17	112
51	16	178
40	5	94
51	19	175
74	7	170

(continued)

x_1	x_2	y
36	12	117
59	13	142
76	16	211

a. Develop an estimated regression equation relating y to x_1. Estimate y if $x_1 = 45$.
b. Develop an estimated regression equation relating y to x_2. Estimate y if $x_2 = 15$.
c. Develop an estimated regression equation relating y to x_1 and x_2. Estimate y if $x_1 = 45$ and $x_2 = 15$.

3. In a regression analysis involving 30 observations, the following estimated regression equation was obtained.

$$\hat{y} = 17.6 + 3.8x_1 - 2.3x_2 + 7.6x_3 + 2.7x_4$$

a. Interpret b_1, b_2, b_3, and b_4 in this estimated regression equation.
b. Estimate y when $x_1 = 10$, $x_2 = 5$, $x_3 = 1$, and $x_4 = 2$.

Applications

4. A shoe store developed the following estimated regression equation relating sales to inventory investment and advertising expenditures.

$$\hat{y} = 25 + 10x_1 + 8x_2$$

where

$$x_1 = \text{inventory investment (\$1000s)}$$
$$x_2 = \text{advertising expenditures (\$1000s)}$$
$$y = \text{sales (\$1000s)}$$

a. Estimate sales resulting from a $15,000 investment in inventory and an advertising budget of $10,000.
b. Interpret b_1 and b_2 in this estimated regression equation.

5. The owner of Showtime Movie Theaters, Inc., would like to estimate weekly gross revenue as a function of advertising expenditures. Historical data for a sample of eight weeks follow.

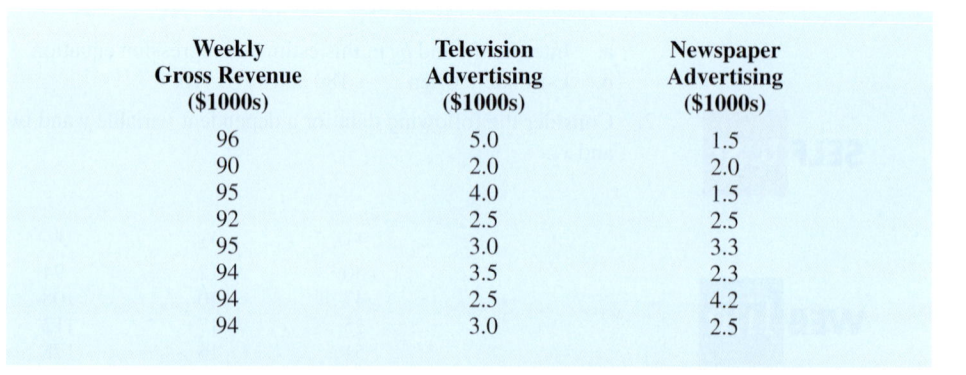

Weekly Gross Revenue ($1000s)	Television Advertising ($1000s)	Newspaper Advertising ($1000s)
96	5.0	1.5
90	2.0	2.0
95	4.0	1.5
92	2.5	2.5
95	3.0	3.3
94	3.5	2.3
94	2.5	4.2
94	3.0	2.5

a. Develop an estimated regression equation with the amount of television advertising as the independent variable.
b. Develop an estimated regression equation with both television advertising and newspaper advertising as the independent variables.

c. Is the estimated regression equation coefficient for television advertising expenditures the same in part (a) and in part (b)? Interpret the coefficient in each case.
d. What is the estimate of the weekly gross revenue for a week when $3500 is spent on television advertising and $1800 is spent on newspaper advertising?

6. Out-of-state tuition and fees at the top graduate schools of business can be very expensive. However, the starting salary and bonus paid to graduates from many of these schools can be very high. The following data show the recruiter assessment score (highest = 5), the out-of-state tuition and fees (rounded to the nearest $1000), and the average starting salary and bonus paid to recent graduates (rounded to the nearest $1000) for a sample of 20 graduate schools of business (*U.S. News & World Report 2009 Edition America's Best Graduate Schools*).

School	Recruiter Score	Tuition & Fees ($)	Salary & Bonus ($)
Arizona State University	3.4	28	98
Babson College	3.5	35	94
Cornell University	3.8	44	119
Georgetown University	3.7	40	109
Georgia Institute of Technology	3.6	30	88
Indiana University—Bloomington	3.7	35	105
Michigan State University	3.2	26	99
Northwestern University	4.3	44	123
Ohio State University	3.1	35	97
Purdue University—West Lafayette	3.6	33	96
Rice University	3.4	36	102
Stanford University	4.4	46	135
University of California—Davis	3.3	35	89
University of Florida	3.5	23	71
University of Iowa	3.1	25	78
University of Minnesota—Twin Cities	3.5	37	100
University of Notre Dame	3.4	36	95
University of Rochester	3.5	38	99
University of Washington	3.4	30	94
University of Wisconsin—Madison	3.3	27	93

a. Develop an estimated regression equation that can be used to predict the starting salary and bonus paid to graduates given the recruiter assessment score.
b. Develop an estimated regression equation that can be used to predict the starting salary and bonus paid to graduates given both the recruiter assessment score and the out-of-state tuition and fees.
c. Is the estimated regression coefficient for the recruiter score the same in part (a) and in part (b)? Interpret the coefficient in each case.
d. Suppose that we randomly select a recent graduate of the University of Virginia graduate school of business. This school has a recruiter assessment score of 4.1 and an out-of-state tuition and fees of $43,000. Estimate the starting salary and bonus for this graduate.

7. *PC World* rated four component characteristics for 10 ultraportable laptop computers: features; performance; design; and price. Each characteristic was rated using a 0–100 point scale. An overall rating, referred to as the *PCW World* Rating, was then developed for each laptop. The following table shows the performance rating, features rating, and the *PCW World* Rating for the 10 laptop computers (PC World website, February 5, 2009).

Model	Performance	Features	*PCW* Rating
Thinkpad X200	77	87	83
VGN-Z598U	97	85	82
U6V	83	80	81
Elitebook 2530P	77	75	78
X360	64	80	78
Thinkpad X300	56	76	78
Ideapad U110	55	81	77
Micro Express JFT2500	76	73	75
Toughbook W7	46	79	73
HP Voodoo Envy133	54	68	72

a. Determine the estimated regression equation that can be used to predict the *PCW World* Rating using the performance rating as the independent variable.
b. Determine the estimated regression equation that can be used to predict the *PCW World* Rating using both the performance rating and the features rating.
c. Predict the *PCW World* Rating for a laptop computer that has a performance rating of 80 and a features rating of 70.

8. Would you expect more reliable and better performing cars to cost more? *Consumer Reports* provided reliability ratings, overall road-test scores, and prices for affordable family sedans, midpriced family sedans, and large sedans (*Consumer Reports*, February 2008). A portion of the data follows. Reliability was rated on a 5-point scale from poor (1) to excellent (5). The road-test score was rated on a 100-point scale, with higher values indicating better performance. The complete data set is contained in the file named Sedans.

Make and Model	Road-Test Score	Reliability	Price ($)
Nissan Altima 2.5 S	85	4	22,705
Honda Accord LX-P	79	4	22,795
Kia Optima EX (4-cyl.)	78	4	22,795
Toyota Camry LE	77	4	21,080
Hyundai Sonata SE	76	3	22,995
.	.	.	.
.	.	.	.
.	.	.	.
Chrysler 300 Touring	60	2	30,255
Dodge Charger SXT	58	4	28,860

a. Develop an estimated regression equation that can be used to predict the price of the car given the reliability rating. Test for significance using $\alpha = .05$.
b. Consider the addition of the independent variable overall road-test score. Develop the estimated regression equation that can be used to predict the price of the car given the road-test score and the reliability rating.
c. Estimate the price for a car with a road-test score of 80 and a reliability rating of 4.

9. Waterskiing and wakeboarding are two popular watersports. Finding a model that best suits your intended needs, whether it is waterskiing, wakeboading, or general boating, can be a difficult task. *WaterSki* magazine did extensive testing for 88 boats and provided a wide variety of information to help consumers select the best boat. A portion of the data they reported for 20 boats with a length of between 20 and 22 feet follows (*WaterSki*, January/February 2006). Beam is the maximum width of the boat in inches, HP is the horsepower of the boat's engine, and TopSpeed is the top speed in miles per hour (mph).

13.2 Least Squares Method

Boats (WEBfile)

Make and Model	Beam	HP	TopSpeed
Calabria Cal Air Pro V-2	100	330	45.3
Correct Craft Air Nautique 210	91	330	47.3
Correct Craft Air Nautique SV-211	93	375	46.9
Correct Craft Ski Nautique 206 Limited	91	330	46.7
Gekko GTR 22	96	375	50.1
Gekko GTS 20	83	375	52.2
Malibu Response LXi	93.5	340	47.2
Malibu Sunsettter LXi	98	400	46
Malibu Sunsetter 21 XTi	98	340	44
Malibu Sunscape 21 LSV	98	400	47.5
Malibu Wakesetter 21 XTi	98	340	44.9
Malibu Wakesetter VLX	98	400	47.3
Malibu vRide	93.5	340	44.5
Malibu Ride XTi	93.5	320	44.5
Mastercraft ProStar 209	96	350	42.5
Mastercraft X-1	90	310	45.8
Mastercraft X-2	94	310	42.8
Mastercraft X-9	96	350	43.2
MB Sports 190 Plus	92	330	45.3
Svfara SVONE	91	330	47.7

a. Using these data, develop an estimated regression equation relating the top speed with the boat's beam and horsepower rating.

b. The Svfara SV609 has a beam of 85 inches and an engine with a 330 horsepower rating. Use the estimated regression equation developed in part (a) to estimate the top speed for the Svfara SV609.

10. The National Basketball Association (NBA) records a variety of statistics for each team. Four of these statistics are the proportion of games won (PCT), the proportion of field goals made by the team (FG%), the proportion of three-point shots made by the team's opponent (Opp 3 Pt%), and the number of turnovers committed by the team's opponent (Opp TO). The following data show the values of these statistics for the 29 teams in the NBA for a portion of the 2004 season (NBA website, January 3, 2004).

NBA (WEBfile)

Team	PCT	FG%	Opp 3 Pt%	Opp TO	Team	PCT	FG%	Opp 3 Pt%	Opp TO
Atlanta	0.265	0.435	0.346	13.206	Minnesota	0.677	0.473	0.348	13.839
Boston	0.471	0.449	0.369	16.176	New Jersey	0.563	0.435	0.338	17.063
Chicago	0.313	0.417	0.372	15.031	New Orleans	0.636	0.421	0.330	16.909
Cleveland	0.303	0.438	0.345	12.515	New York	0.412	0.442	0.330	13.588
Dallas	0.581	0.439	0.332	15.000	Orlando	0.242	0.417	0.360	14.242
Denver	0.606	0.431	0.366	17.818	Philadelphia	0.438	0.428	0.364	16.938
Detroit	0.606	0.423	0.262	15.788	Phoenix	0.364	0.438	0.326	16.515
Golden State	0.452	0.445	0.384	14.290	Portland	0.484	0.447	0.367	12.548
Houston	0.548	0.426	0.324	13.161	Sacramento	0.724	0.466	0.327	15.207
Indiana	0.706	0.428	0.317	15.647	San Antonio	0.688	0.429	0.293	15.344
L.A. Clippers	0.464	0.424	0.326	14.357	Seattle	0.533	0.436	0.350	16.767
L.A. Lakers	0.724	0.465	0.323	16.000	Toronto	0.516	0.424	0.314	14.129
Memphis	0.485	0.432	0.358	17.848	Utah	0.531	0.456	0.368	15.469
Miami	0.424	0.410	0.369	14.970	Washington	0.300	0.411	0.341	16.133
Milwaukee	0.500	0.438	0.349	14.750					

a. Determine the estimated regression equation that can be used to predict the proportion of games won given the proportion of field goals made by the team.

b. Provide an interpretation for the slope of the estimated regression equation developed in part (a).
c. Determine the estimated regression equation that can be used to predict the proportion of games won given the proportion of field goals made by the team, the proportion of three-point shots made by the team's opponent, and the number of turnovers committed by the team's opponent.
d. Discuss the practical implications of the estimated regression equation developed in part (c).
e. Estimate the proportion of games won for a team with the following values for the three independent variables: FG% = .45, Opp 3 Pt% = .34, and Opp TO = 17.

13.3 Multiple Coefficient of Determination

In simple linear regression we showed that the total sum of squares can be partitioned into two components: the sum of squares due to regression and the sum of squares due to error. The same procedure applies to the sum of squares in multiple regression.

RELATIONSHIP AMONG SST, SSR, AND SSE

$$\text{SST} = \text{SSR} + \text{SSE} \tag{13.7}$$

where

$$\text{SST} = \text{total sum of squares} = \Sigma(y_i - \bar{y})^2$$
$$\text{SSR} = \text{sum of squares due to regression} = \Sigma(\hat{y}_i - \bar{y})^2$$
$$\text{SSE} = \text{sum of squares due to error} = \Sigma(y_i - \hat{y}_i)^2$$

Because of the computational difficulty in computing the three sums of squares, we rely on computer packages to determine those values. The Analysis of Variance (ANOVA) part of the Minitab output in Figure 13.4 shows the three values for the Butler Trucking problem with two independent variables: SST = 23.900, SSR = 21.601, and SSE = 2.299. With only one independent variable (number of miles traveled), the Minitab output in Figure 13.3 shows that SST = 23.900, SSR = 15.871, and SSE = 8.029. The value of SST is the same in both cases because it does not depend on $\hat{y}$, but SSR increases and SSE decreases when a second independent variable (number of deliveries) is added. The implication is that the estimated multiple regression equation provides a better fit for the observed data.

In Chapter 12, we used the coefficient of determination, $r^2 = \text{SSR}/\text{SST}$, to measure the goodness of fit for the estimated regression equation. The same concept applies to multiple regression. The term **multiple coefficient of determination** indicates that we are measuring the goodness of fit for the estimated multiple regression equation. The multiple coefficient of determination, denoted R^2, is computed as follows.

MULTIPLE COEFFICIENT OF DETERMINATION

$$R^2 = \frac{\text{SSR}}{\text{SST}} \tag{13.8}$$

The multiple coefficient of determination can be interpreted as the proportion of the variability in the dependent variable that can be explained by the estimated multiple regression equation. Hence, when multiplied by 100, it can be interpreted as the percentage of the variability in y that can be explained by the estimated regression equation.

In the two-independent-variable Butler Trucking example, with SSR = 21.601 and SST = 23.900, we have

$$R^2 = \frac{21.601}{23.900} = .904$$

Therefore, 90.4% of the variability in travel time y is explained by the estimated multiple regression equation with miles traveled and number of deliveries as the independent variables. In Figure 13.4, we see that the multiple coefficient of determination (expressed as a percentage) is also provided by the Minitab output; it is denoted by R-sq = 90.4%.

Figure 13.3 shows that the R-sq value for the estimated regression equation with only one independent variable, number of miles traveled (x_1), is 66.4%. Thus, the percentage of the variability in travel times that is explained by the estimated regression equation increases from 66.4% to 90.4% when number of deliveries is added as a second independent variable. In general, R^2 always increases as independent variables are added to the model.

Many analysts prefer adjusting R^2 for the number of independent variables to avoid overestimating the impact of adding an independent variable on the amount of variability explained by the estimated regression equation. With n denoting the number of observations and p denoting the number of independent variables, the **adjusted multiple coefficient of determination** is computed as follows.

Adding independent variables causes the prediction errors to become smaller, thus reducing the sum of squares due to error, SSE. Because SSR = SST − SSE, when SSE becomes smaller, SSR becomes larger, causing R^2 = SSR/SST to increase.

If a variable is added to the model, R^2 becomes larger even if the variable added is not statistically significant. The adjusted multiple coefficient of determination compensates for the number of independent variables in the model.

ADJUSTED MULTIPLE COEFFICIENT OF DETERMINATION

$$R_a^2 = 1 - (1 - R^2)\frac{n-1}{n-p-1} \tag{13.9}$$

For the Butler Trucking example with $n = 10$ and $p = 2$, we have

$$R_a^2 = 1 - (1 - .904)\frac{10-1}{10-2-1} = .88$$

Thus, after adjusting for the two independent variables, we have an adjusted multiple coefficient of determination of .88. This value (expressed as a percentage) is provided by the Minitab output in Figure 13.4 as R-sq(adj) = 87.6%; the value we calculated differs because we used a rounded value of R^2 in the calculation.

NOTES AND COMMENTS

If the value of R^2 is small and the model contains a large number of independent variables, the adjusted coefficient of determination can take a negative value; in such cases, Minitab sets the adjusted coefficient of determination to zero.

Exercises

Methods

11. In exercise 1, the following estimated regression equation based on 10 observations was presented.

$$\hat{y} = 29.1270 + .5906x_1 + .4980x_2$$

The values of SST and SSR are 6724.125 and 6216.375, respectively.
   a. Find SSE.
   b. Compute R^2.
   c. Compute R_a^2.
   d. Comment on the goodness of fit.

12. **SELF test** In exercise 2, 10 observations were provided for a dependent variable y and two independent variables x_1 and x_2; for these data SST = 15,182.9, and SSR = 14,052.2.
   a. Compute R^2.
   b. Compute R_a^2.
   c. Does the estimated regression equation explain a large amount of the variability in the data? Explain.

13. In exercise 3, the following estimated regression equation based on 30 observations was presented.

$$\hat{y} = 17.6 + 3.8x_1 - 2.3x_2 + 7.6x_3 + 2.7x_4$$

The values of SST and SSR are 1805 and 1760, respectively.
   a. Compute R^2.
   b. Compute R_a^2.
   c. Comment on the goodness of fit.

Applications

14. In exercise 4, the following estimated regression equation relating sales to inventory investment and advertising expenditures was given.

$$\hat{y} = 25 + 10x_1 + 8x_2$$

The data used to develop the model came from a survey of 10 stores; for those data, SST = 16,000 and SSR = 12,000.
   a. For the estimated regression equation given, compute R^2.
   b. Compute R_a^2.
   c. Does the model appear to explain a large amount of variability in the data? Explain.

15. **SELF test** In exercise 5, the owner of Showtime Movie Theaters, Inc., used multiple regression analysis to predict gross revenue (y) as a function of television advertising (x_1) and newspaper advertising (x_2). The estimated regression equation was

$$\hat{y} = 83.2 + 2.29x_1 + 1.30x_2$$

WEB file *Showtime*

The computer solution provided SST = 25.5 and SSR = 23.435.
   a. Compute and interpret R^2 and R_a^2.
   b. When television advertising was the only independent variable, $R^2 = .653$ and $R_a^2 = .595$. Do you prefer the multiple regression results? Explain.

16. In exercise 6, data were given on the recruiter assessment score (highest = 5), the out-of-state tuition and fees (rounded to the nearest $1000), and the average starting salary and bonus

BusinessSchools2

(rounded to the nearest $1000) for a sample of 20 graduate schools of business (*U.S. News & World Report 2009 Edition America's Best Graduate Schools*).
  a. Did the estimated regression equation that uses only the recruiter assessment score as the independent variable to predict the starting salary and bonus provide a good fit? Explain.
  b. Discuss the benefits of using both the recruiter assessment score and the out-of-state tuition and fees to predict the starting salary and bonus.

Boats

17. In exercise 9, an estimated regression equation was developed relating the top speed for a boat to the boat's beam and horsepower rating.
  a. Compute and interpret and R^2 and R_a^2.
  b. Does the estimated regression equation provide a good fit to the data? Explain.

NBA

18. Refer to exercise 10, where data were reported on a variety of statistics for the 29 teams in the National Basketball Association for a portion of the 2004 season (NBA website, January 3, 2004).
  a. In part (c) of exercise 10, an estimated regression equation was developed relating the proportion of games won given the percentage of field goals made by the team, the proportion of three-point shots made by the team's opponent, and the number of turnovers committed by the team's opponent. What are the values of R^2 and R_a^2?
  b. Does the estimated regression equation provide a good fit to the data? Explain.

13.4 Model Assumptions

In Section 13.1 we introduced the following multiple regression model.

MULTIPLE REGRESSION MODEL

$$y = \beta_0 + \beta_1 x_1 + \beta_2 x_2 + \cdots + \beta_p x_p + \epsilon \qquad (13.10)$$

The assumptions about the error term ϵ in the multiple regression model parallel those for the simple linear regression model.

ASSUMPTIONS ABOUT THE ERROR TERM ϵ IN THE MULTIPLE REGRESSION MODEL $y = \beta_0 + \beta_1 x_1 + \cdots + \beta_p x_p + \epsilon$

1. The error term ϵ is a random variable with mean or expected value of zero; that is, $E(\epsilon) = 0$.
   Implication: For given values of $x_1, x_2, \ldots, x_p$, the expected, or average, value of y is given by

   $$E(y) = \beta_0 + \beta_1 x_1 + \beta_2 x_2 + \cdots + \beta_p x_p \qquad (13.11)$$

   Equation (13.11) is the multiple regression equation we introduced in Section 13.1. In this equation, $E(y)$ represents the average of all possible values of y that might occur for the given values of $x_1, x_2, \ldots, x_p$.

2. The variance of ϵ is denoted by σ^2 and is the same for all values of the independent variables $x_1, x_2, \ldots, x_p$.
   Implication: The variance of y about the regression line equals σ^2 and is the same for all values of $x_1, x_2, \ldots, x_p$.

3. The values of ϵ are independent.

Implication: The value of ϵ for a particular set of values for the independent variables is not related to the value of ϵ for any other set of values.

4. The error term ϵ is a normally distributed random variable reflecting the deviation between the y value and the expected value of y given by $\beta_0 + \beta_1 x_1 + \beta_2 x_2 + \cdots + \beta_p x_p$.
Implication: Because $\beta_0, \beta_1, \ldots, \beta_p$ are constants for the given values of $x_1, x_2, \ldots, x_p$, the dependent variable y is also a normally distributed random variable.

To obtain more insight about the form of the relationship given by equation (13.11), consider the following two-independent-variable multiple regression equation.

$$E(y) = \beta_0 + \beta_1 x_1 + \beta_2 x_2$$

The graph of this equation is a plane in three-dimensional space. Figure 13.5 provides an example of such a graph. Note that the value of ϵ shown is the difference between the actual y value and the expected value of y, $E(y)$, when $x_1 = x_1^*$ and $x_2 = x_2^*$.

In regression analysis, the term *response variable* is often used in place of the term *dependent variable*. Furthermore, since the multiple regression equation generates a plane or surface, its graph is called a *response surface*.

13.5 Testing for Significance

In this section we show how to conduct significance tests for a multiple regression relationship. The significance tests we used in simple linear regression were a t test and an F test. In simple linear regression, both tests provide the same conclusion; that is, if the null

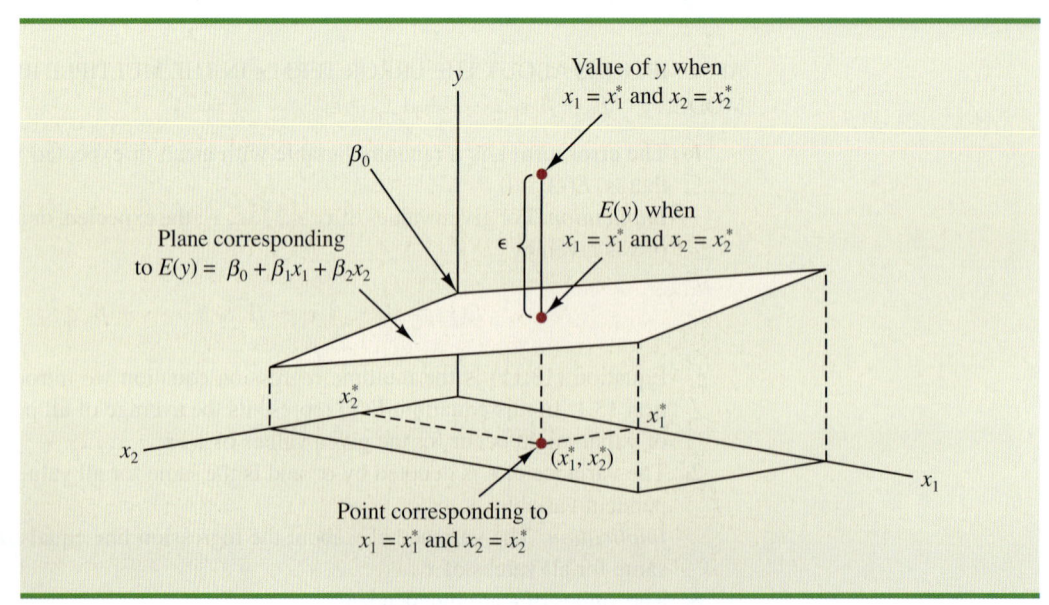

FIGURE 13.5 GRAPH OF THE REGRESSION EQUATION FOR MULTIPLE REGRESSION ANALYSIS WITH TWO INDEPENDENT VARIABLES

hypothesis is rejected, we conclude that $\beta_1 \neq 0$. In multiple regression, the t test and the F test have different purposes.

1. The F test is used to determine whether a significant relationship exists between the dependent variable and the set of all the independent variables; we will refer to the F test as the test for *overall significance*.
2. If the F test shows an overall significance, the t test is used to determine whether each of the individual independent variables is significant. A separate t test is conducted for each of the independent variables in the model; we refer to each of these t tests as a test for *individual significance*.

In the material that follows, we will explain the F test and the t test and apply each to the Butler Trucking Company example.

F Test

The multiple regression model as defined in Section 13.4 is

$$y = \beta_0 + \beta_1 x_1 + \beta_2 x_2 + \cdots + \beta_p x_p + \epsilon$$

The hypotheses for the F test involve the parameters of the multiple regression model.

$H_0: \beta_1 = \beta_2 = \cdots = \beta_p = 0$
$H_a:$ One or more of the parameters is not equal to zero

If H_0 is rejected, the test gives us sufficient statistical evidence to conclude that one or more of the parameters is not equal to zero and that the overall relationship between y and the set of independent variables $x_1, x_2, \ldots, x_p$ is significant. However, if H_0 cannot be rejected, we do not have sufficient evidence to conclude that a significant relationship is present.

Before describing the steps of the F test, we need to review the concept of *mean square*. A mean square is a sum of squares divided by its corresponding degrees of freedom. In the multiple regression case, the total sum of squares has $n - 1$ degrees of freedom, the sum of squares due to regression (SSR) has p degrees of freedom, and the sum of squares due to error has $n - p - 1$ degrees of freedom. Hence, the mean square due to regression (MSR) is SSR/p and the mean square due to error (MSE) is SSE/$(n - p - 1)$.

$$\text{MSR} = \frac{\text{SSR}}{p} \quad \text{(13.12)}$$

and

$$\text{MSE} = \frac{\text{SSE}}{n - p - 1} \quad \text{(13.13)}$$

As discussed in Chapter 12, MSE provides an unbiased estimate of σ^2, the variance of the error term ϵ. If $H_0: \beta_1 = \beta_2 = \cdots = \beta_p = 0$ is true, MSR also provides an unbiased estimate of σ^2, and the value of MSR/MSE should be close to 1. However, if H_0 is false, MSR overestimates σ^2 and the value of MSR/MSE becomes larger. To determine how large the value of MSR/MSE must be to reject H_0, we make use of the fact that if H_0 is true and the assumptions about the multiple regression model are valid, the sampling distribution of MSR/MSE is an F distribution with p degrees of freedom in the numerator and $n - p - 1$ in the denominator. A summary of the F test for significance in multiple regression follows.

> ***F* TEST FOR OVERALL SIGNIFICANCE**
>
> $H_0: \beta_1 = \beta_2 = \cdots = \beta_p = 0$
> H_a: One or more of the parameters is not equal to zero
>
> **TEST STATISTIC**
>
> $$F = \frac{\text{MSR}}{\text{MSE}} \qquad (13.14)$$
>
> **REJECTION RULE**
>
> p-value approach: Reject H_0 if p-value $\leq \alpha$
> Critical value approach: Reject H_0 if $F \geq F_\alpha$
>
> where F_α is based on an F distribution with p degrees of freedom in the numerator and $n - p - 1$ degrees of freedom in the denominator.

Let us apply the F test to the Butler Trucking Company multiple regression problem. With two independent variables, the hypotheses are written as follows.

$$H_0: \beta_1 = \beta_2 = 0$$
$$H_a: \beta_1 \text{ and/or } \beta_2 \text{ is not equal to zero}$$

Figure 13.6 is the Minitab output for the multiple regression model with miles traveled (x_1) and number of deliveries (x_2) as the two independent variables. In the Analysis of Variance (ANOVA) part of the output, we see that MSR = 10.8 and MSE = .328. Using equation (13.14), we obtain the test statistic.

$$F = \frac{10.8}{.328} = 32.9$$

FIGURE 13.6 MINITAB OUTPUT FOR BUTLER TRUCKING WITH TWO INDEPENDENT VARIABLES, MILES TRAVELED (x_1) AND NUMBER OF DELIVERIES (x_2)

```
The regression equation is
Time = - 0.869 + 0.0611 Miles + 0.923 Deliveries

Predictor        Coef     SE Coef        T        p
Constant      -0.8687      0.9515    -0.91    0.392
Miles        0.061135    0.009888     6.18    0.000
Deliveries     0.9234      0.2211     4.18    0.004

S = 0.573142    R-sq = 90.4%    R-sq(adj) = 87.6%

Analysis of Variance

SOURCE            DF        SS        MS        F        p
Regression         2    21.601    10.800    32.88    0.000
Residual Error     7     2.299     0.328
Total              9    23.900
```

13.5 Testing for Significance

TABLE 13.3 ANOVA TABLE FOR A MULTIPLE REGRESSION MODEL WITH p INDEPENDENT VARIABLES

Source	Sum of Squares	Degrees of Freedom	Mean Square	F
Regression	SSR	p	$\text{MSR} = \dfrac{\text{SSR}}{p}$	$F = \dfrac{\text{MSR}}{\text{MSE}}$
Error	SSE	$n - p - 1$	$\text{MSE} = \dfrac{\text{SSE}}{n - p - 1}$	
Total	SST	$n - 1$		

Note that the F value on the Minitab output is $F = 32.88$; the value we calculated differs because we used rounded values for MSR and MSE in the calculation. Using $\alpha = .01$, the p-value $= 0.000$ in the last column of the ANOVA table (Figure 13.6) indicates that we can reject $H_0: \beta_1 = \beta_2 = 0$ because the p-value is less than $\alpha = .01$. Alternatively, Table 4 of Appendix B shows that with two degrees of freedom in the numerator and seven degrees of freedom in the denominator, $F_{.01} = 9.55$. With $32.9 > 9.55$, we reject $H_0: \beta_1 = \beta_2 = 0$ and conclude that a significant relationship is present between travel time y and the two independent variables, miles traveled and number of deliveries.

As noted previously, the mean square error provides an unbiased estimate of σ^2, the variance of the error term ϵ. Referring to Figure 13.6, we see that the estimate of σ^2 is MSE $= .328$. The square root of MSE is the estimate of the standard deviation of the error term. As defined in Section 12.5, this standard deviation is called the standard error of the estimate and is denoted s. Hence, we have $s = \sqrt{\text{MSE}} = \sqrt{.328} = .573$. Note that the value of the standard error of the estimate appears in the Minitab output in Figure 13.6.

Table 13.3 is the general ANOVA table that provides the F test results for a multiple regression model. The value of the F test statistic appears in the last column and can be compared to F_α with p degrees of freedom in the numerator and $n - p - 1$ degrees of freedom in the denominator to make the hypothesis test conclusion. By reviewing the Minitab output for Butler Trucking Company in Figure 13.6, we see that Minitab's ANOVA table contains this information. Moreover, Minitab also provides the p-value corresponding to the F test statistic.

t Test

If the F test shows that the multiple regression relationship is significant, a t test can be conducted to determine the significance of each of the individual parameters. The t test for individual significance follows.

---

t TEST FOR INDIVIDUAL SIGNIFICANCE

For any parameter β_i

$$H_0: \beta_i = 0$$
$$H_a: \beta_i \neq 0$$

TEST STATISTIC

$$t = \frac{b_i}{s_{b_i}} \tag{13.15}$$

> **REJECTION RULE**
>
> p-value approach: Reject H_0 if p-value $\leq \alpha$
> Critical value approach: Reject H_0 if $t \leq -t_{\alpha/2}$ or if $t \geq t_{\alpha/2}$
>
> where $t_{\alpha/2}$ is based on a t distribution with $n - p - 1$ degrees of freedom.

In the test statistic, s_{b_i} is the estimate of the standard deviation of b_i. The value of s_{b_i} will be provided by the computer software package.

Let us conduct the t test for the Butler Trucking regression problem. Refer to the section of Figure 13.6 that shows the Minitab output for the t-ratio calculations. Values of b_1, b_2, s_{b_1}, and s_{b_2} are as follows.

$$b_1 = .061135 \quad s_{b_1} = .009888$$
$$b_2 = .9234 \quad s_{b_2} = .2211$$

Using equation (13.15), we obtain the test statistic for the hypotheses involving parameters β_1 and β_2.

$$t = .061135/.009888 = 6.18$$
$$t = .9234/.2211 = 4.18$$

Note that both of these t-ratio values and the corresponding p-values are provided by the Minitab output in Figure 13.6. Using $\alpha = .01$, the p-values of .000 and .004 on the Minitab output indicate that we can reject $H_0: \beta_1 = 0$ and $H_0: \beta_2 = 0$. Hence, both parameters are statistically significant. Alternatively, Table 2 of Appendix B shows that with $n - p - 1 = 10 - 2 - 1 = 7$ degrees of freedom, $t_{.005} = 3.499$. With $6.18 > 3.499$, we reject $H_0: \beta_1 = 0$. Similarly, with $4.18 > 3.499$, we reject $H_0: \beta_2 = 0$.

Multicollinearity

We use the term *independent variable* in regression analysis to refer to any variable being used to predict or explain the value of the dependent variable. The term does not mean, however, that the independent variables themselves are independent in any statistical sense. On the contrary, most independent variables in a multiple regression problem are correlated to some degree with one another. For example, in the Butler Trucking example involving the two independent variables x_1 (miles traveled) and x_2 (number of deliveries), we could treat the miles traveled as the dependent variable and the number of deliveries as the independent variable to determine whether those two variables are themselves related. We could then compute the sample correlation coefficient $r_{x_1 x_2}$ to determine the extent to which the variables are related. Doing so yields $r_{x_1 x_2} = .16$. Thus, we find some degree of linear association between the two independent variables. In multiple regression analysis, **multicollinearity** refers to the correlation among the independent variables.

To provide a better perspective of the potential problems of multicollinearity, let us consider a modification of the Butler Trucking example. Instead of x_2 being the number of deliveries, let x_2 denote the number of gallons of gasoline consumed. Clearly, x_1 (the miles traveled) and x_2 are related; that is, we know that the number of gallons of gasoline used depends on the number of miles traveled. Hence, we would conclude logically that x_1 and x_2 are highly correlated independent variables.

Assume that we obtain the equation $\hat{y} = b_0 + b_1 x_1 + b_2 x_2$ and find that the F test shows the relationship to be significant. Then suppose we conduct a t test on β_1 to determine whether $\beta_1 \neq 0$, and we cannot reject $H_0: \beta_1 = 0$. Does this result mean that travel time is

13.5 Testing for Significance

not related to miles traveled? Not necessarily. What it probably means is that with x_2 already in the model, x_1 does not make a significant contribution to determining the value of y. This interpretation makes sense in our example; if we know the amount of gasoline consumed, we do not gain much additional information useful in predicting y by knowing the miles traveled. Similarly, a t test might lead us to conclude $\beta_2 = 0$ on the grounds that, with x_1 in the model, knowledge of the amount of gasoline consumed does not add much.

A sample correlation coefficient greater than +.7 or less than −.7 for two independent variables is a rule of thumb warning of potential problems with multicollinearity.

To summarize, in t tests for the significance of individual parameters, the difficulty caused by multicollinearity is that it is possible to conclude that none of the individual parameters are significantly different from zero when an F test on the overall multiple regression equation indicates a significant relationship. This problem is avoided when there is little correlation among the independent variables.

Statisticians have developed several tests for determining whether multicollinearity is high enough to cause problems. According to the rule of thumb test, multicollinearity is a potential problem if the absolute value of the sample correlation coefficient exceeds .7 for any two of the independent variables. The other types of tests are more advanced and beyond the scope of this text.

When the independent variables are highly correlated, it is not possible to determine the separate effect of any particular independent variable on the dependent variable.

If possible, every attempt should be made to avoid including independent variables that are highly correlated. In practice, however, strict adherence to this policy is rarely possible. When decision makers have reason to believe substantial multicollinearity is present, they must realize that separating the effects of the individual independent variables on the dependent variable is difficult.

NOTES AND COMMENTS

Ordinarily, multicollinearity does not affect the way in which we perform our regression analysis or interpret the output from a study. However, when multicollinearity is severe—that is, when two or more of the independent variables are highly correlated with one another—we can have difficulty interpreting the results of t tests on the individual parameters. In addition to the type of problem illustrated in this section, severe cases of multicollinearity have been shown to result in least squares estimates that have the wrong sign. That is, in simulated studies where researchers created the underlying regression model and then applied the least squares technique to develop estimates of β_0, β_1, β_2, and so on, it has been shown that under conditions of high multicollinearity the least squares estimates can have a sign opposite that of the parameter being estimated. For example, β_2 might actually be $+10$ and b_1, its estimate, might turn out to be -2. Thus, little faith can be placed in the individual coefficients if multicollinearity is present to a high degree.

Exercises

Methods

SELF test

19. In exercise 1, the following estimated regression equation based on 10 observations was presented.

$$\hat{y} = 29.1270 + .5906x_1 + .4980x_2$$

Here SST = 6724.125, SSR = 6216.375, s_{b_1} = .0813, and s_{b_2} = .0567.
   a. Compute MSR and MSE.
   b. Compute F and perform the appropriate F test. Use $\alpha = .05$.
   c. Perform a t test for the significance of β_1. Use $\alpha = .05$.
   d. Perform a t test for the significance of β_2. Use $\alpha = .05$.

20. Refer to the data presented in exercise 2. The estimated regression equation for these data is

$$\hat{y} = -18.37 + 2.01x_1 + 4.74x_2$$

Here SST = 15,182.9, SSR = 14,052.2, s_{b_1} = .2471, and s_{b_2} = .9484.
   a. Test for a significant relationship among x_1, x_2, and y. Use $\alpha = .05$.
   b. Is β_1 significant? Use $\alpha = .05$.
   c. Is β_2 significant? Use $\alpha = .05$.

21. The following estimated regression equation was developed for a model involving two independent variables.

$$\hat{y} = 40.7 + 8.63x_1 + 2.71x_2$$

After x_2 was dropped from the model, the least squares method was used to obtain an estimated regression equation involving only x_1 as an independent variable.

$$\hat{y} = 42.0 + 9.01x_1$$

   a. Give an interpretation of the coefficient of x_1 in both models.
   b. Could multicollinearity explain why the coefficient of x_1 differs in the two models? If so, how?

Applications

22. In exercise 4, the following estimated regression equation relating sales to inventory investment and advertising expenditures was given.

$$\hat{y} = 25 + 10x_1 + 8x_2$$

The data used to develop the model came from a survey of 10 stores; for these data SST = 16,000 and SSR = 12,000.
   a. Compute SSE, MSE, and MSR.
   b. Use an F test and a .05 level of significance to determine whether there is a relationship among the variables.

23. Refer to exercise 5.
   a. Use $\alpha = .01$ to test the hypotheses

$$H_0: \beta_1 = \beta_2 = 0$$
$$H_a: \beta_1 \text{ and/or } \beta_2 \text{ is not equal to zero}$$

for the model $y = \beta_0 + \beta_1 x_1 + \beta_2 x_2 + \epsilon$, where

x_1 = television advertising ($1000s)
x_2 = newspaper advertising ($1000s)

   b. Use $\alpha = .05$ to test the significance of β_1. Should x_1 be dropped from the model?
   c. Use $\alpha = .05$ to test the significance of β_2. Should x_2 be dropped from the model?

24. *The Wall Street Journal* conducted a study of basketball spending at top colleges. A portion of the data showing the revenue ($ millions), percentage of wins, and the coach's salary ($ millions) for 39 of the country's top basketball programs follows (*The Wall Street Journal,* March 11–12, 2006).

School	Revenue	% Wins	Salary
Alabama	6.5	61	1.00
Arizona	16.6	63	0.70
Arkansas	11.1	72	0.80
Boston College	3.4	80	0.53
⋮	⋮	⋮	⋮
Washington	5.0	83	0.89
West Virginia	4.9	67	0.70
Wichita State	3.1	75	0.41
Wisconsin	12.0	66	0.70

a. Develop the estimated regression equation that can be used to predict the coach's salary given the revenue generated by the program and the percentage of wins.
b. Use the F test to determine the overall significance of the relationship. What is your conclusion at the .05 level of significance?
c. Use the t test to determine the significance of each independent variable. What is your conclusion at the .05 level of significance?

25. Small cars offer higher fuel efficiency, are easy to maneuver and park, and have generally performed well in road tests. The following data show the overall fuel efficiency rating (miles per gallon), predicted reliability (highest = 5), safety rating (highest = 5), and the price ($) for 20 small cars (*Consumer Reports*, March 2009).

Make & Model	MPG	Reliability	Safety	Price ($)
Chevrolet Aveo LT	25	2	2	16,205
Chevrolet Cobalt LT	24	3	3	17,450
Ford Focus SES	26	4	3	18,490
Ford Focus SES (MT)	29	4	3	17,440
Honda Civic EX	28	5	4	19,610
Honda Civic EX (MT)	31	5	4	18,810
Honda Fit Sport (MT)	33	5	4	16,730
Hyundai Accent GLS	28	4	2	14,230
Hyundai Elantra GLS	27	5	3	17,555
Hyundai Elantra SE	27	5	4	17,980
Kia Spectra EX	25	4	2	16,185
Kia Spectra EX (MT)	28	4	2	15,185
Nissan Versa 1.8 SL	28	3	3	16,675
Nissan Versa 1.8S	27	1	3	16,130
Scion xD	29	5	3	16,620
Scion xD (MT)	34	5	3	15,820
Suzuki SX4 LE	26	5	3	17,378
Toyota Corolla (base, MT)	32	4	4	16,419
Toyota Corolla LE	32	4	4	18,404
Toyota Yaris Hatchback	30	5	3	16,095

a. Determine the estimated regression equation that can be used to predict the price of a small car given its overall fuel efficiency rating, predicted reliability, and safety rating.
b. Use the F test to determine the overall significance of the relationship. What is the conclusion at the .05 level of significance?
c. Use the t test to determine the significance of each independent variable. What is your conclusion at the .05 level of significance?

d. Remove any independent variable that is not significant from the estimated regression equation. What is your recommended estimated regression equation? Compare the value of R^2 for this estimated regression equation with the value of R^2 for the estimated regression equation from part (a). Discuss the differences.

26. In exercise 10 an estimated regression equation was developed relating the proportion of games won given the proportion of field goals made by the team, the proportion of three-point shots made by the team's opponent, and the number of turnovers committed by the team's opponent.
   a. Use the F test to determine the overall significance of the relationship. What is your conclusion at the .05 level of significance?
   b. Use the t test to determine the significance of each independent variable. What is your conclusion at the .05 level of significance?

13.6 Using the Estimated Regression Equation for Estimation and Prediction

The procedures for estimating the mean value of y and predicting an individual value of y in multiple regression are similar to those in regression analysis involving one independent variable. First, recall that in Chapter 12 we showed that the point estimate of the expected value of y for a given value of x was the same as the point estimate of an individual value of y. In both cases, we used $\hat{y} = b_0 + b_1 x$ as the point estimate.

In multiple regression we use the same procedure. That is, we substitute the given values of $x_1, x_2, \ldots, x_p$ into the estimated regression equation and use the corresponding value of $\hat{y}$ as the point estimate. Suppose that for the Butler Trucking example we want to use the estimated regression equation involving x_1 (miles traveled) and x_2 (number of deliveries) to develop two interval estimates:

1. A *confidence interval* of the mean travel time for all trucks that travel 100 miles and make two deliveries
2. A *prediction interval* of the travel time for *one specific* truck that travels 100 miles and makes two deliveries

Using the estimated regression equation $\hat{y} = -.869 + .0611 x_1 + .923 x_2$ with $x_1 = 100$ and $x_2 = 2$, we obtain the following value of $\hat{y}$:

$$\hat{y} = -.869 + .0611(100) + .923(2) = 7.09$$

Hence, the point estimate of travel time in both cases is approximately seven hours.

TABLE 13.4 THE 95% CONFIDENCE AND PREDICTION INTERVALS FOR BUTLER TRUCKING

Value of x_1	Value of x_2	Confidence Interval Lower Limit	Confidence Interval Upper Limit	Prediction Interval Lower Limit	Prediction Interval Upper Limit
50	2	3.146	4.924	2.414	5.656
50	3	4.127	5.789	3.368	6.548
50	4	4.815	6.948	4.157	7.607
100	2	6.258	7.926	5.500	8.683
100	3	7.385	8.645	6.520	9.510
100	4	8.135	9.742	7.362	10.515

13.6 Using the Estimated Regression Equation for Estimation and Prediction

To develop interval estimates for the mean value of y and for an individual value of y, we use a procedure similar to that for regression analysis involving one independent variable. The formulas required are beyond the scope of the text, but computer packages for multiple regression analysis will often provide confidence intervals once the values of $x_1, x_2, \ldots, x_p$ are specified by the user. In Table 13.4 we show the 95% confidence and prediction intervals for the Butler Trucking example for selected values of x_1 and x_2; these values were obtained using Minitab. Note that the interval estimate for an individual value of y is wider than the interval estimate for the expected value of y. This difference simply reflects the fact that for given values of x_1 and x_2 we can estimate the mean travel time for all trucks with more precision than we can predict the travel time for one specific truck.

Exercises

Methods

27. In exercise 1, the following estimated regression equation based on 10 observations was presented.

$$\hat{y} = 29.1270 + .5906x_1 + .4980x_2$$

   a. Develop a point estimate of the mean value of y when $x_1 = 180$ and $x_2 = 310$.
   b. Develop a point estimate for an individual value of y when $x_1 = 180$ and $x_2 = 310$.

28. **SELF test** Refer to the data in exercise 2. The estimated regression equation for those data is

$$\hat{y} = -18.4 + 2.01x_1 + 4.74x_2$$

   a. Develop a 95% confidence interval for the mean value of y when $x_1 = 45$ and $x_2 = 15$.
   b. Develop a 95% prediction interval for y when $x_1 = 45$ and $x_2 = 15$.

Applications

29. **SELF test** In exercise 5, the owner of Showtime Movie Theaters, Inc., used multiple regression analysis to predict gross revenue (y) as a function of television advertising (x_1) and newspaper advertising (x_2). The estimated regression equation was

$$\hat{y} = 83.2 + 2.29x_1 + 1.30x_2$$

   a. What is the gross revenue expected for a week when $3500 is spent on television advertising ($x_1 = 3.5$) and $1800 is spent on newspaper advertising ($x_2 = 1.8$)?
   b. Provide a 95% confidence interval for the mean revenue of all weeks with the expenditures listed in part (a).
   c. Provide a 95% prediction interval for next week's revenue, assuming that the advertising expenditures will be allocated as in part (a).

30. **WEB file** *Boats* — In exercise 9 an estimated regression equation was developed relating the top speed for a boat to the boat's beam and horsepower rating.
   a. Develop a 95% confidence interval for the mean top speed of a boat with a beam of 85 inches and an engine with a 330 horsepower rating.
   b. The Svfara SV609 has a beam of 85 inches and an engine with a 330 horsepower rating. Develop a 95% confidence interval for the mean top speed for the Svfara SV609.

31. The Buyer's Guide section of the website for *Car and Driver* magazine provides reviews and road tests for cars, trucks, SUVs, and vans. The average ratings of overall quality,

vehicle styling, braking, handling, fuel economy, interior comfort, acceleration, dependability, fit and finish, transmission, and ride are summarized for each vehicle using a scale ranging from 1 (worst) to 10 (best). A portion of the data for 14 Sports/GT cars is shown here (Car and Driver website, January 7, 2004).

Sports/GT	Overall	Handling	Dependability	Fit and Finish
Acura 3.2CL	7.80	7.83	8.17	7.67
Acura RSX	9.02	9.46	9.35	8.97
Audi TT	9.00	9.58	8.74	9.38
BMW 3-Series/M3	8.39	9.52	8.39	8.55
Chevrolet Corvette	8.82	9.64	8.54	7.87
Ford Mustang	8.34	8.85	8.70	7.34
Honda Civic Si	8.92	9.31	9.50	7.93
Infiniti G35	8.70	9.34	8.96	8.07
Mazda RX-8	8.58	9.79	8.96	8.12
Mini Cooper	8.76	10.00	8.69	8.33
Mitsubishi Eclipse	8.17	8.95	8.25	7.36
Nissan 350Z	8.07	9.35	7.56	8.21
Porsche 911	9.55	9.91	8.86	9.55
Toyota Celica	8.77	9.29	9.04	7.97

a. Develop an estimated regression equation using handling, dependability, and fit and finish to predict overall quality.
b. Another Sports/GT car rated by *Car and Driver* is the Honda Accord. The ratings for handling, dependability, and fit and finish for the Honda Accord were 8.28, 9.06, and 8.07, respectively. Estimate the overall rating for this car.
c. Provide a 95% confidence interval for overall quality for all sports and GT cars with the characteristics listed in part (b).
d. Provide a 95% prediction interval for overall quality for the Honda Accord described in part (b).
e. The overall rating reported by *Car and Driver* for the Honda Accord was 8.65. How does this rating compare to the estimates you developed in parts (b) and (d)?

13.7 Categorical Independent Variables

The independent variables may be categorical or quantitative.

Thus far, the examples we have considered involved quantitative independent variables such as student population, distance traveled, and number of deliveries. In many situations, however, we must work with **categorical independent variables** such as gender (male, female), method of payment (cash, credit card, check), and so on. The purpose of this section is to show how categorical variables are handled in regression analysis. To illustrate the use and interpretation of a categorical independent variable, we will consider a problem facing the managers of Johnson Filtration, Inc.

An Example: Johnson Filtration, Inc.

Johnson Filtration, Inc., provides maintenance service for water-filtration systems throughout southern Florida. Customers contact Johnson with requests for maintenance service on their water-filtration systems. To estimate the service time and the service cost, Johnson's managers want to predict the repair time necessary for each maintenance request. Hence, repair time in hours is the dependent variable. Repair time is believed to be related to two factors, the number of months since the last maintenance service and the type of repair problem (mechanical or electrical). Data for a sample of 10 service calls are reported in Table 13.5.

TABLE 13.6 DATA FOR THE JOHNSON FILTRATION EXAMPLE WITH TYPE OF REPAIR INDICATED BY A DUMMY VARIABLE ($x_2 = 0$ FOR MECHANICAL; $x_2 = 1$ FOR ELECTRICAL)

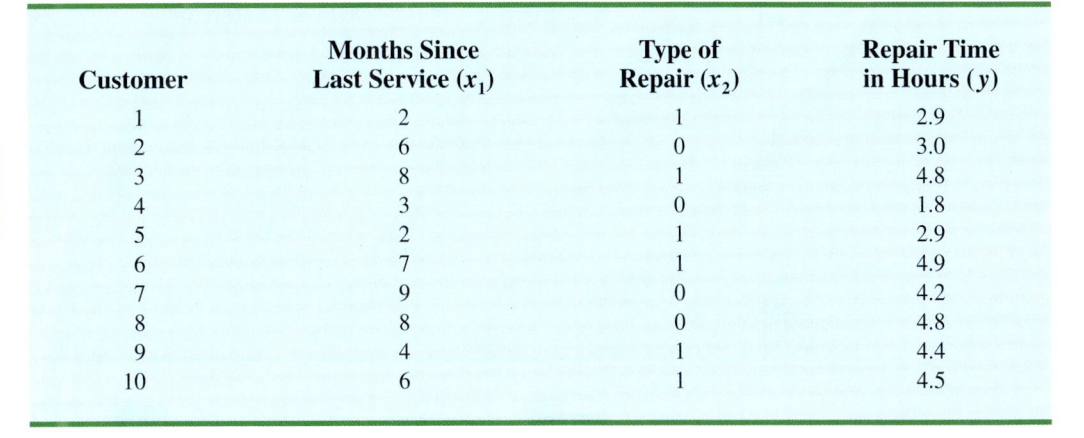

Customer	Months Since Last Service (x_1)	Type of Repair (x_2)	Repair Time in Hours (y)
1	2	1	2.9
2	6	0	3.0
3	8	1	4.8
4	3	0	1.8
5	2	1	2.9
6	7	1	4.9
7	9	0	4.2
8	8	0	4.8
9	4	1	4.4
10	6	1	4.5

In regression analysis x_2 is called a **dummy** or *indicator* **variable**. Using this dummy variable, we can write the multiple regression model as

$$y = \beta_0 + \beta_1 x_1 + \beta_2 x_2 + \epsilon$$

Table 13.6 is the revised data set that includes the values of the dummy variable. Using Minitab and the data in Table 13.6, we can develop estimates of the model parameters. The Minitab output in Figure 13.8 shows that the estimated multiple regression equation is

$$\hat{y} = .93 + .388 x_1 + 1.26 x_2 \qquad (13.17)$$

At the .05 level of significance, the *p*-value of .001 associated with the *F* test ($F = 21.36$) indicates that the regression relationship is significant. The *t* test part of the printout in Figure 13.8 shows that both months since last service (*p*-value = .000) and type of repair

FIGURE 13.8 MINITAB OUTPUT FOR JOHNSON FILTRATION WITH MONTHS SINCE LAST SERVICE (x_1) AND TYPE OF REPAIR (x_2) AS THE INDEPENDENT VARIABLES

In the Minitab output the variable names Months, Type, *and* Time *were entered as the column headings on the worksheet; thus,* $x_1 =$ Months, $x_2 =$ Type, *and* $y =$ Time.

```
The regression equation is
Time = 0.930 + 0.388 Months + 1.26 Type

Predictor      Coef     SE Coef       T       p
Constant     0.9305      0.4670    1.99   0.087
Months       0.38762     0.06257   6.20   0.000
Type         1.2627      0.3141    4.02   0.005

S = 0.459048     R-sq = 85.9%     R-sq(adj) = 81.9%

Analysis of Variance

SOURCE           DF        SS         MS        F       p
Regression        2     9.0009     4.5005   21.36   0.001
Residual Error    7     1.4751     0.2107
Total             9    10.4760
```

13.7 Categorical Independent Variables

TABLE 13.5 DATA FOR THE JOHNSON FILTRATION EXAMPLE

Service Call	Months Since Last Service	Type of Repair	Repair Time in Hours
1	2	electrical	2.9
2	6	mechanical	3.0
3	8	electrical	4.8
4	3	mechanical	1.8
5	2	electrical	2.9
6	7	electrical	4.9
7	9	mechanical	4.2
8	8	mechanical	4.8
9	4	electrical	4.4
10	6	electrical	4.5

Let y denote the repair time in hours and x_1 denote the number of months since the last maintenance service. The regression model that uses only x_1 to predict y is

$$y = \beta_0 + \beta_1 x_1 + \epsilon$$

Using Minitab to develop the estimated regression equation, we obtained the output shown in Figure 13.7. The estimated regression equation is

$$\hat{y} = 2.15 + .304 x_1 \tag{13.16}$$

At the .05 level of significance, the p-value of .016 for the t (or F) test indicates that the number of months since the last service is significantly related to repair time. R-sq = 53.4% indicates that x_1 alone explains 53.4% of the variability in repair time.

To incorporate the type of repair into the regression model, we define the following variable.

$$x_2 = \begin{cases} 0 \text{ if the type of repair is mechanical} \\ 1 \text{ if the type of repair is electrical} \end{cases}$$

FIGURE 13.7 MINITAB OUTPUT FOR JOHNSON FILTRATION WITH MONTHS SINCE LAST SERVICE (x_1) AS THE INDEPENDENT VARIABLE

In the Minitab output the variable names Months and Time were entered as the column headings on the worksheet; thus, $x_1 =$ Months and $y =$ Time.

```
The regression equation is
Time = 2.15 + 0.304 Months

Predictor     Coef    SE Coef     T        p
Constant     2.1473   0.6050    3.55    0.008
Months       0.3041   0.1004    3.03    0.016

S = 0.781022    R-sq = 53.4%    R-sq(adj) = 47.6%

Analysis of Variance

SOURCE            DF        SS         MS       F        p
Regression         1     5.5960     5.5960    9.17    0.016
Residual Error     8     4.8800     0.6100
Total              9    10.4760
```

13.7 Categorical Independent Variables

(p-value = .005) are statistically significant. In addition, R-sq = 85.9% and R-sq(adj) = 81.9% indicate that the estimated regression equation does a good job of explaining the variability in repair times. Thus, equation (13.17) should prove helpful in estimating the repair time necessary for the various service calls.

Interpreting the Parameters

The multiple regression equation for the Johnson Filtration example is

$$E(y) = \beta_0 + \beta_1 x_1 + \beta_2 x_2 \quad (13.18)$$

To understand how to interpret the parameters β_0, β_1, and β_2 when a categorical variable is present, consider the case when $x_2 = 0$ (mechanical repair). Using $E(y \mid \text{mechanical})$ to denote the mean or expected value of repair time *given* a mechanical repair, we have

$$E(y \mid \text{mechanical}) = \beta_0 + \beta_1 x_1 + \beta_2(0) = \beta_0 + \beta_1 x_1 \quad (13.19)$$

Similarly, for an electrical repair ($x_2 = 1$), we have

$$E(y \mid \text{electrical}) = \beta_0 + \beta_1 x_1 + \beta_2(1) = \beta_0 + \beta_1 x_1 + \beta_2 \quad (13.20)$$
$$= (\beta_0 + \beta_2) + \beta_1 x_1$$

Comparing equations (13.19) and (13.20), we see that the mean repair time is a linear function of x_1 for both mechanical and electrical repairs. The slope of both equations is β_1, but the y-intercept differs. The y-intercept is β_0 in equation (13.19) for mechanical repairs and $(\beta_0 + \beta_2)$ in equation (13.20) for electrical repairs. The interpretation of β_2 is that it indicates the difference between the mean repair time for an electrical repair and the mean repair time for a mechanical repair.

If β_2 is positive, the mean repair time for an electrical repair will be greater than that for a mechanical repair; if β_2 is negative, the mean repair time for an electrical repair will be less than that for a mechanical repair. Finally, if $\beta_2 = 0$, there is no difference in the mean repair time between electrical and mechanical repairs and the type of repair is not related to the repair time.

Using the estimated multiple regression equation $\hat{y} = .93 + .388x_1 + 1.26x_2$, we see that .93 is the estimate of β_0 and 1.26 is the estimate of β_2. Thus, when $x_2 = 0$ (mechanical repair)

$$\hat{y} = .93 + .388x_1 \quad (13.21)$$

and when $x_2 = 1$ (electrical repair)

$$\hat{y} = .93 + .388x_1 + 1.26(1) \quad (13.22)$$
$$= 2.19 + .388x_1$$

In effect, the use of a dummy variable for type of repair provides two estimated regression equations that can be used to predict the repair time, one corresponding to mechanical repairs and one corresponding to electrical repairs. In addition, with $b_2 = 1.26$, we learn that, on average, electrical repairs require 1.26 hours longer than mechanical repairs.

Figure 13.9 is the plot of the Johnson data from Table 13.6. Repair time in hours (y) is represented by the vertical axis and months since last service (x_1) is represented by the horizontal axis. A data point for a mechanical repair is indicated by an M and a data point for an electrical repair is indicated by an E. Equations (13.21) and (13.22) are plotted on the

FIGURE 13.9 SCATTER DIAGRAM FOR THE JOHNSON FILTRATION REPAIR DATA FROM TABLE 13.6

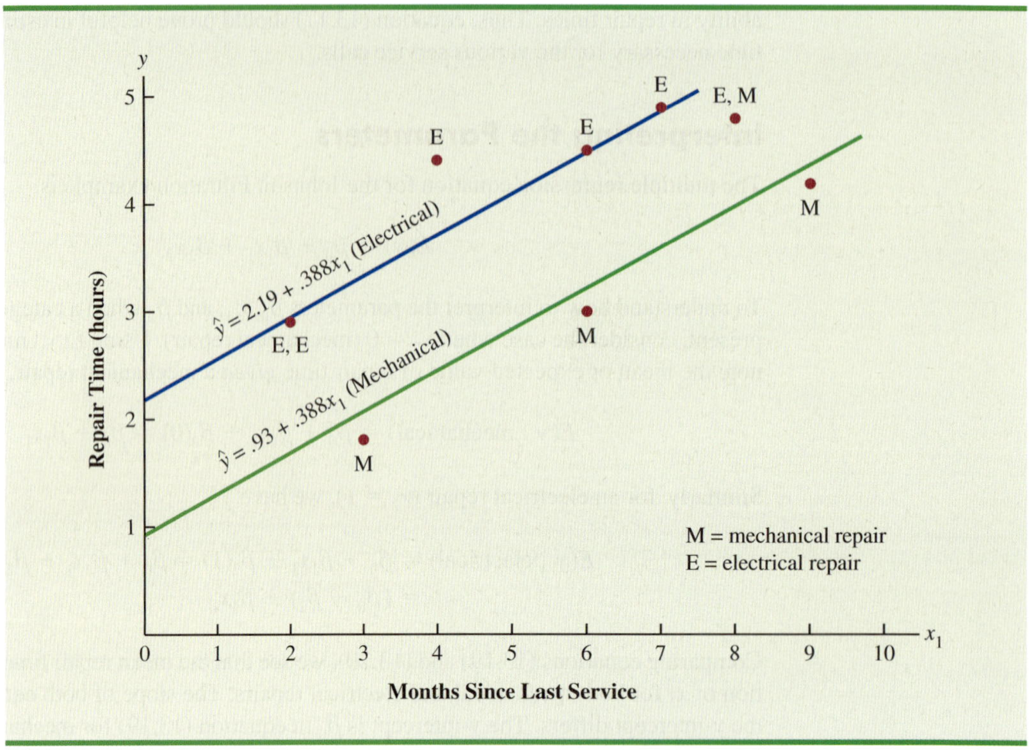

[Scatter diagram showing Repair Time (hours) on y-axis vs Months Since Last Service on x-axis. Two regression lines: $\hat{y} = 2.19 + .388x_1$ (Electrical) and $\hat{y} = .93 + .388x_1$ (Mechanical). M = mechanical repair, E = electrical repair.]

graph to show graphically the two equations that can be used to predict the repair time, one corresponding to mechanical repairs and one corresponding to electrical repairs.

More Complex Categorical Variables

Because the categorical variable for the Johnson Filtration example had two levels (mechanical and electrical), defining a dummy variable with zero indicating a mechanical repair and one indicating an electrical repair was easy. However, when a categorical variable has more than two levels, care must be taken in both defining and interpreting the dummy variables. As we will show, if a categorical variable has k levels, $k - 1$ dummy variables are required, with each dummy variable being coded as 0 or 1.

A categorical variable with k levels must be modeled using k − 1 dummy variables. Care must be taken in defining and interpreting the dummy variables.

For example, suppose a manufacturer of copy machines organized the sales territories for a particular state into three regions: A, B, and C. The managers want to use regression analysis to help predict the number of copiers sold per week. With the number of units sold as the dependent variable, they are considering several independent variables (the number of sales personnel, advertising expenditures, and so on). Suppose the managers believe sales region is also an important factor in predicting the number of copiers sold. Because sales region is a categorical variable with three levels, A, B and C, we will need $3 - 1 = 2$ dummy variables to represent the sales region. Each variable can be coded 0 or 1 as follows.

$$x_1 = \begin{cases} 1 \text{ if sales region B} \\ 0 \text{ otherwise} \end{cases}$$

$$x_2 = \begin{cases} 1 \text{ if sales region C} \\ 0 \text{ otherwise} \end{cases}$$

13.7 Categorical Independent Variables

With this definition, we have the following values of x_1 and x_2.

Region	x_1	x_2
A	0	0
B	1	0
C	0	1

Observations corresponding to region A would be coded $x_1 = 0, x_2 = 0$; observations corresponding to region B would be coded $x_1 = 1, x_2 = 0$; and observations corresponding to region C would be coded $x_1 = 0, x_2 = 1$.

The regression equation relating the expected value of the number of units sold, $E(y)$, to the dummy variables would be written as

$$E(y) = \beta_0 + \beta_1 x_1 + \beta_2 x_2$$

To help us interpret the parameters β_0, β_1, and β_2, consider the following three variations of the regression equation.

$$E(y \mid \text{region A}) = \beta_0 + \beta_1(0) + \beta_2(0) = \beta_0$$
$$E(y \mid \text{region B}) = \beta_0 + \beta_1(1) + \beta_2(0) = \beta_0 + \beta_1$$
$$E(y \mid \text{region C}) = \beta_0 + \beta_1(0) + \beta_2(1) = \beta_0 + \beta_2$$

Thus, β_0 is the mean or expected value of sales for region A; β_1 is the difference between the mean number of units sold in region B and the mean number of units sold in region A; and β_2 is the difference between the mean number of units sold in region C and the mean number of units sold in region A.

Two dummy variables were required because sales region is a categorical variable with three levels. But the assignment of $x_1 = 0, x_2 = 0$ to indicate region A, $x_1 = 1, x_2 = 0$ to indicate region B, and $x_1 = 0, x_2 = 1$ to indicate region C was arbitrary. For example, we could have chosen $x_1 = 1, x_2 = 0$ to indicate region A, $x_1 = 0, x_2 = 0$ to indicate region B, and $x_1 = 0, x_2 = 1$ to indicate region C. In that case, β_1 would have been interpreted as the mean difference between regions A and B and β_2 as the mean difference between regions C and B.

The important point to remember is that when a categorical variable has k levels, $k - 1$ dummy variables are required in the multiple regression analysis. Thus, if the sales region example had a fourth region, labeled D, three dummy variables would be necessary. For example, the three dummy variables can be coded as follows.

$$x_1 = \begin{cases} 1 \text{ if sales region B} \\ 0 \text{ otherwise} \end{cases} \quad x_2 = \begin{cases} 1 \text{ if sales region C} \\ 0 \text{ otherwise} \end{cases} \quad x_3 = \begin{cases} 1 \text{ if sales region D} \\ 0 \text{ otherwise} \end{cases}$$

Exercises

Methods

32. Consider a regression study involving a dependent variable y, a categorical independent variable x_1, and a categorical variable with two levels (level 1 and level 2).
    a. Write a multiple regression equation relating x_1 and the categorical variable to y.
    b. What is the expected value of y corresponding to level 1 of the categorical variable?

c. What is the expected value of y corresponding to level 2 of the categorical variable?
d. Interpret the parameters in your regression equation.

33. Consider a regression study involving a dependent variable y, a quantitative independent variable x_1, and a categorical independent variable with three possible levels (level 1, level 2, and level 3).
   a. How many dummy variables are required to represent the categorical variable?
   b. Write a multiple regression equation relating x_1 and the categorical variable to y.
   c. Interpret the parameters in your regression equation.

Applications

34. Management proposed the following regression model to predict sales at a fast-food outlet.

$$y = \beta_0 + \beta_1 x_1 + \beta_2 x_2 + \beta_3 x_3 + \epsilon$$

where

x_1 = number of competitors within 1 mile
x_2 = population within 1 mile (1000s)
$x_3 = \begin{cases} 1 \text{ if drive-up window present} \\ 0 \text{ otherwise} \end{cases}$
y = sales ($1000s)

The following estimated regression equation was developed after 20 outlets were surveyed.

$$\hat{y} = 10.1 - 4.2x_1 + 6.8x_2 + 15.3x_3$$

a. What is the expected amount of sales attributable to the drive-up window?
b. Predict sales for a store with two competitors, a population of 8000 within 1 mile, and no drive-up window.
c. Predict sales for a store with one competitor, a population of 3000 within 1 mile, and a drive-up window.

35. Refer to the Johnson Filtration problem introduced in this section. Suppose that in addition to information on the number of months since the machine was serviced and whether a mechanical or an electrical repair was necessary, the managers obtained a list showing which repairperson performed the service. The revised data follow.

Repair Time in Hours	Months Since Last Service	Type of Repair	Repairperson
2.9	2	Electrical	Dave Newton
3.0	6	Mechanical	Dave Newton
4.8	8	Electrical	Bob Jones
1.8	3	Mechanical	Dave Newton
2.9	2	Electrical	Dave Newton
4.9	7	Electrical	Bob Jones
4.2	9	Mechanical	Bob Jones
4.8	8	Mechanical	Bob Jones
4.4	4	Electrical	Bob Jones
4.5	6	Electrical	Dave Newton

a. Ignore for now the months since the last maintenance service (x_1) and the repairperson who performed the service. Develop the estimated simple linear regression equation

13.7 Categorical Independent Variables

to predict the repair time (y) given the type of repair (x_2). Recall that $x_2 = 0$ if the type of repair is mechanical and 1 if the type of repair is electrical.

b. Does the equation that you developed in part (a) provide a good fit for the observed data? Explain.

c. Ignore for now the months since the last maintenance service and the type of repair associated with the machine. Develop the estimated simple linear regression equation to predict the repair time given the repairperson who performed the service. Let $x_3 = 0$ if Bob Jones performed the service and $x_3 = 1$ if Dave Newton performed the service.

d. Does the equation that you developed in part (c) provide a good fit for the observed data? Explain.

36. This problem is an extension of the situation described in exercise 35.

   a. Develop the estimated regression equation to predict the repair time given the number of months since the last maintenance service, the type of repair, and the repairperson who performed the service.

   b. At the .05 level of significance, test whether the estimated regression equation developed in part (a) represents a significant relationship between the independent variables and the dependent variable.

   c. Is the addition of the independent variable x_3, the repairperson who performed the service, statistically significant? Use $\alpha = .05$. What explanation can you give for the results observed?

37. The *Consumer Reports* Restaurant Customer Satisfaction Survey is based upon 148,599 visits to full-service restaurant chains (Consumer Reports website, February 11, 2009). Assume the following data are representative of the results reported. The variable Type indicates whether the restaurant is an Italian restaurant or a seafood/steakhouse. Price indicates the average amount paid per person for dinner and drinks, minus the tip. Score reflects diners' overall satisfaction, with higher values indicating greater overall satisfaction. A score of 80 can be interpreted as very satisfied.

Restaurant	Type	Price ($)	Score
Bertucci's	Italian	16	77
Black Angus Steakhouse	Seafood/Steakhouse	24	79
Bonefish Grill	Seafood/Steakhouse	26	85
Bravo! Cucina Italiana	Italian	18	84
Buca di Beppo	Italian	17	81
Bugaboo Creek Steak House	Seafood/Steakhouse	18	77
Carrabba's Italian Grill	Italian	23	86
Charlie Brown's Steakhouse	Seafood/Steakhouse	17	75
Il Fornaio	Italian	28	83
Joe's Crab Shack	Seafood/Steakhouse	15	71
Johnny Carino's Italian	Italian	17	81
Lone Star Steakhouse & Saloon	Seafood/Steakhouse	17	76
LongHorn Steakhouse	Seafood/Steakhouse	19	81
Maggiano's Little Italy	Italian	22	83
McGrath's Fish House	Seafood/Steakhouse	16	81
Olive Garden	Italian	19	81
Outback Steakhouse	Seafood/Steakhouse	20	80
Red Lobster	Seafood/Steakhouse	18	78
Romano's Macaroni Grill	Italian	18	82
The Old Spaghetti Factory	Italian	12	79
Uno Chicago Grill	Italian	16	76

a. Develop the estimated regression equation to show how overall customer satisfaction is related to the independent variable average meal price.

b. At the .05 level of significance, test whether the estimated regression equation developed in part (a) indicates a significant relationship between overall customer satisfaction and average meal price.
c. Develop a dummy variable that will account for the type of restaurant (Italian or seafood/steakhouse).
d. Develop the estimated regression equation to show how overall customer satisfaction is related to the average meal price and the type of restaurant.
e. Is type of restaurant a significant factor in overall customer satisfaction?
f. Estimate the *Consumer Reports* customer satisfaction score for a seafood/steakhouse that has an average meal price of $20. How much would the estimated score have changed for an Italian restaurant?

38. A 10-year study conducted by the American Heart Association provided data on how age, blood pressure, and smoking relate to the risk of strokes. Assume that the following data are from a portion of this study. Risk is interpreted as the probability (times 100) that the patient will have a stroke over the next 10-year period. For the smoking variable, define a dummy variable with 1 indicating a smoker and 0 indicating a nonsmoker.

Risk	Age	Pressure	Smoker
12	57	152	No
24	67	163	No
13	58	155	No
56	86	177	Yes
28	59	196	No
51	76	189	Yes
18	56	155	Yes
31	78	120	No
37	80	135	Yes
15	78	98	No
22	71	152	No
36	70	173	Yes
15	67	135	Yes
48	77	209	Yes
15	60	199	No
36	82	119	Yes
8	66	166	No
34	80	125	Yes
3	62	117	No
37	59	207	Yes

a. Develop an estimated regression equation that relates risk of a stroke to the person's age, blood pressure, and whether the person is a smoker.
b. Is smoking a significant factor in the risk of a stroke? Explain. Use $\alpha = .05$.
c. What is the probability of a stroke over the next 10 years for Art Speen, a 68-year-old smoker who has blood pressure of 175? What action might the physician recommend for this patient?

Summary

In this chapter, we introduced multiple regression analysis as an extension of simple linear regression analysis presented in Chapter 12. Multiple regression analysis enables us to understand how a dependent variable is related to two or more independent variables. The mulitple regression equation $E(y) = \beta_0 + \beta_1 x_1 + \beta_2 x_2 + \cdots + \beta_p x_p$ shows that the mean or expected value of the dependent variable y, denoted $E(y)$, is related to the values of the independent variables $x_1, x_2, \ldots, x_p$. Sample data and the least squares method are used to develop the estimated multiple regression equation $\hat{y} = b_0 + b_1x_1 + b_2x_2 + \ldots + b_px_p$. In effect $b_0, b_1, b_2, \ldots, b_p$ are sample statistics used to estimate the unknown model parameters $\beta_0, \beta_1, \beta_2, \ldots, \beta_p$. Computer printouts were used throughout the chapter to emphasize the fact that statistical software packages are the only realistic means of performing the numerous computations required in multiple regression analysis.

The multiple coefficient of determination was presented as a measure of the goodness of fit of the estimated regression equation. It determines the proportion of the variation of y that can be explained by the estimated regression equation. The adjusted multiple coefficient of determination is a similar measure of goodness of fit that adjusts for the number of independent variables and thus avoids overestimating the impact of adding more independent variables.

An F test and a t test were presented as ways to determine statistically whether the relationship among the variables is significant. The F test is used to determine whether there is a significant overall relationship between the dependent variable and the set of all independent variables. The t test is used to determine whether there is a significant relationship between the dependent variable and an individual independent variable given the other independent variables in the regression model. Correlation among the independent variables, known as multicollinearity, was discussed.

The chapter concluded with a section on how dummy variables can be used to incorporate categorical independent variables into a regression model.

Glossary

Multiple regression analysis Regression analysis involving two or more independent variables.
Multiple regression model The mathematical equation that describes how the dependent variable y is related to the independent variables $x_1, x_2, \ldots, x_p$ and an error term ϵ.
Multiple regression equation The mathematical equation relating the expected value or mean value of the dependent variable to the values of the independent variables; that is, $E(y) = \beta_0 + \beta_1 x_1 + \beta_2 x_2 + \cdots + \beta_p x_p$.
Estimated multiple regression equation The estimate of the multiple regression equation based on sample data and the least squares method; it is $\hat{y} = b_0 + b_1 x_1 + b_2 x_2 + \cdots + b_p x_p$.
Least squares method The method used to develop the estimated regression equation. It minimizes the sum of squared residuals (the deviations between the observed values of the dependent variable, y_i, and the estimated values of the dependent variable, $\hat{y}_i$).
Multiple coefficient of determination A measure of the goodness of fit of the estimated multiple regression equation. It can be interpreted as the proportion of the variability in the dependent variable that is explained by the estimated regression equation.
Adjusted multiple coefficient of determination A measure of the goodness of fit of the estimated multiple regression equation that adjusts for the number of independent variables in the model and thus avoids overestimating the impact of adding more independent variables.
Multicollinearity The term used to describe the correlation among the independent variables.
Categorical independent variable An independent variable with categorical data.
Dummy variable A variable used to model the effect of categorical independent variables. A dummy variable may take only the value zero or one.

Key Formulas

Multiple Regression Model

$$y = \beta_0 + \beta_1 x_1 + \beta_2 x_2 + \cdots + \beta_p x_p + \epsilon \qquad (13.1)$$

Multiple Regression Equation

$$E(y) = \beta_0 + \beta_1 x_1 + \beta_2 x_2 + \cdots + \beta_p x_p \tag{13.2}$$

Estimated Multiple Regression Equation

$$\hat{y} = b_0 + b_1 x_1 + b_2 x_2 + \cdots + b_p x_p \tag{13.3}$$

Least Squares Criterion

$$\min \Sigma(y_i - \hat{y}_i)^2 \tag{13.4}$$

Relationship Among SST, SSR, and SSE

$$\text{SST} = \text{SSR} + \text{SSE} \tag{13.7}$$

Multiple Coefficient of Determination

$$R^2 = \frac{\text{SSR}}{\text{SST}} \tag{13.8}$$

Adjusted Multiple Coefficient of Determination

$$R_a^2 = 1 - (1 - R^2)\frac{n-1}{n-p-1} \tag{13.9}$$

Mean Square Due to Regression

$$\text{MSR} = \frac{\text{SSR}}{p} \tag{13.12}$$

Mean Square Due to Error

$$\text{MSE} = \frac{\text{SSE}}{n-p-1} \tag{13.13}$$

F Test Statistic

$$F = \frac{\text{MSR}}{\text{MSE}} \tag{13.14}$$

t Test Statistic

$$t = \frac{b_i}{s_{b_i}} \tag{13.15}$$

Supplementary Exercises

39. The personnel director for Electronics Associates developed the following estimated regression equation relating an employee's score on a job satisfaction test to his or her length of service and wage rate.

$$\hat{y} = 14.4 - 8.69 x_1 + 13.5 x_2$$

where

$$x_1 = \text{length of service (years)}$$
$$x_2 = \text{wage rate (dollars)}$$
$$y = \text{job satisfaction test score (higher scores indicate greater job satisfaction)}$$

a. Interpret the coefficients in this estimated regression equation.
b. Develop an estimate of the job satisfaction test score for an employee who has four years of service and makes $6.50 per hour.

40. The admissions officer for Clearwater College developed the following estimated regression equation relating the final college GPA (grade point average) to the student's SAT mathematics score and high-school GPA.

$$\hat{y} = -1.41 + .0235x_1 + .00486x_2$$

where

$$x_1 = \text{high-school GPA}$$
$$x_2 = \text{SAT mathematics score}$$
$$y = \text{final college GPA}$$

a. Interpret the coefficients in this estimated regression equation.
b. Estimate the final college GPA for a student who has a high-school average of 84 and a score of 540 on the SAT mathematics test.

41. Recall that in exercise 40, the admissions officer for Clearwater College developed the following estimated regression equation relating final college GPA to the student's SAT mathematics score and high-school GPA.

$$\hat{y} = -1.41 + .0235x_1 + .00486x_2$$

where

$$x_1 = \text{high-school GPA}$$
$$x_2 = \text{SAT mathematics score}$$
$$y = \text{final college GPA}$$

A portion of the Minitab computer output follows.

```
The regression equation is
Y = -1.41 + .0235 X1 + .00486 X2

Predictor          Coef           SE Coef              T
Constant         -1.4053           0.4848          _____
X1              0.023467           0.008666         _____
X2              _____             0.001077         _____

S = 0.1298      R-sq = _____      R-sq(adj) = _____

Analysis of Variance

SOURCE           DF              SS              MS              F
Regression     _____          1.76209         _____          _____
Residual Error _____          _____          _____
Total             9            1.88000
```

a. Complete the missing entries in this output.
b. Use the *F* test and a .05 level of significance to see whether a significant relationship is present.
c. Use the *t* test and $\alpha = .05$ to test $H_0: \beta_1 = 0$ and $H_0: \beta_2 = 0$.
d. Did the estimated regression equation provide a good fit to the data? Explain.

42. Recall that in exercise 39 the personnel director for Electronics Associates developed the following estimated regression equation relating an employee's score on a job satisfaction test to length of service and wage rate.

$$\hat{y} = 14.4 - 8.69x_1 + 13.5x_2$$

where

x_1 = length of service (years)
x_2 = wage rate (dollars)
y = job satisfaction test score (higher scores indicate greater job satisfaction)

A portion of the Minitab computer output follows.

```
The regression equation is
Y = 14.4 - 8.69 X1 + 13.52 X2

Predictor           Coef            SE Coef              T
Constant          14.448              8.191           1.76
X1                                    1.555
X2                13.517              2.085

S = 3.773      R-sq = _____%   R-sq(adj) = _____%

Analysis of Variance

SOURCE             DF              SS              MS              F
Regression          2
Residual Error                  71.17
Total               7           720.0
```

a. Complete the missing entries in this output.
b. Compute *F* and test using $\alpha = .05$ to see whether a significant relationship is present.
c. Did the estimated regression equation provide a good fit to the data? Explain.
d. Use the *t* test and $\alpha = .05$ to test $H_0: \beta_1 = 0$ and $H_0: \beta_2 = 0$.

43. A partial computer output from a regression analysis follows.

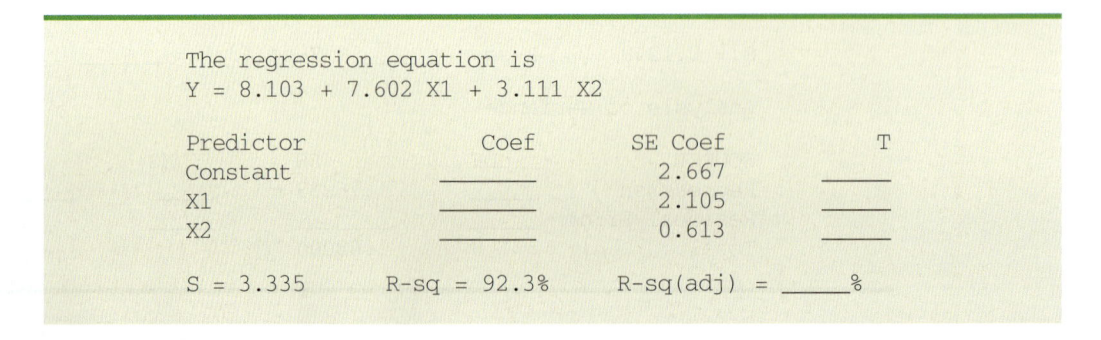

```
The regression equation is
Y = 8.103 + 7.602 X1 + 3.111 X2

Predictor           Coef            SE Coef              T
Constant                             2.667
X1                                   2.105
X2                                   0.613

S = 3.335      R-sq = 92.3%     R-sq(adj) = _____%
```

Supplementary Exercises

```
Analysis of Variance

SOURCE           DF        SS        MS        F
Regression      ____      1612      ____      ____
Residual Error   12       ____      ____
Total           ____      ____
```

a. Compute the missing entries in this output.
b. Use the F test and $\alpha = .05$ to see whether a significant relationship is present.
c. Use the t test and $\alpha = .05$ to test $H_0: \beta_1 = 0$ and $H_0: \beta_2 = 0$.
d. Compute R_a^2.

44. The Tire Rack, America's leading online distributor of tires and wheels, conducts extensive testing to provide customers with products that are right for their vehicle, driving style, and driving conditions. In addition, the Tire Rack maintains an independent consumer survey to help drivers help each other by sharing their long-term tire experiences. The following data show survey ratings (1 to 10 scale with 10 the highest rating) for 18 maximum performance summer tires (Tire Rack website, February 3, 2009). The variable Steering rates the tire's steering responsiveness, Tread Wear rates quickness of wear based on the driver's expectations, and Buy Again rates the driver's overall tire satisfaction and desire to purchase the same tire again.

Tire	Steering	Tread Wear	Buy Again
Goodyear Assurance TripleTred	8.9	8.5	8.1
Michelin HydroEdge	8.9	9.0	8.3
Michelin Harmony	8.3	8.8	8.2
Dunlop SP 60	8.2	8.5	7.9
Goodyear Assurance ComforTred	7.9	7.7	7.1
Yokohama Y372	8.4	8.2	8.9
Yokohama Aegis LS4	7.9	7.0	7.1
Kumho Power Star 758	7.9	7.9	8.3
Goodyear Assurance	7.6	5.8	4.5
Hankook H406	7.8	6.8	6.2
Michelin Energy LX4	7.4	5.7	4.8
Michelin MX4	7.0	6.5	5.3
Michelin Symmetry	6.9	5.7	4.2
Kumho 722	7.2	6.6	5.0
Dunlop SP 40 A/S	6.2	4.2	3.4
Bridgestone Insignia SE200	5.7	5.5	3.6
Goodyear Integrity	5.7	5.4	2.9
Dunlop SP20 FE	5.7	5.0	3.3

a. Develop an estimated regression equation that can be used to predict the Buy Again rating given based on the Steering rating. At the .05 level of significance, test for a significant relationship.
b. Did the estimated regression equation developed in part (a) provide a good fit to the data? Explain.
c. Develop an estimated regression equation that can be used to predict the Buy Again rating given the Steering rating and the Tread Wear rating.
d. Is the addition of the Tread Wear independent variable significant? Use $\alpha = .05$.

45. *Consumer Reports* provided extensive testing and ratings for 24 treadmills. An overall score, based primarily on ease of use, ergonomics, exercise range, and quality, was developed for each treadmill tested. In general, a higher overall score indicates better

performance. The following data show the price, the quality rating, and overall score for the 24 treadmills (*Consumer Reports*, February 2006).

Brand & Model	Price	Quality	Score
Landice L7	2900	Excellent	86
NordicTrack S3000	3500	Very good	85
SportsArt 3110	2900	Excellent	82
Precor	3500	Excellent	81
True Z4 HRC	2300	Excellent	81
Vision Fitness T9500	2000	Excellent	81
Precor M 9.31	3000	Excellent	79
Vision Fitness T9200	1300	Very good	78
Star Trac TR901	3200	Very good	72
Trimline T350HR	1600	Very good	72
Schwinn 820p	1300	Very good	69
Bowflex 7-Series	1500	Excellent	83
NordicTrack S1900	2600	Very good	83
Horizon Fitness PST8	1600	Very good	82
Horizon Fitness 5.2T	1800	Very good	80
Evo by Smooth Fitness FX30	1700	Very good	75
ProForm 1000S	1600	Very good	75
Horizon Fitness CST4.5	1000	Very good	74
Keys Fitness 320t	1200	Very good	73
Smooth Fitness 7.1HR Pro	1600	Very good	73
NordicTrack C2300	1000	Good	70
Spirit Inspire	1400	Very good	70
ProForm 750	1000	Good	67
Image 19.0 R	600	Good	66

a. Use these data to develop an estimated regression equation that could be used to estimate the overall score given the price.
b. Use $\alpha = .05$ to test for overall significance.
c. To incorporate the effect of quality, a categorical variable with three levels, we used two dummy variables: Quality-E and Quality-VG. Each variable was coded 0 or 1 as follows.

$$\text{Quality-E} = \begin{cases} 1 \text{ if quality rating is excellent} \\ 0 \text{ otherwise} \end{cases}$$

$$\text{Quality-VG} = \begin{cases} 1 \text{ if quality rating is very good} \\ 0 \text{ otherwise} \end{cases}$$

Develop an estimated regression equation that could be used to estimate the overall score given the price and the quality rating.
d. For the estimated regression equation developed in part (c), test for overall significance using $\alpha = .10$.
e. For the estimated regression equation developed in part (c), use the *t* test to determine the significance of each independent variable. Use $\alpha = .10$.
f. Develop a standardized residual plot. Does the pattern of the residual plot appear to be reasonable?
g. Do the data contain any outliers or influential observations?
h. Estimate the overall score for a treadmill with a price of $2000 and a good quality rating. How much would the estimate change if the quality rating were very good? Explain.

46. A portion of a data set containing information for 45 mutual funds that are part of the *Morningstar Funds 500* for 2008 follows. The complete data set is available in the file named MutualFunds. The data set includes the following five variables:

Supplementary Exercises

Type: The type of fund, labeled DE (Domestic Equity), IE (International Equity), and FI (Fixed Income).

Net Asset Value ($): The closing price per share on December 31, 2007.

5-Year Average Return (%): The average annual return for the fund over the past five years.

Expense Ratio (%): The percentage of assets deducted each fiscal year for fund expenses.

Morningstar Rank: The risk adjusted star rating for each fund; Morningstar ranks go from a low of 1-Star to a high of 5-Stars.

WEB file

MutualFunds

Fund Name	Fund Type	Net Asset Value ($)	Five-Year Average Return (%)	Expense Ratio (%)	Morningstar Rank
Amer Cent Inc & Growth Inv	DE	28.88	12.39	0.67	2-Star
American Century Intl. Disc	IE	14.37	30.53	1.41	3-Star
American Century Tax-Free Bond	FI	10.73	3.34	0.49	4-Star
American Century Ultra	DE	24.94	10.88	0.99	3-Star
Ariel	DE	46.39	11.32	1.03	2-Star
Artisan Intl Val	IE	25.52	24.95	1.23	3-Star
Artisan Small Cap	DE	16.92	15.67	1.18	3-Star
Baron Asset	DE	50.67	16.77	1.31	5-Star
Brandywine	DE	36.58	18.14	1.08	4-Star

a. Develop an estimated regression equation that can be used to predict the five-year average return given fund type. At the .05 level of significance, test for a significant relationship.

b. Did the estimated regression equation developed in part (a) provide a good fit to the data? Explain.

c. Develop the estimated regression equation that can be used to predict the five-year average return given the type of fund, the net asset value, and the expense ratio. At the .05 level of significance, test for a significant relationship. Do you think any variables should be deleted from the estimated regression equation? Explain.

d. Morningstar Rank is a categorical variable. Because the data set contains only funds with four ranks (2-Star through 5-Star), use the following dummy variables: 3StarRank = 1 for a 3-Star fund, 0 otherwise; 4StarRank = 1 for a 4-Star fund, 0 otherwise; and 5StarRank = 1 for a 5-Star fund, 0 otherwise. Develop an estimated regression equation that can be used to predict the five-year average return given the type of fund, the expense ratio, and the Morningstar Rank. Using $\alpha = .05$, remove any independent variables that are not significant.

e. Use the estimated regression equation developed in part (d) to estimate the five-year average return for a domestic equity fund with an expense ratio of 1.05% and a 3-Star Morningstar Rank.

47. The U.S. Department of Energy's Fuel Economy Guide provides fuel efficiency data for cars and trucks (U.S. Department of Energy website, February 22, 2008). A portion of the data for 311 compact, midsize, and large cars follows. The column labeled Class identifies the size of the car; Compact, Midsize, or Large. The column labeled Displacement shows the engine's displacement in liters. The column labeled Fuel Type shows whether the car uses premium (P) or regular (R) fuel, and the column labeled Hwy MPG shows the fuel efficiency rating for highway driving in terms of miles per gallon. The complete data set is contained in the file named FuelData.

Car	Class	Displacement	Fuel Type	Hwy MPG
1	Compact	3.1	P	25
2	Compact	3.1	P	25
3	Compact	3	P	25
.	.	.	.	.
161	Midsize	2.4	R	30
162	Midsize	2	P	29
.	.	.	.	.
310	Large	3	R	25
311	Large	3	R	25

a. Develop an estimated regression equation that can be used to predict the fuel efficiency for highway driving given the engine's displacement. Test for significance using $\alpha = .05$.
b. Consider the addition of the dummy variables ClassMidsize and ClassLarge. The value of ClassMidsize is 1 if the car is a midsize car and 0 otherwise; the value of ClassLarge is 1 if the car is a large car and 0 otherwise. Thus, for a compact car, the value of ClassMidsize and the value of ClassLarge is 0. Develop the estimated regression equation that can be used to predict the fuel efficiency for highway driving given the engine's displacement and the dummy variables ClassMidsize and ClassLarge.
c. Use $\alpha = .05$ to determine whether the dummy variables added in part (b) are significant.
d. Consider the addition of the dummy variable FuelPremium, where the value of FuelPremium is 1 if the car uses premium fuel and 0 if the car uses regular fuel. Develop the estimated regression equation that can be used to predict the fuel efficiency for highway driving given the engine's displacement, the dummy variables ClassMidsize and ClassLarge, and the dummy variable FuelPremium.
e. For the estimated regression equation developed in part (d), test for overall significance and individual significance using $\alpha = .05$.

Case Problem 1 Consumer Research, Inc.

Consumer Research, Inc., is an independent agency that conducts research on consumer attitudes and behaviors for a variety of firms. In one study, a client asked for an investigation of consumer characteristics that can be used to predict the amount charged by credit card users. Data were collected on annual income, household size, and annual credit card charges for a sample of 50 consumers. The following data are contained in the file named Consumer.

Income ($1000s)	Household Size	Amount Charged ($)	Income ($1000s)	Household Size	Amount Charged ($)
54	3	4016	54	6	5573
30	2	3159	30	1	2583
32	4	5100	48	2	3866
50	5	4742	34	5	3586
31	2	1864	67	4	5037
55	2	4070	50	2	3605
37	1	2731	67	5	5345
40	2	3348	55	6	5370
66	4	4764	52	2	3890

Income ($1000s)	Household Size	Amount Charged ($)	Income ($1000s)	Household Size	Amount Charged ($)
51	3	4110	62	3	4705
25	3	4208	64	2	4157
48	4	4219	22	3	3579
27	1	2477	29	4	3890
33	2	2514	39	2	2972
65	3	4214	35	1	3121
63	4	4965	39	4	4183
42	6	4412	54	3	3730
21	2	2448	23	6	4127
44	1	2995	27	2	2921
37	5	4171	26	7	4603
62	6	5678	61	2	4273
21	3	3623	30	2	3067
55	7	5301	22	4	3074
42	2	3020	46	5	4820
41	7	4828	66	4	5149

Managerial Report

1. Use methods of descriptive statistics to summarize the data. Comment on the findings.
2. Develop estimated regression equations, first using annual income as the independent variable and then using household size as the independent variable. Which variable is the better predictor of annual credit card charges? Discuss your findings.
3. Develop an estimated regression equation with annual income and household size as the independent variables. Discuss your findings.
4. What is the predicted annual credit card charge for a three-person household with an annual income of $40,000?
5. Discuss the need for other independent variables that could be added to the model. What additional variables might be helpful?

Case Problem 2 Alumni Giving

Alumni donations are an important source of revenue for colleges and universities. If administrators could determine the factors that could lead to increases in the percentage of alumni who make a donation, they might be able to implement policies that could lead to increased revenues. Research shows that students who are more satisfied with their contact with teachers are more likely to graduate. As a result, one might suspect that smaller class sizes and lower student-faculty ratios might lead to a higher percentage of satisfied graduates, which in turn might lead to increases in the percentage of alumni who make a donation. Table 13.7 shows data for 48 national universities (*America's Best Colleges,* Year 2000 ed.). The column labeled Graduation Rate is the percentage of students who initially enrolled at the university and graduated. The column labeled % of Classes Under 20 shows the percentage of classes offered with fewer than 20 students. The column labeled Student-Faculty Ratio is the number of students enrolled divided by the total number of faculty. Finally, the column labeled Alumni Giving Rate is the percentage of alumni who made a donation to the university.

Managerial Report

1. Use methods of descriptive statistics to summarize the data.

TABLE 13.7 DATA FOR 48 NATIONAL UNIVERSITIES

	State	Graduation Rate	% of Classes Under 20	Student-Faculty Ratio	Alumni Giving Rate
Boston College	MA	85	39	13	25
Brandeis University	MA	79	68	8	33
Brown University	RI	93	60	8	40
California Institute of Technology	CA	85	65	3	46
Carnegie Mellon University	PA	75	67	10	28
Case Western Reserve Univ.	OH	72	52	8	31
College of William and Mary	VA	89	45	12	27
Columbia University	NY	90	69	7	31
Cornell University	NY	91	72	13	35
Dartmouth College	NH	94	61	10	53
Duke University	NC	92	68	8	45
Emory University	GA	84	65	7	37
Georgetown University	DC	91	54	10	29
Harvard University	MA	97	73	8	46
Johns Hopkins University	MD	89	64	9	27
Lehigh University	PA	81	55	11	40
Massachusetts Inst. of Technology	MA	92	65	6	44
New York University	NY	72	63	13	13
Northwestern University	IL	90	66	8	30
Pennsylvania State Univ.	PA	80	32	19	21
Princeton University	NJ	95	68	5	67
Rice University	TX	92	62	8	40
Stanford University	CA	92	69	7	34
Tufts University	MA	87	67	9	29
Tulane University	LA	72	56	12	17
U. of California—Berkeley	CA	83	58	17	18
U. of California—Davis	CA	74	32	19	7
U. of California—Irvine	CA	74	42	20	9
U. of California—Los Angeles	CA	78	41	18	13
U. of California—San Diego	CA	80	48	19	8
U. of California—Santa Barbara	CA	70	45	20	12
U. of Chicago	IL	84	65	4	36
U. of Florida	FL	67	31	23	19
U. of Illinois—Urbana Champaign	IL	77	29	15	23
U. of Michigan—Ann Arbor	MI	83	51	15	13
U. of North Carolina—Chapel Hill	NC	82	40	16	26
U. of Notre Dame	IN	94	53	13	49
U. of Pennsylvania	PA	90	65	7	41
U. of Rochester	NY	76	63	10	23
U. of Southern California	CA	70	53	13	22
U. of Texas—Austin	TX	66	39	21	13
U. of Virginia	VA	92	44	13	28
U. of Washington	WA	70	37	12	12
U. of Wisconsin—Madison	WI	73	37	13	13
Vanderbilt University	TN	82	68	9	31
Wake Forest University	NC	82	59	11	38
Washington University—St. Louis	MO	86	73	7	33
Yale University	CT	94	77	7	50

2. Develop an estimated regression equation that can be used to predict the alumni giving rate given the number of students who graduate. Discuss your findings.
3. Develop an estimated regression equation that could be used to predict the alumni giving rate using the data provided.
4. What conclusions and recommendations can you derive from your analysis?

Case Problem 3 PGA Tour Statistics

The Professional Golfers Association (PGA) maintains data on performance and earnings for members of the PGA Tour. The top 125 players based on total earnings in PGA Tour events are exempt for the following season. Making the top 125 money list is important because a player who is "exempt" has qualified to be a full-time member of the PGA tour for the following season.

Scoring average is generally considered the most important statistic in terms of success on the PGA Tour. To investigate the effect that variables such as driving distance, driving accuracy, greens in regulation, sand saves, and average putts per round have on average score, year-end performance data for the 125 players who had the highest total earnings in PGA Tour events for 2008 are contained in the file named PGATour (PGA Tour website, 2009). Each row of the data set corresponds to a PGA Tour player, and the data have been sorted based upon total earnings. Descriptions for the variables in the data set follow.

Money: Total earnings in PGA Tour events.

Scoring Average: The average number of strokes per completed round.

DrDist (Driving Distance):   The average number of yards per measured drive. On the PGA Tour driving distance is measured on two holes per round. Care is taken to select two holes which face in opposite directions to counteract the effect of wind. Drives are measured to the point at which they come to rest regardless of whether they are in the fairway or not.

DrAccu (Driving Accuracy):   The percentage of time a tee shot comes to rest in the fairway (regardless of club). Driving accuracy is measured on every hole, excluding par 3s.

GIR (Greens in Regulation):   The percentage of time a player was able to hit the green in regulation. A green is considered hit in regulation if any portion of the ball is touching the putting surface after the GIR stroke has been taken. The GIR stroke is determined by subtracting 2 from par (1st stroke on a par 3, 2nd on a par 4, 3rd on a par 5). In other words, a green is considered hit in regulation if the player has reached the putting surface in par minus two strokes.

Sand Saves: The percentage of time a player was able to get "up and down" once in a greenside sand bunker (regardless of score). "Up and down" indicates it took the player 2 shots or less to put the ball in the hole from a greenside sand bunker.

PPR (Putts Per Round):   The average number of putts per round.

Scrambling: The percentage of time a player missed the green in regulation but still made par or better.

Managerial Report

1. To predict Scoring Average, develop estimated regression equations, first using DrDist as the independent variable and then using DrAccu as the independent variable. Which variable is the better predictor of Scoring Average? Discuss your findings.

2. Develop an estimated regression equation with GIR as the independent variable. Compare your findings with the results obtained using DrDist and DrAccu.
3. Develop an estimated regression equation with GIR and Sand Saves as the independent variables. Discuss your findings.
4. Develop an estimated regression equation with GIR and PPR as the independent variables. Discuss your findings.
5. Develop an estimated regression equation with GIR and Scrambling as the independent variables. Discuss your findings.
6. Compare the results obtained for the estimated regression equations that use GIR and Sand Saves, GIR and PPR, and GIR and Scrambling as the two independent variables. If you had to select one of these two-independent-variable estimated regression equations to predict Scoring Average, which estimated regression equation would you use? Explain.
7. Develop the estimated regression equation that uses GIR, Sand Saves, and PPR to predict Scoring Average. Compare the results to an estimated regression equation that uses GIR, PPR, and Scrambling as the independent variables.
8. Develop an estimated regression equation that uses GIR, Sand Saves, PPR, and Scrambling to predict Scoring Average. Discuss your results.

Case Problem 4 Predicting Winning Percentage for the NFL

The National Football League (NFL) records a variety of performance data for individuals and teams. Some of the year-end performance data for the 2005 season are contained in the file named NFLStats (NFL website). Each row of the data set corresponds to an NFL team, and the teams are ranked by winning percentage. Descriptions for the data follow:

WinPct	Percentage of games won
TakeInt	Takeaway interceptions; the total number of interceptions made by the team's defense
TakeFum	Takeaway fumbles; the total number of fumbles recovered by the team's defense
GiveInt	Giveaway interceptions; the total number of interceptions made by the team's offense
GiveFum	Giveaway fumbles; the total number of fumbles made by the team's offense
DefYds/G	Average number of yards per game given up on defense
RushYds/G	Average number of rushing yards per game
PassYds/G	Average number of passing yards per game
FGPct	Percentage of field goals

Managerial Report

1. Use methods of descriptive statistics to summarize the data. Comment on the findings.
2. Develop an estimated regression equation that can be used to predict WinPct using DefYds/G, RushYds/G, PassYds/G, and FGPct. Discuss your findings.
3. Starting with the estimated regression equation developed in part (2), delete any independent variables that are not significant and develop a new estimated regression equation that can be used to predict WinPct. Use $\alpha = .05$.
4. Some football analysts believe that turnovers are one of the most important factors in determining a team's success. With Takeaways = TakeInt + TakeFum and

Giveaways = GiveInt + GiveFum, let NetDiff = Takeaways − Giveaways. Develop an estimated regression equation that can be used to predict WinPct using NetDiff. Compare your results with the estimated regression equation developed in part (3).

5. Develop an estimated regression equation that can be used to predict WinPct using all the data provided.

Appendix 13.1 Multiple Regression Using Minitab

In Section 13.2 we discussed the computer solution of multiple regression problems by showing Minitab's output for the Butler Trucking Company problem. In this appendix we describe the steps required to generate the Minitab computer solution. First, the data must be entered in a Minitab worksheet. The miles traveled are entered in column C1, the number of deliveries are entered in column C2, and the travel times (hours) are entered in column C3. The variable names Miles, Deliveries, and Time were entered as the column headings on the worksheet. In subsequent steps, we refer to the data by using the variable names Miles, Deliveries, and Time or the column indicators C1, C2, and C3. The following steps describe how to use Minitab to produce the regression results shown in Figure 13.4.

Step 1. Select the **Stat** menu
Step 2. Select the **Regression** menu
Step 3. Choose **Regression**
Step 4. When the **Regression** dialog box appears:
  Enter Time in the **Response** box
  Enter Miles and Deliveries in the **Predictors** box
  Click **OK**

Appendix 13.2 Multiple Regression Using Excel

In Section 13.2 we discussed the computer solution of multiple regression problems by showing Minitab's output for the Butler Trucking Company problem. In this appendix we describe how to use Excel's Regression tool to develop the estimated multiple regression equation for the Butler Trucking problem. Refer to Figure 13.10 as we describe the tasks involved. First, the labels Assignment, Miles, Deliveries, and Time are entered into cells A1:D1 of the worksheet, and the sample data into cells B2:D11. The numbers 1–10 in cells A2:A11 identify each observation.

The following steps describe how to use the Regression tool for the multiple regression analysis.

Step 1. Click the **Data** tab on the Ribbon
Step 2. In the **Analysis** group, click **Data Analysis**
Step 3. Choose **Regression** from the list of Analysis Tools
Step 4. When the Regression dialog box appears:
  Enter D1:D11 in the **Input Y Range** box
  Enter B1:C11 in the **Input X Range** box
  Select **Labels**
  Select **Confidence Level**
  Enter 99 in the **Confidence Level** box
  Select **Output Range**
  Enter A13 in the **Output Range** box (to identify the upper left corner of the section of the worksheet where the output will appear)
  Click **OK**

FIGURE 13.10   EXCEL OUTPUT FOR BUTLER TRUCKING WITH TWO INDEPENDENT VARIABLES

	A	B	C	D	E	F	G	H	I	J
1	Assignment	Miles	Deliveries	Time						
2	1	100	4	9.3						
3	2	50	3	4.8						
4	3	100	4	8.9						
5	4	100	2	6.5						
6	5	50	2	4.2						
7	6	80	2	6.2						
8	7	75	3	7.4						
9	8	65	4	6						
10	9	90	3	7.6						
11	10	90	2	6.1						
12										
13	SUMMARY OUTPUT									
14										
15	*Regression Statistics*									
16	Multiple R	0.9507								
17	R Square	0.9038								
18	Adjusted R Square	0.8763								
19	Standard Error	0.5731								
20	Observations	10								
21										
22	ANOVA									
23		df	SS	MS	F	Significance F				
24	Regression	2	21.6006	10.8003	32.8784	0.0003				
25	Residual	7	2.2994	0.3285						
26	Total	9	23.9							
27										
28		Coefficients	Standard Error	t Stat	P-value	Lower 95%	Upper 95%	Lower 99.0%	Upper 99.0%	
29	Intercept	-0.8687	0.9515	-0.9129	0.3916	-3.1188	1.3813	-4.1986	2.4612	
30	Miles	0.0611	0.0099	6.1824	0.0005	0.0378	0.0845	0.0265	0.0957	
31	Deliveries	0.9234	0.2211	4.1763	0.0042	0.4006	1.4463	0.1496	1.6972	

In the Excel output shown in Figure 13.10 the label for the independent variable x_1 is Miles (see cell A30), and the label for the independent variable x_2 is Deliveries (see cell A31). The estimated regression equation is

$$\hat{y} = -.8687 + .0611x_1 + .9234x_2$$

Note that using Excel's Regression tool for multiple regression is almost the same as using it for simple linear regression. The major difference is that in the multiple regression case a larger range of cells is required in order to identify the independent variables.

Appendix 13.3   Multiple Regression Using StatTools

WEB file
Butler

In this appendix we show how StatTools can be used to perform the regression analysis computations for the Butler Trucking problem. Begin by using the Data Set Manager to create a StatTools data set for these data using the procedure described in the appendix in Chapter 1. The following steps describe how StatTools can be used to provide the regression results.

Appendix 13.3 Multiple Regression Using StatTools

Step 1. Click the **StatTools** tab on the Ribbon
Step 2. In the **Analyses** group, click **Regression and Classification**
Step 3. Choose the **Regression** option
Step 4. When the StatTools - Regression dialog box appears:
    Select **Multiple** in the **Regression Type** box
    In the **Variables** section:
        Click the **Format** button and select **Unstacked**
        In the column labeled **I** select **Miles**
        In the column labeled **I** select **Deliveries**
        In the column labeled **D** select **Time**
  Click **OK**

The regression analysis output will appear in a new worksheet.

The StatTools - Regression dialog box contains a number of more advanced options for developing prediction interval estimates and producing residual plots. The StatTools Help facility provides information on using all of these options.

APPENDIXES

APPENDIX A
References and Bibliography

APPENDIX B
Tables

APPENDIX C
Summation Notation

APPENDIX D
Self-Test Solutions and Answers to Exercises

APPENDIX E
Using Excel Functions

APPENDIX F
Computing p-Values Using Minitab and Excel

Appendix A: References and Bibliography

General

Freedman, D., R. Pisani, and R. Purves. *Statistics,* 4th ed. W. W. Norton, 2007.

Hogg, R. V., A. T. Craig, and J. W. McKean. *Introduction to Mathematical Statistics,* 6th ed. Pearson, 2005.

Hogg, R. V., and E. A. Tanis. *Probability and Statistical Inference,* 8th ed. Pearson, 2008.

Miller, I., and M. Miller. *John E. Freund's Mathematical Statistics,* 7th ed. Pearson, 2004.

Moore, D. S., G. P. McCabe, and B. Craig. *Introduction to the Practice of Statistics,* 6th ed. Freeman, 2009.

Wackerly, D. D., W. Mendenhall, and R. L. Scheaffer. *Mathematical Statistics with Applications,* 7th ed. Cengage Learning, 2008.

Experimental Design

Cochran, W. G., and G. M. Cox. *Experimental Designs,* 2nd ed. Wiley, 1992.

Hicks, C. R., and K. V. Turner. *Fundamental Concepts in the Design of Experiments,* 5th ed. Oxford University Press, 1999.

Montgomery, D. C. *Design and Analysis of Experiments,* 7th ed. Wiley, 2008.

Winer, B. J., K. M. Michels, and D. R. Brown. *Statistical Principles in Experimental Design,* 3rd ed. McGraw-Hill, 1991.

Wu, C. F. Jeff, and M. Hamada. *Experiments: Planning, Analysis, and Parameter Optimization,* 2nd ed. Wiley, 2009.

Probability

Hogg, R. V., and E. A. Tanis. *Probability and Statistical Inference,* 8th ed. Pearson, 2008.

Ross, S. M. *Introduction to Probability Models,* 9th ed. Elsevier, 2006.

Wackerly, D. D., W. Mendenhall, and R. L. Scheaffer. *Mathematical Statistics with Applications,* 7th ed. Cengage Learning, 2008.

Regression Analysis

Chatterjee, S., and A. S. Hadi. *Regression Analysis by Example,* 4th ed. Wiley, 2006.

Draper, N. R., and H. Smith. *Applied Regression Analysis,* 3rd ed. Wiley, 1998.

Graybill, F. A., and H. K. Iyer. *Regression Analysis: Concepts and Applications.* Wadsworth, 1994.

Hosmer, D. W., and S. Lemeshow. *Applied Logistic Regression,* 2nd ed. Wiley, 2000.

Kleinbaum, D. G., L. L. Kupper, and K. E. Muller. *Applied Regression Analysis and Multivariate Methods,* 4th ed. Cengage Learning, 2007.

Neter, J., W. Wasserman, M. H. Kutner, and C. Nashtsheim. *Applied Linear Statistical Models,* 4th ed. McGraw-Hill, 1996.

Mendenhall, M., and T. Sincich. *A Second Course in Statistics: Regression Analysis,* 6th ed. Pearson, 2003.

Sampling

Cochran, W. G. *Sampling Techniques,* 3rd ed. Wiley, 1977.

Hahn, G. J., and W. Meeker. "Assumptions for Statistical Inference," *The American Statistician*, February 1993.

Hansen, M. H., W. N. Hurwitz, W. G. Madow, and M. N. Hanson. *Sample Survey Methods and Theory.* Wiley, 1993.

Kish, L. *Survey Sampling.* Wiley, 2008.

Levy, P. S., and S. Lemeshow. *Sampling of Populations: Methods and Applications,* 4th ed. Wiley, 2008.

Scheaffer, R. L., W. Mendenhall, and L. Ott. *Elementary Survey Sampling,* 6th ed. Cengage Learning, 2005.

Appendix B: Tables

TABLE 1 CUMULATIVE PROBABILITIES FOR THE STANDARD NORMAL DISTRIBUTION

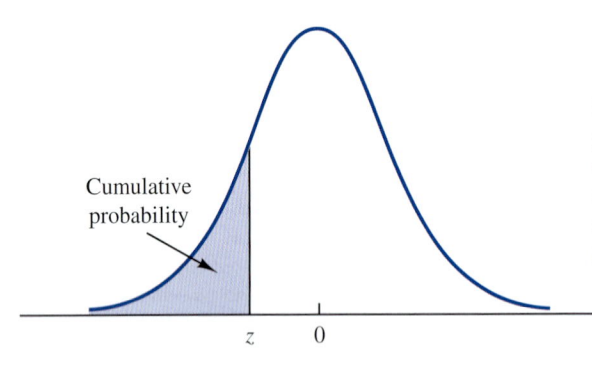

Entries in the table give the area under the curve to the left of the z value. For example, for $z = -.85$, the cumulative probability is .1977.

z	.00	.01	.02	.03	.04	.05	.06	.07	.08	.09
−3.0	.0013	.0013	.0013	.0012	.0012	.0011	.0011	.0011	.0010	.0010
−2.9	.0019	.0018	.0018	.0017	.0016	.0016	.0015	.0015	.0014	.0014
−2.8	.0026	.0025	.0024	.0023	.0023	.0022	.0021	.0021	.0020	.0019
−2.7	.0035	.0034	.0033	.0032	.0031	.0030	.0029	.0028	.0027	.0026
−2.6	.0047	.0045	.0044	.0043	.0041	.0040	.0039	.0038	.0037	.0036
−2.5	.0062	.0060	.0059	.0057	.0055	.0054	.0052	.0051	.0049	.0048
−2.4	.0082	.0080	.0078	.0075	.0073	.0071	.0069	.0068	.0066	.0064
−2.3	.0107	.0104	.0102	.0099	.0096	.0094	.0091	.0089	.0087	.0084
−2.2	.0139	.0136	.0132	.0129	.0125	.0122	.0119	.0116	.0113	.0110
−2.1	.0179	.0174	.0170	.0166	.0162	.0158	.0154	.0150	.0146	.0143
−2.0	.0228	.0222	.0217	.0212	.0207	.0202	.0197	.0192	.0188	.0183
−1.9	.0287	.0281	.0274	.0268	.0262	.0256	.0250	.0244	.0239	.0233
−1.8	.0359	.0351	.0344	.0336	.0329	.0322	.0314	.0307	.0301	.0294
−1.7	.0446	.0436	.0427	.0418	.0409	.0401	.0392	.0384	.0375	.0367
−1.6	.0548	.0537	.0526	.0516	.0505	.0495	.0485	.0475	.0465	.0455
−1.5	.0668	.0655	.0643	.0630	.0618	.0606	.0594	.0582	.0571	.0559
−1.4	.0808	.0793	.0778	.0764	.0749	.0735	.0721	.0708	.0694	.0681
−1.3	.0968	.0951	.0934	.0918	.0901	.0885	.0869	.0853	.0838	.0823
−1.2	.1151	.1131	.1112	.1093	.1075	.1056	.1038	.1020	.1003	.0985
−1.1	.1357	.1335	.1314	.1292	.1271	.1251	.1230	.1210	.1190	.1170
−1.0	.1587	.1562	.1539	.1515	.1492	.1469	.1446	.1423	.1401	.1379
−.9	.1841	.1814	.1788	.1762	.1736	.1711	.1685	.1660	.1635	.1611
−.8	.2119	.2090	.2061	.2033	.2005	.1977	.1949	.1922	.1894	.1867
−.7	.2420	.2389	.2358	.2327	.2296	.2266	.2236	.2206	.2177	.2148
−.6	.2743	.2709	.2676	.2643	.2611	.2578	.2546	.2514	.2483	.2451
−.5	.3085	.3050	.3015	.2981	.2946	.2912	.2877	.2843	.2810	.2776
−.4	.3446	.3409	.3372	.3336	.3300	.3264	.3228	.3192	.3156	.3121
−.3	.3821	.3783	.3745	.3707	.3669	.3632	.3594	.3557	.3520	.3483
−.2	.4207	.4168	.4129	.4090	.4052	.4013	.3974	.3936	.3897	.3859
−.1	.4602	.4562	.4522	.4483	.4443	.4404	.4364	.4325	.4286	.4247
−.0	.5000	.4960	.4920	.4880	.4840	.4801	.4761	.4721	.4681	.4641

TABLE 1 CUMULATIVE PROBABILITIES FOR THE STANDARD NORMAL DISTRIBUTION (*Continued*)

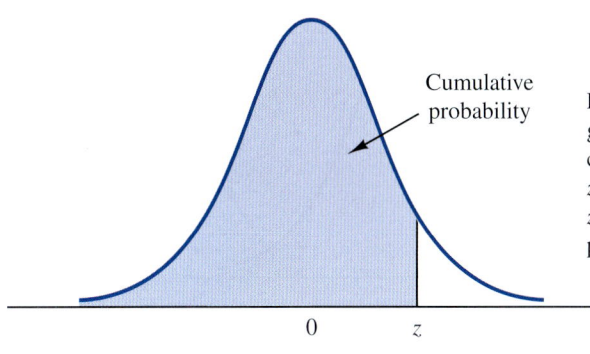

Entries in the table give the area under the curve to the left of the z value. For example, for $z = 1.25$, the cumulative probability is .8944.

z	.00	.01	.02	.03	.04	.05	.06	.07	.08	.09
.0	.5000	.5040	.5080	.5120	.5160	.5199	.5239	.5279	.5319	.5359
.1	.5398	.5438	.5478	.5517	.5557	.5596	.5636	.5675	.5714	.5753
.2	.5793	.5832	.5871	.5910	.5948	.5987	.6026	.6064	.6103	.6141
.3	.6179	.6217	.6255	.6293	.6331	.6368	.6406	.6443	.6480	.6517
.4	.6554	.6591	.6628	.6664	.6700	.6736	.6772	.6808	.6844	.6879
.5	.6915	.6950	.6985	.7019	.7054	.7088	.7123	.7157	.7190	.7224
.6	.7257	.7291	.7324	.7357	.7389	.7422	.7454	.7486	.7517	.7549
.7	.7580	.7611	.7642	.7673	.7704	.7734	.7764	.7794	.7823	.7852
.8	.7881	.7910	.7939	.7967	.7995	.8023	.8051	.8078	.8106	.8133
.9	.8159	.8186	.8212	.8238	.8264	.8289	.8315	.8340	.8365	.8389
1.0	.8413	.8438	.8461	.8485	.8508	.8531	.8554	.8577	.8599	.8621
1.1	.8643	.8665	.8686	.8708	.8729	.8749	.8770	.8790	.8810	.8830
1.2	.8849	.8869	.8888	.8907	.8925	.8944	.8962	.8980	.8997	.9015
1.3	.9032	.9049	.9066	.9082	.9099	.9115	.9131	.9147	.9162	.9177
1.4	.9192	.9207	.9222	.9236	.9251	.9265	.9279	.9292	.9306	.9319
1.5	.9332	.9345	.9357	.9370	.9382	.9394	.9406	.9418	.9429	.9441
1.6	.9452	.9463	.9474	.9484	.9495	.9505	.9515	.9525	.9535	.9545
1.7	.9554	.9564	.9573	.9582	.9591	.9599	.9608	.9616	.9625	.9633
1.8	.9641	.9649	.9656	.9664	.9671	.9678	.9686	.9693	.9699	.9706
1.9	.9713	.9719	.9726	.9732	.9738	.9744	.9750	.9756	.9761	.9767
2.0	.9772	.9778	.9783	.9788	.9793	.9798	.9803	.9808	.9812	.9817
2.1	.9821	.9826	.9830	.9834	.9838	.9842	.9846	.9850	.9854	.9857
2.2	.9861	.9864	.9868	.9871	.9875	.9878	.9881	.9884	.9887	.9890
2.3	.9893	.9896	.9898	.9901	.9904	.9906	.9909	.9911	.9913	.9916
2.4	.9918	.9920	.9922	.9925	.9927	.9929	.9931	.9932	.9934	.9936
2.5	.9938	.9940	.9941	.9943	.9945	.9946	.9948	.9949	.9951	.9952
2.6	.9953	.9955	.9956	.9957	.9959	.9960	.9961	.9962	.9963	.9964
2.7	.9965	.9966	.9967	.9968	.9969	.9970	.9971	.9972	.9973	.9974
2.8	.9974	.9975	.9976	.9977	.9977	.9978	.9979	.9979	.9980	.9981
2.9	.9981	.9982	.9982	.9983	.9984	.9984	.9985	.9985	.9986	.9986
3.0	.9987	.9987	.9987	.9988	.9988	.9989	.9989	.9989	.9990	.9990

TABLE 2 *t* DISTRIBUTION

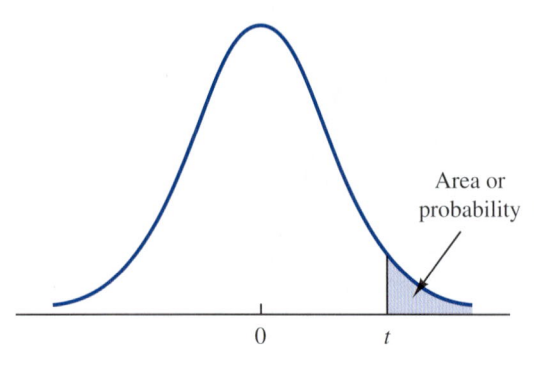

Entries in the table give *t* values for an area or probability in the upper tail of the *t* distribution. For example, with 10 degrees of freedom and a .05 area in the upper tail, $t_{.05} = 1.812$.

Degrees of Freedom	.20	.10	.05	.025	.01	.005
1	1.376	3.078	6.314	12.706	31.821	63.656
2	1.061	1.886	2.920	4.303	6.965	9.925
3	.978	1.638	2.353	3.182	4.541	5.841
4	.941	1.533	2.132	2.776	3.747	4.604
5	.920	1.476	2.015	2.571	3.365	4.032
6	.906	1.440	1.943	2.447	3.143	3.707
7	.896	1.415	1.895	2.365	2.998	3.499
8	.889	1.397	1.860	2.306	2.896	3.355
9	.883	1.383	1.833	2.262	2.821	3.250
10	.879	1.372	1.812	2.228	2.764	3.169
11	.876	1.363	1.796	2.201	2.718	3.106
12	.873	1.356	1.782	2.179	2.681	3.055
13	.870	1.350	1.771	2.160	2.650	3.012
14	.868	1.345	1.761	2.145	2.624	2.977
15	.866	1.341	1.753	2.131	2.602	2.947
16	.865	1.337	1.746	2.120	2.583	2.921
17	.863	1.333	1.740	2.110	2.567	2.898
18	.862	1.330	1.734	2.101	2.552	2.878
19	.861	1.328	1.729	2.093	2.539	2.861
20	.860	1.325	1.725	2.086	2.528	2.845
21	.859	1.323	1.721	2.080	2.518	2.831
22	.858	1.321	1.717	2.074	2.508	2.819
23	.858	1.319	1.714	2.069	2.500	2.807
24	.857	1.318	1.711	2.064	2.492	2.797
25	.856	1.316	1.708	2.060	2.485	2.787
26	.856	1.315	1.706	2.056	2.479	2.779
27	.855	1.314	1.703	2.052	2.473	2.771
28	.855	1.313	1.701	2.048	2.467	2.763
29	.854	1.311	1.699	2.045	2.462	2.756
30	.854	1.310	1.697	2.042	2.457	2.750
31	.853	1.309	1.696	2.040	2.453	2.744
32	.853	1.309	1.694	2.037	2.449	2.738
33	.853	1.308	1.692	2.035	2.445	2.733
34	.852	1.307	1.691	2.032	2.441	2.728

TABLE 2 *t* DISTRIBUTION (*Continued*)

Degrees of Freedom	\.20	\.10	\.05	\.025	\.01	\.005
			Area in Upper Tail			
35	.852	1.306	1.690	2.030	2.438	2.724
36	.852	1.306	1.688	2.028	2.434	2.719
37	.851	1.305	1.687	2.026	2.431	2.715
38	.851	1.304	1.686	2.024	2.429	2.712
39	.851	1.304	1.685	2.023	2.426	2.708
40	.851	1.303	1.684	2.021	2.423	2.704
41	.850	1.303	1.683	2.020	2.421	2.701
42	.850	1.302	1.682	2.018	2.418	2.698
43	.850	1.302	1.681	2.017	2.416	2.695
44	.850	1.301	1.680	2.015	2.414	2.692
45	.850	1.301	1.679	2.014	2.412	2.690
46	.850	1.300	1.679	2.013	2.410	2.687
47	.849	1.300	1.678	2.012	2.408	2.685
48	.849	1.299	1.677	2.011	2.407	2.682
49	.849	1.299	1.677	2.010	2.405	2.680
50	.849	1.299	1.676	2.009	2.403	2.678
51	.849	1.298	1.675	2.008	2.402	2.676
52	.849	1.298	1.675	2.007	2.400	2.674
53	.848	1.298	1.674	2.006	2.399	2.672
54	.848	1.297	1.674	2.005	2.397	2.670
55	.848	1.297	1.673	2.004	2.396	2.668
56	.848	1.297	1.673	2.003	2.395	2.667
57	.848	1.297	1.672	2.002	2.394	2.665
58	.848	1.296	1.672	2.002	2.392	2.663
59	.848	1.296	1.671	2.001	2.391	2.662
60	.848	1.296	1.671	2.000	2.390	2.660
61	.848	1.296	1.670	2.000	2.389	2.659
62	.847	1.295	1.670	1.999	2.388	2.657
63	.847	1.295	1.669	1.998	2.387	2.656
64	.847	1.295	1.669	1.998	2.386	2.655
65	.847	1.295	1.669	1.997	2.385	2.654
66	.847	1.295	1.668	1.997	2.384	2.652
67	.847	1.294	1.668	1.996	2.383	2.651
68	.847	1.294	1.668	1.995	2.382	2.650
69	.847	1.294	1.667	1.995	2.382	2.649
70	.847	1.294	1.667	1.994	2.381	2.648
71	.847	1.294	1.667	1.994	2.380	2.647
72	.847	1.293	1.666	1.993	2.379	2.646
73	.847	1.293	1.666	1.993	2.379	2.645
74	.847	1.293	1.666	1.993	2.378	2.644
75	.846	1.293	1.665	1.992	2.377	2.643
76	.846	1.293	1.665	1.992	2.376	2.642
77	.846	1.293	1.665	1.991	2.376	2.641
78	.846	1.292	1.665	1.991	2.375	2.640
79	.846	1.292	1.664	1.990	2.374	2.639

TABLE 2 *t* DISTRIBUTION (*Continued*)

Degrees of Freedom	\.20	\.10	\.05	\.025	\.01	\.005
80	.846	1.292	1.664	1.990	2.374	2.639
81	.846	1.292	1.664	1.990	2.373	2.638
82	.846	1.292	1.664	1.989	2.373	2.637
83	.846	1.292	1.663	1.989	2.372	2.636
84	.846	1.292	1.663	1.989	2.372	2.636
85	.846	1.292	1.663	1.988	2.371	2.635
86	.846	1.291	1.663	1.988	2.370	2.634
87	.846	1.291	1.663	1.988	2.370	2.634
88	.846	1.291	1.662	1.987	2.369	2.633
89	.846	1.291	1.662	1.987	2.369	2.632
90	.846	1.291	1.662	1.987	2.368	2.632
91	.846	1.291	1.662	1.986	2.368	2.631
92	.846	1.291	1.662	1.986	2.368	2.630
93	.846	1.291	1.661	1.986	2.367	2.630
94	.845	1.291	1.661	1.986	2.367	2.629
95	.845	1.291	1.661	1.985	2.366	2.629
96	.845	1.290	1.661	1.985	2.366	2.628
97	.845	1.290	1.661	1.985	2.365	2.627
98	.845	1.290	1.661	1.984	2.365	2.627
99	.845	1.290	1.660	1.984	2.364	2.626
100	.845	1.290	1.660	1.984	2.364	2.626
∞	.842	1.282	1.645	1.960	2.326	2.576

Area in Upper Tail

Appendix B Tables

TABLE 3 CHI-SQUARE DISTRIBUTION

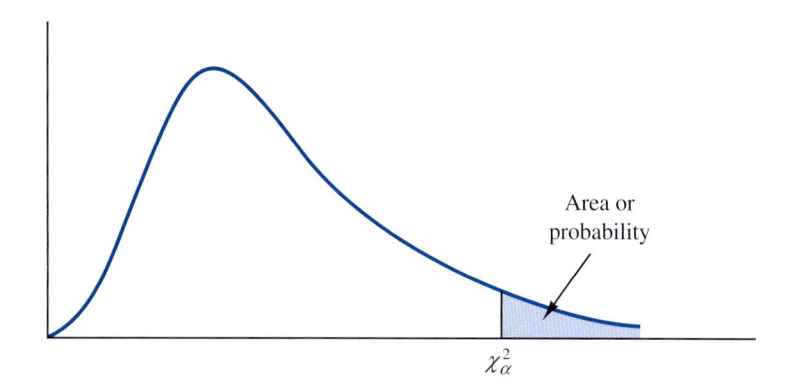

Entries in the table give χ_α^2 values, where α is the area or probability in the upper tail of the chi-square distribution. For example, with 10 degrees of freedom and a .01 area in the upper tail, $\chi_{.01}^2 = 23.209$.

Degrees of Freedom	.995	.99	.975	.95	.90	.10	.05	.025	.01	.005
1	.000	.000	.001	.004	.016	2.706	3.841	5.024	6.635	7.879
2	.010	.020	.051	.103	.211	4.605	5.991	7.378	9.210	10.597
3	.072	.115	.216	.352	.584	6.251	7.815	9.348	11.345	12.838
4	.207	.297	.484	.711	1.064	7.779	9.488	11.143	13.277	14.860
5	.412	.554	.831	1.145	1.610	9.236	11.070	12.832	15.086	16.750
6	.676	.872	1.237	1.635	2.204	10.645	12.592	14.449	16.812	18.548
7	.989	1.239	1.690	2.167	2.833	12.017	14.067	16.013	18.475	20.278
8	1.344	1.647	2.180	2.733	3.490	13.362	15.507	17.535	20.090	21.955
9	1.735	2.088	2.700	3.325	4.168	14.684	16.919	19.023	21.666	23.589
10	2.156	2.558	3.247	3.940	4.865	15.987	18.307	20.483	23.209	25.188
11	2.603	3.053	3.816	4.575	5.578	17.275	19.675	21.920	24.725	26.757
12	3.074	3.571	4.404	5.226	6.304	18.549	21.026	23.337	26.217	28.300
13	3.565	4.107	5.009	5.892	7.041	19.812	22.362	24.736	27.688	29.819
14	4.075	4.660	5.629	6.571	7.790	21.064	23.685	26.119	29.141	31.319
15	4.601	5.229	6.262	7.261	8.547	22.307	24.996	27.488	30.578	32.801
16	5.142	5.812	6.908	7.962	9.312	23.542	26.296	28.845	32.000	34.267
17	5.697	6.408	7.564	8.672	10.085	24.769	27.587	30.191	33.409	35.718
18	6.265	7.015	8.231	9.390	10.865	25.989	28.869	31.526	34.805	37.156
19	6.844	7.633	8.907	10.117	11.651	27.204	30.144	32.852	36.191	38.582
20	7.434	8.260	9.591	10.851	12.443	28.412	31.410	34.170	37.566	39.997
21	8.034	8.897	10.283	11.591	13.240	29.615	32.671	35.479	38.932	41.401
22	8.643	9.542	10.982	12.338	14.041	30.813	33.924	36.781	40.289	42.796
23	9.260	10.196	11.689	13.091	14.848	32.007	35.172	38.076	41.638	44.181
24	9.886	10.856	12.401	13.848	15.659	33.196	36.415	39.364	42.980	45.558
25	10.520	11.524	13.120	14.611	16.473	34.382	37.652	40.646	44.314	46.928
26	11.160	12.198	13.844	15.379	17.292	35.563	38.885	41.923	45.642	48.290
27	11.808	12.878	14.573	16.151	18.114	36.741	40.113	43.195	46.963	49.645
28	12.461	13.565	15.308	16.928	18.939	37.916	41.337	44.461	48.278	50.994
29	13.121	14.256	16.047	17.708	19.768	39.087	42.557	45.722	49.588	52.335

TABLE 3 CHI-SQUARE DISTRIBUTION (*Continued*)

Degrees of Freedom	.995	.99	.975	.95	.90	.10	.05	.025	.01	.005
30	13.787	14.953	16.791	18.493	20.599	40.256	43.773	46.979	50.892	53.672
35	17.192	18.509	20.569	22.465	24.797	46.059	49.802	53.203	57.342	60.275
40	20.707	22.164	24.433	26.509	29.051	51.805	55.758	59.342	63.691	66.766
45	24.311	25.901	28.366	30.612	33.350	57.505	61.656	65.410	69.957	73.166
50	27.991	29.707	32.357	34.764	37.689	63.167	67.505	71.420	76.154	79.490
55	31.735	33.571	36.398	38.958	42.060	68.796	73.311	77.380	82.292	85.749
60	35.534	37.485	40.482	43.188	46.459	74.397	79.082	83.298	88.379	91.952
65	39.383	41.444	44.603	47.450	50.883	79.973	84.821	89.177	94.422	98.105
70	43.275	45.442	48.758	51.739	55.329	85.527	90.531	95.023	100.425	104.215
75	47.206	49.475	52.942	56.054	59.795	91.061	96.217	100.839	106.393	110.285
80	51.172	53.540	57.153	60.391	64.278	96.578	101.879	106.629	112.329	116.321
85	55.170	57.634	61.389	64.749	68.777	102.079	107.522	112.393	118.236	122.324
90	59.196	61.754	65.647	69.126	73.291	107.565	113.145	118.136	124.116	128.299
95	63.250	65.898	69.925	73.520	77.818	113.038	118.752	123.858	129.973	134.247
100	67.328	70.065	74.222	77.929	82.358	118.498	124.342	129.561	135.807	140.170

Area in Upper Tail

TABLE 4 F DISTRIBUTION

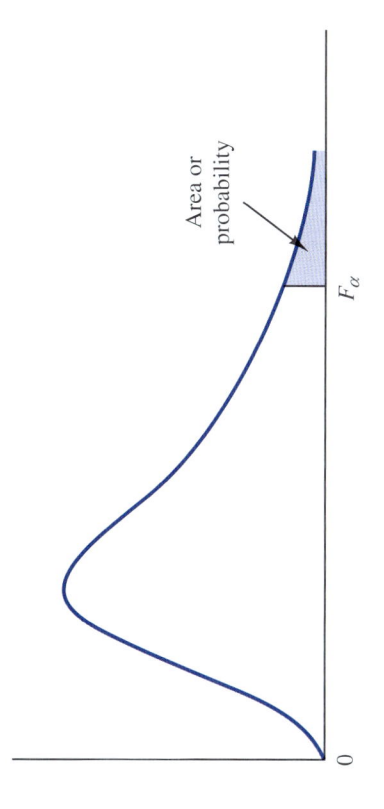

Entries in the table give F_α values, where α is the area or probability in the upper tail of the F distribution. For example, with 4 numerator degrees of freedom, 8 denominator degrees of freedom, and a .05 area in the upper tail, $F_{.05} = 3.84$.

![F distribution curve with shaded upper tail area labeled "Area or probability" at F_α]

Denominator Degrees of Freedom	Area in Upper Tail	1	2	3	4	5	6	7	8	9	10	15	20	25	30	40	60	100	1000
1	.10	39.86	49.50	53.59	55.83	57.24	58.20	58.91	59.44	59.86	60.19	61.22	61.74	62.05	62.26	62.53	62.79	63.01	63.30
	.05	161.45	199.50	215.71	224.58	230.16	233.99	236.77	238.88	240.54	241.88	245.95	248.02	249.26	250.10	251.14	252.20	253.04	254.19
	.025	647.79	799.48	864.15	899.60	921.83	937.11	948.20	956.64	963.28	968.63	984.87	993.08	998.09	1001.40	1005.60	1009.79	1013.16	1017.76
	.01	4052.18	4999.34	5403.53	5624.26	5763.96	5858.95	5928.33	5980.95	6022.40	6055.93	6156.97	6208.66	6239.86	6260.35	6286.43	6312.97	6333.92	6362.80
2	.10	8.53	9.00	9.16	9.24	9.29	9.33	9.35	9.37	9.38	9.39	9.42	9.44	9.45	9.46	9.47	9.47	9.48	9.49
	.05	18.51	19.00	19.16	19.25	19.30	19.33	19.35	19.37	19.38	19.40	19.43	19.45	19.46	19.46	19.47	19.48	19.49	19.49
	.025	38.51	39.00	39.17	39.25	39.30	39.33	39.36	39.37	39.39	39.40	39.43	39.45	39.46	39.46	39.47	39.48	39.49	39.50
	.01	98.50	99.00	99.16	99.25	99.30	99.33	99.36	99.38	99.39	99.40	99.43	99.45	99.46	99.47	99.48	99.48	99.49	99.50
3	.10	5.54	5.46	5.39	5.34	5.31	5.28	5.27	5.25	5.24	5.23	5.20	5.18	5.17	5.17	5.16	5.15	5.14	5.13
	.05	10.13	9.55	9.28	9.12	9.01	8.94	8.89	8.85	8.81	8.79	8.70	8.66	8.63	8.62	8.59	8.57	8.55	8.53
	.025	17.44	16.04	15.44	15.10	14.88	14.73	14.62	14.54	14.47	14.42	14.25	14.17	14.12	14.08	14.04	13.99	13.96	13.91
	.01	34.12	30.82	29.46	28.71	28.24	27.91	27.67	27.49	27.34	27.23	26.87	26.69	26.58	26.50	26.41	26.32	26.24	26.14
4	.10	4.54	4.32	4.19	4.11	4.05	4.01	3.98	3.95	3.94	3.92	3.87	3.84	3.83	3.82	3.80	3.79	3.78	3.76
	.05	7.71	6.94	6.59	6.39	6.26	6.16	6.09	6.04	6.00	5.96	5.86	5.80	5.77	5.75	5.72	5.69	5.66	5.63
	.025	12.22	10.65	9.98	9.60	9.36	9.20	9.07	8.98	8.90	8.84	8.66	8.56	8.50	8.46	8.41	8.36	8.32	8.26
	.01	21.20	18.00	16.69	15.98	15.52	15.21	14.98	14.80	14.66	14.55	14.20	14.02	13.91	13.84	13.75	13.65	13.58	13.47
5	.10	4.06	3.78	3.62	3.52	3.45	3.40	3.37	3.34	3.32	3.30	3.324	3.21	3.19	3.17	3.16	3.14	3.13	3.11
	.05	6.61.	5.79	5.41	5.19	5.05	4.95	4.88	4.82	4.77	4.74	4.62	4.56	4.52	4.50	4.46	4.43	4.41	4.37
	.025	10.01	8.43	7.76	7.39	7.15	6.98	6.85	6.76	6.68	6.62	6.43	6.33	6.27	6.23	6.18	6.12	6.08	6.02
	.01	16.26	13.27	12.06	11.39	10.97	10.67	10.46	10.29	10.16	10.05	9.72	9.55	9.45	9.38	9.29	9.20	9.13	9.03

Numerator Degrees of Freedom

TABLE 4 F DISTRIBUTION (Continued)

Denominator Degrees of Freedom	Area in Upper Tail	\multicolumn{16}{c}{Numerator Degrees of Freedom}
		1	2	3	4	5	6	7	8	9	10	15	20	25	30	40	60	100	1000
6	.10	3.78	3.46	3.29	3.18	3.11	3.05	3.01	2.98	2.96	2.94	2.87	2.84	2.81	2.80	2.78	2.76	2.75	2.72
	.05	5.99	5.14	4.76	4.53	4.39	4.28	4.21	4.15	4.10	4.06	3.94	3.87	3.83	3.81	3.77	3.74	3.71	3.67
	.025	8.81	7.26	6.60	6.23	5.99	5.82	5.70	5.60	5.52	5.46	5.27	5.17	5.11	5.07	5.01	4.96	4.92	4.86
	.01	13.75	10.92	9.78	9.15	8.75	8.47	8.26	8.10	7.98	7.87	7.56	7.40	7.30	7.23	7.14	7.06	6.99	6.89
7	.10	3.59	3.26	3.07	2.96	2.88	2.83	2.78	2.75	2.72	2.70	2.63	2.59	2.57	2.56	2.54	2.51	2.50	2.47
	.05	5.59	4.74	4.35	4.12	3.97	3.87	3.79	3.73	3.68	3.64	3.51	3.44	3.40	3.38	3.34	3.30	3.27	3.23
	.025	8.07	6.54	5.89	5.52	5.29	5.12	4.99	4.90	4.82	4.76	4.57	4.47	4.40	4.36	4.31	4.25	4.21	4.15
	.01	12.25	9.55	8.45	7.85	7.46	7.19	6.99	6.84	6.72	6.62	6.31	6.16	6.06	5.99	5.91	5.82	5.75	5.66
8	.10	3.46	3.11	2.92	2.81	2.73	2.67	2.62	2.59	2.56	2.54	2.46	2.42	2.40	2.38	2.36	2.34	2.32	2.30
	.05	5.32	4.46	4.07	3.84	3.69	3.58	3.50	3.44	3.39	3.35	3.22	3.15	3.11	3.08	3.04	3.01	2.97	2.93
	.025	7.57	6.06	5.42	5.05	4.82	4.65	4.53	4.43	4.36	4.30	4.10	4.00	3.94	3.89	3.84	3.78	3.74	3.68
	.01	11.26	8.65	7.59	7.01	6.63	6.37	6.18	6.03	5.91	5.81	5.52	5.36	5.26	5.20	5.12	5.03	4.96	4.87
9	.10	3.36	3.01	2.81	2.69	2.61	2.55	2.51	2.47	2.44	2.42	2.34	2.30	2.27	2.25	2.23	2.21	2.19	2.16
	.05	5.12	4.26	3.86	3.63	3.48	3.37	3.29	3.23	3.18	3.14	3.01	2.94	2.89	2.86	2.83	2.79	2.76	2.71
	.025	7.21	5.71	5.08	4.72	4.48	4.32	4.20	4.10	4.03	3.96	3.77	3.67	3.60	3.56	3.51	3.45	3.40	3.34
	.01	10.56	8.02	6.99	6.42	6.06	5.80	5.61	5.47	5.35	5.26	4.96	4.81	4.71	4.65	4.57	4.48	4.41	4.32
10	.10	3.29	2.92	2.73	2.61	2.52	2.46	2.41	2.38	2.35	2.32	2.24	2.20	2.17	2.16	2.13	2.11	2.09	2.06
	.05	4.96	4.10	3.71	3.48	3.33	3.22	3.14	3.07	3.02	2.98	2.85	2.77	2.73	2.70	2.66	2.62	2.59	2.54
	.025	6.94	5.46	4.83	4.47	4.24	4.07	3.95	3.85	3.78	3.72	3.52	3.42	3.35	3.31	3.26	3.20	3.15	3.09
	.01	10.04	7.56	6.55	5.99	5.64	5.39	5.20	5.06	4.94	4.85	4.56	4.41	4.31	4.25	4.17	4.08	4.01	3.92
11	.10	3.23	2.86	2.66	2.54	2.45	2.39	2.34	2.30	2.27	2.25	2.17	2.12	2.10	2.08	2.05	2.03	2.01	1.98
	.05	4.84	3.98	3.59	3.36	3.20	3.09	3.01	2.95	2.90	2.85	2.72	2.65	2.60	2.57	2.53	2.49	2.46	2.41
	.025	6.72	5.26	4.63	4.28	4.04	3.88	3.76	3.66	3.59	3.53	3.33	3.23	3.16	3.12	3.06	3.00	2.96	2.89
	.01	9.65	7.21	6.22	5.67	5.32	5.07	4.89	4.74	4.63	4.54	4.25	4.10	4.01	3.94	3.86	3.78	3.71	3.61
12	.10	3.18	2.81	2.61	2.48	2.39	2.33	2.28	2.24	2.21	2.19	2.10	2.06	2.03	2.01	1.99	1.96	1.94	1.91
	.05	4.75	3.89	3.49	3.26	3.11	3.00	2.91	2.85	2.80	2.75	2.62	2.54	2.50	2.47	2.43	2.38	2.35	2.30
	.025	6.55	5.10	4.47	4.12	3.89	3.73	3.61	3.51	3.44	3.37	3.18	3.07	3.01	2.96	2.91	2.85	2.80	2.73
	.01	9.33	6.93	5.95	5.41	5.06	4.82	4.64	4.50	4.39	4.30	4.01	3.86	3.76	3.70	3.62	3.54	3.47	3.37
13	.10	3.14	2.76	2.56	2.43	2.35	2.28	2.23	2.20	2.16	2.14	2.05	2.01	1.98	1.96	1.93	1.90	1.88	1.85
	.05	4.67	3.81	3.41	3.18	3.03	2.92	2.83	2.77	2.71	2.67	2.53	2.46	2.41	2.38	2.34	2.30	2.26	2.21
	.025	6.41	4.97	4.35	4.00	3.77	3.60	3.48	3.39	3.31	3.25	3.05	2.95	2.88	2.84	2.78	2.72	2.67	2.60
	.01	9.07	6.70	5.74	5.21	4.86	4.62	4.44	4.30	4.19	4.10	3.82	3.66	3.57	3.51	3.43	3.34	3.27	3.18
14	.10	3.10	2.73	2.52	2.39	2.31	2.24	2.19	2.15	2.12	2.10	2.01	1.96	1.93	1.91	1.89	1.86	1.83	1.80
	.05	4.60	3.74	3.34	3.11	2.96	2.85	2.76	2.70	2.65	2.60	2.46	2.39	2.34	2.31	2.27	2.22	2.19	2.14
	.025	6.30	4.86	4.24	3.89	3.66	3.50	3.38	3.29	3.21	3.15	2.95	2.84	2.78	2.73	2.67	2.61	2.56	2.50
	.01	8.86	6.51	5.56	5.04	4.69	4.46	4.28	4.14	4.03	3.94	3.66	3.51	3.41	3.35	3.27	3.18	3.11	3.02
15	.10	3.07	2.70	2.49	2.36	2.27	2.21	2.16	2.12	2.09	2.06	1.97	1.92	1.89	1.87	1.85	1.82	1.79	1.76
	.05	4.54	3.68	3.29	3.06	2.90	2.79	2.71	2.64	2.59	2.54	2.40	2.33	2.28	2.25	2.20	2.16	2.12	2.07
	.025	6.20	4.77	4.15	3.80	3.58	3.41	3.29	3.20	3.12	3.06	2.86	2.76	2.69	2.64	2.59	2.52	2.47	2.40
	.01	8.68	6.36	5.42	4.89	4.56	4.32	4.14	4.00	3.89	3.80	3.52	3.37	3.28	3.21	3.13	3.05	2.98	2.88

| Denominator Degrees of Freedom | Area in Upper Tail | \multicolumn{14}{c|}{Numerator Degrees of Freedom} |
		1	2	3	4	5	6	7	8	9	10	15	20	25	30	40	60	100	1000
16	.10	3.05	2.67	2.46	2.33	2.24	2.18	2.13	2.09	2.06	2.03	1.94	1.89	1.86	1.84	1.81	1.78	1.76	1.72
	.05	4.49	3.63	3.24	3.01	2.85	2.74	2.66	2.59	2.54	2.49	2.35	2.28	2.23	2.19	2.15	2.11	2.07	2.02
	.025	6.12	4.69	4.08	3.73	3.50	3.34	3.22	3.12	3.05	2.99	2.79	2.68	2.61	2.57	2.51	2.45	2.40	2.32
	.01	8.53	6.23	5.29	4.77	4.44	4.20	4.03	3.89	3.78	3.69	3.41	3.26	3.16	3.10	3.02	2.93	2.86	2.76
17	.10	3.03	2.64	2.44	2.31	2.22	2.15	2.10	2.06	2.03	2.00	1.91	1.86	1.83	1.81	1.78	1.75	1.73	1.69
	.05	4.45	3.59	3.20	2.96	2.81	2.70	2.61	2.55	2.49	2.45	2.31	2.23	2.18	2.15	2.10	2.06	2.02	1.97
	.025	6.04	4.62	4.01	3.66	3.44	3.28	3.16	3.06	2.98	2.92	2.72	2.62	2.55	2.50	2.44	2.38	2.33	2.26
	.01	8.40	6.11	5.19	4.67	4.34	4.10	3.93	3.79	3.68	3.59	3.31	3.16	3.07	3.00	2.92	2.83	2.76	2.66
18	.10	3.01	2.62	2.42	2.29	2.20	2.13	2.08	2.04	2.00	1.98	1.89	1.84	1.80	1.78	1.75	1.72	1.70	1.66
	.05	4.41	3.55	3.16	2.93	2.77	2.66	2.58	2.51	2.46	2.41	2.27	2.19	2.14	2.11	2.06	2.02	1.98	1.92
	.025	5.98	4.56	3.95	3.61	3.38	3.22	3.10	3.01	2.93	2.87	2.67	2.56	2.49	2.44	2.38	2.32	2.27	2.20
	.01	8.29	6.01	5.09	4.58	4.25	4.01	3.84	3.71	3.60	3.51	3.23	3.08	2.98	2.92	2.84	2.75	2.68	2.58
19	.10	2.99	2.61	2.40	2.27	2.18	2.11	2.06	2.02	1.98	1.96	1.86	1.81	1.78	1.76	1.73	1.70	1.67	1.64
	.05	4.38	3.52	3.13	2.90	2.74	2.63	2.54	2.48	2.42	2.38	2.23	2.16	2.11	2.07	2.03	1.98	1.94	1.88
	.025	5.92	4.51	3.90	3.56	3.33	3.17	3.05	2.96	2.88	2.82	2.62	2.51	2.44	2.39	2.33	2.27	2.22	2.14
	.01	8.18	5.93	5.01	4.50	4.17	3.94	3.77	3.63	3.52	3.43	3.15	3.00	2.91	2.84	2.76	2.67	2.60	2.50
20	.10	2.97	2.59	2.38	2.25	2.16	2.09	2.04	2.00	1.96	1.94	1.84	1.79	1.76	1.74	1.71	1.68	1.65	1.61
	.05	4.35	3.49	3.10	2.87	2.71	2.60	2.51	2.45	2.39	2.35	2.20	2.12	2.07	2.04	1.99	1.95	1.91	1.85
	.025	5.87	4.46	3.86	3.51	3.29	3.13	3.01	2.91	2.84	2.77	2.57	2.46	2.40	2.35	2.29	2.22	2.17	2.09
	.01	8.10	5.85	4.94	4.43	4.10	3.87	3.70	3.56	3.46	3.37	3.09	2.94	2.84	2.78	2.69	2.61	2.54	2.43
21	.10	2.96	2.57	2.36	2.23	2.14	2.08	2.02	1.98	1.95	1.92	1.83	1.78	1.74	1.72	1.69	1.66	1.63	1.59
	.05	4.32	3.47	3.07	2.84	2.68	2.57	2.49	2.42	2.37	2.32	2.18	2.10	2.05	2.01	1.96	1.92	1.88	1.82
	.025	5.83	4.42	3.82	3.48	3.25	3.09	2.97	2.87	2.80	2.73	2.53	2.42	2.36	2.31	2.25	2.18	2.13	2.05
	.01	8.02	5.78	4.87	4.37	4.04	3.81	3.64	3.51	3.40	3.31	3.03	2.88	2.79	2.72	2.64	2.55	2.48	2.37
22	.10	2.95	2.56	2.35	2.22	2.13	2.06	2.01	1.97	1.93	1.90	1.81	1.76	1.73	1.70	1.67	1.64	1.61	1.57
	.05	4.30	3.44	3.05	2.82	2.66	2.55	2.46	2.40	2.34	2.30	2.15	2.07	2.02	1.98	1.94	1.89	1.85	1.79
	.025	5.79	4.38	3.78	3.44	3.22	3.05	2.93	2.84	2.76	2.70	2.50	2.39	2.32	2.27	2.21	2.14	2.09	2.01
	.01	7.95	5.72	4.82	4.31	3.99	3.76	3.59	3.45	3.35	3.26	2.98	2.83	2.73	2.67	2.58	2.50	2.42	2.32
23	.10	2.94	2.55	2.34	2.21	2.11	2.05	1.99	1.95	1.92	1.89	1.80	1.74	1.71	1.69	1.66	1.62	1.59	1.55
	.05	4.28	3.42	3.03	2.80	2.64	2.53	2.44	2.37	2.32	2.27	2.13	2.05	2.00	1.96	1.91	1.86	1.82	1.76
	.025	5.75	4.35	3.75	3.41	3.18	3.02	2.90	2.81	2.73	2.67	2.47	2.36	2.29	2.24	2.18	2.11	2.06	1.98
	.01	7.88	5.66	4.76	4.26	3.94	3.71	3.54	3.41	3.30	3.21	2.93	2.78	2.69	2.62	2.54	2.45	2.37	2.27
24	.10	2.93	2.54	2.33	2.19	2.10	2.04	1.98	1.94	1.91	1.88	1.78	1.73	1.70	1.67	1.64	1.61	1.58	1.54
	.05	4.26	3.40	3.01	2.78	2.62	2.51	2.42	2.36	2.30	2.25	2.11	2.03	1.97	1.94	1.89	1.84	1.80	1.74
	.025	5.72	4.32	3.72	3.38	3.15	2.99	2.87	2.78	2.70	2.64	2.44	2.33	2.26	2.21	2.15	2.08	2.02	1.94
	.01	7.82	5.61	4.72	4.22	3.90	3.67	3.50	3.36	3.26	3.17	2.89	2.74	2.64	2.58	2.49	2.40	2.33	2.22

TABLE 4 F DISTRIBUTION (*Continued*)

Denominator Degrees of Freedom	Area in Upper Tail	1	2	3	4	5	6	7	8	9	10	15	20	25	30	40	60	100	1000
25	.10	2.92	2.53	2.32	2.18	2.09	2.02	1.97	1.93	1.89	1.87	1.77	1.72	1.68	1.66	1.63	1.59	1.56	1.52
	.05	4.24	3.39	2.99	2.76	2.60	2.49	2.40	2.34	2.28	2.24	2.09	2.01	1.96	1.92	1.87	1.82	1.78	1.72
	.025	5.69	4.29	3.69	3.35	3.13	2.97	2.85	2.75	2.68	2.61	2.41	2.30	2.23	2.18	2.12	2.05	2.00	1.91
	.01	7.77	5.57	4.68	4.18	3.85	3.63	3.46	3.32	3.22	3.13	2.85	2.70	2.60	2.54	2.45	2.36	2.29	2.18
26	.10	2.91	2.52	2.31	2.17	2.08	2.01	1.96	1.92	1.88	1.86	1.76	1.71	1.67	1.65	1.61	1.58	1.55	1.51
	.05	4.23	3.37	2.98	2.74	2.59	2.47	2.39	2.32	2.27	2.22	2.07	1.99	1.94	1.90	1.85	1.80	1.76	1.70
	.025	5.66	4.27	3.67	3.33	3.10	2.94	2.82	2.73	2.65	2.59	2.39	2.28	2.21	2.16	2.09	2.03	1.97	1.89
	.01	7.72	5.53	4.64	4.14	3.82	3.59	3.42	3.29	3.18	3.09	2.81	2.66	2.57	2.50	2.42	2.33	2.25	2.14
27	.10	2.90	2.51	2.30	2.17	2.07	2.00	1.95	1.91	1.87	1.85	1.75	1.70	1.66	1.64	1.60	1.57	1.54	1.50
	.05	4.21	3.35	2.96	2.73	2.57	2.46	2.37	2.31	2.25	2.20	2.06	1.97	1.92	1.88	1.84	1.79	1.74	1.68
	.025	5.63	4.24	3.65	3.31	3.08	2.92	2.80	2.71	2.63	2.57	2.36	2.25	2.18	2.13	2.07	2.00	1.94	1.86
	.01	7.68	5.49	4.60	4.11	3.78	3.56	3.39	3.26	3.15	3.06	2.78	2.63	2.54	2.47	2.38	2.29	2.22	2.11
28	.10	2.89	2.50	2.29	2.16	2.06	2.00	1.94	1.90	1.87	1.84	1.74	1.69	1.65	1.63	1.59	1.56	1.53	1.48
	.05	4.20	3.34	2.95	2.71	2.56	2.45	2.36	2.29	2.24	2.19	2.04	1.96	1.91	1.87	1.82	1.77	1.73	1.66
	.025	5.61	4.22	3.63	3.29	3.06	2.90	2.78	2.69	2.61	2.55	2.34	2.23	2.16	2.11	2.05	1.98	1.92	1.84
	.01	7.64	5.45	4.57	4.07	3.75	3.53	3.36	3.23	3.12	3.03	2.75	2.60	2.51	2.44	2.35	2.26	2.19	2.08
29	.10	2.89	2.50	2.28	2.15	2.06	1.99	1.93	1.89	1.86	1.83	1.73	1.68	1.64	1.62	1.58	1.55	1.52	1.47
	.05	4.18	3.33	2.93	2.70	2.55	2.43	2.35	2.28	2.22	2.18	2.03	1.94	1.89	1.85	1.81	1.75	1.71	1.65
	.025	5.59	4.20	3.61	3.27	3.04	2.88	2.76	2.67	2.59	2.53	2.32	2.21	2.14	2.09	2.03	1.96	1.90	1.82
	.01	7.60	5.42	4.54	4.04	3.73	3.50	3.33	3.20	3.09	3.00	2.73	2.57	2.48	2.41	2.33	2.23	2.16	2.05
30	.10	2.88	2.49	2.28	2.14	2.05	1.98	1.93	1.88	1.85	1.82	1.72	1.67	1.63	1.61	1.57	1.54	1.51	1.46
	.05	4.17	3.32	2.92	2.69	2.53	2.42	2.33	2.27	2.21	2.16	2.01	1.93	1.88	1.84	1.79	1.74	1.70	1.63
	.025	5.57	4.18	3.59	3.25	3.03	2.87	2.75	2.65	2.57	2.51	2.31	2.20	2.12	2.07	2.01	1.94	1.88	1.80
	.01	7.56	5.39	4.51	4.02	3.70	3.47	3.30	3.17	3.07	2.98	2.70	2.55	2.45	2.39	2.30	2.21	2.13	2.02
40	.10	2.84	2.44	2.23	2.09	2.00	1.93	1.87	1.83	1.79	1.76	1.66	1.61	1.57	1.54	1.51	1.47	1.43	1.38
	.05	4.08	3.23	2.84	2.61	2.45	2.34	2.25	2.18	2.12	2.08	1.92	1.84	1.78	1.74	1.69	1.64	1.59	1.52
	.025	5.42	4.05	3.46	3.13	2.90	2.74	2.62	2.53	2.45	2.39	2.18	2.07	1.99	1.94	1.88	1.80	1.74	1.65
	.01	7.31	5.18	4.31	3.83	3.51	3.29	3.12	2.99	2.89	2.80	2.52	2.37	2.27	2.20	2.11	2.02	1.94	1.82
60	.10	2.79	2.39	2.18	2.04	1.95	1.87	1.82	1.77	1.74	1.71	1.60	1.54	1.50	1.48	1.44	1.40	1.36	1.30
	.05	4.00	3.15	2.76	2.53	2.37	2.25	2.17	2.10	2.04	1.99	1.84	1.75	1.69	1.65	1.59	1.53	1.48	1.40
	.025	5.29	3.93	3.34	3.01	2.79	2.63	2.51	2.41	2.33	2.27	2.06	1.94	1.87	1.82	1.74	1.67	1.60	1.49
	.01	7.08	4.98	4.13	3.65	3.34	3.12	2.95	2.82	2.72	2.63	2.35	2.20	2.10	2.03	1.94	1.84	1.75	1.62
100	.10	2.76	2.36	2.14	2.00	1.91	1.83	1.78	1.73	1.69	1.66	1.56	1.49	1.45	1.42	1.38	1.34	1.29	1.22
	.05	3.94	3.09	2.70	2.46	2.31	2.19	2.10	2.03	1.97	1.93	1.77	1.68	1.62	1.57	1.52	1.45	1.39	1.30
	.025	5.18	3.83	3.25	2.92	2.70	2.54	2.42	2.32	2.24	2.18	1.97	1.85	1.77	1.71	1.64	1.56	1.48	1.36
	.01	6.90	4.82	3.98	3.51	3.21	2.99	2.82	2.69	2.59	2.50	2.22	2.07	1.97	1.89	1.80	1.69	1.60	1.45
1000	.10	2.71	2.31	2.09	1.95	1.85	1.78	1.72	1.68	1.64	1.61	1.49	1.43	1.38	1.35	1.30	1.25	1.20	1.08
	.05	3.85	3.00	2.61	2.38	2.22	2.11	2.02	1.95	1.89	1.84	1.68	1.58	1.52	1.47	1.41	1.33	1.26	1.11
	.025	5.04	3.70	3.13	2.80	2.58	2.42	2.30	2.20	2.13	2.06	1.85	1.72	1.64	1.58	1.50	1.41	1.32	1.13
	.01	6.66	4.63	3.80	3.34	3.04	2.82	2.66	2.53	2.43	2.34	2.06	1.90	1.79	1.72	1.61	1.50	1.38	1.16

Numerator Degrees of Freedom

TABLE 5 BINOMIAL PROBABILITIES

Entries in the table give the probability of x successes in n trials of a binomial experiment, where p is the probability of a success on one trial. For example, with six trials and $p = .05$, the probability of two successes is .0305.

						p				
n	x	.01	.02	.03	.04	.05	.06	.07	.08	.09
2	0	.9801	.9604	.9409	.9216	.9025	.8836	.8649	.8464	.8281
	1	.0198	.0392	.0582	.0768	.0950	.1128	.1302	.1472	.1638
	2	.0001	.0004	.0009	.0016	.0025	.0036	.0049	.0064	.0081
3	0	.9703	.9412	.9127	.8847	.8574	.8306	.8044	.7787	.7536
	1	.0294	.0576	.0847	.1106	.1354	.1590	.1816	.2031	.2236
	2	.0003	.0012	.0026	.0046	.0071	.0102	.0137	.0177	.0221
	3	.0000	.0000	.0000	.0001	.0001	.0002	.0003	.0005	.0007
4	0	.9606	.9224	.8853	.8493	.8145	.7807	.7481	.7164	.6857
	1	.0388	.0753	.1095	.1416	.1715	.1993	.2252	.2492	.2713
	2	.0006	.0023	.0051	.0088	.0135	.0191	.0254	.0325	.0402
	3	.0000	.0000	.0001	.0002	.0005	.0008	.0013	.0019	.0027
	4	.0000	.0000	.0000	.0000	.0000	.0000	.0000	.0000	.0001
5	0	.9510	.9039	.8587	.8154	.7738	.7339	.6957	.6591	.6240
	1	.0480	.0922	.1328	.1699	.2036	.2342	.2618	.2866	.3086
	2	.0010	.0038	.0082	.0142	.0214	.0299	.0394	.0498	.0610
	3	.0000	.0001	.0003	.0006	.0011	.0019	.0030	.0043	.0060
	4	.0000	.0000	.0000	.0000	.0000	.0001	.0001	.0002	.0003
	5	.0000	.0000	.0000	.0000	.0000	.0000	.0000	.0000	.0000
6	0	.9415	.8858	.8330	.7828	.7351	.6899	.6470	.6064	.5679
	1	.0571	.1085	.1546	.1957	.2321	.2642	.2922	.3164	.3370
	2	.0014	.0055	.0120	.0204	.0305	.0422	.0550	.0688	.0833
	3	.0000	.0002	.0005	.0011	.0021	.0036	.0055	.0080	.0110
	4	.0000	.0000	.0000	.0000	.0001	.0002	.0003	.0005	.0008
	5	.0000	.0000	.0000	.0000	.0000	.0000	.0000	.0000	.0000
	6	.0000	.0000	.0000	.0000	.0000	.0000	.0000	.0000	.0000
7	0	.9321	.8681	.8080	.7514	.6983	.6485	.6017	.5578	.5168
	1	.0659	.1240	.1749	.2192	.2573	.2897	.3170	.3396	.3578
	2	.0020	.0076	.0162	.0274	.0406	.0555	.0716	.0886	.1061
	3	.0000	.0003	.0008	.0019	.0036	.0059	.0090	.0128	.0175
	4	.0000	.0000	.0000	.0001	.0002	.0004	.0007	.0011	.0017
	5	.0000	.0000	.0000	.0000	.0000	.0000	.0000	.0001	.0001
	6	.0000	.0000	.0000	.0000	.0000	.0000	.0000	.0000	.0000
	7	.0000	.0000	.0000	.0000	.0000	.0000	.0000	.0000	.0000
8	0	.9227	.8508	.7837	.7214	.6634	.6096	.5596	.5132	.4703
	1	.0746	.1389	.1939	.2405	.2793	.3113	.3370	.3570	.3721
	2	.0026	.0099	.0210	.0351	.0515	.0695	.0888	.1087	.1288
	3	.0001	.0004	.0013	.0029	.0054	.0089	.0134	.0189	.0255
	4	.0000	.0000	.0001	.0002	.0004	.0007	.0013	.0021	.0031
	5	.0000	.0000	.0000	.0000	.0000	.0000	.0001	.0001	.0002
	6	.0000	.0000	.0000	.0000	.0000	.0000	.0000	.0000	.0000
	7	.0000	.0000	.0000	.0000	.0000	.0000	.0000	.0000	.0000
	8	.0000	.0000	.0000	.0000	.0000	.0000	.0000	.0000	.0000

TABLE 5 BINOMIAL PROBABILITIES (Continued)

						p				
n	x	.01	.02	.03	.04	.05	.06	.07	.08	.09
9	0	.9135	.8337	.7602	.6925	.6302	.5730	.5204	.4722	.4279
	1	.0830	.1531	.2116	.2597	.2985	.3292	.3525	.3695	.3809
	2	.0034	.0125	.0262	.0433	.0629	.0840	.1061	.1285	.1507
	3	.0001	.0006	.0019	.0042	.0077	.0125	.0186	.0261	.0348
	4	.0000	.0000	.0001	.0003	.0006	.0012	.0021	.0034	.0052
	5	.0000	.0000	.0000	.0000	.0000	.0001	.0002	.0003	.0005
	6	.0000	.0000	.0000	.0000	.0000	.0000	.0000	.0000	.0000
	7	.0000	.0000	.0000	.0000	.0000	.0000	.0000	.0000	.0000
	8	.0000	.0000	.0000	.0000	.0000	.0000	.0000	.0000	.0000
	9	.0000	.0000	.0000	.0000	.0000	.0000	.0000	.0000	.0000
10	0	.9044	.8171	.7374	.6648	.5987	.5386	.4840	.4344	.3894
	1	.0914	.1667	.2281	.2770	.3151	.3438	.3643	.3777	.3851
	2	.0042	.0153	.0317	.0519	.0746	.0988	.1234	.1478	.1714
	3	.0001	.0008	.0026	.0058	.0105	.0168	.0248	.0343	.0452
	4	.0000	.0000	.0001	.0004	.0010	.0019	.0033	.0052	.0078
	5	.0000	.0000	.0000	.0000	.0001	.0001	.0003	.0005	.0009
	6	.0000	.0000	.0000	.0000	.0000	.0000	.0000	.0000	.0001
	7	.0000	.0000	.0000	.0000	.0000	.0000	.0000	.0000	.0000
	8	.0000	.0000	.0000	.0000	.0000	.0000	.0000	.0000	.0000
	9	.0000	.0000	.0000	.0000	.0000	.0000	.0000	.0000	.0000
	10	.0000	.0000	.0000	.0000	.0000	.0000	.0000	.0000	.0000
12	0	.8864	.7847	.6938	.6127	.5404	.4759	.4186	.3677	.3225
	1	.1074	.1922	.2575	.3064	.3413	.3645	.3781	.3837	.3827
	2	.0060	.0216	.0438	.0702	.0988	.1280	.1565	.1835	.2082
	3	.0002	.0015	.0045	.0098	.0173	.0272	.0393	.0532	.0686
	4	.0000	.0001	.0003	.0009	.0021	.0039	.0067	.0104	.0153
	5	.0000	.0000	.0000	.0001	.0002	.0004	.0008	.0014	.0024
	6	.0000	.0000	.0000	.0000	.0000	.0000	.0001	.0001	.0003
	7	.0000	.0000	.0000	.0000	.0000	.0000	.0000	.0000	.0000
	8	.0000	.0000	.0000	.0000	.0000	.0000	.0000	.0000	.0000
	9	.0000	.0000	.0000	.0000	.0000	.0000	.0000	.0000	.0000
	10	.0000	.0000	.0000	.0000	.0000	.0000	.0000	.0000	.0000
	11	.0000	.0000	.0000	.0000	.0000	.0000	.0000	.0000	.0000
	12	.0000	.0000	.0000	.0000	.0000	.0000	.0000	.0000	.0000
15	0	.8601	.7386	.6333	.5421	.4633	.3953	.3367	.2863	.2430
	1	.1303	.2261	.2938	.3388	.3658	.3785	.3801	.3734	.3605
	2	.0092	.0323	.0636	.0988	.1348	.1691	.2003	.2273	.2496
	3	.0004	.0029	.0085	.0178	.0307	.0468	.0653	.0857	.1070
	4	.0000	.0002	.0008	.0022	.0049	.0090	.0148	.0223	.0317
	5	.0000	.0000	.0001	.0002	.0006	.0013	.0024	.0043	.0069
	6	.0000	.0000	.0000	.0000	.0000	.0001	.0003	.0006	.0011
	7	.0000	.0000	.0000	.0000	.0000	.0000	.0000	.0001	.0001
	8	.0000	.0000	.0000	.0000	.0000	.0000	.0000	.0000	.0000
	9	.0000	.0000	.0000	.0000	.0000	.0000	.0000	.0000	.0000
	10	.0000	.0000	.0000	.0000	.0000	.0000	.0000	.0000	.0000
	11	.0000	.0000	.0000	.0000	.0000	.0000	.0000	.0000	.0000
	12	.0000	.0000	.0000	.0000	.0000	.0000	.0000	.0000	.0000
	13	.0000	.0000	.0000	.0000	.0000	.0000	.0000	.0000	.0000
	14	.0000	.0000	.0000	.0000	.0000	.0000	.0000	.0000	.0000
	15	.0000	.0000	.0000	.0000	.0000	.0000	.0000	.0000	.0000

TABLE 5 BINOMIAL PROBABILITIES (*Continued*)

n	x	.01	.02	.03	.04	.05	.06	.07	.08	.09
18	0	.8345	.6951	.5780	.4796	.3972	.3283	.2708	.2229	.1831
	1	.1517	.2554	.3217	.3597	.3763	.3772	.3669	.3489	.3260
	2	.0130	.0443	.0846	.1274	.1683	.2047	.2348	.2579	.2741
	3	.0007	.0048	.0140	.0283	.0473	.0697	.0942	.1196	.1446
	4	.0000	.0004	.0016	.0044	.0093	.0167	.0266	.0390	.0536
	5	.0000	.0000	.0001	.0005	.0014	.0030	.0056	.0095	.0148
	6	.0000	.0000	.0000	.0000	.0002	.0004	.0009	.0018	.0032
	7	.0000	.0000	.0000	.0000	.0000	.0000	.0001	.0003	.0005
	8	.0000	.0000	.0000	.0000	.0000	.0000	.0000	.0000	.0001
	9	.0000	.0000	.0000	.0000	.0000	.0000	.0000	.0000	.0000
	10	.0000	.0000	.0000	.0000	.0000	.0000	.0000	.0000	.0000
	11	.0000	.0000	.0000	.0000	.0000	.0000	.0000	.0000	.0000
	12	.0000	.0000	.0000	.0000	.0000	.0000	.0000	.0000	.0000
	13	.0000	.0000	.0000	.0000	.0000	.0000	.0000	.0000	.0000
	14	.0000	.0000	.0000	.0000	.0000	.0000	.0000	.0000	.0000
	15	.0000	.0000	.0000	.0000	.0000	.0000	.0000	.0000	.0000
	16	.0000	.0000	.0000	.0000	.0000	.0000	.0000	.0000	.0000
	17	.0000	.0000	.0000	.0000	.0000	.0000	.0000	.0000	.0000
	18	.0000	.0000	.0000	.0000	.0000	.0000	.0000	.0000	.0000
20	0	.8179	.6676	.5438	.4420	.3585	.2901	.2342	.1887	.1516
	1	.1652	.2725	.3364	.3683	.3774	.3703	.3526	.3282	.3000
	2	.0159	.0528	.0988	.1458	.1887	.2246	.2521	.2711	.2818
	3	.0010	.0065	.0183	.0364	.0596	.0860	.1139	.1414	.1672
	4	.0000	.0006	.0024	.0065	.0133	.0233	.0364	.0523	.0703
	5	.0000	.0000	.0002	.0009	.0022	.0048	.0088	.0145	.0222
	6	.0000	.0000	.0000	.0001	.0003	.0008	.0017	.0032	.0055
	7	.0000	.0000	.0000	.0000	.0000	.0001	.0002	.0005	.0011
	8	.0000	.0000	.0000	.0000	.0000	.0000	.0000	.0001	.0002
	9	.0000	.0000	.0000	.0000	.0000	.0000	.0000	.0000	.0000
	10	.0000	.0000	.0000	.0000	.0000	.0000	.0000	.0000	.0000
	11	.0000	.0000	.0000	.0000	.0000	.0000	.0000	.0000	.0000
	12	.0000	.0000	.0000	.0000	.0000	.0000	.0000	.0000	.0000
	13	.0000	.0000	.0000	.0000	.0000	.0000	.0000	.0000	.0000
	14	.0000	.0000	.0000	.0000	.0000	.0000	.0000	.0000	.0000
	15	.0000	.0000	.0000	.0000	.0000	.0000	.0000	.0000	.0000
	16	.0000	.0000	.0000	.0000	.0000	.0000	.0000	.0000	.0000
	17	.0000	.0000	.0000	.0000	.0000	.0000	.0000	.0000	.0000
	18	.0000	.0000	.0000	.0000	.0000	.0000	.0000	.0000	.0000
	19	.0000	.0000	.0000	.0000	.0000	.0000	.0000	.0000	.0000
	20	.0000	.0000	.0000	.0000	.0000	.0000	.0000	.0000	.0000

TABLE 5 BINOMIAL PROBABILITIES (*Continued*)

						p				
n	*x*	.10	.15	.20	.25	.30	.35	.40	.45	.50
2	0	.8100	.7225	.6400	.5625	.4900	.4225	.3600	.3025	.2500
	1	.1800	.2550	.3200	.3750	.4200	.4550	.4800	.4950	.5000
	2	.0100	.0225	.0400	.0625	.0900	.1225	.1600	.2025	.2500
3	0	.7290	.6141	.5120	.4219	.3430	.2746	.2160	.1664	.1250
	1	.2430	.3251	.3840	.4219	.4410	.4436	.4320	.4084	.3750
	2	.0270	.0574	.0960	.1406	.1890	.2389	.2880	.3341	.3750
	3	.0010	.0034	.0080	.0156	.0270	.0429	.0640	.0911	.1250
4	0	.6561	.5220	.4096	.3164	.2401	.1785	.1296	.0915	.0625
	1	.2916	.3685	.4096	.4219	.4116	.3845	.3456	.2995	.2500
	2	.0486	.0975	.1536	.2109	.2646	.3105	.3456	.3675	.3750
	3	.0036	.0115	.0256	.0469	.0756	.1115	.1536	.2005	.2500
	4	.0001	.0005	.0016	.0039	.0081	.0150	.0256	.0410	.0625
5	0	.5905	.4437	.3277	.2373	.1681	.1160	.0778	.0503	.0312
	1	.3280	.3915	.4096	.3955	.3602	.3124	.2592	.2059	.1562
	2	.0729	.1382	.2048	.2637	.3087	.3364	.3456	.3369	.3125
	3	.0081	.0244	.0512	.0879	.1323	.1811	.2304	.2757	.3125
	4	.0004	.0022	.0064	.0146	.0284	.0488	.0768	.1128	.1562
	5	.0000	.0001	.0003	.0010	.0024	.0053	.0102	.0185	.0312
6	0	.5314	.3771	.2621	.1780	.1176	.0754	.0467	.0277	.0156
	1	.3543	.3993	.3932	.3560	.3025	.2437	.1866	.1359	.0938
	2	.0984	.1762	.2458	.2966	.3241	.3280	.3110	.2780	.2344
	3	.0146	.0415	.0819	.1318	.1852	.2355	.2765	.3032	.3125
	4	.0012	.0055	.0154	.0330	.0595	.0951	.1382	.1861	.2344
	5	.0001	.0004	.0015	.0044	.0102	.0205	.0369	.0609	.0938
	6	.0000	.0000	.0001	.0002	.0007	.0018	.0041	.0083	.0156
7	0	.4783	.3206	.2097	.1335	.0824	.0490	.0280	.0152	.0078
	1	.3720	.3960	.3670	.3115	.2471	.1848	.1306	.0872	.0547
	2	.1240	.2097	.2753	.3115	.3177	.2985	.2613	.2140	.1641
	3	.0230	.0617	.1147	.1730	.2269	.2679	.2903	.2918	.2734
	4	.0026	.0109	.0287	.0577	.0972	.1442	.1935	.2388	.2734
	5	.0002	.0012	.0043	.0115	.0250	.0466	.0774	.1172	.1641
	6	.0000	.0001	.0004	.0013	.0036	.0084	.0172	.0320	.0547
	7	.0000	.0000	.0000	.0001	.0002	.0006	.0016	.0037	.0078
8	0	.4305	.2725	.1678	.1001	.0576	.0319	.0168	.0084	.0039
	1	.3826	.3847	.3355	.2670	.1977	.1373	.0896	.0548	.0312
	2	.1488	.2376	.2936	.3115	.2965	.2587	.2090	.1569	.1094
	3	.0331	.0839	.1468	.2076	.2541	.2786	.2787	.2568	.2188
	4	.0046	.0185	.0459	.0865	.1361	.1875	.2322	.2627	.2734
	5	.0004	.0026	.0092	.0231	.0467	.0808	.1239	.1719	.2188
	6	.0000	.0002	.0011	.0038	.0100	.0217	.0413	.0703	.1094
	7	.0000	.0000	.0001	.0004	.0012	.0033	.0079	.0164	.0313
	8	.0000	.0000	.0000	.0000	.0001	.0002	.0007	.0017	.0039

TABLE 5 BINOMIAL PROBABILITIES (*Continued*)

n	x	.10	.15	.20	.25	.30	.35	.40	.45	.50
9	0	.3874	.2316	.1342	.0751	.0404	.0207	.0101	.0046	.0020
	1	.3874	.3679	.3020	.2253	.1556	.1004	.0605	.0339	.0176
	2	.1722	.2597	.3020	.3003	.2668	.2162	.1612	.1110	.0703
	3	.0446	.1069	.1762	.2336	.2668	.2716	.2508	.2119	.1641
	4	.0074	.0283	.0661	.1168	.1715	.2194	.2508	.2600	.2461
	5	.0008	.0050	.0165	.0389	.0735	.1181	.1672	.2128	.2461
	6	.0001	.0006	.0028	.0087	.0210	.0424	.0743	.1160	.1641
	7	.0000	.0000	.0003	.0012	.0039	.0098	.0212	.0407	.0703
	8	.0000	.0000	.0000	.0001	.0004	.0013	.0035	.0083	.0176
	9	.0000	.0000	.0000	.0000	.0000	.0001	.0003	.0008	.0020
10	0	.3487	.1969	.1074	.0563	.0282	.0135	.0060	.0025	.0010
	1	.3874	.3474	.2684	.1877	.1211	.0725	.0403	.0207	.0098
	2	.1937	.2759	.3020	.2816	.2335	.1757	.1209	.0763	.0439
	3	.0574	.1298	.2013	.2503	.2668	.2522	.2150	.1665	.1172
	4	.0112	.0401	.0881	.1460	.2001	.2377	.2508	.2384	.2051
	5	.0015	.0085	.0264	.0584	.1029	.1536	.2007	.2340	.2461
	6	.0001	.0012	.0055	.0162	.0368	.0689	.1115	.1596	.2051
	7	.0000	.0001	.0008	.0031	.0090	.0212	.0425	.0746	.1172
	8	.0000	.0000	.0001	.0004	.0014	.0043	.0106	.0229	.0439
	9	.0000	.0000	.0000	.0000	.0001	.0005	.0016	.0042	.0098
	10	.0000	.0000	.0000	.0000	.0000	.0000	.0001	.0003	.0010
12	0	.2824	.1422	.0687	.0317	.0138	.0057	.0022	.0008	.0002
	1	.3766	.3012	.2062	.1267	.0712	.0368	.0174	.0075	.0029
	2	.2301	.2924	.2835	.2323	.1678	.1088	.0639	.0339	.0161
	3	.0853	.1720	.2362	.2581	.2397	.1954	.1419	.0923	.0537
	4	.0213	.0683	.1329	.1936	.2311	.2367	.2128	.1700	.1208
	5	.0038	.0193	.0532	.1032	.1585	.2039	.2270	.2225	.1934
	6	.0005	.0040	.0155	.0401	.0792	.1281	.1766	.2124	.2256
	7	.0000	.0006	.0033	.0115	.0291	.0591	.1009	.1489	.1934
	8	.0000	.0001	.0005	.0024	.0078	.0199	.0420	.0762	.1208
	9	.0000	.0000	.0001	.0004	.0015	.0048	.0125	.0277	.0537
	10	.0000	.0000	.0000	.0000	.0002	.0008	.0025	.0068	.0161
	11	.0000	.0000	.0000	.0000	.0000	.0001	.0003	.0010	.0029
	12	.0000	.0000	.0000	.0000	.0000	.0000	.0000	.0001	.0002
15	0	.2059	.0874	.0352	.0134	.0047	.0016	.0005	.0001	.0000
	1	.3432	.2312	.1319	.0668	.0305	.0126	.0047	.0016	.0005
	2	.2669	.2856	.2309	.1559	.0916	.0476	.0219	.0090	.0032
	3	.1285	.2184	.2501	.2252	.1700	.1110	.0634	.0318	.0139
	4	.0428	.1156	.1876	.2252	.2186	.1792	.1268	.0780	.0417
	5	.0105	.0449	.1032	.1651	.2061	.2123	.1859	.1404	.0916
	6	.0019	.0132	.0430	.0917	.1472	.1906	.2066	.1914	.1527
	7	.0003	.0030	.0138	.0393	.0811	.1319	.1771	.2013	.1964
	8	.0000	.0005	.0035	.0131	.0348	.0710	.1181	.1647	.1964
	9	.0000	.0001	.0007	.0034	.0116	.0298	.0612	.1048	.1527
	10	.0000	.0000	.0001	.0007	.0030	.0096	.0245	.0515	.0916
	11	.0000	.0000	.0000	.0001	.0006	.0024	.0074	.0191	.0417
	12	.0000	.0000	.0000	.0000	.0001	.0004	.0016	.0052	.0139
	13	.0000	.0000	.0000	.0000	.0000	.0001	.0003	.0010	.0032
	14	.0000	.0000	.0000	.0000	.0000	.0000	.0000	.0001	.0005
	15	.0000	.0000	.0000	.0000	.0000	.0000	.0000	.0000	.0000

TABLE 5 BINOMIAL PROBABILITIES (*Continued*)

						p				
n	x	.10	.15	.20	.25	.30	.35	.40	.45	.50
18	0	.1501	.0536	.0180	.0056	.0016	.0004	.0001	.0000	.0000
	1	.3002	.1704	.0811	.0338	.0126	.0042	.0012	.0003	.0001
	2	.2835	.2556	.1723	.0958	.0458	.0190	.0069	.0022	.0006
	3	.1680	.2406	.2297	.1704	.1046	.0547	.0246	.0095	.0031
	4	.0700	.1592	.2153	.2130	.1681	.1104	.0614	.0291	.0117
	5	.0218	.0787	.1507	.1988	.2017	.1664	.1146	.0666	.0327
	6	.0052	.0301	.0816	.1436	.1873	.1941	.1655	.1181	.0708
	7	.0010	.0091	.0350	.0820	.1376	.1792	.1892	.1657	.1214
	8	.0002	.0022	.0120	.0376	.0811	.1327	.1734	.1864	.1669
	9	.0000	.0004	.0033	.0139	.0386	.0794	.1284	.1694	.1855
	10	.0000	.0001	.0008	.0042	.0149	.0385	.0771	.1248	.1669
	11	.0000	.0000	.0001	.0010	.0046	.0151	.0374	.0742	.1214
	12	.0000	.0000	.0000	.0002	.0012	.0047	.0145	.0354	.0708
	13	.0000	.0000	.0000	.0000	.0002	.0012	.0045	.0134	.0327
	14	.0000	.0000	.0000	.0000	.0000	.0002	.0011	.0039	.0117
	15	.0000	.0000	.0000	.0000	.0000	.0000	.0002	.0009	.0031
	16	.0000	.0000	.0000	.0000	.0000	.0000	.0000	.0001	.0006
	17	.0000	.0000	.0000	.0000	.0000	.0000	.0000	.0000	.0001
	18	.0000	.0000	.0000	.0000	.0000	.0000	.0000	.0000	.0000
20	0	.1216	.0388	.0115	.0032	.0008	.0002	.0000	.0000	.0000
	1	.2702	.1368	.0576	.0211	.0068	.0020	.0005	.0001	.0000
	2	.2852	.2293	.1369	.0669	.0278	.0100	.0031	.0008	.0002
	3	.1901	.2428	.2054	.1339	.0716	.0323	.0123	.0040	.0011
	4	.0898	.1821	.2182	.1897	.1304	.0738	.0350	.0139	.0046
	5	.0319	.1028	.1746	.2023	.1789	.1272	.0746	.0365	.0148
	6	.0089	.0454	.1091	.1686	.1916	.1712	.1244	.0746	.0370
	7	.0020	.0160	.0545	.1124	.1643	.1844	.1659	.1221	.0739
	8	.0004	.0046	.0222	.0609	.1144	.1614	.1797	.1623	.1201
	9	.0001	.0011	.0074	.0271	.0654	.1158	.1597	.1771	.1602
	10	.0000	.0002	.0020	.0099	.0308	.0686	.1171	.1593	.1762
	11	.0000	.0000	.0005	.0030	.0120	.0336	.0710	.1185	.1602
	12	.0000	.0000	.0001	.0008	.0039	.0136	.0355	.0727	.1201
	13	.0000	.0000	.0000	.0002	.0010	.0045	.0146	.0366	.0739
	14	.0000	.0000	.0000	.0000	.0002	.0012	.0049	.0150	.0370
	15	.0000	.0000	.0000	.0000	.0000	.0003	.0013	.0049	.0148
	16	.0000	.0000	.0000	.0000	.0000	.0000	.0003	.0013	.0046
	17	.0000	.0000	.0000	.0000	.0000	.0000	.0000	.0002	.0011
	18	.0000	.0000	.0000	.0000	.0000	.0000	.0000	.0000	.0002
	19	.0000	.0000	.0000	.0000	.0000	.0000	.0000	.0000	.0000
	20	.0000	.0000	.0000	.0000	.0000	.0000	.0000	.0000	.0000

TABLE 5 BINOMIAL PROBABILITIES (*Continued*)

		\multicolumn{9}{c}{p}
n	x	0.55	0.60	0.65	0.70	0.75	0.80	0.85	0.90	0.95
2	0	0.2025	0.1600	0.1225	0.0900	0.0625	0.0400	0.0225	0.0100	0.0025
	1	0.4950	0.4800	0.4550	0.4200	0.3750	0.3200	0.2550	0.1800	0.0950
	2	0.3025	0.3600	0.4225	0.4900	0.5625	0.6400	0.7225	0.8100	0.9025
3	0	0.0911	0.0640	0.0429	0.0270	0.0156	0.0080	0.0034	0.0010	0.0001
	1	0.3341	0.2880	0.2389	0.1890	0.1406	0.0960	0.0574	0.0270	0.0071
	2	0.4084	0.4320	0.4436	0.4410	0.4219	0.3840	0.3251	0.2430	0.1354
	3	0.1664	0.2160	0.2746	0.3430	0.4219	0.5120	0.6141	0.7290	0.8574
4	0	0.0410	0.0256	0.0150	0.0081	0.0039	0.0016	0.0005	0.0001	0.0000
	1	0.2005	0.1536	0.1115	0.0756	0.0469	0.0256	0.0115	0.0036	0.0005
	2	0.3675	0.3456	0.3105	0.2646	0.2109	0.1536	0.0975	0.0486	0.0135
	3	0.2995	0.3456	0.3845	0.4116	0.4219	0.4096	0.3685	0.2916	0.1715
	4	0.0915	0.1296	0.1785	0.2401	0.3164	0.4096	0.5220	0.6561	0.8145
5	0	0.0185	0.0102	0.0053	0.0024	0.0010	0.0003	0.0001	0.0000	0.0000
	1	0.1128	0.0768	0.0488	0.0284	0.0146	0.0064	0.0022	0.0005	0.0000
	2	0.2757	0.2304	0.1811	0.1323	0.0879	0.0512	0.0244	0.0081	0.0011
	3	0.3369	0.3456	0.3364	0.3087	0.2637	0.2048	0.1382	0.0729	0.0214
	4	0.2059	0.2592	0.3124	0.3601	0.3955	0.4096	0.3915	0.3281	0.2036
	5	0.0503	0.0778	0.1160	0.1681	0.2373	0.3277	0.4437	0.5905	0.7738
6	0	0.0083	0.0041	0.0018	0.0007	0.0002	0.0001	0.0000	0.0000	0.0000
	1	0.0609	0.0369	0.0205	0.0102	0.0044	0.0015	0.0004	0.0001	0.0000
	2	0.1861	0.1382	0.0951	0.0595	0.0330	0.0154	0.0055	0.0012	0.0001
	3	0.3032	0.2765	0.2355	0.1852	0.1318	0.0819	0.0415	0.0146	0.0021
	4	0.2780	0.3110	0.3280	0.3241	0.2966	0.2458	0.1762	0.0984	0.0305
	5	0.1359	0.1866	0.2437	0.3025	0.3560	0.3932	0.3993	0.3543	0.2321
	6	0.0277	0.0467	0.0754	0.1176	0.1780	0.2621	0.3771	0.5314	0.7351
7	0	0.0037	0.0016	0.0006	0.0002	0.0001	0.0000	0.0000	0.0000	0.0000
	1	0.0320	0.0172	0.0084	0.0036	0.0013	0.0004	0.0001	0.0000	0.0000
	2	0.1172	0.0774	0.0466	0.0250	0.0115	0.0043	0.0012	0.0002	0.0000
	3	0.2388	0.1935	0.1442	0.0972	0.0577	0.0287	0.0109	0.0026	0.0002
	4	0.2918	0.2903	0.2679	0.2269	0.1730	0.1147	0.0617	0.0230	0.0036
	5	0.2140	0.2613	0.2985	0.3177	0.3115	0.2753	0.2097	0.1240	0.0406
	6	0.0872	0.1306	0.1848	0.2471	0.3115	0.3670	0.3960	0.3720	0.2573
	7	0.0152	0.0280	0.0490	0.0824	0.1335	0.2097	0.3206	0.4783	0.6983
8	0	0.0017	0.0007	0.0002	0.0001	0.0000	0.0000	0.0000	0.0000	0.0000
	1	0.0164	0.0079	0.0033	0.0012	0.0004	0.0001	0.0000	0.0000	0.0000
	2	0.0703	0.0413	0.0217	0.0100	0.0038	0.0011	0.0002	0.0000	0.0000
	3	0.1719	0.1239	0.0808	0.0467	0.0231	0.0092	0.0026	0.0004	0.0000
	4	0.2627	0.2322	0.1875	0.1361	0.0865	0.0459	0.0185	0.0046	0.0004
	5	0.2568	0.2787	0.2786	0.2541	0.2076	0.1468	0.0839	0.0331	0.0054
	6	0.1569	0.2090	0.2587	0.2965	0.3115	0.2936	0.2376	0.1488	0.0515
	7	0.0548	0.0896	0.1373	0.1977	0.2670	0.3355	0.3847	0.3826	0.2793
	8	0.0084	0.0168	0.0319	0.0576	0.1001	0.1678	0.2725	0.4305	0.6634

TABLE 5 BINOMIAL PROBABILITIES (*Continued*)

						p				
n	x	0.55	0.60	0.65	0.70	0.75	0.80	0.85	0.90	0.95
9	0	0.0008	0.0003	0.0001	0.0000	0.0000	0.0000	0.0000	0.0000	0.0000
	1	0.0083	0.0035	0.0013	0.0004	0.0001	0.0000	0.0000	0.0000	0.0000
	2	0.0407	0.0212	0.0098	0.0039	0.0012	0.0003	0.0000	0.0000	0.0000
	3	0.1160	0.0743	0.0424	0.0210	0.0087	0.0028	0.0006	0.0001	0.0000
	4	0.2128	0.1672	0.1181	0.0735	0.0389	0.0165	0.0050	0.0008	0.0000
	5	0.2600	0.2508	0.2194	0.1715	0.1168	0.0661	0.0283	0.0074	0.0006
	6	0.2119	0.2508	0.2716	0.2668	0.2336	0.1762	0.1069	0.0446	0.0077
	7	0.1110	0.1612	0.2162	0.2668	0.3003	0.3020	0.2597	0.1722	0.0629
	8	0.0339	0.0605	0.1004	0.1556	0.2253	0.3020	0.3679	0.3874	0.2985
	9	0.0046	0.0101	0.0207	0.0404	0.0751	0.1342	0.2316	0.3874	0.6302
10	0	0.0003	0.0001	0.0000	0.0000	0.0000	0.0000	0.0000	0.0000	0.0000
	1	0.0042	0.0016	0.0005	0.0001	0.0000	0.0000	0.0000	0.0000	0.0000
	2	0.0229	0.0106	0.0043	0.0014	0.0004	0.0001	0.0000	0.0000	0.0000
	3	0.0746	0.0425	0.0212	0.0090	0.0031	0.0008	0.0001	0.0000	0.0000
	4	0.1596	0.1115	0.0689	0.0368	0.0162	0.0055	0.0012	0.0001	0.0000
	5	0.2340	0.2007	0.1536	0.1029	0.0584	0.0264	0.0085	0.0015	0.0001
	6	0.2384	0.2508	0.2377	0.2001	0.1460	0.0881	0.0401	0.0112	0.0010
	7	0.1665	0.2150	0.2522	0.2668	0.2503	0.2013	0.1298	0.0574	0.0105
	8	0.0763	0.1209	0.1757	0.2335	0.2816	0.3020	0.2759	0.1937	0.0746
	9	0.0207	0.0403	0.0725	0.1211	0.1877	0.2684	0.3474	0.3874	0.3151
	10	0.0025	0.0060	0.0135	0.0282	0.0563	0.1074	0.1969	0.3487	0.5987
12	0	0.0001	0.0000	0.0000	0.0000	0.0000	0.0000	0.0000	0.0000	0.0000
	1	0.0010	0.0003	0.0001	0.0000	0.0000	0.0000	0.0000	0.0000	0.0000
	2	0.0068	0.0025	0.0008	0.0002	0.0000	0.0000	0.0000	0.0000	0.0000
	3	0.0277	0.0125	0.0048	0.0015	0.0004	0.0001	0.0000	0.0000	0.0000
	4	0.0762	0.0420	0.0199	0.0078	0.0024	0.0005	0.0001	0.0000	0.0000
	5	0.1489	0.1009	0.0591	0.0291	0.0115	0.0033	0.0006	0.0000	0.0000
	6	0.2124	0.1766	0.1281	0.0792	0.0401	0.0155	0.0040	0.0005	0.0000
	7	0.2225	0.2270	0.2039	0.1585	0.1032	0.0532	0.0193	0.0038	0.0002
	8	0.1700	0.2128	0.2367	0.2311	0.1936	0.1329	0.0683	0.0213	0.0021
	9	0.0923	0.1419	0.1954	0.2397	0.2581	0.2362	0.1720	0.0852	0.0173
	10	0.0339	0.0639	0.1088	0.1678	0.2323	0.2835	0.2924	0.2301	0.0988
	11	0.0075	0.0174	0.0368	0.0712	0.1267	0.2062	0.3012	0.3766	0.3413
	12	0.0008	0.0022	0.0057	0.0138	0.0317	0.0687	0.1422	0.2824	0.5404
15	0	0.0000	0.0000	0.0000	0.0000	0.0000	0.0000	0.0000	0.0000	0.0000
	1	0.0001	0.0000	0.0000	0.0000	0.0000	0.0000	0.0000	0.0000	0.0000
	2	0.0010	0.0003	0.0001	0.0000	0.0000	0.0000	0.0000	0.0000	0.0000
	3	0.0052	0.0016	0.0004	0.0001	0.0000	0.0000	0.0000	0.0000	0.0000
	4	0.0191	0.0074	0.0024	0.0006	0.0001	0.0000	0.0000	0.0000	0.0000
	5	0.0515	0.0245	0.0096	0.0030	0.0007	0.0001	0.0000	0.0000	0.0000
	6	0.1048	0.0612	0.0298	0.0116	0.0034	0.0007	0.0001	0.0000	0.0000
	7	0.1647	0.1181	0.0710	0.0348	0.0131	0.0035	0.0005	0.0000	0.0000
	8	0.2013	0.1771	0.1319	0.0811	0.0393	0.0138	0.0030	0.0003	0.0000
	9	0.1914	0.2066	0.1906	0.1472	0.0917	0.0430	0.0132	0.0019	0.0000
	10	0.1404	0.1859	0.2123	0.2061	0.1651	0.1032	0.0449	0.0105	0.0006
	11	0.0780	0.1268	0.1792	0.2186	0.2252	0.1876	0.1156	0.0428	0.0049

TABLE 5 BINOMIAL PROBABILITIES (*Continued*)

						p				
n	*x*	0.55	0.60	0.65	0.70	0.75	0.80	0.85	0.90	0.95
	12	0.0318	0.0634	0.1110	0.1700	0.2252	0.2501	0.2184	0.1285	0.0307
	13	0.0090	0.0219	0.0476	0.0916	0.1559	0.2309	0.2856	0.2669	0.1348
	14	0.0016	0.0047	0.0126	0.0305	0.0668	0.1319	0.2312	0.3432	0.3658
	15	0.0001	0.0005	0.0016	0.0047	0.0134	0.0352	0.0874	0.2059	0.4633
18	0	0.0000	0.0000	0.0000	0.0000	0.0000	0.0000	0.0000	0.0000	0.0000
	1	0.0000	0.0000	0.0000	0.0000	0.0000	0.0000	0.0000	0.0000	0.0000
	2	0.0001	0.0000	0.0000	0.0000	0.0000	0.0000	0.0000	0.0000	0.0000
	3	0.0009	0.0002	0.0000	0.0000	0.0000	0.0000	0.0000	0.0000	0.0000
	4	0.0039	0.0011	0.0002	0.0000	0.0000	0.0000	0.0000	0.0000	0.0000
	5	0.0134	0.0045	0.0012	0.0002	0.0000	0.0000	0.0000	0.0000	0.0000
	6	0.0354	0.0145	0.0047	0.0012	0.0002	0.0000	0.0000	0.0000	0.0000
	7	0.0742	0.0374	0.0151	0.0046	0.0010	0.0001	0.0000	0.0000	0.0000
	8	0.1248	0.0771	0.0385	0.0149	0.0042	0.0008	0.0001	0.0000	0.0000
	9	0.1694	0.1284	0.0794	0.0386	0.0139	0.0033	0.0004	0.0000	0.0000
	10	0.1864	0.1734	0.1327	0.0811	0.0376	0.0120	0.0022	0.0002	0.0000
	11	0.1657	0.1892	0.1792	0.1376	0.0820	0.0350	0.0091	0.0010	0.0000
	12	0.1181	0.1655	0.1941	0.1873	0.1436	0.0816	0.0301	0.0052	0.0002
	13	0.0666	0.1146	0.1664	0.2017	0.1988	0.1507	0.0787	0.0218	0.0014
	14	0.0291	0.0614	0.1104	0.1681	0.2130	0.2153	0.1592	0.0700	0.0093
	15	0.0095	0.0246	0.0547	0.1046	0.1704	0.2297	0.2406	0.1680	0.0473
	16	0.0022	0.0069	0.0190	0.0458	0.0958	0.1723	0.2556	0.2835	0.1683
	17	0.0003	0.0012	0.0042	0.0126	0.0338	0.0811	0.1704	0.3002	0.3763
	18	0.0000	0.0001	0.0004	0.0016	0.0056	0.0180	0.0536	0.1501	0.3972
20	0	0.0000	0.0000	0.0000	0.0000	0.0000	0.0000	0.0000	0.0000	0.0000
	1	0.0000	0.0000	0.0000	0.0000	0.0000	0.0000	0.0000	0.0000	0.0000
	2	0.0000	0.0000	0.0000	0.0000	0.0000	0.0000	0.0000	0.0000	0.0000
	3	0.0002	0.0000	0.0000	0.0000	0.0000	0.0000	0.0000	0.0000	0.0000
	4	0.0013	0.0003	0.0000	0.0000	0.0000	0.0000	0.0000	0.0000	0.0000
	5	0.0049	0.0013	0.0003	0.0000	0.0000	0.0000	0.0000	0.0000	0.0000
	6	0.0150	0.0049	0.0012	0.0002	0.0000	0.0000	0.0000	0.0000	0.0000
	7	0.0366	0.0146	0.0045	0.0010	0.0002	0.0000	0.0000	0.0000	0.0000
	8	0.0727	0.0355	0.0136	0.0039	0.0008	0.0001	0.0000	0.0000	0.0000
	9	0.1185	0.0710	0.0336	0.0120	0.0030	0.0005	0.0000	0.0000	0.0000
	10	0.1593	0.1171	0.0686	0.0308	0.0099	0.0020	0.0002	0.0000	0.0000
	11	0.1771	0.1597	0.1158	0.0654	0.0271	0.0074	0.0011	0.0001	0.0000
	12	0.1623	0.1797	0.1614	0.1144	0.0609	0.0222	0.0046	0.0004	0.0000
	13	0.1221	0.1659	0.1844	0.1643	0.1124	0.0545	0.0160	0.0020	0.0000
	14	0.0746	0.1244	0.1712	0.1916	0.1686	0.1091	0.0454	0.0089	0.0003
	15	0.0365	0.0746	0.1272	0.1789	0.2023	0.1746	0.1028	0.0319	0.0022
	16	0.0139	0.0350	0.0738	0.1304	0.1897	0.2182	0.1821	0.0898	0.0133
	17	0.0040	0.0123	0.0323	0.0716	0.1339	0.2054	0.2428	0.1901	0.0596
	18	0.0008	0.0031	0.0100	0.0278	0.0669	0.1369	0.2293	0.2852	0.1887
	19	0.0001	0.0005	0.0020	0.0068	0.0211	0.0576	0.1368	0.2702	0.3774
	20	0.0000	0.0000	0.0002	0.0008	0.0032	0.0115	0.0388	0.1216	0.3585

TABLE 6 VALUES OF $e^{-\mu}$

μ	$e^{-\mu}$	μ	$e^{-\mu}$	μ	$e^{-\mu}$
.00	1.0000	2.00	.1353	4.00	.0183
.05	.9512	2.05	.1287	4.05	.0174
.10	.9048	2.10	.1225	4.10	.0166
.15	.8607	2.15	.1165	4.15	.0158
.20	.8187	2.20	.1108	4.20	.0150
.25	.7788	2.25	.1054	4.25	.0143
.30	.7408	2.30	.1003	4.30	.0136
.35	.7047	2.35	.0954	4.35	.0129
.40	.6703	2.40	.0907	4.40	.0123
.45	.6376	2.45	.0863	4.45	.0117
.50	.6065	2.50	.0821	4.50	.0111
.55	.5769	2.55	.0781	4.55	.0106
.60	.5488	2.60	.0743	4.60	.0101
.65	.5220	2.65	.0707	4.65	.0096
.70	.4966	2.70	.0672	4.70	.0091
.75	.4724	2.75	.0639	4.75	.0087
.80	.4493	2.80	.0608	4.80	.0082
.85	.4274	2.85	.0578	4.85	.0078
.90	.4066	2.90	.0550	4.90	.0074
.95	.3867	2.95	.0523	4.95	.0071
1.00	.3679	3.00	.0498	5.00	.0067
1.05	.3499	3.05	.0474	6.00	.0025
1.10	.3329	3.10	.0450	7.00	.0009
1.15	.3166	3.15	.0429	8.00	.000335
1.20	.3012	3.20	.0408	9.00	.000123
				10.00	.000045
1.25	.2865	3.25	.0388		
1.30	.2725	3.30	.0369		
1.35	.2592	3.35	.0351		
1.40	.2466	3.40	.0334		
1.45	.2346	3.45	.0317		
1.50	.2231	3.50	.0302		
1.55	.2122	3.55	.0287		
1.60	.2019	3.60	.0273		
1.65	.1920	3.65	.0260		
1.70	.1827	3.70	.0247		
1.75	.1738	3.75	.0235		
1.80	.1653	3.80	.0224		
1.85	.1572	3.85	.0213		
1.90	.1496	3.90	.0202		
1.95	.1423	3.95	.0193		

TABLE 7 POISSON PROBABILITIES

Entries in the table give the probability of x occurrences for a Poisson process with a mean μ. For example, when $\mu = 2.5$, the probability of four occurrences is .1336.

					μ					
x	0.1	0.2	0.3	0.4	0.5	0.6	0.7	0.8	0.9	1.0
0	.9048	.8187	.7408	.6703	.6065	.5488	.4966	.4493	.4066	.3679
1	.0905	.1637	.2222	.2681	.3033	.3293	.3476	.3595	.3659	.3679
2	.0045	.0164	.0333	.0536	.0758	.0988	.1217	.1438	.1647	.1839
3	.0002	.0011	.0033	.0072	.0126	.0198	.0284	.0383	.0494	.0613
4	.0000	.0001	.0002	.0007	.0016	.0030	.0050	.0077	.0111	.0153
5	.0000	.0000	.0000	.0001	.0002	.0004	.0007	.0012	.0020	.0031
6	.0000	.0000	.0000	.0000	.0000	.0000	.0001	.0002	.0003	.0005
7	.0000	.0000	.0000	.0000	.0000	.0000	.0000	.0000	.0000	.0001

					μ					
x	1.1	1.2	1.3	1.4	1.5	1.6	1.7	1.8	1.9	2.0
0	.3329	.3012	.2725	.2466	.2231	.2019	.1827	.1653	.1496	.1353
1	.3662	.3614	.3543	.3452	.3347	.3230	.3106	.2975	.2842	.2707
2	.2014	.2169	.2303	.2417	.2510	.2584	.2640	.2678	.2700	.2707
3	.0738	.0867	.0998	.1128	.1255	.1378	.1496	.1607	.1710	.1804
4	.0203	.0260	.0324	.0395	.0471	.0551	.0636	.0723	.0812	.0902
5	.0045	.0062	.0084	.0111	.0141	.0176	.0216	.0260	.0309	.0361
6	.0008	.0012	.0018	.0026	.0035	.0047	.0061	.0078	.0098	.0120
7	.0001	.0002	.0003	.0005	.0008	.0011	.0015	.0020	.0027	.0034
8	.0000	.0000	.0001	.0001	.0001	.0002	.0003	.0005	.0006	.0009
9	.0000	.0000	.0000	.0000	.0000	.0000	.0001	.0001	.0001	.0002

					μ					
x	2.1	2.2	2.3	2.4	2.5	2.6	2.7	2.8	2.9	3.0
0	.1225	.1108	.1003	.0907	.0821	.0743	.0672	.0608	.0550	.0498
1	.2572	.2438	.2306	.2177	.2052	.1931	.1815	.1703	.1596	.1494
2	.2700	.2681	.2652	.2613	.2565	.2510	.2450	.2384	.2314	.2240
3	.1890	.1966	.2033	.2090	.2138	.2176	.2205	.2225	.2237	.2240
4	.0992	.1082	.1169	.1254	.1336	.1414	.1488	.1557	.1622	.1680
5	.0417	.0476	.0538	.0602	.0668	.0735	.0804	.0872	.0940	.1008
6	.0146	.0174	.0206	.0241	.0278	.0319	.0362	.0407	.0455	.0504
7	.0044	.0055	.0068	.0083	.0099	.0118	.0139	.0163	.0188	.0216
8	.0011	.0015	.0019	.0025	.0031	.0038	.0047	.0057	.0068	.0081
9	.0003	.0004	.0005	.0007	.0009	.0011	.0014	.0018	.0022	.0027
10	.0001	.0001	.0001	.0002	.0002	.0003	.0004	.0005	.0006	.0008
11	.0000	.0000	.0000	.0000	.0000	.0001	.0001	.0001	.0002	.0002
12	.0000	.0000	.0000	.0000	.0000	.0000	.0000	.0000	.0000	.0001

TABLE 7 POISSON PROBABILITIES (*Continued*)

	μ									
x	3.1	3.2	3.3	3.4	3.5	3.6	3.7	3.8	3.9	4.0
0	.0450	.0408	.0369	.0344	.0302	.0273	.0247	.0224	.0202	.0183
1	.1397	.1304	.1217	.1135	.1057	.0984	.0915	.0850	.0789	.0733
2	.2165	.2087	.2008	.1929	.1850	.1771	.1692	.1615	.1539	.1465
3	.2237	.2226	.2209	.2186	.2158	.2125	.2087	.2046	.2001	.1954
4	.1734	.1781	.1823	.1858	.1888	.1912	.1931	.1944	.1951	.1954
5	.1075	.1140	.1203	.1264	.1322	.1377	.1429	.1477	.1522	.1563
6	.0555	.0608	.0662	.0716	.0771	.0826	.0881	.0936	.0989	.1042
7	.0246	.0278	.0312	.0348	.0385	.0425	.0466	.0508	.0551	.0595
8	.0095	.0111	.0129	.0148	.0169	.0191	.0215	.0241	.0269	.0298
9	.0033	.0040	.0047	.0056	.0066	.0076	.0089	.0102	.0116	.0132
10	.0010	.0013	.0016	.0019	.0023	.0028	.0033	.0039	.0045	.0053
11	.0003	.0004	.0005	.0006	.0007	.0009	.0011	.0013	.0016	.0019
12	.0001	.0001	.0001	.0002	.0002	.0003	.0003	.0004	.0005	.0006
13	.0000	.0000	.0000	.0000	.0001	.0001	.0001	.0001	.0002	.0002
14	.0000	.0000	.0000	.0000	.0000	.0000	.0000	.0000	.0000	.0001

	μ									
x	4.1	4.2	4.3	4.4	4.5	4.6	4.7	4.8	4.9	5.0
0	.0166	.0150	.0136	.0123	.0111	.0101	.0091	.0082	.0074	.0067
1	.0679	.0630	.0583	.0540	.0500	.0462	.0427	.0395	.0365	.0337
2	.1393	.1323	.1254	.1188	.1125	.1063	.1005	.0948	.0894	.0842
3	.1904	.1852	.1798	.1743	.1687	.1631	.1574	.1517	.1460	.1404
4	.1951	.1944	.1933	.1917	.1898	.1875	.1849	.1820	.1789	.1755
5	.1600	.1633	.1662	.1687	.1708	.1725	.1738	.1747	.1753	.1755
6	.1093	.1143	.1191	.1237	.1281	.1323	.1362	.1398	.1432	.1462
7	.0640	.0686	.0732	.0778	.0824	.0869	.0914	.0959	.1002	.1044
8	.0328	.0360	.0393	.0428	.0463	.0500	.0537	.0575	.0614	.0653
9	.0150	.0168	.0188	.0209	.0232	.0255	.0280	.0307	.0334	.0363
10	.0061	.0071	.0081	.0092	.0104	.0118	.0132	.0147	.0164	.0181
11	.0023	.0027	.0032	.0037	.0043	.0049	.0056	.0064	.0073	.0082
12	.0008	.0009	.0011	.0014	.0016	.0019	.0022	.0026	.0030	.0034
13	.0002	.0003	.0004	.0005	.0006	.0007	.0008	.0009	.0011	.0013
14	.0001	.0001	.0001	.0001	.0002	.0002	.0003	.0003	.0004	.0005
15	.0000	.0000	.0000	.0000	.0001	.0001	.0001	.0001	.0001	.0002

	μ									
x	5.1	5.2	5.3	5.4	5.5	5.6	5.7	5.8	5.9	6.0
0	.0061	.0055	.0050	.0045	.0041	.0037	.0033	.0030	.0027	.0025
1	.0311	.0287	.0265	.0244	.0225	.0207	.0191	.0176	.0162	.0149
2	.0793	.0746	.0701	.0659	.0618	.0580	.0544	.0509	.0477	.0446
3	.1348	.1293	.1239	.1185	.1133	.1082	.1033	.0985	.0938	.0892
4	.1719	.1681	.1641	.1600	.1558	.1515	.1472	.1428	.1383	.1339

TABLE 7 POISSON PROBABILITIES (*Continued*)

					μ					
x	5.1	5.2	5.3	5.4	5.5	5.6	5.7	5.8	5.9	6.0
5	.1753	.1748	.1740	.1728	.1714	.1697	.1678	.1656	.1632	.1606
6	.1490	.1515	.1537	.1555	.1571	.1587	.1594	.1601	.1605	.1606
7	.1086	.1125	.1163	.1200	.1234	.1267	.1298	.1326	.1353	.1377
8	.0692	.0731	.0771	.0810	.0849	.0887	.0925	.0962	.0998	.1033
9	.0392	.0423	.0454	.0486	.0519	.0552	.0586	.0620	.0654	.0688
10	.0200	.0220	.0241	.0262	.0285	.0309	.0334	.0359	.0386	.0413
11	.0093	.0104	.0116	.0129	.0143	.0157	.0173	.0190	.0207	.0225
12	.0039	.0045	.0051	.0058	.0065	.0073	.0082	.0092	.0102	.0113
13	.0015	.0018	.0021	.0024	.0028	.0032	.0036	.0041	.0046	.0052
14	.0006	.0007	.0008	.0009	.0011	.0013	.0015	.0017	.0019	.0022
15	.0002	.0002	.0003	.0003	.0004	.0005	.0006	.0007	.0008	.0009
16	.0001	.0001	.0001	.0001	.0001	.0002	.0002	.0002	.0003	.0003
17	.0000	.0000	.0000	.0000	.0000	.0001	.0001	.0001	.0001	.0001

					μ					
x	6.1	6.2	6.3	6.4	6.5	6.6	6.7	6.8	6.9	7.0
0	.0022	.0020	.0018	.0017	.0015	.0014	.0012	.0011	.0010	.0009
1	.0137	.0126	.0116	.0106	.0098	.0090	.0082	.0076	.0070	.0064
2	.0417	.0390	.0364	.0340	.0318	.0296	.0276	.0258	.0240	.0223
3	.0848	.0806	.0765	.0726	.0688	.0652	.0617	.0584	.0552	.0521
4	.1294	.1249	.1205	.1162	.1118	.1076	.1034	.0992	.0952	.0912
5	.1579	.1549	.1519	.1487	.1454	.1420	.1385	.1349	.1314	.1277
6	.1605	.1601	.1595	.1586	.1575	.1562	.1546	.1529	.1511	.1490
7	.1399	.1418	.1435	.1450	.1462	.1472	.1480	.1486	.1489	.1490
8	.1066	.1099	.1130	.1160	.1188	.1215	.1240	.1263	.1284	.1304
9	.0723	.0757	.0791	.0825	.0858	.0891	.0923	.0954	.0985	.1014
10	.0441	.0469	.0498	.0528	.0558	.0588	.0618	.0649	.0679	.0710
11	.0245	.0265	.0285	.0307	.0330	.0353	.0377	.0401	.0426	.0452
12	.0124	.0137	.0150	.0164	.0179	.0194	.0210	.0227	.0245	.0264
13	.0058	.0065	.0073	.0081	.0089	.0098	.0108	.0119	.0130	.0142
14	.0025	.0029	.0033	.0037	.0041	.0046	.0052	.0058	.0064	.0071
15	.0010	.0012	.0014	.0016	.0018	.0020	.0023	.0026	.0029	.0033
16	.0004	.0005	.0005	.0006	.0007	.0008	.0010	.0011	.0013	.0014
17	.0001	.0002	.0002	.0002	.0003	.0003	.0004	.0004	.0005	.0006
18	.0000	.0001	.0001	.0001	.0001	.0001	.0001	.0002	.0002	.0002
19	.0000	.0000	.0000	.0000	.0000	.0000	.0000	.0001	.0001	.0001

					μ					
x	7.1	7.2	7.3	7.4	7.5	7.6	7.7	7.8	7.9	8.0
0	.0008	.0007	.0007	.0006	.0006	.0005	.0005	.0004	.0004	.0003
1	.0059	.0054	.0049	.0045	.0041	.0038	.0035	.0032	.0029	.0027
2	.0208	.0194	.0180	.0167	.0156	.0145	.0134	.0125	.0116	.0107
3	.0492	.0464	.0438	.0413	.0389	.0366	.0345	.0324	.0305	.0286
4	.0874	.0836	.0799	.0764	.0729	.0696	.0663	.0632	.0602	.0573

TABLE 7 POISSON PROBABILITIES (*Continued*)

					μ					
x	7.1	7.2	7.3	7.4	7.5	7.6	7.7	7.8	7.9	8.0
5	.1241	.1204	.1167	.1130	.1094	.1057	.1021	.0986	.0951	.0916
6	.1468	.1445	.1420	.1394	.1367	.1339	.1311	.1282	.1252	.1221
7	.1489	.1486	.1481	.1474	.1465	.1454	.1442	.1428	.1413	.1396
8	.1321	.1337	.1351	.1363	.1373	.1382	.1388	.1392	.1395	.1396
9	.1042	.1070	.1096	.1121	.1144	.1167	.1187	.1207	.1224	.1241
10	.0740	.0770	.0800	.0829	.0858	.0887	.0914	.0941	.0967	.0993
11	.0478	.0504	.0531	.0558	.0585	.0613	.0640	.0667	.0695	.0722
12	.0283	.0303	.0323	.0344	.0366	.0388	.0411	.0434	.0457	.0481
13	.0154	.0168	.0181	.0196	.0211	.0227	.0243	.0260	.0278	.0296
14	.0078	.0086	.0095	.0104	.0113	.0123	.0134	.0145	.0157	.0169
15	.0037	.0041	.0046	.0051	.0057	.0062	.0069	.0075	.0083	.0090
16	.0016	.0019	.0021	.0024	.0026	.0030	.0033	.0037	.0041	.0045
17	.0007	.0008	.0009	.0010	.0012	.0013	.0015	.0017	.0019	.0021
18	.0003	.0003	.0004	.0004	.0005	.0006	.0006	.0007	.0008	.0009
19	.0001	.0001	.0001	.0002	.0002	.0002	.0003	.0003	.0003	.0004
20	.0000	.0000	.0001	.0001	.0001	.0001	.0001	.0001	.0001	.0002
21	.0000	.0000	.0000	.0000	.0000	.0000	.0000	.0000	.0001	.0001

					μ					
x	8.1	8.2	8.3	8.4	8.5	8.6	8.7	8.8	8.9	9.0
0	.0003	.0003	.0002	.0002	.0002	.0002	.0002	.0002	.0001	.0001
1	.0025	.0023	.0021	.0019	.0017	.0016	.0014	.0013	.0012	.0011
2	.0100	.0092	.0086	.0079	.0074	.0068	.0063	.0058	.0054	.0050
3	.0269	.0252	.0237	.0222	.0208	.0195	.0183	.0171	.0160	.0150
4	.0544	.0517	.0491	.0466	.0443	.0420	.0398	.0377	.0357	.0337
5	.0882	.0849	.0816	.0784	.0752	.0722	.0692	.0663	.0635	.0607
6	.1191	.1160	.1128	.1097	.1066	.1034	.1003	.0972	.0941	.0911
7	.1378	.1358	.1338	.1317	.1294	.1271	.1247	.1222	.1197	.1171
8	.1395	.1392	.1388	.1382	.1375	.1366	.1356	.1344	.1332	.1318
9	.1256	.1269	.1280	.1290	.1299	.1306	.1311	.1315	.1317	.1318
10	.1017	.1040	.1063	.1084	.1104	.1123	.1140	.1157	.1172	.1186
11	.0749	.0776	.0802	.0828	.0853	.0878	.0902	.0925	.0948	.0970
12	.0505	.0530	.0555	.0579	.0604	.0629	.0654	.0679	.0703	.0728
13	.0315	.0334	.0354	.0374	.0395	.0416	.0438	.0459	.0481	.0504
14	.0182	.0196	.0210	.0225	.0240	.0256	.0272	.0289	.0306	.0324
15	.0098	.0107	.0116	.0126	.0136	.0147	.0158	.0169	.0182	.1094
16	.0050	.0055	.0060	.0066	.0072	.0079	.0086	.0093	.0101	.0109
17	.0024	.0026	.0029	.0033	.0036	.0040	.0044	.0048	.0053	.0058
18	.0011	.0012	.0014	.0015	.0017	.0019	.0021	.0024	.0026	.0029
19	.0005	.0005	.0006	.0007	.0008	.0009	.0010	.0011	.0012	.0014
20	.0002	.0002	.0002	.0003	.0003	.0004	.0004	.0005	.0005	.0006
21	.0001	.0001	.0001	.0001	.0001	.0002	.0002	.0002	.0002	.0003
22	.0000	.0000	.0000	.0000	.0001	.0001	.0001	.0001	.0001	.0001

TABLE 7 POISSON PROBABILITIES (*Continued*)

					μ					
x	9.1	9.2	9.3	9.4	9.5	9.6	9.7	9.8	9.9	10
0	.0001	.0001	.0001	.0001	.0001	.0001	.0001	.0001	.0001	.0000
1	.0010	.0009	.0009	.0008	.0007	.0007	.0006	.0005	.0005	.0005
2	.0046	.0043	.0040	.0037	.0034	.0031	.0029	.0027	.0025	.0023
3	.0140	.0131	.0123	.0115	.0107	.0100	.0093	.0087	.0081	.0076
4	.0319	.0302	.0285	.0269	.0254	.0240	.0226	.0213	.0201	.0189
5	.0581	.0555	.0530	.0506	.0483	.0460	.0439	.0418	.0398	.0378
6	.0881	.0851	.0822	.0793	.0764	.0736	.0709	.0682	.0656	.0631
7	.1145	.1118	.1091	.1064	.1037	.1010	.0982	.0955	.0928	.0901
8	.1302	.1286	.1269	.1251	.1232	.1212	.1191	.1170	.1148	.1126
9	.1317	.1315	.1311	.1306	.1300	.1293	.1284	.1274	.1263	.1251
10	.1198	.1210	.1219	.1228	.1235	.1241	.1245	.1249	.1250	.1251
11	.0991	.1012	.1031	.1049	.1067	.1083	.1098	.1112	.1125	.1137
12	.0752	.0776	.0799	.0822	.0844	.0866	.0888	.0908	.0928	.0948
13	.0526	.0549	.0572	.0594	.0617	.0640	.0662	.0685	.0707	.0729
14	.0342	.0361	.0380	.0399	.0419	.0439	.0459	.0479	.0500	.0521
15	.0208	.0221	.0235	.0250	.0265	.0281	.0297	.0313	.0330	.0347
16	.0118	.0127	.0137	.0147	.0157	.0168	.0180	.0192	.0204	.0217
17	.0063	.0069	.0075	.0081	.0088	.0095	.0103	.0111	.0119	.0128
18	.0032	.0035	.0039	.0042	.0046	.0051	.0055	.0060	.0065	.0071
19	.0015	.0017	.0019	.0021	.0023	.0026	.0028	.0031	.0034	.0037
20	.0007	.0008	.0009	.0010	.0011	.0012	.0014	.0015	.0017	.0019
21	.0003	.0003	.0004	.0004	.0005	.0006	.0006	.0007	.0008	.0009
22	.0001	.0001	.0002	.0002	.0002	.0002	.0003	.0003	.0004	.0004
23	.0000	.0001	.0001	.0001	.0001	.0001	.0001	.0001	.0002	.0002
24	.0000	.0000	.0000	.0000	.0000	.0000	.0000	.0001	.0001	.0001

					μ					
x	11	12	13	14	15	16	17	18	19	20
0	.0000	.0000	.0000	.0000	.0000	.0000	.0000	.0000	.0000	.0000
1	.0002	.0001	.0000	.0000	.0000	.0000	.0000	.0000	.0000	.0000
2	.0010	.0004	.0002	.0001	.0000	.0000	.0000	.0000	.0000	.0000
3	.0037	.0018	.0008	.0004	.0002	.0001	.0000	.0000	.0000	.0000
4	.0102	.0053	.0027	.0013	.0006	.0003	.0001	.0001	.0000	.0000
5	.0224	.0127	.0070	.0037	.0019	.0010	.0005	.0002	.0001	.0001
6	.0411	.0255	.0152	.0087	.0048	.0026	.0014	.0007	.0004	.0002
7	.0646	.0437	.0281	.0174	.0104	.0060	.0034	.0018	.0010	.0005
8	.0888	.0655	.0457	.0304	.0194	.0120	.0072	.0042	.0024	.0013
9	.1085	.0874	.0661	.0473	.0324	.0213	.0135	.0083	.0050	.0029
10	.1194	.1048	.0859	.0663	.0486	.0341	.0230	.0150	.0095	.0058
11	.1194	.1144	.1015	.0844	.0663	.0496	.0355	.0245	.0164	.0106
12	.1094	.1144	.1099	.0984	.0829	.0661	.0504	.0368	.0259	.0176
13	.0926	.1056	.1099	.1060	.0956	.0814	.0658	.0509	.0378	.0271
14	.0728	.0905	.1021	.1060	.1024	.0930	.0800	.0655	.0514	.0387

TABLE 7 POISSON PROBABILITIES (*Continued*)

					μ					
x	11	12	13	14	15	16	17	18	19	20
15	.0534	.0724	.0885	.0989	.1024	.0992	.0906	.0786	.0650	.0516
16	.0367	.0543	.0719	.0866	.0960	.0992	.0963	.0884	.0772	.0646
17	.0237	.0383	.0550	.0713	.0847	.0934	.0963	.0936	.0863	.0760
18	.0145	.0256	.0397	.0554	.0706	.0830	.0909	.0936	.0911	.0844
19	.0084	.0161	.0272	.0409	.0557	.0699	.0814	.0887	.0911	.0888
20	.0046	.0097	.0177	.0286	.0418	.0559	.0692	.0798	.0866	.0888
21	.0024	.0055	.0109	.0191	.0299	.0426	.0560	.0684	.0783	.0846
22	.0012	.0030	.0065	.0121	.0204	.0310	.0433	.0560	.0676	.0769
23	.0006	.0016	.0037	.0074	.0133	.0216	.0320	.0438	.0559	.0669
24	.0003	.0008	.0020	.0043	.0083	.0144	.0226	.0328	.0442	.0557
25	.0001	.0004	.0010	.0024	.0050	.0092	.0154	.0237	.0336	.0446
26	.0000	.0002	.0005	.0013	.0029	.0057	.0101	.0164	.0246	.0343
27	.0000	.0001	.0002	.0007	.0016	.0034	.0063	.0109	.0173	.0254
28	.0000	.0000	.0001	.0003	.0009	.0019	.0038	.0070	.0117	.0181
29	.0000	.0000	.0001	.0002	.0004	.0011	.0023	.0044	.0077	.0125
30	.0000	.0000	.0000	.0001	.0002	.0006	.0013	.0026	.0049	.0083
31	.0000	.0000	.0000	.0000	.0001	.0003	.0007	.0015	.0030	.0054
32	.0000	.0000	.0000	.0000	.0001	.0001	.0004	.0009	.0018	.0034
33	.0000	.0000	.0000	.0000	.0000	.0001	.0002	.0005	.0010	.0020
34	.0000	.0000	.0000	.0000	.0000	.0000	.0001	.0002	.0006	.0012
35	.0000	.0000	.0000	.0000	.0000	.0000	.0000	.0001	.0003	.0007
36	.0000	.0000	.0000	.0000	.0000	.0000	.0000	.0001	.0002	.0004
37	.0000	.0000	.0000	.0000	.0000	.0000	.0000	.0000	.0001	.0002
38	.0000	.0000	.0000	.0000	.0000	.0000	.0000	.0000	.0000	.0001
39	.0000	.0000	.0000	.0000	.0000	.0000	.0000	.0000	.0000	.0001

Appendix C: Summation Notation

Summations

Definition

$$\sum_{i=1}^{n} x_i = x_1 + x_2 + \cdots + x_n \qquad \text{(C.1)}$$

Example for $x_1 = 5, x_2 = 8, x_3 = 14$:

$$\sum_{i=1}^{3} x_i = x_1 + x_2 + x_3$$
$$= 5 + 8 + 14$$
$$= 27$$

Result 1

For a constant c:

$$\sum_{i=1}^{n} c = \underbrace{(c + c + \cdots + c)}_{n \text{ times}} = nc \qquad \text{(C.2)}$$

Example for $c = 5, n = 10$:

$$\sum_{i=1}^{10} 5 = 10(5) = 50$$

Example for $c = \bar{x}$:

$$\sum_{i=1}^{n} \bar{x} = n\bar{x}$$

Result 2

$$\sum_{i=1}^{n} cx_i = cx_1 + cx_2 + \cdots + cx_n$$
$$= c(x_1 + x_2 + \cdots + x_n) = c\sum_{i=1}^{n} x_i \qquad \text{(C.3)}$$

Example for $x_1 = 5, x_2 = 8, x_3 = 14, c = 2$:

$$\sum_{i=1}^{3} 2x_i = 2\sum_{i=1}^{3} x_i = 2(27) = 54$$

Result 3

$$\sum_{i=1}^{n} (ax_i + by_i) = a\sum_{i=1}^{n} x_i + b\sum_{i=1}^{n} y_i \qquad \text{(C.4)}$$

Example for $x_1 = 5$, $x_2 = 8$, $x_3 = 14$, $a = 2$, $y_1 = 7$, $y_2 = 3$, $y_3 = 8$, $b = 4$:

$$\sum_{i=1}^{3}(2x_i + 4y_i) = 2\sum_{i=1}^{3}x_i + 4\sum_{i=1}^{3}y_i$$
$$= 2(27) + 4(18)$$
$$= 54 + 72$$
$$= 126$$

Double Summations

Consider the following data involving the variable x_{ij}, where i is the subscript denoting the row position and j is the subscript denoting the column position:

		Column 1	Column 2	Column 3
Row	1	$x_{11} = 10$	$x_{12} = 8$	$x_{13} = 6$
	2	$x_{21} = 7$	$x_{22} = 4$	$x_{23} = 12$

Definition

$$\sum_{i=1}^{n}\sum_{j=1}^{m} x_{ij} = (x_{11} + x_{12} + \cdots + x_{1m}) + (x_{21} + x_{22} + \cdots + x_{2m})$$
$$+ (x_{31} + x_{32} + \cdots + x_{3m}) + \cdots + (x_{n1} + x_{n2} + \cdots + x_{nm}) \quad \text{(C.5)}$$

Example:

$$\sum_{i=1}^{2}\sum_{j=1}^{3} x_{ij} = x_{11} + x_{12} + x_{13} + x_{21} + x_{22} + x_{23}$$
$$= 10 + 8 + 6 + 7 + 4 + 12$$
$$= 47$$

Definition

$$\sum_{i=1}^{n} x_{ij} = x_{1j} + x_{2j} + \cdots + x_{nj} \quad \text{(C.6)}$$

Example:

$$\sum_{i=1}^{2} x_{i2} = x_{12} + x_{22}$$
$$= 8 + 4$$
$$= 12$$

Shorthand Notation

Sometimes when a summation is for all values of the subscript, we use the following shorthand notations:

$$\sum_{i=1}^{n} x_i = \sum_i x_i \quad \text{(C.7)}$$

$$\sum_{i=1}^{n}\sum_{j=1}^{m} x_{ij} = \sum\sum x_{ij} \quad \text{(C.8)}$$

$$\sum_{i=1}^{n} x_{ij} = \sum_i x_{ij} \quad \text{(C.9)}$$

Appendix D: Self-Test Solutions and Answers to Exercises

Chapter 1

1. a. Quantitative
   b. Categorical
   c. Categorical
   d. Quantitative
   e. Categorical

2. a. 10
   b. 5
   c. Categorical variables: Size and Fuel
      Quantitiative variables: Cylinders, City MPG, and Highway MPG
   d.
Variable	Measurement Scale
Size	Ordinal
Cylinders	Ratio
City MPG	Ratio
Highway MPG	Ratio
Fuel	Nominal

3. a. Average for city driving = 182/10 = 18.2 mpg
   b. Average for highway driving = 261/10 = 26.1 mpg
      On average, the miles per gallon for highway driving is 7.9 mpg greater than for city driving
   c. 3 of 10 or 30% have four-cyclinder engines
   d. 6 of 10 or 60% use regular fuel

5. a. 7
   b. 5
   c. Categorical variables: State, Campus Setting, and NCAA Division
   d. Quantitiative variables: Endowment and Applicants Admitted

7. a. Quantitative; ratio
   b. Categorical; nominal
   c. Categorical; ordinal
   d. Quantitative; ratio
   e. Categorical; nominal

10. a. 1015
    b. Categorical
    c. Percentages
    d. .10(1015) = 101.5; 101 or 102 respondents

11. a. All visitors to Hawaii
    b. Yes
    c. First and fourth questions provide quantitative data
       Second and third questions provide categorical data

13. a. Federal spending ($ trillions)
    b. Quantitative
    c. Time series
    d. Federal spending has increased over time

15. a. Graph with a time series line for each manufacturer
    b. Toyota surpasses General Motors in 2006 to become the leading auto manufacturer
    c. A bar chart would show cross-sectional data for 2007; bar heights would be GM 8.8, Ford 7.9, DC 4.6, and Toyota 9.6

19. a. 43% of managers were bullish or very bullish, and 21% of managers expected health care to be the leading industry over the next 12 months
    b. The average 12-month return estimate is 11.2% for the population of investment managers
    c. The sample average of 2.5 years is an estimate of how long the population of investment managers think it will take to resume sustainable growth

20. a. The population consists of all customers of the chain stores in Charlotte, North Carolina
    b. Some of the ways the grocery store chain could use to collect the data are
       - Customers entering or leaving the store could be surveyed
       - A survey could be mailed to customers who have a shopper's club card
       - Customers could be given a printed survey when they check out
       - Customers could be given a coupon that asks them to complete a brief online survey; if they do, they will receive a 5% discount on their next shopping trip

21. a. 36%
    b. 189
    c. Categorical

25. a. Correct
    b. Incorrect
    c. Correct
    d. Incorrect
    e. Incorrect

Chapter 2

2. a. .20
   b. 40
   c/d.

Class	Frequency	Percent Frequency
A	44	22
B	36	18
C	80	40
D	40	20
Total	200	100

3. a. 360° × 58/120 = 174°
b. 360° × 42/120 = 126°
c.

d.

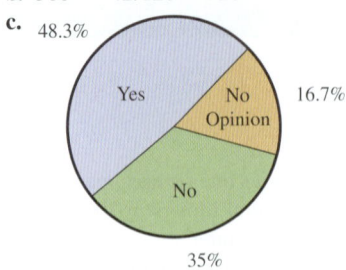

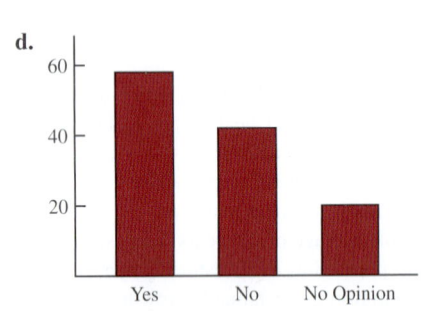

4. a. Categorical
b.

TV Show	Frequency	Percent Frequency
Law & Order	10	20%
CSI	18	36%
Without a Trace	9	18%
Desp Housewives	13	26%
Total:	50	100%

d. *CSI* had the largest viewing audience; *Desperate Housewives* was in second place

6. a.

Network	Frequency	Percent Frequency
ABC	15	30
CBS	17	34
FOX	1	2
NBC	17	34

b. CBS and NBC tied for first; ABC is close with 15

7. a.

Rating	Frequency	Percent Frequency
Excellent	20	40
Very Good	23	46
Good	4	8
Fair	1	2
Poor	2	4
	50	100

Management should be very pleased; 86% of the ratings are Very Good or Excellent

b. Review explanations from the three with Fair or Poor ratings to identify reasons for the low ratings

8. a.

Position	Frequency	Relative Frequency
P	17	.309
H	4	.073
1	5	.091
2	4	.073
3	2	.036
S	5	.091
L	6	.109
C	5	.091
R	7	.127
Totals	55	1.000

b. Pitcher
c. 3rd base
d. Right field
e. Infielders 16 to outfielders 18

10. a/b.

Rating	Frequency	Percent Frequency
Excellent	20	2
Good	101	10
Fair	528	52
Bad	244	24
Terrible	122	12
Total	1015	100

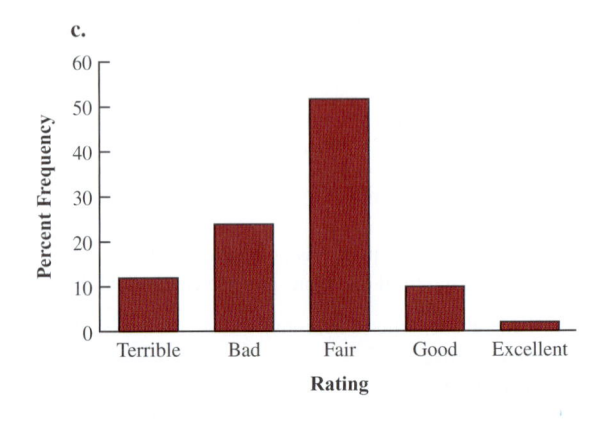

c.

Appendix D Self-Test Solutions and Answers to Exercises   635

d. 36% a bad or a terrible job
   12% a good or excellent job
e. 50% a bad or a terrible job
   4% a good or excellent job
   More pessimism in Spain

12.

Class	Cumulative Frequency	Cumulative Relative Frequency
≤19	10	.20
≤29	24	.48
≤39	41	.82
≤49	48	.96
≤59	50	1.00

14. b/c.

Class	Frequency	Percent Frequency
6.0–7.9	4	20
8.0–9.9	2	10
10.0–11.9	8	40
12.0–13.9	3	15
14.0–15.9	3	15
Totals	20	100

15. a/b.

Waiting Time	Frequency	Relative Frequency
0–4	4	.20
5–9	8	.40
10–14	5	.25
15–19	2	.10
20–24	1	.05
Totals	20	1.00

c/d.

Waiting Time	Cumulative Frequency	Cumulative Relative Frequency
≤4	4	.20
≤9	12	.60
≤14	17	.85
≤19	19	.95
≤24	20	1.00

e. 12/20 = .60

16. a.

Salary	Frequency
150–159	1
160–169	3
170–179	7
180–189	5
190–199	1
200–209	2
210–219	1
Total	20

b.

Salary	Percent Frequency
150–159	5
160–169	15
170–179	35
180–189	25
190–199	5
200–209	10
210–219	5
Total	100

c.

Salary	Cumulative Percent Frequency
Less than or equal to 159	5
Less than or equal to 169	20
Less than or equal to 179	55
Less than or equal to 189	80
Less than or equal to 199	85
Less than or equal to 209	95
Less than or equal to 219	100
Total	100

e. There is skewness to the right
f. 15%

18. a. Lowest $180; highest $2050
b.

Spending	Frequency	Percent Frequency
$0–249	3	12
250–499	6	24
500–749	5	20
750–999	5	20
1000–1249	3	12
1250–1499	1	4
1500–1749	0	0
1750–1999	1	4
2000–2249	1	4
Total	25	100

c. The distribution shows a positive skewness
d. Majority (64%) of consumers spend between $250 and $1000; the middle value is about $750; and two high spenders are above $1750

20. a.

Off-Course Income ($1000s)	Frequency	Percent Frequency
0–4,999	30	60
5,000–9,999	9	18
10,000–14,999	4	8
15,000–19,999	0	0
20,000–24,999	3	6
25,000–29,999	2	4
30,000–34,999	0	0
35,000–39,999	0	0
40,000–44,999	1	2

(*continued*)

Off-Course Income ($1000s)	Frequency	Percent Frequency
45,000–49,999	0	0
Over 50,000	1	2
Total	50	100

c. Off-course income is skewed to the right; only Tiger Woods earns over $50 million

d. The majority (60%) earn less that $5 million; 78% earn less than $10 million; five golfers (10%) earn between $20 million and $30 million; only Tiger Woods and Phil Mickelson earn more than $40 million

22. 5 | 7 8
     6 | 4 5 8
     7 | 0 2 2 5 5 6 8
     8 | 0 2 3 5

23. Leaf unit = .1
     6 | 3
     7 | 5 5 7
     8 | 1 3 4 8
     9 | 3 6
    10 | 0 4 5
    11 | 3

24. Leaf unit = 10
    11 | 6
    12 | 0 2
    13 | 0 6 7
    14 | 2 2 7
    15 | 5
    16 | 0 2 8
    17 | 0 2 3

25. 9 | 8 9
    10 | 2 4 6 6
    11 | 4 5 7 8 8 9
    12 | 2 4 5 7
    13 | 1 2
    14 | 4
    15 | 1

26. a. 6 | 6 7 7
        7 | 2 4 6 7 7 8 9
        8 | 0 0 1 3 7
        9 | 9
       10 | 0 6
       11 | 0
       12 | 1

b. 10 | 0 6 9
   11 | 1 6 9
   12 | 2 5 6
   13 | 0 5 8 8
   14 | 0 6
   15 | 2 5 7
   16 |
   17 |
   18 |
   19 |
   20 |
   21 | 4
   22 | 1

28. a. 2 | 14
        2 | 67
        3 | 011123
        3 | 5677
        4 | 003333344
        4 | 6679
        5 | 00022
        5 | 5679
        6 | 14
        6 | 6
        7 | 2

b. 40–44 with 9
c. 43 with 5
d. 10%; relatively small participation in the race

29. a.

		y		Total
		1	2	
	A	5	0	5
x	B	11	2	13
	C	2	10	12
Total		18	12	30

b.

		y		Total
		1	2	
	A	100.0	0.0	100.0
x	B	84.6	15.4	100.0
	C	16.7	83.3	100.0

c.

		y	
		1	2
	A	27.8	0.0
x	B	61.1	16.7
	C	11.1	83.3
Total		100.0	100.0

d. A values are always in y = 1
B values are most often in y = 1
C values are most often in y = 2

30. a.

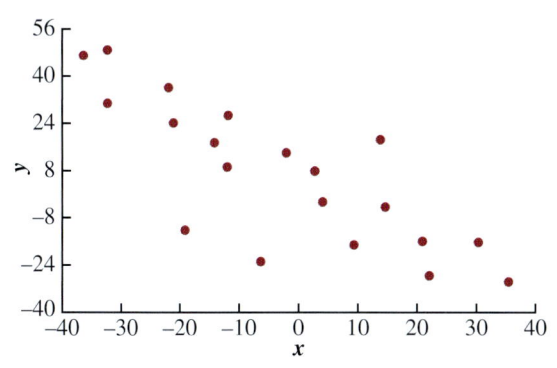

b. A negative relationship between x and y; y decreases as x increases

32. a.

Education Level	Household Income ($1000s)					Total
	Under 25	25.0–49.9	50.0–74.9	75.0–99.9	100 or more	
Not H.S. Graduate	32.10	18.71	9.13	5.26	2.20	13.51
H.S. Graduate	37.52	37.05	33.04	25.73	16.00	29.97
Some College	21.42	28.44	30.74	31.71	24.43	27.21
Bachelor's Degree	6.75	11.33	18.72	25.19	32.26	18.70
Beyond Bach. Deg.	2.21	4.48	8.37	12.11	25.11	10.61
Total	100.00	100.00	100.00	100.00	100.00	100.00

13.51% of the heads of households did not graduate from high school

b. 25.11%, 53.54%
c. Positive relationship between income and education level

34. a.

Fund Type	5-Year Average Return						Total
	0–9.99	10–19.99	20–29.99	30–39.99	40–49.99	50–59.99	
DE	1	25	1	0	0	0	27
FI	9	1	0	0	0	0	10
IE	0	2	3	2	0	1	8
Total	10	28	4	2	0	1	45

b.

5-Year Average Return	Frequency
0–9.99	10
10–19.99	28
20–29.99	4
30–39.99	2
40–49.99	0
50–59.99	1
Total	45

c.

Fund Type	Frequency
DE	27
FI	10
IE	8
Total	45

d. The margin of the crosstabulation shows these frequency distributions
e. Higher returns—International Equity funds
Lower returns—Fixed Income funds

36. b. Higher 5-year returns are associated with higher net asset values

38. a.

Displace	Highway MPG					Total
	15–19	20–24	25–29	30–34	35–39	
1.0–2.9	0	6	72	46	4	128
3.0–4.9	3	56	86	0	0	145
5.0–6.9	23	14	1	0	0	38
Total	26	76	159	46	4	311

b. Higher fuel efficiencies are associated with smaller displacement engines
Lower fuel efficiencies are associated with larger displacement engines
d. Lower fuel efficiencies are associated with larger displacement engines
e. Scatter diagram

40. a.

Division	Frequency	Percent
Buick	10	5
Cadillac	10	5
Chevrolet	122	61
GMC	24	12
Hummer	2	1
Pontiac	18	9
Saab	2	1
Saturn	12	6
Total	200	100

b. Chevrolet, 61%
c. Hummer and Saab, both only 1%
Maintain Chevrolet and GMC

42. a.

SAT Score	Frequency
800–999	1
1000–1199	3
1200–1399	6
1400–1599	10
1600–1799	7
1800–1999	2
2000–2199	1
Total	30

b. Nearly symmetrical
c. 33% of the scores fall between 1400 and 1599
A score below 800 or above 2200 is unusual
The average is near or slightly above 1500

44. a.

Population	Frequency	Percent Frequency
0.0–2.4	17	34
2.5–4.9	12	24
5.0–7.4	9	18
7.5–9.9	4	8
10.0–12.4	3	6
12.5–14.9	1	2
15.0–17.4	1	2
17.5–19.9	1	2
20.0–22.4	0	0
22.5–24.9	1	2
25.0–27.4	0	0
27.5–29.9	0	0
30.0–32.4	0	0
32.5–34.9	0	0
35.0–37.4	1	2
Total	50	100

c. High positive skewness
d. 17 (34%) with population less than 2.5 million
29 (58%) with population less than 5 million
8 (16%) with population greater than 10 million
Largest 35.9 million (California)
Smallest .5 million (Wyoming)

46. a. High Temperatures

```
1 |
2 |
3 | 0
4 | 1 2 2 5
5 | 2 4 5
6 | 0 0 0 1 2 2 5 6 8
7 | 0 7
8 | 4
```

b. Low Temperatures

```
1 | 1
2 | 1 2 6 7 9
3 | 1 5 6 8 9
4 | 0 3 3 6 7
5 | 0 0 4
6 | 5
7 |
8 |
```

c. The most frequent range for high is in 60s (9 of 20) with only one low temperature above 54
High temperatures range mostly from 41 to 68, while low temperatures range mostly from 21 to 47
Low was 11; high was 84

d.

High Temp	Frequency	Low Temp	Frequency
10–19	0	10–19	1
20–29	0	20–29	5
30–39	1	30–39	5
40–49	4	40–49	5
50–59	3	50–59	3
60–69	9	60–69	1
70–79	2	70–79	0
80–89	1	80–89	0
Total	20	Total	20

49. a. Row totals: 247; 54; 82; 121
Column totals: 149; 317; 17; 7; 14

b.

Year	Freq.	Fuel	Freq.
1973 or before	247	Elect.	149
1974–79	54	Nat. Gas	317
1980–86	82	Oil	17
1987–91	121	Propane	7
Total	504	Other	14
		Total	504

c. Crosstabulation of column percentages

Year Constructed	Elect.	Nat. Gas	Oil	Propane	Other
1973 or before	26.9	57.7	70.5	71.4	50.0
1974–1979	16.1	8.2	11.8	28.6	0.0
1980–1986	24.8	12.0	5.9	0.0	42.9
1987–1991	32.2	22.1	11.8	0.0	7.1
Total	100.0	100.0	100.0	100.0	100.0

d. Crosstabulation of row percentages

Year Constructed	Elect.	Nat. Gas	Oil	Propane	Other	Total
1973 or before	16.2	74.1	4.9	2.0	2.8	100.0
1974–1979	44.5	48.1	3.7	3.7	0.0	100.0
1980–1986	45.1	46.4	1.2	0.0	7.3	100.0
1987–1991	39.7	57.8	1.7	0.0	0.8	100.0

50. a.

Level of Support	Percent Frequency
Strongly favor	30.10
Favor more than oppose	34.83
Oppose more than favor	21.13
Strongly oppose	13.94
Total	100.00

Overall favor higher tax = 30.10% + 34.83%
= 64.93%

b. 20.2, 19.5, 20.6, 20.7, 19.0
Roughly 20% per country

c. The crosstabulation with column percentages:

	Country				
Support	Great Britain	Italy	Spain	Germany	United States
Strongly favor	31.00	31.96	45.99	19.98	20.98
Favor more than oppose	34.04	39.04	32.01	36.99	32.06
Oppose more than favor	23.00	17.99	13.98	24.03	26.96
Strongly oppose	11.96	11.01	8.03	18.99	20.00
Total	100.00	100.00	100.00	100.00	100.00

Considering the percentage of respondents who favor the higher tax by either saying "strongly favor" or "favor more than oppose," 65.04%, 71.00%, 78.00%, 56.97%, and 53.04% for the five countries; all show more than 50% support, but all European countries show more support for the tax than the United States; Italy and Spain show the highest level of support

53. b. A positive relationship is demonstrated between market value and stockholders' equity

54. a. Crosstabulation of market value and profit

	Profit ($1000s)				
Market Value ($1000s)	0–300	300–600	600–900	900–1200	Total
0–8000	23	4			27
8000–16,000	4	4	2	2	12
16,000–24,000		2	1	1	4
24,000–32,000		1	2	1	4
32,000–40,000			2	1	3
Total	27	13	6	4	50

b. Crosstabulation of row percentages

	Profit ($1000s)				
Market Value ($1000s)	0–300	300–600	600–900	900–1200	Total
0–8000	85.19	14.81	0.00	0.00	100
8000–16,000	33.33	33.33	16.67	16.67	100
16,000–24,000	0.00	50.00	25.00	25.00	100
24,000–32,000	0.00	25.00	50.00	25.00	100
32,000–40,000	0.00	66.67	33.33	0.00	100

c. A positive relationship is indicated between profit and market value; as profit goes up, market value goes up

Chapter 3

2. 16, 16.5

3. Arrange data in order: 15, 20, 25, 25, 27, 28, 30, 34

$i = \dfrac{20}{100}(8) = 1.6$; round up to position 2

20th percentile = 20

$i = \dfrac{25}{100}(8) = 2$; use positions 2 and 3

25th percentile = $\dfrac{20 + 25}{2} = 22.5$

$i = \dfrac{65}{100}(8) = 5.2$; round up to position 6

65th percentile = 28

$i = \dfrac{75}{100}(8) = 6$; use positions 6 and 7

75th percentile = $\dfrac{28 + 30}{2} = 29$

4. 59.73, 57, 53

6. a. 18.42
b. 6.32
c. 34.3%
d. Reductions of only .65 shots and .9% made shots per game Yes, agree but not dramatically

8. a. $\bar{x} = \dfrac{\Sigma x_i}{n} = \dfrac{3200}{20} = 160$

Order the data from low 100 to high 360

Median: $i = \left(\dfrac{50}{100}\right)20 = 10$; use 10th and 11th positions

Median = $\left(\dfrac{130 + 140}{2}\right) = 135$

Mode = 120 (occur 3 times)

b. $i = \left(\dfrac{25}{100}\right)20 = 5$; use 5th and 6th positions

$Q_1 = \left(\dfrac{115 + 115}{2}\right) = 115$

$i = \left(\dfrac{75}{100}\right)20 = 15$; use 15th and 16th positions

$Q_3 = \left(\dfrac{180 + 195}{2}\right) = 187.5$

c. $i = \left(\dfrac{90}{100}\right)20 = 18$; use 18th and 19th positions

90th percentile = $\left(\dfrac{235 + 255}{2}\right) = 245$

90% of the tax returns cost $245 or less

10. a. .4%, 3.5%
b. 2.3%, 2.5%, 2.7%
c. 2.0%, 2.8%
d. optimistic

12. Disney: 3321, 255.5, 253, 169, 325
Pixar: 3231, 538.5, 505, 363, 631
Pixar films generate approximately twice as much box office revenue per film

14. 16, 4

15. Range = 34 − 15 = 19
Arrange data in order: 15, 20, 25, 25, 27, 28, 30, 34

$i = \dfrac{25}{100}(8) = 2$; $Q_1 = \dfrac{20 + 25}{2} = 22.5$

$i = \frac{75}{100}(8) = 6; Q_3 = \frac{28 + 30}{2} = 29$

IQR $= Q_3 - Q_1 = 29 - 22.5 = 6.5$

$\bar{x} = \frac{\Sigma x_i}{n} = \frac{204}{8} = 25.5$

x_i	$(x_i - \bar{x})$	$(x_i - \bar{x})^2$
27	1.5	2.25
25	−.5	.25
20	−5.5	30.25
15	−10.5	110.25
30	4.5	20.25
34	8.5	72.25
28	2.5	6.25
25	−.5	.25
		242.00

$s^2 = \frac{\Sigma(x_i - \bar{x})^2}{n - 1} = \frac{242}{8 - 1} = 34.57$

$s = \sqrt{34.57} = 5.88$

16. a. Range = 190 − 168 = 22

b. $\bar{x} = \frac{\Sigma x_i}{n} = \frac{1068}{6} = 178$

$s^2 = \frac{\Sigma(x_i - \bar{x})^2}{n - 1}$

$= \frac{4^2 + (-10)^2 + 6^2 + 12^2 + (-8)^2 + (-4)^2}{6 - 1}$

$= \frac{376}{5} = 75.2$

c. $s = \sqrt{75.2} = 8.67$

d. $\frac{s}{\bar{x}}(100) = \frac{8.67}{178}(100\%) = 4.87\%$

18. a. 38, 97, 9.85
b. Eastern shows more variation

20. Dawson: range = 2, $s = .67$
     Clark: range = 8, $s = 2.58$

22. a. 1285, 433
     Freshmen spend more
b. 1720, 352
c. 404, 131.5
d. 367.04, 96.96
e. Freshmen have more variability

24. Quarter-milers: $s = .0564$, Coef. of Var. = 5.8%
     Milers: $s = .1295$, Coef. of Var. = 2.9%

26. .20, 1.50, 0, −.50, −2.20

27. Chebyshev's theorem: at least $(1 - 1/z^2)$

a. $z = \frac{40 - 30}{5} = 2; 1 - \frac{1}{(2)^2} = .75$

b. $z = \frac{45 - 30}{5} = 3; 1 - \frac{1}{(3)^2} = .89$

c. $z = \frac{38 - 30}{5} = 1.6; 1 - \frac{1}{(1.6)^2} = .61$

d. $z = \frac{42 - 30}{5} = 2.4; 1 - \frac{1}{(2.4)^2} = .83$

e. $z = \frac{48 - 30}{5} = 3.6; 1 - \frac{1}{(3.6)^2} = .92$

28. a. 95%
b. Almost all
c. 68%

29. a. $z = 2$ standard deviations

$1 - \frac{1}{z^2} = 1 - \frac{1}{2^2} = \frac{3}{4}$; at least 75%

b. $z = 2.5$ standard deviations

$1 - \frac{1}{z^2} = 1 - \frac{1}{2.5^2} = .84$; at least 84%

c. $z = 2$ standard deviations
Empirical rule: 95%

30. a. 68%
b. 81.5%
c. 2.5%

32. a. −.67
b. 1.50
c. Neither an outlier
d. Yes; $z = 8.25$

34. a. 76.5, 7
b. 16%, 2.5%
c. 12.2, 7.89; no

36. 15, 22.5, 26, 29, 34

38. Arrange data in order: 5, 6, 8, 10, 10, 12, 15, 16, 18

$i = \frac{25}{100}(9) = 2.25$; round up to position 3

$Q_1 = 8$
Median (5th position) = 10

$i = \frac{75}{100}(9) = 6.75$; round up to position 7

$Q_3 = 15$
5-number summary: 5, 8, 10, 15, 18

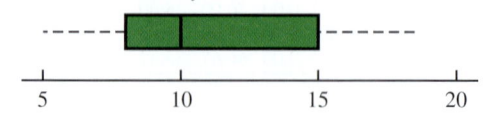

40. a. Subway 29,612 locations
b. Order data smallest to largest
     Median (middle value) = 5889
c. Smallest = 1397; Largest = 29,612

$i = \frac{25}{100}(13) = 3.25$ Round to 4th position

$Q_1 = 4516$

$i = \frac{75}{100}(13) = 9.75$ Round to 10th position

$Q_3 = 10{,}238$
5-number summary: 1397 4516 5889 10,238 29,612

Appendix D Self-Test Solutions and Answers to Exercises   641

   d. IQR = $Q_3 - Q_1$ = 10,238 − 4516 = 5722
      $Q_1 - 1.5(IQR)$ = 4516 − 1.5(5722) = −4067
      $Q_3 + 1.5(IQR)$ = 10,238 + 1.5(5722) = 18,821
      Subway and McDonald's are outliers
   e.

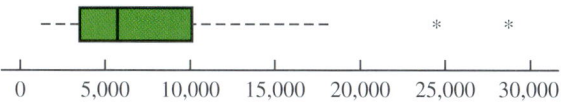

   [box plot from 0 to 30,000 with two outliers marked *]

42. a. 73.5
    b. 68, 71.5, 73.5, 74.5, 77
    c. Limits: 67 and 79; no outliers
    d. 66, 68, 71, 73, 75; 60.5 and 80.5
       63, 65, 66, 67.6, 69; 61.25 and 71.25
       75, 77, 78.5, 79.5, 81; 73.25 and 83.25
       No outliers for any of the services
    e. Verizon is highest rated
       Sprint is lowest rated

44. a. 18.2, 15.35
    b. 11.7, 23.5
    c. 3.4, 11.7, 15.35, 23.5, 41.3
    d. Yes; Alger Small Cap 41.3

45. b. There appears to be a negative linear relationship between x and y
    c.

x_i	y_i	$x_i - \bar{x}$	$y_i - \bar{y}$	$(x_i - \bar{x})(y_i - \bar{y})$
4	50	−4	4	−16
6	50	−2	4	−8
11	40	3	−6	−18
3	60	−5	14	−70
16	30	8	−16	−128
40	230	0	0	−240

    $\bar{x} = 8$; $\bar{y} = 46$

    $$s_{xy} = \frac{\Sigma(x_i - \bar{x})(y_i - \bar{y})}{n - 1} = \frac{-240}{4} = -60$$

    The sample covariance indicates a negative linear association between x and y

    d. $r_{xy} = \dfrac{s_{xy}}{s_x s_y} = \dfrac{-60}{(5.43)(11.40)} = -.969$

    The sample correlation coefficient of −.969 is indicative of a strong negative linear relationship

46. b. There appears to be a positive linear relationship between x and y
    c. $s_{xy} = 26.5$
    d. $r_{xy} = .693$

48. −.91; negative relationship

50. b. .91
    c. Strong positive linear relationship; no

52. a. 3.69
    b. 3.175

53. a.

f_i	M_i	$f_i M_i$
4	5	20
7	10	70
9	15	135
5	20	100
25		325

    $$\bar{x} = \frac{\Sigma f_i M_i}{n} = \frac{325}{25} = 13$$

   b.

f_i	M_i	$(M_i - \bar{x})$	$(M_i - \bar{x})^2$	$f_i(M_i - \bar{x})^2$
4	5	−8	64	256
7	10	−3	9	63
9	15	2	4	36
5	20	7	49	245
25				600

    $$s^2 = \frac{\Sigma f_i(M_i - \bar{x})^2}{n - 1} = \frac{600}{25 - 1} = 25$$
    $$s = \sqrt{25} = 5$$

54. a.

Grade x_i	Weight w_i
4 (A)	9
3 (B)	15
2 (C)	33
1 (D)	3
0 (F)	0
	60 credit hours

    $$\bar{x} = \frac{\Sigma w_i x_i}{\Sigma w_i} = \frac{9(4) + 15(3) + 33(2) + 3(1)}{9 + 15 + 33 + 3}$$
    $$= \frac{150}{60} = 2.5$$

   b. Yes

56. 3.8, 3.7

58. a. 1800, 1351
    b. 387, 1710
    c. 7280, 1323
    d. 3,675,303, 1917
    e. High positive skewness
    f. Using a box plot: 4135 and 7450 are outliers

61. a. 2.3, 1.85
    b. 1.90, 1.38
    c. Altria Group 5%
    d. −.51, below mean
    e. 1.02, above mean
    f. No

62. a. $670
    b. $456
    c. $z = 3$; yes
    d. Save time and prevent a penalty cost

64. a. 215.9
    b. 55%
    c. 175.0, 628.3
    d. 48.8, 175.0, 215.9, 628.3, 2325.0
    e. Yes, any price over 1308.25
    f. 482.1; prefer median

66. a. 364 rooms
    b. $457
    c. −.293; slight negative correlation
       Higher cost per night tends to be associated with smaller hotels

68. a. .268, low or weak positive correlation
    b. Very poor predictor; spring training is practice and does not count toward standings or playoffs

69. a. 60.68
    b. $s^2 = 31.23$; $s = 5.59$

Chapter 4

2. $\binom{6}{3} = \frac{6!}{3!3!} = \frac{6 \cdot 5 \cdot 4 \cdot 3 \cdot 2 \cdot 1}{(3 \cdot 2 \cdot 1)(3 \cdot 2 \cdot 1)} = 20$

   ABC ACE BCD BEF
   ABD ACF BCE CDE
   ABE ADE BCF CDF
   ABF ADF BDE CEF
   ACD AEF BDF DEF

4. b. (H,H,H), (H,H,T), (H,T,H), (H,T,T), (T,H,H), (T,H,T), (T,T,H), (T,T,T)
   c. 1/8

6. $P(E_1) = .40$, $P(E_2) = .26$, $P(E_3) = .34$
   The relative frequency method was used

8. a. 4: Commission Positive—Council Approves
       Commission Positive—Council Disapproves
       Commission Negative—Council Approves
       Commission Negative—Council Disapproves

9. $\binom{50}{4} = \frac{50!}{4!46!} = \frac{50 \cdot 49 \cdot 48 \cdot 47}{4 \cdot 3 \cdot 2 \cdot 1} = 230{,}300$

10. a. Using the table, $P(\text{Debt}) = .94$
    b. Five of the 8 institutions, $P(\text{over } 60\%) = 5/8 = .625$
    c. Two of the 8 institutions, $P(\text{more than } \$30{,}000) = 2/8 = .25$
    d. $P(\text{No debt}) = 1 - P(\text{debt}) = 1 - .72 = .28$
    e. A weighted average with 72% having average debt of $32,980 and 28% having no debt

    $$\text{Average debt per graduate} = \frac{.72(\$32{,}980) + .28(\$0)}{.72 + .28} = \$23{,}746$$

12. a. 3,478,761
    b. 1/3,478,761
    c. 1/146,107,962

14. a. ¼
    b. ½
    c. ¾

15. a. $S = \{\text{ace of clubs, ace of diamonds, ace of hearts, ace of spades}\}$
    b. $S = \{2 \text{ of clubs}, 3 \text{ of clubs}, \ldots, 10 \text{ of clubs}, J \text{ of clubs}, Q \text{ of clubs}, K \text{ of clubs}, A \text{ of clubs}\}$
    c. There are 12; jack, queen, or king in each of the four suits
    d. For (a): $4/52 = 1/13 = .08$
       For (b): $13/52 = 1/4 = .25$
       For (c): $12/52 = .23$

16. a. 36
    c. 1/6
    d. 5/18
    e. No; $P(\text{odd}) = P(\text{even}) = ½$
    f. Classical

17. a. (4, 6), (4, 7), (4, 8)
    b. $.05 + .10 + .15 = .30$
    c. (2, 8), (3, 8), (4, 8)
    d. $.05 + .05 + .15 = .25$
    e. .15

18. a. .0222
    b. .8226
    c. .1048

20. a. .108
    b. .096
    c. .434

22. a. .40, .40, .60
    b. .80, yes
    c. $A^c = \{E_3, E_4, E_5\}$; $C^c = \{E_1, E_4\}$;
       $P(A^c) = .60$; $P(C^c) = .40$
    d. (E_1, E_2, E_5); .60
    e. .80

23. a. $P(A) = P(E_1) + P(E_4) + P(E_6)$
           $= .05 + .25 + .10 = .40$
       $P(B) = P(E_2) + P(E_4) + P(E_7)$
           $= .20 + .25 + .05 = .50$
       $P(C) = P(E_2) + P(E_3) + P(E_5) + P(E_7)$
           $= .20 + .20 + .15 + .05 = .60$
    b. $A \cup B = \{E_1, E_2, E_4, E_6, E_7\}$;
       $P(A \cup B) = P(E_1) + P(E_2) + P(E_4) + P(E_6) + P(E_7)$
           $= .05 + .20 + .25 + .10 + .05$
           $= .65$
    c. $A \cap B = \{E_4\}$; $P(A \cap B) = P(E_4) = .25$
    d. Yes, they are mutually exclusive
    e. $B^c = \{E_1, E_3, E_5, E_6\}$;
       $P(B^c) = P(E_1) + P(E_3) + P(E_5) + P(E_6)$
           $= .05 + .20 + .15 + .10$
           $= .50$

24. a. .05
    b. .70

26. a. .64
    b. .48
    c. .36
    d. .76

Appendix D Self-Test Solutions and Answers to Exercises

28. Let B = rented a car for business reasons
P = rented a car for personal reasons
a. $P(B \cup P) = P(B) + P(P) - P(B \cap P)$
$= .540 + .458 - .300$
$= .698$
b. $P(\text{Neither}) = 1 - .698 = .302$

30. a. $P(A \mid B) = \dfrac{P(A \cap B)}{P(B)} = \dfrac{.40}{.60} = .6667$
b. $P(B \mid A) = \dfrac{P(A \cap B)}{P(A)} = \dfrac{.40}{.50} = .80$
c. No, because $P(A \mid B) \neq P(A)$

32. a.

	Car	Light Truck	Total
U.S.	.1330	.2939	.4269
Non-U.S.	.3478	.2253	.5731
Total	.4808	.5192	1.0000

b. .4269, .5731 Non-U.S. higher
.4808, .5192 Light Truck slightly higher
c. .3115, .6885 Light Truck higher
d. .6909, .3931 Car higher
e. .5661, U.S. higher for Light Trucks

33. a.

	Reason for Applying			
	Quality	Cost/Convenience	Other	Total
Full-time	.218	.204	.039	.461
Part-time	.208	.307	.024	.539
Total	.426	.511	.063	1.000

b. A student is most likely to cite cost or convenience as the first reason (probability = .511); school quality is the reason cited by the second largest number of students (probability = .426)
c. $P(\text{quality} \mid \text{full-time}) = .218/.461 = .473$
d. $P(\text{quality} \mid \text{part-time}) = .208/.539 = .386$
e. For independence, we must have $P(A)P(B) = P(A \cap B)$; from the table
$P(A \cap B) = .218, P(A) = .461, P(B) = .426$
$P(A)P(B) = (.461)(.426) = .196$
Because $P(A)P(B) \neq P(A \cap B)$, the events are not independent

34. a.

	On Time	Late	Total
Southwest	.3336	.0664	.40
US Airways	.2629	.0871	.35
JetBlue	.1753	.0747	.25
Total	.7718	.2282	1.00

b. Southwest (.40)
c. .7718
d. US Airways (.3817); Southwest (.2910)

36. a. .7921
b. .9879
c. .0121
d. .3364, .8236, .1764
Don't foul Jerry Stackhouse

38. a.

Grade	Met Proficiency Standards		Total
	Yes	No	
3	.11196	.05663	0.16858
4	.08271	.08205	0.16476
5	.08517	.07922	0.16439
6	.08588	.07777	0.16365
7	.09671	.07031	0.16702
8	.09618	.07542	0.17159
Total	.55861	.44139	1.00000

b. The column marginal probabilities are .55861 and .441369. The row marginal probabilities are .16858, .16476, .16439, .16365, .16702, and .17159
c. .6641, .5020
d. .2004, .1481

39. a. Yes, because $P(A_1 \cap A_2) = 0$
b. $P(A_1 \cap B) = P(A_1)P(B \mid A_1) = .40(.20) = .08$
$P(A_2 \cap B) = P(A_2)P(B \mid A_2) = .60(.05) = .03$
c. $P(B) = P(A_1 \cap B) + P(A_2 \cap B) = .08 + .03 = .11$
d. $P(A_1 \mid B) = \dfrac{.08}{.11} = .7273$
$P(A_2 \mid B) = \dfrac{.03}{.11} = .2727$

40. a. .10, .20, .09
b. .51
c. .26, .51, .23

42. M = missed payment
D_1 = customer defaults
D_2 = customer does not default
$P(D_1) = .05, P(D_2) = .95, P(M \mid D_2) = .2, P(M \mid D_1) = 1$
a. $P(D_1 \mid M) = \dfrac{P(D_1)P(M \mid D_1)}{P(D_1)P(M \mid D_1) + P(D_2)P(M \mid D_2)}$
$= \dfrac{(.05)(1)}{(.05)(1) + (.95)(.2)}$
$= \dfrac{.05}{.24} = .21$
b. Yes, the probability of default is greater than .20

44. a. .47, .53, .50, .45
b. .4963
c. .4463
d. 47%, 53%

47. a. .40
b. .67

48. b. .2022
c. .4618
d. .4005

51. a.

	Young Adult	Older Adult	Total
Blogger	.0432	.0368	.08
Nonblogger	.2208	.6992	.92
Total	.2640	.7360	1.00

b. .2640
c. .0432
d. .1636

52. a. .76
b. .24

55. a. .25
b. .125
c. .0125
d. .10
e. No

56. a. .2675
b. .3376
c. No
d. .5159
e. .0255, yes

59. a. .49
b. .44
c. .54
d. No
e. Yes

60. a. .60
b. .26
c. .40
d. .74

Chapter 5

1. a. Head, Head (H, H)
Head, Tail (H, T)
Tail, Head (T, H)
Tail, Tail (T, T)
b. x = number of heads on two coin tosses
c.

Outcome	Values of x
(H, H)	2
(H, T)	1
(T, H)	1
(T, T)	0

d. Discrete; 0, 1, and 2

2. a. x = time in minutes to assemble product
b. Any positive value: $x > 0$
c. Continuous

3. Let Y = position is offered
N = position is not offered
a. $S = \{(Y, Y, Y), (Y, Y, N), (Y, N, Y), (Y, N, N), (N, Y, Y), (N, Y, N), (N, N, Y), (N, N, N)\}$
b. Let N = number of offers made; N is a discrete random variable
c.

Experimental Outcome	(Y,Y,Y)	(Y,Y,N)	(Y,N,Y)	(Y,N,N)	(N,Y,Y)	(N,Y,N)	(N,N,Y)	(N,N,N)
Value of N	3	2	2	1	2	1	1	0

4. $x = 0, 1, 2, \ldots, 9$

6. a. 0, 1, 2, . . . , 20; discrete
b. 0, 1, 2, . . . ; discrete
c. 0, 1, 2, . . . , 50; discrete
d. $0 \leq x \leq 8$; continuous
e. $x > 0$; continuous

7. a. $f(x) \geq 0$ for all values of x
$\Sigma f(x) = 1$; therefore, it is a valid probability distribution
b. Probability $x = 30$ is $f(30) = .25$
c. Probability $x \leq 25$ is $f(20) + f(25) = .20 + .15 = .35$
d. Probability $x > 30$ is $f(35) = .40$

8. a.

x	$f(x)$
1	3/20 = .15
2	5/20 = .25
3	8/20 = .40
4	4/20 = .20
Total	1.00

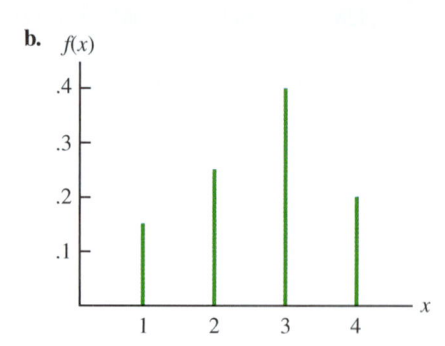

b.

c. $f(x) \geq 0$ for $x = 1, 2, 3, 4$
$\Sigma f(x) = 1$

10. a.

x	1	2	3	4	5
$f(x)$	.05	.09	.03	.42	.41

b.

x	1	2	3	4	5
$f(x)$	.04	.10	.12	.46	.28

c. .83
d. .28
e. Senior executives are more satisfied

12. a. Yes
  b. .15
  c. .10

14. a. .05
  b. .70
  c. .40

16. a.

y	f(y)	yf(y)
2	.20	.4
4	.30	1.2
7	.40	2.8
8	.10	.8
Totals	1.00	5.2

$E(y) = \mu = 5.2$

b.

y	$y - \mu$	$(y - \mu)^2$	f(y)	$(y - \mu)^2 f(y)$
2	-3.20	10.24	.20	2.048
4	-1.20	1.44	.30	.432
7	1.80	3.24	.40	1.296
8	2.80	7.84	.10	.784
			Total	4.560

Var(y) = 4.56
$\sigma = \sqrt{4.56} = 2.14$

18. a/b.

x	f(x)	xf(x)	$x - \mu$	$(x - \mu)^2$	$(x - \mu)^2 f(x)$
0	0.04	0.00	-1.84	3.39	0.12
1	0.34	0.34	-0.84	0.71	0.24
2	0.41	0.82	0.16	0.02	0.01
3	0.18	0.53	1.16	1.34	0.24
4	0.04	0.15	2.16	4.66	0.17
Total	1.00	1.84			0.79
		↑			↑
		E(x)			Var(x)

c/d.

y	f(y)	yf(y)	$y - \mu$	$(y - \mu)^2$	$y - \mu^2 f(y)$
0	0.00	0.00	-2.93	8.58	0.01
1	0.03	0.03	-1.93	3.72	0.12
2	0.23	0.45	-0.93	0.86	0.20
3	0.52	1.55	0.07	0.01	0.00
4	0.22	0.90	1.07	1.15	0.26
Total	1.00	2.93			0.59
		↑			↑
		E(y)			Var(y)

e. The number of bedrooms in owner-occupied houses is greater than in renter-occupied houses; the expected number of bedrooms is 2.93 − 1.84 = 1.09 greater, and the variability in the number of bedrooms is less for the owner-occupied houses

20. a. 430
  b. −90; concern is to protect against the expense of a large loss

22. a. 445
  b. $1250 loss

24. a. Medium: 145; large: 140
  b. Medium: 2725; large: 12,400

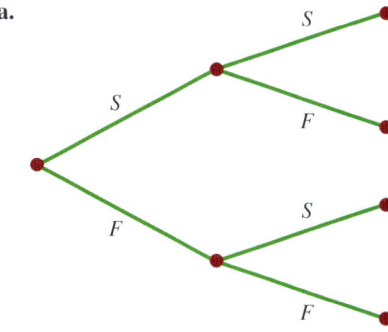

25. a.

b. $f(1) = \binom{2}{1}(.4)^1(.6)^1 = \dfrac{2!}{1!1!}(.4)(.6) = .48$

c. $f(0) = \binom{2}{0}(.4)^0(.6)^2 = \dfrac{2!}{0!2!}(1)(.36) = .36$

d. $f(2) = \binom{2}{2}(.4)^2(.6)^0 = \dfrac{2!}{2!0!}(.16)(.1) = .16$

e. $P(x \geq 1) = f(1) + f(2) = .48 + .16 = .64$
f. $E(x) = np = 2(.4) = .8$
  $Var(x) = np(1-p) = 2(.4)(.6) = .48$
  $\sigma = \sqrt{.48} = .6928$

26. a. .3487
  b. .1937
  c. .9298
  d. .6513
  e. 1
  f. .9, .95

28. a. .2789
  b. .4181
  c. .0733

30. a. Probability of a defective part being produced must be .03 for each part selected; parts must be selected independently
  b. Let D = defective
    G = not defective

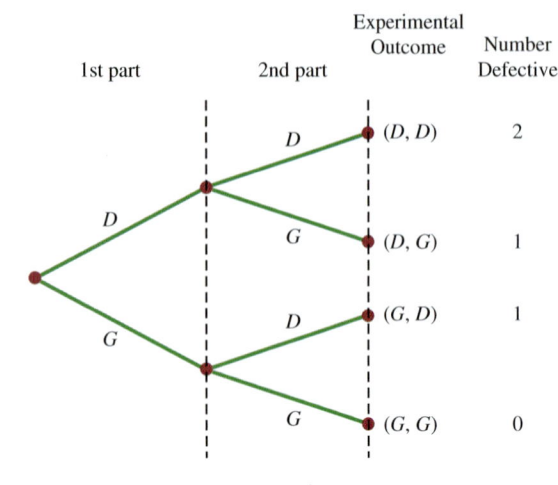

c. Two outcomes result in exactly one defect
d. $P(\text{no defects}) = (.97)(.97) = .9409$
   $P(1 \text{ defect}) = 2(.03)(.97) = .0582$
   $P(2 \text{ defects}) = (.03)(.03) = .0009$

32. a. .90
    b. .99
    c. .999
    d. Yes

34. a. .2262
    b. .8355

36. a. .1897
    b. .9757
    c. $f(12) = .0008$; yes
    d. 5

38. a. $f(x) = \dfrac{3^x e^{-3}}{x!}$
    b. .2241
    c. .1494
    d. .8008

39. a. $f(x) = \dfrac{2^x e^{-2}}{x!}$
    b. $\mu = 6$ for 3 time periods
    c. $f(x) = \dfrac{6^x e^{-6}}{x!}$
    d. $f(2) = \dfrac{2^2 e^{-2}}{2!} = \dfrac{4(.1353)}{2} = .2706$
    e. $f(6) = \dfrac{6^6 e^{-6}}{6!} = .1606$
    f. $f(5) = \dfrac{4^5 e^{-4}}{5!} = .1563$

40. a. .1952
    b. .1048
    c. .0183
    d. .0907

42. a. $f(0) = \dfrac{7^0 e^{-7}}{0!} = e^{-7} = .0009$
    b. probability $= 1 - [f(0) + f(1)]$
       $f(1) = \dfrac{7^1 e^{-7}}{1!} = 7e^{-7} = .0064$
       probability $= 1 - [.0009 + .0064] = .9927$
    c. $\mu = 3.5$
       $f(0) = \dfrac{3.5^0 e^{-3.5}}{0!} = e^{-3.5} = .0302$
       probability $= 1 - f(0) = 1 - .0302 = .9698$
    d.
       probability $= 1 - [f(0) + f(1) + f(2) + f(3) + f(4)]$
       $= 1 - [.0009 + .0064 + .0223 + .0521 + .0912]$
       $= .8271$

44. a. $\mu = 1.25$
    b. .2865
    c. .3581
    d. .3554

46. a. $f(1) = \dfrac{\binom{3}{1}\binom{10-3}{4-1}}{\binom{10}{4}} = \dfrac{\left(\dfrac{3!}{1!2!}\right)\left(\dfrac{7!}{3!4!}\right)}{\dfrac{10!}{4!6!}}$
       $= \dfrac{(3)(35)}{210} = .50$
    b. $f(2) = \dfrac{\binom{3}{2}\binom{10-3}{2-2}}{\binom{10}{2}} = \dfrac{(3)(1)}{45} = .067$
    c. $f(0) = \dfrac{\binom{3}{0}\binom{10-3}{2-0}}{\binom{10}{2}} = \dfrac{(1)(21)}{45} = .4667$
    d. $f(2) = \dfrac{\binom{3}{2}\binom{10-3}{4-2}}{\binom{10}{4}} = \dfrac{(3)(21)}{210} = .30$
    e. $x = 4$ is *greater than* $r = 3$; thus, $f(4) = 0$

48. a. .5250
    b. .8167

50. $N = 60, n = 10$
    a. $r = 20, x = 0$
       $f(0) = \dfrac{\binom{20}{0}\binom{40}{10}}{\binom{60}{10}} = \dfrac{(1)\left(\dfrac{40!}{10!30!}\right)}{\dfrac{60!}{10!50!}}$
       $= \left(\dfrac{40!}{10!30!}\right)\left(\dfrac{10!50!}{60!}\right)$
       $= \dfrac{40 \cdot 39 \cdot 38 \cdot 37 \cdot 36 \cdot 35 \cdot 34 \cdot 33 \cdot 32 \cdot 31}{60 \cdot 59 \cdot 58 \cdot 57 \cdot 56 \cdot 55 \cdot 54 \cdot 53 \cdot 52 \cdot 51}$
       $= .0112$

b. $r = 20, x = 1$

$$f(1) = \frac{\binom{20}{1}\binom{40}{9}}{\binom{60}{10}} = 20\left(\frac{40!}{9!31!}\right)\left(\frac{10!50!}{60!}\right)$$

$$= .0725$$

c. $1 - f(0) - f(1) = 1 - .0112 - .0725 = .9163$

d. Same as the probability that one will be from Hawaii; .0725

52. a. .2917
b. .0083
c. .5250, .1750; 1 bank
d. .7083
e. .90, .49, .70

53. a. 240
b. 12.96
c. 12.96

56. a.

x	1	2	3	4	5
$f(x)$	.24	.21	.10	.21	.24

b. 3.00, 2.34
c. Bonds: $E(x) = 1.36$, $\text{Var}(x) = .23$
Stocks: $E(x) = 4$, $\text{Var}(x) = 1$

57. .1912

60. a. .0596
b. .3585
c. 100
d. 95, 9.75

61. a. .2240
b. .5767

64. a. .9510
b. .0480
c. .0490

65. a. .4667
b. .4667
c. .0667

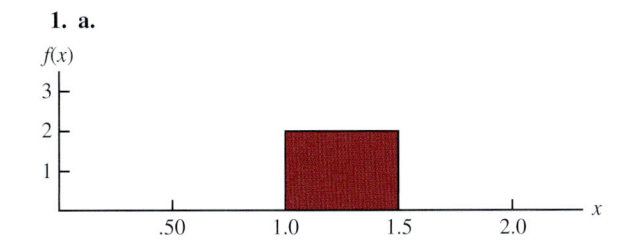

Chapter 6

1. a.

[Graph showing $f(x)$ rectangle of height 2 from $x = .50$ to $x = 1.5$, with x-axis marks at .50, 1.0, 1.5, 2.0]

b. $P(x = 1.25) = 0$; the probability of any single point is zero because the area under the curve above any single point is zero
c. $P(1.0 \leq x \leq 1.25) = 2(.25) = .50$
d. $P(1.20 < x < 1.5) = 2(.30) = .60$

2. b. .50
c. .60
d. 15
e. 8.33

4. a.

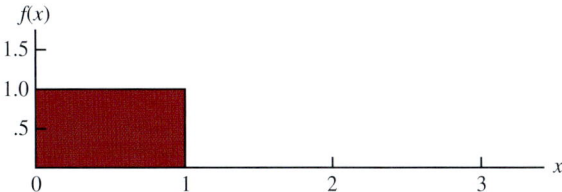

[Graph showing $f(x)$ rectangle of height 1.0 from $x = 0$ to $x = 1$, with x-axis marks at 0, 1, 2, 3 and y-axis marks at .5, 1.0, 1.5]

b. $P(.25 < x < .75) = 1(.50) = .50$
c. $P(x \leq .30) = 1(.30) = .30$
d. $P(x > .60) = 1(.40) = .40$

6. a. .125
b. .50
c. .25

10. a. .9332
b. .8413
c. .0919
d. .4938

12. a. .2967
b. .4418
c. .3300
d. .5910
e. .8849
f. .2389

13. a. $P(-1.98 \leq z \leq .49) = P(z \leq .49) - P(z < -1.98)$
$= .6879 - .0239 = .6640$
b. $P(.52 \leq z \leq 1.22) = P(z \leq 1.22) - P(z < .52)$
$= .8888 - .6985 = .1903$
c. $P(-1.75 \leq z \leq -1.04) = P(z \leq -1.04) - P(z < -1.75) = .1492 - .0401 = .1091$

14. a. $z = 1.96$
b. $z = 1.96$
c. $z = .61$
d. $z = 1.12$
e. $z = .44$
f. $z = .44$

15. a. The z value corresponding to a cumulative probability of .2119 is $z = -.80$
b. Compute $.9030/2 = .4515$; the cumulative probability of $.5000 + .4515 = .9515$ corresponds to $z = 1.66$
c. Compute $.2052/2 = .1026$; z corresponds to a cumulative probability of $.5000 + .1026 = .6026$, so $z = .26$
d. The z value corresponding to a cumulative probability of .9948 is $z = 2.56$
e. The area to the left of z is $1 - .6915 = .3085$, so $z = -.50$

16. a. $z = 2.33$
b. $z = 1.96$
c. $z = 1.645$
d. $z = 1.28$

18. $\mu = 30$ and $\sigma = 8.2$

a. At $x = 40$, $z = \dfrac{40 - 30}{8.2} = 1.22$

$P(z \leq 1.22) = .8888$
$P(x \geq 40) = 1.000 - .8888 = .1112$

b. At $x = 20$, $z = \dfrac{20 - 30}{8.2} = -1.22$

$P(z \leq -1.22) = .1112$
$P(x \leq 20) = .1112$

c. A z value of 1.28 cuts off an area of approximately 10% in the upper tail
$x = 30 + 8.2(1.28)$
$= 40.50$
A stock price of $40.50 or higher will put a company in the top 10%

20. a. .0885
b. 12.51%
c. 93.8 hours or more

22. a. .6553
b. 13.05 hours
c. .9838

24. a. 200, 26.04
b. .2206
c. .1251
d. 242.84 million

26. a. $\mu = np = 100(.20) = 20$
$\sigma^2 = np(1 - p) = 100(.20)(.80) = 16$
$\sigma = \sqrt{16} = 4$

b. Yes, because $np = 20$ and $n(1 - p) = 80$

c. $P(23.5 \leq x \leq 24.5)$

$z = \dfrac{24.5 - 20}{4} = 1.13 \quad P(z \leq 1.13) = .8708$

$z = \dfrac{23.5 - 20}{4} = .88 \quad P(z \leq .88) = .8106$

$P(23.5 \leq x \leq 24.5) = P(.88 \leq z \leq 1.13)$
$= .8708 - .8106 = .0602$

d. $P(17.5 \leq x \leq 22.5)$

$z = \dfrac{22.5 - 20}{4} = .63 \quad P(z \leq .63) = .7357$

$z = \dfrac{17.5 - 20}{4} = -.63 \quad P(z \leq -.63) = .2643$

$P(17.5 \leq x \leq 22.5) = P(-.63 \leq z \leq .63)$
$= .7357 - .2643 = .4714$

e. $P(x \leq 15.5)$

$z = \dfrac{15.5 - 20}{4} = -1.13 \quad P(z \leq -1.13) = .1292$

$P(x \leq 15.5) = P(z \leq -1.13) = .1292$

28. a. $\mu = np = 250(.20) = 50$
b. $\sigma^2 = np(1 - p) = 250(.20)(1 - 20) = 40$
$\sigma = \sqrt{40} = 6.3246$
$P(x < 40) = P(x \leq 39.5)$

$z = \dfrac{x - \mu}{\sigma} = \dfrac{39.5 - 50}{6.3246} = -1.66 \quad \text{Area} = .0485$

$P(x \leq 39.5) = .0485$

c. $P(55 \leq x \leq 60) = P(54.5 \leq x \leq 60.5)$

$z = \dfrac{x - \mu}{\sigma} = \dfrac{54.5 - 50}{6.3246} = .71 \quad \text{Area} = .7611$

$z = \dfrac{x - \mu}{\sigma} = \dfrac{60.5 - 50}{6.3246} = 1.66 \quad \text{Area} = .9515$

$P(54.5 \leq x \leq 60.5) = .9515 - .7611 = .1904$

d. $P(x \geq 70) = P(x \geq 69.5)$

$z = \dfrac{x - \mu}{\sigma} = \dfrac{69.5 - 50}{6.3246} = 3.08 \quad \text{Area} = .9990$

$P(x \geq 69.5) = 1 - .9990 = .0010$

30. a. 220
b. .0392
c. .8962

32. a. .5276
b. .3935
c. .4724
d. .1341

33. a. $P(x \leq x_0) = 1 - e^{-x_0/3}$
b. $P(x \leq 2) = 1 - e^{-2/3} = 1 - .5134 = .4866$
c. $P(x \geq 3) = 1 - P(x \leq 3) = 1 - (1 - e^{-3/3})$
$= e^{-1} = .3679$
d. $P(x \leq 5) = 1 - e^{-5/3} = 1 - .1889 = .8111$
e. $P(2 \leq x \leq 5) = P(x \leq 5) - P(x \leq 2)$
$= .8111 - .4866 = .3245$

34. a. .5624
b. .1915
c. .2461
d. .2259

35. a.

b. $P(x \leq 12) = 1 - e^{-12/12} = 1 - .3679 = .6321$
c. $P(x \leq 6) = 1 - e^{-6/12} = 1 - .6065 = .3935$
d. $P(x \geq 30) = 1 - P(x < 30)$
$= 1 - (1 - e^{-30/12})$
$= .0821$

36. a. .3935
b. .2386
c. .1353

38. a. $f(x) = 5.5e^{-5.5x}$
  b. .2528
  c. .6002

39. a. 2 minutes
  b. .2212
  c. .3935
  d. .0821

42. a. $3780 or less
  b. 19.22%
  c. $8167.50

43. a. 1/7 minute
  b. $7e^{-7x}$
  c. .0009
  d. .2466

46. a. 3229
  b. .2244
  c. $12,383 or more

47. a. Lose $240
  b. .1788
  c. .3557
  d. .0594

49. $\mu = 19.23$ ounces

50. a. .0228
  b. $50

54. a. 38.3%
  b. 3.59% better, 96.41% worse
  c. 38.21%

Chapter 7

1. a. AB, AC, AD, AE, BC, BD, BE, CD, CE, DE
  b. With 10 samples, each has a 1/10 probability
  c. E and C because 8 and 0 do not apply; 5 identifies E; 7 does not apply; 5 is skipped because E is already in the sample; 3 identifies C; 2 is not needed because the sample of size 2 is complete

2. 22, 147, 229, 289

3. 459, 147, 385, 113, 340, 401, 215, 2, 33, 348

4. a. Bell South, LSI Logic, General Electric
  b. 120

6. 2782, 493, 825, 1807, 289

8. ExxonMobil, Chevron, Travelers, Microsoft, Pfizer, and Intel

10. a. finite; **b.** infinite; **c.** infinite; **d.** finite; **e.** infinite

11. a. $\bar{x} = \dfrac{\Sigma x_i}{n} = \dfrac{54}{6} = 9$

  b. $s = \sqrt{\dfrac{\Sigma(x_i - \bar{x})^2}{n - 1}}$

  $\Sigma(x_i - \bar{x})^2 = (-4)^2 + (-1)^2 + 1^2 + (-2)^2 + 1^2 + 5^2 = 48$

  $s = \sqrt{\dfrac{48}{6 - 1}} = 3.1$

12. a. .50
  b. .3667

13. a. $\bar{x} = \dfrac{\Sigma x_i}{n} = \dfrac{465}{5} = 93$

  b.

x_i	$(x_i - \bar{x})$	$(x_i - \bar{x})^2$
94	+1	1
100	+7	49
85	−8	64
94	+1	1
92	−1	1
Totals 465	0	116

  $s = \sqrt{\dfrac{\Sigma(x_i - \bar{x})^2}{n - 1}} = \sqrt{\dfrac{116}{4}} = 5.39$

14. a. .45
  b. .15
  c. .45

16. a. .10
  b. 20
  c. .72

18. a. 200
  b. 5
  c. Normal with $E(\bar{x}) = 200$ and $\sigma_{\bar{x}} = 5$
  d. The probability distribution of $\bar{x}$

19. a. The sampling distribution is normal with
  $E(\bar{x}) = \mu = 200$
  $\sigma_{\bar{x}} = \sigma/\sqrt{n} = 50/\sqrt{100} = 5$
  For ±5, $195 \leq \bar{x} \leq 205$
  Using the standard normal probability table:
  At $\bar{x} = 205$, $z = \dfrac{\bar{x} - \mu}{\sigma_{\bar{x}}} = \dfrac{5}{5} = 1$
  $P(z \leq 1) = .8413$
  At $\bar{x} = 195$, $z = \dfrac{\bar{x} - \mu}{\sigma_{\bar{x}}} = \dfrac{-5}{5} = -1$
  $P(z < -1) = .1587$
  $P(195 \leq \bar{x} \leq 205) = .8413 - .1587 = .6826$

  b. For ±10, $190 \leq \bar{x} \leq 210$
  Using the standard normal probability table:
  At $\bar{x} = 210$, $z = \dfrac{\bar{x} - \mu}{\sigma_{\bar{x}}} = \dfrac{10}{5} = 2$
  $P(z \leq 2) = .9772$
  At $\bar{x} = 190$, $z = \dfrac{\bar{x} - \mu}{\sigma_{\bar{x}}} = \dfrac{-10}{5} = -2$
  $P(z < -2) = .0228$
  $P(190 \leq \bar{x} \leq 210) = .9722 - .0228 = .9544$

20. 3.54, 2.50, 2.04, 1.77
$\sigma_{\bar{x}}$ decreases as n increases

22. a. Normal with $E(\bar{x}) = 51{,}800$ and $\sigma_{\bar{x}} = 516.40$
  b. $\sigma_{\bar{x}}$ decreases to 365.15
  c. $\sigma_{\bar{x}}$ decreases as n increases

23. a.

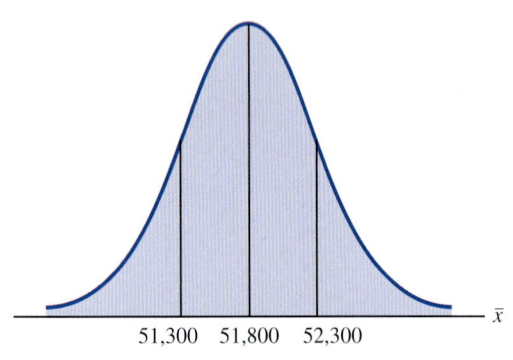

$$\sigma_{\bar{x}} = \frac{\sigma}{\sqrt{n}} = \frac{4000}{\sqrt{60}} = 516.40$$

At $\bar{x} = 52{,}300$, $z = \dfrac{52{,}300 - 51{,}800}{516.40} = .97$

$P(\bar{x} \leq 52{,}300) = P(z \leq .97) = .8340$

At $\bar{x} = 51{,}300$, $z = \dfrac{51{,}300 - 51{,}800}{516.40} = -.97$

$P(\bar{x} < 51{,}300) = P(z < -.97) = .1660$

$P(51{,}300 \leq \bar{x} \leq 52{,}300) = .8340 - .1660 = .6680$

b. $\sigma_{\bar{x}} = \dfrac{\sigma}{\sqrt{n}} = \dfrac{4000}{\sqrt{120}} = 365.15$

At $\bar{x} = 52{,}300$, $z = \dfrac{52{,}300 - 51{,}800}{365.15} = 1.37$

$P(\bar{x} \leq 52{,}300) = P(z \leq 1.37) = .9147$

At $\bar{x} = 51{,}300$, $z = \dfrac{51{,}300 - 51{,}800}{365.15} = -1.37$

$P(\bar{x} < 51{,}300) = P(z < -1.37) = .0853$

$P(51{,}300 \leq \bar{x} \leq 52{,}300) = .9147 - .0853 = .8294$

24. a. Normal with $E(\bar{x}) = 17.5$ and $\sigma_{\bar{x}} = .57$
  b. .9198
  c. .6212

26. a. .4246, .5284, .6922, .9586
  b. Higher probability the sample mean will be close to population mean

28. a. Normal with $E(\bar{x}) = 95$ and $\sigma_{\bar{x}} = 2.56$
  b. .7580
  c. .8502
  d. Part (c), larger sample size

30. a. $n/N = .01$; no
  b. 1.29, 1.30; little difference
  c. .8764

32. a. $E(\bar{p}) = .40$

$$\sigma_{\bar{p}} = \sqrt{\frac{p(1-p)}{n}} = \sqrt{\frac{(.40)(.60)}{200}} = .0346$$

Within $\pm .03$ means $.37 \leq \bar{p} \leq .43$

$$z = \frac{\bar{p} - p}{\sigma_{\bar{p}}} = \frac{.03}{.0346} = .87$$

$P(.37 \leq \bar{p} \leq .43) = P(-.87 \leq z \leq .87)$
$\qquad = .8078 - .1922$
$\qquad = .6156$

b. $z = \dfrac{\bar{p} - p}{\sigma_{\bar{p}}} = \dfrac{.05}{.0346} = 1.44$

$P(.35 \leq \bar{p} \leq .45) = P(-1.44 \leq z \leq 1.44)$
$\qquad = .9251 - .0749$
$\qquad = .8502$

34. a. .6156
  b. .7814
  c. .9488
  d. .9942
  e. Higher probability with larger n

35. a.

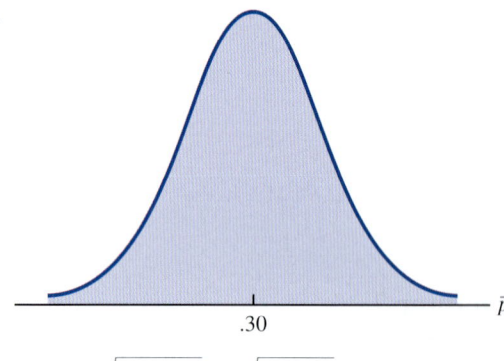

$$\sigma_{\bar{p}} = \sqrt{\frac{p(1-p)}{n}} = \sqrt{\frac{.30(.70)}{100}} = .0458$$

The normal distribution is appropriate because $np = 100(.30) = 30$ and $n(1-p) = 100(.70) = 70$ are both greater than 5.

b. $P(.20 \leq \bar{p} \leq .40) = ?$

$$z = \frac{.40 - .30}{.0458} = 2.18$$

$P(.20 \leq \bar{p} \leq .40) = P(-2.18 \leq z \leq 2.18)$
$\qquad = .9854 - .0146$
$\qquad = .9708$

c. $P(.25 \leq \bar{p} \leq .35) = ?$

$$z = \frac{.35 - .30}{.0458} = 1.09$$

$P(.25 \leq \bar{p} \leq .35) = P(-1.09 \leq z \leq 1.09)$
$\qquad = .8621 - .1379$
$\qquad = .7242$

36. a. Normal with $E(\bar{p}) = .66$ and $\sigma_{\bar{p}} = .0273$
    b. .8584
    c. .9606
    d. Yes, standard error is smaller in part (c)
    e. .9616, the probability is larger because the increased sample size reduces the standard error
38. a. Normal with $E(\bar{p}) = .56$ and $\sigma_{\bar{p}} = .0248$
    b. .5820
    c. .8926
40. a. Normal with $E(\bar{p}) = .76$ and $\sigma_{\bar{p}} = .0214$
    b. .8384
    c. .9452
42. a. 48
    b. Normal, $E(\bar{p}) = .25$, $\sigma_{\bar{p}} = .0625$
    c. .2119
44. a. .8882
    b. .0233
46. a. Normal with $E(\bar{p}) = .28$ and $\sigma_{\bar{p}} = .0290$
    b. .8324
    c. .5098
47. a. 625
    b. .7888
49. a. 955
    b. .50
    c. .7062
    d. .8230
51. a. Normal with $E(\bar{x}) = 115.50$ and $\sigma_{\bar{x}} = 5.53$
    b. .9298
    c. $z = -2.80$, .0026
53. 122, 99, 25, 55, 115, 102, 61

Chapter 8

2. Use $\bar{x} \pm z_{\alpha/2}(\sigma/\sqrt{n})$
   a. $32 \pm 1.645(6/\sqrt{50})$
      32 ± 1.4; 30.6 to 33.4
   b. $32 \pm 1.96(6/\sqrt{50})$
      32 ± 1.66; 30.34 to 33.66
   c. $32 \pm 2.576(6/\sqrt{50})$
      32 ± 2.19; 29.81 to 34.19
4. 54
5. a. $1.96\sigma/\sqrt{n} = 1.96(5/\sqrt{49}) = 1.40$
   b. 24.80 ± 1.40; 23.40 to 26.20
6. 8.1 to 8.9
8. a. Population is at least approximately normal
   b. 3.1
   c. 4.1
10. a. $113,638 to $124,672
    b. $112,581 to $125,729
    c. $110,515 to $127,795
    d. Width increases as confidence level increases

12. a. 2.179
    b. -1.676
    c. 2.457
    d. -1.708 and 1.708
    e. -2.014 and 2.014
13. a. $\bar{x} = \dfrac{\Sigma x_i}{n} = \dfrac{80}{8} = 10$
    b. $s = \sqrt{\dfrac{\Sigma(x_i - \bar{x})^2}{n-1}} = \sqrt{\dfrac{84}{7}} = 3.464$
    c. $t_{.025}\left(\dfrac{s}{\sqrt{n}}\right) = 2.365\left(\dfrac{3.46}{\sqrt{8}}\right) = 2.9$
    d. $\bar{x} \pm t_{.025}\left(\dfrac{s}{\sqrt{n}}\right)$
       10 ± 2.9 (7.1 to 12.9)
14. a. 21.5 to 23.5
    b. 21.3 to 23.7
    c. 20.9 to 24.1
    d. A larger margin of error and a wider interval
15. $\bar{x} \pm t_{\alpha/2}(s/\sqrt{n})$
    90% confidence: $df = 64$ and $t_{.05} = 1.669$
    $19.5 \pm 1.669\left(\dfrac{5.2}{\sqrt{65}}\right)$
    19.5 ± 1.08 or (18.42 to 20.58)
    95% confidence: $df = 64$ and $t_{.025} = 1.998$
    $19.5 \pm 1.998\left(\dfrac{5.2}{\sqrt{65}}\right)$
    19.5 ± 1.29 or (18.21 to 20.79)
16. a. 1.69
    b. 47.31 to 50.69
    c. Fewer hours and higher cost for United
18. a. 22 weeks
    b. 3.8020
    c. 18.20 to 25.80
    d. Larger n next time
20. $\bar{x} = 22$; 21.48 to 22.52
22. a. $9,269 to $12,541
    b. 1523
    c. 4,748,714, $34 million
24. a. Planning value of $\sigma = \dfrac{\text{Range}}{4} = \dfrac{36}{4} = 9$
    b. $n = \dfrac{z_{.025}^2 \sigma^2}{E^2} = \dfrac{(1.96)^2(9)^2}{(3)^2} = 34.57$; use $n = 35$
    c. $n = \dfrac{(1.96)^2(9)^2}{(2)^2} = 77.79$; use $n = 78$
25. a. Use $n = \dfrac{z_{\alpha/2}^2 \sigma^2}{E^2}$
       $n = \dfrac{(1.96)^2(6.84)^2}{(1.5)^2} = 79.88$; use $n = 80$
    b. $n = \dfrac{(1.645)^2(6.84)^2}{(2)^2} = 31.65$; use $n = 32$

26. a. 18
    b. 35
    c. 97
28. a. 328
    b. 465
    c. 803
    d. n gets larger; no to 99% confidence
30. 81
31. a. $\bar{p} = \dfrac{100}{400} = .25$
    b. $\sqrt{\dfrac{\bar{p}(1-\bar{p})}{n}} = \sqrt{\dfrac{.25(.75)}{400}} = .0217$
    c. $\bar{p} \pm z_{.025}\sqrt{\dfrac{\bar{p}(1-\bar{p})}{n}}$
       $.25 \pm 1.96(.0217)$
       $.25 \pm .0424$; .2076 to .2924
32. a. .6733 to .7267
    b. .6682 to .7318
34. 1068
35. a. $\bar{p} = \dfrac{1760}{2000} = .88$
    b. Margin of error
       $z_{.05}\sqrt{\dfrac{\bar{p}(1-\bar{p})}{n}} = 1.645\sqrt{\dfrac{.88(1-.88)}{2000}} = .0120$
    c. Confidence interval
       $.88 \pm .0120$ or .868 to .892
    d. Margin of error
       $z_{.05}\sqrt{\dfrac{\bar{p}(1-\bar{p})}{n}} = 1.96\sqrt{\dfrac{.88(1-.88)}{2000}} = .0142$
       95% confidence interval
       $.88 \pm .0142$ or .8658 to .8942
36. a. .23
    b. .1716 to .2884
38. a. .1790
    b. .0738, .5682 to .7158
    c. 354
39. a. $n = \dfrac{z_{.025}^2 p^*(1-p^*)}{E^2} = \dfrac{(1.96)^2(.156)(1-.156)}{(.03)^2}$
       $= 562$
    b. $n = \dfrac{z_{.005}^2 p^*(1-p^*)}{E^2} = \dfrac{(2.576)^2(.156)(1-.156)}{(.03)^2}$
       $= 970.77$; use 971
40. .0346 (.4854 to .5546)
42. a. .0442
    b. 601, 1068, 2401, 9604
45. a. 122
    b. $1751 to $1995
    c. $172, 316 million
    d. Less than $1873
46. a. 4.00
    b. $29.77 to $37.77

48. a. 14 minutes
    b. 13.38 to 14.62
    c. 32 per day
    d. Staff reduction
50. a. .3101
    b. .2898 to .3304
    c. 8219; no, this sample size is unnecessarily large
51. a. 1267
    b. 1509
55. a. .8273
    b. .7957 to .8589
56. 176
57. a. .5420
    b. .0508
    c. .4912 to .5928
59. 37

Chapter 9

2. a. $H_0: \mu \leq 14$
      $H_a: \mu > 14$
   b. No evidence that the new plan increases sales
   c. The research hypothesis $\mu > 14$ is supported; the new plan increases sales
4. a. $H_0: \mu \geq 220$
      $H_a: \mu < 220$
5. a. Rejecting $H_0: \mu \leq 56.2$ when it is true
   b. Accepting $H_0: \mu \leq 56.2$ when it is false
6. a. $H_0: \mu \leq 1$
      $H_a: \mu > 1$
   b. Claiming $\mu > 1$ when it is not true
   c. Claiming $\mu \leq 1$ when it is not true
8. a. $H_0: \mu \geq 220$
      $H_a: \mu < 220$
   b. Claiming $\mu < 220$ when it is not true
   c. Claiming $\mu \geq 220$ when it is not true
10. a. $z = \dfrac{\bar{x} - \mu_0}{\sigma/\sqrt{n}} = \dfrac{26.4 - 25}{6/\sqrt{40}} = 1.48$
    b. Using normal table with $z = 1.48$: p-value = $1.0000 - .9306 = .0694$
    c. p-value $> .01$, do not reject H_0
    d. Reject H_0 if $z \geq 2.33$
       $1.48 < 2.33$, do not reject H_0
11. a. $z = \dfrac{\bar{x} - \mu_0}{\sigma/\sqrt{n}} = \dfrac{14.15 - 15}{3/\sqrt{50}} = -2.00$
    b. p-value = $2(.0228) = .0456$
    c. p-value $\leq .05$, reject H_0
    d. Reject H_0 if $z \leq -1.96$ or $z \geq 1.96$
       $-2.00 \leq -1.96$, reject H_0
12. a. .1056; do not reject H_0
    b. .0062; reject H_0
    c. ≈ 0; reject H_0
    d. .7967; do not reject H_0

14. a. .3844; do not reject H_0
   b. .0074; reject H_0
   c. .0836; do not reject H_0

15. a. $H_0: \mu \geq 1056$
    $H_a: \mu < 1056$
   b. $z = \dfrac{\bar{x} - \mu_0}{\sigma/\sqrt{n}} = \dfrac{910 - 1056}{1600/\sqrt{400}} = -1.83$
   p-value = .0336
   c. p-value $\leq .05$, reject H_0; the mean refund of "last-minute" filers is less than $1056
   d. Reject H_0 if $z \leq -1.645$
   $-1.83 \leq -1.645$; reject H_0

16. a. $H_0: \mu \leq 3173$
    $H_a: \mu > 3173$
   b. .0207
   c. Reject H_0, conclude mean credit card balance for undergraduate student has increased

18. a. $H_0: \mu = 4.1$
    $H_a: \mu \neq 4.1$
   b. -2.21, .0272
   c. Reject H_0; return for Mid-Cap Growth Funds differs from that for U.S. Diversified Funds

20. a. $H_0: \mu \geq 32.79$
    $H_a: \mu < 32.79$
   b. -2.73
   c. .0032
   d. Reject H_0; conclude the mean monthly Internet bill is less in the southern state

22. a. $H_0: \mu = 8$
    $H_a: \mu \neq 8$
   b. .1706
   c. Do not reject H_0; we cannot conclude the mean waiting time differs from 8 minutes
   d. 7.83 to 8.97; yes

24. a. $t = \dfrac{\bar{x} - \mu_0}{s/\sqrt{n}} = \dfrac{17 - 18}{4.5/\sqrt{48}} = -1.54$
   b. Degrees of freedom = $n - 1 = 47$
   Area in lower tail is between .05 and .10
   p-value (two-tail) is between .10 and .20
   Exact p-value = .1303
   c. p-value $> .05$; do not reject H_0
   d. With $df = 47$, $t_{.025} = 2.012$
   Reject H_0 if $t \leq -2.012$ or $t \geq 2.012$
   $t = -1.54$; do not reject H_0

26. a. Between .02 and .05; exact p-value = .0397; reject H_0
   b. Between .01 and .02; exact p-value = .0125; reject H_0
   c. Between .10 and .20; exact p-value = .1285; do not reject H_0

27. a. $H_0: \mu \geq 238$
    $H_a: \mu < 238$
   b. $t = \dfrac{\bar{x} - \mu_0}{s/\sqrt{n}} = \dfrac{231 - 238}{80/\sqrt{100}} = -.88$
   Degrees of freedom = $n - 1 = 99$
   p-value is between .10 and .20
   Exact p-value = .1905
   c. p-value $> .05$; do not reject H_0
   Cannot conclude mean weekly benefit in Virginia is less than the national mean
   d. $df = 99$, $t_{.05} = -1.66$
   Reject H_0 if $t \leq -1.66$
   $-.88 > -1.66$; do not reject H_0

28. a. $H_0: \mu \geq 9$
    $H_a: \mu < 9$
   b. Between .005 and .01
   Exact p-value = .0072
   c. Reject H_0; mean tenure of a CEO is less than 9 years

30. a. $H_0: \mu = 600$
    $H_a: \mu \neq 600$
   b. Between .20 and .40
   Exact p-value = .2491
   c. Do not reject H_0; cannot conclude there has been a change in mean CNN viewing audience
   d. A larger sample size

32. a. $H_0: \mu = 10{,}192$
    $H_a: \mu \neq 10{,}192$
   b. Between .02 and .05
   Exact p-value = .0304
   c. Reject H_0; mean price at dealership differs from national mean price

34. a. $H_0: \mu = 2$
    $H_a: \mu \neq 2$
   b. 2.2
   c. .52
   d. Between .20 and .40
   Exact p-value = .2535
   e. Do not reject H_0; no reason to change from 2 hours for cost estimating

36. a. $z = \dfrac{\bar{p} - p_0}{\sqrt{\dfrac{p_0(1 - p_0)}{n}}} = \dfrac{.68 - .75}{\sqrt{\dfrac{.75(1 - .75)}{300}}} = -2.80$
   p-value = .0026
   p-value $\leq .05$; reject H_0
   b. $z = \dfrac{.72 - .75}{\sqrt{\dfrac{.75(1 - .75)}{300}}} = -1.20$
   p-value = .1151
   p-value $> .05$; do not reject H_0
   c. $z = \dfrac{.70 - .75}{\sqrt{\dfrac{.75(1 - .75)}{300}}} = -2.00$
   p-value = .0228
   p-value $\leq .05$; reject H_0
   d. $z = \dfrac{.77 - .75}{\sqrt{\dfrac{.75(1 - .75)}{300}}} = .80$
   p-value = .7881
   p-value $> .05$; do not reject H_0

38. a. $H_0: p = .64$
$H_a: p \neq .64$

b. $\bar{p} = 52/100 = .52$

$$z = \frac{\bar{p} - p_0}{\sqrt{\frac{p_0(1-p_0)}{n}}} = \frac{.52 - .64}{\sqrt{\frac{.64(1-.64)}{100}}} = -2.50$$

p-value $= 2(.0062) = .0124$

c. p-value $\leq .05$; reject H_0
Proportion differs from the reported .64

d. Yes, because $\bar{p} = .52$ indicates that fewer believe the supermarket brand is as good as the name brand

40. a. 21
b. .0436
c. Conclude the number of business owners providing gifts has decreased, .0436

42. a. $\bar{p} = .15$
b. .0718 to .2282
c. The return rate for the Houston store is different than the national average

44. a. $H_0: p \leq .51$
$H_a: p > .51$

b. $\bar{p} = .58$, p-value $= .0026$
c. Reject H_0; people working the night shift get drowsy more often

47. a. $H_0: \mu \leq 119{,}155$
$H_a: \mu > 119{,}155$

b. .0047
c. Reject H_0; mean annual income for theatergoers in Bay Area is higher

49. $t = 2.26$
p-value between .01 and .025
Exact p-value $= .0155$
Reject H_0; mean cost is greater than \$125,000

50. $t = -1.05$
p-value between .20 and .40
Exact p-value $= .2999$
Do not reject H_0; there is no evidence to conclude that the age at which women had their first child has changed

52. a. $H_0: \mu = 16$
$H_a: \mu \neq 16$
b. .0286; reject H_0
Readjust line
c. .2186; do not reject H_0
Continue operation
d. $z = 2.19$; reject H_0
$z = -1.23$; do not reject H_0
Yes, same conclusion

55. $H_0: p \geq .90$
$H_a: p < .90$
p-value $= .0808$
Do not reject H_0; claim of at least 90% cannot be rejected

57. a. $H_0: p \leq .80$
$H_a: p > .80$
b. .84
c. .0418

d. Reject H_0; more than 80% of customers are satisfied with service of home agents

59. a. Yes
b. Cannot conclude that over 75% of travelers approve; mandatory use is not recommended

Chapter 10

1. a. $\bar{x}_1 - \bar{x}_2 = 13.6 - 11.6 = 2$

b. $z_{\alpha/2} = z_{.05} = 1.645$

$$\bar{x}_1 - \bar{x}_2 \pm 1.645 \sqrt{\frac{\sigma_1^2}{n_1} + \frac{\sigma_2^2}{n_2}}$$

$$2 \pm 1.645 \sqrt{\frac{(2.2)^2}{50} + \frac{(3)^2}{35}}$$

$2 \pm .98 \quad (1.02 \text{ to } 2.98)$

c. $z_{\alpha/2} = z_{.05} = 1.96$

$$2 \pm 1.96 \sqrt{\frac{(2.2)^2}{50} + \frac{(3)^2}{35}}$$

$2 \pm 1.17 \ (.83 \text{ to } 3.17)$

2. a. $z = \dfrac{(\bar{x}_1 - \bar{x}_2) - D_0}{\sqrt{\dfrac{\sigma_1^2}{n_1} + \dfrac{\sigma_2^2}{n_2}}} = \dfrac{(25.2 - 22.8) - 0}{\sqrt{\dfrac{(5.2)^2}{40} + \dfrac{(6)^2}{50}}} = 2.03$

b. p-value $= 1.0000 - .9788 = .0212$
c. p-value $\leq .05$; reject H_0

4. a. $\bar{x}_1 - \bar{x}_2 = 85.36 - 81.40 = 3.96$

b. $z_{.025}\sqrt{\dfrac{\sigma_1^2}{n_1} + \dfrac{\sigma_2^2}{n_2}} = 1.96\sqrt{\dfrac{(4.55)^2}{37} + \dfrac{(3.97)^2}{44}} = 1.88$

c. $3.96 \pm 1.88 \ (2.08 \text{ to } 5.84)$

6. p-value $= .0351$
Reject H_0; mean price in Atlanta lower than mean price in Houston

8. a. Reject H_0; customer service has improved for Rite Aid
b. Do not reject H_0; the difference is not statistically significant
c. p-value $= .0336$; reject H_0; customer service has improved for Expedia
d. 1.80
e. The increase for J.C. Penney is not statistically significant

9. a. $\bar{x}_1 - \bar{x}_2 = 22.5 - 20.1 = 2.4$

b. $df = \dfrac{\left(\dfrac{s_1^2}{n_1} + \dfrac{s_2^2}{n_2}\right)^2}{\dfrac{1}{n_1 - 1}\left(\dfrac{s_1^2}{n_1}\right)^2 + \dfrac{1}{n_2 - 1}\left(\dfrac{s_2^2}{n_2}\right)^2}$

$$= \dfrac{\left(\dfrac{2.5^2}{20} + \dfrac{4.8^2}{30}\right)^2}{\dfrac{1}{19}\left(\dfrac{2.5^2}{20}\right)^2 + \dfrac{1}{29}\left(\dfrac{4.8^2}{30}\right)^2} = 45.8$$

c. $df = 45$, $t_{.025} = 2.014$

$t_{.025}\sqrt{\dfrac{s_1^2}{n_1} + \dfrac{s_2^2}{n_2}} = 2.014\sqrt{\dfrac{2.5^2}{20} + \dfrac{4.8^2}{30}} = 2.1$

d. $2.4 \pm 2.1 \ (.3 \text{ to } 4.5)$

Appendix D   Self-Test Solutions and Answers to Exercises

10. a. $t = \dfrac{(\bar{x}_1 - \bar{x}_2) - 0}{\sqrt{\dfrac{s_1^2}{n_1} + \dfrac{s_2^2}{n_2}}} = \dfrac{(13.6 - 10.1) - 0}{\sqrt{\dfrac{5.2^2}{35} + \dfrac{8.5^2}{40}}} = 2.18$

b. $df = \dfrac{\left(\dfrac{s_1^2}{n_1} + \dfrac{s_2^2}{n_2}\right)^2}{\dfrac{1}{n_1 - 1}\left(\dfrac{s_1^2}{n_1}\right)^2 + \dfrac{1}{n_2 - 1}\left(\dfrac{s_2^2}{n_2}\right)^2}$

$= \dfrac{\left(\dfrac{5.2^2}{35} + \dfrac{8.5^2}{40}\right)^2}{\dfrac{1}{34}\left(\dfrac{5.2^2}{35}\right)^2 + \dfrac{1}{39}\left(\dfrac{8.5^2}{40}\right)^2} = 65.7$

Use $df = 65$

c. $df = 65$, area in tail is between .01 and .025; two-tailed p-value is between .02 and .05
Exact p-value = .0329

d. p-value $\leq$ .05; reject H_0

12. a. $\bar{x}_1 - \bar{x}_2 = 22.5 - 18.6 = 3.9$ miles

b. $df = \dfrac{\left(\dfrac{s_1^2}{n_1} + \dfrac{s_2^2}{n_2}\right)^2}{\dfrac{1}{n_1 - 1}\left(\dfrac{s_1^2}{n_1}\right)^2 + \dfrac{1}{n_2 - 1}\left(\dfrac{s_2^2}{n_2}\right)^2}$

$= \dfrac{\left(\dfrac{8.4^2}{50} + \dfrac{7.4^2}{40}\right)^2}{\dfrac{1}{49}\left(\dfrac{8.4^2}{50}\right)^2 + \dfrac{1}{39}\left(\dfrac{7.4^2}{40}\right)^2} = 87.1$

Use $df = 87$, $t_{.025} = 1.988$

$3.9 \pm 1.988\sqrt{\dfrac{8.4^2}{50} + \dfrac{7.4^2}{40}}$

3.9 ± 3.3 (.6 to 7.2)

14. a. $H_0: \mu_1 - \mu_2 \geq 0$
$H_a: \mu_1 - \mu_2 < 0$

b. -2.41

c. Using t table, p-value is between .005 and .01
Exact p-value = .009

d. Reject H_0; nursing salaries are lower in Tampa

16. a. $H_0: \mu_1 - \mu_2 \leq 0$
$H_a: \mu_1 - \mu_2 > 0$

b. 38

c. $t = 1.80$, $df = 25$
Using t table, p-value is between .025 and .05
Exact p-value = .0420

d. Reject H_0; conclude higher mean score if college grad

18. a. $H_0: \mu_1 - \mu_2 \geq 120$
$H_a: \mu_1 - \mu_2 < 120$

b. -2.10
Using t table, p-value is between .01 and .025
Exact p-value = .0195

c. 32 to 118

d. Larger sample size

19. a. 1, 2, 0, 0, 2

b. $\bar{d} = \Sigma d_i/n = 5/5 = 1$

c. $s_d = \sqrt{\dfrac{\Sigma(d_i - \bar{d})^2}{n - 1}} = \sqrt{\dfrac{4}{5 - 1}} = 1$

d. $t = \dfrac{\bar{d} - \mu}{s_d/\sqrt{n}} = \dfrac{1 - 0}{1/\sqrt{5}} = 2.24$

$df = n - 1 = 4$
Using t table, p-value is between .025 and .05
Exact p-value = .0443
p-value $\leq$ .05; reject H_0

20. a. 3, -1, 3, 5, 3, 0, 1

b. 2

c. 2.08

d. 2

e. .07 to 3.93

21. $H_0: \mu_d \leq 0$
$H_a: \mu_d > 0$
$\bar{d} = .625$
$s_d = 1.30$
$t = \dfrac{\bar{d} - \mu_d}{s_d/\sqrt{n}} = \dfrac{.625 - 0}{1.30/\sqrt{8}} = 1.36$

$df = n - 1 = 7$
Using t table, p-value is between .10 and .20
Exact p-value = .1080
p-value $>$ .05; do not reject H_0; cannot conclude commercial improves mean potential to purchase

22. $.10 to $.32; earnings have increased

24. $t = 1.32$
Using t table, p-value is greater than .10
Exact p-value = .1142
Do not reject H_0; cannot conclude airfares from Dayton are higher

26. a. $t = -1.42$
Using t table, p-value is between .10 and .20
Exact p-value = .1718
Do not reject H_0; no difference in mean scores

b. -1.05

c. 1.28; yes

27. a. $\bar{\bar{x}} = (156 + 142 + 134)/3 = 144$

$\text{SSTR} = \sum_{j=1}^{k} n_j(\bar{x}_j - \bar{\bar{x}})^2$

$= 6(156 - 144)^2 + 6(142 - 144)^2 + 6(134 - 144)^2$
$= 1488$

b. $\text{MSTR} = \dfrac{\text{SSTR}}{k - 1} = \dfrac{1488}{2} = 744$

c. $s_1^2 = 164.4$, $s_2^2 = 131.2$, $s_3^2 = 110.4$

$\text{SSE} = \sum_{j=1}^{k} (n_j - 1)s_j^2$

$= 5(164.4) + 5(131.2) + 5(110.4)$
$= 2030$

d. $\text{MSE} = \dfrac{\text{SSE}}{n_T - k} = \dfrac{2030}{18 - 3} = 135.3$

e.

Source of Variation	Sum of Squares	Degrees of Freedom	Mean Square	F	p-value
Treatments	1488	2	744	5.50	.0162
Error	2030	15	135.3		
Total	3518	17			

f. $F = \dfrac{\text{MSTR}}{\text{MSE}} = \dfrac{744}{135.3} = 5.50$

From the F table (2 numerator degrees of freedom and 15 denominator), p-value is between .01 and .025

Using Excel or Minitab, the p-value corresponding to $F = 5.50$ is .0162

Because p-value $\leq \alpha = .05$, we reject the hypothesis that the means for the three treatments are equal

28.

Source of Variation	Sum of Squares	Degrees of Freedom	Mean Square	F	p-value
Treatments	300	4	75	14.07	.0000
Error	160	30	5.33		
Total	460	34			

30.

Source of Variation	Sum of Squares	Degrees of Freedom	Mean Square	F	p-value
Treatments	150	2	75	4.80	.0233
Error	250	16	15.63		
Total	400	18			

Reject H_0 because p-value $\leq \alpha = .05$

32. Because p-value = .0082 is less than $\alpha = .05$, we reject the null hypothesis that the means of the three treatments are equal

34. $\bar{\bar{x}} = (79 + 74 + 66)/3 = 73$

$\text{SSTR} = \sum_{j=1}^{k} n_j(\bar{x}_j - \bar{\bar{x}})^2 = 6(79 - 73)^2 + 6(74 - 73)^2$
$\qquad\qquad + 6(66 - 73)^2 = 516$

$\text{MSTR} = \dfrac{\text{SSTR}}{k - 1} = \dfrac{516}{2} = 258$

$s_1^2 = 34 \quad s_2^2 = 20 \quad s_3^2 = 32$

$\text{SSE} = \sum_{j=1}^{k}(n_j - 1)s_j^2 = 5(34) + 5(20) + 5(32) = 430$

$\text{MSE} = \dfrac{\text{SSE}}{n_T - k} = \dfrac{430}{18 - 3} = 28.67$

$F = \dfrac{\text{MSTR}}{\text{MSE}} = \dfrac{258}{28.67} = 9.00$

Source of Variation	Sum of Squares	Degrees of Freedom	Mean Square	F	p-value
Treatments	516	2	258	9.00	.003
Error	430	15	28.67		
Total	946	17			

Using F table (2 numerator degrees of freedom and 15 denominator), p-value is less than .01

Using Excel or Minitab, the p-value corresponding to $F = 9.00$ is .003

Because p-value $\leq \alpha = .05$, we reject the null hypothesis that the means for the three plants are equal; in other words, analysis of variance supports the conclusion that the population mean examination scores at the three NCP plants are not equal

36. p-value = .0000
Because p-value $\leq \alpha = .05$, we reject the null hypothesis that the means for the three groups are equal

38. p-value = .0038
Because p-value $\leq \alpha = .05$, we reject the null hypothesis that the mean meal prices are the same for the three types of restaurants

39. $H_0: \mu_1 - \mu_2 = 0$
$H_a: \mu_1 - \mu_2 \neq 0$
$z = 2.79$
p-value = .0052
Reject H_0; the mean checkout times differ

42. a. $H_0: \mu_1 - \mu_2 \leq 0$
$H_a: \mu_1 - \mu_2 > 0$
b. $t = .60 \ df = 57$
Using t table, p-value is greater than .20
Exact p-value = .2754
Do not reject H_0; cannot conclude load mutual funds have a higher return

44. a. A decline of $2.45
b. 2.45 ± 2.15 (.30 to 4.60)
c. 8% decrease
d. $23.93

46. Difference is significant; p-value = .046

48. Not significant; p-value = .2455

Chapter 11

1. a. $\bar{p}_1 - \bar{p}_2 = .48 - .36 = .12$

b. $\bar{p}_1 - \bar{p}_2 \pm z_{.05}\sqrt{\dfrac{\bar{p}_1(1 - \bar{p}_1)}{n_1} + \dfrac{\bar{p}_2(1 - \bar{p}_2)}{n_2}}$

$.12 \pm 1.645\sqrt{\dfrac{.48(1 - .48)}{400} + \dfrac{.36(1 - .36)}{300}}$

$.12 \pm .0614$ (.0586 to .1814)

c. $.12 \pm 1.96\sqrt{\dfrac{.48(1-.48)}{400} + \dfrac{.36(1-.36)}{300}}$

$.12 \pm .0731$ (.0469 to .1931)

2. a. .2333
  b. .1498
  c. Do not reject H_0; cannot conclude population proportions differ

3. a. $\bar{p} = \dfrac{n_1\bar{p}_1 + n_2\bar{p}_2}{n_1 + n_2} = \dfrac{200(.22) + 300(.16)}{200 + 300} = .1840$

$z = \dfrac{\bar{p}_1 - \bar{p}_2}{\sqrt{\bar{p}(1-\bar{p})\left(\dfrac{1}{n_1} + \dfrac{1}{n_2}\right)}}$

$= \dfrac{.22 - .16}{\sqrt{.1840(1-.1840)\left(\dfrac{1}{200} + \dfrac{1}{300}\right)}} = 1.70$

p-value $= 1.0000 - .9554 = .0446$

  b. p-value $\leq .05$; reject H_0; conclude p_1 is greater than p_2

4. a. .64; .58; professional
  b. .06; professional 6% more
  c. .02 to .10
  from 2% to 10% more

6. a. .64
  b. .45
  c. $.19 \pm .0813$ (.1087 to .2713)

8. a. $H_0: p_1 - p_2 = 0$
  $H_a: p_1 - p_2 \neq 0$
  b. .28
  c. .26
  d. .3078
  Do not reject H_0; cannot conclude airports differ in proportion delayed

10. a. $H_0: p_1 - p_2 = 0$
  $H_a: p_1 - p_2 \neq 0$
  b. .13
  c. p-value $= .0404$
  d. Reject H_0; there is a significant difference between the younger and older age groups

11. a. Expected frequencies: $e_1 = 200(.40) = 80$
  $e_2 = 200(.40) = 80$
  $e_3 = 200(.20) = 40$
  Actual frequencies: $f_1 = 60, f_2 = 120, f_3 = 20$

$\chi^2 = \dfrac{(60-80)^2}{80} + \dfrac{(120-80)^2}{80} + \dfrac{(20-40)^2}{40}$

$= \dfrac{400}{80} + \dfrac{1600}{80} + \dfrac{400}{40}$

$= 5 + 20 + 10 = 35$

Degrees of freedom: $k - 1 = 2$
$\chi^2 = 35$ shows p-value is less than .005
p-value $\leq .01$; reject H_0; the proportions are not .40, .40, and .20

  b. Reject H_0 if $\chi^2 \geq 9.210$
  $\chi^2 = 35$; reject H_0

12. $\chi^2 = 15.33, df = 3$
  p-value less than .005
  Reject H_0; the proportions are not all .25

13. $H_0: p_{ABC} = .29, p_{CBS} = .28, p_{NBC} = .25, p_{IND} = .18$
  H_a: The proportions are not
  $p_{ABC} = .29, p_{CBS} = .28, p_{NBC} = .25, p_{IND} = .18$
  Expected frequencies: $300(.29) = 87, 300(.28) = 84$
  $300(.25) = 75, 300(.18) = 54$
  $e_1 = 87, e_2 = 84, e_3 = 75, e_4 = 54$
  Actual frequencies: $f_1 = 95, f_2 = 70, f_3 = 89, f_4 = 46$

$\chi^2 = \dfrac{(95-87)^2}{87} + \dfrac{(70-84)^2}{84} + \dfrac{(89-75)^2}{75}$

$+ \dfrac{(46-54)^2}{54} = 6.87$

Degrees of freedom: $k - 1 = 3$
$\chi^2 = 6.87$, p-value between .05 and .10
Do not reject H_0; cannot conclude that the audience proportions have changed

14. $\chi^2 = 29.51, df = 5$
  p-value is less than .005
  Reject H_0; the percentages differ from those reported by the company

16. a. $\chi^2 = 12.21, df = 3$
  p-value is between .005 and .01
  Conclude difference for 2003
  b. 21%, 30%, 15%, 34%
  Increased use of debit card
  c. 51%

18. $\chi^2 = 16.31, df = 3$
  p-value less than .005
  Reject H_0; ratings differ, with telephone service slightly better

19. H_0: The column variable is independent of the row variable
  H_a: The column variable is not independent of the row variable

Expected frequencies:

	A	B	C
P	28.5	39.9	45.6
Q	21.5	30.1	34.4

$\chi^2 = \dfrac{(20-28.5)^2}{28.5} + \dfrac{(44-39.9)^2}{39.9} + \dfrac{(50-45.6)^2}{45.6}$

$+ \dfrac{(30-21.5)^2}{21.5} + \dfrac{(26-30.1)^2}{30.1} + \dfrac{(30-34.4)^2}{34.4}$

$= 7.86$

Degrees of freedom: $(2-1)(3-1) = 2$
$\chi^2 = 7.86$, p-value between .01 and .025
Reject H_0; column variable and row variable are not independent

20. $\chi^2 = 19.77$, $df = 4$
p-value less than .005
Reject H_0; column variable and row variable are not independent

21. H_0: Type of ticket purchased is independent of the type of flight
H_a: Type of ticket purchased is not independent of the type of flight

Expected frequencies:

$e_{11} = 35.59$ $\quad e_{12} = 15.41$
$e_{21} = 150.73$ $\quad e_{22} = 65.27$
$e_{31} = 455.68$ $\quad e_{32} = 197.32$

Ticket	Flight	Observed Frequency (f_i)	Expected Frequency (e_i)	$(f_i - e_i)^2/e_i$
First	Domestic	29	35.59	1.22
First	International	22	15.41	2.82
Business	Domestic	95	150.73	20.61
Business	International	121	65.27	47.59
Full-fare	Domestic	518	455.68	8.52
Full-fare	International	135	197.32	19.68
Totals		920		$\chi^2 = 100.43$

Degrees of freedom: $(3-1)(2-1) = 2$
$\chi^2 = 100.43$, p-value is less than .005
Reject H_0; type of ticket is not independent of type of flight

22. a. $\chi^2 = 7.95$, $df = 3$
p-value is between .025 and .05
Reject H_0; method of payment is not independent of age group
b. 18 to 24 use most

24. a. $\chi^2 = 8.47$; p-value is between .025 and .05
Reject H_0; intent to purchase again is not independent of the automobile
b. Accord 77, Camry 71, Taurus 62, Impala 57
c. Impala and Taurus below, Accord and Camry above; Accord and Camry have greater owner satisfaction, which may help future market share

26. a. 6446
b. $\chi^2 = 425.4$; p-value = 0
Reject H_0; attitude toward nuclear power is not independent of country
c. Italy (58%), Spain (32%)

28. $\chi^2 = 3.01$, $df = 2$
p-value is greater than .10
Do not reject H_0; couples working is independent of location; 63.3%

30. $\chi^2 = 42.53$, $df = 4$
p-value is less than .005
Reject H_0; conclude job satisfaction differs

32. a. .35 and .47
b. $.12 \pm .1037$ (.0163 to .2237)

c. Yes, we would expect occupancy rates to be higher

34. $\chi^2 = 23.37$, $df = 3$
p-value is less than .005
Reject H_0; employment status is not independent of region

35. $\chi^2 = 4.64$, $df = 2$
p-value between .05 and .10
Do not reject H_0; cannot conclude market shares have changed

37. $\chi^2 = 8.04$, $df = 3$
p-value between .025 and .05
Reject H_0; potentials are not the same for each sales territory

38. a. 71%, 22%, slower preferred
b. $\chi^2 = 2.99$, $df = 2$
p-value greater than .10
Do not reject H_0; cannot conclude men and women differ in preference

42. $\chi^2 = 7.75$, $df = 3$
p-value is between .05 and .10
Do not reject H_0; cannot conclude office vacancies differ by metropolitan area

43. a. p-value ≈ 0, reject H_0
b. .0468 to .1332

Chapter 12

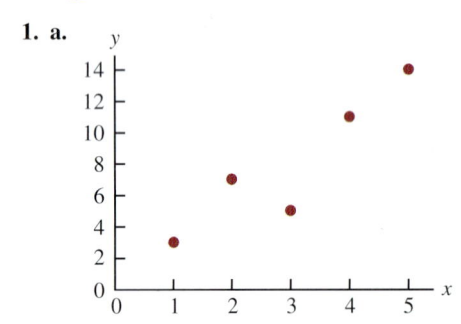

1. a.

b. There appears to be a positive linear relationship between x and y
c. Many different straight lines can be drawn to provide a linear approximation of the relationship between x and y; in part (d) we will determine the equation of a straight line that "best" represents the relationship according to the least squares criterion
d. Summations needed to compute the slope and y-intercept:

$$\bar{x} = \frac{\Sigma x_i}{n} = \frac{15}{5} = 3, \quad \bar{y} = \frac{\Sigma y_i}{n} = \frac{40}{5} = 8,$$

$\Sigma(x_i - \bar{x})(y_i - \bar{y}) = 26, \quad \Sigma(x_i - \bar{x})^2 = 10$

$$b_1 = \frac{\Sigma(x_i - \bar{x})(y_i - \bar{y})}{\Sigma(x_i - \bar{x})^2} = \frac{26}{10} = 2.6$$

$b_0 = \bar{y} - b_1\bar{x} = 8 - (2.6)(3) = 0.2$
$\hat{y} = 0.2 - 2.6x$

e. $\hat{y} = .2 + 2.6x = .2 + 2.6(4) = 10.6$

2. b. There appears to be a negative linear relationship between x and y
d. $\hat{y} = 68 - 3x$
e. 38

4. a.

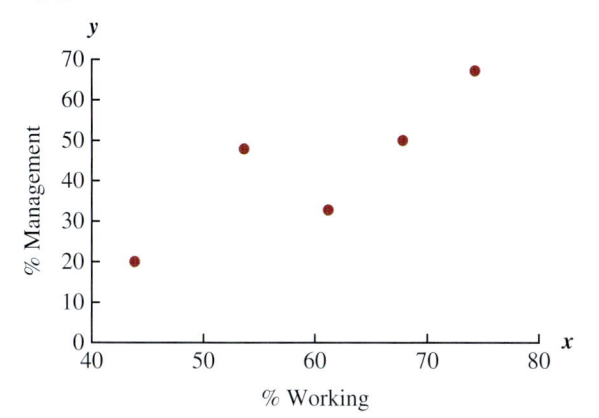

x_i	y_i	$\hat{y}_i$	$y_i - \hat{y}_i$	$(y_i - \hat{y}_i)^2$	$y_i - \bar{y}$	$(y_i - \bar{y})^2$
1	3	2.8	.2	.04	−5	25
2	7	5.4	1.6	2.56	−1	1
3	5	8.0	−3.0	9.00	−3	9
4	11	10.6	.4	.16	3	9
5	14	13.2	.8	.64	6	36
				SSE = 12.40		SST = 80

SSR = SST − SSE = 80 − 12.4 = 67.6

b. $r^2 = \dfrac{SSR}{SST} = \dfrac{67.6}{80} = .845$

The least squares line provided a good fit; 84.5% of the variability in y has been explained by the least squares line

c. $r_{xy} = \sqrt{.845} = +.9192$

16. a. SSE = 230, SST = 1850, SSR = 1620
b. $r^2 = .876$
c. $r_{xy} = -.936$

18. a. The estimated regression equation and the mean for the dependent variable:
$\hat{y} = 1790.5 + 581.1x$, $\bar{y} = 3650$
The sum of squares due to error and the total sum of squares:
SSE = $\Sigma(y_i - \hat{y}_i)^2 = 85,135.14$
SST = $\Sigma(y_i - \bar{y})^2 = 335,000$
Thus, SSR = SST − SSE
    = 335,000 − 85,135.14 = 249,864.86

b. $r^2 = \dfrac{SSR}{SST} = \dfrac{249,864.86}{335,000} = .746$

The least squares line accounted for 74.6% of the total sum of squares

c. $r_{xy} = \sqrt{.746} = +.8637$

20. a. $\hat{y} = 12.0169 + .0127x$
b. $r^2 = .4503$
c. 53

22. a. .77
b. Yes
c. $r_{xy} = +.88$, strong

23. a. $s^2 = MSE = \dfrac{SSE}{n-2} = \dfrac{12.4}{3} = 4.133$
b. $s = \sqrt{MSE} = \sqrt{4.133} = 2.033$
c. $\Sigma(x_i - \bar{x})^2 = 10$

$$s_{b_1} = \dfrac{s}{\sqrt{\Sigma(x_i - \bar{x})^2}} = \dfrac{2.033}{\sqrt{10}} = .643$$

d. $t = \dfrac{b_1 - \beta_1}{s_{b_1}} = \dfrac{2.6 - 0}{.643} = 4.044$

From the t table (3 degrees of freedom), area in tail is between .01 and .025
p-value is between .02 and .05
Using Excel or Minitab, the p-value corresponding to t = 4.04 is .0272
Because p-value ≤ α, we reject $H_0: \beta_1 = 0$

---

4. a.

% Working

b. There appears to be a positive linear relationship between the precentage of women working in the five companies (x) and the percentage of management jobs held by women in each company (y)

c. Many different straight lines can be drawn to provide a linear approximation of the relationship between x and y; in part (d) we will determine the equation of a straight line that "best" represents the relationship according to the least squares criterion

d. Summations needed to compute the slope and y-intercept:

$\bar{x} = \dfrac{\Sigma x_i}{n} = \dfrac{300}{5} = 60$, $\bar{y} = \dfrac{\Sigma y_i}{n} = \dfrac{215}{5} = 43$,

$\Sigma(x_i - \bar{x})(y_i - \bar{y}) = 624$, $\Sigma(x_i - \bar{x})^2 = 480$

$b_1 = \dfrac{\Sigma(x_i - \bar{x})(y_i - \bar{y})}{\Sigma(x_i - \bar{x})^2} = \dfrac{624}{480} = 103$

$b_0 = \bar{y} - b_1\bar{x} = 43 - 1.3(60) = -35$

$\hat{y} = -35 + 1.3x$

e. $\hat{y} = -35 + 1.3x = -35 + 1.3(60) = 43\%$

6. c. $\hat{y} = 8.9412 - .02633x$
e. 6.3 or approximately $6300

8. c. $\hat{y} = 359.2668 - 5.2772x$
d. 4003 or $4,003,000

10. c. $\hat{y} = -6,745.44 + 149.29x$
d. 4003 or $4,003,000

12. c. $\hat{y} = -8129.4439 + 22.4443x$
d. $8704

14. c. $\hat{y} = 37.1217 + .51758x$
d. 73

15. a. $\hat{y}_i = .2 + 2.6x_i$ and $\bar{y} = 8$

e. MSR = $\frac{SSR}{1}$ = 67.6

$F = \frac{MSR}{MSE} = \frac{67.6}{4.133} = 16.36$

From the F table (1 numerator degree of freedom and 3 denominator), p-value is between .025 and .05

Using Excel or Minitab, the p-value corresponding to $F = 16.36$ is .0272

Because p-value $\leq \alpha$, we reject $H_0: \beta_1 = 0$

Source of Variation	Sum of Squares	Degrees of Freedom	Mean Square	F	p-value
Regression	67.6	1	67.6	16.36	.0272
Error	12.4	3	4.133		
Total	80	4			

24. a. 76.6667
   b. 8.7560
   c. .6526
   d. Significant; p-value = .0193
   e. Significant; p-value = .0193

26. a. $s^2 = MSE = \frac{SSE}{n-2} = \frac{85,135.14}{4} = 21,283.79$

$s = \sqrt{MSE} = \sqrt{21,283.79} = 145.89$

$\Sigma(x_i - \bar{x})^2 = .74$

$s_{b_1} = \frac{s}{\sqrt{\Sigma(x_i - \bar{x})^2}} = \frac{145.89}{\sqrt{.74}} = 169.59$

$t = \frac{b_1 - \beta_1}{s_{b_1}} = \frac{581.08 - 0}{169.59} = 3.43$

From the t table (4 degrees of freedom), area in tail is between .01 and .025

p-value is between .02 and .05

Using Excel or Minitab, the p-value corresponding to $t = 3.43$ is .0266

Because p-value $\leq \alpha$, we reject $H_0: \beta_1 = 0$

b. MSR = $\frac{SSR}{1} = \frac{249,864.86}{1} = 249,864.86$

$F = \frac{MSR}{MSE} = \frac{249,864.86}{21,283.79} = 11.74$

From the F table (1 numerator degree of freedom and 4 denominator), p-value is between .025 and .05

Using Excel or Minitab, the p-value corresponding to $F = 11.74$ is .0266

Because p-value $\leq \alpha$, we reject $H_0: \beta_1 = 0$

c.

Source of Variation	Sum of Squares	Degrees of Freedom	Mean Square	F	p-value
Regression	249,864.86	1	249,864.86	11.74	.0266
Error	85,135.14	4	21,283.79		
Total	335,000	5			

28. They are related; p-value = .000
30. Significant; p-value = .0042
32. a. $s = 2.033$

$\bar{x} = 3, \Sigma(x_i - \bar{x})^2 = 10$

$s_{\hat{y}_p} = s\sqrt{\frac{1}{n} + \frac{(x_p - \bar{x})^2}{\Sigma(x_i - \bar{x})^2}}$

$= 2.033\sqrt{\frac{1}{5} + \frac{(4-3)^2}{10}} = 1.11$

b. $\hat{y} = .2 + 2.6x = .2 + 2.6(4) = 10.6$

$\hat{y}_p \pm t_{\alpha/2} s_{\hat{y}_p}$

$10.6 \pm 3.182(1.11)$

10.6 ± 3.53, or 7.07 to 14.13

c. $s_{ind} = s\sqrt{1 + \frac{1}{n} + \frac{(x_p - \bar{x})^2}{\Sigma(x_i - \bar{x})^2}}$

$= 2.033\sqrt{1 + \frac{1}{5} + \frac{(4-3)^2}{10}} = 2.32$

d. $\hat{y}_p \pm t_{\alpha/2} s_{ind}$

$10.6 \pm 3.182(2.32)$

10.6 ± 7.38, or 3.22 to 17.98

34. Confidence interval: 8.65 to 21.15
Prediction interval: -4.50 to 41.30

35. a. $s = 145.89, \bar{x} = 3.2, \Sigma(x_i - \bar{x})^2 = .74$

$\hat{y} = 1790.5 + 581.1x = 1790.5 + 581.1(3)$

$= 3533.8$

$s_{\hat{y}_p} = s\sqrt{\frac{1}{n} + \frac{(x_p - \bar{x})^2}{\Sigma(x_i - \bar{x})^2}}$

$= 145.89\sqrt{\frac{1}{6} + \frac{(3 - 3.2)^2}{.74}} = 68.54$

$\hat{y}_p \pm t_{\alpha/2} s_{\hat{y}_p}$

$3533.8 \pm 2.776(68.54)$

3533.8 ± 190.27, or \$3343.53 to \$3724.07

b. $s_{ind} = s\sqrt{1 + \frac{1}{n} + \frac{(x_p - \bar{x})^2}{\Sigma(x_i - \bar{x})^2}}$

$= 145.89\sqrt{1 + \frac{1}{6} + \frac{(3 - 3.2)^2}{.74}} = 161.19$

$\hat{y}_p \pm t_{\alpha/2} s_{ind}$

$3533.8 \pm 2.776(161.19)$

3533.8 ± 447.46, or \$3086.34 to \$3981.26

36. a. \$201
   b. 167.25 to 234.65
   c. 108.75 to 293.15
38. a. \$5046.67
   b. \$3815.10 to \$6278.24
   c. Not out of line
40. a. 9
   b. $\hat{y} = 20.0 + 7.21x$
   c. 1.3626
   d. SSE = SST − SSR = 51,984.1 − 41,587.3 = 10,396.8
   MSE = 10,396.8/7 = 1485.3

$F = \frac{MSR}{MSE} = \frac{41,587.3}{1485.3} = 28.0$

From the F table (1 numerator degree of freedom and 7 denominator), p-value is less than .01

Using Excel or Minitab, the p-value corresponding to $F = 28.0$ is .0011

Because p-value $\leq \alpha = .05$, we reject $H_0: \beta_1 = 0$

e. $\hat{y} = 20.0 + 7.21(50) = 380.5$, or $380,500

42. a. $\hat{y} = 80.0 + 50.0x$
    b. 30
    c. Significant; p-value = .000
    d. $680,000

44. b. Yes
    c. $\hat{y} = 2044.38 - 28.35$ weight
    d. Significant; p-value = .000
    e. .774; a good fit

45. a. $\bar{x} = \dfrac{\Sigma x_i}{n} = \dfrac{70}{5} = 14$, $\bar{y} = \dfrac{\Sigma y_i}{n} = \dfrac{76}{5} = 15.2$,
    $\Sigma(x_i - \bar{x})(y_i - \bar{y}) = 200$, $\Sigma(x_i - \bar{x})^2 = 126$
    $b_1 = \dfrac{\Sigma(x_i - \bar{x})(y_i - \bar{y})}{\Sigma(x_i - \bar{x})^2} = \dfrac{200}{126} = 1.5873$
    $b_0 = \bar{y} - b_1\bar{x} = 15.2 - (1.5873)(14) = -7.0222$
    $\hat{y} = -7.02 + 1.59x$

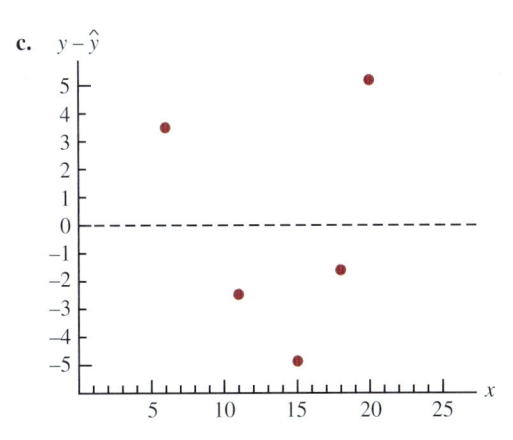

    b.

x_i	y_i	$\hat{y}_i$	$y_i - \hat{y}_i$
6	6	2.52	3.48
11	8	10.47	-2.47
15	12	16.83	-4.83
18	20	21.60	-1.60
20	30	24.78	5.22

    c. $y - \hat{y}$ plot

    With only five observations, it is difficult to determine whether the assumptions are satisfied; however, the plot does suggest curvature in the residuals, which would indicate that the error term assumptions are not satisfied; the scatter diagram for these data also indicates that the underlying relationship between x and y may be curvilinear

d. $s^2 = 23.78$
   $h_i = \dfrac{1}{n} + \dfrac{(x_i - \bar{x})^2}{\Sigma(x_i - \bar{x})^2}$
   $= \dfrac{1}{5} + \dfrac{(x_i - 14)^2}{126}$

x_i	h_i	$s_{y_i - \hat{y}_i}$	$y_i - \hat{y}_i$	Standardized Residuals
6	.7079	2.64	3.48	1.32
11	.2714	4.16	-2.47	-.59
15	.2079	4.34	-4.83	-1.11
18	.3270	4.00	-1.60	-.40
20	.4857	3.50	5.22	1.49

   e. The plot of the standardized residuals against $\hat{y}$ has the same shape as the original residual plot; as stated in part (c), the curvature observed indicates that the assumptions regarding the error term may not be satisfied

46. a. $\hat{y} = 2.32 + .64x$
    b. No; the variance appears to increase for larger values of x

47. a. Let x = advertising expenditures and y = revenue
       $\hat{y} = 29.4 + 1.55x$
    b. SST = 1002, SSE = 310.28, SSR = 691.72
       $MSR = \dfrac{SSR}{1} = 691.72$
       $MSE = \dfrac{SSE}{n - 2} = \dfrac{310.28}{5} = 62.0554$
       $F = \dfrac{MSR}{MSE} = \dfrac{691.72}{62.0554} = 11.15$
       From the F table (1 numerator degree of freedom and 5 denominator), p-value is between .01 and .025
       Using Excel or Minitab, p-value = .0206
       Because p-value $\leq \alpha = .05$, we conclude that the two variables are related

    c.

x_i	y_i	$\hat{y}_i = 29.40 + 1.55x_i$	$y_i - \hat{y}_i$
1	19	30.95	-11.95
2	32	32.50	-.50
4	44	35.60	8.40
6	40	38.70	1.30
10	52	44.90	7.10
14	53	51.10	1.90
20	54	60.40	-6.40

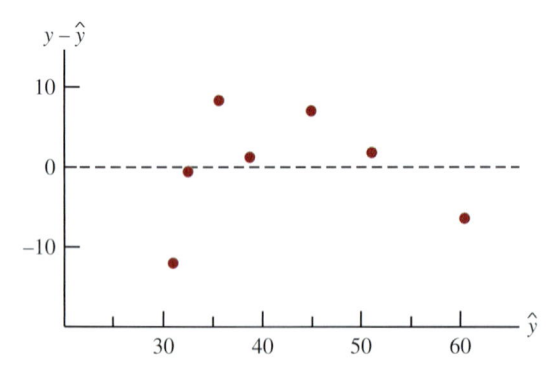

d. The residual plot leads us to question the assumption of a linear relationship between x and y; even though the relationship is significant at the $\alpha = .05$ level, it would be extremely dangerous to extrapolate beyond the range of the data

48. b. Yes
50. a. $\hat{y} = 9.26 + .711x$
    b. Significant; p-value = .001
    c. $r^2 = .744$; good fit
    d. $13.53
52. b. GR(%) = 25.4 + .285 RR(%)
    c. Significant; p-value = .000
    d. No; $r^2 = .449$
    e. Yes
    f. Yes
54. a. $\hat{y} = 22.2 - .148x$
    b. Significant relationship; p-value = .028
    c. Good fit; $r^2 = .739$
    d. 12.294 to 17.271
56. a. $\hat{y} = 220 + 132x$
    b. Significant; p-value = .000
    c. $r^2 = .873$; very good fit
    d. $559.50 to $933.90
58. a. Market beta = .95
    b. Significant; p-value = .029
    c. $r^2 = .470$; not a good fit
    d. Xerox has a higher risk
60. b. There appears to be a positive linear relationship between the two variables
    c. $\hat{y} = 9.37 + 1.2875$ Top Five (%)
    d. Significant; p-value = .000
    e. $r^2 = .741$; good fit
    f. $r_{xy} = .86$

Chapter 13

2. a. The estimated regression equation is
   $\hat{y} = 45.06 + 1.94x_1$
   An estimate of y when $x_1 = 45$ is
   $\hat{y} = 45.06 + 1.94(45) = 132.36$
   b. The estimated regression equation is
   $\hat{y} = 85.22 + 4.32x_2$
   An estimate of y when $x_2 = 15$ is
   $\hat{y} = 85.22 + 4.32(15) = 150.02$
   c. The estimated regression equation is
   $\hat{y} = -18.37 + 2.01x_1 + 4.74x_2$
   An estimate of y when $x_1 = 45$ and $x_2 = 15$ is
   $\hat{y} = -18.37 + 2.01(45) + 4.74(15) = 143.18$
4. a. $255,000
5. a. The Minitab output is shown in Figure D13.5a
   b. The Minitab output is shown in Figure D13.5b
   c. It is 1.60 in part (a) and 2.29 in part (b); in part (a) the coefficient is an estimate of the change in revenue due to a one-unit change in television advertising expenditures; in part (b) it represents an estimate of the change in revenue due to a one-unit change in television advertising expenditures when the amount of newspaper advertising is held constant
   d. Revenue = 83.2 + 2.29(3.5) + 1.30(1.8) = 93.56 or $93,560
6. a. Salary & Bonus ($1000s) = $-20.0 + 33.7$ Recruiter Score
   b. Salary & Bonus ($1000s) = $0.9 + 14.6$ Recruiter Score + 1.37 Tuition & Fees ($1000s)
   c. No
   d. 119.56 or approximately $120,000
8. a. $\hat{y} = 31054 + 1328.7$ Reliability
   b. $\hat{y} = 21313 + 136.69$ Score $- 1446.3$ Reliability
   c. $26,643
10. a. PCT = $-1.22 + 3.96$ FG%
    b. Increase of 1% in FG% will increase PCT by .04
    c. PCT = $-1.23 + 4.82$ FG% $- 2.59$ Opp 3 Pt% + .0344 Opp TO
    d. Increase FG%; decrease Opp 3 Pt%; increase Opp TO
    e. .638
12. a. $R^2 = \dfrac{\text{SSR}}{\text{SST}} = \dfrac{14{,}052.2}{15{,}182.9} = .926$
    b. $R_a^2 = 1 - (1 - R^2)\dfrac{n-1}{n-p-1}$
       $= 1 - (1 - .926)\dfrac{10-1}{10-2-1} = .905$
    c. Yes; after adjusting for the number of independent variables in the model, we see that 90.5% of the variability in y has been accounted for
14. a. .75   b. .68
15. a. $R^2 = \dfrac{\text{SSR}}{\text{SST}} = \dfrac{23.435}{25.5} = .919$
    $R_a^2 = 1 - (1 - R^2)\dfrac{n-1}{n-p-1}$
    $= 1 - (1 - .919)\dfrac{8-1}{8-2-1} = .887$

FIGURE D13.5a

The regression equation is
Revenue = 88.6 + 1.60 TVAdv

Predictor	Coef	SE Coef	T	p
Constant	88.638	1.582	56.02	0.000
TVAdv	1.6039	0.4778	3.36	0.015

S = 1.215   R-sq = 65.3%   R-sq(adj) = 59.5%

Analysis of Variance

SOURCE	DF	SS	MS	F	p
Regression	1	16.640	16.640	11.27	0.015
Residual Error	6	8.860	1.477		
Total	7	25.500			

FIGURE D13.5b

The regression equation is
Revenue = 83.2 + 2.29 TVAdv + 1.30 NewsAdv

Predictor	Coef	SE Coef	T	p
Constant	83.230	1.574	52.88	0.000
TVAdv	2.2902	0.3041	7.53	0.001
NewsAdv	1.3010	0.3207	4.06	0.010

S = 0.6426   R-sq = 91.9%   R-sq(adj) = 88.7%

Analysis of Variance

SOURCE	DF	SS	MS	F	p
Regression	2	23.435	11.718	28.38	0.002
Residual Error	5	2.065	0.413		
Total	7	25.500			

    b. Multiple regression analysis is preferred because both R^2 and R_a^2 show an increased percentage of the variability of y explained when both independent variables are used

16. a. $r^2 = .613$, not too bad a fit
    b. Better fit with multiple regression, $R^2 = .793$ and $R_a^2 = .768$

18. a. $R^2 = .564$, $R_a^2 = .511$
    b. The fit is not very good

19. a. $\text{MSR} = \dfrac{\text{SSR}}{p} = \dfrac{6216.375}{2} = 3108.188$

    $\text{MSE} = \dfrac{\text{SSE}}{n - p - 1} = \dfrac{507.75}{10 - 2 - 1} = 72.536$

    b. $F = \dfrac{\text{MSR}}{\text{MSE}} = \dfrac{3108.188}{72.536} = 42.85$

From the F table (2 numerator degrees of freedom and 7 denominator), p-value is less than .01

Using Excel or Minitab, the p-value corresponding to $F = 42.85$ is .0001

Because p-value $\leq \alpha$, the overall model is significant

    c. $t = \dfrac{b_1}{s_{b_1}} = \dfrac{.5906}{.0813} = 7.26$

    p-value $= .0002$

    Because p-value $\leq \alpha$, β_1 is significant

    d. $t = \dfrac{b_2}{s_{b_2}} = \dfrac{.4980}{.0567} = 8.78$

    p-value $= .0001$

    Because p-value $\leq \alpha$, β_2 is significant

20. a. Significant; p-value $= .000$
    b. Significant; p-value $= .000$
    c. Significant; p-value $= .002$

22. a. SSE $= 4000$, $s^2 = 571.43$,
       MSR $= 6000$
    b. Significant; p-value $= .008$

23. a. $F = 28.38$
       p-value $= .002$
       Because p-value $\leq \alpha$, there is a significant relationship
    b. $t = 7.53$
       p-value $= .001$
       Because p-value $\leq \alpha$, β_1 is significant and x_1 should not be dropped from the model
    c. $t = 4.06$
       p-value $= .010$
       Because p-value $\leq \alpha$, β_2 is significant and x_2 should not be dropped from the model

24. a. $\hat{y} = -.682 + .0498$ Revenue $+ .0147$ % Wins
    b. Significant; p-value $= .001$
    c. Revenue is significant; p-value $= .001$
       %Wins is significant; p-value $= .025$

26. a. Significant; p-value $= .000$
    b. All significant; p-values are all $< \alpha = .05$

28. a. Using Minitab, the 95% confidence interval is 132.16 to 154.16
    b. Using Minitab, the 95% prediction interval is 111.13 at 175.18

29. a. See Minitab output in Figure D13.5b.
       $\hat{y} = 83.23 + 2.29(3.5) + 1.30(1.8) = 93.555$ or $93,555
    b. Minitab results: 92.840 to 94.335, or $92,840 to $94,335
    c. Minitab results: 91.774 to 95.401, or $91,774 to $95,401

30. a. 46.758 to 50.646
    b. 44.815 to 52.589

32. a. $E(y) = \beta_0 + \beta_1 x_1 + \beta_2 x_2$
       where $x_2 = \begin{cases} 0 \text{ if level 1} \\ 1 \text{ if level 2} \end{cases}$
    b. $E(y) = \beta_0 + \beta_1 x_1 + \beta_2(0) = \beta_0 + \beta_1 x_1$
    c. $E(y) = \beta_0 + \beta_1 x_1 + \beta_2(1) = \beta_0 + \beta_1 x_1 + \beta_2$
    d. $\beta_2 = E(y \mid \text{level 2}) - E(y \mid \text{level 1})$
       β_1 is the change in $E(y)$ for a 1-unit change in x_1 holding x_2 constant

34. a. $15,300
    b. $\hat{y} = 10.1 - 4.2(2) + 6.8(8) + 15.3(0) = 56.1$
       Sales prediction: $56,100
    c. $\hat{y} = 10.1 - 4.2(1) + 6.8(3) + 15.3(1) = 41.6$
       Sales prediction: $41,600

36. a. $\hat{y} = 1.86 + 0.291$ Months $+ 1.10$ Type $- 0.609$ Person
    b. Significant; p-value $= .002$
    c. Person is not significant; p-value $= .167$

38. a. $\hat{y} = -91.8 + 1.08$ Age $+ .252$ Pressure $+ 8.74$ Smoker
    b. Significant; p-value $= .01$
    c. 95% prediction interval is 21.35 to 47.18 or a probability of .2135 to .4718; quit smoking and begin some type of treatment to reduce his blood pressure

39. b. 67.39

41. a. $\hat{y} = -1.41 + .0235 x_1 + .00486 x_2$
    b. Significant; p-value $= .0001$
    c. Both significant
    d. $R^2 = .937$; $R_a^2 = 9.19$; good fit

44. a. Buy Again $= -7.522 + 1.8151$ Steering
    b. Yes
    c. Buy Again $= -5.388 + .6899$ Steering $+ .9113$ Treadwear
    d. Significant; p-value $= .001$

46. a. $\hat{y} = 4.9090 + 10.4658$ FundDE $+ 21.6823$ FundIE
    b. $R^2 = .6144$; reasonably good fit
    c. $\hat{y} = 1.1899 + 6.8969$ FundDE $+ 17.6800$ FundIE $+ 0.0265$ Net Asset Value ($) $+ 6.4564$ Expense Ratio (%)
       Net Asset Value ($) is not significant and can be deleted
    d. $\hat{y} = -4.6074 + 8.1713$ FundDE $+ 19.5194$ FundIE $+ 5.5197$ Expense Ratio (%) $+ 5.9237$ 3StarRank $+ 8.2367$ 4StarRank $+ 6.6241$ 5StarRank
    e. 15.28%

Appendix E: Using Excel Functions

Excel provides a wealth of functions for data management and statistical analysis. If we know what function is needed, and how to use it, we can simply enter the function into the appropriate worksheet cell. However, if we are not sure what functions are available to accomplish a task or are not sure how to use a particular function, Excel can provide assistance.

The functions used in this text are available to both Excel 2007 and Excel 2010 users, and all the procedures described here for inserting functions are the same in both Excel 2007 and Excel 2010. In Excel 2010, some new functions have been added and some new functions have replaced functions available in Excel 2007. However, all the Excel 2007 functions continue to be available to Excel 2010 users for compatibility purposes.

Finding the Right Excel Function

To identify the functions available in Excel, click the **Formulas** tab on the Ribbon. In the Function Library group, click **Insert Function.** Alternatively, click the f_x button on the formula bar. Either approach provides the **Insert Function** dialog box shown in Figure E.1.

The **Search for a function** box at the top of the Insert Function dialog box enables us to type a brief description of what we want to do. After doing so and clicking **Go**, Excel will

FIGURE E.1 INSERT FUNCTION DIALOG BOX

Note to Excel 2010 users: The BETADIST and BETAINV functions shown here have been replaced by the BETA.DIST and BETA.INV functions.

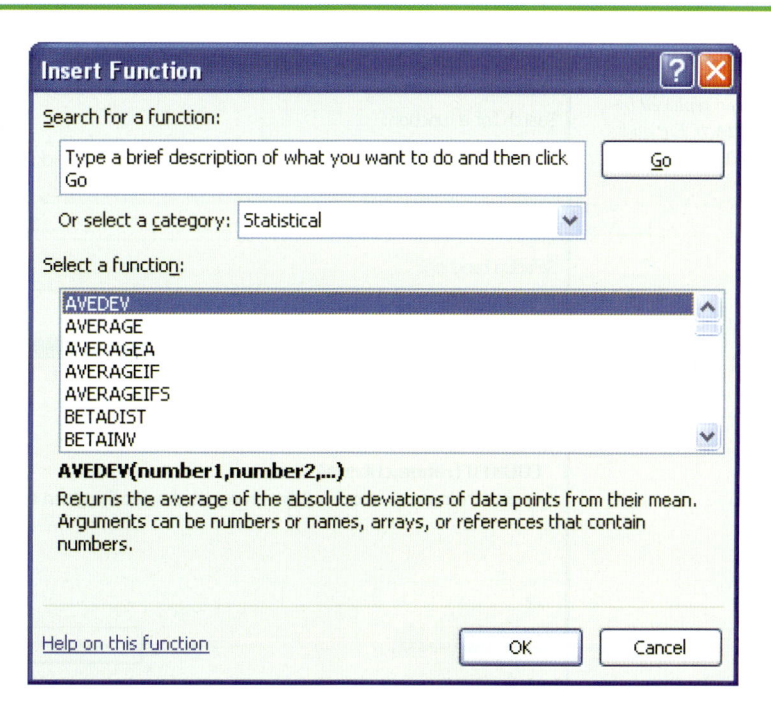

search for and display, in the **Select a function** box, the functions that may accomplish our task. In many situations, however, we may want to browse through an entire category of functions to see what is available. For this task, the **Or select a category** box is helpful. It contains a drop-down list of several categories of functions provided by Excel. Figure E.1 shows that we selected the **Statistical** category. As a result, Excel's statistical functions appear in alphabetic order in the Select a function box. We see the AVEDEV function listed first, followed by the AVERAGE function, and so on.

The AVEDEV function is highlighted in Figure E.1, indicating it is the function currently selected. The proper syntax for the function and a brief description of the function appear below the Select a function box. We can scroll through the list in the Select a function box to display the syntax and a brief description for each of the statistical functions available. For instance, scrolling down farther, we select the COUNTIF function, as shown in Figure E.2. Note that COUNTIF is now highlighted, and that immediately below the Select a function box we see **COUNTIF(range,criteria)**, which indicates that the COUNTIF function contains two arguments, range and criteria. In addition, we see that the description of the COUNTIF function is "Counts the number of cells within a range that meet the given condition."

If the function selected (highlighted) is the one we want to use, we click **OK**; the **Function Arguments** dialog box then appears. The Function Arguments dialog box for the COUNTIF function is shown in Figure E.3. This dialog box assists in creating the appropriate arguments for the function selected. When finished entering the arguments, we click **OK**; Excel then inserts the function into a worksheet cell.

FIGURE E.2 DESCRIPTION OF THE COUNTIF FUNCTION IN THE INSERT FUNCTION DIALOG BOX

Note to Excel 2010 users: The COVAR function shown here has been replaced by the COVARIANCE.P and COVARIANCE.S functions.

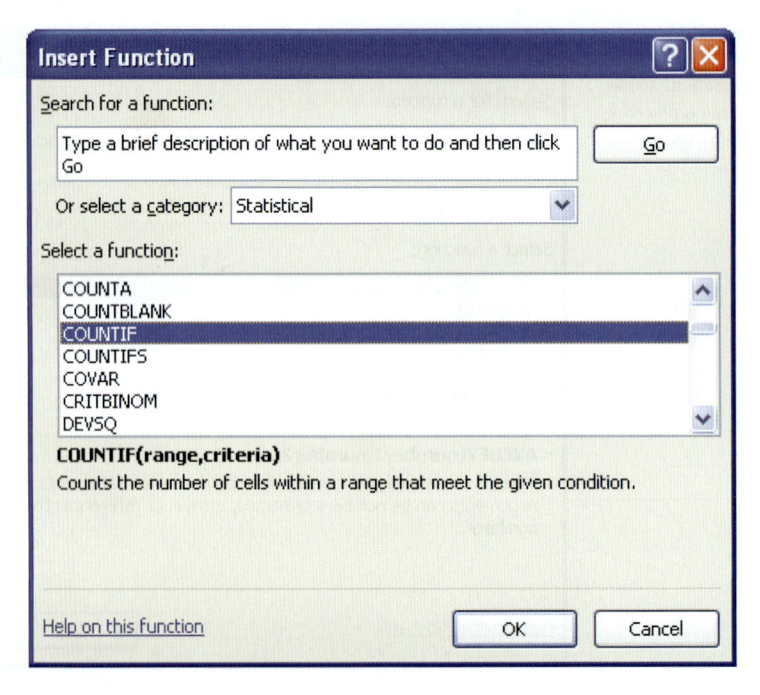

Inserting a Function into a Worksheet Cell

We will now show how to use the Insert Function and Function Arguments dialog boxes to select a function, develop its arguments, and insert the function into a worksheet cell.

In Appendix 2.2, we used Excel's COUNTIF function to construct a frequency distribution for soft drink purchases. Figure E.4 displays an Excel worksheet containing the soft

FIGURE E.3 FUNCTION ARGUMENTS DIALOG BOX FOR THE COUNTIF FUNCTION

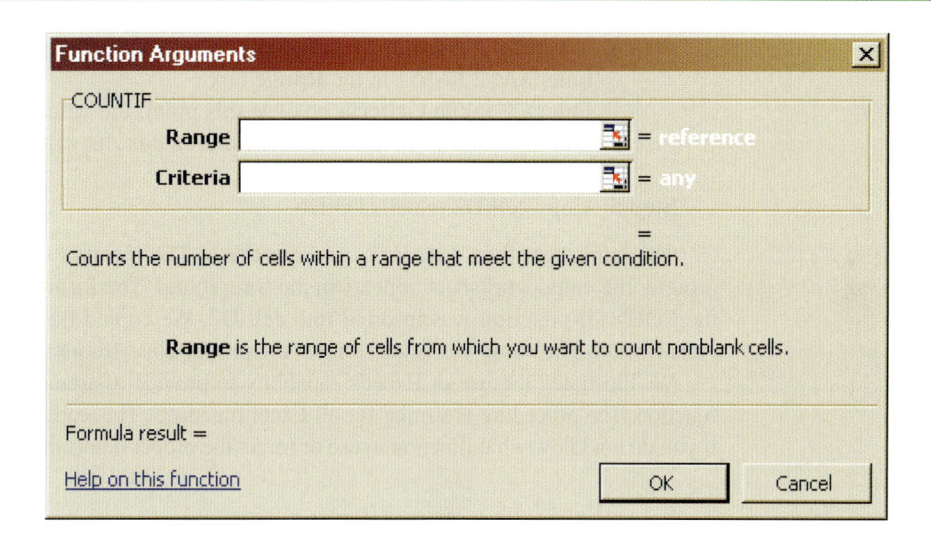

FIGURE E.4 EXCEL WORKSHEET WITH SOFT DRINK DATA AND LABELS FOR THE FREQUENCY DISTRIBUTION WE WOULD LIKE TO CONSTRUCT

WEB file

SoftDrink

Note: Rows 11–44 are hidden.

	A	B	C	D	E
1	Brand Purchased		Soft Drink	Frequency	
2	Coke Classic		Coke Classic		
3	Diet Coke		Diet Coke		
4	Pepsi		Dr. Pepper		
5	Diet Coke		Pepsi		
6	Coke Classic		Sprite		
7	Coke Classic				
8	Dr. Pepper				
9	Diet Coke				
10	Pepsi				
45	Pepsi				
46	Pepsi				
47	Pepsi				
48	Coke Classic				
49	Dr. Pepper				
50	Pepsi				
51	Sprite				
52					

drink data and labels for the frequency distribution we would like to construct. We see that the frequency of Coke Classic purchases will go into cell D2, the frequency of Diet Coke purchases will go into cell D3, and so on. Suppose we want to use the COUNTIF function to compute the frequencies for these cells and would like some assistance from Excel.

Step 1. Select cell D2
Step 2. Click f_x on the formula bar (or click the **Formulas** tab on the Ribbon and click **Insert Function** in the Function Library group)
Step 3. When the **Insert Function** dialog box appears:
Select **Statistical** in the **Or select a category box**
Select **COUNTIF** in the **Select a function box**
Click **OK**
Step 4. When the **Function Arguments** box appears (see Figure E.5):
Enter $A$2:$A$51 in the **Range** box
Enter C2 in the **Criteria** box (At this point, the value of the function will appear on the next-to-last line of the dialog box. Its value is 19.)
Click **OK**
Step 5. Copy cell D2 to cells D3:D6

The worksheet then appears as in Figure E.6. The formula worksheet is in the background; the value worksheet appears in the foreground. The formula worksheet shows that the COUNTIF function was inserted into cell D2. We copied the contents of cell D2 into cells D3:D6. The value worksheet shows the proper class frequencies as computed.

We illustrated the use of Excel's capability to provide assistance in using the COUNTIF function. The procedure is similar for all Excel functions. This capability is especially helpful if you do not know what function to use or forget the proper name and/or syntax for a function.

FIGURE E.5 COMPLETED FUNCTION ARGUMENTS DIALOG BOX FOR THE COUNTIF FUNCTION

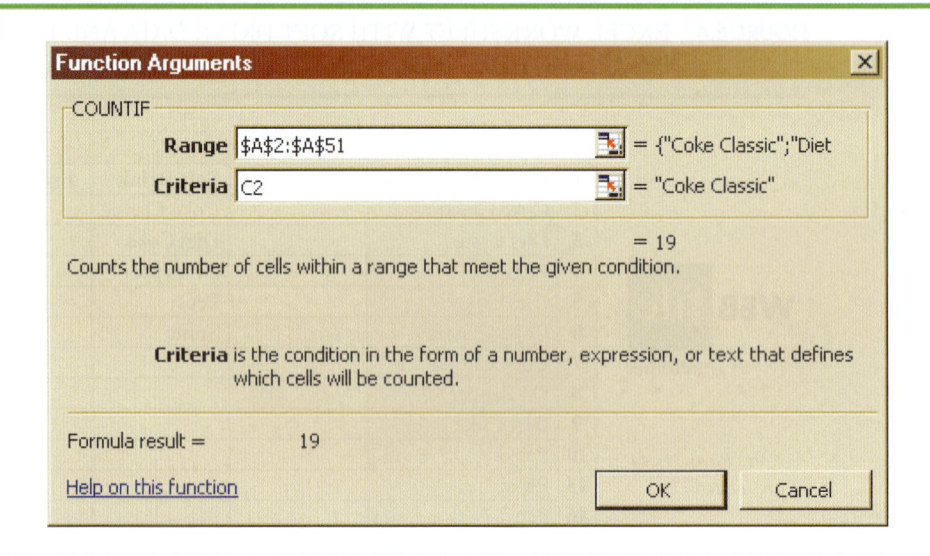

Appendix E Using Excel Functions

FIGURE E.6 EXCEL WORKSHEET SHOWING THE USE OF EXCEL'S COUNTIF FUNCTION TO CONSTRUCT A FREQUENCY DISTRIBUTION

	A	B	C	D	E
1	Brand Purchased		Soft Drink	Frequency	
2	Coke Classic		Coke Classic	=COUNTIF($A$2:$A$51,C2)	
3	Diet Coke		Diet Coke	=COUNTIF($A$2:$A$51,C3)	
4	Pepsi		Dr. Pepper	=COUNTIF($A$2:$A$51,C4)	
5	Diet Coke		Pepsi	=COUNTIF($A$2:$A$51,C5)	
6	Coke Classic		Sprite	=COUNTIF($A$2:$A$51,C6)	
7	Coke Classic				
8	Dr. Pepper				
9	Diet Coke				
10	Pepsi				
45	Pepsi				
46	Pepsi				
47	Pepsi				
48	Coke Classic				
49	Dr. Pepper				
50	Pepsi				
51	Sprite				
52					

Note: Rows 11–44 are hidden.

	A	B	C	D	E
1	Brand Purchased		Soft Drink	Frequency	
2	Coke Classic		Coke Classic	19	
3	Diet Coke		Diet Coke	8	
4	Pepsi		Dr. Pepper	5	
5	Diet Coke		Pepsi	13	
6	Coke Classic		Sprite	5	
7	Coke Classic				
8	Dr. Pepper				
9	Diet Coke				
10	Pepsi				
45	Pepsi				
46	Pepsi				
47	Pepsi				
48	Coke Classic				
49	Dr. Pepper				
50	Pepsi				
51	Sprite				
52					

Appendix F: Computing *p*-Values Using Minitab and Excel

Here we describe how Minitab and Excel can be used to compute *p*-values for the z, t, χ^2, and F statistics that are used in hypothesis tests. As discussed in the text, only approximate *p*-values for the t, χ^2, and F statistics can be obtained by using tables. This appendix is helpful to a person who has computed the test statistic by hand, or by other means, and wishes to use computer software to compute the exact *p*-value.

Using Minitab

Minitab can be used to provide the cumulative probability associated with the z, t, χ^2, and F test statistics, so the lower tail *p*-value is obtained directly. The upper tail *p*-value is computed by subtracting the lower tail *p*-value from 1. The two-tailed *p*-value is obtained by doubling the smaller of the lower and upper tail *p*-values.

The z test statistic We use the Hilltop Coffee lower tail hypothesis test in Section 9.3 as an illustration; the value of the test statistic is $z = -2.67$. The Minitab steps used to compute the cumulative probability corresponding to $z = -2.67$ follow.

    Step 1. Select the **Calc** menu
    Step 2. Choose **Probability Distributions**
    Step 3. Choose **Normal**
    Step 4. When the Normal Distribution dialog box appears:
        Select **Cumulative probability**
        Enter 0 in the **Mean** box
        Enter 1 in the **Standard deviation** box
        Select **Input Constant**
        Enter -2.67 in the **Input Constant** box
        Click **OK**

Minitab provides the cumulative probability of .0038. This cumulative probability is the lower tail *p*-value used for the Hilltop Coffee hypothesis test.

For an upper tail test, the *p*-value is computed from the cumulative probability provided by Minitab as follows:

$$p\text{-value} = 1 - \text{cumulative probability}$$

For instance, the upper tail *p*-value corresponding to a test statistic of $z = -2.67$ is $1 - .0038 = .9962$. The two-tailed *p*-value corresponding to a test statistic of $z = -2.67$ is two times the minimum of the upper and lower tail *p*-values; that is, the two-tailed *p*-value corresponding to $z = -2.67$ is $2(.0038) = .0076$.

The t test statistic We use the Heathrow Airport example from Section 9.4 as an illustration; the value of the test statistic is $t = 1.84$ with 59 degrees of freedom. The Minitab steps used to compute the cumulative probability corresponding to $t = 1.84$ follow.

Step 1. Select the **Calc** menu
Step 2. Choose **Probability Distributions**
Step 3. Choose **t**
Step 4. When the t Distribution dialog box appears:
    Select **Cumulative probability**
    Enter 59 in the **Degrees of freedom** box
    Select **Input Constant**
    Enter 1.84 in the **Input Constant** box
    Click **OK**

Minitab provides a cumulative probability of .9646, and hence the lower tail p-value = .9646. The Heathrow Airport example is an upper tail test; the upper tail p-value is 1 − .9646 = .0354. In the case of a two-tailed test, we would use the minimum of .9646 and .0354 to compute p-value = 2(.0354) = .0708.

The χ^2 test statistic Suppose we are conducting an upper tail test and the value of the test statistic is χ^2 = 28.18 with 23 degrees of freedom. The Minitab steps used to compute the cumulative probability corresponding to χ^2 = 28.18 follow.

Step 1. Select the **Calc** menu
Step 2. Choose **Probability Distributions**
Step 3. Choose **Chi-Square**
Step 4. When the Chi-Square Distribution dialog box appears:
    Select **Cumulative probability**
    Enter 23 in the **Degrees of freedom** box
    Select **Input Constant**
    Enter 28.18 in the **Input Constant** box
    Click **OK**

Minitab provides a cumulative probability of .7909, which is the lower tail p-value. The upper tail p-value = 1 − the cumulative probability, or 1 − .7909 = .2091. The two-tailed p-value is two times the minimum of the lower and upper tail p-values. Thus, the two-tailed p-value is 2(.2091) = .4182. We are conducting an upper tail test, so we use p-value = .2091.

The F test statistic Suppose we are conducting a two-tailed test and the test statistic is F = 2.40 with 25 numerator degrees of freedom and 15 denominator degrees of freedom. The Minitab steps to compute the cumulative probability corresponding to F = 2.40 follow.

Step 1. Select the **Calc** menu
Step 2. Choose **Probability Distributions**
Step 3. Choose **F**
Step 4. When the F Distribution dialog box appears:
    Select **Cumulative probability**
    Enter 25 in the **Numerator degrees of freedom** box
    Enter 15 in the **Denominator degrees of freedom** box
    Select **Input Constant**
    Enter 2.40 in the **Input Constant** box
    Click **OK**

Minitab provides the cumulative probability and hence a lower tail p-value = .9594. The upper tail p-value is 1 − .9594 = .0406. Because we are conducting a two-tailed test, the minimum of .9594 and .0406 is used to compute p-value = 2(.0406) = .0812.

Appendix F Computing p-Values Using Minitab and Excel

Using Excel

Excel functions and formulas can be used to compute p-values associated with the z, t, χ^2, and F test statistics. We provide a template in the data file entitled p-Value for use in computing these p-values. Using the template, it is only necessary to enter the value of the test statistic and, if necessary, the appropriate degrees of freedom. Refer to Figure F.1 as we describe how the template is used. For users interested in the Excel functions and formulas being used, just click on the appropriate cell in the template.

The z test statistic We use the Hilltop Coffee lower tail hypothesis test in Section 9.3 as an illustration; the value of the test statistic is $z = -2.67$. To use the p-value template for this hypothesis test, simply enter -2.67 into cell B6 (see Figure F.1). After doing so, p-values for all three types of hypothesis tests will appear. For Hilltop Coffee, we would use the lower tail p-value = .0038 in cell B9. For an upper tail test, we would use the p-value in cell B10, and for a two-tailed test we would use the p-value in cell B11.

The t test statistic We use the Heathrow Airport example from Section 9.4 as an illustration; the value of the test statistic is $t = 1.84$ with 59 degrees of freedom. To use the p-value template for this hypothesis test, enter 1.84 into cell E6 and enter 59 into cell E7 (see Figure F.1). After doing so, p-values for all three types of hypothesis tests will appear.

FIGURE F.1   EXCEL WORKSHEET FOR COMPUTING p-VALUES

	A	B	C	D	E
1	Computing p-Values				
2					
3					
4	Using the Test Statistic z			Using the Test Statistic t	
5					
6	Enter z -->	−2.67		Enter t -->	1.84
7				df -->	59
8					
9	p-value (Lower Tail)	0.0038		p-value (Lower Tail)	0.9646
10	p-value (Upper Tail)	0.9962		p-value (Upper Tail)	0.0354
11	p-value (Two Tail)	0.0076		p-value (Two Tail)	0.0708
12					
13					
14					
15					
16	Using the Test Statistic Chi Square			Using the Test Statistic F	
17					
18	Enter Chi Square -->	28.18		Enter F -->	2.40
19	df -->	23		Numerator df -->	25
20				Denominator df -->	15
21					
22	p-value (Lower Tail)	0.7909		p-value (Lower Tail)	0.9594
23	p-value (Upper Tail)	0.2091		p-value (Upper Tail)	0.0406
24	p-value (Two Tail)	0.4181		p-value (Two Tail)	0.0812

The Heathrow Airport example involves an upper tail test, so we would use the upper tail p-value $= .0354$ provided in cell E10 for the hypothesis test.

The χ^2 test statistic Suppose we are conducting an upper tail test and the value of the test statistic is $\chi^2 = 28.18$ with 23 degrees of freedom. To use the p-value template for this hypothesis test, enter 28.18 into cell B18 and enter 23 into cell B19 (see Figure F.1). After doing so, p-values for all three types of hypothesis tests will appear. We are conducting an upper tail test, so we would use the upper tail p-value $= .2091$ provided in cell B23 for the hypothesis test.

The F test statistic Suppose we are conducting a two-tailed test and the test statistic is $F = 2.40$ with 25 numerator degrees of freedom and 15 denominator degrees of freedom. To use the p-value template for this hypothesis test, enter 2.40 into cell E18, enter 25 into cell E19, and enter 15 into cell E20 (see Figure F.1). After doing so, p-values for all three types of hypothesis tests will appear. We are conducting a two-tailed test, so we would use the two-tailed p-value $= .0812$ provided in cell E24 for the hypothesis test.

Index

A

Accounting, statistical applications, 3
Addition law
  formula, 186
  mutually exclusive events, 168
  probability and, 165–166
Adjusted multiple coefficient of determination,
    multiple regression analysis, 565, 588
Alliance Data Systems, 484
Alternative hypothesis. *See also* Hypothesis tests
  defined, 345–349
  simple linear regression, *t* test, 509–510
  summary of forms for, 348–349
American Statistical Association, ethical guidelines,
    statistical practices, 18–19
Analysis of variance (ANOVA)
  assumptions, 417
  completely randomized design, 420–428
  computer results, 425–426
  conceptual overview, 417–419
  data collection, 416–417
  equality of *k* population means, observational
      study, 427–428
  Excel applications, 550
  experimental design and, 414–419
  Minitab applications, 442
  multiple regression analysis, 564–565, 570–571
  simple linear regression, 512–513
  StatTools applications, 446–447
  table, 424–425, 512–513, 571
Approximate class width, 40–41, 66
Assumptions
  analysis of variance, 417
  multiple regression models, 567–568
  null hypothesis, challenge to, 347–349
  residual analysis validation, 527–531
  simple linear regression models, 506–508

B

Bar charts, categorical data, 34–35
Basic assignment for assigning probability, 155–157
Bayes, Thomas, 178
Bayes' theorem, 178–183
  formula, 186
  probability, subjective method of probability
      assignment, 157
Bell-shaped distribution, 106–107
Bernoulli, Jakob, 208
Bernoulli process, 208

Between-treatments estimates
  analysis of variance, 418–419
  population variance, 421–422
Bimodal data, 90
Binomial probability distribution, 207–215
Binomial probability function, 212–215
  formulas, 226
Box plot
  exploratory data analysis, 110–111
  Minitab applications, 112, 143
  StatTools applications, 147
Business
  descriptive statistics in, 32–33
  sampling applications, 266–267
  statistical applications in, 2–4
Business Week, 2

C

Categorical data, 7
  bar and pie charts, 34–35
  descriptive statistics, 33
  frequency distribution, 33–34
  relative frequency/percent frequency
      distributions, 34
Categorical independent variable, multiple
    regression analysis, 581–583
Categorical variable, 7
Cause-and-effect relationship, 415
Census, defined, 15
Central limit theorem, 281–283
Chebyshev's theorem, 105–106
Chi-square distribution, goodness of fit test,
    458–460
Citibank, 194
Classes
  number of, 40
  width of, 40–41
Classical method, probability assignment,
    155–158
Class limits, 40–41
Class midpoint, 41
Cluster sampling, 295–296
Coefficient of determination
  formula, 535
  simple linear regression, 499–503
Coefficient of variation, 100
  formula, 133
Colgate-Palmolive Company, 32
Combinations, counting rule, 154, 185
Complement of A, 165–166

Complement of events, probability and, 164–165, 186
Completely randomized design, 415–416
  analysis of variance, 420–428
  analysis of variance results, 442
  ANOVA table, 424–425
Computers
  analysis of variance results, 425–426
  simple linear regression, 523–524
  simple random sample generation, 268–269
  statistical analysis and, 17
Conditional probability, 171–175
  formula, 186
  multiplication law, 174–175
Confidence coefficient
  population mean, σ known, 309–310
  simple linear regression, β_1 confidence interval, 510–511
Confidence interval
  estimated regression equation, 517–521
  formula, 536
  Minitab applications, population proportions, 477–478
  multiple regression analysis, 576–577
  population mean, σ known, 309–310
  population proportion, 325–327
  sample size determination, 321–323
  simple linear regression, β_1, 510–511
  StatTools applications, 480–482
Confidence level, population mean, σ known, 309–310
Contingency table, test of independence, 463–467, 472
Continuous random variables, 196
Convenience sampling, 296–297
Correlation coefficient, 120–122
  Excel applications, 145
  formulas, 133–134
  Minitab applications, 143
  simple linear regression, 502–503
  StatTools applications, 147
Counting rules
  combinations, 154
  formulas, 185
  permutations, 154–155
  probability and, 151–155
Covariance, 116–120
  Excel applications, 145
  formulas, 133
  Minitab applications, 143
  StatTools applications, 147
Critical value
  F distribution, population variance, 424
  one-tailed tests, population mean σ known, 356–358
  two-tailed tests, population mean σ known, 360–363
Cross-sectional data, 7–8
Crosstabulation
  applications, 61–64
  descriptive statistics, 54–60

Simpson's paradox, 57–58
  test of independence, 463–467
Cumulative distributions, quantitative data, 44
Customer opinion questionnaire, 12

D

Data acquisition and collection
  analysis of variance, 416–417
  errors in, 13
Data components
  categorical and quantitative data, 7
  cross-sectional and time series data, 7–8
  grouped data, 126–127
  scales of measurement, 6–7
  statistics, 5–9
Data mining, 17–18
Data set, 5
Data sources, 10–13
Data warehousing, 17–18
Degree of belief, 156
Degrees of freedom
  between-treatments estimates, population variance, 421–422
  chi-square distribution, 458–460
  F distribution, population variance, 424
  population mean, σ unknown, 313–318, 432
  population means, σ_1 and σ_2 unknown, 402–403, 432
  simple linear regression, β_1 confidence interval, 510–511
Dependent variable
  defined, 485
  regression model, 506–508
  residual plot against, 531
Descriptive statistics, 13–15
  business applications, 32–33
  categorical data, 33–36
  crosstabulations, 54–60
  Excel applications, 143–146
  exploratory data analysis, 49–54
  Minitab applications, 141–143
  numerical measures, 87–147
  quantitative data, 33, 39–45
  scatter diagram, 58–60
  StatTools applications, 146–147
Deviation about the mean, 98. *See also* Standard deviation
Difference data, matched samples, 410–411
Discrete probability distributions, 197–200
  Excel, 230–231
  Minitab, 230
Discrete random variables, 195
  expected value, 202–203
  formulas, 226
  variance, 203–204
Discrete uniform probability function, 226
Distance intervals, Poisson probability distribution, 220
Distribution, shape of, 103–104
Dot plot, quantitative data, 41–42
Dummy variable, 580–583

E

Economics, statistical applications in, 2–4
Elements, defined, 5
Empirical rule, 106–107
Equality
    F distribution, population variance, 424
    k population means, observational study, 427–428, 433
Errors
    in data acquisition, 13
    type I/type II errors, hypothesis tests, 349–351
Error term
    multiple regression models, 567–568
    regression model, 506–508
Estimated multiple regression equation, 554–555, 576–577, 588
Estimated regression equation
    defined, 486–487
    estimation and prediction, 517–521
    Excel applications, 548–550
    formulas, 535–536
Ethical guidelines, statistical practices, 18–19
Events
    complement of, 164–165
    independent events, 174
    probability and, 160–162
Excel (Microsoft)
    analysis of variance results, 425–426
    descriptive statistics in, 143–146
    discrete probability distributions, 230–231
    goodness of fit and independence tests, 479–480
    hypothesis testing, 386–390, 404–405
    inferences about two populations, 442–443
    interval estimations, 339–341
    multiple regression analysis, 599–600
    population proportions, hypothesis testing, 386–390
    random sampling applications, 302
    regression analysis applications, 548–550
    σ known applications, 339–340, 386–388
    σ unknown applications, 340, 388
    test of independence, 466–467
Expected frequencies, test of independence, 463–467, 472
Expected value, 202–203
    binomial probability distribution, 214–215
    formulas, 226–227
    sample distribution of p, 289–290, 299
    sample distribution of x, 279–280, 299
Experimental design
    analysis of variance and, 414–419
    data collection, 416–417
Experimental outcomes, formulas, 226
Experimental statistical studies, 11–13
Experimental units, 415
Experiments
    binomial probability distribution, 207–215
    probability and, 150–155
Exploratory data analysis
    box plot, 111–112
    five-number summary, 110–111
    stem-and-leaf display, 49–54

F

Factor, 415
Factorial combinations, counting rule, 154
Finance, statistical applications, 4
Finite population correction factor, 280–281
Fisher, Ronald Aylmer, 415
Five-number summary, 110–111
Food and Drug Administration (FDA) (U.S.), 393
Food Lion, 305
Frame, defined, 267
Frequency distribution
    categorical data, 33–34
    crosstabulation, 56–60
    cumulative, 44
    grouped data, 127–128
    quantitative data, 39–41
F test
    multiple regression analysis, 569–571, 588
    residual analysis validation, 528–531
    simple linear regression, 511–513

G

Galton, Francis (Sir), 485
Gauss, Carl Friedrich, 490
Goodness of fit test
    Excel applications, 479–480
    Minitab applications, 478–479
    multinomial population proportion, 457–460, 472
    regression model, 506–508
Gosset, William Sealy, 312
Government agencies, as data source, 11
Graphical data summary, 61–65. *See also* Descriptive statistics
    binomial probability function, 213–215
    discrete probability distribution, 199–200
Grouped data, 126–127
    formulas, 134

H

Histograms
    population mean, σ unknown, 317–318
    quantitative data, 42–43
    relative frequency, simple random samples, 278–279
    skewness of distribution, 104
    variability measurements, 96–97
Hypergeometric probability distribution, 221–223
    formula, 227
Hypergeometric probability function, 222–223
    formula, 227
Hypothesis tests
    alternative hypothesis, 346–349
    analysis of variance, 417–419
    completely randomized design, 420–428
    equality of k population means, observational study, 427–428
    Excel applications, 386–390
    experimental design, 416–417

Index   677

Hypothesis tests (*continued*)
  F distribution, population variance, 423–424
  inferential difference, μ_1 and μ_2, 397–398, 403–405, 432
  interval estimation, 362–363
  matched samples, 410–411
  Minitab applications, 385–386, 477–478
  null hypothesis, 346–349
  population means, σ_1 and σ_2 known, 397–398
  population means, σ_1 and σ_2 unknown, 403–405
  population mean σ known, 352–363
  population mean σ unknown, 367–370
  population proportion, 373–375, 452–453
  proportions of multinomial population, 456–460, 472
  StatTools applications, 391, 445, 482
  type I and type II errors, 349–351

I

Independence, test of, 463–467, 472, 478–479
  Excel applications, 479–480
  Minitab applications, 478–479
  StatTools, 482
Independent events
  conditional probability, 174
  multiplication law, 186
Independent random sample
  population means, σ_1 and σ_2 known, 394–398, 432
  population means, σ_1 and σ_2 unknown, 402–403, 432
Independent sample design, population means, 409–411
Independent variables
  defined, 485
  multiple coefficient of determination, 565
  multiple regression analysis, 558–559, 568–573, 578–583
  regression model, 506–508
  residual plot against, 529–531
Indicator variable, multiple regression analysis, 580–583
Individual value of *y*
  formula, 536
  prediction interval, 519–521
Inferences
  difference between two population proportions, 450–453
  Excel applications, 442–443
  Minitab applications, 440–441
  population means, matched samples, 409–411
  population means, σ_1 and σ_2 known, 394–398
  population means, σ_1 and σ_2 unknown, 401–405
  StatTools applications, 444–446
Internal company records, as data source, 10
International Paper, 553
Interquartile range, 97–98
  formula, 133
Intersection of A and B, 166–168
Interval estimation

  computer-based ANOVA, 426
  defined, 305
  difference between two populations, 450–452, 471
  estimated regression equation, 517–518
  Excel applications, 339–341
  hypothesis tests and, 362–363
  Minitab applications, 338
  multiple regression analysis, 576–577
  population mean, sample size, 322–323
  population mean, σ known, 306–310
  population mean, σ unknown, 313–316
  population means, σ_1 and σ_2 known, 394–398, 432
  population means, σ_1 and σ_2 unknown, 401–403, 432
  population proportion, 325–327, 331
  sample size for, 331
  StatTools applications, 341–343, 445
  summary, 318–319
Interval scale, 6
*i*th residual
  formula, 536
  residual analysis validation, 527–531
  simple linear regression, 499–503

J

John Morrell & Company, 345
Joint probability, 172
Judgment sampling, 297

K

k levels, categorical variables, multiple regression analysis, 582–583

L

Least squares criterion
  multiple regression analysis, 555–559, 588
  simple linear regression, 488–492, 535
Length intervals, Poisson probability distribution, 220
Levels of significance, type I/type II errors, hypothesis tests, 350–351
Linear approximation, nonlinear relationships, 513–514
Linear regression. *See* Simple linear regression
Location measures, 88–93
  mean, 88–89
  median, 89–90
  mode, 90
  percentiles, 91–92
  quartiles, 92–93
Lower class limit, frequency distribution, 40–41
Lower tail test, population mean σ known, 357–358

M

Marginal probability, 172–173
Margin of error
  confidence interval, estimated regression equation, 518–521

Margin of error (*continued*)
 defined, 305–306
 population mean, sample size, 321–323
 population mean, σ known, 306–310
 population mean, σ unknown, 313–316
 population proportion, 289–293, 324–327
 prediction interval, 519–521
 simple linear regression, β_1 confidence interval, 510–511
Marketing, statistical applications, 4
Matched samples
 Excel applications, 443
 population means, 409–411, 441
 StatTools applications, 446
Meadwestvaco Corporation, 266
Mean
 Excel applications, 144
 numerical measures, 88–89
 population mean, 89, 133
 sample mean formula, 133
 weighted mean, 125–126
 z-scores, 105
Mean square due to error (MSE)
 computer-based ANOVA, 426
 F distribution, population variance, 423–424
 formula, 535, 588
 F test, simple linear regression, 512–513
 multiple regression analysis, 569–571
 significance test, simple linear regression, 508–509
 within-treatments estimates, population variance, 422, 433
Mean square due to treatments (MSTR)
 between-treatments estimates, population variance, 421–422, 433
 F distribution, population variance, 423–424
Mean square regression (MSR)
 formula, 536, 588
 F test, simple linear regression, 511–513
 multiple regression analysis, 569–571
Mean value, confidence interval, estimated regression equation, 518–521
Measurement, scales of, 6–7
Media, statistics in, 2–3
Median
 50th percentile as, 92
 Excel applications, 144
 location measurement, 89–90
Microsoft Excel. *See* Excel (Microsoft)
Minimum sample size, population mean, 322–323
Minitab
 analysis of variance results, 425–426, 442
 binomial probability experiments, 215
 box plots, 112, 143
 confidence interval, population mean, σ unknown, 315–316
 covariance and correlation, 143
 descriptive statistics, 141–143
 discrete probability distributions, 230
 goodness of fit and independence testing, 478–479
 hypothesis testing, 385–386, 404–405

 interval estimates using, 338–339
 matched samples, population means, 441
 multiple regression analysis, 557–559, 570–571, 579–583, 599
 population proportion inferences, 477–478
 random sampling applications, 301
 regression analysis applications, 547–548
 simple linear regression applications, 523–524
 test of independence, 466–467
 two-population inferences, 440–441
Mode
 Excel applications, 144
 location measurement, 90
Multicollinearity, multiple regression analysis, 572–573
Multimodal data, 90
Multinomial population, proportions of, hypothesis testing, 456–460
Multiple coefficient of determination, 564–565, 588
Multiple regression
 categorical independent variables, 578–583
 coefficient interpretations, 558
 defined, 554
 estimated regression equation, 554–555, 576–577, 588
 Excel applications, 599–600
 formulas, 587–588
 F test, 569–571
 least squares method, 555–559
 Minitab applications, 557–559, 570–571, 579–583, 599
 model, 554–555, 567–568
 multicollinearity, 572–573
 multiple coefficient of determination, 564–565
 regression equation, 554–555
 StatTools applications, 600–601
 test for significance, 568–573
 t test, 571–572
Multiple-step experiments, probability, 151–152
Multiplication law
 formula, 186
 independent events, 174–175
Mutually exclusive events, probability and, 168

N

95% confidence interval, multiple regression analysis, 576–577
Nominal scale, 6
Nonlinear relationship, linear approximation, 513–514
Nonprobability sampling, 296–297
Normal probability distribution. *See* Probability; Probability distribution
Null hypothesis
 analysis of variance, 417–419
 defined, 345–349
 equality of *k* population means, observational study, 427–428
 experimental design, 416–417
 one-tailed tests, population mean σ known, 353–358

Null hypothesis (*continued*)
    population mean, μ_1 and μ_2, 397–398
    population proportion, 452–453
    simple linear regression, *t* test, 509–510
    summary of forms for, 348–349
    test of independence, 463–467
Numerical measures
    association between two variables, 116–122
    Chebyshev's theorem, 105–106
    distribution shape, 103–104
    empirical rule, 106–107
    exploratory data analysis, 110–112
    grouped data, 126–128
    of location, 88–93
    outlier detection, 107
    variability, 96–100
    weighted mean, 125–126
    z-scores, 104–105

O

Observation
    defined, 6
    independent, analysis of variance, 417
Observational statistical studies, 11–13
    equality of *k* population means, 427–428
    experimental design, 415
Observed frequencies, test of independence, 463–467
Oceanwide Seafood, 149
Ogive graph, quantitative data, 44–45
One-tailed tests
    critical value approach, 356–358
    goodness of fit test, 458–460
    population mean σ known, 352–358
    population mean σ unknown, 367–368
    p-value approach, 354–356
    test of independence, 463–467
Ordinal scale, 6
Outlier detection, 107
Overall sample mean
    analysis of variance, 418–419, 433
    completely randomized design, analysis of variance, 420–428

P

Parameters
    multiple regression analysis, 581–583
    population characteristics, 268
Pareto, Vilfredo, 35
Pareto chart, 35
Partitioning, analysis of variance table, 425, 433
Pearson, Karl, 485
Pearson product moment, correlation coefficient, 120–121, 133–134
Percent frequency distribution
    audit time data, 40–41
    categorical data, 34
    crosstabulation, 56–60
    cumulative, 44
    quantitative data, 41

Percentile, 91–92
Perfect positive linear relationship, scatter diagram, 121
Permutations, counting rule, 154–155, 185
Pie charts, categorical data, 35–36
Planning value
    population proportion, 326–327
    population standard deviation, 322–323
Point estimator/point estimation
    difference between two populations, 450–452, 471
    estimated regression equation, 517
    margin of error, 305–306
    numerical measures, 88
    population mean, σ unknown, 315–316, 431
    population means, σ_1 and σ_2 known, 395–398, 431
    sampling procedures, 273–275
    simple linear regression, β_1 confidence interval, 510–511
Poisson, Siméon, 218
Poisson distribution, 218–220
Poisson probability function, 218–220
    formula, 227
Pooled estimates/pooled estimator
    analysis of variance, 419
    population proportion, 452–453, 472
Pooled StDev, computer-based ANOVA, 426
Population
    covariance, 117–118, 133
    infinite sample population, 270–271
    normal distribution, 281
    parameters, numerical measures, 88
    proportion, 324–327, 331
    sampled population, 267
    statistical inference, 15
    without normal distribution, 281–283
Population mean(s)
    basic properties, 284–285
    equality of, (*k* population means), 427–428
    Excel applications, 339–341, 442–443
    formula for, 133–134
    inferential difference, μ_1 and μ_2, 394–398
    location measures, 89
    matched samples, inferences about, 409–411
    Minitab applications, 338–339, 385–386, 440–441
    sample size, 321–323
    σ known, 306–310, 331, 352–363, 385–388, 394–398
    σ unknown, 312–319, 331, 367–370, 385–386, 388, 391, 401–405
    StatTools applications, 341–343, 391, 444–446
Population proportion
    difference between two populations, inferences about, 450–453
    Excel applications, 340–341, 388–390
    hypothesis tests, 373–375
    Minitab applications, 338–339, 386, 477–478
    multinomial population, hypothesis testing, 456–460
    sample size formula, 331
    StatTools applications, 343, 391, 480–482
Population variance, 98–99
    between-treatments estimates, 421–422
    formulas, 133–134

Index

Population variance (*continued*)
   F test, 423–424
   within-treatments estimates, 422
Posterior probability, 178–183
Prediction interval
   estimated regression equation, 517
   formula, 536
   multiple regression analysis, 576–577
Prior probability, 178–183
Probability
   addition law, 165–168
   assignment of, 155–157
   Bayes' theorem, 178–182
   complement of events, 164–165
   conditional probability, 171–175
   counting rules, combinations, and permutations, 151–155
   defined, 150
   events and, 160–162
   experiments, 150–151
   independent events, 174
   multiplication law, 174–175
   relative frequency and, 157–158
   sample distribution of p, 292–293
   sample mean, population, 285–286
   tabular approach, 182
Probability density function, 197–200
Probability distribution
   binomial, 207–215
   discrete probability distributions, 197–200
   hypergeometric probability, 221–223
   Poisson probability, 218–220
   population mean, σ unknown, 312–318
Probability sampling, 296–297
Production, statistical applications, 4
p-values
   chi-square distribution, goodness-of-fit test, 458–460
   F distribution, population variance, 423–424
   one-tailed tests, population mean σ known, 354–356
   population mean, μ_1 and μ_2, 398, 404–405
   population proportion, hypothesis tests, 374–375, 453
   simple random samples, 278–279
   two-tailed tests, population mean σ known, 359–363

Q

Qualitative variables, 568–573
Quantitative data, 7. *See also* Categorical data
   cumulative distribution, 44
   descriptive statistics, 33, 39–45
   dot plot, 41–42
   frequency distribution, 39–41
   histogram, 42–43
   ogive, 44–45
   relative frequence/percent frequency distribution, 41
Quantitative variable, 7
Quartiles, 92–93

R

Random experiments, probability, 158
Randomization, experimental design, 415
Random numbers, simple random sample, 268–269
Random samples
   Excel applications, 302
   infinite population, 270–271
   Minitab applications, 301
   simple, 268–270
   StatTools applications, 302–303
   stratified random sampling, 295
Random variables, 194–196
   standard deviation, 203–204, 266
Range, 97
Ratio scale, 6–7
Regression analysis. *See also* Multiple regression analysis; Simple linear regression
   computer solutions, 523–524
   Excel applications, 548–550
   limitations of, 513–514
   Minitab applications, 547–548
   model assumptions, 506–508
   residual analysis validation, 527–531
   StatTools, 550–551
Regression coefficients, multiple regression analysis, 558–559
Regression equation
   estimated, 486–487, 517–521
   formulas, 535–536
   multiple regression analysis, 554–555, 568–571, 576–577
   simple linear regression, 485–486
Regression model
   assumptions, 506–508
   multiple regression analysis, 554–555, 567–568
   simple linear regression, 485–486
Rejection rule
   F distribution, population variance, 424
   F test, simple linear regression, 512–513
   multiple regression analysis, 570–572
Relative frequency distribution
   audit time data, 40–41
   categorical data, 34
   crosstabulation, 56–60
   cumulative, 44
   formula, 66
   probability assignment, 156–158
   quantitative data, 41
Replication, experimental design, 416
Research hypothesis, 346–349
Residual analysis, model assumptions, 527–531
Residual plot
   dependent variable, 531
   independent variable x, 529–531
Response variable, 415, 417

S

Sample correlation coefficient, 120–122
   formula, 535

Sample correlation coefficient (*continued*)
  formulas, 133–134
  multiple regression analysis, 573
Sample covariance, 116–120
Sample mean, 88–89
  analysis of variance, 418–419, 432
  completely randomized design, analysis of variance, 420–428
  experimental design, 416–417
  formulas, 133–134
  grouped data, 126–127
  population mean, σ known, 306–310
  probability, simple random population sample, 285–286
Sample points
  events as, 160–162
  probability experiments, 150–151
Sample population, 267
Sample/sampling
  business applications, 266–267
  cluster sampling, 295–296
  convenience sampling, 296–297
  correlation coefficient, 122
  covariance, 116–118, 133
  distribution, simple linear regression, *t* test, 509–510
  distribution of *p*, 289–293, 324–327
  distribution of *x*, 278–286, 358–363
  distributions, 276–277, 417–419, 423–424
  finite population, 268–269
  infinite population, 270–271
  judgment sampling, 297
  mean, formulas, 133–134
  point estimation, 273–275
  selection, 268–271
  simple random sample, 268–270
  statistical inference, 15
  stratified random sampling, 295
  systematic sampling, 296
  with/without replacement, 269–270
Sample size
  determination, 321–323
  interval estimate, population mean, 331
  matched samples, 410–411
  population mean, σ unknown, 315–316
  population proportion, 326–327
  sampling distribution of *x*, 285–286
  StatTools applications, 343
Sample space, probability experiments, 150
Sample statistics, numerical measures, 88
Sample survey, 15
Sample variance, 98–99
  completely randomized design, analysis of variance, 420–428, 432
  formulas, 133–134
  grouped data, 126–127
Scales of measurement, 6–7
Scatter diagrams
  correlation coefficient, 121
  covariance measurements, 118–120
  descriptive statistics, 58–60
  multiple regression analysis, 556–559, 582–583
  simple linear regression, 488–492

σ known
  Excel applications, 339–340, 386–388
  formula, 331
  hypothesis tests, 352–363
  inferential difference, σ_1 and σ_2 known, 394–398
  Minitab applications, 338, 385
  population mean, 306–310, 331, 352–363, 385–388, 394–498
  two-tailed test, 358–363
Sigma squared (σ^2) estimation
  formula, 535
  simple linear regression, 508–509
σ unknown, 331
  Excel applications, 340, 388
  hypothesis tests, 367–370
  margin of error and interval estimate, 313–316
  Minitab applications, 338, 385–386
  one-tailed tests, 367–368
  population mean, 312–319, 331, 367–370, 385–386, 388, 391, 401–405
  sample size, 316–318
  StatTools applications, 341–342, 391
  two-tailed tests, 368–370
Significance test
  multiple regression analysis, 568–573
  simple linear regression, 508–514
Simple linear regression
  ANOVA table, 512–513
  coefficient of determination, 499–503
  computer solution, 523–524
  confidence for β_1, 510–511
  confidence interval, mean value of *y*, 518–519
  correlation coefficient, 502–503
  defined, 485
  estimated regression equation, 486–487, 517–521
  estimation process, 487
  Excel applications, 548–550
  formulas, 535–536
  F test, 511–512
  interval estimation, 517
  least squares method, 488–492
  Minitab applications, 547–548
  model, 485–487, 506–508, 535
  point estimation, 517
  prediction interval, individual value of *y*, 519–521
  regression equation, 485–486
  regression model, 485–486
  residual analysis, 527–531
  residual plot against *x*, 529–531
  residual plot against *y*, 531
  sigma squared (σ^2) estimation, 508–509
  significance testing, 508–514
  StatTools, 550–551
  t test, 509
Simple random sample, 268–270
  completely randomized design, analysis of variance, 420–428
  population mean, σ known, 306–310
  sample size, 285–286
  sampling distribution of *x*, 283–285
Simpson's paradox, crosstabulation, 57–58
Single-factor experiment, 415

Skewness levels
  distribution shape, 103–104
  histogram, 43
Small Fry Design, 87
Squared deviation, 98–99
Standard deviation
  discrete random variable variance, 204
  Excel applications, 144
  formula, 133
  sample distribution of p, 290, 299
  sample distribution of x, 280–281, 299
  simple linear regression, formula, 536
  simple linear regression, t test, 509–510
  variability measurement, 100
  z-scores, 105
Standard error
  difference between two populations, 450–452, 471
  of estimate, formula, 536
  of estimate, significance test, simple linear regression, 508–509
  of mean, 133, 281, 290–291
  point estimator, $\chi_1 - \chi_2$ known, 395–398, 432
  of proportion, 290
Standardized value, z-scores, 104–105
Statistical inference, 15–16
Statistical modeling, data mining, 18
Statistical studies, as data source, 11–13
Statistics
  business/economics applications, 2–4
  computers and, 17
  data components, 5–9
  data mining, 17–18
  data sources, 10–13
  defined, 3
  descriptive statistics, 13–15
  ethical practice guidelines, 18–19
StatTools
  analysis of variance results, 425–426, 446–447
  descriptive statistics applications, 146–147
  hypothesis testing, 391, 445
  Interval estimate applications, 341–343
  interval estimation, 445
  matched samples, 446
  multiple regression analysis, 600–601
  population inferences, 444–446
  population proportion applications, 480–482
  random sampling applications, 302–303
  regression analysis applications, 550–551
  test of independence with, 482
Stem-and-leaf display
  applications, 53–54
  exploratory data analysis, 49–54
  methods, 53
Stratified random sampling, 295
Subjective method of probability assignment, 156–158
Summation sign, mean, 88
Sum of squares due to error (SSE)
  formula, 535
  multiple regression analysis, 564–565, 569–571
  significance test, simple linear regression, 508–509
  simple linear regression, 499–503
  within-treatments estimates, population variance, 422, 433
Sum of squares due to regression (SSR)
  formula, 535
  multiple regression analysis, 564–565, 569–571
  simple linear regression, 500–503
Sum of squares due to treatments (SSTR), between-treatments estimates, population variance, 421–422, 433
Sum of squares of deviations, simple linear regression, 489–492
Systematic sampling, 296

T

Tabular data summary, 61–65. *See also* Descriptive statistics
  Bayes' theorem, 182
  binomial probabilities, 213–215
Tail distribution, population mean, σ unknown, 313–318
Test statistic
  equality of k population means, observational study, 427–428, 433
  F distribution, population variance, 423–424
  F test, 512–513, 588
  goodness of fit test, 458–460, 472
  hypothesis tests, μ_1 and μ_2, 397–398, 403–405, 432
  hypothesis tests, population mean σ unknown, 379, 403–405
  matched samples, 410–411
  multiple regression analysis, 570–571
  one-tailed tests, population mean σ known, 353–358
  population mean σ known, 397–398
  population mean σ unknown, 403–405
  population proportion, hypothesis tests, 374–375, 379, 452–453
  test of independence, 463–467
Time intervals, Poisson probability distribution, 218–220
Time series data, 7–9
Total sum of squares (SST)
  ANOVA table, 424–425, 433
  formula, 535
  multiple regression analysis, 564–565
  simple linear regression, 500–503
Treatments, defined, 415
Tree diagrams
  Bayes' theorem example, 180–182
  binomial probability distribution, 210–215
  multiple-step experiments, 152–155
Trendlines, 58–60
Trimmed mean, 93
t test
  formula, 536
  matched samples, 410–411
  multiple regression analysis, 571–572, 588
  population mean, σ unknown, 313–318, 402–403
  population mean σ known, 397–398
  residual analysis validation, 528–531

t test (*continued*)
   μ unknown, two-tailed tests, 369–370
   simple linear regression, 509–510
Two-tailed test
   population mean, μ_1 and μ_2, 398
   population mean σ known, 358–363
   population mean σ unknown, 368–370
Type I/type II errors, hypothesis tests, 349–351

U

Union of A and B, defined, 165–166
United Way, 449
Upper class limit
   cumulative distributions, 44
   frequency distribution, 40–41
Upper tail test
   F distribution, population variance, 424
   goodness of fit test, 458–460
   population mean σ known, 357–358
   test of independence, 463–467

V

Variability measures, 96–100
   coefficient of variation, 100
   interquartile range, 97–98
   range, 97
   standard deviation, 100
   variance, 98–100

Variables
   categorical, 7
   defined, 5
   dependent variable, 485, 506–508, 531
   independent variables, 485, 506–508, 529–531, 558–559, 565, 568–573, 578–583
   indicator variable, 580–583
   quantitative, 7
Variance, 98–100
   binomial probability distribution, 214–215, 227
   discrete random variables, 203–204, 226
   Excel applications, 144
Venn diagram
   addition law, 165–166
   complement of events, 164–165
   conditional probability, 173–175

W

Weighted mean, 125–126
   formula, 134
Within-treatments estimates
   analysis of variance, 419
   population variance, 422

Z

z-scores, 104–105
   formula, 133

CUMULATIVE PROBABILITIES FOR THE STANDARD NORMAL DISTRIBUTION

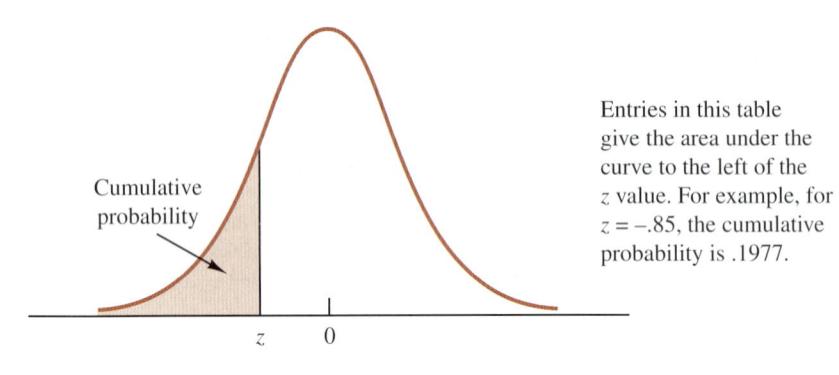

Entries in this table give the area under the curve to the left of the z value. For example, for $z = -.85$, the cumulative probability is .1977.

z	.00	.01	.02	.03	.04	.05	.06	.07	.08	.09
−3.0	.0013	.0013	.0013	.0012	.0012	.0011	.0011	.0011	.0010	.0010
−2.9	.0019	.0018	.0018	.0017	.0016	.0016	.0015	.0015	.0014	.0014
−2.8	.0026	.0025	.0024	.0023	.0023	.0022	.0021	.0021	.0020	.0019
−2.7	.0035	.0034	.0033	.0032	.0031	.0030	.0029	.0028	.0027	.0026
−2.6	.0047	.0045	.0044	.0043	.0041	.0040	.0039	.0038	.0037	.0036
−2.5	.0062	.0060	.0059	.0057	.0055	.0054	.0052	.0051	.0049	.0048
−2.4	.0082	.0080	.0078	.0075	.0073	.0071	.0069	.0068	.0066	.0064
−2.3	.0107	.0104	.0102	.0099	.0096	.0094	.0091	.0089	.0087	.0084
−2.2	.0139	.0136	.0132	.0129	.0125	.0122	.0119	.0116	.0113	.0110
−2.1	.0179	.0174	.0170	.0166	.0162	.0158	.0154	.0150	.0146	.0143
−2.0	.0228	.0222	.0217	.0212	.0207	.0202	.0197	.0192	.0188	.0183
−1.9	.0287	.0281	.0274	.0268	.0262	.0256	.0250	.0244	.0239	.0233
−1.8	.0359	.0351	.0344	.0336	.0329	.0322	.0314	.0307	.0301	.0294
−1.7	.0446	.0436	.0427	.0418	.0409	.0401	.0392	.0384	.0375	.0367
−1.6	.0548	.0537	.0526	.0516	.0505	.0495	.0485	.0475	.0465	.0455
−1.5	.0668	.0655	.0643	.0630	.0618	.0606	.0594	.0582	.0571	.0559
−1.4	.0808	.0793	.0778	.0764	.0749	.0735	.0721	.0708	.0694	.0681
−1.3	.0968	.0951	.0934	.0918	.0901	.0885	.0869	.0853	.0838	.0823
−1.2	.1151	.1131	.1112	.1093	.1075	.1056	.1038	.1020	.1003	.0985
−1.1	.1357	.1335	.1314	.1292	.1271	.1251	.1230	.1210	.1190	.1170
−1.0	.1587	.1562	.1539	.1515	.1492	.1469	.1446	.1423	.1401	.1379
−.9	.1841	.1814	.1788	.1762	.1736	.1711	.1685	.1660	.1635	.1611
−.8	.2119	.2090	.2061	.2033	.2005	.1977	.1949	.1922	.1894	.1867
−.7	.2420	.2389	.2358	.2327	.2296	.2266	.2236	.2206	.2177	.2148
−.6	.2743	.2709	.2676	.2643	.2611	.2578	.2546	.2514	.2483	.2451
−.5	.3085	.3050	.3015	.2981	.2946	.2912	.2877	.2843	.2810	.2776
−.4	.3446	.3409	.3372	.3336	.3300	.3264	.3228	.3192	.3156	.3121
−.3	.3821	.3783	.3745	.3707	.3669	.3632	.3594	.3557	.3520	.3483
−.2	.4207	.4168	.4129	.4090	.4052	.4013	.3974	.3936	.3897	.3859
−.1	.4602	.4562	.4522	.4483	.4443	.4404	.4364	.4325	.4286	.4247
−.0	.5000	.4960	.4920	.4880	.4840	.4801	.4761	.4721	.4681	.4641

CUMULATIVE PROBABILITIES FOR THE STANDARD NORMAL DISTRIBUTION

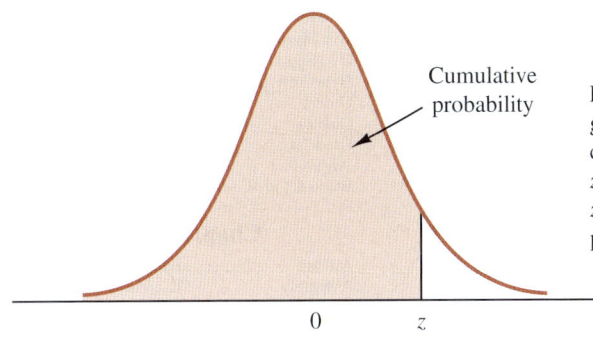

Entries in the table give the area under the curve to the left of the z value. For example, for $z = 1.25$, the cumulative probability is .8944.

z	.00	.01	.02	.03	.04	.05	.06	.07	.08	.09
.0	.5000	.5040	.5080	.5120	.5160	.5199	.5239	.5279	.5319	.5359
.1	.5398	.5438	.5478	.5517	.5557	.5596	.5636	.5675	.5714	.5753
.2	.5793	.5832	.5871	.5910	.5948	.5987	.6026	.6064	.6103	.6141
.3	.6179	.6217	.6255	.6293	.6331	.6368	.6406	.6443	.6480	.6517
.4	.6554	.6591	.6628	.6664	.6700	.6736	.6772	.6808	.6844	.6879
.5	.6915	.6950	.6985	.7019	.7054	.7088	.7123	.7157	.7190	.7224
.6	.7257	.7291	.7324	.7357	.7389	.7422	.7454	.7486	.7517	.7549
.7	.7580	.7611	.7642	.7673	.7704	.7734	.7764	.7794	.7823	.7852
.8	.7881	.7910	.7939	.7967	.7995	.8023	.8051	.8078	.8106	.8133
.9	.8159	.8186	.8212	.8238	.8264	.8289	.8315	.8340	.8365	.8389
1.0	.8413	.8438	.8461	.8485	.8508	.8531	.8554	.8577	.8599	.8621
1.1	.8643	.8665	.8686	.8708	.8729	.8749	.8770	.8790	.8810	.8830
1.2	.8849	.8869	.8888	.8907	.8925	.8944	.8962	.8980	.8997	.9015
1.3	.9032	.9049	.9066	.9082	.9099	.9115	.9131	.9147	.9162	.9177
1.4	.9192	.9207	.9222	.9236	.9251	.9265	.9279	.9292	.9306	.9319
1.5	.9332	.9345	.9357	.9370	.9382	.9394	.9406	.9418	.9429	.9441
1.6	.9452	.9463	.9474	.9484	.9495	.9505	.9515	.9525	.9535	.9545
1.7	.9554	.9564	.9573	.9582	.9591	.9599	.9608	.9616	.9625	.9633
1.8	.9641	.9649	.9656	.9664	.9671	.9678	.9686	.9693	.9699	.9706
1.9	.9713	.9719	.9726	.9732	.9738	.9744	.9750	.9756	.9761	.9767
2.0	.9772	.9778	.9783	.9788	.9793	.9798	.9803	.9808	.9812	.9817
2.1	.9821	.9826	.9830	.9834	.9838	.9842	.9846	.9850	.9854	.9857
2.2	.9861	.9864	.9868	.9871	.9875	.9878	.9881	.9884	.9887	.9890
2.3	.9893	.9896	.9898	.9901	.9904	.9906	.9909	.9911	.9913	.9916
2.4	.9918	.9920	.9922	.9925	.9927	.9929	.9931	.9932	.9934	.9936
2.5	.9938	.9940	.9941	.9943	.9945	.9946	.9948	.9949	.9951	.9952
2.6	.9953	.9955	.9956	.9957	.9959	.9960	.9961	.9962	.9963	.9964
2.7	.9965	.9966	.9967	.9968	.9969	.9970	.9971	.9972	.9973	.9974
2.8	.9974	.9975	.9976	.9977	.9977	.9978	.9979	.9979	.9980	.9981
2.9	.9981	.9982	.9982	.9983	.9984	.9984	.9985	.9985	.9986	.9986
3.0	.9987	.9987	.9987	.9988	.9988	.9989	.9989	.9989	.9990	.9990

Fundamentals of Business Statistics 6e WEBfiles

Chapter 1
Morningstar	Table 1.1
Norris	Table 1.5
Shadow02	Exercise 25

Chapter 2
AirSurvey	Exercise 7
ApTest	Table 2.8
Audit	Table 2.4
BestTV	Exercise 4
Careers	Exercise 26
CityTemp	Exercise 46
Computer	Exercise 21
Crosstab	Exercise 29
DJIAPrices	Exercise 17
DYield	Exercise 41
FedBank	Exercise 10
Fortune	Exercise 51
Frequency	Exercise 11
FuelData08	Exercise 37
GMSales	Exercise 40
Holiday	Exercise 18
LargeCorp	Exercise 19
LivingArea	Exercise 9
Major	Exercise 39
Marathon	Exercise 28
Movies	Case Problem 2
MutualFunds	Exercise 34
Names	Exercise 5
Networks	Exercise 6
NewSAT	Exercise 42
OffCourse	Exercise 20
PelicanStores	Case Problem 1
Population	Exercise 44
Restaurant	Table 2.9
Scatter	Exercise 30
SoftDrink	Table 2.1
Stereo	Table 2.12
SuperBowl	Exercise 43

Chapter 3
3Points	Exercise 6
Ages	Exercise 59
BackToSchool	Exercise 22
BowlGames	Exercise 47
CellService	Exercise 42
Disney	Exercise 12
Economy	Exercise 10
FairValue	Exercise 67
Franchise	Exercise 40
Homes	Exercise 64
Hotels	Exercise 5
Housing	Exercise 49
MajorSalary	Figure 3.7
MLBSalaries	Exercise 43
Movies	Case Problem 2
Mutual	Exercise 44
NCAA	Exercise 34
NFLTickets	Exercise 35
PelicanStores	Case Problem 1
Penalty	Exercise 62
PovertyLevel	Exercise 65
Runners	Exercise 41
Shoppers	Case Problem 3
SpringTraining	Exercise 68
StartSalary	Table 3.1
Stereo	Table 3.6
StockMarket	Exercise 50
TaxCost	Exercise 8
Travel	Exercise 66
Visa	Exercise 58
WorldTemp	Exercise 51

Chapter 4
Judge	Case Problem

Chapter 6
Volume	Exercise 24

Chapter 7
EAI	Section 7.1
MetAreas	Appendix 7.1, 7.2, & 7.3
MutualFund	Exercise 14

Chapter 8
Alcohol	Exercise 21
Auto	Case Problem 3
Flights	Exercise 48
GulfProp	Case Problem 2
Interval p	Appendix 8.2
JobSatisfaction	Exercise 37
JobSearch	Exercise 18
Lloyd's	Section 8.1
Miami	Exercise 17
NewBalance	Table 8.3
Nielsen	Exercise 6
NYSEStocks	Exercise 47
Professional	Case Problem 1
Program	Exercise 20
Scheer	Table 8.4
Standing	Exercise 49
TaxReturn	Exercise 9
TeeTimes	Section 8.4
TicketSales	Exercise 22

Chapter 9
AgeGroup	Exercise 39
AirRating	Section 9.4
Bayview	Case Problem 2
Coffee	Section 9.3
Diamonds	Exercise 29
Drowsy	Exercise 44
Eagle	Exercise 43
FirstBirth	Exercise 50
Fowle	Exercise 21
Gasoline	Exercise 53
GolfTest	Section 9.3
Hyp Sigma Known	Appendix 9.2
Hyp Sigma Unknown	Appendix 9.2
Hypothesis p	Appendix 9.2
Orders	Section 9.4
Quality	Case Problem 1
UsedCars	Exercise 32
WeeklyPay	Exercise 51
WomenGolf	Section 9.5

Chapter 10
AirFare	Exercise 24
Assembly	Exercise 48
AudJudg	Exercise 36
Browsing	Exercise 49
Cargo	Exercise 13
CheckAcct	Section 10.2
Chemitech	Table 10.3
Earnings2005	Exercise 22
ExamScores	Section 10.1
Funds	Exercise 46
Golf	Case Problem 1
GolfScores	Exercise 26
GrandStrand	Exercise 38
HomePrices	Exercise 41
Hotel	Exercise 6
Matched	Table 10.2
Medical1	Case Problem 2
Medical2	Case Problem 2
Mutual	Exercise 42
NCP	Table 10.6
Paint	Exercise 37
PriceChange	Exercise 44
RandomDesign	Exercise 32
RentalVacancy	Exercise 47
SalesSalary	Case Problem 3
SAT	Exercise 18
SatisJob	Exercise 45
SATMath	Exercise 16
SoftwareTest	Table 10.1
TVRadio	Exercise 25

Chapter 11
Albers	Appendix 11.5
FitTest	Appendix 11.2
Independence	Appendix 11.2
NYReform	Case Problem
Occupancy	Exercise 32
TaxPrep	Section 11.1

Chapter 12
Absent	Exercise 55
AgeCost	Exercise 56
Alumni	Case Problem 3
Armand's	Table 12.1
Beta	Case Problem 1
BusinessSchools	Exercise 43
DigitalCameras	Exercise 27
Ellipticals	Exercises 5, 22, & 30
ExecSalary	Exercise 10
HomePrices	Exercise 49
HondaAccord	Exercise 6
HoursPts	Exercise 57
IPO	Exercise 50
IRSAudit	Exercise 59
Jensen	Exercise 53
JetSki	Exercise 12
JobSat	Exercise 60
Laptop	Exercise 14
MktBeta	Exercise 58
OnlineEdu	Exercise 52
PGATour	Case Problem 4
PlasmaTV	Exercise 20
RaceHelmets	Exercise 44
Safety	Case Problem 2
Sales	Exercises 7 & 19
SleepingBags	Exercises 8, 28, & 36
SportyCars	Exercise 11
Stocks500	Exercise 51
Suitcases	Exercise 9

Chapter 13
Alumni	Case Problem 2
Basketball	Exercise 24
Boats	Exercises 9, 17, & 30
BusinessSchools2	Exercises 6 & 16
Butler	Tables 13.1 & 13.2
Consumer	Case Problem 1
Exer2	Exercise 2
FuelData	Exercise 47
Johnson	Table 13.6
Laptop	Exercise 7
MutualFunds	Exercise 46
NBA	Exercises 10, 18, & 26
NFLStats	Case Problem 4
PGATour	Case Problem 3
Repair	Exercise 35
RestaurantRatings	Exercise 37
Sedans	Exercise 8
Showtime	Exercises 5 & 15
SmallCars	Exercise 25
SportsCar	Exercise 31
Stroke	Exercise 38
TireRack	Exercise 44
Treadmills	Exercise 45

Appendix F
p-Value	Appendix F